Measurement in Fluid Mechanics

Second Edition

Thoroughly revised and expanded, the new edition of this established textbook equips readers with a robust and practical understanding of experimental fluid mechanics. Enhanced features include improved support for students with emphasis on pedagogical instruction and self-learning, end-of-chapter summaries, 127 examples, 165 problems, refined illustrations, as well as new coverage of techniques in digital photography, frequency analysis of signals and the measurement of forces. It describes comprehensively classical and modern methods for flow visualisation and measuring flow rate, pressure, velocity, temperature, concentration, forces and wall shear stress, alongside supporting material on system response, measurement uncertainty, signal analysis, data analysis, optics, laboratory apparatus and laboratory practice. With enhanced instructor resources, including lecture slides, additional problems, laboratory support materials and online solutions, this is the ideal textbook for senior undergraduate and graduate students studying experimental fluid mechanics and is also suitable for an introductory measurements laboratory. Moreover, it is a valuable resource for practising engineers and scientists in this area.

Stavros Tavoularis is a Professor of Mechanical Engineering at the University of Ottawa, where he has taught undergraduate and graduate courses in experimental fluid mechanics for more than 40 years. Among other distinctions, he is a Fellow of the American Physical Society and a Fellow of the Engineering Institute of Canada.

Jovan Nedić is an Associate Professor in Mechanical Engineering at McGill University, where his teaching focuses on measurement techniques and statistical methods in undergraduate fluid mechanics. He was awarded the 2023 Wighton Fellowship for innovative, distinctive, and exceptional undergraduate laboratory course instruction in Engineering.

Measurement in Fluid Mechanics

Second Edition

STAVROS TAVOULARIS
University of Ottawa, Ottawa, Canada

JOVAN NEDIĆ
McGill University, Montreal, Canada

Shaftesbury Road, Cambridge CB2 8EA, United Kingdom

One Liberty Plaza, 20th Floor, New York, NY 10006, USA

477 Williamstown Road, Port Melbourne, VIC 3207, Australia

314–321, 3rd Floor, Plot 3, Splendor Forum, Jasola District Centre, New Delhi – 110025, India

103 Penang Road, #05–06/07, Visioncrest Commercial, Singapore 238467

Cambridge University Press is part of Cambridge University Press & Assessment, a department of the University of Cambridge.

We share the University's mission to contribute to society through the pursuit of education, learning and research at the highest international levels of excellence.

www.cambridge.org
Information on this title: www.cambridge.org/highereducation/isbn/9781009343626

DOI: 10.1017/9781009343657

First published 2005
Reprinted 2006
First paperback edition 2009
Second edition published 2024

A catalogue record for this publication is available from the British Library

A Cataloging-in-Publication data record for this book is available from the Library of Congress

ISBN 978-1-009-34362-6 Hardback

Additional resources for this publication at www.cambridge.org/Tavoularis_Nedic

To Sofia, Christina, Jason, Christopher, Sophia, Thomas and Benjamin

STAVROS TAVOULARIS

To Charlie and Mays

JOVAN NEDIĆ

Contents

Preface

This book is the thoroughly revised and expanded second edition of a monograph by the first author, having the same title and published in 2005. The main objective of both editions is to introduce students and other researchers into the multi-faceted world of experimental fluid mechanics. As in the first edition, we have organised the material in two parts: Background (Chapters 1–8) and Measurement Techniques (Chapters 9–15). In the first part, we have addressed the possible needs of readers with diverse backgrounds and skill levels by reviewing introductory material from several other fields, while also referring to specialised sources for an in-depth learning in such fields. In the second part, we have reviewed instrumentation and methodology that are used most commonly in fluid mechanics research and in the development of many technologies. Many statements in the preface to the first edition are largely valid for this edition as well. The main differences between the two editions may be summarised as follows:

- All material has been updated to reflect recent developments in instrumentation and experimental methodology, focusing on current research activities.
- The book's scope has been enlarged with the introduction of additional topics, including the use of digital cameras and the measurement of forces and moments.
- Parts of the book are presented so that they are suitable for undergraduate student instruction and self-learning.
- Much of the material has been reorganised and presented in a sequence that is easier to follow; for example, concepts related to the frequency analysis of signals are covered in a separate chapter.
- A multitude of examples are included in a format that distinguishes them from the main text; these examples illustrate and, in many cases, expand the concepts presented elsewhere. As much as possible, we have explained our rationale in answering the posed questions, identified possible pitfalls, and included detailed mathematical and numerical calculations, as well as self-explanatory sketches and plots. Some of these examples address research needs, whereas others are meant to elucidate introductory concepts.
- A number of new sketches and other illustrations have been provided and the quality of most figures from the previous edition has been improved.
- A Chapter Digest has been included at the end of each chapter, aimed at helping the reader assimilate definitions and distinguish differences between concepts.

- Additional problems have been included at the end of each chapter and the clarity of some previous problems has been improved.

A main intended use of this edition, like the first one, is as a textbook for an introductory graduate course in experimental fluid mechanics. A selection of topics covered in this book, as well as the accompanying examples, would also be suitable for an undergraduate course on experimental techniques or a laboratory-oriented undergraduate course. We further hope that the present book will be of use outside the classroom, serving as a first reference to engineers and scientists engaged in experimental work, primarily in fluid mechanics, but also in many other related fields.

While working on this edition, we benefited from suggestions and comments by several of our colleagues and students, to which we are thankful.

Part I

Background

1 Flow Properties and Basic Principles

Before being able to measure a flow property, it is necessary to understand its nature and its relationship to other properties. Furthermore, to make proper use of a measuring instrument, one must be thoroughly familiar with the principles of its operation, which usually involves concepts and relationships from several different fields. In the present chapter, we shall review the basic principles of fluid mechanics, and identify the properties of interest and their groupings in dimensionless form. This is meant to be a refresher of familiar concepts, as well as to identify a possible need for more in-depth reviews of fluid mechanics [2, 4, 10], thermodynamics [6, 9] and heat transfer [3]. Background material from system dynamics, signal analysis and optics will be reviewed separately, in later chapters.

1.1 Forces, Stresses and the Continuum Hypothesis

All material objects are subjected to external forces, which are of two types, body forces and surface forces. *Body forces* act on the bulk of the object from a distance and are proportional to its mass; the most common examples are gravitational and electromagnetic forces. *Surface forces* are exerted on the surface of the object by other objects in contact with it; they generally increase with increasing contact area.

Any surface force acting upon an elementary surface section of an object can be decomposed into a *normal* component, with a direction normal to the local tangent plane, and a *tangential* or *shear* component, with a direction parallel to the local tangent plane (see Fig. 1.1).

The *stress* at a point of an object is defined as the corresponding surface force per unit area; consequently, there are two types of stresses, *normal stresses* and *shear stresses.*

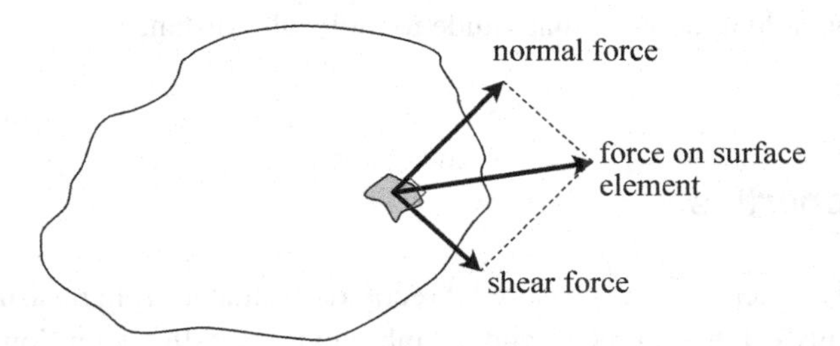

Figure 1.1 Force on a surface element and its decomposition to normal and shear components.

With respect to a Cartesian coordinate system, all stresses acting on three planes normal to the three axes form a second order Cartesian tensor, which has nine components, only six of which (three normal stresses and three shear stresses) are independent. In classical fluid mechanics, the (static) *pressure* is defined as the average normal stress along any three orthogonal directions.

According to the classical definition, a *fluid* is a material which cannot withstand a shear stress when at rest; in other words, a fluid under a shear stress will always be in internal motion, or deformation. Fluids are easily deformable materials and take the shape of any container in which they are contained. They are further distinguished into *liquids*, which have a relatively high density and require an extremely large change of normal stresses for a change of their volume, and *gases*, which have a relatively low density and can easily change their volume. Unlike liquids, gases tend to occupy the entire available volume of their container. Besides 'simple' liquids and gases, there is a number of materials, which, although not satisfying all properties of classical fluids, exhibit fluid-like properties. Examples include viscoelastic materials, which may sustain a certain amount of shear stress without being set in motion but behave like fluids when the shear stress exceeds a certain level, and plasmas, which form when gases are exposed to extremely high temperatures, in which case their molecules are dissociated into free atoms.

In most applications within the scope of conventional fluid mechanics, the phenomena of interest are characterised by scales which are far larger than the distances between molecules. Then, the flow properties are defined as statistical averages over a volume that contains a very large number of molecules. In such cases, one need not be concerned about individual molecular or atomic motions and masses and, instead, invoke the *continuum hypothesis*, by which any property of the fluid is assumed to have a continuous distribution within the volume of the fluid. Thus, one may define the *local* value of the property as the limit of the volume-averaged value of this property as the volume collapses towards a mathematical point. One may refer to a *fluid element*, or *fluid particle*, as a material entity that has an arbitrarily small volume, in which case its properties are considered to be uniform within this volume. In multiphase flows of immiscible fluids, the continuum hypothesis applies within each individual fluid, while some properties may be considered as discontinuous at the interface. Obviously, there are also situations where the continuum hypothesis does not apply at all. An example is the case of rarefied gases, in which the distances between gas molecules are relatively large, and for which one must account for individual molecules and their motions. Another example is the case of nanofluids, in which the phenomena of interest have length scales that are comparable in order of magnitude to molecular distances.

1.2 Measurable Properties

A property of a fluid element can be measured directly or estimated from measurements of other properties only if it has a precise and unambiguous scientific definition, associated with a measurement procedure. The following list identifies measurable properties of common interest classified in four general classes.

- *Material properties*:

mass
density
specific volume
viscosity
thermal conductivity
molecular diffusivity
specific heat under constant pressure
specific heat under constant volume
specific heat ratio
gas constant
bulk modulus of elasticity
coefficient of thermal expansion
electric conductivity
surface tension
index of refraction
fluorescence

- *Kinematic properties*, namely, properties which describe the motion of a fluid without consideration of applied forces:

position
displacement
velocity
volume flow rate
mass flow rate
acceleration
vorticity
strain rate
angular position
angular displacement
angular velocity
angular acceleration
momentum
angular momentum

- *Dynamic properties*, namely, properties related to the applied forces:

force
pressure (mechanical – Section 9.1)
stress
torque

- *Thermodynamic properties*, namely, properties related to heat and work:

temperature
pressure (thermodynamic – Section 9.1)
internal energy
enthalpy
entropy
heat flux
work
energy

Material properties are usually not the subject of experimental fluid mechanics, as their values have, in many cases, been measured independently and can be found in handbooks [5] or other sources. However, if a material property is unknown or overly sensitive to the particular experimental conditions, its value may have to be determined either as part of the overall experiment or by a specific experimental investigation.

1.3 Flow Velocity and Velocity Fields

A position in space is specified in terms of its coordinates x_i, $i = 1, 2, 3$, with respect to a Cartesian coordinate system. At any time t, this position is occupied by some fluid element, assumed to maintain its mass within an infinitesimally small volume. With the fluid considered as a continuum, one may define the flow velocity at a given position and a given time as the velocity of a fluid element that occupies that position at that time. One is also interested in defining the *velocity field*, which consists of the velocities of all fluid elements that comprise a material system. Thus, it becomes necessary to distinguish the fluid element in question from any other fluid element. For clarity, we shall specify as X_i the coordinates of the fluid element that occupies position x_i at time t. This element moves along its trajectory, indicated by the dashed line in Fig. 1.2. At time $t + \delta t$, the same element will have the coordinates $X_i + \delta X_i$. Then, the *flow velocity* is defined as

$$V_i = \lim_{\delta t \to 0} \frac{\delta X_i}{\delta t} = \frac{\mathrm{d}X_i}{\mathrm{d}t} . \tag{1.1}$$

One approach to identify the fluid element is by specifying its initial coordinates X_{0i}, namely its position coordinates at the origin of time t_0. Then, its coordinates at any time t are functions only of the initial coordinates and t, namely $X_i = X_i(X_{0i}, t)$. The velocity field may also be specified as a function of these two variables, as $V_i = V_i(X_{0i}, t)$. This approach is known as the *material* or *Lagrangian* description of flow motion. Because identifying ('tagging') individual fluid elements is not usually practical, it is also customary to express the velocity field in terms of a fixed position with respect to the coordinate system, and, of course, time; this approach is known as the *spatial* or *Eulerian* description . To avoid confusion when differentiating, the velocity field according to the Eulerian description is denoted by a different symbol, as $U_i(x_i, t)$; it is understood, nevertheless, that, at all positions and for all times, the definition of flow velocity is unique and that

$$V_i(X_{0i}, t) = U_i(x_i, t) . \tag{1.2}$$

Following the Lagrangian description, the fluid element *acceleration* is defined as

$$\alpha_i = \frac{\mathrm{d}V_i}{\mathrm{d}t} . \tag{1.3}$$

However, if one follows the Eulerian description, one has to account for changes of the

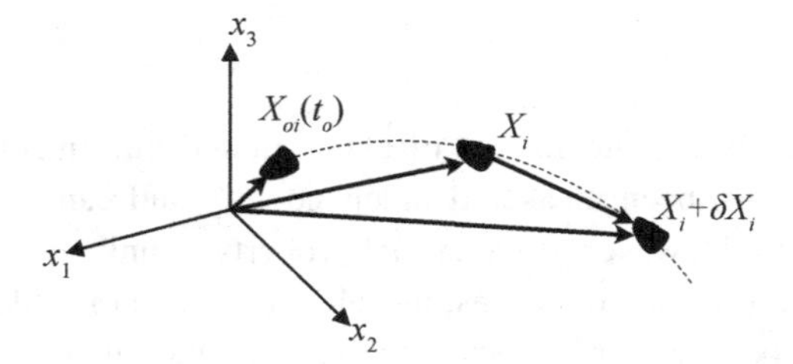

Figure 1.2 Fluid element positions.

velocity from one location of the particle to the next and the fluid acceleration becomes

$$\alpha_i = \frac{\partial U_i}{\partial t} + U_1 \frac{\partial U_i}{\partial x_1} + U_2 \frac{\partial U_i}{\partial x_2} + U_3 \frac{\partial U_i}{\partial x_3} \,, \tag{1.4}$$

where the first term on the right hand side is called the *local acceleration* and the remaining terms are collectively called the *convective acceleration.* The right hand side of the previous equation is called the *material* or *substantial derivative* of U_i and is commonly denoted as DU_i/Dt.

Fluid deformation under the influence of stresses is described by the *rate of strain* or *rate of deformation tensor.*

$$e_{ij} = \frac{1}{2}\left(\frac{\partial U_i}{\partial x_j} + \frac{\partial U_j}{\partial x_i}\right) \,, \quad i, j = 1, 2, 3 \,. \tag{1.5}$$

Example 1.1 Express the velocity vector and its components in cylindrical coordinates.

Answer

The coordinate transformation from the Cartesian coordinate system (x_1, x_2, x_3) to the cylindrical coordinate system (r, θ, z) is shown in Fig. 1.3. The unit vectors in the Cartesian coordinate system are $\vec{i}, \vec{j}, \vec{k}$ and in the cylindrical coordinate system they are $\vec{e_r}, \vec{e_\theta}, \vec{e_z}$. From geometrical considerations, we can get the coordinates of a point in the cylindrical coordinate system as

$$\begin{aligned} r &= \sqrt{x_1^2 + x_2^2} \,, \\ \theta &= \tan \frac{x_2}{x_1} \,, \quad 0 \leq \theta < 2\pi \,, \\ z &= x_3 \,. \end{aligned} \tag{1.6}$$

Similar geometrical considerations give us the velocity components in cylindrical

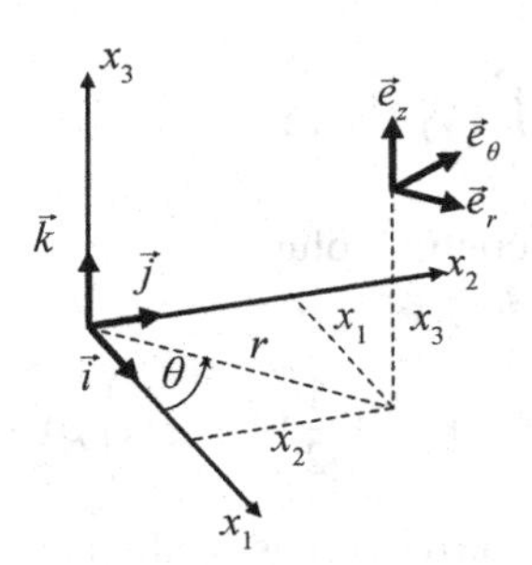

Figure 1.3 Cartesian and cylindrical coordinate systems.

coordinates as

$$U_r = U_1 \cos\theta + V_1 \sin\theta\,,$$
$$U_\theta = -U_1 \sin\theta + V_2 \cos\theta\,, \qquad (1.7)$$
$$U_z = U_3\,.$$

1.4 Mathematical Description of Flows

The physical relationships among various flow properties are represented by mathematical expressions, which constitute algebraic, differential, integral or integro-differential equations. Such relationships include *axiomatic principles*, which are generally accepted as natural laws subjected to experimental verification, and *empirical relationships*, which range from semi-theoretical concepts, based on physical arguments with some experimental input, to purely empirical expressions, obtained by statistical curve fitting to experimental results.

Conventional fluid mechanics and thermodynamics are based on four basic axiomatic principles: the conservation of mass, the momentum principle, and the first and second laws of thermodynamics. When applied to a *closed fluid system*, these principles result in a set of integral relationships that describe the fluid motion. However, it is usually more convenient to apply these principles to the contents of a fixed *control volume*, in which case one may derive the following four basic relationships [4]:

- The *conservation of mass equation* or *continuity equation*:

$$\frac{\partial}{\partial t}\int_{\text{CV}} \rho \mathrm{d}\mathcal{V} + \int_{\text{CS}} \rho \vec{V} \cdot \mathrm{d}\vec{A} = 0\,, \qquad (1.8)$$

  where ρ is the fluid density, $\vec{V}$ is the velocity vector, CV is the control volume, $\mathrm{d}\mathcal{V}$ is the volume of a fluid particle in the CV, CS is the control surface, which encloses the control volume, and $\mathrm{d}\vec{A} = \vec{n}\mathrm{d}A$ ($\mathrm{d}A$ is an element of the CS and $\vec{n}$ is the local, outwards, unit normal vector). For non-reacting multi-component flows, one may formulate similar equations expressing the *conservation of species*.
- The *momentum equation* (*Newton's second law*), which, for an inertial coordinate system, can be written as

$$\vec{F} = \frac{\partial}{\partial t}\int_{\text{CV}} \vec{V}\rho \mathrm{d}\mathcal{V} + \int_{\text{CS}} \vec{V}\rho\vec{V} \cdot \mathrm{d}\vec{A}\,, \qquad (1.9)$$

  where $\vec{F}$ is the net external force acting on the control volume.
- The *energy equation* (*first law of thermodynamics*):

$$\dot{Q} - \dot{W} = \frac{\partial}{\partial t}\int_{\text{CV}} \left(u + \frac{1}{2}V^2 + gx_3\right) \rho \mathrm{d}\mathcal{V} + \int_{\text{CS}} \left(u + \frac{p}{\rho} + \frac{1}{2}V^2 + gx_3\right) \rho\vec{V} \cdot \mathrm{d}\vec{A}\,, \qquad (1.10)$$

  where $\dot{Q}$ is the rate of heat transfer from the surroundings to the control volume,

$\dot{W}$ is the mechanical power produced by moving solid components, shear stresses acting on the boundary or electromagnetic forces, but not normal stresses acting on the control volume (the latter have been included on the right hand side), u is the specific internal energy, p is the pressure, x_3 is an upwards vertical axis and g is the gravitational acceleration.

- The *second law of thermodynamics*:

$$\frac{\partial}{\partial t}\int_{\mathrm{CV}} s\rho \mathrm{d}\mathcal{V} + \int_{\mathrm{CS}} s\rho\vec{V}\cdot \mathrm{d}\vec{A} \geqslant \int_{\mathrm{CV}} \frac{1}{T}\left(\frac{\dot{Q}}{A}\right)\mathrm{d}A\,, \tag{1.11}$$

where s is the specific entropy, T is the absolute temperature and $\dot{Q}/A$ is the heat flux into the CV through $\mathrm{d}A$.

By letting the control volume vanish towards a point, one may convert the previous equations into differential forms (see examples in the next section). These must be complemented by initial and boundary conditions. The usual conditions employed at a solid boundary in conventional fluid mechanics are the *no penetration* condition, which specifies that the relative velocity normal to the contact surface vanishes, and the *no slip* condition, which specifies that the relative velocity tangential to the contact surface vanishes. In two-phase flows, the pressure difference Δp across the interface is related to the surface tension σ by the expression

$$\Delta p = \sigma\left(\frac{1}{R_1} + \frac{1}{R_2}\right)\,, \tag{1.12}$$

where R_1 and R_2 are the principal radii of curvature; obviously, $\Delta p = 0$ across plane interfaces. In liquid–gas flows, it is customary to neglect the shear stress on the liquid side of the interface (*stress-free boundary*).

Certain types of flow require the use of additional axiomatic principles, as, for example:

- the *laws of chemical reaction* for reactive flows (e.g., $H_2 + \frac{1}{2}O_2 \rightleftarrows H_2O$);
- *magnetohydrodynamic laws* for electrically conductive fluids in magnetic fields; an example is *Maxwell's equations*.

Among the common types of empirical relationships, one may mention the following:

- *Equations of state* or *constitutive relationships*, relating various thermodynamic properties of a material. The simplest example of an equation of state is the ideal gas law

$$\frac{p}{\rho} = RT\,, \tag{1.13}$$

where R is the corresponding gas constant.
- *Stress–strain relationships*, also called constitutive relationships. Most common is a linear relationship between the stress tensor and the rate of strain tensor, which defines a *Newtonian fluid* and has a single material parameter, the viscosity μ. Different stress–strain relationships have also been proposed to describe the motion of *non-Newtonian fluids*.

- *Turbulence models*, which are relationships among various statistical properties of turbulent flows. A widely used turbulence model is the *gradient transport model*, by which the turbulent shear stress $-\rho\overline{u_1u_2}$ is assumed to be proportional to the mean shear $\partial\overline{U}_1/\partial x_2$; this model requires a means to determine the proportionality coefficient μ_T, called the *eddy viscosity*.

Example 1.2 Starting from the momentum equation, show that the drag force F_D (namely the streamwise component of the force) acting on a two-dimensional body in a steady, incompressible, uniform stream, can be obtained from flow velocity measurements. Explain all assumptions and simplifications made in the derivation.

Answer

Friction between the fluid and the object wall creates a low-speed wake downstream of the object, where the local velocity $U_1(x_1, x_2)$ is lower than the free stream velocity $U_{1\infty}$ and so there is a loss of the fluid momentum, which is equal to the external force. We wish to apply the momentum equation (Eq. (1.9)) over a control volume (CV) that contains the object. The CV must be chosen so that the mass flow rate through its upstream side (namely, the momentum of the uniform stream) is equal to the mass flow rate through its downstream side (namely, the momentum of the wake), as shown in Fig. 1.4. If a CV with parallel top and bottom sides were used instead, integration over its downstream side would miss the momentum of the fluid that exits the CV from the top and bottom.

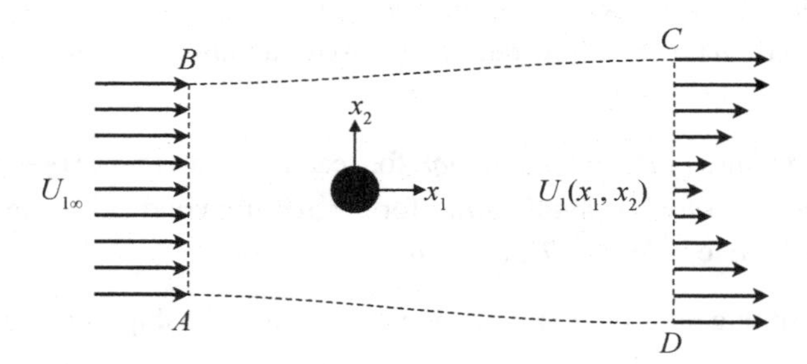

Figure 1.4 Control volume for flow around a two-dimensional object.

The downstream side of the CV must extend sufficiently far from the plane of symmetry for the difference between the wake velocity and the free stream velocity to be negligible. We can determine the appropriate height h of the upstream side by applying the mass conservation (Eq. (1.8)). For a steady, incompressible flow, this equation is simplified to

$$\int_{\text{CS}} \vec{V} \cdot \mathrm{d}\vec{A} = 0\,,$$

from which one gets

$$\int_A^B U_{1\infty}\, b\, \mathrm{d}x_2 = \int_D^C U_1(x_1, x_2)\, b\, \mathrm{d}x_2\,, \tag{1.14}$$

where b is the depth of the CV in the x_3 direction, equal to the length of the object. Then,

$$h = \frac{1}{U_{1\infty}} \int_D^C U_1(x_1, x_2)\, \mathrm{d}x_2\,.$$

Multiplying both sides of Eq. (1.14) by $U_{1\infty}$, we get

$$\int_A^B U_{1\infty}^2 \mathrm{d}x_2 = \int_D^C U_{1\infty} U_1(x_1, x_2)\, \mathrm{d}x_2\,.$$

We note that the pressure in the wake far from the object is uniform and equal to the upstream pressure, because the flow velocity in the wake is nearly parallel to the x_1 axis. Thus, the pressure force on this CV vanishes. Then, the momentum equation in the x_1 direction, simplified for steady, incompressible flow, becomes

$$\begin{aligned} F_D &= \rho b \int_A^B U_{1\infty}^2 \mathrm{d}x_2 - \rho b \int_D^C U_1^2(x_1, x_2)\, \mathrm{d}x_2 \\ &= \rho b \int_D^C U_{1\infty} U_1(x_1, x_2)\, \mathrm{d}x_2 - \rho b \int_D^C U_1^2(x_1, x_2)\, \mathrm{d}x_2 \\ &= \rho b \int_D^C U_1(x_1, x_2)\, [U_{1\infty} - U_1(x_1, x_2)]\, \mathrm{d}x_2\,. \end{aligned}$$

Measurements of the wake velocity can be obtained by any local velocity measurement method, for example by traversing a Pitot-static tube (to be discussed in a later chapter).

1.5 The Choice of Analytical Approach

Although accurate mathematical models of fluid flow are available, it is generally advisable, and even necessary, to employ simplifications, whenever possible and to the greatest possible extent. This strategy has been applied extensively to the analysis of measuring instrument operation. As a rule, one should strive to use the simplest possible analytical model that permits the desired measurement with an acceptable uncertainty. Of course, the differences between an approximate and a more 'exact' method must be analysed, and, if found to be excessive, one must either apply appropriate corrections, or abandon the approximate method in favour of a more accurate one.

Among the important effects, which may or may not be accounted for in a particular type of analysis, are deformation, friction, compressibility and turbulence; additional effects complicating the analysis may be present under specific circumstances. Accordingly, one may distinguish the following theoretical approaches used commonly to describe fluid motion.

1.5.1 Fluid Statics

When a fluid is at rest, or in rigid body motion, it does not deform and, therefore, it cannot sustain shear stresses. In fluid statics, the three normal stresses are equal in magnitude to each other and to the pressure. Static fluids are subjected to gravity, which causes the development of hydrostatic pressure and of the buoyancy force. Static fluid analysis is useful in analysing the performance of many instruments, notably those of manometers, barometers and certain types of pressure transducers. On the other hand, the use of static fluid analysis in situations where there are fluids in motion could lead to substantial errors. For example, a static analysis of a liquid manometer subjected to pressure fluctuations would result in erroneously low readings of pressure differences.

1.5.2 Inviscid Incompressible Flows

The simplest mathematical model of fluid flow is one which neglects the effects of friction and compressibility. Continuity imposes the requirement of conservation of volume of an incompressible fluid. In differential form, this can be expressed as

$$\frac{\partial U_1}{\partial x_1} + \frac{\partial U_2}{\partial x_2} + \frac{\partial U_3}{\partial x_3} = 0 . \tag{1.15}$$

When friction is neglected, turbulence must also be disregarded. The differential momentum equation for an *inviscid incompressible* fluid is known as the *Euler* equation

$$\frac{DU_i}{Dt} = g_i - \frac{1}{\rho}\frac{\partial p}{\partial x_i} . \tag{1.16}$$

Together, the continuity equation and the Euler equation form a closed system, which is sufficient for the determination of the fluid velocity and pressure. Integration of Euler's equation along a streamline leads to the simple and frequently used (as well as misused), steady-flow *Bernoulli*'s equation, which is an algebraic expression relating velocity magnitude and pressure as

$$p + \frac{1}{2}\rho U^2 + \rho g x_3 = \text{const.} , \tag{1.17}$$

where x_3 is a vertical upward elevation. The flow analysis can be further simplified by the assumption of *irrotationality*, namely the vanishing of *vorticity*

$$\vec{\zeta} = \text{curl}\vec{U} \equiv \nabla \times \vec{U} \tag{1.18}$$

everywhere in the flow domain, with the possible exception of isolated singularities. Such a flow is called *potential* and is described by the *velocity potential*, which satisfies Laplace's equation. In potential flow, Bernoulli's equation can be applied not only along a streamline but also from one streamline to another. Potential flow analysis is a common approach in aerodynamics. It is also used to explain the operation of several simple instruments. For example, the measurement of velocity with the use of immersed pressure tubes routinely employs the use of the steady-flow Bernoulli's equation; this approach is acceptable in many wind-tunnel applications, in which the effects of friction are known to be smaller than the measurement uncertainty, but it introduces large errors in measurements near walls, where friction effects are important, or in high-speed flows, where compressibility must be accounted for.

Example 1.3 A wind tunnel has a two-dimensional contraction with an area ratio of 6:1, leading to the test section. Two wall pressure taps are located at the start and the end of the contraction respectively, as shown in Fig. 1.5. Using a differential pressure transducer, we obtain a pressure difference of 1.50 kPa. Determine the air speed at the end of the contraction. Neglect friction effects.

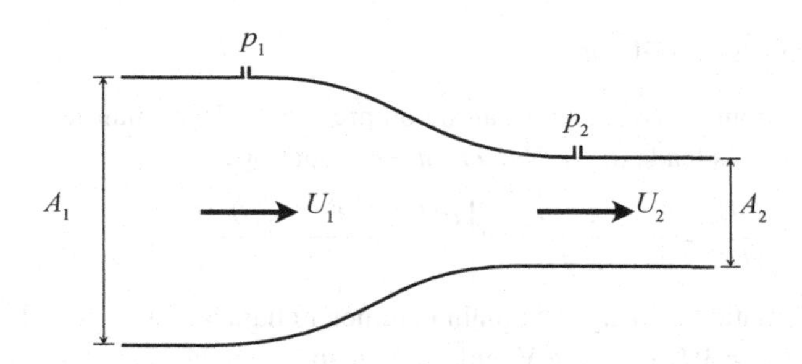

Figure 1.5 Flow through a wind tunnel contraction.

Answer

The contraction ratio for wind tunnels, the design of which will be expanded upon in Chapter 8, is defined as $C = A_1/A_2$, where A_1, A_2 are, respectively, the upstream and downstream cross-sectional areas. For simplicity, we will neglect the effects of friction and will assume that the flow velocity is uniform in each cross section of the wind tunnel. Using conservation of mass (Eq. (1.8)), we can relate the areas at the start and end of the contraction to the respective velocities as

$$\rho U_1 A_1 = \rho U_2 A_2 .$$

The flow through the contraction of the wind tunnel is steady by design and may be approximated as incompressible. Considering the relatively small size of the facility, we may further neglect the elevation pressure differences and assume that pressure is uniform within each cross section. Then, following any streamline along the contraction, we may apply Bernoulli's equation (Eq. (1.17)) to relate the pressure and the flow velocity as

$$p_1 + \tfrac{1}{2}\rho U_1^2 = p_2 + \tfrac{1}{2}\rho U_2^2 .$$

Combining the two previous expressions, one obtains

$$U_2^2 = \frac{2(p_1 - p_2)}{\rho}\left[1 - \left(\tfrac{A_2}{A_1}\right)^2\right] = \frac{2(p_1 - p_2)}{\rho}\left[1 - \left(\tfrac{1}{C}\right)^2\right].$$

Inserting the given values, and assuming a standard atmospheric air density of $\rho =$ 1.225 kg/m^3, one obtains a test section velocity of $U_2 = 48.8$ m/s. One could obviously improve the accuracy of the measurement by using an air density value that corresponds to the barometric pressure, temperature and humidity of the room where the wind tunnel is located. If the contraction ratio is sufficiently large (e.g., $C \geq 10$), one may assume that $(1/C)^2$ is negligible and compute the test section velocity as

$$U_2 \approx \sqrt{\frac{2(p_1 - p_2)}{\rho}}.$$

For the present contraction, the simplified expression gives $U_2 = 49.5$ m/s, which is 1% larger than the more accurate value obtained previously.

1.5.3 Viscous Incompressible Flows

Application of Newton's second law to an incompressible, Newtonian fluid with constant material properties leads to the *Navier–Stokes* equations

$$\frac{DU_i}{Dt} = g_i - \frac{1}{\rho}\frac{\partial p}{\partial x_i} + \nu\left[\frac{\partial^2 U_i}{\partial x_1^2} + \frac{\partial^2 U_i}{\partial x_2^2} + \frac{\partial^2 U_i}{\partial x_3^2}\right], \tag{1.19}$$

where ν is the kinematic viscosity. The main parameter characterising such flows is the *Reynolds number* $\mathrm{Re} = VL/\nu$, where V and L are, respectively a characteristic velocity and a characteristic length. Re represents the ratio of inertia forces (proportional to the term on the left-hand side) and viscous forces (proportional to the last term on the right-hand side). In the limiting case of vanishing Re, inertial effects become negligible and the Navier–Stokes equations are reduced to the *Stokes* equations [8, 10], which contain no non-linear terms. The form of applicable solutions and the measuring system response in low-Re flows are radically different from those at higher Re. On the other extreme, as $\mathrm{Re} \to \infty$, one may at first glance anticipate that viscous effects would become negligible. However, this limit contradicts the physical fact that, as the Reynolds number increases, a flow would become increasingly unstable and eventually turbulent. At large Re, one must not only include friction in the analysis, but also account for turbulence effects. The infinite-Re limit also leads to an analytical singularity, because the order of the Navier–Stokes equations changes from second to first.

1.5.4 Compressible Flows

In compressible flow, density variation is significant. Density changes are related to pressure changes and are also accompanied by temperature changes. Unlike incompressible flow, compressible flow is not divergence-free, and the corresponding differential continuity equation includes the unknown variable density. Besides velocity and pressure, compressible flow models contain two additional unknowns, density and temperature, and, therefore, require two additional equations. One independent differential equation (the *energy* equation) is provided by the first law of thermodynamics, and a second relationship is provided by an equation of state. To simplify the analysis of compressible flow, friction is often neglected. *Isentropic flow*, namely a reversible and adiabatic change of state, is commonly used as an approximation in compressible flow instrumentation. Allowing for density changes permits the modelling of propagation of weak disturbances with a finite speed, which, for isentropic flows, is called the *speed of sound*. The ratio of fluid velocity to the speed of sound, called the *Mach number*, M, is a dimensionless parameter that serves as a measure of the importance of compressibility. It can be viewed as representing the ratio of inertia forces to elastic forces. For air, compressibility effects become significant when $M > 0.3$. When $M > 1$ (supersonic flow), dramatic and sudden changes of fluid velocity and thermodynamic properties may occur across a *shock wave*, which is an irreversible, and therefore non-isentropic, process.

1.5.5 Turbulent Flows

Turbulence is a state of motion in which flow properties, including velocity, pressure and vorticity, vary rapidly and randomly in space x_1, x_2, x_3 and time t. A turbulent flow is always three-dimensional and unsteady, but, under certain conditions, its statistical properties at a point in space would be independent of time, in which case the turbulence is called *stationary*. The velocity $U_i(x_1, x_2, x_3, t)$, $i = 1, 2, 3$ in a stationary turbulent velocity field is called the *instantaneous velocity*. It can be decomposed into a time-independent *average* (or *mean*) $\overline{U}_i$ and a time-dependent *fluctuation* u_i as

$$U_i = \overline{U}_i + u_i \, . \tag{1.20}$$

This process is called *Reynolds decomposition*. The average is calculated by integrating the instantaneous velocity over a sufficiently long time interval T as

$$\overline{U}_i = \lim_{T \to \infty} \frac{1}{T} \int_0^T U_i \mathrm{d}t \, . \tag{1.21}$$

By definition, the fluctuation has zero average ($\overline{u}_i = 0$) and, therefore, the strength of turbulence cannot be described by $\overline{u}_i$. Instead, it is described by the variances $\overline{u_1^2}, \overline{u_2^2}, \overline{u_3^2}$ or the corresponding standard deviations, such as $u_1' = \sqrt{\overline{u_1^2}}$, which are always positive. These concepts are illustrated in Fig. 1.6.

Statistical equations for the means and fluctuations have been formulated but lead to systems containing unknowns whose number exceeds the number of available equations and cannot be solved without the use of additional relationships (turbulence models). Turbulence affects the operation of even the simplest instruments. In some cases, relatively simple corrections for turbulence effects have been devised. An important parameter is the *turbulence intensity*, that is, the ratio $u_1'/\overline{U}_1$, however, information on the time and length scales of turbulent motions is often required in the correction method. The measurement of turbulent properties themselves can be achieved with the use of special instrumentation, having particularly refined spatial, temporal and amplitude resolutions. Special methods, including *phase averaging* and *conditional sampling*, have been developed to resolve quasi-deterministic turbulence patterns, known as *coherent structures*.

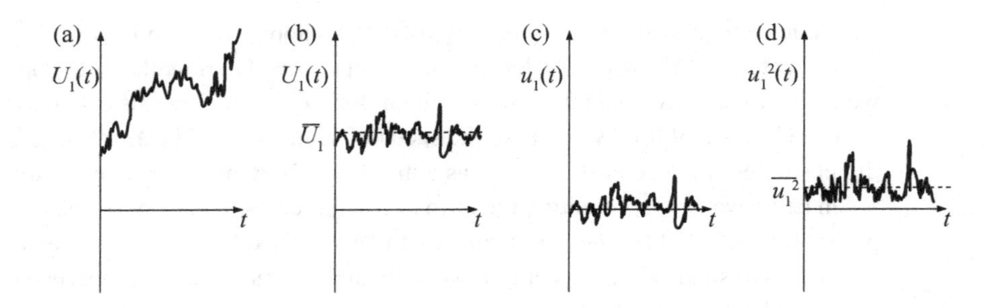

Figure 1.6 Examples of a turbulent velocity component: (a) instantaneous velocity in non-stationary turbulence; (b) instantaneous velocity and time average in stationary turbulence; (c) velocity fluctuation in stationary turbulence; and (d) squared velocity fluctuation and variance in stationary turbulence.

1.5.6 Complex Flows

Fluids and flows are sometimes complicated by particular conditions which necessitate the use of specialised analytical models and methods as well as special instrumentation, calibration and measuring procedures. Of particular interest are the following cases:

- *Multi-phase* flows, including liquid–solid, gas–solid, gas–liquid and liquid–liquid flows, and flows with phase change.
- Flows with *chemical reaction*, including combustion. These involve changes of composition, accompanied by release or absorption of chemical energy and changes of temperature.
- *Low density* (*rarefied*) gas flows, for which the continuum hypothesis is inappropriate and whose measurement relies on molecular and atomic phenomena.
- Flows of *non-Newtonian* fluids, having non-linear stress–strain rate relationships (e.g., , polymer flows and flow of blood, which is a suspension of red blood cells, white blood cells and platelets).
- *Magnetohydrodynamic* flows, namely flows of electrically conductive fluids in magnetic fields.

1.6 Similarity

1.6.1 Standard Dimensions and Units

Measurable properties may be either *primary*, namely independent of each other, or *secondary*, namely related to other properties *via* their definition or a basic principle. International conventions regulate the choice of a sufficient set or primary properties (called *dimensions*) and appropriate standard amounts (called *units*) that may be used as scales for expressing magnitudes. Table 1.1 lists the seven primary properties and corresponding base units that have been adopted by the SI (système international d'unités) System of Units. All other dimensions are related to these seven base dimensions through dimensional analysis of various mathematical relationships among primary and secondary properties.

Table 1.1 Primary properties and corresponding base units in the SI system

Property	Unit name	Unit symbol	Dimension symbol
Length	meter	m	L
Mass	kilogram	kg	M
Time	second	s	T
Temperature	kelvin	K	Θ
Electric current	ampere	A	I
Amount of substance	mole	mol	N
Luminous intensity	candela	cd	J

Example 1.4 Determine the dimensions of force in the SI system. Also express the unit of force in terms of base units.

Answer

One may derive the dimension of force F by using Newton's second law, which equates a force with a mass m and its acceleration a (this is a vector relationship, but here we consider equality of the magnitudes):

$$F = ma\,.$$

Mass is a primary property and has a dimension M. Acceleration is the rate of change of the velocity, which in turn is the rate of change of the distance from the origin. Thus, the dimension of acceleration can be found as $[a] = \mathrm{LT}^{-2}$. Thus, the dimension of force is related to the primary dimensions of mass, length and time as

$$[F] = [m][a] = \mathrm{MLT}^{-2}\,.$$

Newton's second law can also relate the unit of force, called appropriately one newton (1 N), to the base units as $1\ \mathrm{N} = 1\ \mathrm{kg\ m\ s}^{-2}$.

1.6.2 Similarity and Non-dimensionalisation

Although measurement may occasionally be performed directly on the actual system of interest under actual operating conditions, more often than not, the measurement is performed either on a model of the system or on the actual system but under modified and controlled conditions, which may differ from the actual operating conditions. The concept of *similarity* (also known as *similitude*) permits the application of information obtained from studies using models to the actual systems they represent. An essential requirement is that an actual system and its experimental model be *geometrically similar*;

this means that they should have the same shape, so that the ratios of all corresponding dimensions are the same. In addition, the flows in the model and the actual system must be *kinematically similar*; this requires that the velocities at corresponding positions have the same directions and a constant ratio of their magnitudes. For complete similarity, the flows must also be *dynamically similar*; this requires that the corresponding forces have the same directions and a constant ratio of their magnitudes. When two flows are dynamically similar, results collected in one may be used for the prediction of corresponding properties in the other. This usually requires an intermediate step, called *scaling*, that is, resizing of the values of the different properties. Measurable flow properties are normally *dimensional*, having their magnitudes expressed in terms of the corresponding units. *Non-dimensionalisation* of the results is the process of converting measured properties into dimensionless numbers by dividing their magnitudes by combinations of appropriate powers of chosen scales. Besides the economy in presentation, non-dimensionalisation produces results which are independent of the unit system and serves as a guide for the selection of optimal geometrical and operating conditions. Similarity and non-dimensionalisation permit the derivation of scaling laws and the design of models.

1.6.3 Common Dimensionless Parameters

A large number of dimensionless groups of various flow properties have already been identified in previous analyses of problems in fluid mechanics, heat and mass transfer and related fields. In many cases, these groups appear naturally, when all terms in the governing equations are converted to dimensionless forms by dividing all properties by appropriate scales, a process often called *similarity analysis*.

Example 1.5 As an example of similarity analysis and the construction of dimensionless groups, derive a dimensionless form of the Navier–Stokes equations for incompressible flow with a constant viscosity in a gravitational field and identify dimensionless groups as coefficients of the different terms in these equations.

Answer

We start with the dimensional Navier–Stokes equations (Eq. (1.19)). Assume that a length scale L and a velocity scale V_o can be identified, for example as a characteristic dimension of the boundary and an average or free-stream velocity. Then, one may non-dimensionalise all distances and velocity components by dividing them by the corresponding scales, while time is made dimensionless by the time scale L/V_o. Assume further that pressure variation is non-dimensionalised by a reference value p_{ref}, while the gravitational acceleration vector is non-dimensionalised by its magnitude. Denoting dimensionless variables by asterisks, one may then express the Navier–Stokes equations in dimensionless form as

$$\frac{D^* U_i^*}{D^* t^*} = \frac{gL}{V_o^2} g_i^* - \frac{p_{ref}}{\rho V_o^2} \frac{\partial p^*}{\partial x_i^*} + \frac{\nu}{V_o L} \left(\frac{\partial^2 U_i^*}{\partial x_1^{*2}} + \frac{\partial^2 U_i^*}{\partial x_2^{*2}} + \frac{\partial^2 U_i^*}{\partial x_3^{*2}} \right) , \quad i = 1, 2, 3 . \tag{1.22}$$

The dimensionless coefficients of the three terms on the right-hand side have been identified as the squared reciprocal of the *Froude number*, the *Euler number* and the reciprocal of the *Reynolds number* for the system (definitions of these numbers are given further in this section). The left-hand side of Eq. (1.22) is the dimensionless acceleration of a fluid element; brought to the right-hand side, it may be considered as representing a fictitious dimensionless 'inertia force', which balances the sum of external forces. The three terms on the right-hand side represent, respectively, the dimensionless gravitational, pressure and viscous forces. Thus, the previously identified dimensionless groups may be considered as representing the relative magnitudes of external forces by comparison to the inertia force.

Among the most common dimensionless groups that are encountered in experimental fluid mechanics are the following:

- **Reynolds number:**

$$\mathrm{Re} = \frac{\rho V L}{\mu} \equiv \frac{V L}{\nu}, \tag{1.23}$$

where V is a characteristic velocity, L is a characteristic length, μ is the dynamic viscosity and $\nu = \mu/\rho$ is the kinematic viscosity; Re represents the ratio of inertia forces to viscous forces.

- **Mach number:**

$$\mathrm{M} = \frac{V}{c}, \tag{1.24}$$

where c is the speed of sound; M represents the ratio of inertia forces and elastic forces (compressibility effect); it is noted that, in Section 1.6, the symbol M has also been used to denote the dimension of mass.

- **Pressure coefficient (Euler number):**

$$C_p(\equiv \mathrm{Eu}) = \frac{p - p_{ref}}{\frac{1}{2}\rho V^2}, \tag{1.25}$$

where p_{ref} is a reference pressure; it represents the ratio of pressure forces and inertia forces; some sources define Eu as one half of the previously defined pressure coefficient.

- **Drag coefficient:**

$$C_D = \frac{F_D}{\frac{1}{2}\rho A V^2}, \tag{1.26}$$

where F_D is the drag force and A is the frontal (i.e., normal to the free stream) area of the immersed object for bluff objects or the planform (i.e., projected on a plane parallel to the free stream) area for streamlined objects; C_D represents the ratio of drag forces and inertia forces.

- **Lift coefficient:**

$$C_L = \frac{F_L}{\frac{1}{2}\rho A V^2}, \tag{1.27}$$

where F_L is the lift force and A is the planform area; C_L represents the ratio of lift forces and inertia forces.

- **Prandtl number:**

$$\mathrm{Pr} = \frac{\nu}{\gamma} = \frac{c_p \mu}{k} , \tag{1.28}$$

where c_p is the specific heat under constant pressure, k is the thermal conductivity and $\gamma = k/(\rho c_p)$ is the thermal diffusivity; Pr represents the ratio of the rates of diffusion of momentum and heat due to molecular motions.

- **Schmidt number:**

$$\mathrm{Sc} = \frac{\nu}{\gamma_c} , \tag{1.29}$$

where γ_c is the molecular diffusivity of a species in a fluid mixture; Sc represents the ratio of the rates of diffusion of momentum and mass in the fluid.

- **Froude number:**

$$\mathrm{Fr} = \frac{V}{\sqrt{gL}} , \tag{1.30}$$

where L is a characteristic length; Fr^2 represents the ratio of inertia forces and gravitational forces and applies mainly to liquid flows with a free surface.

- **Weber number:**

$$\mathrm{We} = \frac{\rho V^2 L}{\sigma} , \tag{1.31}$$

where σ is the surface tension; We represents the ratio of inertia forces and surface tension forces on interfaces between two immiscible fluids.

- **Capillary number**:

$$\mathrm{Ca} = \frac{\mu V}{\sigma} = \frac{\mathrm{We}}{\mathrm{Re}} ; \tag{1.32}$$

Ca represents the ratio of viscous forces and surface tension forces on interfaces between two immiscible fluids.

- **Cavitation number** (for liquids):

$$\sigma_c = \frac{p - p_v}{\frac{1}{2}\rho V^2} , \tag{1.33}$$

where p_v is the vapour pressure; σ_c applies to liquids.

- **Nusselt number:**

$$\mathrm{Nu} = \frac{hL}{k} , \tag{1.34}$$

where h is the overall heat transfer coefficient and k is the thermal conductivity of the fluid; Nu represents the ratio of total and conductive heat transfer rates in a fluid.

- **Biot number:**

$$\mathrm{Bi} = \frac{hL}{k_s} , \tag{1.35}$$

where h is the overall heat transfer coefficient from a solid surface to a fluid and k_s is the thermal conductivity of the solid; Bi represents the ratio of heat transfer rates to the surrounding fluid and the solid interior.

- **Péclet number:**

$$\mathrm{Pe} = \frac{VL}{\gamma} = \mathrm{Re}\,\mathrm{Pr}\,; \tag{1.36}$$

Pe represents the ratio of heat convection and heat conduction.

- **Grashof number:**

$$\mathrm{Gr} = \frac{\alpha g L^3 \Delta T}{\nu^2}\,, \tag{1.37}$$

where α is the thermal expansion coefficient and ΔT is a temperature difference; Gr represents the ratio of buoyancy forces and viscous forces in flows with free thermal convection.

- **Rayleigh number:**

$$\mathrm{Ra} = \frac{\alpha g L^3 \Delta T}{\nu\gamma} = \mathrm{Gr}\,\mathrm{Pr}\,; \tag{1.38}$$

Ra is used in flows with thermal convection.

- **Marangoni number:**

$$\mathrm{Ma} = \frac{\frac{\partial\sigma}{\partial c}\frac{\partial c}{\partial x}L^2}{\mu\gamma_c} \quad \text{or} \quad \mathrm{Ma} = \frac{\frac{\partial\sigma}{\partial T}\frac{\partial T}{\partial x}L^2}{\mu\gamma}\,; \tag{1.39}$$

the first definition of Ma applies to flows with convection induced by a surface tension gradient due to a concentration gradient, whereas the second definition of Ma applies to flows with convection induced by a surface tension gradient due to a temperature gradient.

- **Richardson number:**

$$\mathrm{Ri} = -\frac{g\rho/x_3}{\rho(V/x_3)^2} = -\frac{g x_3}{V^2} = -\mathrm{Fr}^{-\frac{1}{2}}\,, \tag{1.40}$$

where x_3 is a vertical upwards axis; Ri represents the ratio of potential energy and kinetic energy in stratified flows.

- **Taylor number:**

$$\mathrm{Ta} = \frac{\Omega^2 L^4}{\nu^2}\,, \tag{1.41}$$

where Ω is the rotation rate; Ta applies to rotating flows.

- **Rossby number:**

$$\mathrm{Ro} = \frac{V}{\Omega L}\,; \tag{1.42}$$

Ro represents the ratio of inertia and Coriolis forces and applies to rotating flows.

- **Strouhal number:**

$$\mathrm{St} = \frac{fL}{V}\,, \tag{1.43}$$

where f is the frequency of a periodic phenomenon; St applies to periodic vortex shedding from bluff objects and other periodic phenomena.

- **Womersley number:**

$$\mathrm{Wo}\ (\text{or}\ \alpha) = \frac{d}{2}\sqrt{\frac{\omega\rho}{\mu}}, \tag{1.44}$$

where $\omega = 2\pi f$ is the angular frequency of a periodic flow in a tube with a diameter d; Wo is a measure of the ratio of unsteady 'inertia' forces and viscous forces, namely, a kind of unsteady Reynolds number.

- **Dean number:**

$$\mathrm{De} = \mathrm{Re}\sqrt{\frac{r_{tube}}{r_{curv}}}, \tag{1.45}$$

where r_{tube} is the radius of a curved tube and r_{curv} is the radius of curvature of the tube axis; De represents the relative strength of centripetal forces, which, under certain conditions, generate secondary flows.

- **Knudsen number:**

$$\mathrm{Kn} = \frac{\lambda}{L}, \tag{1.46}$$

where λ is the mean free path of molecules in a gas.

A large number of additional dimensionless parameters have been identified in specialised areas [7].

Example 1.6 A fire engine carries a long telescopic ladder, which may be extended vertically to reach an upper floor of a building (see Fig. 1.7). The average frontal width of the ladder is $L_l = 0.5$ m and its span is $b_l = 10$ m. We need to estimate the drag force F_{Dl} on the ladder and the vortex shedding frequency f_l from it, when it is operated during a storm with a wind speed of 80 km/hr. Because of the unusual shape of the ladder, we cannot rely on available information for circular cylinders, square beams and other simple objects. We have, however, available a water tunnel facility, which has a square test section with a side length $w = 0.50$ m and a maximum speed of 0.50 m/s and which is equipped with means to perform flow visualisation and local velocity measurements.

Answer

First, we need to design a model for use in the water tunnel. The model should obviously be geometrically similar to the ladder, but we still need to determine its size. The area of the test section occupied by the model must be small enough for test section wall effects not to be significant. If we assume that the maximum allowable blockage ratio is 5%, we find that the maximum acceptable frontal width of the model is $L_{m,\max} = 0.05 \times 0.5$ m $= 0.025$ m. The span of this model for geometric similarity with the ladder is found to be 0.5 m, which just fits in the water tunnel test section. To reduce wall effects on the flow around the free end of the model, we shall make the model a little shorter; let's choose $b_m = 0.45$ m, in which case the model width will be chosen as $L_m = 0.0225$ m (see Fig. 1.7). It is important to note that, for the present aspect ratio of $b_l/L_l = 20$, the drag coefficient on the ladder would likely be significantly lower than the 2D value and so a

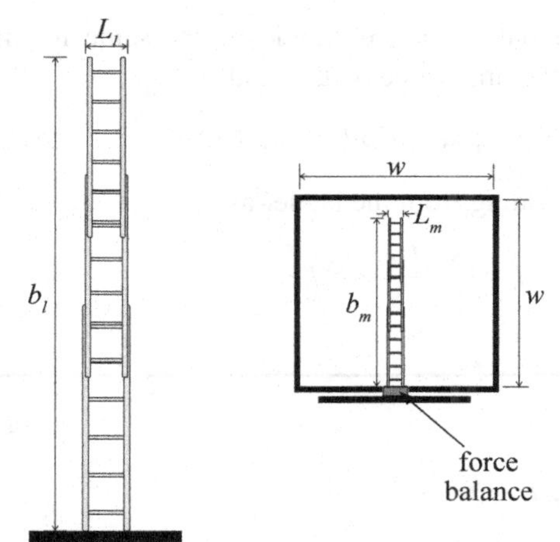

Figure 1.7 Sketches of the ladder (left) and its model inside the test section (right); not to scale.

model with a smaller aspect ratio would produce a scaling error. For dynamic similarity, we need to match the Reynolds numbers of the ladder and the model. From tables, we find the kinematic viscosities of air and water at room temperature to be, respectively, $\nu_{air} \approx 15 \times 10^{-6}$ m^2/s and $\nu_{wat} \approx 1 \times 10^{-6}$ m^2/s, and the corresponding densities to be $\rho_{air} = 1.225$ kg/m^3 and $\rho_{wat} = 1{,}000$ kg/m^3. The speeds of operation are, respectively, $U_l = 80$ km/hr $= 22.22$ m/s and $U_m = U_{mmax} = 0.5$ m/s. The ladder Reynolds number is $\text{Re}_l = U_l L_l/\nu_{air} \approx 741{,}000$. The model Reynolds number is $\text{Re}_m = U_m L_m/\nu_{wat} = 11{,}250$. These results mean that we cannot match the two Reynolds numbers. It is well known, however, that flow patterns behind bluff objects with sharp edges (i.e., fixed separation points) are insensitive to Re for Re greater than a few thousand. Therefore, even with this difference of conditions, we may expect dynamic similarity to occur fairly well, which means that the ladder and model drag coefficients and Strouhal numbers would be approximately matched.

So, we proceed with fabricating the model and conducting water tunnel tests. We use the momentum loss method (Example 1.2) to determine the drag force on the model and analysis of videos of vortex shedding visualised by dye injection to determine the frequency of vortex shedding. It is expected that the flow would not be two-dimensional and that the momentum loss and the vortex shedding frequency would not be uniform along the model. We may, however, approximate the average values of these properties by those measured at mid-height of the model. These measurements gave the model drag as $F_{Dm} = 2.02$ N and the model shedding frequency as $f_m = 3.3$ Hz. From these, we may calculate the drag coefficient and the Strouhal number for the model as, respectively,

$$C_{Dm} = \frac{F_{Dm}}{\frac{1}{2}\rho_{wat} L_m b_m U_m^2} = 1.6$$

and

$$\text{St}_m = \frac{f_m L_m}{U_m} = 0.15\,.$$

The values of these dimensionless groups should be the same for the fire engine ladder. Then we can calculate the drag force on the ladder as

$$F_{Dl} = C_{Dm}\tfrac{1}{2}\rho_{air}L_l b_l U_l^2 = 2,420 \text{ N}$$

and the frequency of vortex shedding from the ladder as

$$f_l = \frac{\text{St}_m U_l}{L_l} = 6.7 \text{ Hz} .$$

1.7 Patterns of Fluid Motion

Many flow visualisation and measurement methods provide images or other records of the fluid or of transported admixtures, which contain information on the fluid motion or the variation of some flow property. For a correct interpretation of the observed patterns, it is necessary to understand their relationship to the actual flow characteristics. Among the simplest types of visual patterns that can be obtained are the following:

- The *instantaneous position* of a visible fluid particle or marker as, for example, provided by a still camera with a short exposure time. To illustrate this concept, Fig. 1.8a shows a small particle source, which releases, or marks, fluid particles at consecutive time instants, separated by a small time increment δt. The figure shows representative instantaneous positions of three particles released in an unsteady flow at three different times t_0 (dark grey circles), $t_0 + \delta t$ (light grey circles), and $t_0 + 2\delta t$ (open circles). In steady flows, all positions of all particles originating from the same source would coincide with those of other particles at different times. Even in unsteady flows, it is possible for two or more particles originating from the same source to have coinciding instantaneous positions at different times.
- A *pathline*, namely the locus of positions of a fluid particle during a time interval; this would, for example, be recorded by a still camera using a relatively long exposure time; short pathlines provide a measure of the local velocity direction and magnitude. Examples of pathlines, which pass through the instantaneous positions of three particles released from a 'point' source, are shown in Fig. 1.8a.
- A *streakline*, namely the locus of all fluid particles which have passed through a fixed position in the fluid during some time interval; this can be achieved by introducing continuously fluid markers at some point and taking a short-exposure photograph of the flow. Representative images of a streakline in an unsteady flow, recorded at different instants, are shown in Fig. 1.8b. In a steady flow, all such images would coincide.
- A *timeline*, namely the locus, at a given time, of all fluid particles which formed a continuous line in the fluid at some previous time; this can be achieved by introducing the markers through a line source, which is active over short time intervals, separated by time intervals of inactivity. Examples of timelines are shown in Fig. 1.8c, in which

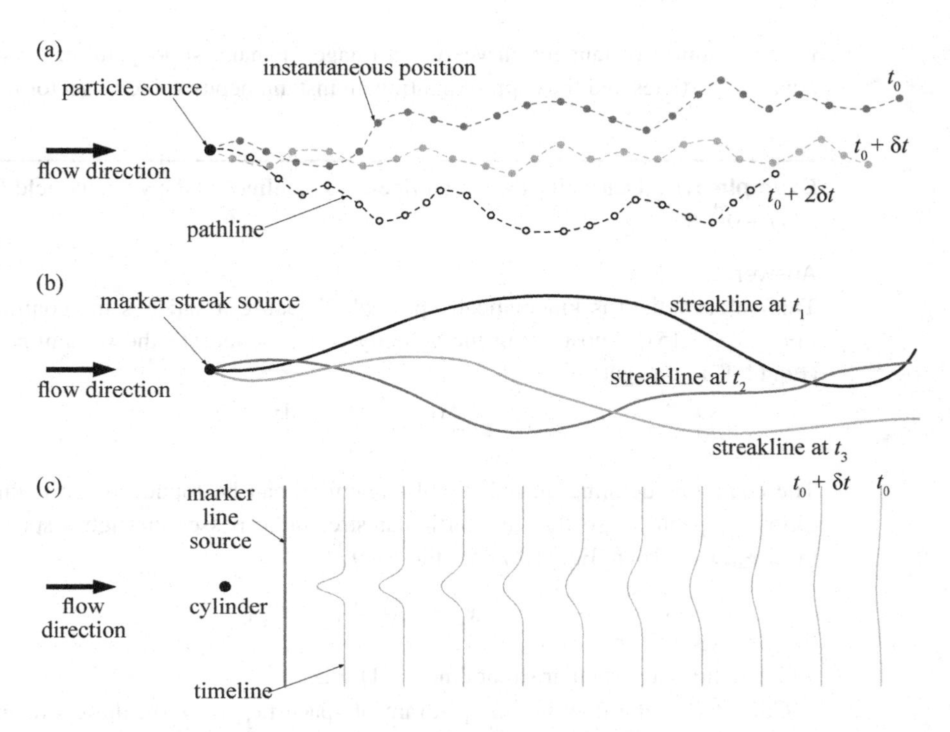

Figure 1.8 (a) Representative instantaneous positions and pathlines of three particles released at different times from a point source in an unsteady flow; (b) three representative snapshots of a streakline of markers released from a point source in an unsteady flow; (c) a snapshot of timelines of markers released at different previous instants from a line source past a cylinder.

a marker line source (e.g., a smoke wire in an air flow or a wire producing hydrogen bubbles in water flow) has been positioned downstream of a cylinder with its axis perpendicular to the cylinder axis. These timelines illustrate clearly the development of the cylinder wake.

- Combinations of the previous patterns, for example, time-streaklines, or photographic exposures of markers over intermediate time intervals.

It is essential to understand that all previously mentioned patterns are distinct, in general, from the instantaneous *streamlines*, namely lines in the fluid, which are tangent at all their points to the local velocity vector. In a Cartesian coordinate system, a streamline is defined by the following pair of equations:

$$\frac{dx_1}{U_1} = \frac{dx_2}{U_2} = \frac{dx_3}{U_3} \,. \tag{1.47}$$

In steady flows, pathlines, streaklines and streamlines coincide, whereas, in unsteady flows, they may be vastly different. In stationary turbulent flows of relatively low intensity, these lines may approximately coincide on the average, but their instantaneous features are distinct. A set of streamlines in unsteady flows may be reconstructed by fitting a family of tangent curves to an image of many short pathlines generated by adjacent particles and thus approximating an instantaneous velocity vector map.

Example 1.7 Determine the streamlines and pathlines of the velocity field $\vec{U} = ax_1\vec{i} - ax_2\vec{j} + 0\vec{k}$.

Answer

This velocity field is kinematically possible, because it satisfies the continuity equation (Eq. (1.15)). Substituting the velocity components into the streamline equations (Eq. (1.47)), one gets

$$\frac{dx_1}{ax_1} = -\frac{dx_2}{ax_2} = \frac{dx_3}{0}.$$

The equations defining an individual streamline can be found once a boundary condition is specified, so that the particular streamline passes through a specified point (x_{10}, x_{20}, x_{30}). Then, Eq. (1.47) has the solution

$$x_2 = \frac{x_{10}x_{20}}{x_1}, \quad x_3 = x_{30}$$

which defines a particular streamline in 3D space.

Considering the flow in one quadrant of space $(x_1, x_2 > 0)$, these streamlines correspond to flow in a corner, which is illustrated in Fig. 1.9a. One may note that the streamlines in this particular problem are independent of the coefficient a and so they are time-independent, regardless of whether this coefficient depends on time or not.

If the flow were steady (a = const.), pathlines would coincide with streamlines. If, however, a were time dependent, the pathlines would also be time dependent and their coordinates could be found by integrating the velocity components over time in a La-

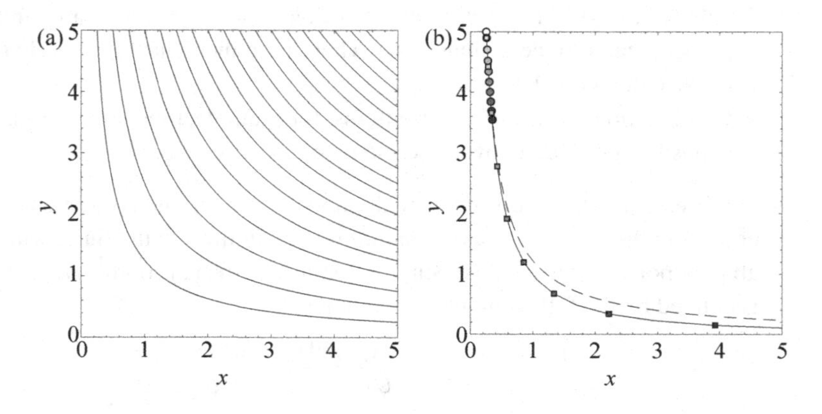

Figure 1.9 (a) A set of streamlines for the flow discussed in Example 1.7; (b) the streamline passing through the open circle (dashed line) and the pathline passing through the same point at the origin of time (solid line) for the case with $a = t/\tau$, $\tau = 0.02$ s. Consecutive particle positions along the pathline are spaced by $\delta t = 0.04$ s.

grangian frame of reference to compute the particle positions at different times. Consider, for example, a case with $a = t/\tau$, where τ is a constant with a dimension of time. Then, the pathline that passes through a particular point would be different from the streamline that passes through the same point, as shown in Fig. 1.9b. In this example, the position of the particle at each time instant $t_i = t_0 + i\delta t, i = 1, 2, ...$ is found from its previous position at time instant t_{i-1} as

$$X_{1,i} \approx X_{1,i-1} + V_{1,i-1}\delta t = X_{1,i-1} + \tfrac{t_{i-1}}{\tau}X_{1,i-1}\delta t = X_{1,i-1}(1 + \tfrac{t_{i-1}}{\tau}\delta t)\,,$$
$$X_{2,i} \approx X_{2,i-1} + V_{2,i-1}\delta t = X_{2,i-1} + \tfrac{t_{i-1}}{\tau}X_{2,i-1}\delta t = X_{2,i-1}(1 - \tfrac{t_{i-1}}{\tau}\delta t)\,,$$
$$X_{3,i} = X_{30}\,.$$

Chapter Digest

1.1 Explain the differences between a fluid and a solid and the differences between a liquid and a gas.

1.2 Discuss the different types of forces.

1.3 Define stresses and pressure; under which conditions are all normal stresses equal? When is pressure equal, in magnitude, to a particular stress and not the average normal stress?

1.4 What is the continuum hypothesis, why do we make it and under which conditions is it sufficiently accurate?

1.5 Discuss, from a measurement perspective, how the kinematic properties discussed in Section 1.2 are related to each other. Which properties would you need to measure directly in order to determine each of the listed kinematic properties?

1.6 Distinguish between the Eulerian and Lagrangian frames of reference. Which frame of reference applies to (a) the velocity field around an aircraft model mounted on a fixed sting in a water tunnel and measured point-by point by a velocity probe, sequentially traversed to different positions, and (b) the velocity measured by a probe that is attached to a model that is towed through the water tunnel test section?

1.7 Why is, in most cases, the control volume formulation of fluid motion preferable to the closed system formulation?

1.8 For which types of problems would the differential formulation be preferable to the control volume formulation? How are these two formulations related to each other?

1.9 Express the components of the rate of strain tensor in cylindrical coordinates.

1.10 Write the van der Waals equation of state for a gas and compare it briefly to the ideal gas law.

1.11 Define a viscoelastic material and give an example of a constitutive equation for such a material.

1.12 At first glance, one might think that potential flow analysis has limited practical use, yet it is widely used in aerodynamics, in particular for determining the lift on aerofoils [1]. Identify the crucial assumption made (Kutta condition) and discuss why this approach works.

1.13 Explain whether Bernoulli's equation may be used (a) in air flow entering the engine of a cruising jet airplane, (b) in flow around swimming bacteria and (c) in the gas plume rising from a smokestack.

1.14 Which are the SI units of pressure, energy and power and how are they related to base units?

1.15 Occasionally one has to work in the British gravitational system, particularly when it comes to pressure and lengths. Make sure you can easily convert from one to the other, for example, convert a length measured in feet and inches to millimeters (1 in = 25.4 mm and 1 ft = 12 in).

1.16 Why are the radian (rad) and the steradian (sr) not classified as base SI units?

1.17 How is the surface pressure coefficient around a circular cylinder in uniform flow related to its drag coefficient? What are the assumptions involved? Is this approach appropriate when the Reynolds number is (a) 1, (b) 10^4 or (c) 10^6?

1.18 How do the drag coefficients on a circular cylinder and a sphere vary with Reynolds number and what is responsible for visible qualitative changes of this dependence?

1.19 How do the the lift and drag coefficients on an aerofoil change with angle of attack?

1.20 From tables, find the values of the Prandtl number for (a) air at standard temperature and pressure; (b) water at standard atmospheric pressure and a temperature of (i) 4°C, (ii) 20°C and (iii) 80°C; and (c) mercury at room conditions. Comment on these numbers.

1.21 Give typical values of the Strouhal number for vortex shedding from (i) circular and (ii) square cylinders at a Reynolds number of 5×10^4.

1.22 A common misunderstanding of students is that a flow with a large Mach number would also have a large Reynolds number. To illustrate the fallacy of this intuitive presumption, estimate roughly and discuss the Mach and Reynolds numbers of (a) a cruising blue whale (*Balaenoptera musculus*) and (b) cosmic dust entering the upper layers of the earth's atmosphere.

1.23 What is a streamline and what is a pathline? If you have a complete set of all fluid particle positions at all times in a time dependent flow, would you be able to determine the corresponding streamlines? How would you do this?

Problems

1.1 A spacecraft orbiting the moon returns to the earth, reentering its atmosphere and eventually landing on its surface. Describe the applicability of the continuum hypothesis to the motion of air in the proximity of the spacecraft during this process.

1.2 Helium-filled balloons are released carrying instruments measuring temperature, humidity and other properties of the air and a transmitter to transmit this information to the base. Explain whether the information collected by these balloons would be in an Eulerian frame of reference, a Lagrangian one or a different type.

1.3 Assume that gas bubbles rise very slowly in a glass of beer. Explain how you would estimate the pressure inside the bubbles by measuring their size.

1.4 Consider Couette flow, in which the velocity vector is parallel to the x_1 axis and its magnitude changes linearly along the x_2 axis. Determine the components of the rate of strain tensor and the vorticity vector and discuss them comparatively.

1.5 The wings of butterflies are covered with scales, the height of which is typically 0.1 mm. Based on simple analysis, it is estimated that the relative air velocity just over the scales is about 15 mm/s. For a visual/optical study of the aerodynamic characteristics of these scales, it is advisable to use a large-scale model, having scales at least 10 mm high. Is it possible to do these tests in a wind-tunnel? A water tunnel? If so, describe and discuss the conditions and possible problems arising. Can you describe alternative types of experiment that would be more suitable for this research? Hints: consider using different fluids and/or tow the model.

1.6 In order to scale pumps, we need to match appropriate dimensionless groups. First, we need to define length, time and mass scales. Determine such scales in terms of the diameter D of the pump impeller, the angular velocity ω of the impeller and the fluid density ρ. Other properties of the pump that require to be scaled include the volume flow rate Q through the device, the viscosity μ of the fluid, the head of the pump $H = \Delta p/\rho g$ (Δp is the pressure rise across the pump; note that, although it has a dimension of length, H represents energy per unit weight and must be scaled accordingly) and the external power $\mathcal{P}$ (brake horse power) used to drive the pump impeller. Derive appropriate dimensionless groups for all these properties. Also define an efficiency η for the pump.

1.7 Consider a uniform stream past a 2D plate that is normal to it. Suddenly, the plate starts rotating about its axis of symmetry. Sketch simultaneous pathlines and streamlines at different angles of rotation. Would pathlines coincide with streamlines?

1.8 A very large flat plate is immersed in a fluid and oscillates periodically in the x_1 direction with a velocity $V_o \cos \omega t$, while remaining parallel to the x_1–x_2 plane. An exact solution of the system of the continuity equation and the Navier–Stokes equation in the x_1 direction is available for this problem. The wall motion induces a periodic motion of the surrounding fluid in a direction that is also parallel to the x_1 axis and having a velocity magnitude

$$V(x_3, t) = V_o e^{-x_3/\delta} \cos(\omega t - x_3/\delta) ,$$

where x_3 is the direction normal to the plate and $\delta = \sqrt{2\nu/\omega}$ is the *Stokes length.* Sketch typical streamlines for this flow. Then, sketch pathlines for particles at distances δ, 2δ and 5δ from the plate. Further sketch the time evolution of timelines generated by a stationary wire stretched in the x_3 direction, while neglecting gravitational effects on the markers. Explain how you can use this method to measure the viscosity of the fluid.

References

[1] J.D. Anderson. *Fundamentals of Aerodynamics (6th Edition).* McGraw-Hill, New York, 2017.

[2] G.K. Batchelor. *An Introduction to Fluid Dynamics.* Cambridge University Press, Cambridge, UK, 1970.

[3] T.L. Bergman, A.S. Lavine, F.P. Incropera, and D.P. DeWitt. *Fundamentals of Heat and Mass Transfer (8th Edition)*. Wiley., New York, 2017.

[4] R.W. Fox, A.T. McDonald, and J.W. Mitchell. *Fox and McDonald's Introduction to Fluid Mechanics (10th Edition)*. Wiley, New York, 2019.

[5] D.R. Lide. *CRC Handbook of Chemistry and Physics (78th Edition)*. The Chemical Rubber Co., Cleveland, OH, 1997.

[6] M.J. Moran, H.N. Shapiro, D.D. Boettner, and M.B. Bailey. *Fundamentals of Engineering Thermodynamics, 9th Edition*. Wiley, New York, 2018.

[7] S.P. Parker. *Fluid Mechanics Source Book*. McGraw Hill, New York, 1987.

[8] L. Rosenhead. *Laminar Boundary Layers*. Oxford University Press, Oxford, 1963.

[9] R.E. Sonntag, C. Borgnakke, and G.J. Van Wylen. *Fundamentals of Themodynamics (5th Edition)*. Wiley, New York, 1998.

[10] F.M. White. *Viscous Fluid Flow (3rd Edition)*. McGraw-Hill, New York, 2005.

2 Measuring Systems

This chapter contains a review of definitions and concepts that apply to all types of measuring systems. In addition, given the prevalence of electronic components used within the measurement process, this chapter also provides a review of electronic components and basic circuit analysis. This is meant to assist in the understanding of the generic operations and steps that are involved in any measurement process. Some general discussion of the static and dynamic responses of common types of measuring systems will also be provided. More thorough coverage of these topics may be found in texts dealing with measuring systems [1, 4, 7].

2.1 Fundamentals of Measuring Systems

A fluid mechanics experiment is conducted by one or more experimenters, who measure properties of the flow. Thus, in any fluid mechanics experiment one may identify three essential and distinct systems:

- The *physical system*, consisting of one or more flowing fluids, the flow-producing apparatus and, possibly, test models and other objects that are part of the experiment. It is the properties of the physical system, referred to as the *measurands* that we wish to measure using a measuring system.
- The *measuring system*, consisting of a number of interconnected components, including sensors, electric and electronic circuits, data acquisition and processing devices and software.
- The *experimenters*, who plan, execute and interpret the measurements.

It is obvious that, for an accurate measurement, all three systems must be functional and compatible with each other.

2.1.1 Inputs and Outputs

A measuring system, and each of its components, have one or more *inputs* and one or more *outputs* (Fig. 2.1). The output of each component may represent an input to another component. Each input or output corresponds to a physical property, as, for example, a displacement or an electric voltage, commonly referred to as a *signal*. The relationship

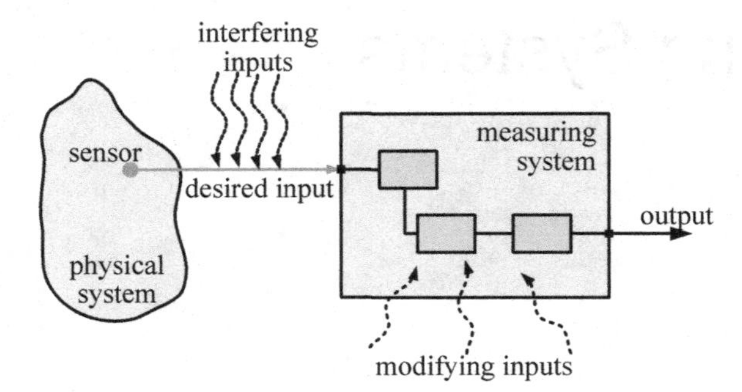

Figure 2.1 Inputs and outputs of a measuring system.

between the value of an input and the value of an output is called the *response* of that component to the particular input.

The flow properties that the measuring system is intended to respond to and measure are the *desired* inputs. In addition, a measuring system is also subjected to *undesirable* inputs, which can be further classified into *interfering* and *modifying* inputs. An interfering input is a property to which the system is unintentionally sensitive, whereas a modifying input is a property that modifies the response to a particular desired input. Undesirable inputs may introduce a *measurement error*, defined as the difference between the measured and the true value of the measurand (see Section 6.1.1). Undesirable inputs may even produce an output in the absence of desired inputs.

Example 2.1 A velocity sensor (e.g., a hot-wire anemometer) is used to measure the local air velocity in a jet issuing through a nozzle into the laboratory. A sketch of the experimental setting is shown in Fig. 2.2. The jet is exposed to a cross-draft, generated by air discharged through a ventilation outlet on the ceiling and the room temperature is controlled in the range between 18°C and 22°C by a thermostat. Identify and categorise the different inputs in this experiment.

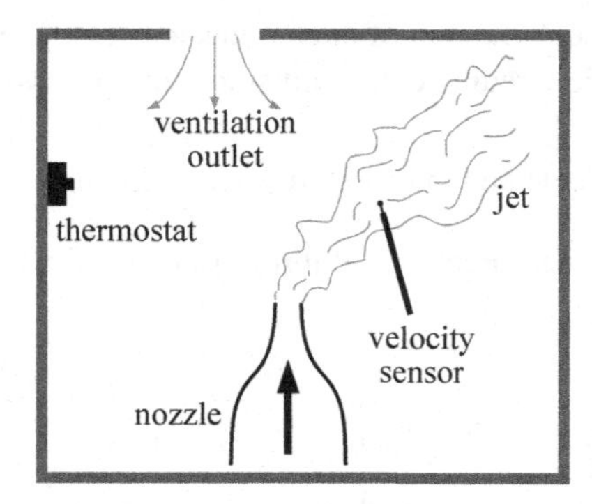

Figure 2.2 Interfering and modifying inputs for a jet velocity measurement.

Answer

The desired input is obviously the jet flow velocity. The draft of air produced by the ventilation system acts as an interfering input, because it distorts the jet's velocity field. A change in room temperature, due to the same ventilation system, might also have an interfering effect, as it may affect the operation of the jet. However, the main undesirable effect of the room temperature change is that it modifies the response of the electronic circuit that powers the sensor, as well as causing deviations from the sensor's calibrated response. Thus, the room temperature change acts as both a modifying and an interfering input to the velocity measurement.

2.1.2 Distortion, Loading and Cross-talk

The physical presence or the operation of instruments often introduce undesirable influences to the physical system or to the measurement process. Among the most common adverse influences are *flow distortion*, *loading* and *instrument cross-talk.*

Any instrument inserted in a physical system would cause a certain amount of distortion of measurands, as the result of flow blockage, streamline displacement, vortex shedding, instability, phase change, transition to turbulence, change of turbulence structure, shock waves or other distorting phenomena. Such effects may be local or distributed over some volume of the fluid, in which case they would depend on the local flow conditions and would be difficult to correct for. Measuring devices, which need to be introduced into the physical system in order to measure a measurand, are called *intrusive*. For example, the velocity sensor in Example 2.1, which has to be placed within the jet in order to measure the velocity of the jet, is an intrusive device, as the flow velocity in the vicinity of the sensor would be different from that in the absence of the sensor. This sensor is actually a modifying input to the velocity measurement. Measurement methods that do not require the use of intrusive devices are called *non-intrusive.* Most optical measurement methods are considered to be non-intrusive. When one must necessarily use an intrusive device, one should ensure that the design of this device and its insertion mechanism is such as to minimise distortion of measurands.

Loading of a physical system occurs when a measuring component extracts significant power from it, thus altering the values of measurands. A common form of loading occurs in electric and electronic circuits. For example, the connection of an ammeter in series with some electric circuit would change the measured current, unless the ammeter resistance its negligible compared to the resistance of the circuit. Another example of loading occurrence is the measurement of a voltage difference between two nodes in an electric circuit by connecting a voltmeter to these nodes with probes. In this case, it is necessary for the *impedance* of the voltmeter–probe combination to be much larger than the impedance of the electric circuit across these nodes. If this is not the case, a significant current will flow through the probes and the voltmeter and the measured voltage will be incorrect. Methods for analysing electric circuits and determining their impedance are discussed in Section 2.2. As an example of loading of a mechanical system, we can mention the use of a turbine flow meter to measure flow rate through a pipe, which may introduce sufficient resistance to the flow, thus reducing the flow rate. Although, in some cases, it may be possible to compensate analytically for the effects of loading, it is obviously preferable to avoid or minimise such effects by proper design or choice of instrumentation.

Instrument cross-talk occurs when a measuring component may be coupled in its operation with another measuring component, such that the output of one may act as an undesirable input to the other. Electric and electronic circuits are particularly prone to cross-talk, as one influences another through the generation of electric currents or electromagnetic fields. Instrumentation in fluid systems may also be subjected to thermal cross-talk, as for example in the mutual interference of the wakes of closely positioned heated thermal anemometers, and to mechanical cross-talk, commonly in the form of coupled vibrations. To ensure accurate measurement, any significant cross-talk must be identified and removed.

2.1.3 Reducing Undesirable Effects

Undesirable inputs cause measurement errors and may produce an output even in the absence of desired inputs. For accurate measurement, the experimenter must identify all undesirable inputs and estimate their effects, which, if found significant, must be accounted for. It goes without saying that the preferred approach would be to redesign a component that is sensitive to undesirable inputs in a way that it becomes insensitive to them.

Example 2.2 A mechanical flow meter has been calibrated to measure flow rate from the deformation of a certain component. An error is produced when the flow meter is exposed to changing temperature due to deformation caused by thermal expansion of this component. Is there a simple way to reduce or eliminate this error?

Answer

Use a material that has a very small thermal expansion coefficient within the range of interest to reduce drastically the temperature sensitivity of this component. For example, the nickel-iron alloy Invar has a linear coefficient of thermal expansion that is equal to 1.2×10^{-6} K^{-1}, which is roughly one tenth of the value for steel and one twentieth of the value for aluminium.

An effective way to reduce undesirable effects is the use of *compensation*. By this term, we understand the deliberate introduction of additional interfering or modifying inputs, which may partly or entirely cancel the original undesirable effects.

Example 2.3 Music produced by an orchestra has tones spanning a wide range of frequencies. When replayed by speakers and headphones, this music tends to miss the strength of high pitch (treble) and low pitch (bass) tones, as a result of speaker limitations. How can one restore the tone balance of replayed music to its original state?

Answer

Many audio amplifiers have been designed as to permit the user to amplify treble and/or bass tones with a higher gain than middle-range tones. This compensates for the speaker-related attenuation of tones in the extreme frequency ranges.

Undesirable effects and measurement errors, if known, can be removed by applying an analytical correction to the output of an experiment. This approach clearly requires knowledge of both the values of undesirable inputs and the dependence of the measuring system response to such inputs. Examples of correction methods specific to particular measuring techniques will be given in following chapters.

2.1.4 Filtering

In many experiments, the values of measurands are fluctuating in time. A time-dependent function may be transformed, for example, through a Fourier series, Fourier transform, Laplace transform or wavelet transform, to an equivalent function that depends on frequency (see Chapter 5). Thus, one may equivalently analyse the characteristics of a time-dependent property in either the *time domain* or the *frequency domain*. Any operation on the property in the time domain will also affect its equivalent property in the frequency domain and vice versa.

Undesirable inputs may introduce time-dependent disturbances, which could be concentrated at one or more discrete frequencies (e.g., vibration of a wall) or extend over a wide range of frequencies (e.g., electronic noise). A common way to reduce or eliminate the effects of such undesirable inputs within a range of frequencies is the use of *filters*. A filter may be applied directly to an input or an output of the measuring system or one of its components.

In terms of physical operation, filters may be of several different types:

- **Electrical/electronic filters:** These are applied to electric signals (see Section 3.2.2).
- **Mechanical filters:** These are designed to filter motion or forces. An example is the use of shock absorbers to reduce vibrations of an apparatus.
- **Thermal filters:** These are designed to reduce or eliminate temperature fluctuations. The simplest thermal filter is thermal insulation.
- **Electromagnetic filters:** These are designed to remove the interfering effects of electric and magnetic fields. The containment of instrumentation within a grounded metallic shield is an effective method of such filtering.

2.1.5 Modes and Functions of Measuring System Components

Each component of a measuring system may operate in an analogue, discrete (digital, binary, etc.) or hybrid mode. It may be *passive*, if the energy necessary for producing its output is supplied by the input, or *active*, if this energy is supplied mainly by an external excitation source.

In order to understand the detailed operation of a measuring system, it is instructional to identify the function of each component. The main functions that take place during measurement are usually the following:

- **Sensing:** This is the first step in the measuring process. The measurand excites a sensor, which produces an output, which could be a mechanical property (e.g., a displacement or rotation angle), a material property (e.g., an electric resistance), a voltage, or some other property.
- **Conversion:** This is the process by which the sensor output is converted into a form that is convenient for further processing. In most cases, the latter is a signal.
- **Conditioning:** This is the process by which the converted output is modified in order to ensure that its magnitude is suitable for data collection. Common types of conditioning are amplification and filtering, which may be applied to a signal, a displacement or other type of output.
- **Transmission:** This includes the step(s) of transferring signals or other information from one component to another.
- **Presentation and storage**: This is the final act of measurement, during which the output is displayed or stored. In most measurement systems that produce signals, the signal is discretised and recorded on storage devices.

One may note that each component of a measuring system may contribute to more than one of the above functions and that each function may be performed by more than one components.

Example 2.4 Identify the different measurement components of a liquid-in-glass thermometer.

Answer

Sensing: The liquid inside the bulb at the end of the thermometer senses a change in temperature of the surrounding fluid, which is the measurand.
Conversion: The volume of the liquid inside the thermometer will expand when its temperature increases; thus, a change in temperature is converted into a change in volume.
Conditioning: The height of the expanded volume of liquid is amplified by forcing the liquid trough a narrow tube, thus making it easier to see the volume change.
Presentation: The height of the liquid column is measured by comparison to lines drawn on the tube and labeled in units of temperature. The correspondence between liquid column height and temperature is established via a *calibration*. A classic thermometer calibration consisted of marking the column heights at the freezing and boiling temperatures of water and drawing equally spaced lines at equally spaced intermediate temperatures.

Example 2.5 Identify the different measurement components of the strain measuring system shown in Fig. 2.3. In this system, the measurand (desired input) is the elongation of the shaft under tension and the sensing element is a strain gauge.

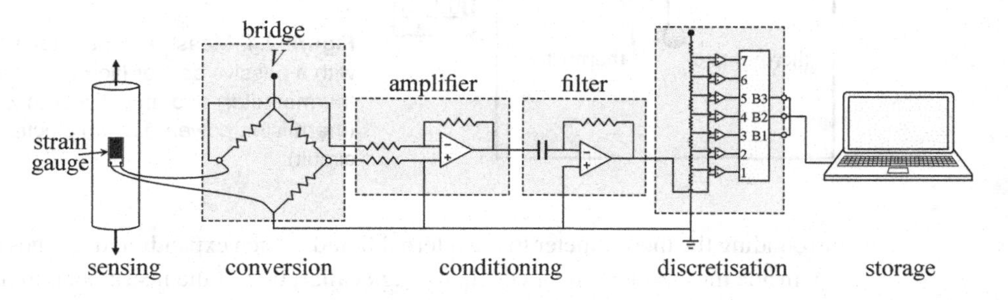

Figure 2.3 Functions of the components of a strain measuring system.

Answer

Sensing: The strain gauge is the sensor, whose electric resistance changes, when it gets elongated.

Conversion: The output of the sensor (electric resistance) is converted to an electric signal by a Wheatstone bridge, which contains the strain gauge in one of its arms. The bridge output is noisy and has a very small value, which makes it hard to measure.

Conditioning: A substantial amount of noise in the bridge signal is filtered out by a low-pass filter and the signal is amplified by an electronic amplifier.

Transmission: Cables are used to transmit the signal from one component to the next.

Presentation and storage: The conditioned signal is discretised using an analogue-to-digital converter, and transmitted to a computer, where it is further conditioned by software and stored on magnetic storage devices. The signal itself and other properties that are calculated from the signal can be displayed on the monitor or printed, either in real time from a live signal or at a later time from a stored signal.

Note that a distinction is made between a *sensor* and a *transducer*. A sensor is a device that responds to a change in the measurand to produce an output, which could be a signal or other property. A transducer is a measuring device that contains a sensor but also converts the sensor's output into an electric signal.

Example 2.6 The temperature of a liquid inside a container is measured using a liquid-in-glass thermometer and a thermistor, as illustrated in Fig. 2.4. Discuss whether these are active or passive sensors or transducers.

Answer

The liquid-in-glass thermometer is a passive sensor: heat is transferred from the liquid

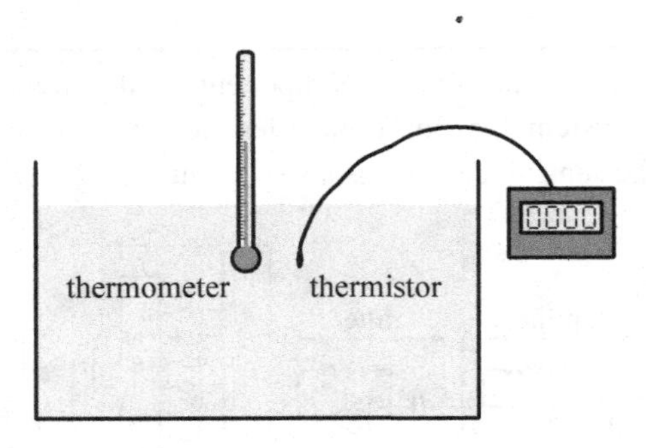

Figure 2.4 Measurement of temperature with a passive sensor (liquid-in-glass thermometer) and an active transducer (thermistor, powered by an electronic circuit).

surrounding the thermometer to the internal liquid, which expands and reaches an elevation inside the capillary tube, which, through calibration of the instrument, indicates the temperature. In contrast, the thermistor-based measuring system is an active transducer: the thermistor itself is a sensor, which is supplied with a current by an electronic circuit that produces a signal, the value of which depends on the thermistor's electric resistance and is converted to temperature with the use of a calibration relationship.

2.2 Fundamentals of Electric and Electronic Circuits

The usual distinction between the terms *electric* and *electronic* is that electric pertains to the flow of electric charges, whereas electronic pertains to the control of this flow.

2.2.1 Basic Properties

In the vast majority of cases, all aspects of the measurement process are executed using electronic components and electronic circuits. Before proceeding with the analysis of electric and electronic circuits, it seems advisable to review their basic properties [3, 11]. All properties of electric circuits and components are ultimately related to the *electric charge* of a single proton, considered as positive, or a single electron, equal in magnitude to that of a proton but considered as negative. The value of an elementary electric charge is 1.6×10^{-19} C (coulomb). The movement of an electric charge q along a path is called *electric current* i, defined as

$$i = \frac{\mathrm{d}q}{\mathrm{d}t}\,. \tag{2.1}$$

In electric circuits that include metallic conductors, the moving charges consist of electrons. The current is considered to be positive, when its conventional direction is opposite to that of electron movement. The unit of electric current is the ampere (A) and it is one of the base units of the SI system. The work, per unit charge, done on an electric charge moving from one position to another is called *voltage* V. The voltage unit is the volt (V). An ideal *current source* (Fig. 2.5a) is a power source that provides a current that is independent of the voltage across it. An ideal *voltage source* is a power source

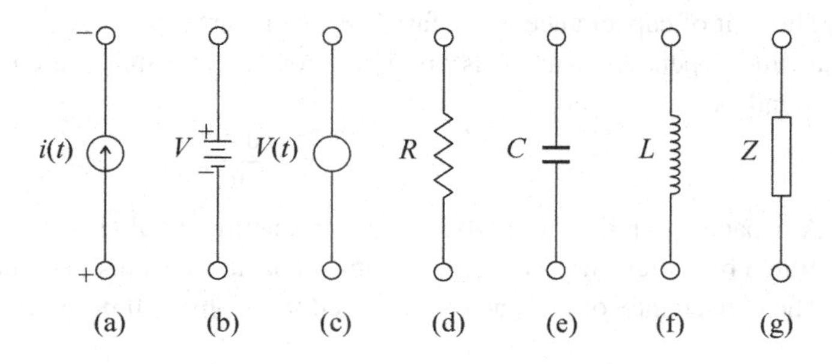

Figure 2.5 Elementary power sources and passive elements of linear electric circuits: (a) current source, (b) DC voltage source, (c) AC voltage source, (d) resistor, (e) capacitor, (f) inductor and (g) impedance element.

that provides a voltage that is independent of the current through it. Voltage sources could be either *DC* (direct current; Fig. 2.5b) or *AC* (alternating current; Fig. 2.5c).

Linear electric circuits have three common passive elements: the resistor, having a resistance R, the capacitor, having a capacitance C, and the inductor, having an inductance L. These elements are illustrated symbolically in Fig. 2.5d-f.

Resistor: A device through which the current is proportional to the voltage across it according to *Ohm's law*,

$$R = \frac{V}{i} \,. \tag{2.2}$$

The unit of resistance is the ohm (Ω). A (pure) resistor is assumed to have a lumped, constant resistance and no capacitance or inductance. The inverse of resistance, $G = 1/R$, is called *conductance*; its unit is the siemens (S). The resistance of any electrically conductive material (e.g., a metallic wire) is related to its physical properties via *Pouillet's law*,

$$R = \rho \frac{L}{A} \,, \tag{2.3}$$

where ρ is the *electric resistivity* (a measure of how resistive the material is to the flow of electric charges), L is the length of the material and A is its cross-sectional area. Metals, in general, have a low electric resistivity ($\rho \approx 10^{-8}$–10^{-7} Ω m), air has a very high electric resistivity ($\rho \approx 10^{14}$ Ω m) and silicon has an intermediate value ($\rho \approx 10^{2}$ Ω m). Resistors with fixed values are available as single devices or as arrays of multiple resistors with nearly equal resistances. A different type of resistor is the *potentiometer*, the resistance of which can be changed within a range by turning a knob or screw. When a current passes through a resistor, it consumes electric power, which is equal to Ri^2 and is converted irreversibly to heat in a process that is called *Joule heating*.

Capacitor: A device that can store an electric charge q when a voltage V is applied across it. Its *capacitance*, assumed to be pure, lumped and constant, is defined as

$$C = \frac{q}{V} \,. \tag{2.4}$$

The unit of capacitance is the farad (F); its inverse, $S = 1/C$, is called *elastance*. When a time-dependent voltage is applied across a capacitor, the current through it can be found as

$$i = C\frac{\mathrm{d}V}{\mathrm{d}t} . \tag{2.5}$$

A capacitor typically consists of two conducting parallel plates, called *electrodes*, separated by a dielectric material. In some capacitor designs, the plates are rolled together. The capacitance of a capacitor is related to its physical properties by the expression

$$C = \varepsilon_0 \varepsilon_r \frac{A}{d} , \tag{2.6}$$

where $\varepsilon_0 = 8.854 \times 10^{-12}$ F/m is the *dielectric constant*, ε_r is the *relative static permittivity*, whose value depends on the dielectric material between the two electrodes ($\varepsilon_r \approx 1$ for air and 11.68 for silicon), A is the area of each electrode and d is the separation distance between the electrodes.

Inductor: A device, typically a coiled metallic wire, that produces a *magnetic flux* Ψ, when a current passes through it. Its *self-inductance,* or simply *inductance* (assumed to be pure, lumped and constant), is defined as

$$L = \frac{\mathrm{d}\Psi}{\mathrm{d}i} . \tag{2.7}$$

The unit of inductance is the henry (H). *Faraday's law* connects the voltage to the magnetic flux as

$$V = \frac{\mathrm{d}\Psi}{\mathrm{d}t} , \tag{2.8}$$

which leads to the voltage–current relationship for an inductor as

$$V = L\frac{\mathrm{d}i}{\mathrm{d}t} . \tag{2.9}$$

When more than one inductors are connected in a circuit, such that their magnetic fields interfere with each other, their voltage–current relationships are coupled through a property M called *mutual inductance*; its unit is also H.

Impedance: Linear electric circuits may be conveniently analysed with the use of Laplace transform (Section 2.4.7). Let $\mathbf{V}(s)$ and $\mathbf{I}(s)$ be, respectively, the Laplace transforms of the voltage across and the current through a passive element. The *impedance* (Fig. 2.5g) of this element is defined as

$$Z(s) = \frac{\mathbf{V}(s)}{\mathbf{I}(s)} . \tag{2.10}$$

The impedance of an element or a passive electric circuit is, generally, complex, with the exception of purely resistive circuits, the impedance of which is real. The unit of impedance, like that of resistance, is the ohm. The inverse of impedance $Y(s) = 1/Z(s)$ is called the *admittance*. The impedances of resistors, capacitors and inductors are, respectively,

$$Z = R\,, \quad Z = 1/(sC) \quad \text{and}\, Z = sL\,. \tag{2.11}$$

Substituting these expressions into Eq. (2.10), one gets the voltage–current relationships for resistors, capacitors, and inductors in the complex domain, also referred to as the *Laplace s-domain*, as, respectively,

$$\mathbf{V}(s) = \mathbf{I}(s)R,\ \mathbf{V}(s) = \mathbf{I}(s)/Cs \text{ and } \mathbf{V}(s) = \mathbf{I}(s)Ls\,. \tag{2.12}$$

The impedances of elements connected in series add up, that is, $Z = Z_1 + Z_2$, whereas, when elements are connected in parallel, it is their admittances that add up, that is, $Y = Y_1 + Y_2$. In all previous expressions, the complex frequency s may be replaced by $j\omega$.

Example 2.7 Compute the impedance of the electric circuit shown in Fig. 2.6.

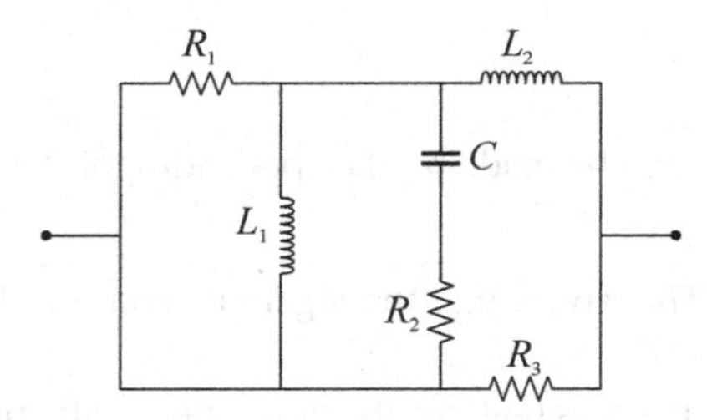

Figure 2.6 An electric circuit consisting of pure passive elements.

Answer

The impedances of the different elements in this circuit will be denoted as $Z_1 = R_1$, $Z_2 = sL_1$, $Z_3 = 1/(sC)$, $Z_4 = R_2$, $Z_5 = sL_2$ and $Z_6 = R_3$. To distinguish easily the in-series and parallel connections of the different elements, we redraw the circuit as shown in Fig. 2.7.

Now, it is straightforward to derive the impedance Z_t of the entire circuit in a step-by-step process, during which we sequentially find the impedance of elements enclosed

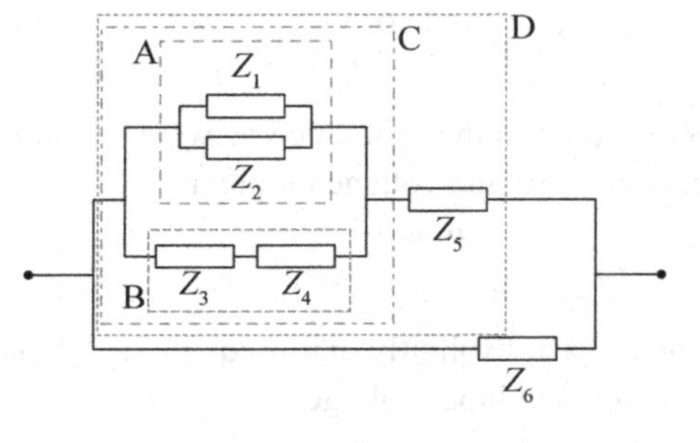

Figure 2.7 The same circuit as the one shown in Fig. 2.6, drawn in a way that makes obvious the in-series and parallel connections of different elements.

within the ellipses sketched in Fig. 2.7 as

$$\frac{1}{Z_A} = \frac{1}{Z_1} + \frac{1}{Z_2},$$

$$Z_B = Z_3 + Z_4,$$

$$\frac{1}{Z_C} = \frac{1}{Z_A} + \frac{1}{Z_B},$$

$$Z_D = Z_C + Z_5,$$

$$\frac{1}{Z_t} = \frac{1}{Z_D} + \frac{1}{Z_6}.$$

2.2.2 Principles of Circuit Analysis

An analysis of simple electric circuits can be made by the application of the following two principles:

- **Kirchhoff's current (node) law:** This states that the algebraic sum of all currents entering or leaving a node equals zero.
- **Kirchhoff's voltage (loop) law:** This states that the algebraic sum of all voltage rises and drops around any closed path (loop) equals zero.

It is easy to show that Kirchhoff's laws apply equally well on the real and the complex domains.

Example 2.8 Determine expressions for the voltage outputs of (a) a voltage divider (Fig. 2.8a) and (b) an RC circuit (Fig. 2.8b).

Answer

(a) **Voltage divider:** Kirchhoff's current law requires that the current across resistor R_1 must be equal to the current across resistor R_2, that is, $i_1 = i_2$. Using Ohm's law, we can obtain the voltage drop across both resistors as

$$\frac{V_i - V_o}{R_1} = \frac{V_o - 0}{R_2},$$

noting that the voltage across R_2 is equal to the difference between V_o and the ground. Re-arranging the previous expression gives the voltage output as

$$V_o = V_i \frac{R_2}{R_1 + R_2}. \tag{2.13}$$

We may observe that time does not appear explicitly in this equation and that the output voltage is adjusted instantly to a change in input voltage.

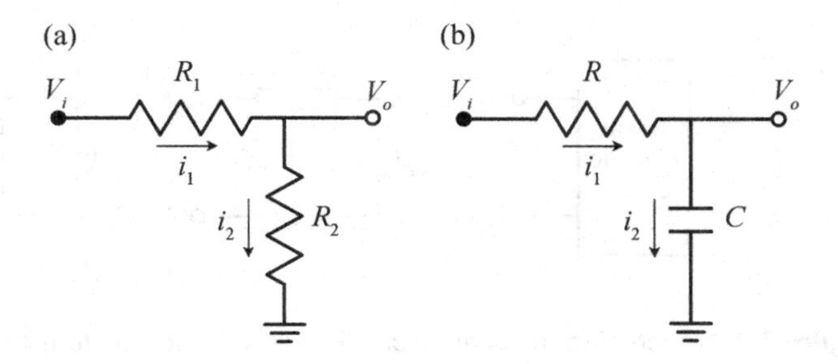

Figure 2.8 Two simple electric circuits: (a) a voltage divider, in which the output voltage is read across a resistor R_2, and (b) an RC circuit, in which the output voltage is read across a capacitor.

(b) **RC circuit:** As in (a), application of Kirchhoff's current law gives $i_1 = i_2$. The current across the resistor is given by Ohm's law, whereas the current across the capacitor is given by Eq. (2.5) as $i_2 = C\mathrm{d}V_o/\mathrm{d}t$. Hence, application of Kirchhoff's current law gives the expression

$$\frac{V_i - V_o}{R} = C\frac{\mathrm{d}V_o}{\mathrm{d}t} ,$$

which may be rearranged to the form

$$RC\frac{\mathrm{d}V_o}{\mathrm{d}t} + V_o = V_i .$$

We note that the output of a passive RC circuit is time-dependent and can be found by solving a first-order ordinary differential equation.

Any linear electric circuit, which contains an arbitrary number of power sources and passive elements, can be represented in a simple form as a single power source and a single impedance element. This can be accomplished in two different ways, as stated in the following two theorems [3, 6]:

- **Thévenin's theorem**, which states that any electric circuit having two free terminals A_1 and A_2 and comprising of current sources, voltage sources, resistors, capacitors and/or inductors (Fig. 2.9a) can be transformed into an equivalent circuit (Fig. 2.9b), consisting of an *equivalent voltage source* V_{oc} in series with an *equivalent impedance* Z_{eq}. The value V_{oc} is equal to the voltage across the terminals A_1 and A_2 in the original circuit and is called the *open-circuit voltage*. The equivalent impedance Z_{eq} can be found by short-circuiting all voltage sources and leaving open-circuited all current sources in the original circuit and then combining all passive components.
- **Norton's theorem**, which states that any electric circuit having two free terminals A_1 and A_2 and comprising of current sources, voltage sources, resistors, capacitors and/or inductors (Fig. 2.9a) can be transformed into an equivalent circuit (Fig. 2.9c), consisting of an *equivalent current source* i_{sc} in parallel with an *equivalent impedance*

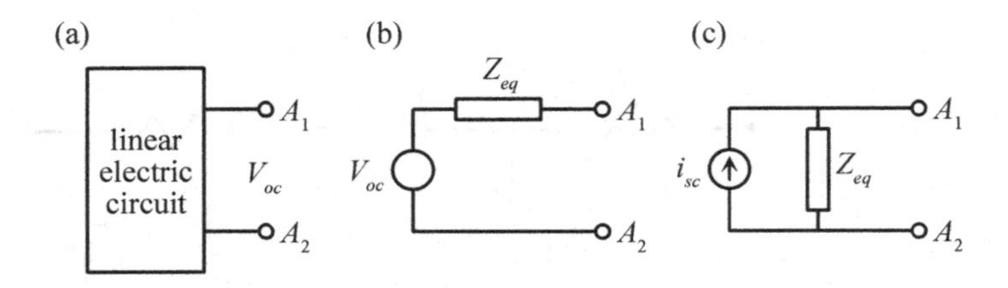

Figure 2.9 Sketch of (a) a linear electric circuit with two free terminals, (b) its Thévenin-equivalent circuit and (c) its Norton-equivalent circuit.

Z_{eq}. The value i_{sc} is equal to the current that would flow between the terminals A_1 and A_2 if they were short-circuited in the original circuit and is called the *short-circuit current.* The Norton-equivalent impedance is the same one as the Thévenin-equivalent impedance and it is also called the *output impedance* of the circuit as seen between the terminals A_1 and A_2. It is related to the Laplace transforms of V_{oc} and i_{sc} as

$$Z_{eq}(s) = \frac{\mathbf{V}_{oc}(s)}{\mathbf{I}_{sc}(s)} \,. \tag{2.14}$$

Example 2.9 Determine the Thévenin- and Norton-equivalent circuits of the voltage divider shown in Fig. 2.8a.

Answer

To help visualise the procedure, we have re-drawn the voltage divider circuit in Fig. 2.10a. The Thévenin-equivalent voltage V_{oc} is equal to the voltage output V_o, which was derived previously in Example 2.8 as

$$V_o = V_i \frac{R_2}{R_1 + R_2} \,.$$

To find the equivalent impedance, we short the power source and notice that the cir-

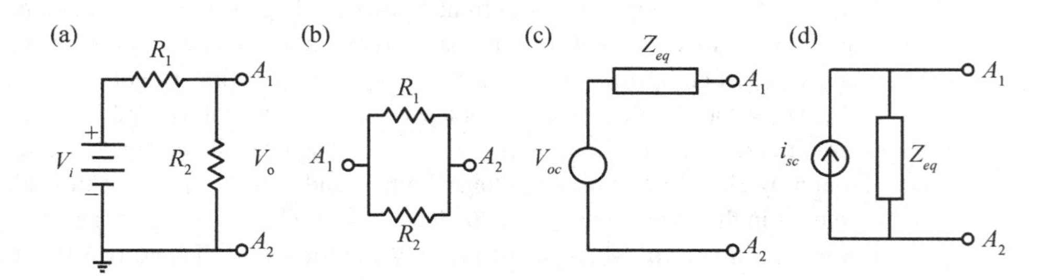

Figure 2.10 (a) A voltage divider; (b) the same circuit with its power source shorted; (c) the Thévenin-equivalent circuit; and (d) the Norton-equivalent circuit.

cuit consists of a pair of resistors connected in parallel (Fig. 2.10b). The Thévenin-equivalent impedance, which in this case is purely resistive, is, therefore,

$$Z_{eq} = \left(\frac{1}{R_1} + \frac{1}{R_2}\right)^{-1} = \frac{R_1 R_2}{R_1 + R_2} .$$

The short-circuit current for the Norton-equivalent circuit is found as the ratio of the Thévenin-equivalent voltage and the equivalent impedance, namely as

$$i_{sc} = \frac{V_{oc}}{Z_{eq}} = V_i \frac{R_2}{R_1 + R_2} \frac{R_1 + R_2}{R_1 R_2} = \frac{V_i}{R_1} .$$

The Thévenin- and Norton-equivalent circuits are shown in Fig. 2.10c and d, respectively.

2.2.3 The Wheatstone Bridge

The Wheatstone bridge is one of the most important electrical circuits in the context of measurements. Its original function was to determine an unknown resistance; however, from a measurement point of view, its function is to produce a voltage output when there is a change in the resistance of one of its arms, let's say that of the resistor R_1. As we shall discuss in following chapters, this resistor can be a sensor, whose resistance changes as a result of an applied strain, pressure, temperature, or other property. Towards this purpose, the sensor is connected in the bridge configuration with three other resistors whose resistances R_2, R_3 and R_4 are known and one of which, let's say R_2, is adjustable. The ratio k_B of the resistances on each *branch of the bridge* is known as the *bridge ratio*; the bridge ratios of the two branches in the particular circuit shown in Fig. 2.11 are $k_{B1} = R_1/R_3$ and $k_{B2} = R_2/R_4$. The bridge is connected to a DC voltage source, and a sensitive voltmeter (e.g., a galvanometer) is connected across the terminals A_1 and A_2 to read the voltage output $V_o = V_{A_1} - V_{A_2}$. This voltage output can be found by applying Kirchhoff's current law on the two sides of the bridge. When applied to the left-hand branch, this law gives

$$\frac{V - V_{A_1}}{R_1} = \frac{V_{A_1} - 0}{R_3} , \tag{2.15}$$

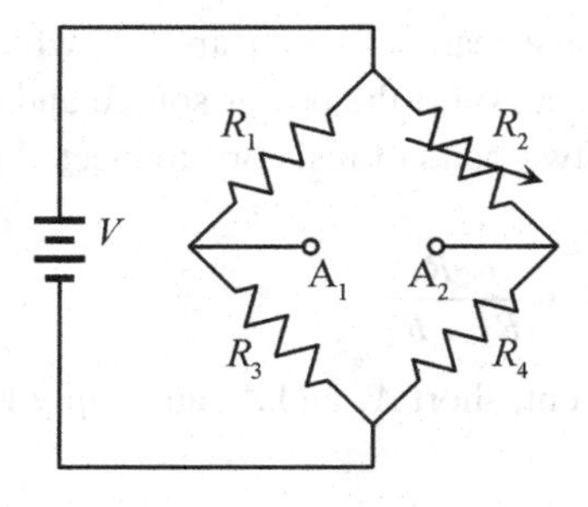

Figure 2.11 A resistive Wheatstone bridge.

which can be solved for V_{A_1} as

$$V_{A_1} = V\frac{R_3}{R_1 + R_3} \,. \tag{2.16}$$

A similar analysis on the right-hand branch of the bridge gives

$$V_{A_2} = V\frac{R_4}{R_2 + R_4} \,. \tag{2.17}$$

The bridge voltage output is the difference between the previous voltages, namely,

$$V_o = V_{A_1} - V_{A_2} = V\left(\frac{R_3}{R_1 + R_3} - \frac{R_4}{R_2 + R_4}\right) \,. \tag{2.18}$$

By adjusting the variable resistor, one may set the voltage shown by the voltmeter to zero, in which case we say that the bridge is balanced. The condition for a balanced bridge is

$$\frac{R_1}{R_3} = \frac{R_2}{R_4} \,, \tag{2.19}$$

from which one can readily find R_1. We therefore see that a necessary condition for the bridge to be balanced is that the bridge ratios for both branches are equal, that is, $k_{B1} = k_{B2}$. Note that the values of corresponding resistors in the two branches need not be the same in order to have a balanced bridge, as the only requirement is that the ratios of resistors are the same. For example, the choice of resistors $R_1 = 1$ kΩ, $R_2 = 10$ Ω, $R_3 = 100$ Ω, and $R_4 = 1$ Ω would give a bridge ratio $k_B = 10$ and would result in a balanced bridge, which would produce a zero voltage output. The bridge would become imbalanced and produce a non-zero voltage output, if resistors R_2, R_3, R_4 maintained their values at bridge balance, but the value of R_1 changed, for example, due to an external stimulus in a sensor. Thus, following appropriate calibration, the bridge voltage output could used to measure the value of the property to which the sensor is sensitive. The use of Wheatstone bridges as measurement devices shall be built upon in Part 2. We further note that one may generalise the design of a Wheatstone bridge with the addition of capacitors and/or inductors in one or both branches of the bridge.

Example 2.10 Determine the Thévenin- and Norton-equivalent circuits of a resistive Wheatstone bridge, connected to a dry cell (Fig. 2.11).

Answer

The Thévenin-equivalent voltage is the voltage read across A_1 and A_2, which is given in Eq. (2.18). To find the equivalent impedance, short the power source and notice that the circuit consists of a series connection of two pairs of resistors connected in parallel. Then,

$$Z_{eq} = \frac{R_1 R_3}{R_1 + R_3} + \frac{R_2 R_4}{R_2 + R_4} \,.$$

Finally, to determine the short-circuit current, short A_1 and A_2 and apply Kirchhoff's laws to derive

$$i_{sc} = V \frac{R_2R_3 - R_1R_4}{R_1R_2R_3 + R_1R_2R_4 + R_1R_3R_4 + R_2R_3R_4} .$$

An alternative analysis can be made with the use of the Laplace transform. First, assume that V_{oc} and i_{sc} are step functions, as necessary for their Laplace transforms to exist. Then, one may easily find that

$$\mathbf{V}_{oc}(s) = \frac{V_{oc}}{s}$$

and

$$\mathbf{I}_{sc}(s) = \frac{i_{sc}}{s} ,$$

and verify that, indeed, $Z_{eq} = \mathbf{V}_{oc}(s)/\mathbf{I}_{sc}(s)$.

2.2.4 Operational Amplifiers

A serious disadvantage of passive signal conditioning circuits is that, when connected to another circuit, they might alter the voltages and currents through it, a phenomenon identified previously as loading (Section 2.1.2). Loading effects may be circumvented by proper selection of the circuit components, following a process called *impedance matching*. Although of historical significance in metrology, impedance matching is both inconvenient and restrictive. Instead, it is a common approach to essentially decouple electric circuits form each other with the use of active devices, powered by external power supplies. The basic building block for common signal conditioning operations is the *operational amplifier*, usually abbreviated to *op-amp* [5, 8, 9, 10]. This is an electronic device having a high internal gain (namely, a strong amplification of the input voltage), a high input impedance, a low output impedance and connected in a feedback mode. The equivalent circuit of an op-amp is sketched in Fig. 2.12a. It can be seen that an op-amp has two inputs, referred to as *inverting input* and *non-inverting input*, and an *output*; the corresponding voltages will be denoted as V_-, V_+ and V_o. Power is provided by a *power supply*, usually a dual supply, with voltages $\pm V_s$ and a common ground. Among the most important op-amp specifications are:

- *open-loop gain A*, typically of the order of 10^5 or higher;
- *input impedance* Z_i, typically 10^5 to 10^{11} Ω;
- *output impedance* Z_o, typically 1 to 10 Ω;
- *voltage offset* V_{os}, typically ± 1 mV;
- *bias currents* i_- and i_+, typically 10^{-14} to 10^{-6}A;
- *slew rate*, which indicates how fast the op-amp responds to a change in input voltage; typical values are in the range of V/μs;
- *common-mode-rejection-ratio* (see Section 3.2.1), typically in the range of 60 to 120 dB.

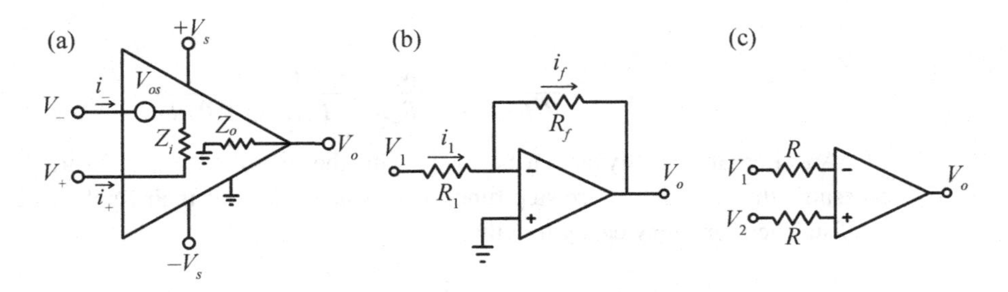

Figure 2.12 Sketches of (a) a simplified op-amp circuit, (b) an op-amp in an inverting amplifier configuration and (c) an op-amp in a comparator configuration.

- *Gain-bandwidth product*, which is the product of the low-frequency gain and the 3 dB high-frequency cut-off (see Section 3.2.2); the higher the gain is set, the lower the bandwidth of the amplifier would be; a high gain-bandwidth product op-amp would be required for applications involving low-level or high-frequency signals.

Because of their large input, and small output, impedance, op-amps normally exert negligible loading on circuits connected to their inputs and outputs. When placed within specific electrical circuits, op-amps can be used to execute mathematical operations on voltage input signal(s), for example, inversion, addition, subtraction, multiplication, integration, and differentiation, as well as amplifying the resulting values. A simplified analysis of op-amp operation within a given electrical circuit may be based on the assumption of an *ideal op-amp*, namely a device that has an infinite open-loop gain, an infinite input impedance and a zero output impedance [9]. Then, one may assume that

$$V_- \approx V_+ \,,\, i_- \approx 0 \text{ and } i_+ \approx 0 \,. \tag{2.20}$$

Example 2.11 Determine the voltage output of the simple op-amp configuration shown in Fig. 2.12b.

Answer

In this circuit, the non-inverting input has been grounded and the inverting input has been connected to a voltage source V_1 through a resistor R_1 and to the output through the *feedback resistor* R_f. Assuming that this is an ideal op-amp, we take its input impedance to be infinite, from which we infer that there is zero current flowing between the inverting and non-inverting inputs, while, at the same time, the voltages at these inputs are identical. Kirchhoff's current law at the inverting input gives $i_1 + i_f = 0$, which is equivalent to $V_1/R_1 + V_o/R_f = 0$, or

$$V_o = -\frac{R_f}{R_1} V_1 \,.$$

Thus, this device is an *inverting amplifier*: it inverts the sign of the input voltage signal and amplifies its magnitude by a factor of R_f/R_1, called the *amplification gain*.

The output voltage of an op-amp cannot exceed the range of the power supply voltages. If driven towards higher positive or negative values, V_o will reach a *saturation value* $\pm V_{sat}$, which cannot be exceeded in magnitude. For the typical power supply voltage of ±15 V, the saturation voltage would be approximately ±13 V. One must be careful to avoid saturation, especially when dealing with fluctuating input voltages, in which case the output voltage would be clipped if driven to values exceeding V_{sat} in magnitude. In fact, the op-amp may be configured as a *comparator*, providing an output that is either $+V_{sat}$, if $V_1 < V_2$, or $-V_{sat}$, if $V_2 < V_1$ (Fig. 2.12c). In the hypothetical case of $V_1 = V_2$, the comparator output would ideally be 0, but there is zero probability that two physical properties would have exactly equal values and so this case can be ignored. Similarly, when $V_1 \approx V_2$, there would be a range of voltages for which the comparator output would be intermediate between $-V_{sat}$ and $+V_{sat}$, but, because the op-amp open-loop gain is extremely large, this range would be extremely narrow and can also be ignored for practical purposes.

2.3 Static Response of Measuring Systems

2.3.1 Static Response and Static Calibration

The operation of a measuring system is called *static*, when all its inputs and outputs are either constant or vary very slowly with time. The relationship that provides the value of the input as a function of the system's output during static operation is called the *static response* of the measuring system. If there is a physical law that connects the input and output of the measuring system, the static response may be derived entirely theoretically. More commonly, however, the input–output relationship is based on static calibration and contains adjustable coefficients, the optimal values of which are determined by curve fitting (Section 4.6.2). A measuring system, operating under static conditions, is called *linear*, if its static response can be represented by a straight line within its entire range of operation. In practice, all measuring systems will become non-linear if the input magnitude becomes sufficiently large; for example, a metallic spring that behaves linearly for small extensions will become non-linear when its extension exceeds the elastic limit. Conversely, any continuous non-linear static response may be linearised within a relatively narrow input range (Section 4.6).

Static calibration is preferably performed separately for each desired input. In case of multiple inputs, all inputs should be kept fixed, except for one, which is given successively different values within a range, causing the output to reach corresponding values. The values of the input are measured independently by some other instrument, whose response is known and which serves as a *standard*; therefore, the accuracy of a calibration depends on the accuracy of the instruments used as standards. Sometimes, it is impossible or difficult to maintain all but one inputs constant, because changes in the desired input may trigger changes in other desired or undesirable inputs; in such cases, all changing inputs must be measured during calibration and the different effects must be identified and separated, with the use of, for example, available theoretical relationships among the various inputs. If possible, all interfering and modifying inputs during calibration should be maintained at the same levels as those during the actual measurement. Undesirable inputs may distort the static response by introducing a *zero drift*, namely a parallel shift of the primary calibration curve, or a *sensitivity drift*, namely a change in the slope of the primary calibration curve.

As an example of the two types of static response, consider the measurement of the pressure difference $p_1 - p_2$ in a gas (input) using a U-tube manometer and a variable reluctance pressure transducer (Fig. 2.13). In the first case, the input is related to the difference Δh in elevation of the two columns of the liquid in the manometer (output) by the hydrostatic law $p_1 - p_2 \approx \rho_l g \Delta h$, where ρ_l is the density of the liquid inside the manometer and g is the gravitational acceleration. In the second case, the same input has been correlated to the electric voltage output V by the empirical expression $p_1 - p_2 \approx \alpha V + \beta$, where α and β are constants determined by calibration of the transducer, for instance vs. a U-tube manometer (see Example 2.12).

Example 2.12 Outline the calibration method of a pressure transducer using a U-tube manometer as the *standard* (both instruments are shown in Fig. 2.13). Consider that we have a pressure transducer with a digital display, which shows values between 0.000 and 0.999 V with an increment of 0.001 V.

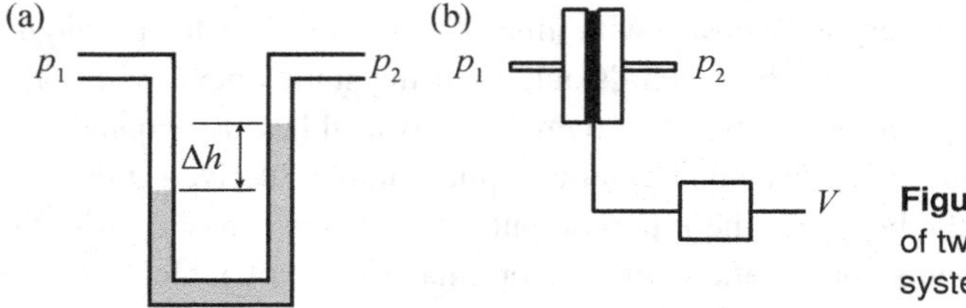

Figure 2.13 Static responses of two pressure measuring systems.

Answer

To calibrate the pressure transducer, we need to record the values of its output voltage V for a number of applied pressure difference values $p_2 - p_1$, which are measured by some other instrument that is deemed to be sufficiently accurate and serves as the *standard* of the calibration. A convenient standard for pressure measurement is the U-tube manometer. One can fabricate an inexpensive U-tube manometer by using transparent, flexible, plastic tubing and distilled water, preferably mixed with some fluorescent dye for better visibility and a small amount of detergent to reduce its surface tension so that it has a relatively clean free surface. To extend the range of measurable pressure differences, one may mount a long piece of tubing vertically on the laboratory wall. To generate a range of pressure differences, one may leave one side of the manometer and one side of the pressure transducer open to the atmosphere, while connecting the other side of both devices to a small sealed vessel or tube, filled with air and pressurised with the use of a bicycle pump. The calibration consists of applying a series of pressure differences to both devices and recording the corresponding pairs of values of the column elevation difference Δh and the transducer voltage output V. The pressure difference is determined as $p_2 - p_1 \approx \rho_{water} g \Delta h$. It is good experimental practice not to apply the pressure differences in a sequence with monotonically increasing or decreasing values, but in a randomly arranged one. This will eliminate a possible bias introduced by the fact that manometer readings for a given pressure that are taken in ascending order may be slightly different from those taken in descending order; although this bias will appear in the data as scatter, it can be easily eliminated by statistical means, such as linear least squares fitting.

A typical set of calibration data has been plotted in Fig. 2.14. One may notice that the transducer output is not zero when the applied pressure difference is zero, but has a *zero offset*. Many transducers allow for the removal of this offset, which may actually drift from one day to the next. Another important observation is that V increases as $p_2 - p_1$ increases, and that this increase is, upon visual inspection, approximately linear up to pressures of about 20 kPa. If we assume a linear fit of the form

$$p_2 - p_1 \approx \alpha V + \beta ,$$

we observe that there is significantly more deviation from the fitted line when the fit is applied to the entire measurement range (dashed line), than to values in the range $0 \leq p_2 - p_1 \leq 20$ kPa (solid line), in which case the empirical constants are $\alpha = 39{,}290$ and $\beta = -3{,}877$. An extrapolation of this expression to higher pressure differences (dotted line) exposes clearly the non-linearity of the pressure–voltage relationship in that range.

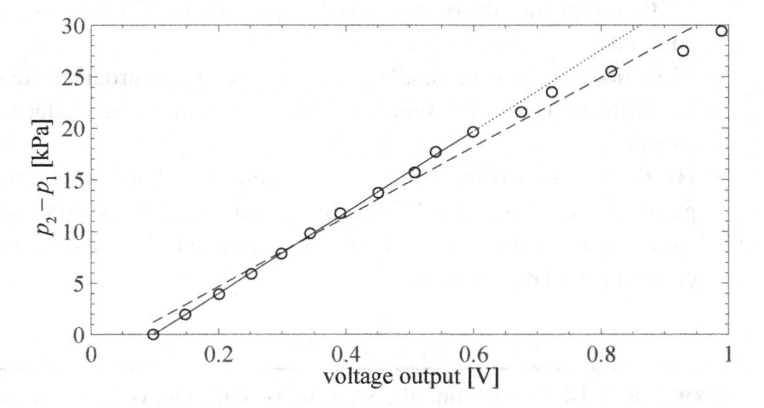

Figure 2.14 Measured voltage output of a pressure transducer vs. pressure difference, measured with a U-tube manometer; symbols represent measurements and the solid line is a linear least squares fit to the data in the range $0 \leq p_2 - p_1 \leq 20$ kPa.

2.3.2 Static Performance Characteristics

The static performance of a measuring system is characterised by several parameters, as follows:

- **Static sensitivity:** This is the slope of an input–output relationship. In other words, if x is the input of a measuring system and y is its output, then the static sensitivity is defined as $\mathrm{d}y/\mathrm{d}x$. The reciprocal $\mathrm{d}x/\mathrm{d}y$ of the static sensitivity is called the *inverse sensitivity*. Obviously, the static sensitivity of a linear system is constant over its operating range, while non-linear systems have a *local* sensitivity, which varies over their input range. In case of multiple inputs, it is understood that the static sensitivity with respect to a particular input is specified for particular fixed values of the other desired inputs.
- **Scale readability:** This property refers exclusively to analogue instruments. It is the lowest change in the output that can be recognised by an observer. Obviously, this is a subjective property and should not be confused with the accuracy of the instrument.
- **Range:** The range of a measuring system consists of the input values between a lower limit and an upper limit, for which the system is designed to operate properly.
- **Span:** This is the algebraic difference of the upper and lower limits of the range.
- **Full-scale output:** This is the algebraic difference between the output values measured with maximum and minimum input values applied.
- **Dynamic range:** This is the ratio of largest to smallest values of the input that the system can measure. The larger its dynamic range is, the higher the quality of a measuring system is considered to be.
- **Linearity:** This property, which is actually a measure of non-linearity, is the maximum deviation of the actual response from the straight line fitted to calibration measurements. Manufacturers specify linearity as a percentage of the instrument's reading, typically denoted as RDG, a percentage of full scale, denoted as FS, or a combination of the two.
- **Threshold:** This is the smallest input level that will produce a detectable output.
- **Resolution:** This is the smallest input change that will produce a detectable output change.
- **Hysteresis:** This is the difference between the output value corresponding to an input value that was reached, while being increased from smaller values and the output value corresponding to the same input value, but this time reached while being decreased from larger values.

Example 2.13 Determine the static sensitivity and other static performance characteristics of the pressure transducer, whose static calibration was discussed in Example 2.12.

Answer

In the following, we shall consider only the linear range of operation of the transducer, namely, the pressure difference range $0 \leq p_2 - p_1 \leq 20$ kPa, which corresponds to the voltage output range $0.1 \text{ V} \leq V \leq 0.6$ V. We will further assume that the transducer output is completely free of noise, an assumption that may not be met in practice.

- The static sensitivity is, by definition, $S = \mathrm{d}V / \mathrm{d}(p_2 - p_1) = 25$ mV/kPa.
- The operating range is 0–20 kPa, hence the input span is 20 kPa.

- The full-scale output is 0.6 V – 0.1 V = 0.5 V.
- The smallest value of the input that the transducer can detect is one that gives an output of 0.001 V, which corresponds to $0.001\ \text{V}/S = 0.04$ kPa, whereas the largest value of the input the transducer can measure accurately is 20 kPa. Therefore, the dynamic range is $20/0.04 = 500$.
- The threshold and the resolution are ideally both equal to 0.04 kPa. In practice, noise and other disturbances would likely increase these values.

Example 2.14 Consider the voltage divider, shown in Fig. 2.8a. Replace resistor R_2 by a resistance thermometer, namely a sensor the electric resistance R of which changes with its temperature T, as shown in Fig. 2.15, and explore the suitability of this circuit as a simple temperature measurement system. The resistance of a metallic temperature sensor over a relatively narrow range of temperatures can be modelled by the linear relationship $R(T) = R_0 + \alpha_0(T - T_0)$, where R_0 is the sensor resistance at the reference temperature T_0 and α_0 is the thermal resistivity coefficient at T_0; α_0 depends on the material of the sensor (e.g., $\alpha_0 = 0.385\ \Omega/\text{K}$ for a particular platinum sensor). Determine and discuss the sensitivity of this measurement system. Can this sensitivity be maximised for a given sensor? Can this system be configured so that it has a linear response?

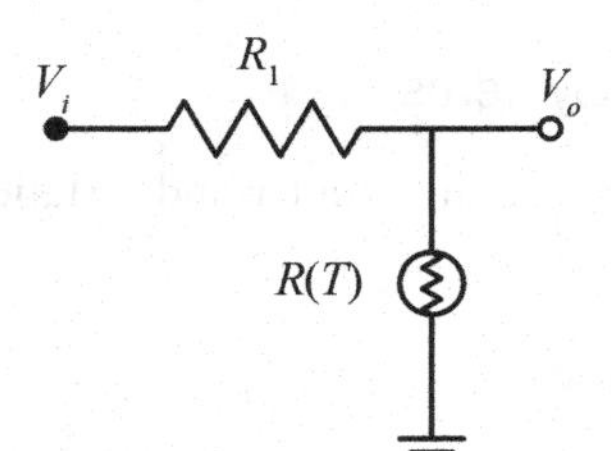

Figure 2.15 A simple circuit that may be used as a temperature measurement system.

Answer

As shown in Example 2.8, the voltage output is

$$V_o = V_i \frac{R(T)}{R_1 + R(T)} = V_i \frac{R_0 + \alpha_0(T - T_0)}{R_1 + R_0 + \alpha_0(T - T_0)} .$$

The sensitivity of this measurement system can be found as

$$S = \frac{\mathrm{d}V_o}{\mathrm{d}T} = V_i \frac{\alpha_0 T R_1}{[R_1 + R_0 + \alpha_0(T - T_0)]^2} .$$

This sensitivity depends non-linearly on the value of the temperature, which is the unknown measurand. Because of this, this circuit is not a convenient temperature measurement system. For example, as will be shown in the next section, the average voltage would not be related to the average temperature by the expression which relates the instantaneous values V_o and T. Even so, let us examine the sensitivity further. It is obvious that, to increase the sensitivity, one should increase the reference voltage V_i, noting that the reference values R_0 and α_0 are properties of the sensor and cannot be changed. Nevertheless, the range of the data acquisition system that may be used and safety considerations dictate an upper limit for V_i. The sensitivity can also be maximised by optimising the value of resistor R_1. Its optimal value, which can be found by solving the expression $\mathrm{d}S/\mathrm{d}R_1 = 0$ and ensuring that $\mathrm{d}^2S/\mathrm{d}R_1^2 < 0$, is $R_1 = R$, is also a function of the unknown and variable temperature.

Fortunately, the general drawback of this system can be overcome by a simple observation: if the fixed resistor R_1 is chosen such that $R_1 \gg R$ (e.g., if $R_0 \approx 100\ \Omega$, one could select $R_1 = 100\ \mathrm{k}\Omega$), then the response of the system becomes

$$V_o \approx \frac{V_i}{R_1} R(T) = \frac{V_i}{R_1} [R_0 + \alpha_0 (T - T_0)] ,$$

which is linear and has the known sensitivity

$$S = \frac{\mathrm{d}V_o}{\mathrm{d}T} = \frac{V_i \alpha_0}{R_1} .$$

Such measuring systems have been used in past fluid mechanics research, especially before the advent of low-noise electronics. The disadvantage of this approach is the large power loss on the *ballast resistor* R_1, which would be of concern, if dry cells were used as the power source.

2.3.3 Effects of Input Fluctuations in Non-linear Systems

Let us consider a measuring system with an input x, an output y and a static response that is described by the expression

$$y = f(x) . \tag{2.21}$$

In many experimental settings, the input may be fluctuating periodically or randomly in time, but sufficiently slowly for the instantaneous value $y(t)$ of the output to be related to the instantaneous value $x(t)$ of the input by the previous static response expression. One may define the *time-averages* of the input and output over a time interval T as, respectively,

$$\overline{x} = \frac{1}{T} \int_0^T x(t)\mathrm{d}t , \tag{2.22}$$

$$\overline{y} = \frac{1}{T} \int_0^T y(t)\mathrm{d}t . \tag{2.23}$$

Moreover, one may define the input *fluctuations* as

$$\tilde{x}(t) = x(t) - \overline{x} , \tag{2.24}$$

from which we note that the time-average of the fluctuations vanishes, that is, $\overline{\tilde{x}} = 0$.

When the response of the measuring system is linear, namely, if

$$y(x) = ax + b\,, \tag{2.25}$$

where a, b are constants, linearity of the integration operation ensures that the time-average of the output can be found from the time-average of the input with the use of the static response expression, that is,

$$\overline{y} = f(\overline{x}) = a\overline{x} + b\,. \tag{2.26}$$

This is convenient, because the averaging process would not introduce any error in addition to the unavoidable error in the determination of the static response. Thus, one can easily determine the time-averaged input from the time-averaged output, obtained by visually averaging a fluctuating reading, or with the use of an averaging voltmeter, or by post-processing a discretised record. When, however, the measuring system is non-linear, $\overline{y} \neq f(\overline{x})$ and so an error would occur, if the equality were used. Several popular instruments used in fluid mechanics research, including Pitot-static tubes, hot-wire anemometers and thermistors, are distinctly non-linear and thus subjected to fluctuation-induced errors. This error can be positive or negative, depending on the form of the static response. The way this error is generated will be illustrated by the following example.

Example 2.15 The static response of a measuring system is found to be

$$y = \alpha x^n\,,$$

where α and n are constants fitted to static calibration data. Consider that this system is subjected to a time-dependent input that fluctuates slowly enough for the static response to apply. Determine a relationship between the time-averaged input $\overline{x}$ and the time-averaged output $\overline{y}$.

Answer

The time-averaged output can be computed as

$$\begin{aligned}\overline{y} = \alpha\overline{x^n} &= \alpha\frac{1}{T}\int_0^T x^n(t)\,\mathrm{d}t \\ &= \alpha\frac{1}{T}\int_0^T (\overline{x} + \tilde{x})^n\,\mathrm{d}t \\ &= \alpha\overline{x}^n\frac{1}{T}\int_0^T \left(1 + \frac{\tilde{x}}{\overline{x}}\right)^n \mathrm{d}t\,.\end{aligned}$$

Assuming that the fluctuations are small ($|\tilde{x}| \ll |\overline{x}|$), one may use the binomial theorem to approximate the integrand as $1 + n\tilde{x}/\overline{x} + \frac{1}{2}n(n-1)(\tilde{x}/\overline{x})^2$ and then perform the integration, taking into consideration that $\overline{\tilde{x}} = 0$. This produces an approximate relationship between $\overline{x}$ and $\overline{y}$ as

$$\overline{y} \approx \alpha\overline{x}^n\left[1 + \frac{n(n-1)}{2}\frac{\overline{\tilde{x}^2}}{\overline{x}^2}\right],$$

where

$$\overline{\tilde{x}^2} = \frac{1}{T}\int_0^T \tilde{x}^2(t)\mathrm{d}t\ .$$

For the power-law response of this example, $\overline{y} < \alpha\overline{x}^n$, if $0 < n < 1$, and $\overline{y} > \alpha\overline{x}^n$, if $n < 0$ or $n > 1$; if $n = 1$, $\overline{y} = \alpha\overline{x}^n$ (linear system), while the case $n = 0$ is trivial.

To demonstrate an application of this discussion, consider the case of Pitot-static tube. The input is the flow velocity, that is, $x(t) = U(t)$ and the output is a pressure difference Δp, which, from Bernoulli's equation, is equal to the dynamic pressure, that is, $y(t) = \Delta p = \frac{1}{2}\rho U^2$. In the notation of this example, $\alpha = \frac{1}{2}\rho$ and $n = 2$ and so

$$\overline{\Delta p} \approx \tfrac{1}{2}\rho\overline{U}^2\left(1 + \frac{\overline{\tilde{U}^2}}{\overline{U}^2}\right) > \tfrac{1}{2}\rho\overline{U}^2\ .$$

The true mean velocity is therefore

$$\overline{U} = \left(\frac{2\overline{\Delta p}}{\rho} - \overline{\tilde{U}^2}\right)^{1/2}\ .$$

If we do not account for the effect of velocity fluctuations in our calculations, we would obtain an *apparent time-averaged velocity*

$$\overline{U}_{app} = \sqrt{2\overline{\Delta p}/\rho} > \overline{U}\ ,$$

which means that the Pitot-static tube would always overestimate the average flow velocity in the presence of fluctuations.

A non-linear static response distorts not only the relationship between time-averaged input and output, but also the relationships between any other statistical properties of the input and the output. It also makes the waveform of the output look different from the input waveform, a distortion which may have adverse consequences on the understanding of a physical process. This will be illustrated by the following example.

Example 2.16 Consider a Pitot-static tube inserted in a pulsatile flow of water, which has sinusoidal flow oscillations that are known to have an amplitude equal to $0.4\overline{U}$, where $\overline{U}$ is the true, but unknown, time-averaged velocity. The time-averaged pressure difference read by the Pitot-static tube is $\overline{\Delta p} = 5.00$ kPa. Determine $\overline{U}$ and plot the fluctuating velocity and pressure difference vs. time. The density of water is $\rho = 998\ \mathrm{kg/m^3}$.

Answer

The instantaneous flow velocity is given by

$$U(t) = \overline{U}[1 + 0.4\sin(2\pi t/T)]\ ,$$

where T is the period of oscillation. Then, the instantaneous pressure difference read by

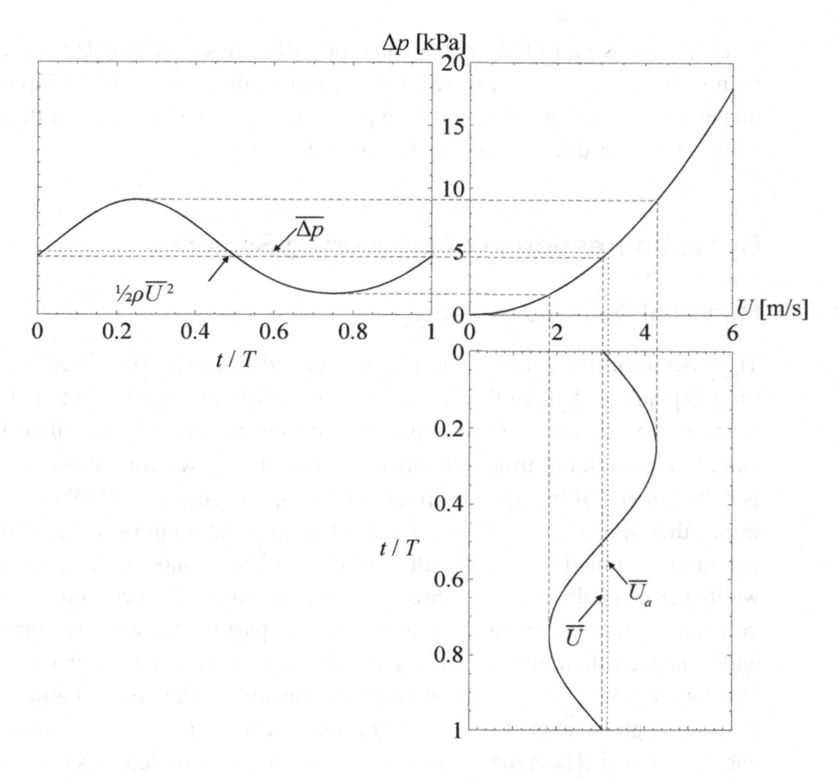

Figure 2.16 Plot of the pressure difference indicated by the Pitot-static tube vs. the flow velocity.

the tube will be

$$\Delta p(t) = \tfrac{1}{2}\rho\overline{U}^2[1 + 0.4\sin(2\pi t/T)]^2\,,$$

which, when integrated over one period, gives the time-averaged pressure as $\overline{\Delta p} = 0.54\rho\overline{U}^2$. Then, the time-average flow velocity can be found as

$$\overline{U} = \sqrt{1.85\overline{\Delta p}/\rho} = 3.05\ \mathrm{m/s}\,,$$

whereas the apparent velocity is

$$\overline{U}_{app} = \sqrt{2\overline{\Delta p}/\rho} = 3.17\ \mathrm{m/s} > \overline{U}\,.$$

Now, we can plot the waveforms of the input and output of this measuring system in a form that connects them via the static response expression. As shown in Fig. 2.16, the pressure difference waveform is distinctly non-sinusoidal, exhibiting an asymmetric amplitude about the mean.

In closing, one should point out that errors due to a non-linear response appear only when averages and other statistical properties are calculated from the output values. If, however, one inverts Eq. (2.21) to compute the instantaneous input value for each instantaneous output value and then performs statistical analysis of the computed input values, no error due to non-linearity would arise.

2.4 Dynamic Response of Measuring Systems

2.4.1 Models of Dynamic Systems

The operation of a measuring system is called *dynamic*, if at least one of its inputs is time dependent. Unlike the static response, which is usually specified by an algebraic relationship, the *dynamic response* of a system is generally specified by a differential equation, containing time derivatives. A measuring system, or one of its components, is called *linear*, if its input–output relationship is a linear algebraic or differential equation, otherwise it is called *non-linear*. Because the solution of non-linear differential equations is much more difficult than that of linear ones, it is customary to linearise, whenever possible, these systems. Also for simplicity, it is common to examine certain non-linear phenomena, such as hysteresis, as part of the static response of the system and to neglect their effects in the determination of the dynamic characteristics.

Like electrical circuits, measuring systems are modelled by mathematical equations describing physical principles, for example, electric circuit laws, momentum balance or energy balance. The dynamic response of such a system depends on the type of input(s) the system receives, as well as the nature of the system. When an output depends on several inputs, the full mathematical model of the system would likely be a set of non-linear, coupled, partial differential equations. This may be simplified by considering the time variation of the output y, when only one input x changes, further assuming that the system response can be approximated by a single, linear, ordinary differential equation with constant coefficients, namely,

$$a_n \frac{\mathrm{d}^n y}{\mathrm{d}t^n} + a_{n-1} \frac{\mathrm{d}^{n-1} y}{\mathrm{d}t^{n-1}} + \cdots + a_1 \frac{\mathrm{d}y}{\mathrm{d}t} + a_0 y = b_m \frac{\mathrm{d}^m x}{\mathrm{d}t^m} + b_{m-1} \frac{\mathrm{d}^{m-1} x}{\mathrm{d}t^{m-1}} + \cdots + b_1 \frac{\mathrm{d}x}{\mathrm{d}t} + b_0 x \, , \quad n \geq m \, . \tag{2.27}$$

Such a model is called *linear time-invariant* (LTI), so that it can be distinguished from models with time-dependent coefficients. It turns out that the nature of the input to many measuring systems is such that the coefficients on the right-hand side of Eq. (2.27) vanish, with the exception of b_0. This means that the mathematical equation of measuring systems usually takes the simpler form

$$a_n \frac{\mathrm{d}^n y}{\mathrm{d}t^n} + a_{n-1} \frac{\mathrm{d}^{n-1} y}{\mathrm{d}t^{n-1}} + \cdots + a_1 \frac{\mathrm{d}y}{\mathrm{d}t} + a_0 y = b_0 x \, . \tag{2.28}$$

If the system operates under static conditions, all derivatives vanish and both previous equations are reduced to

$$a_0 y = b_0 x \, . \tag{2.29}$$

The ratio $K = b_0/a_0$ is called the *static sensitivity* of the system. It is important to emphasise that, for linear systems, the static sensitivity is constant.

In the following, we shall discuss the simplest and most common linear systems found in measurement practice, which are the zero-, first- and second-order systems.

Zero-order systems: These have an algebraic response, namely,

$$y = Kx \,. \tag{2.30}$$

A zero-order system is characterised by a single parameter, the *static sensitivity* K, which has the same dimensions as the ratio y/x. The dynamic response of zero-order systems is independent of time, as the output remains proportional to the input at all times. Examples of zero-order systems are an electric resistor (Fig. 2.17a) and an elastic spring. Most physical systems, however, even those normally described by zero-order models, will show some time dependence when subjected to fast-changing inputs. For example, an elastic spring, normally considered to have a negligible mass, will show effects of inertia when the rate of change of the applied force exceeds some limit; a more accurate representation of this spring would be by a second-order system.

First-order systems: These have a response

$$\tau \frac{dy}{dt} + y = Kx \,. \tag{2.31}$$

A first-order system is characterised by two parameters, the static sensitivity K and the *time constant* τ, which has dimensions of time. Examples are the common thermometer (Fig. 2.17b) and a resistor-capacitor (RC) electric circuit (see Example 2.8), the time constant of which is $\tau = RC$.

Second-order systems: These have have a response

$$\frac{d^2y}{dt^2} + 2\zeta\omega_n \frac{dy}{dt} + \omega_n^2 y = K\omega_n^2 x \,. \tag{2.32}$$

A second-order system is characterised by three parameters, the static sensitivity K, the *undamped natural frequency* ω_n (measured in rad/s) and the (dimensionless) *damping ratio* ζ. A second-order system is called *undamped* when $\zeta = 0$, *underdamped* when $0 < \zeta < 1$, *critically damped* when $\zeta = 1$ and *overdamped* when $1 < \zeta$. The parameter

$$\omega_d = \omega_n \sqrt{1 - \zeta^2} \,, \tag{2.33}$$

defined for underdamped systems only, is called the *damped natural frequency*. Damping is generally due to a mechanism that dissipates energy and opposes some action; mechanical friction is the most common damping mechanism. Because energy dissipation is always present in real systems, perfectly undamped systems do not exist; the closest approximation to an undamped system is a *lightly damped* system, having $0 < \zeta \ll 1$. Examples of measuring systems that can be represented approximately by a second-order model are the liquid manometer (Fig. 2.17c), in which damping is caused by friction between the liquid and the glass tube wall, and force transducers, in which the damping is due to the stiffness of the transducer itself.

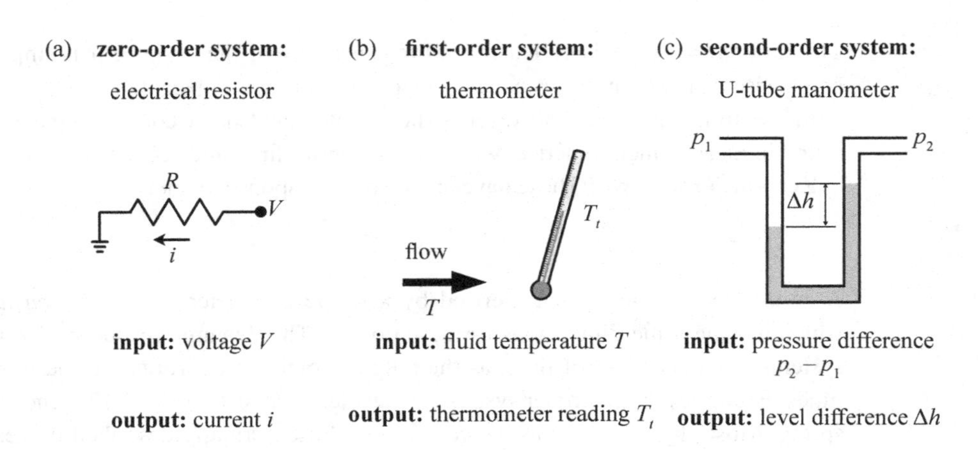

Figure 2.17 Examples of measuring systems whose dynamic response may be approximated by zero-, first- and second-order models.

Example 2.17 Develop dynamic models of a diaphragm-type pressure transducer.

Answer

Many common types of pressure transducers contain a metallic diaphragm, which separates two chambers. When two different pressures are applied on the two chambers, the diaphragm deforms. The pressure difference Δp, which is the input to the measuring system, creates the force $F = \Delta p A$, where A is the area of the diaphragm. When a steady force is applied within the transducer's range of operation, the diaphragm deforms elastically, such that the centre of the diaphragm is displaced by Δx from its equilibrium position, following a relationship of the type $F = k\Delta x$, where k is the elastic constant of the diaphragm. The deformation of the diaphragm changes some electric property (e.g., a capacitance), which creates a change in the electric voltage V of a circuit, which is the output of the transducer. We may assume that, within the transducer's range of operation, $V \propto \Delta x$.

The static response of this system would be simply

$$V = K\Delta p\,,$$

where K is the static sensitivity of the transducer. Now, consider that the transducer is subjected to a time-dependent pressure difference. If we neglect the inertia of the diaphragm, the inertia of the displaced fluid within the transducer and any frictional losses, the deformation of the diaphragm would be proportional to the applied force, and so the system can be represented by a zero-order model, which is identical to the previously mentioned static response.

Next, let us take into account the inertia of the diaphragm, which is assumed to have a lumped mass m. Time-dependent deformation is accompanied by a change of momentum $m\,\mathrm{d}^2(\Delta x)/\mathrm{d}t^2$, and so Newton's second law on the diaphragm may be written in the form

$$\Delta pA - k\Delta x = m\frac{\mathrm{d}^2(\Delta x)}{\mathrm{d}t^2}\,.$$

This expression leads to an undamped second-order model for the voltage output V.

For a more realistic model, let us assume that friction in the fluid creates a force that opposes the motion of the diaphragm and that this force is proportional to the velocity $\mathrm{d}(\Delta x)/\mathrm{d}t$ of the diaphragm. With the additional frictional force, the transducer model becomes a damped second-order type. Depending on the geometry and the material properties of the diaphragm and the fluid, this model could be underdamped, critically damped or overdamped.

2.4.2 Idealised Types of Dynamic Input

With the exception of zero-order systems, the output of all other systems exposed to a time dependent input depends not only on the current value of the input but also on its history. For the description of dynamic response, it is customary to consider a few idealised types of input, as approximate models of more realistic input types.

Step input: A nearly abrupt change of the input from its current value to another value may be idealised mathematically by a discontinuous value change. An elementary value change is the *step function* $AU(t)$, $A =$ const., where the *unit-step*, or *Heaviside, function* is defined as

$$U(t) = \begin{cases} 0\,, & \text{for } t < 0\,, \\ 1\,, & \text{for } 0 \leq t\,, \end{cases} \tag{2.34}$$

and is hence dimensionless. The constant A is referred to as the *loading constant*. Depending on whether A is positive or negative, the change would be upwards or downwards. Note that for step inputs, the dimensions of A should be the same as those of the input variable x. For example, consider a scale that measures the weight of a box with a mass equal to 30 kg; the sudden increase of the load on the scale by the addition of a second box, having a mass of 20 kg, can be modelled as a step input with a loading constant $A = 20$ kg.

Impulse input: A sudden, impulsive application of a different value of the input, lasting only briefly before it returns to the original level, may be idealised as proportional to the *unit-impulse function* or *Dirac's delta function*, defined as

$$\delta(t) = 0 \text{ for } t \neq 0\,,\ \delta(t) \to \infty \text{ for } t \to 0\,,\ \int_{-\infty}^{\infty} \delta(t)\,\mathrm{d}t = 1\,. \tag{2.35}$$

It is noted that, although $\delta(t)$ has an infinite value at 0, its integral is finite and equal to 1. Dirac's function has the additional property that, for any continuous function $g(t)$,

$$\int_{-\infty}^{\infty} \delta(t - t_o)g(t)\,\mathrm{d}t = g(t_o)\,. \tag{2.36}$$

Thus, an *impulse function* is defined as $A\delta(t)$, $A =$ const., where A is positive or negative and has the same dimensions as the input. For the balance example, an impulse input would correspond to a sudden loading and unloading the scale by the second box; in this case, we also have $A = 20$ kg.

Ramp input: A gradual increase or decrease of the input may be modelled by a mathematical function. An idealised monotonic change is the *ramp function* $Ar(t)$, $A =$ const., which is proportional to the *unit-slope ramp function*

$$r(t) = \begin{cases} 0, & \text{for } t < 0, \\ t, & \text{for } 0 \leq t. \end{cases} \tag{2.37}$$

Once more, depending on whether A is positive or negative, the change would be upwards or downwards. For ramp dynamic inputs, the dimensions of the loading constant are those of the input x divided by time. Revisiting the balance example, assume that the second box is an open vessel and that we pour water into it from a hose at a constant mass flow rate of 0.1 kg/s; in this case, $A = 0.1$ kg/s.

The three types of idealised inputs that were defined so far in this section are illustrated in Fig. 2.18.

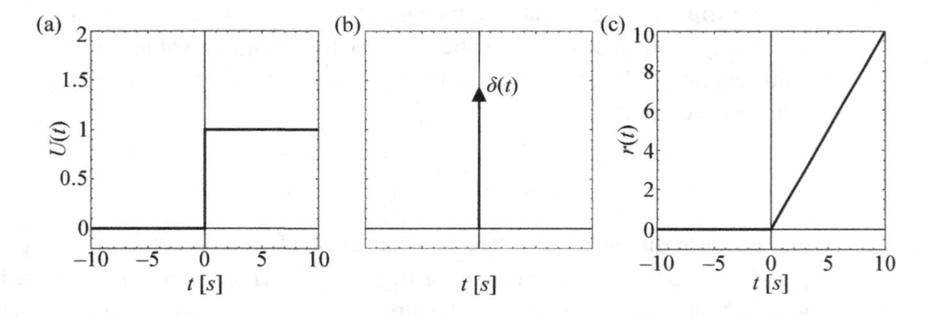

Figure 2.18 (a) The Heaviside function; (b) Dirac's delta function (the arrowhead indicates extent to infinity); and (c) the unit-slope ramp function.

Example 2.18 Describe how step, ramp, and impulse inputs can be applied to a thermometer in order to determine its dynamic response.

Answer

A step input can most readily be applied by letting the thermometer rest in a cup of water at room temperature for a period of time and then dipping it into a pot with boiling water. An impulse input can be replicated by momentarily dipping the thermometer into boiling water and then immediately taking it back out. Finally, a ramp input can be replicated by placing the thermometer into a pot of water that, starting at room temperature, is steadily heated up to boiling.

Sinusoidal input: A sinusoidal input is defined as

$$x(t) = A\sin(\omega t) \equiv A\sin(2\pi f t) \equiv A\sin\left(\frac{2\pi t}{T}\right), \tag{2.38}$$

where ω is the *angular frequency*, measured in rad/s, f is the *cyclic frequency*, measured in cycles/s (Hz) and T is the period. These three properties are equivalent, because they are related as

$$\omega = 2\pi f = \frac{2\pi}{T}. \tag{2.39}$$

One must be warned, however, that both ω and f are often referred to casually as simply *frequency*.

Periodic input: A function $x(t)$ is called *periodic* with a period T, if $x(t) = x(t + nT)$, where n is any integer number. A periodic input to a measuring system can be decomposed into a *Fourier series* [2], as

$$x(t) = \frac{a_0}{2} + \sum_{n=1}^{\infty}\left[a_n\cos\left(2\pi n\frac{t}{T}\right) + b_n\sin\left(2\pi n\frac{t}{T}\right)\right], \tag{2.40}$$

where the *Fourier coefficients* are determined as

$$a_n = \frac{2}{T}\int_{-T/2}^{T/2} x(t)\cos\left(2\pi n\frac{t}{T}\right)\mathrm{d}t, n = 0, 1, 2, \ldots, \tag{2.41}$$

$$b_n = \frac{2}{T}\int_{-T/2}^{T/2} x(t)\sin\left(2\pi n\frac{t}{T}\right)\mathrm{d}t, n = 1, 2, 3, \ldots. \tag{2.42}$$

The only requirement for this representation is that $x(t)$ satisfies the *Dirichlet conditions*, which is always the case for measurement data.

Using a familiar trigonometric identity, one may express the term inside the square brackets in Eq. (2.40) as

$$A_n\cos\left(2\pi n\frac{t}{T} - \varphi_n\right), \quad A_n = \sqrt{a_n^2 + b_n^2}, \quad \varphi_n = \arccos\left(\frac{a_n}{\sqrt{a_n^2 + b_n^2}}\right), \tag{2.43}$$

which implies that a_n, b_n are the Cartesian components of a two-dimensional vector, which, in polar coordinates, has an amplitude A_n and a phase φ_n. Therefore, any periodic input can be considered as the sum of sinusoidal inputs. The first term in the Fourier series, Eq. (2.40), is a constant, equal to $a_0/2$, and is sometimes referred to as the *zeroth harmonic*. The second term (namely, the sum of the two terms within the square brackets in Eq. (2.40) for $n = 1$) oscillates with the *fundamental frequency* $f_1 = 1/T$ (or $\omega_1 = 2\pi/T$) and is called the *first harmonic*. The following terms oscillate with the *harmonic frequencies* $f_i = i/T, i = 2, 3, \ldots$ and are called *second*, *third*, etc. *harmonics*. A periodic function $x(t)$ may be approximated as the sum of a finite number N of terms in the Fourier series, with the lowest acceptable N depending on the waveform shape of the function and the level of desired accuracy.

Example 2.19 Expand a square wave and a triangle wave into Fourier series.

Answer

The square wave takes two values, A and $-A$, each persisting over half of each period, namely,

$$x(t) = \begin{cases} A\,, & \text{for } 0 < t/T \leq 0.5\,, \\ -A\,, & \text{for } 0.5 < t/T \leq 1\,. \end{cases} \tag{2.44}$$

Its Fourier series expansion contains only odd harmonics and can be found as

$$\begin{aligned} x(t) &= \frac{4A}{\pi}\sum_{n=1}^{\infty}\frac{\sin[(2n-1)\omega_0 t)]}{2n-1} \\ &= \frac{4A}{\pi}\left[\sin(\omega t) + \frac{1}{3}\sin(2\omega t) + \frac{1}{5}\sin(5\omega t) + \cdots\right]. \end{aligned} \tag{2.45}$$

The triangle wave consists of ramps of finite length within each period, as

$$x(t) = \begin{cases} At\,, & \text{for } 0 \leq t/T \leq 0.25\,, \\ -At\,, & \text{for } 0.25 \leq t/T \leq 0.75\,, \\ At, & \text{for } 0.75 \leq t/T \leq 1\,. \end{cases} \tag{2.46}$$

Its Fourier series expansion also consists of odd harmonics only, however, these har-

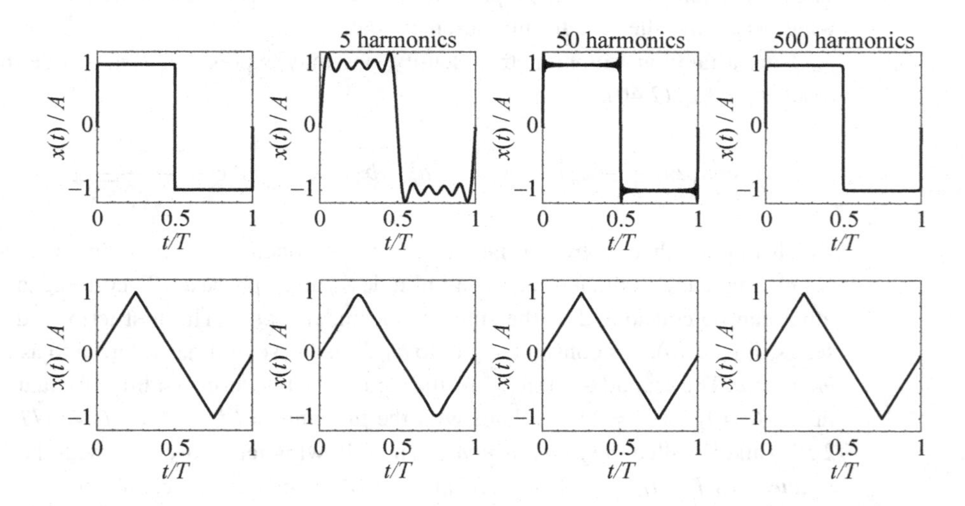

Figure 2.19 The idealised square wave (top row) and triangle wave (bottom row), and the corresponding waves composed of the first five, 50 and 500 odd harmonics of the Fourier series.

monics alternate in sign, as

$$x(t) = \frac{8A}{\pi^2}\sum_{n=1}^{\infty}(-1)^{(n-1)/2}n^{-2}\sin(n\omega_0 t)$$
$$= \frac{8A}{\pi^2}\left[\sin(\omega_0 t) - \frac{1}{9}\sin(3\omega_0 t) + \frac{1}{25}\sin(5\omega_0 t) + \cdots\right]. \tag{2.47}$$

Both waves, along with the corresponding truncated Fourier series for the first five, 50, and 500 odd-number harmonics are shown in Fig. 2.19.

We may note that the truncated Fourier series of the triangle wave converges within a few harmonics, in contrast to the square wave, which shows ripples even when 50 odd harmonics were used.

2.4.3 Steady-State Response

Given an input function, one can, in principle, determine the output of a linear time-invariant system by solving the corresponding differential equation, in our case usually Eq. (2.28). As will be shown in Sections 2.4.4 and 2.4.5, such solutions often consist of two parts: a *transient part*, which vanishes after sufficiently long time, and a *steady-state* part, which persists as $t \to \infty$ [7]. The *steady-state response* of the system corresponds to the steady-state part of the solution. Understanding the behaviour of the output during both the transient state and the steady-state, as well as determining the duration of the transient state, are essential from a measurement point of view.

The objective of taking measurements is to determine the current value of a measurand (input) from the measured value of the output of the measuring system. Under static operation of a measuring system, the input can be computed from the output by solving the algebraic static-response expression. For example, under static operation of a linear system, the input is easily computed from the output as $x(t) = y(t)/K$. Under dynamic conditions, however, the input value that is computed from the corresponding static-response expression would, in general, differ from the actual current input value. We will call this difference the *dynamic error* $\varepsilon_d(t)$. For a linear system, the dynamic error is defined as

$$\varepsilon_d(t) = x(t) - \frac{1}{K}y(t). \tag{2.48}$$

Note that systems control theory defines the error as $Kx(t) - y(t)$, which is different from the error $\varepsilon_d(t)$ associated with a measurement, as defined here. The dynamic error depends on the dynamic response expression of the measuring system (for a linear system, its order), as well as the type of input that the system receives. Like the solution of the corresponding dynamic response expression, the dynamic error may consist of a *transient dynamic error* and a *steady-state dynamic error* ε_{ds}, which is defined as

$$\varepsilon_{ds}(t) = \lim_{t\to\infty}\varepsilon_d(t). \tag{2.49}$$

The steady-state output of a linear system that is subjected to a sinusoidal input $x(t) = A\sin(\omega t)$ is $y(t) = B\sin(\omega t + \varphi)$, where φ is the *phase difference* or *phase shift*; thus, $y(t)$

would also be sinusoidal, with the same frequency as $x(t)$ but out of phase with it. The *amplitude ratio* B/A and the phase difference φ depend on the input frequency ω and the order of the system. The relationships describing B/A and φ as functions of frequency are referred to as the *frequency response* of the system and are presented in analytical and/or graphical forms. The amplitude ratio function is often presented in logarithmic form as a *gain*

$$G_{dB}(\omega) = 20\log_{10}(|B/A|)\,, \tag{2.50}$$

which is measured in decibels (dB).

A common type of frequency response plots are the *Bode plots*, or *Bode diagrams*, which show G_{dB} and φ on linear axes vs. the frequency on logarithmic axes [11]. The frequency axis in a Bode plot is divided into *decades*, namely ranges of frequency between end points having a ratio 1:10 (e.g., between 1 kHz and 10 kHz) or *octaves*, namely ranges of frequency between end points having a ratio 1:2 (e.g., between 1 kHz and 2 kHz).

Many measuring systems have an approximately constant amplitude ratio over a range of frequencies, which is called the *measurement bandwidth* of the system, outside of which the amplitude ratio drops, as depicted in Fig. 2.20. Note that a measuring system will normally revert to its static operation, if it is subjected to a steady input for sufficiently long time. Letting $\omega \to 0$, one can obtain the static gain as $G_{dB}(0) = 20\log_{10}(|K|)$. Let $(B/A)_b$ be the approximately constant,'nominal' amplitude ratio within the bandwidth. Many measuring systems also have an amplitude ratio, which drops at high frequencies only, in which case the lower limit of their bandwidth is considered to be 0 Hz and $(B/A)_b = K$. The upper bound for the bandwidth is taken to be the *cut-off frequency* ω_c, defined as the frequency at which the amplitude ratio

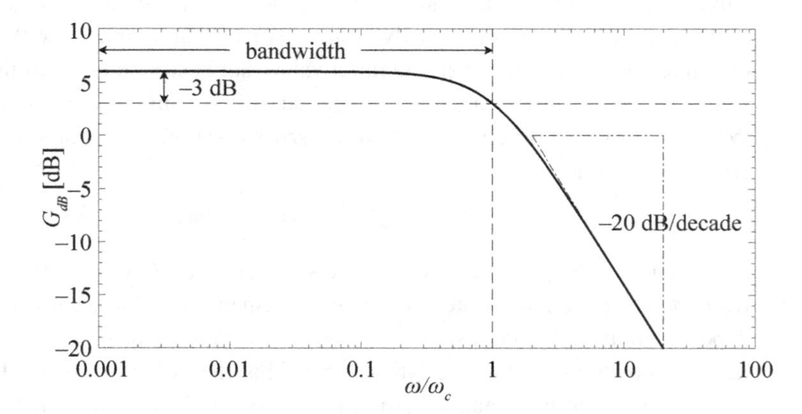

Figure 2.20 Sample frequency response of a first-order measuring system, with a static sensitivity of $K = 2$, hence a static gain of $G_{dB} = 20\log_{10}(2) \approx 6$ dB. The cut-off frequency ω_c is defined as the frequency at which the gain has decreased by approximately 3 dB, which in this case occurs when the gain is at approximately 3 dB. A roll-off of −20 dB/decade is observed as $\omega \to \infty$.

drops to a value $(B/A)_b/\sqrt{2} \approx 0.707(B/A)_b$. The difference between the maximum gain within the bandwidth and the gain at the cut-off frequency is

$$20\log_{10}(|(B/A)_b|) - 20\log_{10}\left(|(B/A)_b/\sqrt{2}|\right) = 20\log_{10}\left(\sqrt{2}\right) \approx -3\text{ dB}\,. \tag{2.51}$$

Thus, a cut-off frequency can be found on a Bode plot as a value at which the gain drops by 3 dB from its nominal bandwidth value. Another property of interest for measuring systems is the *roll-off*, defined as the slope of the gain function in a Bode plot as $\omega \to \infty$. As will be proved in Section 2.4.7, the roll-off is equal to $-20n$ dB/decade or $-6n$ dB/octave, where n is the order of the system. Therefore, a first-order measuring system will have a roll-off of -20 dB/decade, whilst a second-order measuring system will have a roll-off of -40 dB/decade.

The Bode plots can be used to determine the steady-state output of any linear system subjected to any periodic input. First, the periodic input is decomposed into a Fourier series and then the steady-state output is reconstructed as the sum of the outputs that would be produced, if individual sinusoidal inputs, corresponding to each term of the input Fourier series, were applied to the system. This is a consequence of the *superposition principle*, which applies only to linear systems. It is evident then that the steady-state output of a linear system subjected to a periodic input would also be periodic with the same period.

The dynamic response of a linear measurement system depends, in general, on the order of the system, the type of dynamic input the system receives and the initial conditions. In the next two subsections, we shall discuss the response of first- and second-order systems to the previously presented idealised inputs.

2.4.4 Dynamic Response of First-Order systems

The response of first-order systems, which are described by Eq. (2.31), to the idealised types of input mentioned in Section 2.4.2 is as follows.

Step response: Assume a step-change of the input from a constant level $x(t) = x_0$ for $t \leq t_0$ to another constant level $x(t) = x_0 + A$ for $t > t_0$. The output is found by solving Eq. (2.31) as

$$\frac{y(t) - y_0}{KA} = 1 - e^{-(t-t_0)/\tau}\,, \tag{2.52}$$

where $y(t) = y_0 = Kx_0$ for $t \leq t_0$. This equation, plotted in Fig. 2.21, describes the transient output of a first-order system to a step-change input. The steady-state output, found by letting $t \to \infty$, is simply $[y(t) - y_0]/(KA) = 1$. Accordingly, the dynamic error for $t > t_0$ is

$$\begin{aligned}\varepsilon_d(t) &= x(t) - y(t)/K \\ &= x_0 + A - y_0/K - A\left[1 - e^{-(t-t_0)/\tau}\right] \\ &= Ae^{-(t-t_0)/\tau}\end{aligned} \tag{2.53}$$

and, hence, the steady-state error is

$$\varepsilon_{ds}(t) \equiv \lim_{t\to\infty} \varepsilon_d(t) = 0\,. \tag{2.54}$$

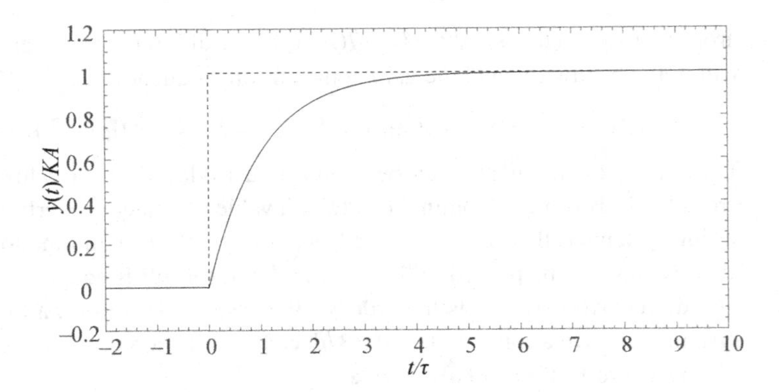

Figure 2.21 Step response of a first-order system. For simplicity, we assumed that $y(t) = 0$ for $t \leq 0$.

Therefore, a first-order system subjected to a step input has no steady-state error. At finite times, however, the error would be non-zero. The relative dynamic error ε_d/A is 37% after time τ, 13.5% after time 2τ, 5% after time 3τ, and 0.7% after time 5τ. It is customary to consider that the error is negligible for $t \geq 5\tau$, with the value 5τ referred to as the *settling time*. We also note that, at a given time t, the dynamic error decreases as the time constant decreases; in other words, the dynamic response of systems with small τ is better than that of systems with larger τ.

Impulse response: Let an impulse input be of the form $x(t) = x_0 + A\delta(t - t_0)$, where x_0 is a constant. The output for $t > t_0$ can be found by solving Eq. (2.31) as

$$\frac{y(t) - y_0}{KA} = \frac{1}{\tau} e^{-(t-t_0)/\tau} , \tag{2.55}$$

where $y_0 = Kx_0$. The dynamic error for $t > t_0$ becomes

$$\begin{aligned} \varepsilon_d(t) &= x(t) - y(t)/K \\ &= x_0 - \frac{A}{\tau} e^{-(t-t_0)/\tau} - \frac{y_0}{K} \\ &= -\frac{A}{\tau} e^{-(t-t_0)/\tau} . \end{aligned} \tag{2.56}$$

The previous relationships show that a system with a small time constant, although preferable in terms of its step response, would be subjected to large-amplitude spikes when exposed to impulsive inputs. As in the case of a step input, the error following an impulse decreases with time and vanishes when $t \gg \tau$, such that the steady-state error becomes $\varepsilon_{ds}(t) = 0$.

Ramp response: Let a constant input $x(t) = x_0$ for $t \leq t_0$ be followed by a ramp of the form $x(t) = x_0 + A(t - t_0)$ for $t > t_0$. By solving Eq. (2.31), we find the output for $t > t_0$ as

$$\frac{y(t) - y_0}{KA} = (t - t_0) - \tau[1 - e^{-(t-t_0)/\tau}] . \tag{2.57}$$

The dynamic error is, therefore,

$$\begin{aligned}\varepsilon_d &= x(t) - y(t)/K \\ &= x_0 + A(t - t_0) - y_0/K - A[t - t_0 - \tau(1 - e^{-(t-t_0)/\tau})] \\ &= A\tau[1 - e^{-(t-t_0)/\tau}] ,\end{aligned} \tag{2.58}$$

and the steady-state error becomes $\varepsilon_{ds}(t) = A\tau$. It is interesting to note that a first-order system will always be in error when subjected to a ramp-type input. Even after a very long time, the error would not vanish, but it would increase asymptotically to the constant value $A\tau$.

Frequency response: The steady-state amplitude ratio and phase change of a first-order measurement system, subjected to a sinusoidal input, are

$$\frac{B}{A} = \frac{K}{\sqrt{\omega^2\tau^2 + 1}} \tag{2.59}$$

and

$$\varphi = -\arctan \omega\tau . \tag{2.60}$$

Eqs. (2.59) and (2.60) show that, as $\omega \to 0$, $B/A \to K$ and $\varphi \to 0$, whereas, as $\omega \to \infty$, $B/A \to K/(\omega\tau)$ and $\varphi \to -\pi/2$. Thus, a first-order measuring system acts as a low-pass filter; filters are further discussed in Section 3.2.2. If a periodic input with a wide range of frequencies is applied to it, a lower harmonic of its output will have a higher amplitude ratio than a higher harmonic and the Fourier series of the output will necessarily be truncated. Moreover, the phase of lower input harmonic will be shifted less at the output than the phase of a higher harmonic. The consequence of this non-uniform phase shift of different components of the steady-state periodic output is that that the waveform of the output signal may look very different from the waveform of the input signal. The amplitude ratio and the phase change of first-order systems are plotted in linear axes in Fig. 2.22 and in the form of Bode plots in Fig. 2.23. It is evident that the

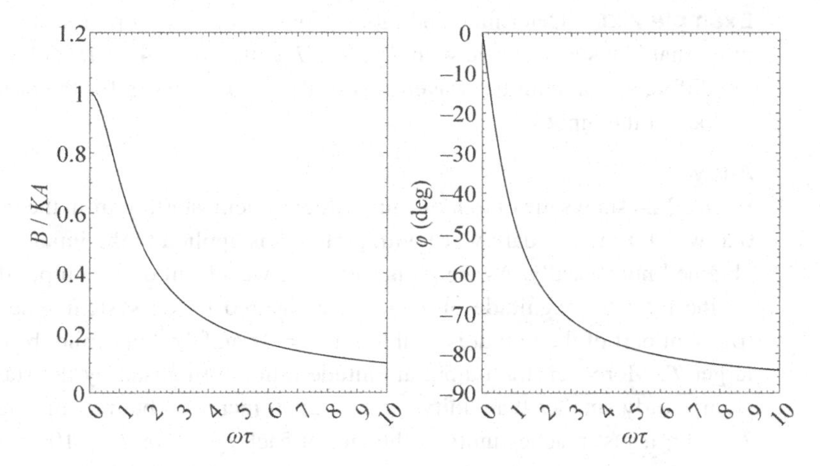

Figure 2.22 Normalised frequency response of a first-order system.

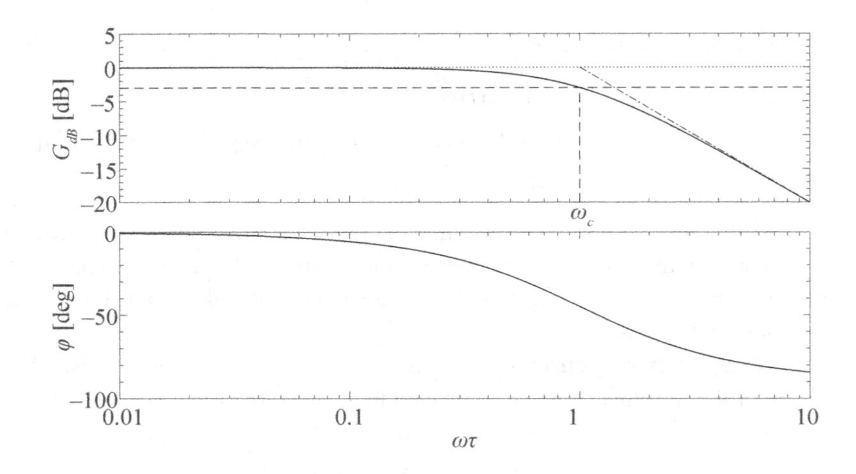

Figure 2.23 Bode plot for a first-order system with a static sensitivity $K = 1$.

amplitude ratio decreases monotonically with increasing frequency. Expressed in dB, the static gain of a first-order system is $G_{dB}(0) = 20\log_{10}(|K|)$, which for $K = 1$ gives $G_{dB}(0) = 0$. The bandwidth of a first-order system extends over the range of frequencies $0 \le \omega \le \omega_c$. Setting $B/A = K/\sqrt{2}$ in Eq. (2.59) provides the cut-off frequency as

$$\omega_c = 1/\tau \,. \tag{2.61}$$

Therefore, the bandwidth of a first-order measurement system is inversely proportional to its time-constant; the smaller the time-constant, the larger the operational bandwidth. It is also easy to show that the roll-off of a first-order system is −20 dB/decade or, equivalently, −6 dB/octave.

Example 2.20 Determine and plot the output of a first-order system subjected to an input that is a square wave with a period T equal to $2\tau, 4\tau, 10\tau$ and 20τ. Comment on the differences among the waveforms of the output signals for the cases with different periods of the input.

Answer

Figure 2.24 shows the output of a first-order system, starting from the initial value $y(t) = 0$ at which time a square wave with period T is applied to the input. This solution was obtained numerically. As the plots show, the waveform of the output depends strongly on the relative magnitude of T, when compared to the system time constant τ. The transient part of the output is visible for $T = 2\tau$ and 4τ, but cannot be distinguished for larger T. Moreover, the output amplitude ratio, normalised by the static sensitivity, is significantly smaller than unity for $T = 2\tau$, it increases but remains less than unity for $T = 4\tau$, it just reaches unity at the end of each cycle for $T = 10\tau$ and has near-unity values for a significant part of each cycle for $T = 20\tau$. It is only for the latter case that the output starts to resemble a square wave.

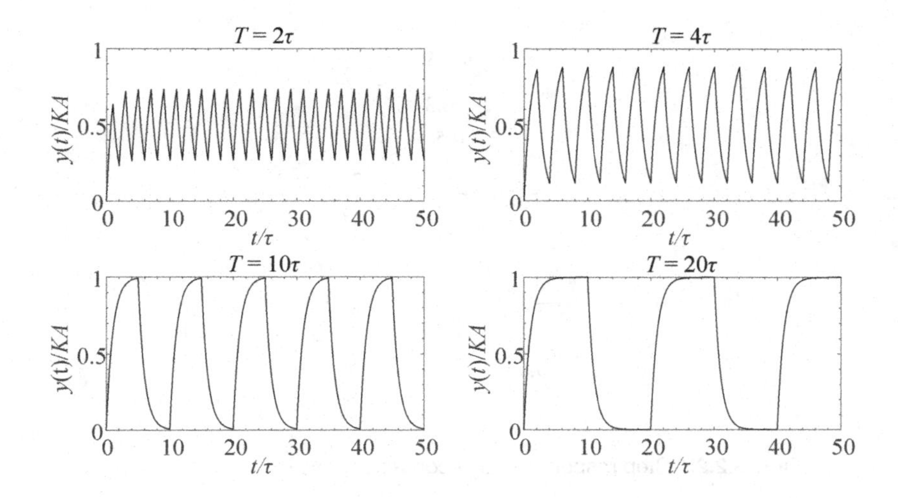

Figure 2.24 The output of a first-order measuring system with a time constant τ, following application of a square-wave input with a period T and an amplitude A at time $t = 0$ under the initial condition $y(0)/(KA) = 0$.

2.4.5 Dynamic Response of Second-Order Systems

This section summarises the mathematical expressions that describe the dynamic response of second-order systems to the idealised types of input mentioned in Section 2.4.2. It is important to note that these expressions take different functional forms for undamped, underdamped, critically damped and overdamped systems.

Step response: For simplicity, consider a step change in input from $x_0 = 0, t \leq 0$ to $A, t > 0$. The output of a second-order system following this step-like input change is

$$\frac{y(t)}{KA} = \begin{cases} 2\sin(\omega_n t)\,, & \text{for } \zeta = 0\,, \\ 1 - \frac{e^{-\zeta\omega_n t}}{\sqrt{1-\zeta^2}}\sin\left(\omega_n t\sqrt{1-\zeta^2} + \arcsin\sqrt{1-\zeta^2}\right), & \text{for } 0 < \zeta < 1\,, \\ 1 - (1+\omega_n t)\,e^{-\omega_n t}\,, & \text{for } \zeta = 1\,, \\ 1 - \dfrac{\zeta + \sqrt{\zeta^2-1}}{2\sqrt{\zeta^2-1}} e^{\left(-\zeta + \sqrt{\zeta^2-1}\right)\omega_n t} & \\ \qquad + \dfrac{\zeta - \sqrt{\zeta^2-1}}{2\sqrt{\zeta^2-1}} e^{\left(-\zeta - \sqrt{\zeta^2-1}\right)\omega_n t}, & \text{for } 1 < \zeta\,. \end{cases} \tag{2.62}$$

These expressions are plotted in Fig. 2.25. As Eq. (2.62) and Fig. 2.25 show, the output of overdamped and critically damped systems increases monotonically towards its steady-state level, whereas the output of underdamped systems oscillates about the

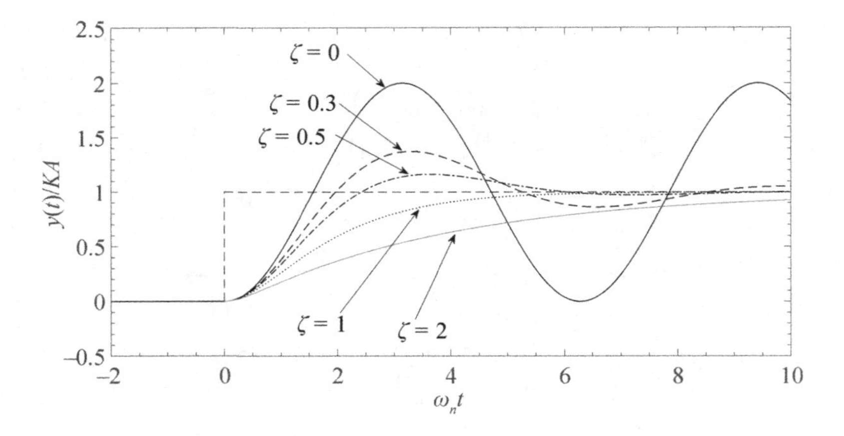

Figure 2.25 Step response of a second-order system.

steady-state level with diminishing amplitude in each cycle but eventually settles at that level. Lightly damped systems ($\zeta \ll 1$) undergo large-amplitude oscillations, which persist over a long time and obscure a measurement. Undamped systems ($\zeta = 0$) would oscillate with a large, undiminished amplitude indefinitely.As mentioned in Section 2.4.1, however, undamped systems do not exist in reality, because all physical systems contain some dissipative mechanism, like friction, which opposes the action of the system, thus introducing damping of oscillations. The output of an underdamped ($0 < \zeta < 1$) system subjected to a step-input would oscillate with a diminishing amplitude, which would be enclosed between two *envelope curves* with equations

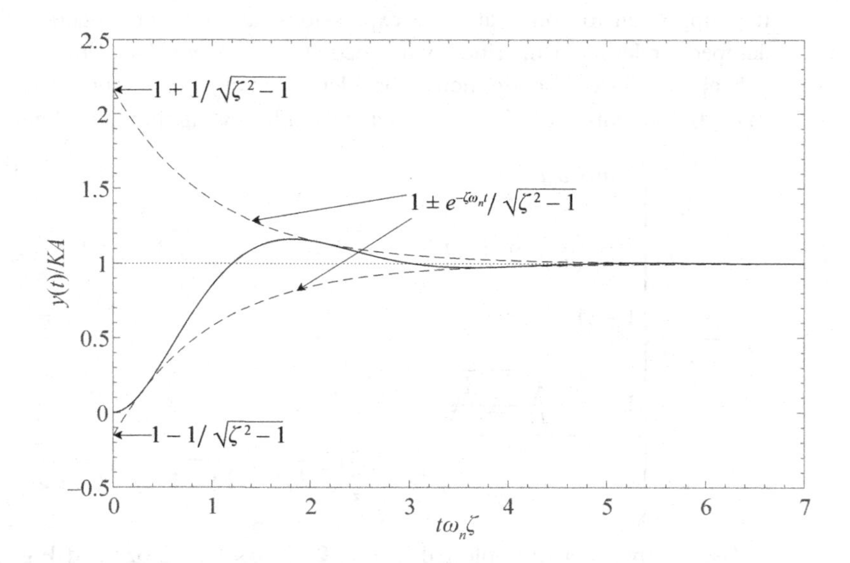

Figure 2.26 Normalised output (solid line) of an underdamped second-order system with $\zeta = 0.5$, subjected to a step-input, and its two envelope curves (dashed lines).

$$1 \pm \frac{e^{-\zeta\omega_n t}}{\sqrt{1-\zeta^2}}, \tag{2.63}$$

as shown in Fig. 2.26. The envelope curves show that the upper bound of the absolute error of an underdamped system following a step-change in the input would be $e^{-\zeta\omega_n t}/\sqrt{1-\zeta^2}$, which, at first glance, is reminiscent of the corresponding error of a first-order measuring system, if one considers the coefficient $1/(\zeta\omega_n)$ as equivalent to the time constant. A more careful comparison of the two errors, however, shows that, while the first-order-system error at a given time depends only on the time constant, the error for underdamped systems not only depends on $1/(\zeta\omega_n)$, but it is also divided by $\sqrt{1-\zeta^2}$.

Example 2.21 The dynamic response of a second-order system is determined by two parameters, the natural frequency ω_n and the damping ratio ζ. In practice, however, it is convenient to specify the dynamic response of a measuring system by the time it takes to approach its steady state. Suppliers of instrumentation, such as pressure transducers and load-cells, often specify a 'response time', albeit without explaining the significance of this term. In this example, we will determine the *settling time* T_s of second-order systems, defined as the time it takes for the output error to be within a certain percentage ε_p of the steady-state output step increase. For consistency with first-order systems, for which the settling time was defined such as $\varepsilon_p = e^{-5} = 0.0067 \equiv 0.67\%$, and noticing the presence of an exponential function within the expressions in Eq. (2.62) for $0 < \zeta$, we may also define the settling time for second-order systems to be the time for which $\varepsilon_p = 0.67\%$.

Answer

It is evident that undamped systems will never settle. For underdamped systems, setting the tolerable error, as approximated by its upper bound that is computed from Eq. (2.63), to our prescribed limit of $\varepsilon_p = e^{-5}$, gives a settling time, in normalised form, of

$$T_s\omega_n \approx \frac{-\ln\left(\varepsilon_p\sqrt{1-\zeta^2}\right)}{\zeta}, \quad \text{for } 0 < \zeta < 1. \tag{2.64}$$

Similarly, for critically damped systems, i.e., for $\zeta = 1$, one can numerically find the normalised time when the error is within 0.67% as $T_s\omega_n \approx 7.09$, by solving the relevant equation from Eq. (2.62). For overdamped systems, the presence of two exponential terms in the step-response equation does not permit an exact analytical solution, but the product $T_s\omega_n$ can be computed numerically by setting the error in the last expression in Eq. (2.62) to $\varepsilon_p = e^{-5}$ and solving this expression for different values of ζ. The product $T_s\omega_n$ for a wide range of ζ has been plotted in Fig. 2.27. It seems instructive, however, to derive an approximate analytical solution for overdamped systems as well, by noting that the second exponential term in the last expression of Eq. (2.62) is negligible by comparison to the first exponential term when $\zeta \gg 1$, and hence this expression can be approximated as

$$\frac{y(t)}{KA} \approx 1 - e^{-\left(\zeta - \sqrt{\zeta^2-1}\right)\omega_n t} \text{ for } 1 \ll \zeta. \tag{2.65}$$

The normalised settling time, under such conditions, can therefore be approximated as

$$T_s\omega_n \approx \frac{-\ln \varepsilon_p}{\zeta - \sqrt{\zeta^2 - 1}}, \quad \text{for } 1 \ll \zeta\,. \tag{2.66}$$

As shown in Fig. 2.27, this estimate of $T_s\omega_n$ is very close to the numerical solution of the exact equation, not only for $\zeta \gg 1$, but also for all values of $\zeta > 1$, except those in the immediate vicinity of 1. Figure 2.27 makes it clear that the natural frequency alone is not sufficient for determining, even roughly, the settling time of a second-order system, because the latter is also sensitive to the value of the damping ratio.

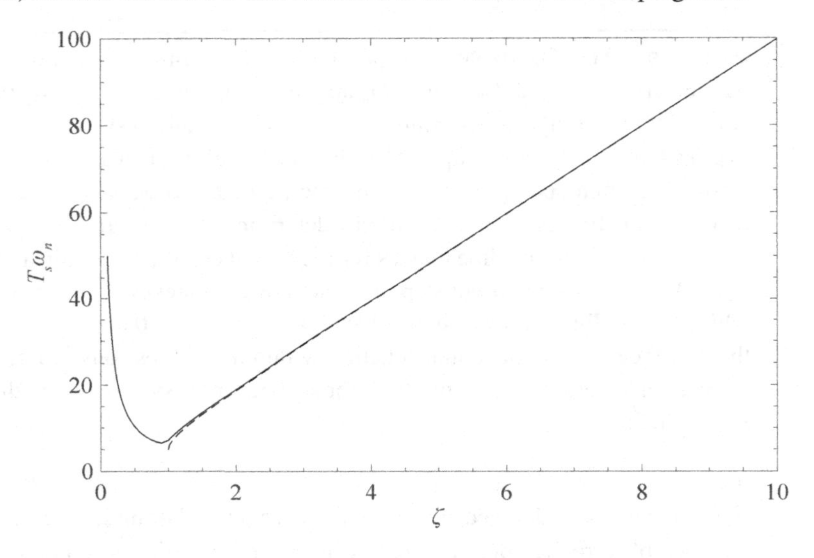

Figure 2.27 Normalised settling time of second-order systems (solid line) and its approximation for $\zeta \gg 1$ (dashed line).

Impulse response: The output of a second-order system, following an impulsive input at $t = 0$, is

$$\frac{y(t)}{KA} = \begin{cases} \omega_n \sin(\omega_n t)\,, & \text{for } \zeta = 0 \\ \frac{\omega_n e^{-\zeta\omega_n t}}{\sqrt{1-\zeta^2}} \sin\left(\sqrt{1-\zeta^2}\,\omega_n t\right), & \text{for } 0 < \zeta < 1 \\ \omega_n^2 t e^{-\omega_n t}, & \text{for } \zeta = 1 \\ \frac{\omega_n}{2\sqrt{\zeta^2-1}} \left[e^{\left(-\zeta+\sqrt{\zeta^2-1}\right)\omega_n t} - e^{\left(-\zeta-\sqrt{\zeta^2-1}\right)\omega_n t}\right], & \text{for } 1 < \zeta. \end{cases} \tag{2.67}$$

Similarly to their step response, critically damped and overdamped systems have a non-oscillatory impulse response, while underdamped systems have an oscillatory response. The steady-state error for an impulse-type input, obtained by letting $\omega_n t \to \infty$, vanishes, except for $\zeta = 0$, in which case the error oscillates indefinitely.

Ramp response: The output of a second-order system with a ramp-like input that starts at $t = 0$ and vanishes for $t \leq 0$ is

$$\frac{y(t)}{KA} = \begin{cases} t - \frac{1}{\omega_n} \sin(\omega_n t)\,, & \text{for } \zeta = 0\,, \\ t - \frac{2\zeta}{\omega_n} + \frac{e^{-\zeta\omega_n t}}{\omega_n \sqrt{1-\zeta^2}} \sin\left(\sqrt{1-\zeta^2}\omega_n t + \arctan\frac{2\zeta\sqrt{1-\zeta^2}}{2\zeta^2-1}\right), & \text{for } 0 < \zeta < 1\,, \\ t(1+e^{-\omega_n t}) - \frac{2}{\omega_n}(1-e^{-\omega_n t})\,, & \text{for } \zeta = 1\,, \\ t - \frac{2\zeta}{\omega_n} - \frac{2\zeta^2 - 1 - 2\zeta\sqrt{\zeta^2-1}}{2\omega_n\sqrt{\zeta^2-1}} e^{\left(-\zeta+\sqrt{\zeta^2-1}\right)\omega_n t} \\ \quad + \frac{-2\zeta^2 + 1 - 2\zeta\sqrt{\zeta^2-1}}{2\omega_n\sqrt{\zeta^2-1}} e^{\left(-\zeta-\sqrt{\zeta^2-1}\right)\omega_n t}, & \text{for } 1 < \zeta\,. \end{cases} \tag{2.68}$$

Similarly to their step and impulse responses, critically damped and overdamped systems have a non-oscillatory ramp response, whereas underdamped systems have an oscillatory response. The steady-state error for a ramp-type input, obtained by letting $\omega_n t \to \infty$, is equal to the constant $2\zeta/\omega_n$, except for $\zeta = 0$, in which case the error oscillates indefinitely.

Frequency response: The steady-state amplitude ratio and phase shift for a second-order system subjected to a sinusoidal input are

$$\frac{B}{A} = \frac{K}{\sqrt{[1-(\omega/\omega_n)^2]^2 + 4\zeta^2(\omega/\omega_n)^2}}\,, \tag{2.69}$$

$$\varphi = -\arctan\frac{2\zeta\omega/\omega_n}{1-(\omega/\omega_n)^2}\,. \tag{2.70}$$

The response and Bode plots for second-order systems are shown in Fig. 2.28 and Fig. 2.29, respectively. As $\omega/\omega_n \to \infty$, $B/A \to 0$ and $\varphi \to -\pi$. Undamped systems have an infinite output amplitude when $\omega = \omega_n$, a condition referred to as *resonance*. Underdamped systems with $0 < \zeta < \sqrt{2}/2$ have a peak $B/(KA) = 1/\left(2\zeta\sqrt{1-\zeta^2}\right)$ in their output amplitude at the *resonant frequency*

$$\omega_r = \omega_n\sqrt{1-2\zeta^2}\,, \tag{2.71}$$

whereas underdamped systems with $\zeta \geq \sqrt{2}/2$ have no resonant peak, but, instead, their amplitude ratio decreases monotonically with increasing frequency. It is noted that ω_r is smaller than the damped natural frequency, defined previously as $\omega_d = \omega_n\sqrt{1-\zeta^2}$, which is the frequency of free oscillation of underdamped systems. In a manner reminiscent of first-order measuring systems, critically damped and overdamped second-order measuring systems, as well as underdamped ones with $\zeta \geq \sqrt{2}/2$, act like a low-pass filter and have diminishing output amplitudes, when subjected to sinusoidal inputs of increasing frequency. The cut-off frequency of second-order systems can be easily found as

$$\omega_c = \omega_n \left\{ \left(1 - 2\zeta^2\right) + \left[\left(1 - 2\zeta^2\right)^2 + 1 \right]^{1/2} \right\}^{1/2} . \tag{2.72}$$

This expression shows that, for a fixed ω_n, the bandwidth of a second-order system decreases with increasing ζ. Moreover, as shown in ?? and Fig. 2.29, it is only for cases with $\zeta \geq \sqrt{2}/2$ that the gain does not exceed the static level within this bandwidth, whereas, as ζ drops below $\sqrt{2}/2$, the gain increasingly exceeds the static level within part of the bandwidth. The value $\zeta = \sqrt{2}/2 \approx 0.7$ is recommended for many measuring systems, because it produces the widest measurement bandwidth that is free of resonance and overshooting of the output. Regardless of the value of ζ, it is clear that the frequency response of second-order systems depends to a great extent upon the natural frequency of the system.

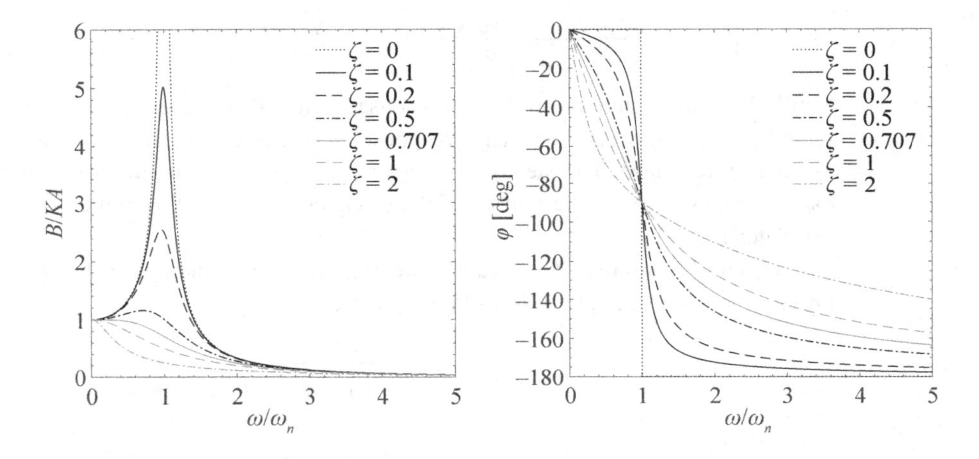

Figure 2.28 Frequency response of second-order systems.

Example 2.22 Determine the relationship between the settling time of a second-order system subjected to a step input, as described in Example 2.21, and its cut-off frequency.

Answer

The product $T_s\omega_c$, calculated from the product $T_s\omega_n$, which was plotted Fig. 2.27, and Eq. (2.72), has been plotted against ζ in Fig. 2.30. It is evident that, for underdamped systems, T_s cannot be determined solely from the value of ω_c, because it is also sensitive to the value of ζ. For overdamped systems, however, $T_s\omega_c \approx 5$, irrespectively of ζ. Considering that, for first-order systems, $T_s/\tau = 5$, one may view $1/\omega_c$ for overdamped second-order systems as equivalent to the time constant of a first-order system.

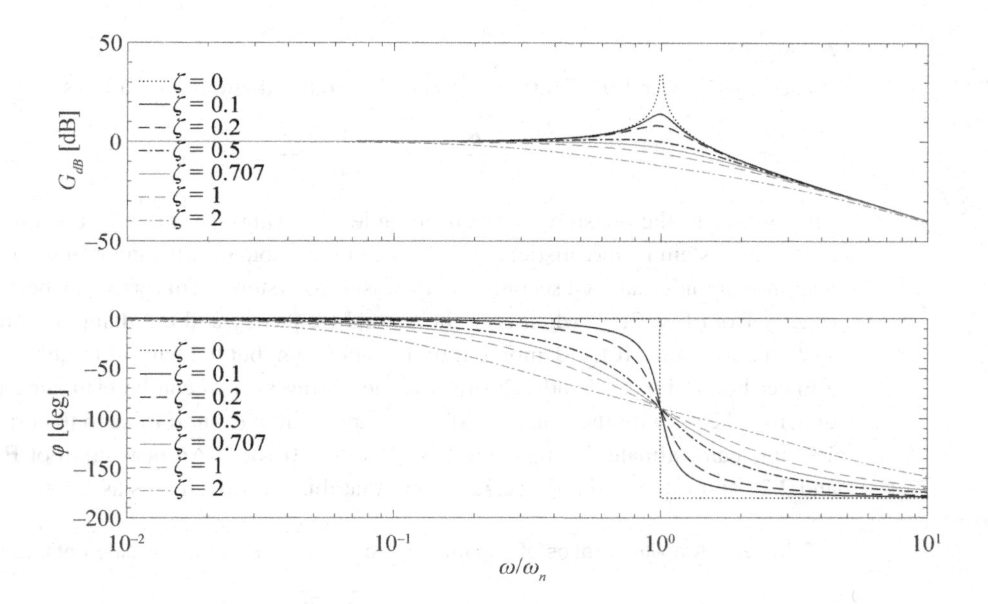

Figure 2.29 Bode plot for second-order systems with $K = 1$.

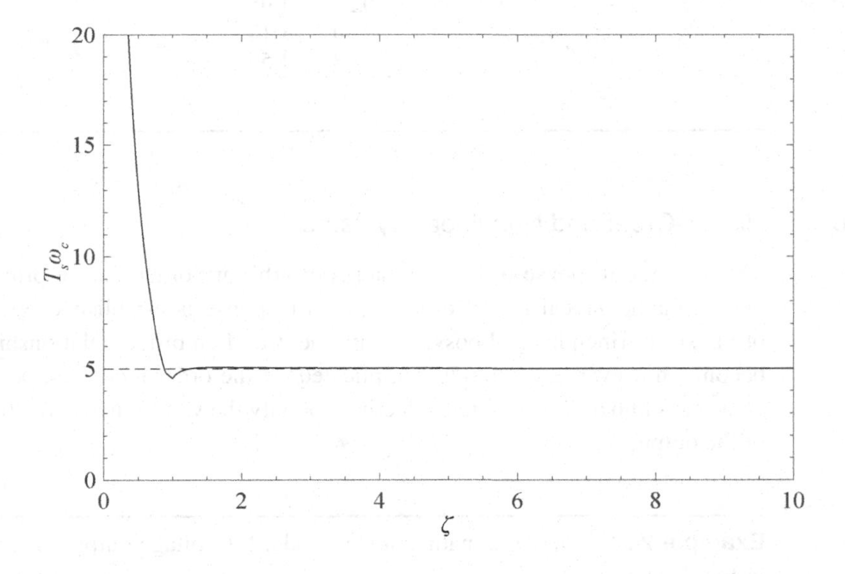

Figure 2.30 Settling time of second-order systems, normalised by the cut-off frequency.

Example 2.23 Consider an undamped second-order measurement device. Determine the frequencies of an input signal for which the output amplitude B, normalised by the static sensitivity K, would be larger than the input amplitude A by 1%, 5%, 10%, and 50%.

Answer

By setting $\zeta = 0$ in Eq. (2.69), one gets the normalised amplitude ratio as

$$\frac{B}{AK} = \frac{1}{[1 - (\omega/\omega_n)^2]}.$$

The answer to the question is given in Table 2.1. Although a purely undamped measurement system is unrealistic, it provides an upper bound of the error in the frequency response of underdamped second-order measuring systems. This error can be calculated exactly from Fig. 2.28, when both the natural frequency and the damping ratio of the system are known. If the damping ratio is not known, but the natural frequency is, then a lower bound for the bandwidth for the measuring system can be estimated from the undamped case with the same ω_n. If, for example, it is desired to keep the error below 1%, one can estimate this bandwidth as $0 \leq \omega < 0.10\omega_n$. An inspection of Figs. 2.28 and 2.29 and Eqs. (2.71) and (2.72) shows that this error decreases as ζ is increased.

Table 2.1 Amplitude ratios of undamped second-order systems for different frequencies

ω/ω_n	B/AK
0.10	1.01
0.22	1.05
0.30	1.10
0.58	1.50

2.4.6 Higher-Order and Non-linear Systems

The dynamic response of linear systems of arbitrary order can, in principle, be found once a mathematical model of the system response is established by the application of physical principles and possibly with the use of empirical relationships. The results become, however, exceedingly complicated, as the order increases, because of the increasing number of parameters affecting not only the value but also the functional shape of the output.

Example 2.24 Derive a mathematical model for voltage output of the circuit shown in Fig. 2.31.

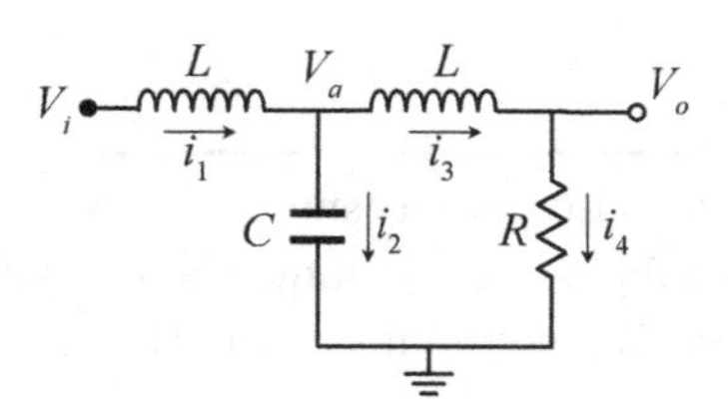

Figure 2.31 A resistor, inductor, capacitor circuit.

Answer

To derive a mathematical model, we first write the voltage–current relationships for all passive elements as

$$V_i - V_a = L\frac{\mathrm{d}i_1}{\mathrm{d}t},$$
$$V_a - V_o = L\frac{\mathrm{d}i_3}{\mathrm{d}t},$$
$$V_o - 0 = Ri_4,$$
$$i_2 = C\frac{\mathrm{d}(V_a - 0)}{\mathrm{d}t}.$$

Then, we write Kirchhoff's current law for the two internal nodes of the circuit as

$$i_1 = i_2 + i_3,$$

$$i_3 = i_4.$$

Our objective is to find a relationship between the output V_o and the input V_i, which may contain the passive elements L, C, R, but no other unknown. The set of six previous equations contains, besides V_i and V_o, the five unknowns V_a, i_1, i_2, i_3, i_4. These five unknowns can be eliminated by substitution, which can follow different equivalent paths.

The currents can be computed as

$$i_3 = i_4 = \frac{1}{R}V_o,$$

$$i_2 = C\frac{\mathrm{d}V_a}{\mathrm{d}t} = C\frac{\mathrm{d}V_o}{\mathrm{d}t} + CL\frac{\mathrm{d}^2 i_3}{\mathrm{d}t^2} = C\frac{\mathrm{d}V_o}{\mathrm{d}t} + \frac{CL}{R}\frac{\mathrm{d}^2 V_o}{\mathrm{d}t^2},$$

$$i_1 = i_2 + i_3 = C\frac{\mathrm{d}V_o}{\mathrm{d}t} + \frac{CL}{R}\frac{\mathrm{d}^2 V_o}{\mathrm{d}t^2} + \frac{1}{R}V_o.$$

Finally, eliminating V_a, we get

$$V_i = V_a + L\frac{\mathrm{d}i_1}{\mathrm{d}t} = V_o + L\frac{\mathrm{d}i_3}{\mathrm{d}t} + L\frac{\mathrm{d}i_1}{\mathrm{d}t},$$

whence we get the mathematical model of this circuit, written in the form of Eq. (2.27), as

$$\frac{L^2C}{R}\frac{\mathrm{d}^3 V_o}{\mathrm{d}t^3} + LC\frac{\mathrm{d}^2 V_o}{\mathrm{d}t^2} + \frac{2L}{R}\frac{\mathrm{d}V_o}{\mathrm{d}t} + V_o = V_i. \tag{2.73}$$

This equation demonstrates that this circuit is a third-order LTI system. The only non-zero coefficients of the various terms in this equation are $a_3 = L^2C/R$, $a_2 = LC$, $a_1 = 2L/R$, $a_0 = 1$ and $b_0 = 1$. The output of this circuit, when subjected to a specified input, can be found by solving Eq. (2.73). Then, one can determine other properties, for example, the steady-state amplitude ratio following the application of a sinusoidal input. Although such solution may be possible mathematically and it is certainly possible numerically, one can obtain many useful results much easier with the use of the Laplace transform, which will be discussed in the next section.

The dynamic response of non-linear systems can be found analytically only for a few special cases, which leaves numerical analysis as the only practical approach. For these reasons, the use of higher-order and non-linear models is rather rare in the measurement field.

2.4.7 Dynamic Analysis Using Laplace Transform

The *Laplace transform* of any time-dependent property $f(t)$, such that $f(t) = 0$ for $t < 0$, is defined as [3, 4, 7]

$$\mathcal{L}[f(t)] = \mathbf{F}(s) = \int_0^{\infty} f(t)e^{-st}\mathrm{d}t\,, \tag{2.74}$$

where $s = j\omega$ is a complex variable with a dimension of frequency, $j = \sqrt{-1}$ is the imaginary unit and the transformed properties are denoted by bold capital characters. The Laplace transform is unique and reversible. Given $\mathbf{F}(s)$, one may compute the original time-dependent property $f(t)$ using the *inverse Laplace transform*

$$f(t) = \mathcal{L}^{-1}\left[\mathbf{F}(s)\right]\,. \tag{2.75}$$

The reader is reminded that, unlike integrals of real functions, the value of which depends uniquely on the limits of integration, integrals of complex functions depend on the integration path in the complex variable plane. Consequently, expressions derived by integrating real functions do not, in general, apply to complex integrals. To evaluate a complex integral, one should first express it as the sum of a real and an imaginary part, carry out the two integrations separately, and then sum the two parts to determine a unique expression for the complex integral.

Laplace transform pairs $f(t), \mathbf{F}(s)$ have been tabulated in many sources [e.g., 7]. An important property of the Laplace transform, which makes it very attractive for use with linear, time-invariant systems, is the *differentiation property*

$$\mathcal{L}\left[\frac{\mathrm{d}^n f(t)}{\mathrm{d}t^n}\right] = s^n\mathbf{F}(s) - \left(s^{n-1}\frac{\mathrm{d}^{n-1}f(0)}{\mathrm{d}t^{n-1}} + \cdots + s\frac{\mathrm{d}f(0)}{\mathrm{d}t} + f(0)\right)\,. \tag{2.76}$$

This property allows the Laplace transform of any derivative of $f(t)$ to be computed from the Laplace transform of $f(t)$ itself and a set of initial conditions.

Consider a linear, time-invariant system with an input $x(t)$ and an output $y(t)$ and let $\mathbf{X}(s)$ and $\mathbf{Y}(s)$ be the corresponding Laplace transforms. The *Laplace transfer function* of this system is defined as

$$\mathbf{H}(s) = \frac{\mathbf{Y}(s)}{\mathbf{X}(s)}\,, \tag{2.77}$$

under the assumption that all initial conditions are equal to zero. Applying the Laplace transform to both sides of the general mathematical model of such systems (Eq. (2.27)) and using the differentiation property on both sides, under the assumption that the initial values of the input and all its derivatives are equal to zero, readily provides an expression for the Laplace transfer function as

$$\mathbf{H}(s) = \frac{b_m s^m + b_{m-1} s^{m-1} + \cdots + b_1 s + b_0}{a_n s^n + a_{n-1} s^{n-1} + \cdots + a_1 s + a_0} . \tag{2.78}$$

For systems described by the simpler model of Eq. (2.28), the Laplace transfer function is simplified to

$$\mathbf{H}(s) = \frac{b_0}{a_n s^n + a_{n-1} s^{n-1} + \cdots + a_1 s + a_0} . \tag{2.79}$$

The Laplace transfer function defines uniquely the relationship between the output and the input of the measurement system. The order of the system is equal to the highest power of s found on the denominator of Eq. (2.78). If the differential equation connecting input and output is known in the time domain, then the transfer function can be easily computed from Eq. (2.78). Then, one may compute the output in the complex frequency domain as

$$\mathbf{Y}(s) = \mathbf{H}(s)\mathbf{X}(s) , \tag{2.80}$$

from which, given the input function and taking an inverse Laplace transform, one can readily compute the output in the time domain as

$$y(t) = \mathcal{L}^{-1} [\mathbf{Y}(s)] . \tag{2.81}$$

The Laplace transforms of some idealised types of input are listed in Table 2.2. When applied to Eqs. (2.31) and (2.32), Eq. (2.78) provides the Laplace transfer function of a first-order measuring system as

$$\mathbf{H}(s) = \frac{K}{1 + \tau s} \tag{2.82}$$

and that of a second-order measuring system as

$$\mathbf{H}(s) = \frac{K\omega_n^2}{s^2 + 2\zeta\omega_n s + \omega_n^2} . \tag{2.83}$$

Table 2.2 Laplace transforms of some idealised dynamic inputs

Input type	$x(t)$	$\mathbf{X}(s)$
Impulse	$A\delta(t)$	A
Step	A	A/s
Ramp	At	A/s^2
Sinusoidal	$A \sin \omega t$	$A\omega/\left(s^2 + \omega^2\right)$

Example 2.25 Apply the Laplace transform method to determine the output of a first-order system subjected to an impulse-type input.

Answer

Substituting Eq. (2.82) and $\mathbf{X}(s)$ from Table 2.2 into Eq. (2.80), one gets

$$\mathbf{Y}(s) = \mathbf{H}(s)\mathbf{X}(s) = \frac{AK}{\tau s + 1} .$$

By taking the inverse Laplace transform of this expression, one gets the output of a first-order system subjected to an impulse-type input as

$$y(t) = \mathcal{L}^{-1}\left[\mathbf{Y}(s)\right] = \frac{AK}{\tau} e^{-t/\tau} ,$$

which is the same as the previously derived Eq. (2.55).

In systems analysis, it is customary, instead of the complex variable s, to use the real frequency ω, defined such as $s = j\omega$. Then, instead of the term Laplace transfer function, one may use the term *sinusoidal transfer function* $\mathbf{H}(j\omega)$. As a complex function, $\mathbf{H}(j\omega)$ (or $\mathbf{H}(s)$) is composed of a real part and an imaginary part, namely,

$$\mathbf{H}(j\omega) = \mathrm{Re}\,\mathbf{H}(j\omega) + j\,\mathrm{Im}\,\mathbf{H}(j\omega) . \tag{2.84}$$

In polar form, $\mathbf{H}(j\omega)$ may be written as

$$\mathbf{H}(j\omega) = |\mathbf{H}(j\omega)|\, e^{j\varphi(j\omega)} , \tag{2.85}$$

where its *magnitude* and *argument* (*phase*) are, respectively, defined as

$$|\mathbf{H}(j\omega)| = \sqrt{[\mathrm{Re}\,\mathbf{H}(j\omega)]^2 + [\mathrm{Im}\,\mathbf{H}(j\omega)]^2} \tag{2.86}$$

and

$$\varphi(j\omega) = \arctan \frac{\mathrm{Im}\,\mathbf{H}(j\omega)}{\mathrm{Re}\,\mathbf{H}(j\omega)} . \tag{2.87}$$

Example 2.26 Derive the magnitude of the Laplace transfer function for a third-order system.

Answer

Eq. (2.79) gives the Laplace transfer function for a third-order system as

$$\mathbf{H}(s) = \frac{b_0}{a_3 s^3 + a_2 s^2 + a_1 s + a_0} ,$$

which, after s is replaced by $j\omega$, takes the form

$$\mathbf{H}(j\omega) = \frac{b_0}{-ja_3\omega^3 - a_2\omega^2 + ja_1\omega + a_0}$$
$$= \frac{b_0}{(a_0 - a_2\omega^2) + j\omega(a_1 - a_3\omega^2)}$$
$$= \frac{b_0[(a_0 - a_2\omega^2) - j\omega(a_1 - a_3\omega^2)]}{[(a_0 - a_2\omega^2) + j\omega(a_1 - a_3\omega^2)][(a_0 - a_2\omega^2) - j\omega(a_1 - a_3\omega^2)]}$$
$$= \frac{b_0[(a_0 - a_2\omega^2)]}{(a_0 - a_2\omega^2)^2 + \omega^2(a_1 - a_3\omega^2)^2} + j\frac{b_0[-\omega(a_1 - a_3\omega^2)]}{(a_0 - a_2\omega^2)^2 + \omega^2(a_1 - a_3\omega^2)^2} .$$

The magnitude of the Laplace transfer function, as defined in Eq. (2.86), is therefore

$$|\mathbf{H}(j\omega)| = \left[\frac{b_0^2}{(a_0 - a_2\omega^2)^2 + \omega^2(a_1 - a_3\omega^2)^2}\right]^{1/2} .$$

Note that, letting $a_3 = 0$, one may simplify this expression so that it applies to second-order systems, in which case it takes the form of Eq. (2.69). Further letting also $a_2 = 0$, one may obtain the Laplace transfer function magnitude for second-order systems, which takes the form of Eq. (2.59). Finally, letting also $a_1 = 0$, one obtains the Laplace transfer function for zero-order systems, which is equal to the static sensitivity K.

One of the useful properties of the sinusoidal transfer function is that its magnitude and argument are equal to the amplitude ratio and phase change in the steady-state frequency response of a linear, time-invariant system of arbitrary order. Thus, the gain in the Bode plot can be determined as

$$G_{dB}(\omega) = 20\log_{10}(|\mathbf{H}(j\omega)|) . \tag{2.88}$$

The magnitude and the argument of $\mathbf{H}(j\omega)$ take particularly simple forms in the two limiting cases. As $\omega \to 0$, $|\mathbf{H}(j\omega)| \to |K|$ and $\varphi(j\omega) \to 0$ for any linear system. As $\omega \to \infty$, $|\mathbf{H}(j\omega)| \to |b_0/a_n|/\omega^n$ and $\varphi(j\omega) \to -\pi n/2$ for any n-order LTI system that is described by the simplified model of Eq. (2.28). From these expressions, one may easily determine the roll-off of an n-order system as $-20n$ dB/decade or $-6n$ dB/octave.

Another useful characteristic of LTI systems that can be provided by the sinusoidal transfer function is the *corner frequency* ω_{cf}, defined as the frequency at the intersection between the asymptotes to the amplitude ratio expressions at the two limits. It is elementary to determine the corner frequency as

$$\omega_{cf} = (a_0/a_n)^{1/n} . \tag{2.89}$$

For first-order systems, this becomes $\omega_{cf} = 1/\tau$, whereas for second-order systems, it becomes $\omega_{cf} = \omega_n$. These results show that the corner frequency, which is relatively easy to determine experimentally with the use of a frequency response test (see Section 2.4.8), provides important characteristics of the measuring system.

Example 2.27 Determine the steady-state frequency response of the circuit shown in Fig. 2.31 (Example 2.24) and again in Fig. 2.32a by using the Laplace transform. Draw Bode plots for $R = 1\ \text{k}\Omega$, $L = 1$ mH and $C = 10$ pF.

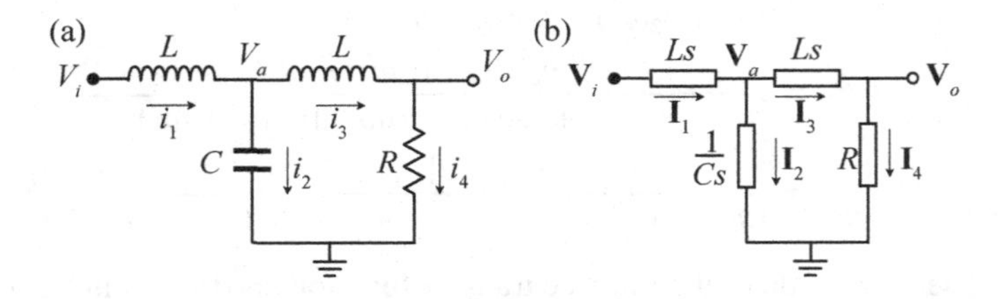

Figure 2.32 A resistor–inductor–capacitor circuit in (a) real space and (b) complex space.

Answer

In Example 2.24, we derived a mathematical model of this circuit (Eq. (2.73)), which shows that this circuit is a third-order LTI system. Then, applying Eq. (2.79), we find the Laplace transfer function as

$$\mathbf{H}(s) = \frac{1}{\frac{L^2C}{R}s^3 + LCs^2 + \frac{2L}{R}s + 1} .$$

An alternative derivation of the Laplace transfer function can be made by considering the same circuit in the complex frequency domain, as shown in Fig. 2.32b. This is possible because the voltage–current relationships for circuit elements, Kirchhoff's laws and the Laplace transform are all linear operations. First, we replace all circuit elements by the corresponding impedances. Then, we apply Kirchhoff's current law in the complex frequency domain as

$$\mathbf{I}_1(s) = \mathbf{I}_2(s) + \mathbf{I}_3(s)$$

and

$$\mathbf{I}_3(s) = \mathbf{I}_4(s) .$$

Substituting the currents by the ratios of the corresponding voltages and impedances ($\mathbf{I}(s) = \mathbf{V}(s)/\mathbf{Z}(s)$), we get

$$\frac{\mathbf{V}_i(s) - \mathbf{V}_a(s)}{Ls} = \mathbf{V}_a(s)Cs + \frac{\mathbf{V}_a(s) - \mathbf{V}_o(s)}{Ls}$$

and

$$\frac{\mathbf{V}_a(s) - \mathbf{V}_o(s)}{Ls} = \frac{\mathbf{V}_o(s)}{R} .$$

Finally, eliminating $\mathbf{V}_a(s)$ and rearranging the terms, we get the Laplace transfer function once more as

$$\mathbf{H}(s) \equiv \frac{\mathbf{V_o}(s)}{\mathbf{V_i}(s)} = \frac{1}{\frac{L^2C}{R}s^3 + LCs^2 + \frac{2L}{R}s + 1} ,$$

which, upon substitution of the given values for the circuit elements R, L, C, becomes

$$\mathbf{H}(s) = \frac{1}{1 \times 10^{-20}s^3 + 1 \times 10^{-14}s^2 + 2 \times 10^{-6}s + 1}.$$

Comparing this expression to the general form of the transfer function (Eq. (2.79)), we find that $b_0 = 1, a_3 = 1 \times 10^{-20}, a_2 = 1 \times 10^{-14}, a_1 = 2 \times 10^{-6}$ and $a_0 = 1$. We further note that the static sensitivity of this system is $K = 1$, or, equivalently, its low frequency gain is $G_{dB}(0) = 0$. Therefore, this circuit neither amplifies nor attenuates low frequency input signals. On the other hand, as $\omega \to \infty$, $|\mathrm{H}(s)| \to 0$, and therefore $G_{dB}(\omega) \to -\infty$, which implies that this circuit attenuates high frequency input signals and so it acts like a low-pass filter. Substituting the given values into the Laplace transfer function for a third-order system, which was derived in Example 2.26, setting the right-hand side in this expression to $1/\sqrt{2}$ and solving it numerically for ω, we get the cut-off frequency as $\omega_c = 4.98 \times 10^5$ rad/s. Eq. (2.89) further provides the corner frequency as $\omega_{cf} = (a_0/a_n)^{1/n} = \left(1/1 \times 10^{-20}\right)^{1/3} = 4.6 \times 10^6$ rad/s. Given that the circuit is a third-order system, one may assert that, as $\omega \to \infty$, the roll-off would be -60 dB/decade and the phase shift would tend to $-3\pi/2$. The Bode plots for this circuit are shown in Fig. 2.33. As one might have expected, these plots are quite different from those of any first- or second-order system. In fact, the appearance of these plots would be drastically different for different combinations of values of the circuit elements. It is noted that the design and specifications of this circuit are entirely arbitrary and void of any physical significance. The most prominent features of these Bode plots are a distinct peak in gain and a nearly stepwise drop in phase change at $\omega = 1.4 \times 10^7$ rad/s.

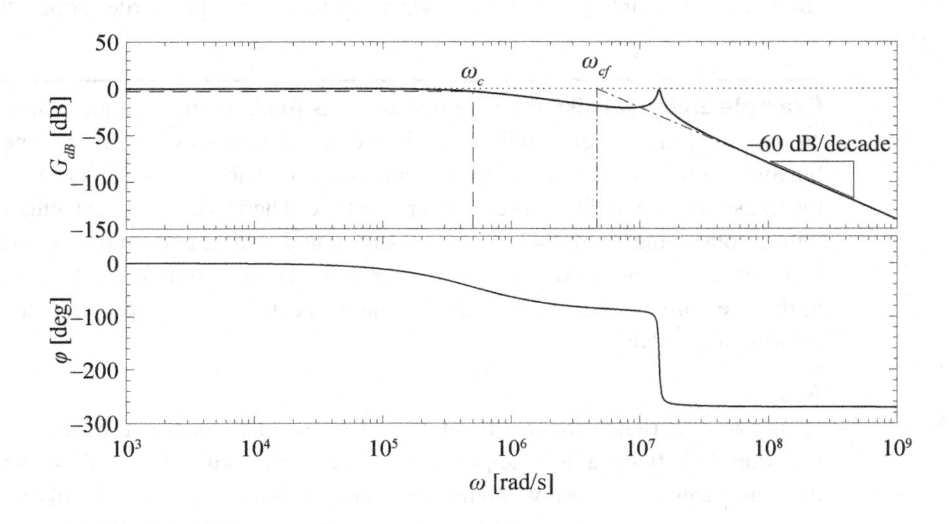

Figure 2.33 Bode plots for the circuit shown in Figs. 2.31 and 2.32.

2.4.8 Experimental Determination of the Dynamic Response

The dynamic response of a measuring system can, in general, be determined by *dynamic calibration*, which is achieved by measuring the time-dependent output, while the system is subjected to a controlled and known time-dependent input. It will be generally assumed that the static characteristics of the measuring system have been determined in advance by static calibration. Obviously, dynamic calibration would be easier for systems that are known to be linear and of low order than for high-order systems, non-linear systems and systems for which there is no mathematical model. The most common types of dynamic tests are the *step-input test*, the *square-wave test* and the *frequency test*.

Step-input test: For a step-input test, the input to the measuring system or device is changed from one constant value to another, and its output is recorded over time at a sufficiently fast rate for the dynamic response to be observable and over a sufficiently long time interval for the steady-state output to be reached. Let y_0 be the output value before the step-change of the input is applied and y_∞ be the steady-state output value. If it is known that the system is of first order, one can determine the time-constant τ as the negative slope of the line fitted by linear least squares to the data, plotted in the form

$$t = -\tau \ln\left[1 - \frac{y(t) - y_0}{y_\infty - y_0}\right]. \tag{2.90}$$

If the system is known to be of second order, the test will first demonstrate whether the system is overdamped, in which case the output will approach its steady-state monotonically, or underdamped, in which case the output will be oscillatory before reaching its steady-state. Then, one may employ curve-fitting algorithms to determine the values of ζ and ω_n for the appropriate set of equations from Eq. (2.62). The data from this test would be hard to interpret if the measuring system is of higher order or non-linear.

Example 2.28 A bulky pressure transducer is found in the laboratory and is considered for measuring sequentially multiple pressure differences in a wind tunnel, in combination with a fast-acting, programmable scanning valve, which connects one port of the transducer to a different pressure tap at a time. There is a significant length of plastic tubing connecting the transducer to the valve and the valve to each pressure tap. Devise and implement a procedure for the experimental determination of the dynamic response of this pressure measurement system, so that you can select an appropriate time-step for the scanning valve.

Answer

From past experience and the literature, we expect that this pressure measurement system would likely be, at least approximately, of second-order. The static sensitivity K of the transducer can be easily obtained by static calibration against a U-tube manometer, as described in Example 2.12. Therefore, we need to devise tests that will provide the natural frequency ω_n and damping ratio ζ of the transducer-valve-tubing system. The simplest test to determine the dynamic response of this system would be a step-input test. For such a test, the input, that is, the pressure difference Δp applied to the sys-

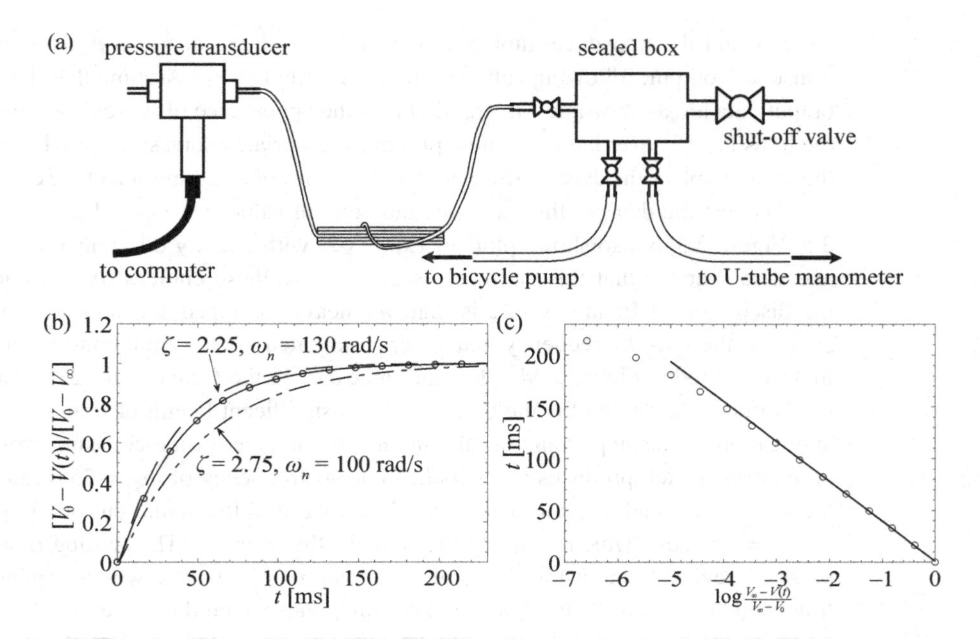

Figure 2.34 (a) Sketch of a simple apparatus for performing static and dynamic calibrations of pressure measurement systems. (b) Measurements (○) of the pressure transducer output, following a step-change in the input; the solid line is a plot of Eq. (2.62) for overdamped systems with $\omega_n = 125$ rad/s and $\zeta = 2.5$, which best fits the data; the two other lines show the same equation with slightly different values of ω_n and ζ for comparison. (c) The same data, plotted in the form of Eq. (2.90), to examine whether the pressure transducer can be represented by a first-order model.

tem, must be changed abruptly from one value to another, while its output, that is, the voltage V provided by the electric circuit that operates the transducer, is recorded at a sufficiently fast rate with the use of a data acquisition board and a computer or a storage oscilloscope. Apparatus on which to perform such tests would not be commonly found in small laboratories and so one may need to improvise. In the following, we will describe a simple apparatus that can be easily put together and a possible scenario of the tests.

Fabricate a pressure calibrator, as shown in Fig. 2.34a. This device consists of a small sealed box, equipped with four ports, connected to shut-off ball valves, which can be quickly opened or closed completely by a quarter-turn of their handle. Connect one port of the transducer to the tubing and then to one valve of the box, while leaving the second port of the transducer open to the atmospheric pressure p_∞. Connect the second valve to a U-tube manometer and the third valve to a bicycle pump, which will be used to pressurise the box to a pressure p_0, ensuring that this value is within the range of the device, as determined by static calibration. Finally, leave the free port of the fourth valve (preferably a large one) open to the atmospheric pressure p_∞.

The procedure for the test is as follows. First, pressurise the box to a pressure p_0 and record the transducer output voltage as V_0. Then, open the free valve quickly to expose the box and the transducer/tubing system to atmospheric pressure p_∞, for which the transducer output, following sufficient time, is settled at V_∞. Assume that the recorded output data are as shown in Fig. 2.34b. From the appearance of these data, observe that the transducer is overdamped. Then, plot the appropriate expression from Eq. (2.62) on the same graph as the data for different combinations of ω_n, ζ and select the combination that fits best the data. In this example, the optimal values are $\omega_n = 125$ rad/s and $\zeta = 2.5$. Figure 2.34b also shows plots of Eq. (2.62) with slightly different values of ω_n, ζ and demonstrates that this approach is sensitive to these choices. A consequence of the discussion in Example 2.22 is that, for heavily damped second-order measuring systems, the cut-off frequency can be approximated by the time-constant of a fitted first-order model. Figure 2.34c plots the same data in the form of Eq. (2.90) and shows that a first-order model fits the data well for a significant length of time following the application of the step-change in the input. The obtained 'time-constant' from this fit is 38.6 ms, which produces an estimated cut-off frequency of $\omega_c = 25.9$ rad/s. Given the values of ζ and ω_n, and using Eq. (2.72), we find the actual cut-off frequency to be $\omega_c = 26$ rad/s, which is nearly the same as the estimate. The settling time for this pressure device, found using Eq. (2.66), is therefore $T_s \approx 0.19$ s, which implies that the time-step between port changes on the scanning valve should be at least 0.2 s.

Square-wave test: During a square-wave test, the input is made to switch periodically from one level to another, which is a process consisting of a sequence of step changes. If it is known that the system under test is of first order, the square-wave test will easily provide the value of the static sensitivity and the time constant. If the system is known to be of second order and its output during a square-wave test is found to be oscillatory, it would become evident that the system is underdamped. Then, the undamped natural frequency and the damping ratio can be found as the values in Eq. (2.62) that provide the best fit to the output waveform. Determination of these parameters for critically damped and overdamped systems is possible, but more difficult, due to the reduced sensitivity of this approach. The determination of the parameters of higher-order systems from a single square wave test is even more complicated or impossible.

Frequency test: During a frequency test, the input is made to oscillate sinusoidally, if possible exactly, or at least approximately. The frequency test is relatively easy to interpret and applies equally to linear systems of arbitrary order. It consists of exposing the system to a sinusoidal input of a constant amplitude and varying frequency. The amplitude ratio B/A is measured and plotted against frequency to characterise the system. Again, if it is known that the measuring system is of first order, the procedure is straightforward: (a) measure the amplitude ratio for a wide range of frequencies; (b) determine the static sensitivity K as the amplitude ratio at very low frequencies; and(c) determine the corner frequency ω_{cf}, from which one can get the cut-off frequency $\omega_c = \omega_{cf}$ and the time constant $\tau = 1/\omega_c$. If it is known that the system is of second order, the procedure is as follows: (a) measure the amplitude ratio for a wide range of frequencies; (b) determine the static sensitivity K as the amplitude ratio at very low frequencies; (c) determine the corner frequency ω_{cf}, which would be equal to ω_n; and (d) plot the gain

$G_{dB}(\omega/\omega_n)$ vs. ω/ω_n in a Bode plot and compare it to curves for different values of the damping ratio ζ, from which one can find the most appropriate ζ.

Dynamic calibration of an electric circuit can be performed conveniently with the use of a simple periodic input voltage (e.g., a square wave or a sinusoidal wave), supplied by an analogue multi-function generator or a digital-to-analogue converter connected to a computer. The dynamic response of velocity, stress, pressure, temperature and composition sensors and measuring systems can be determined by exposing them to a fluid under conditions that change in a step-like, impulse-like or periodic manner. A variety of configurations for such tests have been suggested in the past. A binary field of velocity, temperature or pressure may be approximately created with the use of a perforated disk rotating in front of a fluid jet. An essentially step-wise change in these parameters may also be achieved by bursting a diaphragm separating two chambers containing fluids under different pressures. Dynamic testing of fast-response transducers is sometimes done in shock tubes, which produce travelling shock waves across which there are drastic changes in flow properties. Local heating can be induced by pulsed or chopped laser beams. And sensors can be exposed to known relative velocity fields, while mounted on linear actuators, towing devices and rotating or oscillating arms.

Example 2.29 We need to determine the dynamic response of the pressure transducer described in Example 2.28, but without the tubing. We found that the step-change in pressure produced by the apparatus used was not sudden enough to allow an accurate determination of ω_n and ζ. Devise and implement a procedure for a frequency test of this transducer, which will, hopefully, be suitable for this purpose.

Answer

As with the step-input test, the goal of the frequency test is to determine the values of ζ and ω_n, as well as the bandwidth of the measurement device and its performance across that bandwidth. Hence, our aim is to also determine the corner frequency ω_c.

To begin with, we need to select a device that produces approximately sinusoidal pressure fluctuations in the frequency range of interest and a second device that can be used as the dynamic calibration standard. A readily available device that is capable of producing pressure waves is a loudspeaker and a convenient standard for this dynamic calibration is a microphone. Among the loudspeaker choices, we may consider a woofer, which typically operates in the range 20–2,000 Hz, and a tweeter, which operates in the range 2–20 kHz. A standard microphone typically has a nearly, but not exactly, uniform gain in the range 20–20,000 Hz and we may assume that we use a device for which the manufacturer has provided a Bode plot. It is noted that a microphone cannot replace a pressure transducer, because the former only measures pressure fluctuations and not the pressure value. Thus, we put together a dynamic calibration apparatus, consisting of a sealed and sound-insulated box with one or two loudspeakers mounted on one wall, while the pressure transducer with its tubing and the microphone are connected to the opposite wall, as shown in Fig. 2.35a. An issue that needs some attention is the amplitude of the pressure fluctuations that serve as input to the transducer. We cannot

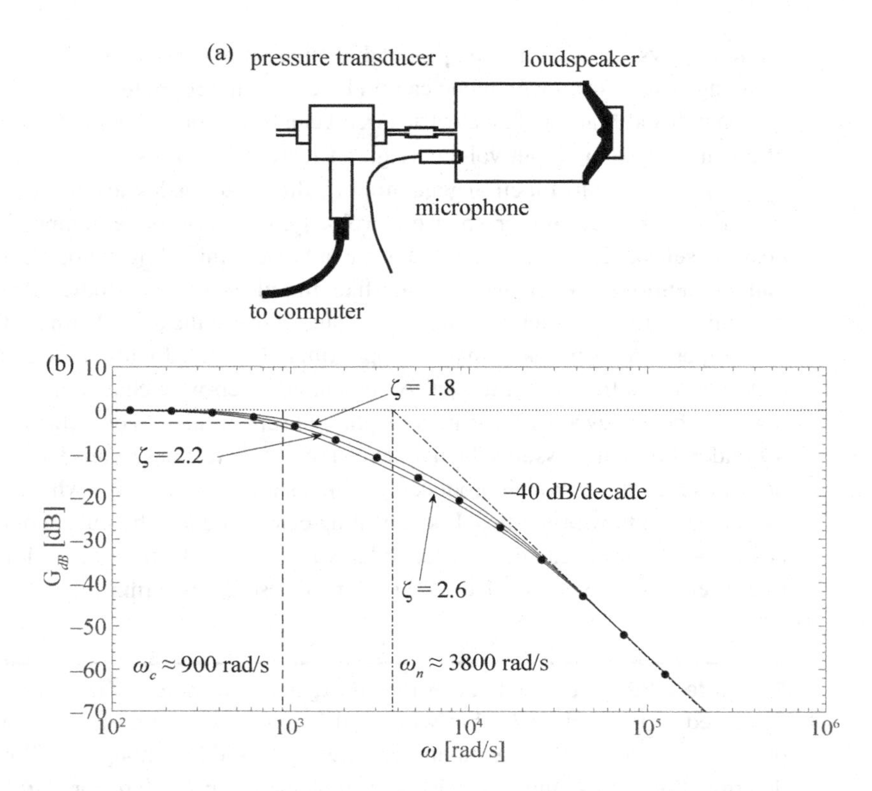

Figure 2.35 (a) Sketch of a simple apparatus for performing a frequency test of pressure transducer. (b) A Bode plot showing the determination of the corner frequency, the cut-off frequency and the best-fitted damping ratio, in this case $\zeta \approx 2.2$.

know this amplitude, because the microphone Bode plot is on a relative scale. We can, however, use the microphone Bode plot to adjust the amplitude of the input signal that drives the loudspeaker to ensure that this amplitude remains constant as we change the frequency. Then, we can normalise the amplitude of the transducer output voltage by its value at the low-frequency plateau, if such a plateau exists in the measurements or can be determined reliably by a moderate extrapolation of the data.

We adjust the amplitude of the sinusoidal signal fed into the loudspeaker at each frequency and record the amplitude of the signal measured by the pressure transducer. For the low-frequency range, and for this particular transducer, we find that the amplitude measured by the transducer is roughly constant. We use this value to normalise all amplitudes and produce a gain vs. frequency plot, which is shown in Fig. 2.35b. Note that the plot is constructed using an apparent static sensitivity $K_{app} = 1$, because we do not know the true amplitude of the pressure fluctuations from which to determine the true static sensitivity K. Nevertheless, the true value of K does not matter for our purposes, as a change in the value of this parameter will only shift the data points in Fig. 2.35b up or down, without affecting the determination of the corner frequency. Assuming that the device is a second-order system, we expect its roll-off to be −40 dB/decade.

Fitting this gradient to the high-frequency data, we find the corner frequency to be $\omega_{cf} = 3{,}800$ rad/s. This is also the value of the undamped natural frequency, as, for second-order systems, $\omega_{cf} = \omega_n$. To determine the damping ratio ζ, we plot Eq. (2.79) for different ζ and select the value $\zeta \approx 2.2$, which fits the data better than other values. Finally, substituting the obtained values of ζ and ω_n in Eq. (2.72), we obtain the cut-off frequency to be $\omega_c \approx 900$ rad/s.

Chapter Digest

2.1 Which electric property has a base unit in the SI system?

2.2 Which are the three pure passive elements in electric circuits? How is impedance related to the properties of each of these elements?

2.3 In your opinion, is circuit analysis based on Kirchhoff's laws equivalent to analysis with the use of Laplace transform? Do you see any advantages or disadvantages of one or the other?

2.4 Which is the major advantage of an active circuit by comparison to a passive one with the same nominal action? Do you see any disadvantage of an active circuit?

2.5 Explain the difference between a modifying input and an interfering input. Is it possible for an input to be both modifying and interfering? Provide an example that is different from Example 2.1.

2.6 A load cell, which is a device that is used for measuring forces, is known to have a second-order response. Describe tests suitable for determining the values of the corresponding parameters.

2.7 What is a Wheatstone bridge and why is it used so commonly in instrumentation and measurement?

2.8 Explain the difference between filtering out an undesirable input and compensating for it.

2.9 Why should the input impedance of an op-amp be very large and why should its output impedance be very small?

2.10 How does a comparator work?

2.11 Determine the sensitivity of the simple op-amp configuration shown in Fig. 2.12b.

2.12 Consider a pressure transducer that will be used for the measurement of wall pressure of an aircraft model in the wind tunnel. Explain how you would perform a static calibration and a dynamic calibration of this transducer. For both cases, sketch the calibration setup, identifying all instruments to be used and their functions.

2.13 A manometer at atmospheric pressure is suddenly connected to a high-pressure air cylinder. What can you conclude about the type of dynamic model that is appropriate for this manometer, if its reading: (i) monotonically increases to a value and (ii) oscillates before it settles to a value. Which kind would you prefer to use and why?

2.14 A first-order system and an overdamped second-order system have both a monotonically increasing output, when subjected to a step change in input. Discuss how you would distinguish one from the other.

2.15 The input of a first-order electric circuit that has a time constant $\tau = 1$ s is connected to a square-wave generator. Sketch the expected output of the circuit on an oscilloscope, when the fundamental frequency f_o of the square wave is set to (a) 0.1 Hz, (b) 1 Hz and (c) 10 Hz.

2.16 In what type of system that will be subjected to a fluctuating input can one use static calibration to establish the relationship between the average input and the average output values?

2.17 Explain the difference between a system operating under static conditions and one operating under steady state conditions.

2.18 Sketch roughly the impulse responses of (a) a zero-order system, (b) a first-order system, (c) an underdamped second-order system, (d) a critically damped second-order system and (e) an overdamped second-order system.

2.19 The Laplace transform is suitable for the analysis of linear systems. Express your thoughts on the possible use of Laplace transform for non-linear systems.

2.20 Prove that the necessary condition for a second-order measuring system not to have any resonance is $\zeta \geq 1/\sqrt{2}$.

Problems

2.1 To deliver an online lecture, a professor uses a microphone connected to a personal computer and a student listens to the lecture through headphones connected to another personal computer. We are only concerned about the audio communication. (a) Identify the desired input and the output of this system. (b) Identify and discuss as many as possible interfering and modifying inputs to this system. (c) Identify as many as possible components of this system and the function of each component.

2.2 A resistance temperature detector (RTD) with a time constant $\tau = 2$ s was used to measure the temperature of coffee contained in three thermos flasks, to be referred to as A, B and C. After the RTD was left in each flask for 60 s, it read temperatures of 30°C for A, 60°C for B and 80°C for C. Then, we repeated the measurements, except this time we first left the RTD in A for 60 s, then moved it quickly to B and, after 3 s, we moved it quickly to C, taking a reading after another 3 s. Determine the dynamic error. Describe all the assumptions that you have made and discuss how realistic each assumption is and how deviation from each assumption might influence the dynamic error. Based on heat transfer analysis, one can model the temperature reading T of an RTD that is submerged in a liquid with temperature T_c by the first-order differential equation $\tau \mathrm{d}T/\mathrm{d}t + T = T_c$. Recommend a minimum time for the RTD to remain in flasks B and C, if you wish the final dynamic error not to exceed 0.5°C.

2.3 A temperature measuring system consists of a sensor, which is a resistance temperature detector (RTD), and an electronic circuit, which has been adjusted by static calibration to produce a voltage output V, the numerical value of which, in volts, is equal to the RTD temperature T_c, in degrees Celsius. It is given that the time constant of the RTD is $\tau = 6$ s. The RTD is immersed for a long time in a kettle containing water with a constant initial temperature $T_0 = 20$°C. At some instant, taken to be the origin of time ($t = 0$ s), the kettle starts getting heated by a stove at a rate of 20°C/min up to the boiling temperature $T_b = 100°C$, at which it is maintained for some time. Let $T(t)$ be the water temperature, which is assumed to be uniform within the entire kettle.

(a) Identify the input and the output of this temperature measuring system. You may use T_c instead of V.

(b) Determine the static sensitivity K of this system.

(c) Derive an equation for the dynamic response of this measuring system. Stating any assumptions that you make, simplify the Energy Equation (First Law of Thermodynamics), for the case for which heat transfer from the water to the RTD takes place only by convection with a constant heat transfer coefficient h (discuss whether h would actually be constant during the entire process). Show that the dynamic response is described by the linear, first-order, ordinary differential equation

$$\tau \frac{\mathrm{d}T_c(t)}{\mathrm{d}t} + T_c(t) = T(t) ,$$

where τ is a time constant. Express the time constant in terms of geometrical and thermophysical properties of the water and/or the RTD.

(d) Describe the water temperature $T(t)$ as a function of time and plot this function. Determine the time at which the water in the kettle would start boiling.

(e) Determine the specific forms that the dynamic response equation takes (i) before the water starts getting heated; (ii) during the time interval between the instant the water starts getting heated and the instant it starts boiling; (iii) after the water has started boiling.

(f) Solve the differential equations obtained in the previous question to determine the temperature T_c, indicated by the RTD, and the measurement error, namely the difference $T - T_c$, during the three different time intervals. Plot these functions in the same graph as the plot of T. Discuss the value of the measurement error before the start of heating, during heating and after the water has started boiling.

Note: You may assume that, at the start of boiling, a step input is applied, having an amplitude equal to the difference between the boiling temperature and the temperature indicated by the RTD at that instant.

2.4 Consider a U-tube manometer, containing water as the working liquid and used for measuring pressure differences in air flows. (a) Identify the desired input and the output of this measuring system. (b) Derive the ideal response of this system, assuming that air density is negligible by comparison to water density. (c) Identify and discuss as many as possible interfering and modifying inputs to this measuring system. (d) Derive the response of this system, if it is tilted with respect to the vertical direction (i) about a horizontal axis that is on the plane defined by the axes of the two glass tubes of the manometer and (ii) about a horizontal axis that is perpendicular to the previous plane; denote the distance between the axes of the glass tubes as w. (e) Determine the error in pressure difference measurement, if we use the manometer response at ground conditions for a manometer placed in a stationary balloon, 3,000 m above ground. (f) Explain the problems that may arise, if we use the response at indoor conditions for a manometer located outdoors throughout the year.

2.5 Consider a U-tube manometer, containing water as the working liquid and used for measuring pressure differences in air flows. Derive the response of this manometer, if it is transported on a vehicle moving along a horizontal road with a constant acceleration, when the plane defined by the axes of the two glass tubes, which are separated by a distance w, is parallel to the direction of motion. Hint: Simplify the Navier–Stokes equations and solve them to find the column elevation difference, when the applied pressure difference is zero.

2.6 A platinum resistance temperature detector (RTD) has a resistance of 100 Ω at a temperature of $T = 0°C$. It is connected to one arm of a Wheatstone Bridge circuit, which is supplied with a voltage of $V = 1$ V. The remaining arms of the bridge have resistors of 100 kΩ. The voltmeter used to read the voltage across the bridge has an input span of 0–0.99 V. Determine the static sensitivity and other static performance characteristics of the RTD.

2.7 A thermometer, initially at room temperature $T_0 = 22°C$, is suddenly dipped into boiling water, which has a temperature $T_b = 100°C$. The temperature T that is indicated by the thermometer at different times t, rounded to the closest degree, is shown in the following table. For your solution, consider that the temperature readings are rounded and that both temperature and time readings may contain some small random errors.

t [s]	0	1	2	3	4	5	6	7	8	9	10
T [°C]	22	56	75	86	92	96	98	99	99	100	100

(a) Determine the time constant τ of this thermometer.

(b) Determine the temperature error $\varepsilon = T_b - T$, when the thermometer has been dipped in the boiling water for times equal to $\tau, 2\tau$ and 5τ.

(c) Determine the bandwidth of this thermometer. At which frequency would the steady-state error be less than 1%, and how does this frequency relate to the time constant?

2.8 Plot the impulse response and the ramp response, both as amplitude ratios and phases, of first-order and second-order systems, the latter ones for damping ratio values $\zeta = 0, 0.2, 0.5, \sqrt{2}/2, 1$ and 2. Discuss the appearance of these plots and compare qualitatively the responses of first-order systems and those of second-order systems with different damping ratios.

2.9 In order to perform a dynamic calibration, a thermometer is attached to an oscillating arm such that it is immersed alternately in one or another of two air jets with different temperatures. The oscillation period is 8 s and it may be assumed that the thermometer is exposed to a sinusoidal temperature field with a maximum of 330 K and a minimum of 300 K. The reading of the thermometer fluctuates between a maximum of 325 K and a minimum of 305 K. Assuming that the thermometer is a first-order system, determine its time constant. Also determine the difference between the time that the thermometer is exposed to the highest temperature and the time that the thermometer indicates the highest temperature (first find the phase shift).

2.10 After analysing the dynamic response of a second-order measuring system, the following transfer function was determined:

$$\mathbf{H}(s) = \frac{50{,}000}{s^2 + 438s + 33{,}320},$$

with s in rad/s. Draw the Bode plot of the measuring system, as well as its dynamic response due to a step-, impulse and ramp input with $A = 2$. Determine the settling time and bandwidth of this device. How would the results change, if the damping coefficient were halved? Compare the maximum measurement errors of the systems that have these two damping coefficients (within their bandwidth).

2.11 The sinusoidal transfer function of a third-order low-pass Butterworth filter is

$$\mathbf{H}(j\omega) = \frac{K}{\left(\frac{j\omega}{\omega_c} + 1\right)\left(-\frac{\omega^2}{\omega_c^2} + \frac{j\omega}{\omega_c} + 1\right)}.$$

(a) Write a differential equation that describes the response of this system.

(b) Prove that K is the static sensitivity and ω_c is the -3 dB high-frequency cut-off.

(c) Make Bode plots of this filter vs. the frequency ratio ω/ω_c, assuming that $K = 1$ and ensuring that the phase plot is continuous and indicates a negative phase shift.

References

[1] T.G. Beckwith, R.D. Marangoni, and J.H. Lienhard. *Mechanical Measurements (6th Edition)*. Pearson, Upper Saddle River, NJ, 2007.

[2] W.H. Beyer. *CRC Handbook of Mathematical Sciences (6th Edition)*. CRC Press, Boca Raton, FL, 1987.

[3] A.R. Cohen. *Linear Circuits and Systems*. Regents Publishing Company, Inc., New York, 1965.

[4] E.O. Doebelin and D.N. Manik. *Doebelin's Measurement Systems (SIE) (7th Edition)*. McGraw-Hill, New York, 2019.

[5] R.A. Gayakwad. *Op-Amps and Linear Integrated Circuits (3rd Edition)*. Prentice Hall, Englewood Cliffs, NJ, 1993.

[6] J. Millman and C.H. Halkias. *Integrated Electronics: Analog and Digital Circuits and Systems*. McGraw-Hill, New York, 1972.

[7] K. Ogata. *Modern Control Engineering (5th Edition)*. Prentice Hall, Englewood Cliffs, NJ, 2010.

[8] M. Sayer and A. Mansingh. *Measurement, Instrumentation and Experiment Design in Physics and Engineering*. Prentice-Hall of India, New Delhi, 2000.

[9] D.H. Sheingold, editor. *Analog-Digital Conversion Handbook*. Analog Devices, Inc., Norwood, MA, 1972.

[10] W.D. Stanley. *Operational Amplifiers with Linear Integrated Circuits (3rd Edition)*. Merrill, New York, 1994.

[11] R.E. Thomas and A.J. Rosa. *The Analysis and Design of Linear Circuits (2nd Edition)*. Prentice Hall, Upper Saddle River, NJ, 1998.

3 Signal Conditioning and Discretisation

A signal is a quantitative description of the dependence of a property upon time and/or location. The inputs and outputs of measuring systems are signals and the goal of measurement is to collect these signals, such that they can be analysed and, if applicable, compared to theoretical predictions and models. Analogue (namely, continuous) electric signals, provided by electric and electronic circuits, constitute the most common form of inputs and outputs of measuring systems and their components. In many cases, these signals have a level, form or frequency content that are not suitable for observation, recording or processing and need to be modified, a process generally referred to as signal conditioning. While much of signal conditioning as well as observation and recording of signals are still done by means of analogue instrumentation, the overwhelming trend is to transform the signals into discrete time series and then further condition and process them with the use of digital computers. This chapter summarises the definitions and background necessary for understanding the operation of common devices and the procedures used for signal conditioning and discretisation of analogue signals.

3.1 Properties of Electric Signals

Consider a time-dependent electric signal $V(t)$, measured over a time interval ΔT. Then, one can define the *time average* (or *mean*) of this signal, to be denoted by an overline, as

$$\overline{V} = \frac{1}{\Delta T} \int_0^{\Delta T} V(t)\,\mathrm{d}t\,. \tag{3.1}$$

One may then subtract the mean from the signal to define the signal fluctuation, to be denoted by a tilde, as

$$\tilde{V}(t) = V(t) - \overline{V}\,. \tag{3.2}$$

It is evident that the mean of the fluctuations vanishes and so this property cannot be used for any purpose. The strength of a signal that is always positive or always negative can be expressed by its mean $\overline{V}$. A more general measure of signal strength is its *root mean square* (*rms*) value, defined as

$$V_{rms} = \left[\frac{1}{\Delta T} \int_0^{\Delta T} V(t)^2\mathrm{d}t\right]^{1/2}\,. \tag{3.3}$$

A measure of the strength of signal fluctuations is its *standard deviation*

$$\sigma_V = \left[\frac{1}{\Delta T} \int_0^{\Delta T} \tilde{V}(t)^2 \mathrm{d}t \right]^{1/2} . \tag{3.4}$$

Assume that, over the interval ΔT, a signal fluctuates between a maximum value V_{max} and a minimum value V_{min}. Then, one can define the *peak-to-peak voltage difference* (or *signal span*) as

$$V_{pp} = V_{max} - V_{min} . \tag{3.5}$$

Example 3.1 Determine the mean $\overline{V}$, the rms voltage V_{rms}, the standard deviation σ_V and the peak-to-peak voltage difference V_{pp} of the signal $V(t) = [2 + 4\sin(2\pi t/T)]$ V, shown in Fig. 3.1, over one period T.

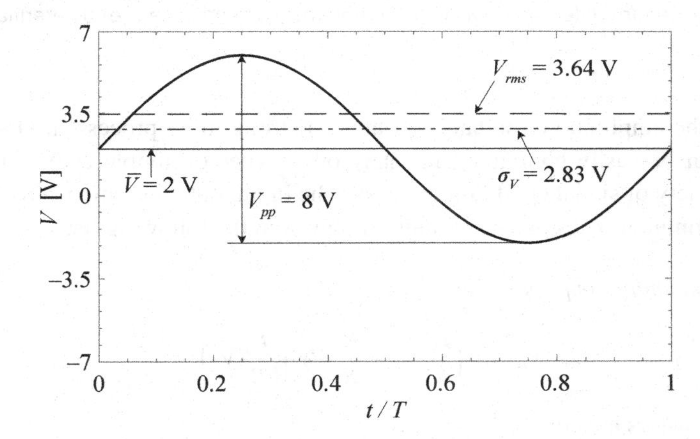

Figure 3.1 A periodic signal and its mean, rms value and peak-to-peak value.

Answer

From Eq. (3.1), we can easily find the mean voltage as $\overline{V} = 2$ V. The voltage fluctuation is $\tilde{V}(t) = 4\sin(2\pi t/T)$ V and the peak-to-peak voltage difference is $V_{pp} = 8$ V. Finally, from Eq. (3.3) we get the rms voltage as $V_{rms} = 3.46$ V and from Eq. (3.4) we get the voltage standard deviation as $\sigma_V = 2.83$ V.

3.2 Analogue Signal Conditioning

3.2.1 Analogue Signal Operations

The most common analogue conditioning operation is the amplification of weak signals, typically signals with voltages of the order of microvolts or even millivolts. This can be achieved with the use of the inverting amplifier (Fig. 2.12b), by selecting $R_f/R_1 > 1$. If $R_f/R_1 < 1$, the same device can be used as an *inverting attenuator*, for the purpose

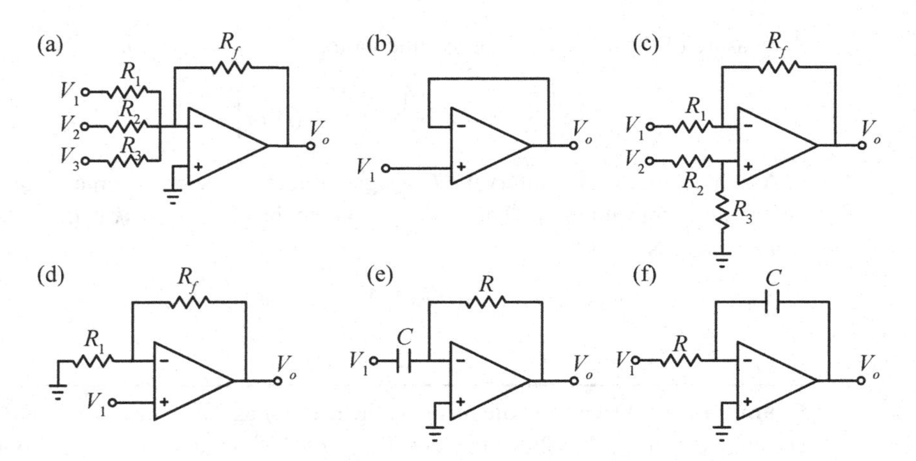

Figure 3.2 Simple active signal conditioning circuits using operational amplifiers: (a) adder, (b) voltage follower, (c) difference amplifier, (d) non-inverting amplifier, (e) differentiator and (f) integrator.

of reducing the amplitude of signals that are too strong to be processed. Operational amplifiers can be easily configured in many other types of simple active circuits to perform a variety of signal conditioning tasks. The input–output relationships for some of the most common types of such circuits are given in the following list:

- *Adder*, or *summing amplifier* (Fig. 3.2a):

$$V_o = -\left(\frac{R_f}{R_1}V_1 + \frac{R_f}{R_2}V_2 + \frac{R_f}{R_3}V_3\right) ; \tag{3.6}$$

  when all resistors are equal,

$$V_o = -(V_1 + V_2 + V_3) . \tag{3.7}$$

  If R_1, R_2 and R_3 are set to some fixed fraction of R_f, that is, $R_1 = R_2 = R_3 = R_f/g$, then

$$V_o = -g(V_1 + V_2 + V_3) \tag{3.8}$$

  where g is referred to as the *amplification gain*. Therefore this circuit will first sum the voltages together, and then amplify the signal by a factor of g.
- *Voltage follower*, or *buffer* (Fig. 3.2b):

$$V_o = V_1 ; \tag{3.9}$$

  a buffer is inserted between circuits that would load each other if connected directly.
- *Difference amplifier* (Fig. 3.2c):

$$V_o = \frac{R_f R_3}{R_2 + R_3}\left(\frac{1}{R_f} + \frac{1}{R_1}\right)V_2 - \frac{R_f}{R_1}V_1 ; \tag{3.10}$$

when all resistors are equal, this circuit becomes a *subtractor*:

$$V_o = V_2 - V_1 \,. \tag{3.11}$$

As for the summing amplifier, the resistors could be selected in a way that they produce an amplification gain g. In such case, the circuit will take the difference between the voltages and then amplify the output by a factor g, so that $V_o = g(V_2 - V_1)$.

- *Non-inverting amplifier* (Fig. 3.2d):

$$V_o = \left(1 + \frac{R_f}{R_1}\right) V_1 \equiv g V_1 \,. \tag{3.12}$$

- *Differentiator* (Fig. 3.2e):

$$V_o = -RC\frac{\mathrm{d}V_1}{\mathrm{d}t} \,. \tag{3.13}$$

Note that, when the input V_1 of the differentiator is a function of time, its output would also be time-dependent. When the input is constant, the output would vanish.

- *Integrator* (Fig. 3.2f):

$$V_o(t) = -\frac{1}{RC}\int_{t_0}^{t_0+\Delta t} V_1(t)\mathrm{d}t \,. \tag{3.14}$$

Hence, the integrator output would always be time dependent.

Example 3.2 A velocity measurement device produces a voltage output, the numerical value of which is equal to the flow speed in m/s, but we want the voltage output to be in mph (miles per hour). Suggest an operational amplifier circuit that can perform this conversion and select its component values.

Answer

To convert m/s to mph, we need to multiply the former value by 2.237. This can be achieved with the use of the non-inverting amplifier shown in Fig. 3.2d, set up such that it has a gain $g = 1 + R_f/R_1 = 2.237$, which means that $R_f = 1.127R_1$. If we select $R_1 = 10$ kΩ, then we must select $R_f = 10.127$ kΩ. Considering that it would be very unlikely to find an off-the-shelf fixed resistor with a resistance of 10.127 kΩ, we should instead use a potentiometer with a range between 0 and 20 kΩ and then adjust it to the required resistance value. The fixed resistor choice of 10 kΩ is a convenient one in terms of availability, as well as sufficiently large not to incur significant power losses.

Before the advent of digital data acquisition and processing, a variety of analogue signal conditioning operations were routinely performed with the use of dedicated devices, including time delays, linearisers, multipliers and dividers, correlators and spectrum analysers. These operations are more conveniently performed by software on discretised signals (Section 3.3.1). Even so, specialised integrated circuits, called *function modules*, are available to perform multiplications, divisions, roots, powers, logarithms, trigonometric functions and other mathematical operations [7], for cases in which it is preferable to condition and process signals by analogue rather than digital means.

Particular care must be taken in the processing of electric signals at very low levels (e.g., of the order of μV), such as often produced by strain gauges, thermocouples and other transducers. Such signals must be first amplified to an intermediate level before they can be further processed by ordinary circuitry. In such cases, the amplifier noise as well as interference voltages from various sources would be amplified as well. For an improved signal-to-noise ratio, it is necessary to use dedicated, very low-noise *preamplifiers*. *Instrumentation amplifiers* are particularly suitable as preamplifiers. These are specialised integrated circuits containing op-amps but having a committed response, for example the amplification of a voltage difference with a gain either fixed or selectable by means of a jumper or a single resistor. Instrumentation amplifiers have extremely low internal noise and drift and a moderate gain range. An important quality is their ability to remove a common interference voltage from a differential signal, called the *common-mode rejection* (CMR – Section 6.5). A related parameter is the *common-mode rejection ratio* (CMRR), defined as the ratio of the common-mode voltage and the common-mode error voltage, which is referred to the input. CMR is usually expressed in terms of CMRR as [8]

$$\mathrm{CMR}_{dB} = 10 \log_{10} \mathrm{CMRR} \,. \tag{3.15}$$

Instrumentation amplifiers normally have a large CMR, typically exceeding 60 dB. When very large common-mode voltages are present or for safety purposes (e.g., in medical instrumentation, in which circuits may come in contact with humans), it may be necessary to avoid physical contact between two circuits altogether. This may be achieved by transmitting fluctuating signals through transformers or optical couplings. Devices that perform such tasks are known as *isolation amplifiers*.

3.2.2 Analogue Filters

Filters are used commonly in measuring systems to remove or reduce undesirable effects, particularly noise, from a time-dependent signal. In this section, we discuss *analogue filters*, which are electric and electronic circuits connected to measuring circuits for the purpose of conditioning voltage signals. Besides these, there are also *digital filters* (Section 3.3.5), which are applied to recorded time series. Although analogue filters can be replaced by digital ones for many purposes, they are necessary components of nearly all circuits before these are discretised (Section 3.3.2).

According to the frequency range of the fluctuations that they remove, both analogue and digital filters are classified as in the following list. Typical Bode plots of such filters are shown in Fig. 3.3.

- *No-pass* filters, which remove all fluctuations, permitting only a steady component, if any at all.
- *Low-pass* filters, which remove fluctuations with frequencies that are higher than a cut-off value.

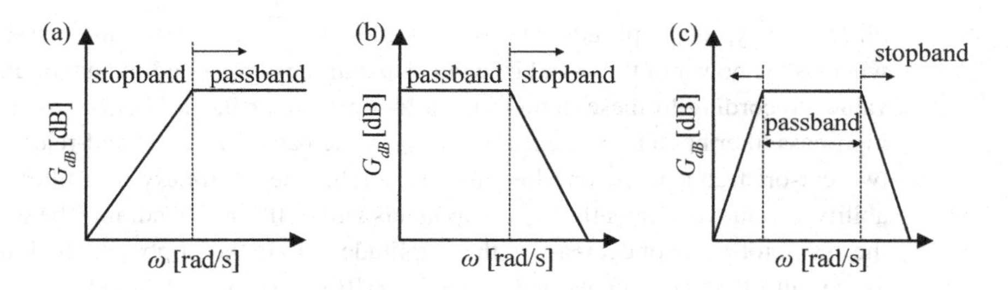

Figure 3.3 Idealised Bode plots for a (a) high-pass filter, (b) low-pass filter and (c) band-pass filter. The passband and stopband regions are also shown.

- *High-pass* filters, which remove fluctuations with frequencies that are lower than a cut-off value.
- *Band-pass* filters, which remove all fluctuations except those with frequencies within a certain band.
- *Band-reject* filters, which remove all fluctuations with frequencies within a certain band.

Example 3.3 The AC coupling function on an oscilloscope allows the user to see only the fluctuations of the signal. For analogue oscilloscopes, this is achieved by using an analogue filter. What type of filter should be used for this function and which frequency settings of this filter should be selected?

Answer

We know that the voltage signal is composed of a mean ('DC component') and a fluctuation ('AC component'), that is, $V(t) = \overline{V} + \tilde{V}$. We therefore have to remove the mean, which has a zero frequency. This can be achieved with a high-pass filter with a frequency cut-off that is set as closely as possible to 0, so as to remove the mean but as little as possible of fluctuations at non-zero frequencies.

A convenient presentation of filter performance is in the form of Bode plots. A filter is designed to have one or more frequency ranges, called the *passband*, over which the gain $G_{dB}(\omega)$ does not drop significantly and one or more frequency ranges, called the *stopband*, over which the gain drops significantly. A gain of 0 dB means that the input and output amplitudes at a particular frequency are equal. A positive gain, in dB, means that the output amplitude is larger than the input amplitude, whereas a negative gain means the opposite. The gain within the passband is not necessarily close to zero, but to another value, called the *maximum passband value*. The width of the passband frequency range is the *bandwidth* of the filter.

An important characteristic of a filter is its *cut-off frequency*, which is the frequency at which the gain is reduced by 3 dB from its maximum passband value. At the cut-off frequency, the amplitude ratio is reduced to $1/\sqrt{2}$ of its maximum passband value, whereas the power of the signal is reduced to half the corresponding maximum passband value. According to these definitions, a low-pass filter has a *high-frequency cut-off*, a high-pass filter has a *low-frequency cut-off*, while band-pass and band-reject filters have two cut-off frequencies, one low and one high. The *sharpness* of a filter, namely its ability to remove efficiently any components within the stopband, may be quantified by the asymptotic rate of decrease of the amplitude ratio in the stopband, previously termed as the roll-off and expressed in dB/decade or dB/octave. The rate of change of the phase can be expressed as degrees per decade or degrees per octave.

Analogue filters are classified as *passive*, which are circuits consisting of exclusively passive components, or *active*, which in addition include operational amplifiers or other active components.

The analysis of analogue filters is more conveniently conducted in the complex domain, or *Laplace s-domain*, namely by analysing the transfer function of the filter. As discussed in Section 2.4.7, the order of the filter would be equal to the highest power of s in the denominator of the transfer function. The performance of the filter can be determined by analysing the magnitude of the transfer function as a function of the input frequency.

Example 3.4 Two electric circuits are shown in Fig. 3.4. Specify whether they are passive or active. Then, show that both of these circuits act as first-order low-pass filters and determine their cut-off frequencies and other characteristics.

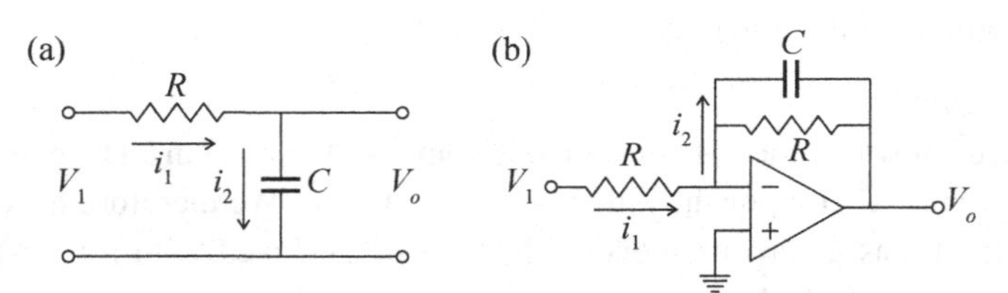

Figure 3.4 Sketches of (a) a passive first-order low-pass filter and (b) an active first-order low-pass filter.

Answer

The circuit shown in shown in Fig. 3.4a is a passive RC circuit, which has already been analysed in Example 2.8. We therefore know that it satisfies the first-order, time invariant, ordinary differential equation

$$RC\frac{\mathrm{d}V_o}{\mathrm{d}t} + V_o = V_1 \ ,$$

and, thus, has the response of a first-order system with a static sensitivity $K = 1$, a time constant $\tau = RC$ and a transfer function

$$\mathbf{H}(j\omega) = \frac{1}{1 + jRC\omega} \ .$$

We can see that, when $\omega \to 0$, $\mathbf{H}(j\omega) \to 1$, and so we note that it allows low frequencies to pass, that is, it is a low-pass filter. The cut-off frequency can be found by setting the magnitude of the transfer function to $1/\sqrt{2}$, which gives $\omega_c = 1/(RC)$ (or $f_c = 1/(2\pi RC)$), a high-frequency ($\omega \gg \omega_c$) roll-off of -20 dB/decade and a high-frequency phase shift of $-\pi/2$. The low-frequency ($\omega \to 0$) asymptote is $|\mathbf{H}(j\omega)| = 1$, while the high-frequency ($\omega \to \infty$) asymptote is $|\mathbf{H}(j\omega)| = 1/RC\omega$. The two asymptotes intersect at the corner frequency $\omega_{cf} = 1/(RC)$, which, in this case, is equal to the cut-off frequency ω_c.

Now, consider the circuit shown in Fig. 3.4b. This circuit is obviously active, because it contains an operational amplifier, which has an external power source. The transfer function of this circuit can be obtained by applying the circuit laws in the s-domain. The resistor and capacitor in the feedback branch of the op-amp are connected in parallel and can be replaced by the impedance $Z_2 = R/(1 + RCs)$. For an ideal op-amp, the voltages V_- at the inverting input and V_+ at the non-inverting input would be equal. In this circuit, $V_+ = 0$ and so also $V_- = 0$. Moreover, for an ideal op-amp, the currents into both the non-inverting and inverting inputs would be zero. Then, Kirchhoff's current law requires that $i_1 = i_2$ (see Fig. 3.4b), from which we get $\mathbf{I}_1(s) = \mathbf{I}_2(s)$, or

$$\frac{\mathbf{V}_1 - \mathbf{V}_-}{R} = \frac{\mathbf{V}_- - \mathbf{V}_o}{Z_2} \quad \Leftrightarrow \quad \frac{\mathbf{V}_1 - 0}{R} = \frac{(0 - \mathbf{V}_o)(1 + RCs)}{R}.$$

Rearranging the latter expression, we get the circuit transfer function as

$$\mathbf{H}(s) = \frac{\mathbf{V}_o(s)}{\mathbf{V}_1(s)} = \frac{-1}{1 + RCs}.$$

This is the transfer function of an inverting first-order system and so this circuit is also a first-order low-pass filter. The active filter has a response identical to that of the passive filter, with the exception of a negative sign in its transfer function.

Active filters have the general advantage over passive ones that they do not load other circuits. This allows them to be *cascaded*, namely, connected in series, to modify or enhance the filtering action. For example, one may easily construct a first-order band-pass or band-reject filter by cascading a first-order low-pass filter and a first-order high-pass filter; similarly, one may construct a higher-order filter by cascading a series of identical lower-order filters.

A variety of active filter designs are available for different operations. In general, one would prefer to use a filter with a steep roll-off ('sharp' filter). As was the case for measuring systems, the roll-off of a filter increases with increasing filter order, which, however, also increases the complexity and cost of the circuit. Another adverse effect of increasing the filter order is an accompanying increase in the phase shift. As a consequence, higher-order analogue filters would distort the waveform of a signal more than lower-order ones.

Example 3.5 Construct a sharp low-pass filter by cascading first-order, active, low-pass filters.

Answer

Connect in series n identical first-order, active, low-pass filters, as shown in Fig. 3.4b, so that the output of each serves as the input to the next one. The transfer function of the composite filter can be found by multiplying the transfer functions of its parts, namely, as

$$\mathbf{H}(j\omega) = \frac{1}{(1 + jRC\omega)^n},$$

which provides an amplitude ratio of

$$|\mathbf{H}(j\omega)| = \frac{1}{\left[\sqrt{1 + (RC\omega)^2}\right]^n},$$

a cut-off frequency of

$$\omega_c = \frac{\sqrt{2^{1/n} - 1}}{RC}, \tag{3.16}$$

a corner frequency of

$$\omega_{cf} = \frac{1}{RC}$$

and a roll-off of $-20n$ dB/decade. It is obvious that the cascade filter is much sharper than each of its parts. Nevertheless, the cascade filter has a serious disadvantage: as Eq. (3.16) and Fig. 3.5 show, its cut-off frequency, and thus its bandwidth, decrease with an increasing number of stages. This means that, if one desires to have a specific value of ω_c, one would need to know the number of stages before selecting the resistors

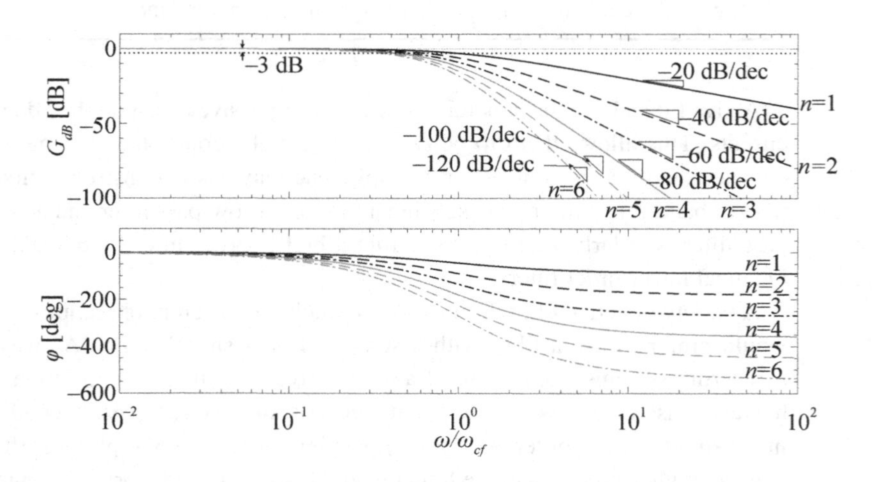

Figure 3.5 Bode plot for a first-order filter and cascade filters up to sixth order.

and capacitors in the circuits. On the other hand, the corner frequency is independent of the filter order and such that $\omega_{cf} \geq \omega_c$ (the equality holds for the first-order filter only).

When designing a filter, one strives to approximate, as closely as possible, the performance of the corresponding *ideal filter*, which has a uniform gain within its passband and an infinite roll-off at the cut-off frequency. The most popular designs are based on the Butterworth polynomials, which approximate step functions. Low-pass Butterworth filters are designed to have transfer functions given by

$$\mathbf{H}(s) = \frac{K}{q_n(s)}, \tag{3.17}$$

where the constant K is the static gain of the filter, n is the order of the filter and $q_n(s)$ are the Butterworth polynomials, which are listed, up to fourth order, in Table 3.1. Note that the complex frequency in these polynomials has been conveniently normalised by the corresponding cut-off frequency $\omega_c = 1/RC$. It is straightforward to prove that the amplitude ratio of these filters is

$$|\mathbf{H}(j\omega)| = \frac{|K|}{\sqrt{1 + (\omega/\omega_c)^{2n}}}. \tag{3.18}$$

Noticing that the amplitude ratio has a low-frequency asymptote $|\mathbf{H}(j\omega)| = |K|$ and a high-frequency asymptote $|\mathbf{H}(j\omega)| = |K| \,/\, (\omega/\omega_c)^n$, which intersect at a corner frequency that is always equal to the cut-off frequency, one can show that the bandwidth of a Butterworth filter is independent of the filter order. Like first-order cascade filters, Butterworth filters have a high-frequency roll-off of $-20n$ dB/decade. However, if one compares two filters of the same order and with the same ω_c, one will find that the Butterworth filter will attenuate its output at a given frequency within the stopband ($\omega > \omega_c$) significantly more than the corresponding cascade filter would.

Table 3.1 Butterworth polynomials

n	$q_n(s)$
1	$s/\omega_c + 1$
2	$(s/\omega_c)^2 + 1.414s/\omega_c + 1$
3	$(s/\omega_c + 1)\left[(s/\omega_c)^2 + s/\omega_c + 1\right]$
4	$\left[(s/\omega_c)^2 + 0.7654s/\omega_c + 1\right]\left[(s/\omega_c)^2 + 1.8478s/\omega_c + 1\right]$

Example 3.6 A fourth-order low-pass Butterworth filter is shown in Fig. 3.6 [4]. Determine its bandwidth and compare this to the bandwidth of a fourth-order cascade filter with the same cut-off frequency.

Answer

This filter consists of two second-order stages in series, the transfer function for each of which can be found using the methodology shown in the previous chapter. The static

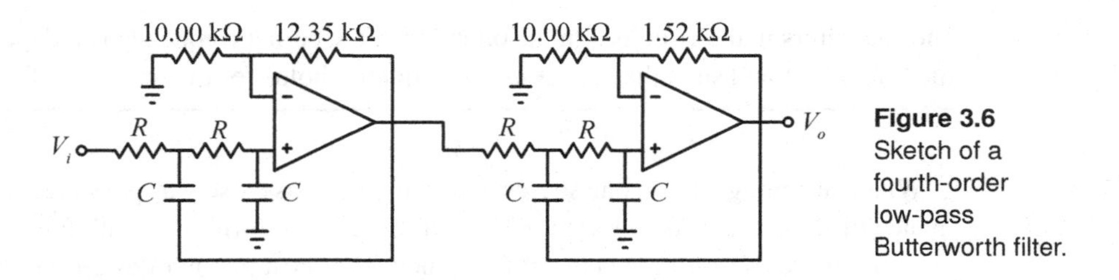

Figure 3.6 Sketch of a fourth-order low-pass Butterworth filter.

sensitivity for the first stage is $K_1 = (12.35 + 10)/10 = 2.235$, whilst for the second stage it is $K_2 = (1.52 + 10)/10 = 1.152$. The overall static sensitivity of the filter is $K = K_1 K_2 = 2.575$. The overall transfer function for this filter is

$$\mathbf{H}(s) = \frac{K_1 K_2}{[(s/\omega_c)^2 + (3 - K_1)(s/\omega_c) + 1][(s/\omega_c)^2 + (3 - K_2)(s/\omega_c) + 1]} .$$

We therefore note that this particular Butterworth filter has a static gain of $G_{dB} = 20 \log_{10}(|K|) \approx 8.2$ dB. The cut-off frequency is $\omega_c = 1/(RC)$ and, thus, it depends on the choices of resistor and capacitor values; the gain at that frequency is $G_{dB}(\omega_c) = 20 \log_{10}(|K|) - 3 \approx 5.2$. As a fourth-order system, it has a high-frequency roll-off equal to -80 dB/decade. The Bode plots for this filter, as well as for a fourth-order cascade filter with the same cut-off frequency, are shown in Fig. 3.7, which, among other features, also illustrates that the Butterworth filter is much sharper than the cascade filter.

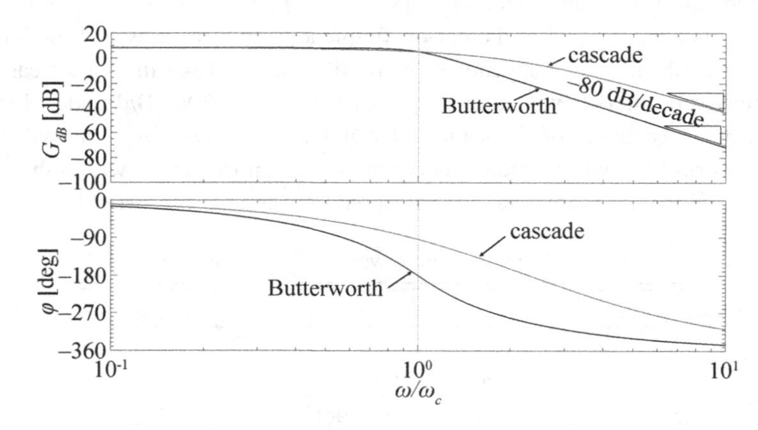

Figure 3.7 Bode plots for the fourth-order Butterworth filter shown in Fig. 3.6, and a fourth-order cascade filter with the same cut-off frequency.

Example 3.7 A sinusoidal force is applied to a load cell during an experiment in a wind tunnel. The corresponding voltage signal $V(t)$, measured in volts and plotted vs. time t, measured in seconds (see Fig. 3.8), can be fitted well by the function $V(t) = [-1.5 + 1.9 \sin(2\pi t) + 0.1 \sin(200\pi t)]$. Thus, this signal consists of three components: (a) a constant (DC) component equal to -1.5 V; (b) a main sinusoidal component, which presumably corresponds to the force on the load cell, with a frequency $\omega_1 = 2\pi$ rad/s and an amplitude 1.9 V; and (c) a secondary sinusoidal component with the much higher frequency $\omega_2 = 200\pi = 100\omega_1$ and the much smaller amplitude 0.1 V, which may be attributed to interference from an unknown source. Select an appropriate Butterworth filter to eliminate the interference in the signal. How do the rms voltage and the standard deviation of the voltage change after the application of the filter?

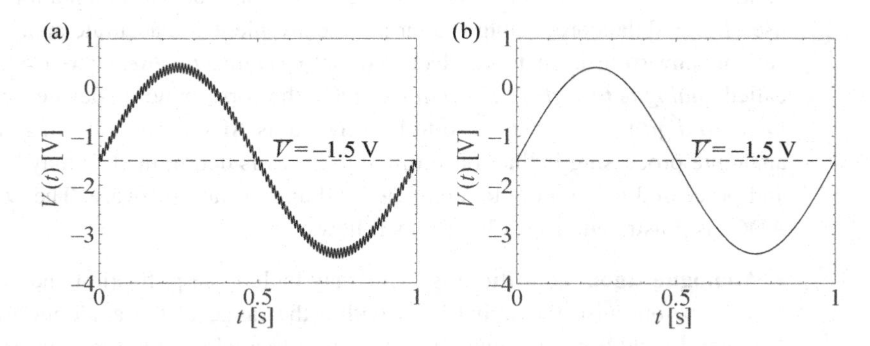

Figure 3.8 One period of the voltage signal received from a load cell (a) before applying the low-pass filter and (b) after applying a low-pass filter with a cut-off frequency $\omega_c = 50\pi$ rad/s.

Answer

We note that the two sinusoidal components are in phase and that the signal is periodic with a period $T = 1$ s. We need to remove the high-frequency component and keep the low-frequency component, a task that can be achieved by a low-pass filter. It is noted that a band-pass filter is inappropriate, because it would also filter out the mean, which is needed in this measurement. Next, we need to choose the order of the low-pass filter. As discussed previously, the higher the order of a filter is, the more it would attenuate the high-frequency component of the signal. A promising choice seems to be the fourth-order Butterworth filter. Finally, we need to select the cut-off frequency ω_c. Because the difference between the frequencies of the two components is very large, we can choose any intermediate value, as long as it is not close to either of the two frequencies of the sinusoidal signal components. For example, it seems appropriate to use any frequency in the range between $2\omega_1$ and $0.5\omega_2$, namely, to choose ω_c between 4π and 100π rad/s. Let us then use $\omega_c = 50\pi$ rad/s. As shown in Fig. 3.8, the unfiltered signal has strong evidence of interference, whereas the filtered signal is interference-free, while also preserving both the DC component and the amplitude and waveform of the main sinusoidal component.

Using Eq. (3.3), we find the rms voltage of the original signal to be $V_{rms} = 2.015$ V and the rms voltage of the filtered signal to be slightly smaller and equal to 2.014 V. The standard deviations of the original and filtered signals, found using Eq. (3.4), were, respectively, $\sigma_V = 1.345$ V and 1.344 V.

3.3 Discretisation of Analogue Signals

3.3.1 Analogue to Digital Conversion

Most measuring instruments and transducers produce continuous, or *analogue*, electric signals, which are generally time-dependent. To some degree, analogue signals can be conditioned, observed and stored by exclusively analogue instrumentation; however, the use of digital data processing is a far more convenient and accurate approach. The process of converting an analogue electric signal $x(t)$ into a *time series* x_i, $i = 0, \pm 1, \pm 2, \ldots$ is called *analogue to digital conversion* (ADC); the corresponding devices are called *analogue to digital converters* (also to be denoted as ADC). The advantage of ADC over analogue processing is that, once the time series is determined, it may be conditioned and processed using a digital computer and appropriate software. The typical steps in ADC, as illustrated in Fig. 3.9, are as follows:

- **Analogue signal conditioning:** This may include amplification and/or offsetting in order to optimise the signal level within the range of the analogue to digital converter. In addition, analogue filtering is best to apply just before conversion, in order to remove noise and other undesirable effects most effectively. Low-pass filtering is usually applied at this stage to prevent *aliasing* (see Section 3.3.2).
- **Multiplexing:** For multi-channel conversion, a *multiplexer* is used to select the particular channel that will be discretised at a given time and to establish the sequence of channels to be discretised; a variety of multiplexing methods are available [6].
- **Sample-and-holding:** The selected signal is fed to a *sample-and-hold circuit*, which continuously tracks ('sample' state) its input voltage and, when it receives a control signal, it captures the current voltage value by storing it in a capacitor ('hold' state); thus the voltage is maintained essentially constant over a time interval that is long enough for the conversion to be completed.
- **Conversion:** There are several ADC methods, including the successive approximation method, the counter-comparator method and the dual-ramp method. In all these cases, the ADC produces a discrete value which approximates the voltage stored in the sample-and-hold circuit. For flash ADC (or direct-conversion ADC), however, so named due to the their fast conversion rates, there is usually no need for a sample-and-hold circuit. The most common code in which the discrete value is stored is the binary code, in which a number is represented as a series of successive powers of 2,

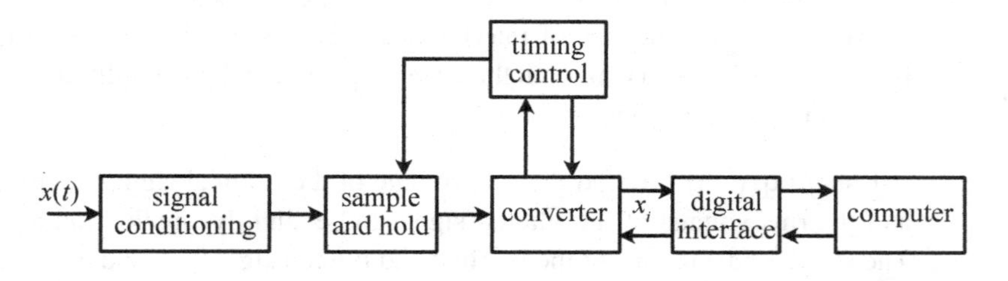

Figure 3.9 Typical steps in single-channel analogue to digital conversion.

called the *bits*, with each bit having a value of 0 or 1. Each ADC is characterised by the number n of bits of its output. An n-bit ADC will produce a discrete output that is equal to one among 2^n integer numbers. The smallest number that can be represented in binary code is 0 and the largest number is equal to $2^n - 1$. For example, the largest number that can be represented by a 2-bit ADC is $2^2 - 1 = 3$.

- **Storage:** A *digital interface* transfers the discrete values to the computer memory or a storage device; when high-speed ADC is required, the discretised signals are stored in intermediate *buffers*, which are periodically emptied to their permanent destination.

Example 3.8 Express the decimal number 13 in 3-bit, 4-bit and 5-bit binary forms.

Answer

The largest decimal numbers that can be represented in these three binary forms are, respectively, $2^3 - 1 = 7$, $2^4 - 1 = 15$ and $2^5 - 1 = 31$. Hence, the number 13 cannot be expressed in 3-bit binary form. To determine the representation of 13 in the other two binary forms, we first express this number as the sum of decimal numbers, each of which is equal to a different power of 2 or zero. Then, we rewrite each of the terms of this sum as 0 or 1 times 2 to some power, still in decimal form. The binary form of this number is a sequence of the coefficients 0 or 1 of all terms, with the smallest power corresponding to the rightmost digit. The step-by-step procedure for the 4-bit binary form is

$$\begin{aligned} 13 &= 8 + 4 + 0 + 1 \\ &= \left(1 \times 2^3\right) + \left(1 \times 2^2\right) + \left(0 \times 2^1\right) + \left(1 \times 2^0\right) \\ &= 1101|_{bin} \,. \end{aligned}$$

For the 5-bit binary form, the same procedure provides the binary number $01101|_{bin}$.

A 3-bit analogue to digital converter: An illustration of a flash ADC is shown in Fig. 3.10a [1, 3, 12], in which a 3-bit ADC is used to discretise the analogue voltage V_{in}. The device contains eight comparators (Section 2.2.4), connected in parallel. The non-inverting inputs of all comparators are connected to the input voltage V_{in}. The inverting input of each comparator is connected to an intermediate point in a voltage divider, which is connected to two constant reference voltages V_+ and V_-, with $V_+ > V_-$. The *span* of the voltage divider, which may also be considered as the *nominal span* of the ADC, is

$$V_S = V_+ - V_- \,. \tag{3.19}$$

The inverting input of the lowest comparator can be found as $V_- + (0.5/8)(V_+ - V_-) = (0.5V_+ + 7.5V_-)/8$. In a similar manner, we can find that the inverting inputs of the eight comparators take, respectively, the values $(0.5V_+ + 7.5V_-)/8$, $(1.5V_+ + 6.5V_-)/8$, $(2.5V_+ + 5.5V_-)/8$, $(3.5V_+ + 4.5V_-)/8$, $(4.5V_+ + 3.5V_-)/8$, $(5.5V_+ + 2.5V_-)/8$, $(6.5V_+ +$

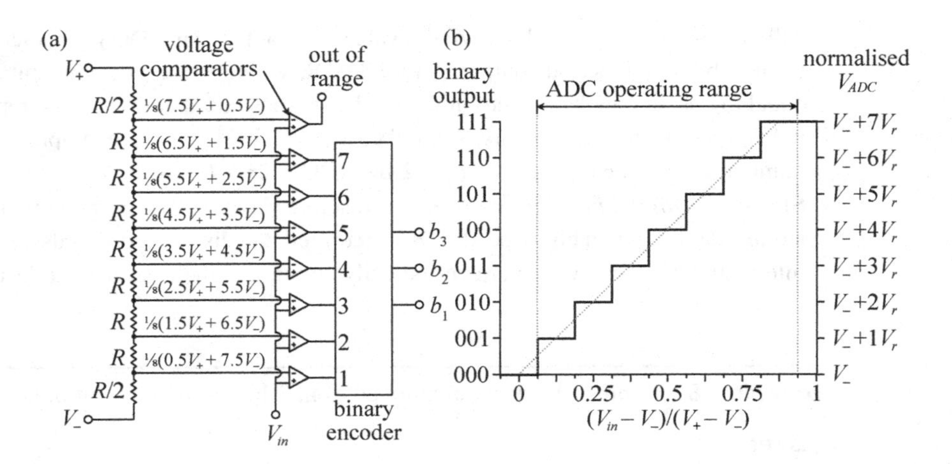

Figure 3.10 (a) Conversion of an analogue voltage V_{in} to a binary number $b_2b_1b_o|_{bin}$ and a discrete voltage V_{ADC} using a flash ADC, comprised of eight voltage comparators and a 3-bit binary encoder; (b) plot of the ideal 3-bit binary encoder output (left axis) and the normalised ADC discrete voltage output $(V_{ADC} - V_-)/(V_+ - V_-)$ (right axis) vs. the normalised analogue voltage input.

$1.5V_-)/8$ and $(7.5V_+ + 0.5V_-)/8$. The output of each comparator will be either $-V_{sat}$, if V_{in} is smaller than the voltage connected to the corresponding inverting input, or $+V_{sat}$, if V_{in} is greater than the voltage connected to the corresponding inverting input (see Section 2.2.4). The output of each of the first seven comparators is fed to a separate input of a *binary encoder*, which produces a 3-bit binary output $b_2b_1b_o|_{bin}$. Each bit will take the value 0 or 1, depending on whether the corresponding comparator output is $-V_{sat}$ or $+V_{sat}$. The relationship between the binary output of the ADC and the normalised voltage input is shown in Fig. 3.10b. The output of the binary encoder will be one among the eight integer numbers between 0 and 7, inclusive, for any value of the voltage input V_{in}. For example, if $(4.5V_+ + 3.5V_-)/8 < V_{in} < (5.5V_+ + 2.5V_-)/8$, the encoder will provide the output $101|_{bin} = 5$. If $V_{in} < (0.5V_+ + 7.5V_-)/8$, the output will be $000|_{bin} = 0$, which means that a binary output of 0 corresponds not only to input values in the range $(-0.5V_+ + 6.5V_-)/8 < V_{in} < (0.5V_+ + 7.5V_-)/8$, but also to any lower value, in which case the discretisation would be erroneous. This type of error is known as *saturation*. Therefore, an output of $000|_{bin} = 0$ indicates that the analogue voltage input is below the input range of the ADC and so the ADC output is not usable. Saturation may also have occurred if the binary output were 7, which would be the case when $V_{in} > (6.5V_+ + 0.5V_-)/8$. For such values, the ADC takes advantage of the top (eighth) comparator, which, unlike the seven other comparators, does not contribute to the binary output, but, instead, provides an output that indicates whether V_{in} is smaller or larger than $(7.5V_+ + 0.5V_-)/8$. Thus, the ADC detects the occurrence of saturation only when the top comparator is activated, which means that $V_{in} > (7.5V_+ + 0.5V_-)/8$.

ADC specifications and errors: Following the previous discussion of the flash ADC operation, it is evident that the *effective input range* of an ADC would be narrower than its *nominal input range*, which is from V_- to V_+. For example, for the 3-bit ADC, the effective input range is from $(0.5V_+ + 7.5V_-)/8$ to $(7.5V_+ + 0.5V_-)/8$. As a consequence, the *effective span* of an n-bit ADC is not V_S, but

$$V_{SADC} = \frac{2^n - 1}{2^n} V_S \ . \tag{3.20}$$

Note that the effective range and span of the ADC approach asymptotically the nominal values as the number of bits increase. For example, the effective span and range for a 3-bit ADC with $V_+ = 8$ V and $V_- = 0$ V are $V_{SADC} = 7$ V and $0.5 \text{ V} < V_{in} < 7.5$ V, respectively, but, when the number of bits is increased to 12, these properties take the values $V_{SADC} = 7.998$ V and $0.977 \times 10^{-3} \text{ V} < V_{in} < 7.999$ V.

The usable values of the binary output $b_{n-1}, ..., b_0|_{bin}$ of an n-bit ADC are the integers between 1 and $2^n - 1$, with 0 being reserved for informing the user that the signal is saturated at the lowest input level. Therefore, the number of usable output values that can be produced by this ADC is $2^n - 1$. The *resolution* of this ADC is

$$V_r = V_{SADC}/(2^n - 1) = V_S/2^n \ . \tag{3.21}$$

An operation that remains to be performed is to convert the binary output into a discrete voltage value V_{ADC}, which will be part of the time series corresponding to the analogue input voltage V_{in}. This operation can be done by software installed in the ADC or computer used. It is noted, however, that each value of V_{ADC} corresponds not to a single value of V_{in}, but to a range of input values. The difference

$$V_q = V_{in} - V_{ADC} \tag{3.22}$$

is called the *quantisation error*. To minimise this error, the discretised voltage from the ADC is determined as

$$V_{ADC} = b_{n-1}, ..., b_0|_{bin} \ V_r + V_- \ , \tag{3.23}$$

which leads to the following bounds of the quantisation error within the ADC operating range

$$-V_r/2 \leq V_q \leq V_r/2 \ . \tag{3.24}$$

Consider now that we apply a ramp-type input to the ADC, namely, an analogue signal that, during a given time interval, changes linearly in time. The discrete output voltage will have a staircase pattern, as shown in Fig. 3.11a, and a sawtooth-type error, bounded by $\pm V_r/2$, as shown in Fig. 3.11b. This time-dependent error may be considered as noise and it is referred to as *quantisation noise*. As a measure of the strength of this noise, we can use its standard deviation σ_{V_q} (see Section 3.1), which can be calculated from Eq. (3.4) as [3]

$$\sigma_{V_q} = V_r/\sqrt{12} \ . \tag{3.25}$$

In practice, this noise may be considerably larger due to various other errors in the ADC process.

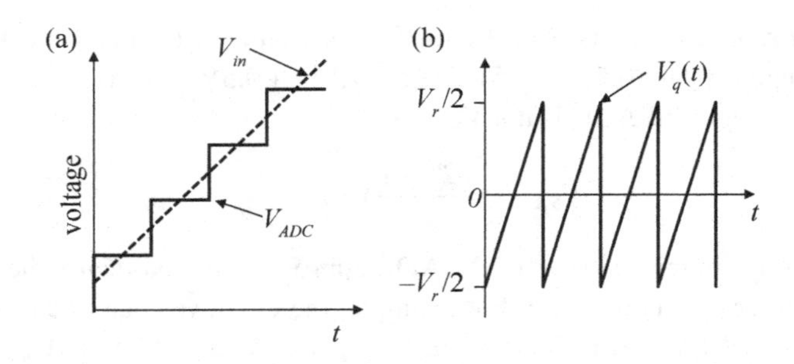

Figure 3.11 (a) Comparison between an analogue voltage that increases linearly in time and the corresponding digitised signal; (b) the corresponding time dependent quantisation error.

Example 3.9 A 3-bit ADC with a nominal input range of 0–4 V is used to measure an analogue voltage, the value of which was measured independently by a voltmeter to be 3.24 V. Determine the digitised reading, the maximum quantisation error and the quantisation noise.

Answer

The nominal span of this ADC is $V_S = 4$ V and so its resolution is $V_r = V_S/2^3 = 0.5$ V and its maximum quantisation error is 0.25 V. This ADC will provide only one of seven usable different output voltages V_{ADC}, which will be 0.5, 1, 1.5, 2, 2.5, 3 and 3.5 V. The effective operating range of this ADC is 0.25–3.75 V. The voltages at the inverting inputs of comparators 6 and 7 are, respectively, 2.75 V and 3.25 V. Consequently, for $V_{in} = 3.24$ V, the binary output will be $110|_{bin} = 6$, because the seventh comparator will not be triggered by this voltage, whereas the sixth comparator will. The ADC output will be $V_{ADC} = 110|_{bin} V_r + V_- = 3.0$ V. Accounting for uncertainty due to the quantisation error, the ADC output may be presented as $V_{ADC} = 3.0 \pm 0.25$ V, which includes the input voltage of 3.24 V. Note that, if the input voltage were increased to 3.26 V, then the digitised voltage would become $V_{ADC} = 3.5 \pm 0.25$ V, which includes 3.26. For this input voltage, the eighth comparator will not be triggered, which indicates that the converted voltage output is within the effective range of the ADC. The quantisation noise is found from Eq. (3.25) to be $\sigma_{V_q} = 0.14$ V.

To reduce the error in V_{ADC}, one needs to either increase the number of bits of the ADC (which would likely increase the cost of the device and reduce the maximum discretisation rate) or decrease the span V_S, while possibly shifting V_- so that the input voltage remains within the range of the ADC. Alternatively, one may also amplify, and offset, when needed, the input signal. For example, let us say that the input analogue voltage fluctuates between 2.5 and 3.5 V. It is evident that setting the ADC range between 0 and 5 V would result in a resolution $5/2^n$ V, whereas setting the ADC range between 0 and 10 V would result in the larger resolution value of $10/2^n$ V. Keeping the input range fixed, but increasing the number of bits in the converter, would result into a smaller resolution value. Let us now assume that the ADC range and the number of bits are fixed. To reduce the relative discretisation error, we may offset and amplify the input signal, so that its fluctuations occupy a larger part of the ADC range.

When selecting an appropriate ADC for a particular measurement, one should not only strive to have the highest possible resolution, so as to minimise quantisation error and noise, but also to ensure that the operating range of the ADC is wider than the range of the signal. In many cases, the range of the signal that is produced by a sensor or other measuring system component is within the operating range of the ADC and the ADC resolution is adequate for the needs of the measurement. There are cases, however, for which part or all of the signal is outside the ADC range or the signal fluctuations are too weak to be discretised accurately. In such cases, the signal must be conditioned before being inputted to the ADC. This signal conditioning generally includes a combination of offsetting to bring the signal range within the ADC range and amplification to strengthen the input signal fluctuations. It is also possible that the signal fluctuations are too strong and would need to be attenuated before being inputted to the ADC. Of course, signal conditioning operations would introduce additional noise and uncertainty and, for this reason, they should only be used when necessary.

Example 3.10 We want to discretise the low-pass filtered signal from the load cell discussed in Example 3.7, which is given as

$$V(t) = -1.5 + 1.9\sin(2\pi t)\,,$$

with voltage in volts and time in seconds. Two ADC are at our disposal, the one discussed in Example 3.9 ('ADC-1') and a 4-bit ADC with an input range of 0–8 V ('ADC-2'). Design a signal conditioner, if possible, by adapting one of the circuits shown in Fig. 3.2, to enable the measurement, clearly justifying your choice. Plot the discretised voltage signal for both ADC and compare standard deviation voltage of the converted signal over one period.

Answer

Let us first determine the characteristics of the two ADC. From Example 3.9, we know that the resolution of ADC-1 is $V_{r1} = V_{S\,1}/2^3 = 0.5$ V, whilst, for ADC-2, we have $V_{r2} = V_{S\,2}/2^4 = 8/2^4 = 0.5$ V. Therefore, both ADC have the same resolution $V_r = 0.5$ V, maximum quantisation error $V_{qmax} = \pm 0.5 V_r = \pm 0.25$ V and quantisation noise $\sigma_{V_q} = V_r / \sqrt{12} \approx 0.14$ V.

The properties of the input signal are as follows: the minimum value is -3.4 V; the maximum value is 0.4 V; the peak-to-peak value is $V_{pp} = 3.8$ V; and the mean is $\overline{V} = -1.5$ V. By comparing these values with the characteristics of the two ADC, we see that the signal must be conditioned, because part of it is outside the range of either ADC. First of all, it is evident that the signal must be offset so that its minimum value is larger than the minimum output of either ADC. To determine whether amplification or attenuation is also required, we note that the signal peak-to-peak value is larger than the effective span of $V_{SADC} = 3.5$ V for ADC-1. The signal is therefore guaranteed to be saturated, and therefore has to be attenuated before being digitised. On the other hand, the signal peak-to-peak value is significantly smaller than the effective span of ADC-2, but large compared to the ADC resolution, which means that no amplification or attenuation is required, if ADC-2 is used. In view of the previous considerations, we choose to proceed with ADC-2. In order to bring the signal into the range of ADC-2, one must offset it by adding a fixed voltage to it, ensuring that the lowest voltage of the analogue signal is within the effective range of ADC-2, which is 0.25–7.75 V. Best practice, if applying an offset is required, is to set it such that the *median* (i.e., the average of the largest and smallest values) of the signal is roughly equal to the mid-point of the ADC range. In this case, the median of the analogue signal is equal to its mean $\overline{V} = -1.5$ V, and so, if the original signal were to be fed directly into ADC-2, it should be offset by adding 5.5 V to it, so that its mean would become equal to the midpoint 4 V of the ADC range. Suitable circuits for offsetting the input signal are the summing amplifier (Fig. 3.2a) and the difference amplifier (Fig. 3.2c). We shall proceed with the use of the summing amplifier, which is somewhat simpler than the difference amplifier, leaving the use of the latter as an exercise. Because there is no need for amplification or attenuation, we will set the gain of the amplifier to 1 by choosing equal values of all resistors. In setting the voltage offset, we must consider that the summing amplifier also inverts the input, so that the mean of the inverted analogue signal would become +1.5 V. Consequently, the voltage offset should be +2.5 V. To perform the offsetting operation, we connect the load cell signal to the V_1 terminal of the summing amplifier, a constant voltage of −2.5 V to V_2 and leave V_3 unconnected. If ADC-1 were used instead, the same arguments lead us to connect a constant voltage of −0.5 V to the V_2 terminal of the summing amplifier in order to offset the mean of the analogue signal to the mid-point 2 V of the ADC-1 range.

The filtered signal from the load cell and the signals discretised by both ADC are plotted in Fig. 3.12. The actual outputs of the ADC are discrete time series, but, in

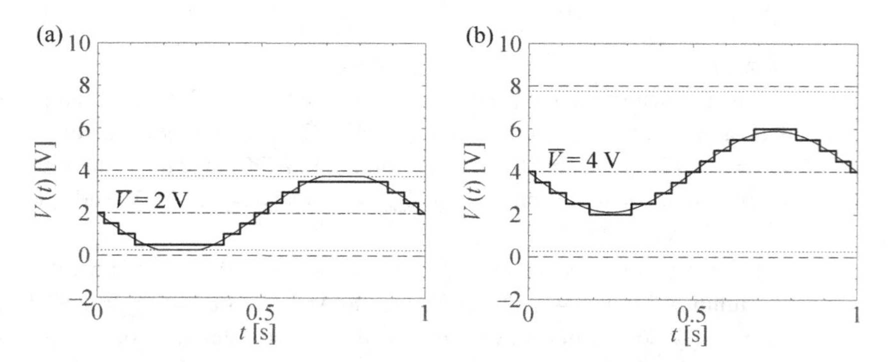

Figure 3.12 Thick solid lines show the filtered analogue signal received from a load cell (a) after it was offset by −0.5 V, inverted and discretised using ADC-1 and (b) after it was offset by −2.5 V, inverted and discretised using ADC-2. Thin lines show the same signal as it would have been discretised by an ADC with a resolution equal to zero. Dashed lines mark the nominal operating range of each ADC and the dotted lines mark the effective operating range of the ADC.

this figure, we plot them as continuous functions, under the approximation that the discrete values are very closely spaced in time (the significance of sampling time will be discussed in the next section). To illustrate more clearly the possible occurrence of saturation, we have also plotted the hypothetical output of an ideal ADC with a resolution equal to zero (i.e., an infinite number of bits). It is evident that the ADC-1 output is saturated during part of the period, whereas the ADC-2 output is not. To recover the mean of the input signal, we must first invert the mean of the ADC output and then subtract the offset from this value. Although saturation of the ADC-1 output occurred during some time intervals, because it was symmetric, it did not contaminate the value of the discretised signal mean, which was equal to the value −1.5 V of the analogue input. This serves as evidence that the mean is not a sensitive indicator of discretisation error, or even the occurrence of saturation. As errors due to different sources may be positive of negative, and so they may be added to or subtracted from the correct value, one should avoid generalising anecdotal observations. In this particular example, the voltage standard deviations were 1.210 V for ADC-1 and 1.382 V for ADC-2, which were on either side of the analogue voltage value of 1.344 V.

Multi-channel ADC: An important specification for ADC systems is the *maximum conversion rate* per channel, usually provided in samples per second, or, equivalently, in Hz. In multi-channel ADC systems, the overall conversion rate is equal to the maximum conversion rate divided by the number of discretised channels. In multi-channel applications, conversion of different channels can be made in either of two ways:

- *Sequentially*: all channels are connected to a multiplexer, which selects channels one at a time and feeds them to a single sample-and-hold; after the signal is held, it is fed to the converter which finishes the conversion before the next channel is fed to the same sample-and-hold; this method requires a single sample-and-hold and is simpler than the next one, but suffers the disadvantage of discretising samples of different channels at different times.
- *Simultaneously*: each channel is fed to a separate sample-and-hold and a timing circuit triggers all holdings at the same time; when sampling is completed, all sample-and-hold outputs are fed to a multiplexer, which selects one at a time and sends them to the converter. Ideally, this method would provide simultaneous samples. In practice, however, the voltages stored in the sample-and-hold capacitors would leak, causing errors. For this reason, this method requires higher-quality, more expensive components than the previous one. Dedicated converters for each channel may be used to improve the performance of ADC.

ADC programming: Programming of the computer for data acquisition and processing can be done in text-based programming languages, such as assembly language, Python, MATLAB, C^{++} or FORTRAN. A convenient approach would be to use one of several commercial software packages, which employ a graphical representation of the intermediate steps. A widely used package is LabVIEW, which uses the graphical programming language G to create programs in block diagram form. LabVIEW is a general-purpose programming system but also includes libraries of functions and development tools designed specifically for data acquisition and control. LabVIEW programs are called *virtual instruments* (VIs), because their appearance and operation imitate those of actual instruments. VIs have an interactive user interface, called the front panel, because it simulates the front panel of an actual instrument (e.g., a voltmeter), displaying images of knobs, push-buttons, switches, graphs etc. The VI receives instructions from a block diagram, created by the programmer in G-language.

3.3.2 Sampling Theorem

Consider an analogue (i.e., continuous) time-dependent electric signal $x(t)$ and a series of equally spaced discrete samples (*time series*) $x_i, i = 0, \pm 1, \pm 2, ...$, such that $x_i = x(i\Delta t)$ (Fig. 3.13a), where Δt is the *sampling time*, corresponding to the *sampling rate* $\omega_s = 2\pi/\Delta t$ (or $f_s = 1/\Delta t$). In general, there is an infinite number of analogue signals that correspond to the same time series. However, the following theorem permits a reversible, at least ideally, representation of analogue signals by a time series [5].

Sampling theorem: If the continuous signal $x(t)$ is *band-limited*, namely if $\mathbf{X}(j\omega) = 0$ for $|\omega| > \omega_h$, then it can be uniquely determined by the time series $x_i = x(i\Delta t), i = 0, \pm 1, \pm 2, ...$, if Δt is chosen such that $\omega_s = 2\pi/\Delta t > 2\omega_h$ (*Nyquist criterion*). One may reconstruct the signal $x(t)$ exactly by generating a train of impulses with amplitudes equal to the sample values and low-pass filtering it with an ideal low-pass filter having a gain Δt and a cut-off frequency that is greater than ω_h and smaller than $\omega_s - \omega_h$.

Nyquist frequency: The frequency $2\omega_h$ is called the *Nyquist rate*, whereas the frequency ω_h itself is called the *Nyquist frequency*. The Nyquist criterion states that exact equality of the sampling and the Nyquist rates is insufficient for the sampling theorem to hold and so one would need to choose a value of ω_s that is somewhat larger than $2\omega_h$. Let us discuss a practical situation, in which we have an analogue signal which, besides its physical fluctuations, also contains white noise extending to infinite frequencies. From considerations about the frequency content of physical fluctuations in the signal, we decided to sample it at a sampling rate of ω_s. In order to avoid aliasing, by which noise power will be shifted to lower frequencies, we should first ensure that there is no power in the sampled signal fluctuations at frequencies that are equal to or larger than

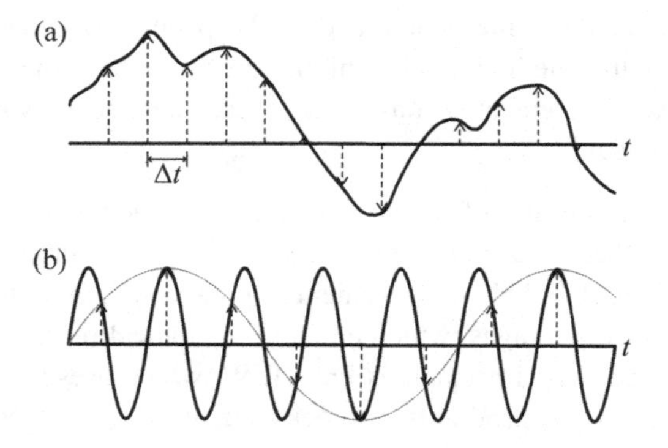

Figure 3.13 (a) Sketch showing an analogue voltage (solid line) and corresponding equally spaced discrete samples (dashed lines); (b) sketch of a sinusoidal analogue signal with a frequency ω_o (solid line), corresponding discrete samples (dashed lines) obtained at a rate $\omega_s < 2\omega_o$ and the perceived analogue signal (dotted line), which, due to aliasing, has a frequency $\omega_s - \omega_o$.

$\omega_s/2$. The only way to achieve this is by low-pass filtering the signal with an analogue filter having a cut-off frequency $\omega_c < \omega_s/2$, for example, $\omega_c = 0.4\omega_s$.

It is emphasised that it is not possible to eliminate aliasing with the use of digital filters or other data processing applied after signal discretisation and so analogue filtering is a necessary part of any aliasing-free discretisation process. The choice of filter type and ω_h depends on the frequency content of the particular signal, the level and frequency content of the noise and the specifications of the available instrumentation. To prevent any confusion in selecting ω_c and ω_s, we suggest the following sequence of actions:

- Based on physical considerations for the measurand frequency content, identify the highest frequency ω_h of interest in the signal.
- Set the low-pass filter cut-off frequency as $\omega_c > \omega_h$.
- Select a sampling rate so that $\omega_s > 2\omega_c$; for example, select $\omega_s = 2.5\omega_c$.

Aliasing: When the Nyquist criterion is not satisfied, the analogue signal may not be reconstructed by manipulations of the time series. In particular, high-frequency components of the analogue signal would appear as lower-frequency components, thus permanently changing the signal composition in the frequency domain. This phenomenon is called *aliasing*; it is illustrated in Fig. 3.13b, in which the original signal $x(t) = \sin(\omega_o t)$, sampled at the rate $\omega_s < 2\omega_o$, appears to be $x'(t) = \sin[(\omega_s - \omega_o)t] \neq x(t)$, that is, the perceived frequency from the digitisation process would be lower than the true frequency of the analogue signal. It must be emphasised that, once aliasing occurs, it becomes permanent and cannot be removed by post-processing of data.

Example 3.11 Consider the periodic analogue signal

$$x(t) = 2\sin(2\pi t) + \cos(8\pi t) \ ,$$

where the values of time t are in seconds and the values of $x(t)$ are in volts. Discuss how aliasing would affect the time series produced by discretising this signal using different sampling rates ω_s.

Answer

The lowest frequency of fluctuations in this periodic signal is $\omega_0 = 2\pi$ rad/s and so the period is $T = 2\pi/\omega_0 = 1$ s. The Nyquist frequency, namely, the highest frequency of fluctuations in this signal, is $\omega_h = 4\omega_0 = 8\pi$ rad/s. Let us assume that we discretise this signal using a sampling frequency ω_s over a single period T, so that data are collected from $t = 0$ to $t = T$, inclusive. The total number of data points within one period is, therefore, $N = T\omega_s/(2\pi) + 1$. The sampling theorem states that, to prevent aliasing, the sampling frequency must be chosen as $\omega_s > 2\omega_h = 16\pi$ rad/s. Figure 3.14 shows the analogue signal and two discrete times series, one obtained when the Nyquist criterion is not satisfied (a) and another when it is (b). It is clear that, in the first case, the time series is aliased and connection of the discrete values by splines results in a vastly different waveform, in which the low-frequency component appears to have a larger amplitude

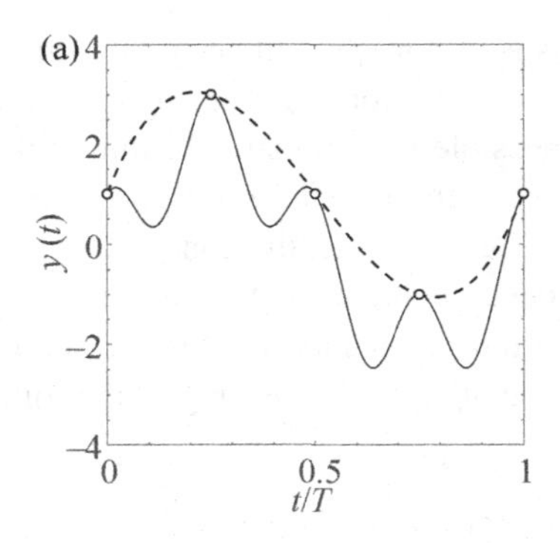

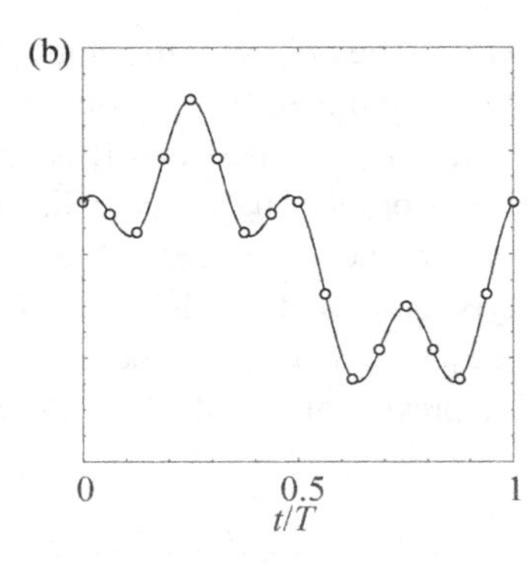

Figure 3.14 Plots showing a periodic analogue signal (solid line) and the corresponding discrete time series (symbols), obtained by using a sampling rate (a) $\omega_s = \omega_h$ and (b) $\omega_s = 4\omega_h$; the dashed line connects the discrete values by splines.

than the one in the analogue signal. On the other hand, in the second case, the time series faithfully reproduces the signal waveform.

The sampling theorem also applies to discrete-time signals [5]. In such cases, a closely spaced discrete-time signal is represented by another that only retains part of the original values, spaced at equal time intervals. This process, which is the inverse of *interpolation*, is known as *decimation*.

3.3.3 Mean and Standard Deviation of a Time Series

Consider a analogue signal $x(t)$, which is sampled over a time interval $0 \leq t \leq \Delta T$, with ΔT to be referred to as the *record length*. The mean $\overline{x}$ and the standard deviation σ_x of this signal can be determined from Eqs. (3.1) and (3.4), respectively. Discretising the analogue signal at a rate of $\omega_s = 2\pi/\Delta t$, leads to a time series with N discrete values within the time interval from 0 to ΔT, inclusive. Hence the analogue signal is represented by the discrete *time series* $x_i = x(i\Delta t), i = 0, 1, 2, ..., N - 1$. The mean and the standard deviation of this time series are, respectively, defined as

$$\overline{x_i} = \frac{1}{N} \sum_{i=0}^{N-1} x_i \tag{3.26}$$

and

$$s_{x_i} = \left[\frac{1}{N-1} \sum_{i=0}^{N-1} (x_i - \overline{x_i})^2 \right]^{1/2} . \tag{3.27}$$

To begin with, we can state that, in general, the occurrence of aliasing does not by itself introduce an error in the values of the signal mean and standard deviation, because aliasing does not filter out the activity of high-frequency fluctuations, but ‘folds it over’, so to speak, to low-frequency fluctuations. Even so, an important issue is to identify the conditions under which the time series average and standard deviation would be approximately equal to those of the analogue signal. It is apparent that this may not be the case if the number of samples N is very small, because accurate averaging requires a sufficiently large number of data points. The question that remains, however, is whether having a large N is sufficient for the equality to hold. The answer to this question depends on whether the discrete values in the recorded time series are generally independent of each other or not. An example for which this is not the case is a periodic signal, the period of which happens to be an integral multiple of the sampling time: samples obtained one period apart would have identical values and so increasing the record length would not increase the number of independent values from which we can calculate accurate averages. Therefore, we would need a large number of samples within at least one period. Because, in general, we may not know in advance whether the signal is periodic and the value of its period, it is safer to use a record length that is sure to include many periods. Similarly, for signals that are not periodic but have one or more periodic components, one must ensure that the record length is sufficiently long to include many occurrences of the component with the lowest frequency. For signals that are perfectly random and have no periodic component, it would be sufficient for the number of samples to be large. At this point, we will present some representative examples, following which we will return to the discussion of record length.

Example 3.12 Consider the sinusoidal signal

$$x(t) = \sin(2\pi t) \ ,$$

where time t is measured in seconds. Compare its average and standard deviation to those of corresponding time series having different numbers of data points per period.

Answer

It is given that this signal is periodic with a period $T = 1$ s. Because this signal is periodic, all its statistical properties can be calculated from values in a single period, which allows us to set $\Delta T = T$. Let us assume that we discretise this signal using a sampling frequency ω_s, selected so that the period T is an integral multiple of the sampling time $\Delta t = 2\pi/\omega_s$. Therefore, the total number of data points in the range $0 \le t \le T$ is $N = (T/\Delta t) + 1$ or, equivalently, the sampling time is $\Delta t = T/(N-1)$. It is evident that the mean of this analogue signal is zero. Its standard deviation is

$$\sigma_x = \left[\frac{1}{T}\int_0^T x^2(t)\mathrm{d}t\right]^{1/2} = \left[\frac{1}{T}\int_0^T \sin^2(\omega_0 t)\,\mathrm{d}t\right]^{1/2} = \frac{1}{\sqrt{2}} \ .$$

The mean and the standard deviation of a corresponding time series with N values over one period are, respectively,

$$\overline{x_i} = \frac{1}{N}\sum_{i=0}^{N-1} x_i = \frac{1}{N}\sum_{i=0}^{N-1} \sin\left(\frac{2\pi i}{N-1}\right)$$

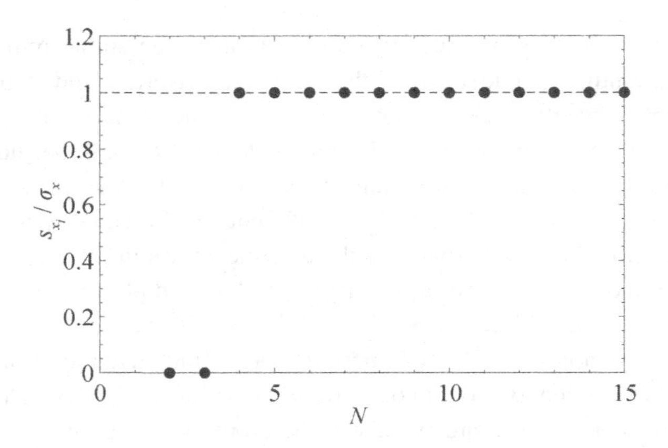

Figure 3.15 The dependence of the standard deviation of a time series corresponding to a periodic analogue signal. normalised by the standard deviation of the analogue signal, vs. the number of samples per period.

and

$$s_{x_i} = \left[\frac{1}{N-1}\sum_{i=0}^{N-1}(x_i - \overline{x_i})^2\right]^{1/2} = \left\{\frac{1}{N-1}\sum_{i=0}^{N-1}\left[\sin\left(\frac{2\pi i}{N-1}\right) - \overline{x_i}\right]^2\right\}^{1/2} .$$

Because of the anti-symmetry of the sinusoidal function and the fact that the time series includes values at $t = 0$ and $t = T$, the mean $\overline{x_i}$ of this time series will vanish, irrespective of the number of data points N. As Fig. 3.15 shows, the values of the two standard deviations σ_x, s_{x_i} differ drastically for $N = 2$ and 3, but are equal for $N \geq 4$. Therefore, the minimum sampling rate that is appropriate for this sinusoidal signal and the present purposes is $\omega_{s,min} = 6\pi$ rad/s.

Example 3.13 Consider the analogue signal

$$x(t) = \sin(\omega_0 t) + \cos(\omega_1 t) ,$$

where $\omega_0 = 2\pi$ rad/s and $\omega_1 = 100\omega_0$. How are the mean and standard deviation of this analogue signal affected when aliasing is present?

Answer

Because ω_1 is an integral multiple of ω_0, the signal is periodic with a period $T = 2\pi/\omega_0 = 1$ s and, therefore, all its statistical properties can be calculated from values in a single period, that is, $\Delta T = T$. It is evident that the mean $\overline{x}$ of this signal is zero. Its standard deviation is

$$\sigma_x = \left[\frac{1}{T}\int_0^T x^2(t)\mathrm{d}t\right]^{1/2} = \left[\frac{1}{T}\int_0^T \{\sin(\omega_0 t) + \cos(\omega_1 t)\}^2\mathrm{d}t\right]^{1/2} = 1 .$$

Now, let us discretise this signal to obtain a time series with N samples per period. The sampling rate is $\omega_s = 2\pi(N-1)/T$. The mean and the standard deviation of this time series are, respectively,

$$\overline{x_i} = \frac{1}{N}\sum_{i=0}^{N-1} x_i = \frac{1}{N}\sum_{i=0}^{N-1}\left\{\sin\left(\frac{2\pi i}{N-1}\right) + \cos\left(100\frac{2\pi i}{N-1}\right)\right\}$$

and

$$\begin{aligned} s_{x_i} &= \left[\frac{1}{N-1}\sum_{i=0}^{N-1}(x_i - \overline{x_i})^2\right]^{1/2} \\ &= \left\{\frac{1}{N-1}\sum_{i=1}^{N-1}\left[\sin\left(\frac{2\pi i}{N-1}\right) + \cos\left(100\frac{2\pi i}{N-1}\right) - \overline{x_i}\right]^2\right\}^{1/2}. \end{aligned}$$

We now examine how $\overline{x_i}$ and s_{x_i} vary, as ω_s (and, hence, N) changes relative to the highest frequency in this signal, namely ω_1. The time series mean and the ratio of standard deviations are shown in Fig. 3.16, which shows that the sampling rate value affects both of these parameters and that the time series properties tend towards the corresponding analogue signal properties, as $\omega_s/\omega_1 \to \infty$. Application of the Nyquist criterion to this signal infers that, in order to prevent aliasing, one must choose $\omega_s > 2\omega_1$. Figure 3.16 shows that, when aliasing is present ($\omega_s \leq 2\omega_1$), the time series mean and standard deviation differ significantly from the analogue signal values. Even for the borderline value $\omega_s = 2\omega_1$, we find a weak affect ($\overline{x_i} \approx 0.005$ and $s_{x_i} \approx 1.002$). Nevertheless, one is warned not to generalise this observation to all types of signals. Aliasing has such a noticeable effect on this signal, because the signal is periodic. Random signals, to be discussed in later chapters, would not be affected by such issues, when the Nyquist criterion is satisfied.

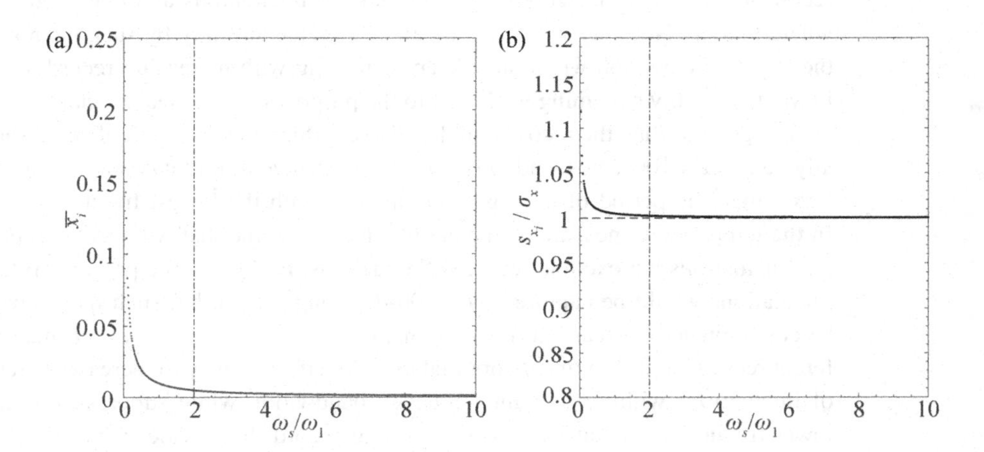

Figure 3.16 (a) The mean of the time series and (b) the ratio of the standard deviations, both plotted vs. the sampling rate, normalised by the highest frequency in the signal.

3.3.4 Record Length

The statistical and frequency analyses of signals will be discussed in following chapters, but it seems necessary to summarise a few relevant points here. As mentioned previously, the choice of an appropriate record length ΔT is very important for calculating properties of a time series that approximate accurately the corresponding properties of an analogue signal. Because of the diversity of analogue signals, this choice cannot be based on a universal rule, but it depends on the physical system that is being measured and the phenomena that affect the value of the measurand. Some signals may contain one or more approximately periodic components, which have frequencies that are much lower than the highest frequency of interest in the signal. The same or other signals may contain distinct and identifiable patterns, which may not be periodic but recur with some degree of regularity; we may call these *quasi-periodic* components, keeping in mind that this is not an established term. Such components could be of interest to the measurement and need to be represented in the time series. On the other hand, a signal may also contain slowly changing components, which are the result of undesirable inputs or phenomena that are not of interest in the measurement; including such components in the time series may distort the measurement. In summary, one must strive to specify a record length that captures phenomena of interest and, as much as possible, excludes phenomena and influences that introduce undesirable effects. As an example, consider the measurement of temperature at some outdoor location. The temperature value will fluctuate continuously as a result of multiple influences. Nevertheless, some temperature fluctuations occur with a degree of regularity. In order of decreasing average frequency, we may list the following fluctuations: (a) diurnal fluctuations, as nights are generally cooler than days; (b) seasonal fluctuations, as winters are generally cooler than summers; (c) fluctuations due to local phenomena, such as El Niño, which recur every few years; and (d) fluctuations due to the onset of ice ages and other climate changes, which may recur every few thousand years. The point to be made here is that increasing the record length would progressively capture such fluctuations and one needs to decide where to stop. Not only would the mean be affected strongly by such phenomena, but the standard deviation would also keep increasing with increasing record length, with its value possibly becoming irrelevant to the purposes of the measurement.

As a general rule, the record length ΔT over which the data are collected should ideally be at least 100 times, and only under very stringent limitations as low as 20 times, larger than the period of the signal component with the lowest frequency of interest. In the temperature measurement example, let us assume that we need to capture diurnal fluctuations but exclude seasonal fluctuations. Following the previous rule, diurnal fluctuations would be captured by a 100-day long record, but such record would also be contaminated by seasonal fluctuations. In such cases, one can experiment with different record lengths to find an optimal one. An effective way to increase representation of fluctuations within a certain frequency bandwidth, while suppressing undesirable lower-frequency influences is to take a long record, but divide it into shorter *blocks* of data, then calculate averages over each block and finally average the corresponding block values. This is called *ensemble averaging* (see also Section 4.9.3). Back to the temperature example, let us assume that we need to estimate the standard deviation of the temperature fluctuations in a temperate region due to time-of-day effects. If we use a record that is 100 days long, seasonal effects will be relatively strong, affecting the average and, consequently, resulting in an erroneously large value of the standard deviation. The result would be substantially improved, if we used several 20-day blocks and ensemble-averaged them.

A question that arises is how to find the lowest frequency of interest in a signal. In some cases, such frequency can be estimated from theoretical arguments or empirical familiarity with a topic. Examples of this type include the frequency of pendulum oscillation and the vortex shedding frequency behind a cylinder. If this approach is not possible, one may perform preliminary tests for different values of ΔT and select a value in a range within which the calculated block means and standard deviations are insensitive to ΔT. A final point is that choosing a record length that is much longer than the period of the component with the lowest frequency of interest has an additional advantage: calculating averages requires one to divide the corresponding sums or integrals by the record length and so any contribution of the signal that occupies only a fraction of one period would vanish as $\Delta T \to \infty$.

3.3.5 Digital Filters

Digital filters of various types can be applied to recorded time series in a manner similar to the application of analogue filters to analogue signals, but with some important advantages, to be discussed in the following [2, 9, 10]. The output of analogue filters has a frequency-dependent amplitude change, compared to the input, but also a phase shift, which depends on frequency in a non-linear manner. Thus, analogue filters would distort the waveform shape of a signal containing various frequency components. As shown in Example 3.11, if an analogue signal consisting of the sum of two sinusoidal terms with different frequencies were low-pass filtered by an analogue filter, not only would the amplitudes of the two terms change by different proportions, but also the appearance (waveform) of the signal would change as a result of phase shifts of the two terms the would not be in proportion to their frequencies. In some applications, it is essential to preserve the waveform shape and so a different type of filter would be required. Digital filters are more flexible than analogue ones, because they can use parts of or the entire time series at the time of application of the filter. They can also use past values of the output and may also be applied in multiple passes in forward or reverse direction.

One may then wonder why use analogue filters at all. It must be emphasised that a low-pass analogue filter is an indispensable component of any measuring system that measures a broadband measurand and the only means to prevent aliasing. A time series that is obtained from an unfiltered analogue signal would be subjected to aliasing, which cannot be reversed by post-processing with the use of digital filters. As will be discussed in later chapters, electronic noise is wideband and, in general, would likely have energy at frequencies that are higher than the estimated highest frequency of interest in the measurand (namely, the highest frequency that we do not wish to suppress significantly in the time series). To prevent high-frequency noise, and possibly other types of high-frequency interference, from aliasing the analogue signal, we need to use a low-pass analogue filter with a cut-off frequency that is determined by the Nyquist criterion. Once this is done, one may additionally use a digital filter to further reduce the bandwidth of the time series, at the possible expense of filtering out some physical fluctuations that are present in the measurand, albeit without aliasing risk. Another advantage of analogue filters is that they act in real time, unlike digital filters that are applied to recorded time series. To apply a digital filter, one needs to wait until the entire time series is recorded, whereas, with ongoing analogue filtering, one may compute tentative statistical properties as the measurement proceeds and update them in essentially real time.

Digital filters applied in the time domain are generally divided into two categories:

- *non-recursive* or *finite impulse response (FFR) filters*, where the current output is not an explicit function of past output values;
- *recursive* or *infinite impulse response (IIR) filters*, where the current output is a function of past output values, in addition to the input values.

There is a great variety of digital filters for general use as well as filters that have been optimised for specific purposes [e.g., 13]. In the following, we shall only discuss an example of a non-recursive digital filter, which introduces zero phase shift at all frequencies and, thus, causes no waveform distortion of the time series [11]. For the phase shift to vanish, the transfer function $\mathbf{H}(j\omega)$ of the filter must be real and so we will use the notation $H(\omega)$ to denote it. Let $x_n, n = 1, 2, ...,$ be a time series, used as input to the filter, sampled at a time step Δt. Then, the output of the filter can be written as

$$y_m = b_0 x_m + \sum_{n=1}^{\infty} b_n \left(x_{m-n} + x_{m+n} \right) , \tag{3.28}$$

where b_0 and b_n are the cosine Fourier coefficients of the filter transfer function (Section 5.1), namely,

$$b_0 = \frac{\Delta t}{\pi} \int_0^{\frac{\Delta t}{\pi}} H(\omega) \mathrm{d}\omega , \tag{3.29}$$

$$b_n = \frac{\Delta t}{\pi} \int_0^{\frac{\Delta t}{\pi}} H(\omega) \cos\left(n\omega\Delta t\right) \mathrm{d}\omega , \tag{3.30}$$

such that

$$H(\omega) = b_0 + 2 \sum_{n=1}^{\infty} b_n \cos\left(n\omega\Delta t\right) . \tag{3.31}$$

Let us now assume that the desired filter has an ideal band-pass response with a zero phase shift, with the desired transfer function given by

$$H(\omega) = \begin{cases} 1, & \omega_1 \leq \omega \leq \omega_2 , \\ 0, & \text{otherwise} , \end{cases} \tag{3.32}$$

for which the Fourier coefficients become

$$b_0 = \frac{(\omega_2 - \omega_1)\,\Delta t}{\pi}\,, \tag{3.33}$$

$$b_n = \frac{\sin(n\omega_2\Delta t) - \sin(n\omega_1\Delta t)}{n\pi}\,. \tag{3.34}$$

A complication arises when the Fourier series is truncated to a finite number of terms N, because this process generates oscillations in the transfer function estimate, known as the *Gibbs phenomenon*. To reduce the amplitude of such oscillations, one may employ the use of a 'window function' w_n, such that

$$y_m = b_0 x_m + \sum_{n=1}^{N} b_n w_n \left(x_{m-n} + x_{m+n}\right)\,, \tag{3.35}$$

$$H(\omega) \approx b_0 + 2\sum_{n=1}^{N} b_n w_n \cos(n\omega\Delta t)\,. \tag{3.36}$$

An example of a suitable window function is the *Hanning window*

$$w_n = \frac{1}{2}\left(1 + \cos\frac{n\pi}{N}\right)\,. \tag{3.37}$$

The sharpness of the filter increases with the number of terms N, but so is the computational time required for the application of the filter.

It is important to emphasise that digital filters cannot be used to remove aliasing caused by an insufficient sampling frequency. To prevent aliasing, low-pass analogue filtering must be applied before discretisation with a cut-off that must be less than $\omega_2/2\pi$ and the sampling interval time Δt must be less than π/ω_2.

An alternative approach that also results in zero phase shift is to apply a digital filter to a time series twice. First, to apply the filter in a forward direction (namely successively applied to input values acquired in increasing time sequence) and afterwards to apply the same filter to the output of the first pass in reverse direction. The amplitude ratio of the transfer function of the dual filter would be equal to the square of the single-pass transfer function amplitude ratio.

Besides the time domain, digital filters can be also applied in the frequency domain [2]. This would require first to transform the time series to the frequency domain through a Fast Fourier transform (Section 5.1), then to multiply this by the transfer function of the desired filter and finally to perform an inverse FFT in order to recover the filtered time series. This procedure introduces no phase shift but is computationally more intensive than the time domain analysis.

Chapter Digest

3.1 What type of operational amplifier circuit should be connected to the two terminals of a Wheatstone bridge?

3.2 Is it possible to construct a first-order band-pass filter?

3.3 If we want to construct an active, second-order band-pass filter by connecting low-pass and high-pass filters, what is the order of these filters and how should they be connected?

3.4 If we amplify a sinusoidal signal, what would happen to V_{rms} and to σ_V? What if we normalise these properties with $\overline{V}$?

3.5 Discuss the type of circuit you would need to offset a signal.

3.6 If you want to remove the mean (DC component) of a signal, why is it more practical to use AC coupling (namely, a high-pass filter) instead of applying an offset?

3.7 The output of a buffer is equal to its input; so, what is the use of this device?

3.8 How would you choose the time constant *RC* of a differentiator and that of an integrator?

3.9 What is the specific use of a preamplifier? Why would a general-purpose operational amplifier not be suitable as a preamplifier?

3.10 When can a passive low-pass filter not replace an active low-pass filter?

3.11 What is a cascade filter? What major disadvantage does a cascade filter have? Suggest a better type of filter and explain your reasons.

3.12 Do you see a risk that would arise, if you set up the frequency cut-off of a low-pass filter that is applied to a noisy sensor to a very small value?

3.13 You observe that the frequency spectrum of a signal has a spike at 60 Hz, which in a North American laboratory is the frequency of single-phase electric power. Discuss any considerations that you would make when attempting to remove this spike with the use of a band-pass filter.

3.14 How would you discretise multiple signals in a way that you can determine the instantaneous value of their sum?

3.15 Assume that you wish to digitise a signal that represents a turbulent velocity, which has fluctuations with frequencies up to 10 kHz. Describe the devices that you would use and specify their settings.

3.16 What is aliasing and how can it be avoided?

3.17 When can a digital low-pass filter not replace an analogue low-pass filter?

3.18 Is there any benefit in filtering a signal before discretisation, if the sum of the amplitudes of its interference and its noise is smaller than the resolution of the ADC?

Problems

3.1 Applying Kirchhoff's laws and assuming ideal operational amplifier operation, derive algebraic or differential equations for the circuits shown in Fig. 3.2. Then determine the transfer function for each circuit; in cases of multiple inputs, consider one input at a time, connecting all other inputs to the ground.

3.2 Derive Eq. (3.18) and find an expression for the cut-off frequency of Butterworth filters.

3.3 Derive expressions for the standard deviations of square, triangle, and sawtooth signals, and compare them to the standard deviation of a sinusoidal signal with the same amplitude. Which one has the lowest value?

3.4 Consider three analogue to digital converters with the same input range from −5 to 5 V, but 8-, 12- and 16-bit conversions, respectively. Determine their quantisation uncertainties.

3.5 Derive an expression for the gain g of a difference amplifier, such that $V_o = g(V_2 - V_1)$.

3.6 Derive the transfer function of a second-order active band-pass filter and determine the expression for the cut-off frequencies.
3.7 We wish to offset the signal described in Example 3.1 by (i) −0.5 V, (ii) −1 V and (iii) −2 V. Design a single circuit that can perform all three operations.
3.8 We wish to double the input range of an ADC, but want to maintain the same resolution. How many more bits do we need?
3.9 Design a non-inverting summing amplifier and derive an expression for its amplification gain. Determine the settings of this circuit, if it is used in Example 3.10.
3.10 The specifications of available ADC are listed in the following table. A temperature sensor produces a sinusoidal voltage with no DC offset and a standard deviation $\sigma_V = 1.3$ V. Which device would you use? Justify your answer in detail.

Device	Input range [V]	Number of bits
1	±1	16
2	±2	8
3	0–2	16
4	0–4	24
5	±2	12
6	±8	14

3.11 A noisy velocity sensor is used to measure the period of a pulsatile flow, which we expect to be in the vicinity of 15 ms. Would you filter the output of this sensor? If so, select the filter settings. If you discretise this signal, select the ADC sampling rate for this experiment. Clearly justify your selections.
3.12 Consider a sinusoidal signal with a period $T = 1$ s, having superimposed random fluctuations, which can be obtained by linearly connecting random numbers that are generated by appropriate software and are spaced by time increments of 1 ms. Adjust the random number amplitude so that the standard deviation of the random numbers is equal to the standard deviation of the sinusoidal signal. Apply a first-order low-pass analogue filter to this signal in the Laplace domain. Use the inverse Laplace transform to reconstruct the filtered signal in the time domain. Plot the unfiltered and filtered signals and the means and standard deviations of the filtered signal for different filter cut-off frequencies. Discuss how the analogue filter cut-off frequency affects the signal and suggest an optimal cut-off frequency value for this signal.
3.13 The voltage output of a Wheatstone bridge, measured in volts, is given as $V(t) = 0.25[1.2 + \sin(5t) + 0.6\sin(32t)]$, where the time t is measured in seconds. We wish to discretise the voltage using an ADC with an input range of $\pm 5V$. Design a signal conditioner that would allow us to do this. What should the sampling rate be?
3.14 The flow behind a cylinder is known to experience a periodic phenomenon known as vortex shedding, which occurs at a Strouhal number of 0.2. We place a cylinder with a diameter $D = 150$ mm in a wind tunnel and wish to conduct experiments over a range of Reynolds numbers $1 \times 10^5 \leq Re \leq 5 \times 10^5$. We have two velocity sensors, one with a cut-off frequency of 50 Hz and another with a cut-off frequency of 100 Hz. Which device should we use, and to which values should the sampling rate and record length be set?

References

[1] T.G. Beckwith, R.D. Marangoni, and J.H. Lienhard. *Mechanical Measurements (6th Edition).* Pearson, Upper Saddle River, NJ, 2007.

[2] J.S. Bendat and A.G. Piersol. *Random Data Analysis and Measurement Procedures (4th Edition).* Wiley, New York, 2010.

[3] T.C. Carusone, D. Johns, and K. Martin. *Analog Integrated Circuit Design (2nd Edition).* Wiley, Hoboken, NJ, 2011.

[4] J. Millman and C.H. Halkias. *Integrated Electronics: Analog and Digital Circuits and Systems.* McGraw-Hill, New York, 1972.

[5] A.V. Oppenheim and A.S. Willsky. *Signals and Systems (2nd Edition).* Prentice Hall, Upper Saddle River, NJ, 1996.

[6] D.H. Sheingold, editor. *Analog-Digital Conversion Handbook.* Analog Devices, Inc., Norwood, MA, 1972.

[7] D.H. Sheingold, editor. *Nonlinear Circuits Handbook.* Analog Devices, Inc., Norwood, MA, 1976.

[8] D.H. Sheingold, editor. *Transducer Interfacing Handbook.* Analog Devices, Inc., Norwood, MA, 1980.

[9] D.J. Stearns. *Digital Signal Processing with Examples in MATLAB.* CRC Press, Boca Raton, FL, 2004.

[10] S.D. Stearns. *Digital Signal Analysis.* Hayden, Rochelle Park, NJ, 1975.

[11] S. Tavoularis and S. Corrsin. Experiments in nearly homogeneous turbulent shear flow with a uniform mean temperature gradient, part 2: The fine structure. *J. Fluid Mech.*, 104:349–367, 1981.

[12] R.J. van de Plassche. *Integrated Analog-to-Digital and Digital-To-Analog Converters.* Kluwer Academic Publishers, Dordrecht, 1994.

[13] A. Williams and F. Taylor. *Electronic Filter Design Handbook (4th Edition).* McGraw-Hill, New York, 2006.

4 Statistical Analysis of Measurements

This chapter is a review of the fundamentals of probability, random variables and random processes. It also outlines some common statistical operations and tests that can be applied to measurements to help describe and understand their properties.

4.1 Probability

The term *probability* is used colloquially to express one's belief that something may or may not happen or have happened. For example, one may assess that 'the probability that it will rain this afternoon is high', based on observation of the morning sky, combined with one's experience and intuition. Clearly, this notion is subjective and non-quantifiable (weather forecasting, which specifies the probability of precipitation etc. as a percentage, is presumably based on extensive databases and elaborate models).

The classical scientific definition of probability is based on the notion of *events*, namely the possible outcomes of a process. According to this definition, probability is the a priori ratio of favourable to total number of alternative events, assuming that all events are equally likely to occur. The gross inaccuracy of this approach can be illustrated by considering as events the readings of a thermometer, rounded to the closest marking on its scale. If the thermometer scale has 40 markings, the classical definition would lead to all probabilities of all events being equal to 1/41, which is obviously wrong.

Probability has alternatively been defined as the relative frequency of an event. In this approach, the probability of future events can be predicted from past records of the same process, under the implicit assumption that all past and future events would be subjected to the same external conditions and that there are no particular circumstances that may significantly affect a specific event but not the others. An example showing that this approach may also lead to errors is the prediction of the probability that an expecting mother will bear twins as the relative frequency of twin births with respect to total births in a certain country over a certain period of years. This estimate disregards the effect of genetic background of the parents, the possible use of fertility medication, and other factors, which may substantially affect the conception of twins.

The modern theory of probability is a branch of measure theory. According to this theory, probability is defined axiomatically, thus avoiding any assumptions or reference to past events. Consider the set of all possible events $\mathcal{A}_i, i = 1, 2, ...$ in a repeatable experiment. Two events are called *mutually exclusive*, if it is impossible for both to occur at the same trial. The *compound event* $\mathcal{A}_i + \mathcal{A}_j, i \neq j$, is the event that occurs when either $\mathcal{A}_i$ or $\mathcal{A}_j$ or both occur at a given trial. Finally, let $\mathcal{S}$ denote the *certain event*, namely the event that occurs in every trial. Then, the probability of an event $\mathcal{A}_i$ is defined as a number $P(\mathcal{A}_i)$, which obeys the following three postulates:

- $P(\mathcal{A}_i) \geq 0$;
- $P(\mathcal{S}) = 1$;
- If $\mathcal{A}_i$ and $\mathcal{A}_j, i \neq j$ are mutually exclusive, then $P\left(\mathcal{A}_i + \mathcal{A}_j\right) = P(\mathcal{A}_i) + P\left(\mathcal{A}_j\right)$.

Thus, the probability of an event is simply a non-negative real number, which is less than or equal to one. This definition does not provide any means by which to estimate the probability value.

Example 4.1 Discuss the probability of heads and tails in coin tossing.

Answer

Consider an experiment by which a coin is tossed in the air and falls on a smooth horizontal surface. Assume that the only two possible events are heads and tails; any other outcome, such as the coin standing on its edge or never reaching the surface for some strange reason, are completely disregarded.

According to the classical definition, the probability of heads and the probability of tails should be both equal to 0.5. This is based on the facts that there are only two events and the collective probability of all events should be equal to 1. This prediction would be approximately valid for a fair coin, but not for one that has been fixed to give preference to one side.

According to the relative frequency definition, one would have to use a relatively large database of past repetitions of this experiment. If, for instance, among 1000 past coin tosses, heads came up 535 times, then one would say that the probability of heads in a future tossing would be 0.535 and the probability of tails would be 0.465. This prediction is based on the assumption that the conditions during past and future coin tosses would be identical. One may, however, imagine that there are ways to bias future events, such as tampering with the coin, developing a favourable tossing technique and so on.

Finally, consider the axiomatic definition of probability. The coin-tossing experiment has two mutually exclusive events, to be referred to as $\mathcal{A}_1$ (heads) and $\mathcal{A}_2$ (tails). The event $\mathcal{S} = \mathcal{A}_1 + \mathcal{A}_2$ is a certain event and so $P(\mathcal{S}) = 1$. If the probabilities of the two events are denoted as $P(\mathcal{A}_1) = p$ and $P(\mathcal{A}_2) = q$, respectively, the only requirements of the axiomatic definition are that $0 \leq p, q \leq 1$ and $p + q = 1$. This definition does not provide estimates of p and q.

4.2 Random Variables

4.2.1 Definition of a Random Variable

A *random property* is commonly understood to be one that has unpredictable values. For example, each time we toss a coin, we will get 'heads' or 'tails' without being able to predict which one will occur.

In probability theory, however, the term *real random variable* does not apply to an arbitrary random property, but only to properties the values of which are real numbers and, therefore, have no dimensions and units. In the coin-tossing experiment, for instance, the events heads and tails do not constitute a real random variable. To make a real random variable, we need to assign a real number to each of these events, let's say 0 to heads and 1 to tails; any other real numbers would be equally acceptable as values of the random variable. This random variable and any other that may only take discrete values are called *discrete random variables*.

A different type of random variables are the *continuous random variables*, namely, ones the values of which extend continuously over an interval or even the entire real axis. As an example of a continuous random variable, consider an experiment for which we repeatedly throw a coin towards a wall that is 10 m away and measure the distance x in metres between the coin edge and the wall. A continuous random variable would be the numerical value of x, under the assumption that it is measured with an ideal measuring tape, which can resolve any distance, however small it is. One should distinguish between the random value of a measurand, which has dimensions and units (e.g., distance measured in metres), and the corresponding real random variable, which has only numerical values.

It is also possible for a random variable to be *mixed*, namely to take both discrete and continuous values. We can easily make up an example of a mixed random variable by modifying a continuous random variable. For the coin throwing experiment, we can define a mixed random variable consisting of (a) a continuous part, which is the value of x in m, when $1 \text{ m} \leq x \leq 9 \text{ m}$, and (b) two discrete values, which are the number 10, when $9 \text{ m} < x$ and the number 0, when $x < 1 \text{ m}$.

According to its formal definition, a *real random variable* $\boldsymbol{x}$ is a real function of the events ζ, such that the set $\{\boldsymbol{x}(\zeta) \leq x\}$ is an event for any real number x and $P\{\boldsymbol{x} = \infty\} = P\{\boldsymbol{x} = -\infty\} = 0$. In this notation, bold-faced characters, such as $\boldsymbol{x}$, denote the random variable, whereas roman characters, such as x, denote the real values that this random variable takes.

4.2.2 Statistical Properties of a Random Variable

The *distribution function* (or *cumulative distribution function*) of a random variable $\boldsymbol{x}$ is defined as

$$F_x(x) = P\{\boldsymbol{x} \leq x\} . \tag{4.1}$$

For the sake of simplicity, a roman character appears in the subscript of F, but it is noted that a bold-faced character would have been more appropriate for this subscript, as it indicates the random variable of interest and not its value. The distribution function has the following properties:

$$F_x(-\infty) = 0\,, \tag{4.2}$$

$$F_x(+\infty) = 1\,, \tag{4.3}$$

$$F_x(x_1) \le F_x(x_2)\,,\ \text{if } x_1 < x_2\,. \tag{4.4}$$

The *probability density function* (*pdf*) $f_x(x)$ of a random variable $\boldsymbol{x}$ is the derivative of its distribution function, that is,

$$f_x(x) = \frac{\mathrm{d}F_x(x)}{\mathrm{d}x}\,, \tag{4.5}$$

which may be inverted to give

$$F_x(x) = \int_{-\infty}^{x} f_x(x')\mathrm{d}x'\,. \tag{4.6}$$

The pdf of a continuous random variable, which would have a continuous $F_x(x)$, requires no explanation. The distribution function of a discrete random variable would be the sum of step functions and its pdf may be expressed as the sum of *Dirac's delta functions* (Eq. (2.35)).

Example 4.2 Describe the distribution function and the pdf of a random variable that corresponds to the coin-tossing experiment.

Answer

Let us define the random variable to take the value of -1 for a heads event and the value of $+1$ for a tails event and assume that the coin is fair, so that the probabilities of both events are equal to 0.5. This is a *bimodal* discrete random variable. Its distribution function is a step function, namely,

$$F_x(x) = \begin{cases} 0, & \text{for } x < -1\,, \\ 0.5, & \text{for } -1 \le x < 1\,, \\ 1, & \text{for } 1 \le x\,. \end{cases}$$

Its pdf is

$$f_x(x) = \frac{1}{2}\delta(-1) + \frac{1}{2}\delta(+1)\,.$$

These two functions are illustrated in Fig. 4.1.

The *mean* (or *average* or *expectation* or *expected value*) of a random variable $\boldsymbol{x}$, to be denoted as μ_x or by enclosing the random variable symbol in angle brackets as $\langle\boldsymbol{x}\rangle$, is defined as

$$\mu_x \equiv \langle\boldsymbol{x}\rangle = \int_{-\infty}^{\infty} x f_x(x)\mathrm{d}x\,. \tag{4.7}$$

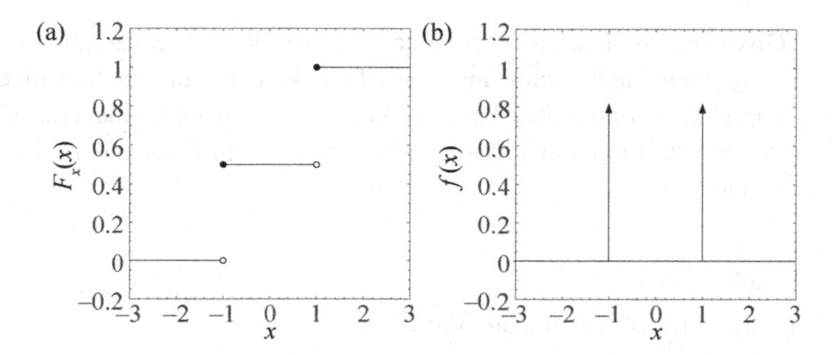

Figure 4.1 (a) Distribution function and (b) pdf of a bimodal random variable; white and black circles indicate, respectively, the non-inclusion and inclusion of boundary points in a discontinuous function.

The *variance* of a random variable $\boldsymbol{x}$ is defined as

$$\sigma_x^2 = \left\langle (\boldsymbol{x} - \langle \boldsymbol{x} \rangle)^2 \right\rangle = \int_{-\infty}^{\infty} (x - \langle \boldsymbol{x} \rangle)^2 \, f_x(x) \mathrm{d}x \, . \tag{4.8}$$

The positive square root σ_x of the variance is called the *standard deviation*. The terms variance and standard deviation are sometimes used alternatively with the terms *mean squared value* and *root-mean-squared (rms) value*, respectively. Such practice is discouraged, because the definitions of the former two terms require removal of the mean, whereas the definitions of the latter two terms retain the mean.

In general, a *moment of nth order* of a random variable is defined as

$$\langle \boldsymbol{x}^n \rangle = \int_{-\infty}^{\infty} x^n f_x(x) \mathrm{d}x \, , \tag{4.9}$$

whereas a *central moment of nth order* is defined as

$$\langle (\boldsymbol{x} - \langle \boldsymbol{x} \rangle)^n \rangle = \int_{-\infty}^{\infty} (x - \langle \boldsymbol{x} \rangle)^n \, f_x(x) \mathrm{d}x \, . \tag{4.10}$$

Central moments are often denoted by the symbol μ_n, where n is the moment order. We therefore note that $\mu_0 = 1$, $\mu_1 = 0$ and $\mu_2 = \sigma_x^2$. One is warned against confusing the first central moment μ_1 with the mean μ_x, which is is the first moment, namely, not a central one. Two other commonly used statistical properties are the *skewness factor* and the *flatness factor*, respectively defined as

$$S = \frac{\mu_3}{\sigma_x^3} \tag{4.11}$$

and

$$F = \frac{\mu_4}{\sigma_x^4} \, . \tag{4.12}$$

The flatness factor is sometimes presented as the *kurtosis*

$$K = \frac{\mu_4}{\sigma_x^4} - 3 \, . \tag{4.13}$$

Given the distribution function or the pdf of a random variable, one may compute all its moments and central moments. Reversely, the distribution function or the pdf of a random variable may be computed only if all its moments are known. Therefore, in general, a random variable cannot be described statistically by a finite number of moments.

4.2.3 Functions of a Random Variable

A task that sometimes arises during a measurement process is to determine the pdf of a property that is specified as a function of a random variable with a known pdf. This can be done with the use of the probabilistic relationship between the two variables, as will be illustrated by the following example.

Example 4.3 Let $\boldsymbol{\theta}$ be a random variable, representing an angle which is uniformly distributed in the interval $[-\pi, \pi]$, such that its pdf is

$$f_\theta(\boldsymbol{\theta}) = \begin{cases} \frac{1}{2\pi}, & -\pi \leq \boldsymbol{\theta} \leq \pi , \\ 0 , & \pi < |\boldsymbol{\theta}| . \end{cases}$$

The pdf and the distribution function of this random variable are plotted in Fig. 4.2. Determine the pdf of the dependent random variable

$$\boldsymbol{x}(\boldsymbol{\theta}) = a \cos \boldsymbol{\theta} ,$$

where a is a positive constant number.

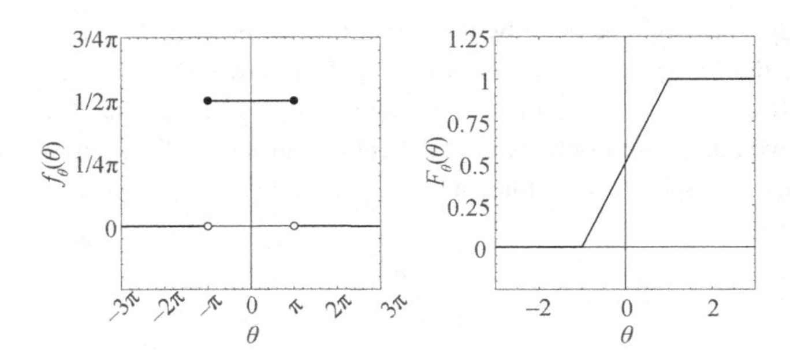

Figure 4.2 Pdf and distribution function of a random variable that is uniformly distributed within an interval. White and black circles, indicate, respectively, the non-inclusion and inclusion of boundary points of a discontinuous function.

Answer

For any value $\boldsymbol{\theta} = \theta$, one gets the value $\boldsymbol{x}(\boldsymbol{\theta}) = x = a\cos\theta$, which is equivalent to $\theta = \cos^{-1}(x/a)$. Let us first compute $F_x(x)$ and $f_x(x)$. Starting with the definition of

$F_x(x)$, one gets

$$F_x(x) = P\{\boldsymbol{x} \leq x\} = \begin{cases} 0\,, & x \leq -a\,, \\ P\{-\pi < \boldsymbol{\theta} < -\theta\} + P\{\theta < \boldsymbol{\theta} < \pi\}\,, & -a < x < a\,, \\ 1\,, & a \leq x \end{cases}$$

$$= \begin{cases} 0\,, & x \leq -a\,, \\ \frac{-\theta-(-\pi)}{2\pi} + \frac{\pi-\theta}{2\pi}\,, & -a < x < a\,, \\ 1\,, & a \leq x \end{cases}$$

$$= \begin{cases} 0\,, & x \leq -a\,, \\ 1 - \frac{1}{\pi}\cos^{-1}(x/a)\,, & -a < x < a\,, \\ 1\,, & a \leq x\,. \end{cases}$$

Then,

$$f_x(x) = \frac{\mathrm{d}F_x(x)}{\mathrm{d}x} = \begin{cases} 0\,, & x \leq -a\,, \\ \frac{1}{\pi}\frac{1}{\sqrt{a^2-x^2}}\,, & -a < x < a\,, \\ 0, & a \leq x\,. \end{cases}$$

The distribution function and the pdf of $\boldsymbol{x}(\boldsymbol{\theta})$ are illustrated in Fig. 4.3. Notice that, although $f_x(x) \to \infty$, as $|x| \to a$, this is not a concern, because $f_x(a) = f_x(-a) = 0$. Note also that $f_x(a)$ is an even function and so the mean of $\boldsymbol{x}$, as well as all its odd-order moments, would vanish. One may proceed in this manner to compute additional statistical properties of $\boldsymbol{x}(\boldsymbol{\theta})$ through application of the corresponding definitions.

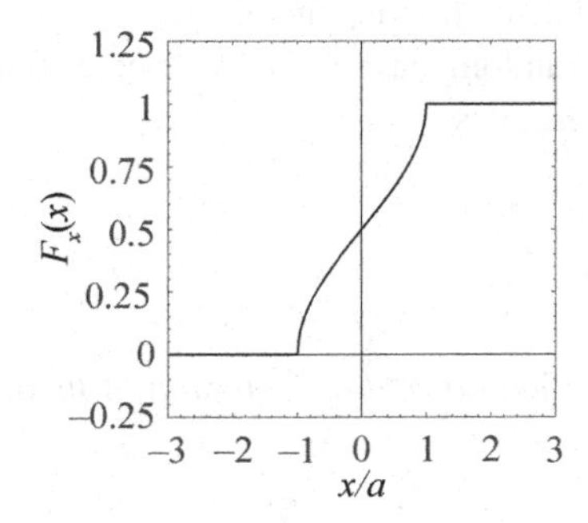

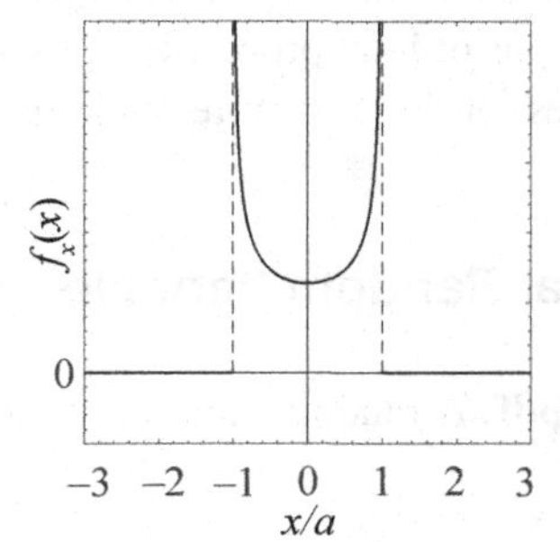

Figure 4.3 Distribution function and pdf of a sinusoidal random variable with a uniformly distributed phase.

4.2.4 Multiple Random Variables

Now consider two random variables $\boldsymbol{x}$ and $\boldsymbol{y}$. Their *joint distribution function* is defined as

$$F_{xy}(x, y) = P\{\boldsymbol{x} \leq x \text{ and } \boldsymbol{y} \leq y\}\,. \tag{4.14}$$

It has the properties

$$F_{xy}(-\infty, y) = F_{xy}(x, -\infty) = 0 , \tag{4.15}$$

$$F_{xy}(+\infty, +\infty) = 1 . \tag{4.16}$$

The distribution functions F_x and F_y of the individual random variables are called *marginal* and satisfy the relationships

$$F_{xy}(x, \infty) = F_x(x) , \tag{4.17}$$

$$F_{xy}(\infty, y) = F_y(y) . \tag{4.18}$$

The *joint probability density function (jpdf)* of $\boldsymbol{x}$ and $\boldsymbol{y}$ is defined as

$$f_{xy}(x, y) = \frac{\partial^2 F_{xy}(x, y)}{\partial x \partial y} . \tag{4.19}$$

Two random variables are *statistically independent*, if

$$f_{xy}(x, y) = f_x(x) f_y(y) . \tag{4.20}$$

An indicator of possible interdependence between two random variables is their *correlation coefficient*

$$\rho = \frac{\langle (\boldsymbol{x} - \langle \boldsymbol{x} \rangle)(\boldsymbol{y} - \langle \boldsymbol{y} \rangle) \rangle}{\sigma_x \sigma_y} = \frac{\int_{-\infty}^{\infty} \int_{-\infty}^{\infty} (x - \langle \boldsymbol{x} \rangle)(y - \langle \boldsymbol{y} \rangle) f_{xy}(x, y) \, \mathrm{d}x \, \mathrm{d}y}{\sigma_x \sigma_y} , \tag{4.21}$$

which spans the range $[-1, 1]$. Two random variables are called *uncorrelated* if $\rho = 0$, otherwise they are called *correlated*, in which case, they are also statistically dependent on each other. On the other hand, two uncorrelated random variables do not necessarily have to be statistically independent. A value $|\rho| = 1$ is proof that the two variables are proportional to each other and not simply statistically dependent.

The definitions of joint properties of two random variables may be extended to define joint properties for three or more random variables.

4.3 The Normal Random Variable

The normal pdf: A random variable $\boldsymbol{x}$ is called *normal* or *Gaussian*, if its pdf is given by

$$f_x(x) = \frac{1}{\sqrt{2\pi}\sigma_x} e^{-\frac{(x-\mu_x)^2}{2\sigma_x^2}} . \tag{4.22}$$

One may notice that, in this pdf, the random variable appears in a *standardised form*, which entails subtracting the mean μ_x from all values and dividing the difference by the standard deviation σ_x. Thus, instead of a value x, one may use the corresponding *z-score*, defined as

$$z = \frac{x - \mu_x}{\sigma_x} . \tag{4.23}$$

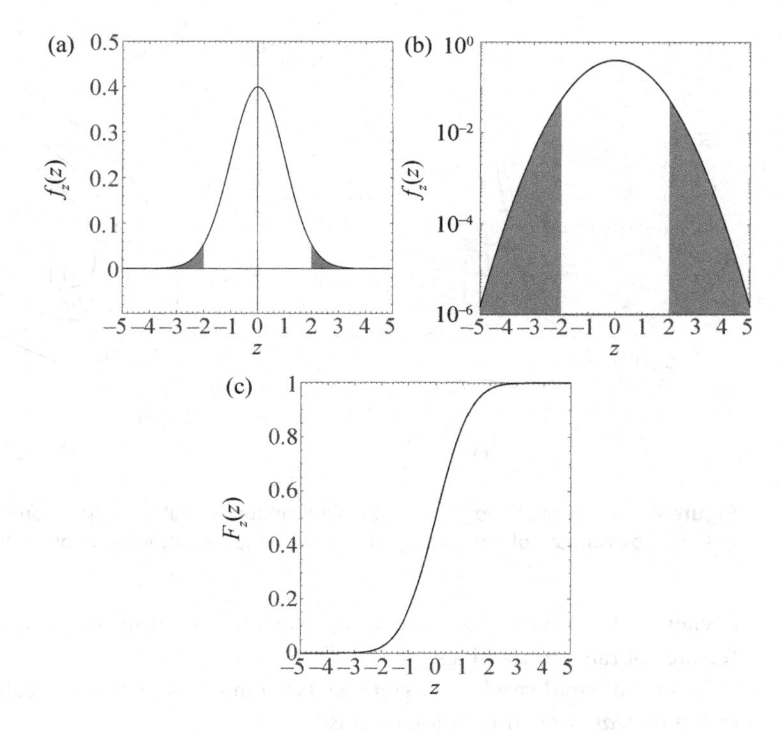

Figure 4.4 (a) Pdf of a standardised Gaussian random variable in linear coordinates, with each shaded area (on either tail) containing approximately 2.5% of the probability; (b) the same pdf in semi-logarithmic coordinates; (c) distribution function of the same random variable.

The pdf of a *standardised normal random variable* is, therefore,

$$f_z(z) = \frac{1}{\sqrt{2\pi}} e^{-z^2/2} . \tag{4.24}$$

This pdf and the corresponding distribution function $F_z(z)$ are illustrated in Fig. 4.4. Values of these functions may be found in many handbooks and textbooks of probability [3, 4], as well as in many internet sources. The normal random variable is an exceptional one, because its pdf depends only on two parameters, namely, its mean and its standard deviation. Therefore, a normal random variable would be completely characterised statistically, if its mean and variance were known. One is also reminded that all moments of a random variable can be determined from its pdf, and so the moments of a normal random variable can be expressed as functions of its mean and standard deviation alone. It is easy to show that all odd central moments of a Gaussian random variable vanish ($\mu_{2i+1} = 0$, i integer) and that its even central moments can be expressed as

$$\mu_{2i} = 1 \cdot 3 \cdot 5 \cdot \ \cdots \ \cdot (2i-1)\, \sigma_x^{2i} , \ i \text{ integer} . \tag{4.25}$$

Both the skewness factor and the kurtosis of Gaussian random variables vanish and

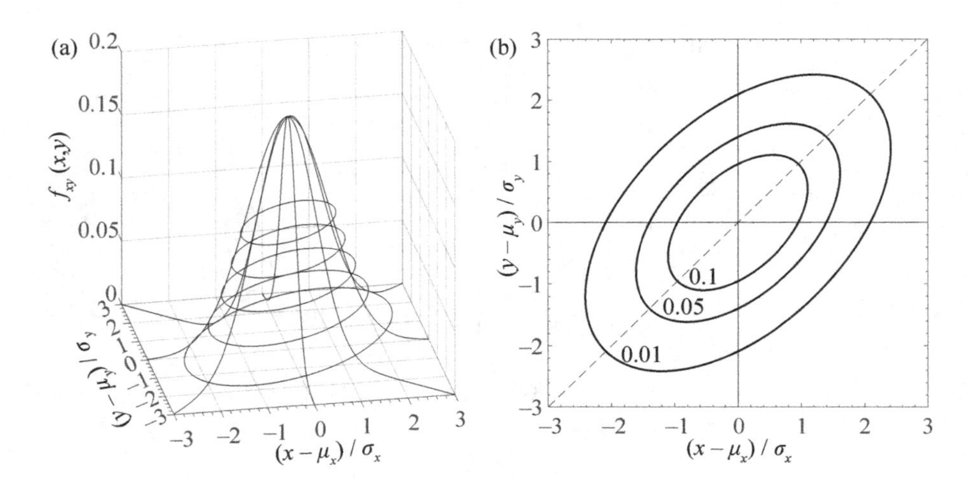

Figure 4.5 (a) Jpdf of two jointly Gaussian random variables with a correlation coefficient of 0.5; (b) iso-contours of the same jpdf with its value indicated on each contour.

deviations from these reference values are often used to detect systematic statistical features of random variables.

Jointly normal random variables: Two random variables are called *jointly normal* or *jointly Gaussian*, if their joint pdf is

$$f_{xy}(x,y) = \frac{1}{2\pi\sigma_x\sigma_y\sqrt{1-\rho^2}} e^{-\frac{1}{2(1-\rho^2)}\left[\frac{(x-\langle x\rangle)^2}{\sigma_x^2} - \frac{2\rho(x-\langle x\rangle)(y-\langle y\rangle)}{\sigma_x\sigma_y} + \frac{(y-\langle y\rangle)^2}{\sigma_y^2}\right]}. \tag{4.26}$$

Uncorrelated jointly Gaussian variables are also statistically independent. A jointly Gaussian pdf in a 3D plot and a contour plot is shown in Fig. 4.5. Notice that the iso-probability contours are ellipses, whose axis ratio depends on ρ; if $\rho = 0$, these contours would be circles and, if $\rho = \pm 1$, all contours would collapse on a diagonal line.

95% confidence interval: The Gaussian random variable is a continuous function over the entire real axis, which means that its pdf has always positive values, even as $x \to \pm\infty$. Physical properties, however, have bounded values and so the pdf of very large (positive or negative) values of physical random variables is expected to vanish. A common practice in engineering applications is to truncate the range of the normal pdf at its two tails. Typically, one neglects approximately 2.5% of the total probability on either tail (Fig. 4.4a) by setting the bounds of the random variable range as

$$\mu_x - 2\sigma_x < x < \mu_x + 2\sigma_x\,. \tag{4.27}$$

Equation (4.27) will be used interchangeably with the notation

$$x = \mu_x \pm 2\sigma_x\,. \tag{4.28}$$

This means that, in practice, one often only considers values that, collectively, would have an approximately 95% probability to occur, if they belonged to a normal random variable. The range of these values is called the *95% confidence interval*. To be more precise, the 95% confidence interval corresponds to the range $\mu_x \pm 1.96\sigma_x$, but, instead, it is Eq. (4.28) that is used universally. It should also be noted that, in some applications, different confidence intervals, for example, 90% or 99%, are used instead of the 95% confidence interval.

The prevalence of normality: Random variables encountered in many science and engineering areas are frequently assumed to be normal. Justification for this practice is provided by the *central limit theorem* [9], which states that the pdf of the sum of a large number of independent random variables will approach the Gaussian pdf as the number of variables increases. This situation will arise when the value of the random property of interest is the result of a large number of many small influences, which cannot be predicted and controlled. When multiple ('repeat') measurements of the same property are made, the values may not be identical, even if all conditions are seemingly the same and one might expect these values to be identical. The reasons for such variation could be *systematic*, such as a sudden change in the room temperature or in the power line voltage that affect the operation of the flow apparatus or a measuring instrument. Systematic effects should, as much as possible, be detected and removed or corrected for. In general, however, at least part of the observed differences in repeat values would be *random*, namely attributable to many unspecifiable, undetectable and uncorrectable small effects. To determine whether significant systematic effects are present in a measuring process, one needs to to assess the 'randomness' of repeat measurements, or, equivalently, to determine whether the process is under *statistical control*. In such cases, one would expect that the measured values would likely have a normal distribution. To determine whether this is the case, it is advisable to perform a *normality test*, to be discussed in Section 4.5.

4.4 The Mean of Repeat Measurements

Population and set statistics: Consider a normal random variable $\boldsymbol{x}$ with a mean μ_x and a standard deviation σ_x, to be referred to as *population mean* and *population standard deviation*, respectively. The term population implies an infinite, or, at least, very large, number of values. Next, consider a set of values, for example, 'repeat measurements', $x_i, i = 1, 2, ..., N$, taken randomly from the population of $\boldsymbol{x}$ and having a *set mean* μ_{xN} and a *set standard deviation* σ_{xN}. In general, μ_{xN} and σ_{xN} for each set of N values would be different from those for a different set of N values of the same population and would, therefore, be random variables as well. According to the *law of large numbers* in probability theory, as $N \to \infty$, $\mu_{xN} \to \mu_x$ and $\sigma_{xN} \to \sigma_x$. The central limit theorem dictates that μ_{xN} is a normal random variable with a mean equal to μ_x and a standard deviation $\sigma_{\mu_{xN}} = \sigma_x / \sqrt{N}$. Note that $\sigma_{\mu_{xN}}$ depends on both σ_x and N. Equivalently, the random variable $(\mu_{xN} - \mu_x)/(\sigma_x/\sqrt{N})$ is a standardised normal random variable with a zero mean and a standard deviation equal to 1. In many applications, we do not know the population mean and standard deviation but we may know the statistical properties of a set of N values that belong to that population.

Student's *t* distribution: A random variable **t** is said to follow *Student's t statistical*

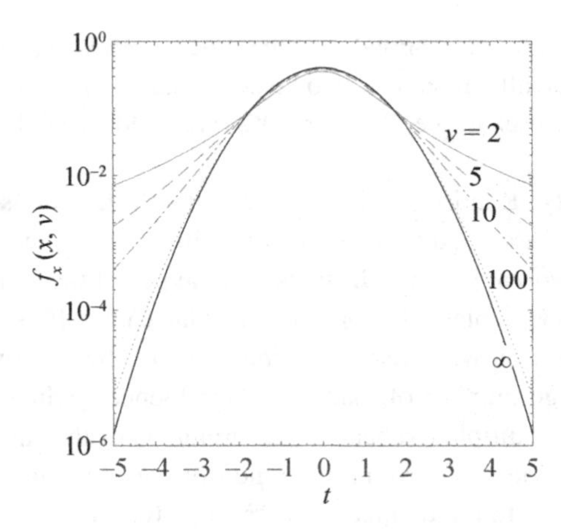

Figure 4.6 Probability density function of a random variable that follows the Student t-distribution for $\nu = 2, 5, 10, 100$, and ∞; the latter of these corresponds to a normal distribution; note that these properties are plotted in semi-logarithmic axes.

distribution with ν degrees of freedom, if its pdf is

$$f_t(t, \nu) = \frac{\Gamma\left(\frac{\nu+1}{2}\right)}{\sqrt{\nu\pi}\Gamma\left(\frac{\nu}{2}\right)} \left(1 + \frac{t^2}{\nu}\right)^{-\frac{\nu+1}{2}} . \tag{4.29}$$

As shown in Fig. 4.6, the tails of this pdf are heavier than those of the Gaussian pdf, and become heavier, as ν decreases; as $\nu \to \infty$, however, this pdf tends towards a standardised Gaussian one.

Now, consider the mean μ_{xN} of N random values belonging to a normal population. This mean may be standardised in the form of the *t-score*

$$t = \frac{\mu_{xN} - \mu_x}{\sigma_{xN}/\sqrt{N}} . \tag{4.30}$$

It has been shown [5] that the random variable **t** has a Student's t statistical distribution with a mean μ_x, a standard deviation $\sigma_{xN}/\sqrt{N}$ and $\nu = N - 1$ degrees of freedom.

95% confidence interval: From a set of N values of a random variable **x**, we can determine μ_{xN} and σ_{xN}. Then, under the assumption that this set of values is part of a normal population, Eqs. (4.29) and (4.30) allow us to determine the interval within which the population mean μ_x lies with a specified probability. Setting, as commonly done, this probability at 95%, we may determine t_ν, which is the value of t that specifies the bounds of the interval that contains 95% of the probability. Then, the probability that μ_x will be within the interval

$$\mu_{xN} - \frac{t_\nu \sigma_{xN}}{\sqrt{N}} < \mu_x < \mu_{xN} + \frac{t_\nu \sigma_{xN}}{\sqrt{N}} \tag{4.31}$$

would be 0.95, which is equivalent to saying that we are 95% confident that μ_x will be within that interval. The values of t_ν for different values of ν are listed in Table 4.1. For example, for $N = 2$, the number of degrees of freedom is $\nu = 1$ and $t_\nu = 12.706$, which demonstrates that the 95% confidence interval of the average of two values is much wider than that of a Gaussian random variable.

The mean of 'large' data sets: The central limit theorem stipulates that, as $N \to \infty$, the statistical distribution of the mean μ_{xN} of each data set will tend towards a normal

Table 4.1 t-scores at 95% confidence interval for various numbers of degrees of freedom ν

ν	1	2	3	4	5	6	7	8	9	10
t_ν	12.706	4.303	3.182	2.776	2.571	2.447	2.365	2.306	2.262	2.228
ν	11	12	13	14	15	16	17	18	19	20
t_ν	2.201	2.179	2.160	2.145	2.131	2.120	2.110	2.101	2.093	2.086
ν	21	22	23	24	25	26	27	28	29	30
t_ν	2.080	2.074	2.069	2.064	2.060	2.056	2.052	2.048	2.045	2.042

distribution with a variance

$$\sigma^2_{\mu_{xN}} \approx \frac{\sigma^2_x}{N} \approx \frac{\sigma^2_{xN}}{N} \, . \tag{4.32}$$

Therefore, if N is sufficiently large, we may assess that, for 95% of the values of μ_{xN}, the population mean μ_x will be within the approximate interval

$$\mu_{xN} - \frac{2\sigma_{xN}}{\sqrt{N}} < \mu_x < \mu_{xN} + \frac{2\sigma_{xN}}{\sqrt{N}} \, . \tag{4.33}$$

The lowest value of N for one to use Eq. (4.33) instead of Eq. (4.31) depends on the requirements of each application. For many purposes, one may use $N = 29$ as a threshold, as this is the lowest value of N for which t_ν may be rounded to 2.0, a number equal to the corresponding Gaussian z-score, when rounded in the same way. Nevertheless, keep in mind that, as Fig. 4.6 shows, differences between the normal pdf and the Student t pdf are visible for N as large as 100.

Nominal value vs. mean: Before showing an example, it is important to distinguish the mean that is determined from a population or a set of repeat values from a *nominal value* that is provided by a manufacturer. Although this nominal value is sometimes representative of the population average, it is generally understood to be just a product name, rounded for convenience. For certain types of products, the nominal value could be very different from the actual value. For example, in North America, pipes having a nominal size of 1 in, as specified in manufacturer catalogues, actually have an outer diameter of 1.315 in and an inner diameter that is specified separately in terms of the *pipe schedule*.

Example 4.4 In a measurements class with $N = 100$ students, the course instructor provides each student with an operational amplifier circuit bought from a manufacturer, who states that the nominal amplification gain of the amplifier circuit is $g = 10$. Following the instructor's request, the students determined the gain for each circuit and the results are shown in Fig. 4.7. The average gain for all 100 amplifiers is $\mu_{gN} = 10.23$ and the standard deviation is $\sigma_{gN} = 0.79$. The 100 amplifiers were then separated into five groups, each having $n = 20$ circuits. The average and the standard deviation for each of these groups are also determined and are shown in Fig. 4.7.

Using the measurements of this class, determine the 95% confidence interval of the mean gain of the amplifiers provided by this manufacturer, which may be assumed to be a normally distributed population. Compare values obtained using all 100 circuits and

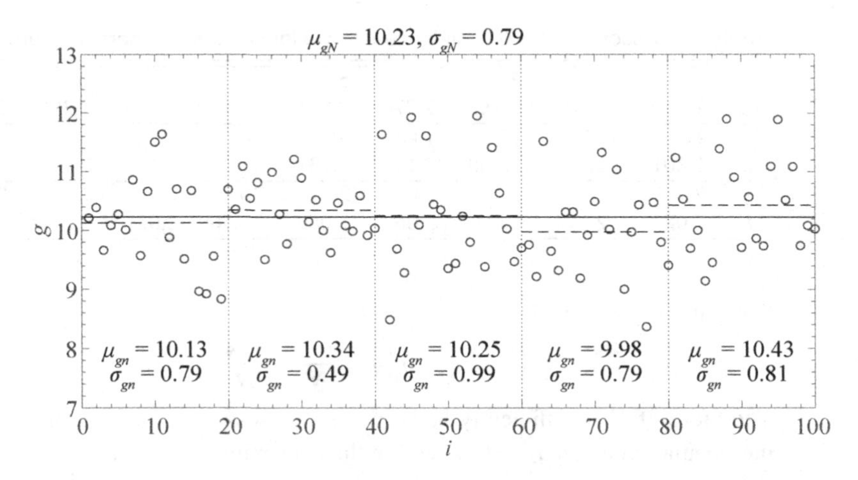

Figure 4.7 Measured gains of 100 amplifiers, split into five groups of 20 amplifiers each. The mean and the standard deviation of the entire set of 100 amplifiers, as well as those of each group of 20 amplifiers are also shown.

values obtained from each group of 20 circuits. Compare the previous estimates with the nominal gain. Which gain value should each student use for their calculations?

Answer

We first note that the averages of groups, calculated using $n = 20$ amplifiers, are different from the average of the $N = 100$ amplifiers and from each other. For this particular example, we find the maximum difference between two group means to be $10.43 - 9.98 = 0.45$. We also note that the average gains of the 100 amplifiers or any group of 20 amplifiers are different from the nominal value provided by the manufacturer.

First, let us consider the entire set of $N = 100$ values. This number is large enough for Eq. (4.33) to apply. Considering that $2\sigma_{gN}/\sqrt{N} = 0.16$, we find that the 95% confidence interval of the manufacturer's average μ_g will be from 10.07 to 10.39, that is, $\mu_g = 10.23 \pm 0.16$. We note that, in this particular case, the nominal value provided by the manufacturer falls outside the 95% confidence interval.

Next, we estimate the population mean from measurements in each group of $n = 20$ values. As this number is smaller than 28, we may not use Eq. (4.33), but we should use Eq. (4.31). For each group, $\nu = n - 1 = 19$, for which Table 4.1 gives $t_\nu = 2.093$. Then, we may determine the 95% confidence interval of the gain population mean from each of the five data groups as

$$\mu_g = \mu_{gn} \pm \frac{t_\nu \sigma_{gn}}{\sqrt{n}} = \begin{cases} 10.13 \pm 0.37\,, & \text{for } i = 1, ..., 20\,, \\ 10.34 \pm 0.23\,, & \text{for } i = 21, ..., 40\,, \\ 10.25 \pm 0.47\,, & \text{for } i = 41, ..., 60\,, \\ 9.98 \pm 0.37\,, & \text{for } i = 61, ..., 80\,, \\ 10.43 \pm 0.38\,, & \text{for } i = 81, ..., 100\,. \end{cases}$$

Having obtained different estimates of the circuit gain, it remains to choose the most appropriate one in any following circuit analysis. Obviously, if the actual gain of the specific amplifier that will be analysed is known from direct measurement, this value should be used in all subsequent calculations. Let us, however, follow a scenario by which gain measurements collected by one class are provided to a following class, which is asked to use the same 100 circuits, but without specific identification of each unit. In such a scenario, all students should use the average gain that was calculated for the 100 amplifiers available to the class, that is, 10.23, as this is a more accurate representative value than the nominal value stated by the manufacturer. If, however, each of the groups of 20 amplifiers is clearly marked and labelled, then each student should use the average gain for the corresponding group, as this would provide a more accurate estimate for this particular group. The nominal gain of 10 that was specified in the manufacturer's catalogue should be used for a specific unit as a last resort, if there is no other information about this unit and no possibility to perform gain measurements.

4.5 Normality Tests

A normality test is a standard procedure of assessing randomness, which entails comparing the statistical properties of a set of values of a random variable to those of a normal random variable. A positive normality test provides some reassurance, though not conclusive evidence, that an experiment is under statistical control and not subjected to strong systematic influences, such as electronic drift caused by room temperature changes.

4.5.1 Standardising the Data

Before performing a normality test, it is convenient to standardise the values of the random variable, so that they can be compared statistically to the standardised normal random variable. To do so, one must take the following actions:

- Let $x_i, i = 1, 2, ..., N$ be a set of values of a random variable.
- Compute the mean of these values as

$$\mu_{xN} = \frac{1}{N} \sum_{i=1}^{N} x_i \,, \tag{4.34}$$

and their variance as

$$\sigma_{xN}^2 = \frac{1}{N-1} \sum_{i=1}^{N} (x_i - \mu_{xN})^2 \,. \tag{4.35}$$

- Standardise the values to obtain their z-scores

$$z_i = \frac{x_i - \mu_{xN}}{\sigma_{xN}} \,. \tag{4.36}$$

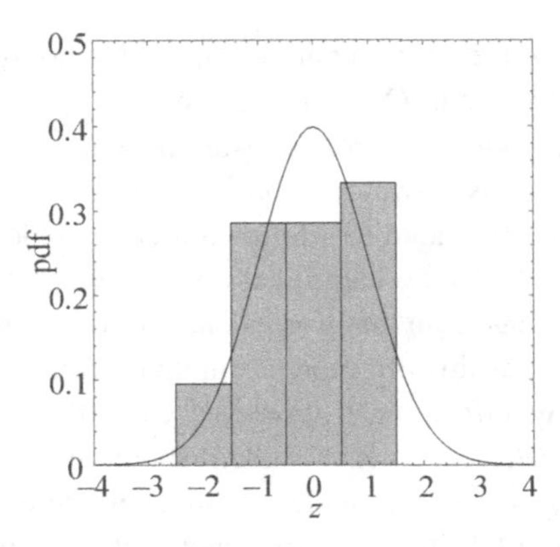

Figure 4.8 Histogram of standardised pressure measurements with a superimposed standardised normal pdf (solid line).

4.5.2 Histogram Inspection

The *histogram* (which is an approximation to a pdf) is a bar graph, in which the height of each bar corresponds to the *data count* (namely, the number of values) within each of the bins, into which we separate these values. Histograms are not unique, and their shape depends upon the number of bins we choose, as well as the boundaries of the bins. Histograms may be created automatically by different software packages. A simple normality test consists of plotting a histogram of the standardised values on the same graph as the standardised Gaussian pdf and compare the two by visual inspection. It is evident that the results of this test would be subjective. The confidence to such assessment of normality would be low for relatively small N, but it would increase for $N \gg 1$.

Example 4.5 Measurements of pressure difference in kPa at two taps along a pipe were taken by a pressure transducer at intervals of 30 s from each other. The numerical values of these measurements, which constitute a random variable, are as follows: 13, 10, 12, 11, 12, 12, 13, 10, 16, 17, 16, 18, 17, 16, 19, 15, 17, 16, 19, 15, 18. Inspect the histogram of these data and discuss whether the experimental conditions under which these measurements were taken are likely to be under statistical control.

Answer

In preparation for this analysis, we note that the number of repeat values x_i, $i = 1, 2, ..., N$ is $N = 21$ and we compute the mean and the standard deviation of these measurements as, respectively, $\mu_{xN} = 14.9$ and $\sigma_{xN} = 2.9$. Then, we standardise the measured values as $z_i = (x_i - \mu_{xN})/\sigma_{xN}$. We are going to plot the histogram of the standardised values, rather than the pressure values, so that we can compare it with the pdf of a standardised normal random variable.

To construct the histogram, we decided to divide the real axis into seven bins, with all bins having a width equal to 1, except the two that are furthest away from the mean, which are wider. Hence, the bin boundaries are set as $(-\infty, -2.5), [-2.5, -1.5), [-1.5, -0.5), [-0.5, 0.5], (0.5, 1.5], (1.5, 2.5], (2.5, \infty)$. Note that a larger number of bins would be more appropriate, if N were significantly larger than 20, and that a choice of progressively widening bins with distance from the mean might be preferable. We then computed the percentage of values of z_i within each bin and plotted them vs. z, together with the normal pdf, as shown in Fig. 4.8. Our policy for values that would happen to be on the boundary between two bins was to count them in the bin that is closer to the mean, as indicated by a square bracket in the bin boundary definition. Figure 4.8 shows that the histogram shape is quite different from that of the normal pdf. In particular, unlike the Gaussian pdf, the histogram is strongly asymmetric about the mean. In this case, we must conclude that the repeat measurements do not have an approximately normal distribution and that the experiment may not be under statistical control. A more detailed analysis of this dataset is required to expose the reasons for non-normality.

4.5.3 The Chi-Square Test

A statistical method, which determines the probability that the observed set of data belongs to a random population with a particular pdf, is the *Pearson's* chi-square (χ^2) *goodness of fit test*. We will use this method as a quantitative normality test, by determining the probability that the collected set of repeat values belongs to a population that has a normal statistical distribution. This test is based on the assumptions that the data values are random and independent of each other and that the number of values is sufficiently large, namely $N \geq 20$.

A random variable $\boldsymbol{x}$ that has a χ^2 statistical distribution with ν degrees of freedom is the sum of the squares of ν independent, standardised, normal random variables. The probability density function of this random variable is found to be

$$f_{\chi^2}(x;\nu) = \begin{cases} \frac{x^{\nu/2-1}e^{-x/2}}{2^{\nu/2}\Gamma(\nu/2)}, & x > 0, \\ 0, & x \leq 0, \end{cases} \tag{4.37}$$

where $\Gamma(\nu/2)$ is the gamma function. The cumulative distribution function

$$F_{\chi^2}(x;\nu) = \int_0^x f_{\chi^2}(x;\nu)\mathrm{d}x \tag{4.38}$$

is the probability that a value of this random variable is less than or equal to x.

The χ^2 normality test may be performed as follows:

- Let $x_i, i = 1, 2, ..., N$ be a set of random values. Determine the mean μ_{xN} and the standard deviation σ_{xN} of the given data set. Standardise the measured values as $z_i = (x_i - \mu_{xN})/\sigma_{xN}$.
- Determine the number m of bins, within which we will distribute the N data values, as we did to create a histogram. If $20 \leq N \leq 40$, select m by truncating the number $N/5$ to the integer that is equal to or just below this value; for example, for $N = 23$, for which $N/5 = 4.6$, select $m = 4$ bins, which gives an average of $N/m = 5.75$ data values in each bin. If $N > 40$, select m by truncating the value $1.87(N-1)^{0.4}$, given by the *Kendall–Stuart grouping guide*, to the closest integer; for example, for $N = 45$, select $m = 8$ bins, for which $N/m = 5.6$, whereas, for $N = 46$, select $m = 9$ bins, for which $N/m = 5.1$.

- Determine the boundaries of the m bins. Note that the bins need not be of equal width and that the boundaries are not uniquely determined. Each bin has a lower edge z_{jl} and an upper edge z_{ju}, where $j = 1, 2, ..., m$. It is evident that $z_{1l} \rightarrow -\infty, z_{mu} \rightarrow \infty$ and $z_{ju} = z_{(j+1)l}, j = 1, 2, ..., m - 1$.
- Determine the *observed count* $O_j, j = 1, 2, ..., m$ in each bin, which is defined as the number of data values in the bin. If a value happens to be on the boundary between two bins, count it in the bin that is closer to the mean.
- Determine the *expected count* E_i in each bin (for a normal distribution) as

$$E_j = [F(z_{ju}) - F(z_{jl})]N \,, \tag{4.39}$$

  where $F(z_{ju})$ is the cumulative distribution function of the standardised normal random variable, namely the probability that a value is in the range from $-\infty$ to z_{ju}. It is also evident that $F(z_{1l}) = 0$ and $F(z_{mu}) = 1$.
- If the expected count in any bin is less than 5, then change the bin boundaries, until all bins contain at least five values. As a guide, we recommend the following approach, which results in all bins having an expected count $E_j = N/m \geq 5$: starting from $-\infty$, select the boundaries of all bins such that

$$F(z_{ju}) - F(z_{jl}) = 1/m \,.$$

  For example, if the number of values is $N = 20$, we find that we need $m = 4$ bins. If we set the bin boundaries to $-\infty, -0.67, 0, 0.67, \infty$, we get an expected count of 5 for all bins. If N is increased to 50, we find the number of bins as $m = 9$, which provides an expected count of 5.6 for all bins, if the bin boundaries are set to $-\infty, -1.22, -0.76, -0.43, -0.14, 0.14, 0.43, 0.76, 1.22, \infty$. Note that, if m is odd, the mean will be in the middle of the central bin, whereas, if m is even, there will be two central bins, separated by the mean as one of their boundaries.
- Determine the chi-square value of the given data set as

$$\chi^2 = \sum_{j=1}^{m} \frac{(O_j - E_j)^2}{E_j} \,. \tag{4.40}$$

- Determine the number of degrees of freedom as $\nu = m - 3$.
- Choose a *significance level* α and determine the *critical chi-square value* χ_c^2 that corresponds to $F_{\chi^2}(\chi_c^2; \nu) = 1 - \alpha$, where the cumulative distribution function $F_{\chi^2}(\chi_c^2; \nu)$ represents the probability that a set of values with ν degrees of freedom that are drawn randomly from a normal population would have a χ^2 value that satisfies $\chi^2 \leq \chi_c^2$. A significance level $\alpha = 0.05$ is deemed to be appropriate for assessing the normality of typical engineering data.
- Conclude that a given data set is drawn from a normal population at a significance level α, if $\chi^2 \leq \chi_c^2$. If $\chi^2 > \chi_c^2$, you may conclude that this data set is likely not normal. For example, consider a set of repeat values that has $\nu = 2$ and $\chi^2 = 0.20$ and determine whether this data set is normal at a 0.10 significance level. The critical chi-square value that gives $F_{\chi^2}(\chi_c^2; \nu) = 0.90$ is $\chi_c^2 = 4.61$. As this data set has $\chi^2 < \chi_c^2$, we conclude that it is normal. By the same logic, we would conclude that a different set of data with $\nu = 2$ and $\chi^2 = 6.0$ would not be normal at a 0.10 significance level.

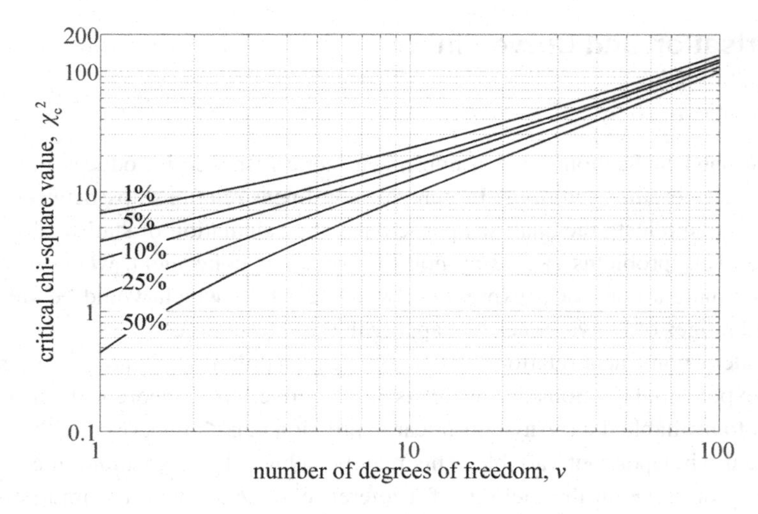

Figure 4.9 The critical chi-square value at different significance levels α, for chi-square normality tests.

The chi-square test of normality can be performed with the use of scientific software packages. To assist the reader, we provide Fig. 4.9, from which one can easily assess the normality of different data sets at different significance levels. Unfortunately, the chi-square test does not provide any clue for the form and the cause of non-normality, when such is observed.

Example 4.6 Perform a chi-square normality test at a 0.05 significance level for the data given in Example 4.5.

Answer

To perform the χ^2 test, following the guidelines listed previously, we first allotted the 21 standardised values z_i into $m = 4$ bins. The bin boundaries, and the observed and expected counts, in each bin are shown in Table 4.2. Using Eq. (4.40), we calculated the chi-square value as $\chi^2 = 2.81$. The number of degrees of freedom is $\nu = m - 3 = 1$. The critical chi-square value for $\nu = m - 3 = 1$ and a 0.05 significance level is $\chi_c^2 = 3.82$. As $\chi^2 < \chi_c^2$, we must conclude that this data set is drawn from a normal distribution. It is noted, however, that, although this set of data passes the chi-square normality test, the confidence in this result is rather limited, because ν is very small and χ^2 is only 26% smaller than χ_c^2. Consistently, the histogram in Example 4.5 also failed to provide a strong indication that the sample is normal.

Table 4.2 Values of variables for the χ^2 test

j	z_{jl}	z_{ju}	O_j	E_j
1	$-\infty$	-0.67	6	5.3
2	-0.67	0	2	5.3
3	0	0.67	7	5.3
4	0.67	∞	6	5.3

4.6 Linearisation and Curve Fitting

4.6.1 Linearisation

As mentioned in Section 2.3.3, a non-linear static response introduces measurement errors under certain conditions. In general, non-linear relationships between different properties complicate the analysis, particularly the calculation of statistical properties. To circumvent problems caused by non-linearity, it is convenient, whenever possible, to approximate a non-linear expression by a linear one, which would be valid within specified ranges of the variables that appear in this expression.

Consider a non-linear relationship $y = f(x)$ (e.g., a polynomial of order different from 1 or an exponential relationship), which is either derived from theoretical considerations or fitted to available data. Any non-linear expression can be linearised within a limited range of the independent variable. This may be achieved by expanding the expression into a Taylor series in the vicinity of a reference value x_o (e.g., the mid-point of the range of x values) and retaining only the zero-order and first-order terms, namely,

$$y(x) \approx f(x_o) + \frac{\mathrm{d}f(x_o)}{\mathrm{d}x}(x - x_o)\,. \tag{4.41}$$

Although generally useful, in practice, linearisation becomes a futile exercise when the range of values of x is sufficiently large for non-linear effects to become significant. The maximum error due to linearisation can be computed by comparing the values of the linearised expression with those of the original expression at the boundaries of the linearisation range or at other appropriate points, if the expression is not a monotonic function.

Example 4.7 Linearise the cubic function $y(x) = ax^3 + bx + c$ about reference points $x_o = 1$ and $x_o = 0$.

Answer

For the reference point $x_o = 1$, Eq. (4.41) provides the linearised expression

$$y(x) \approx (3a + b)x - 2a + c\,.$$

For the reference point $x_o = 0$, the linearised expression becomes

$$y(x) \approx bx + c\,.$$

Each linearised expression has its own range of approximate validity and its own error. For example, let's assume that $a = b = c = 1$ and consider the intermediate point $x = 0.5$. The two linearised expressions give, respectively, $y = 1$ and $y = 1.5$, which are both lower than the value $y = 1.625$ provided by the cubic expression, but the error in the first of the two linearised expressions is much larger than the error in the second one.

Now consider a multi-variate function $y = f(x_1, x_2, ...)$, which depends on multiple independent variables $x_1, x_2, ...$ (e.g., an output that depends on multiple inputs). One may linearise this function about a *reference point* $(x_{1o}, x_{2o}, ...)$, with respect to one or more inputs, by expanding it into a multi-variate Taylor series and retaining the zero-order term and all first-order terms, namely,

$$y(x_1, x_2, ...) \approx f(x_{1o}, x_{2o}, ...) + \sum \frac{\partial f(x_{1o}, x_{2o}, ...)}{\partial x_i}(x_i - x_{io}) \,. \tag{4.42}$$

Example 4.8 Linearise the expression $y(x_1, x_2) = x_1^2 e^{x_2}$ in the vicinity of the reference point $x_1 = 1, \ x_2 = 1$.

Answer

Equation (4.42) provides the following linearised expression

$$y(x_1, x_2) \approx e\,[2(x_1 - 1) + 2(x_2 - 1) + 1] \,.$$

4.6.2 Curve Fitting

Scientific software packages generally have curve-fitting capabilities and usually provide an indication of the *goodness of fit*, namely how much the measurements depart statistically from the fitted line, as well as of the uncertainty in the estimated coefficients of the fitted expressions. In selecting the type of expression to be fitted to a set of data that contain random fluctuations (so-called *scatter*), one must judiciously compromise between the smoothness of a fitted curve and the goodness of fit to a specific sample of measurements. In particular, it is advisable to fit only expressions having a number of adjustable coefficients that is significantly smaller than the number of data points available. Otherwise, one may obtain an expression with an excellent goodness of fit to a specific data sample, but wild fluctuations in the intervals between consecutive points, which is obviously an unacceptable situation.

Consider a set of pairs of values of two measured properties $x_i, \ y_i, \ i = 1, 2, ..., N$. For instance, these can be values of the input and output of a measuring system, measured during static calibration and used to determine the static response of the measuring system. If we plot each value of y_i vs. the corresponding value of x_i, we will often notice that the points on the plot do not follow a smooth line but seem to fluctuate about some mental mean line that follows the data in trend. It is even possible that there are more than one values of y_i for the same value of x_i. Experimental values usually contain random errors, which are unknown and cannot be corrected. When dealing with such results, it is common practice to represent them by a continuous analytical expression $Y(x)$.

The curve-fitting procedure will be elaborated for the simple case of fitting the data with the 'best' straight line $Y(x) = ax+b$. The constants a and b can be determined from the condition that the sum of the squared differences between each measured value y_i

and its corresponding fitted value Y_i is minimised, that is,

$$\frac{\partial \sum (Y_i - y_i)^2}{\partial a} = 0 , \quad \frac{\partial \sum (Y_i - y_i)^2}{\partial b} = 0 . \tag{4.43}$$

All summations in the previous and following expressions are from 1 to N. The resulting expression is called a *linear least squares fit* (llsf). For the straight line case, the two constants are

$$a = \frac{N \sum x_i y_i - \sum x_i \sum y_i}{N \sum x_i^2 - (\sum x_i)^2} , \quad b = \frac{\sum y_i \sum x_i^2 - \sum x_i y_i \sum x_i}{N \sum x_i^2 - (\sum x_i)^2} . \tag{4.44}$$

Two common measures for the goodness of fit are the *coefficient of determination*

$$r^2 = 1 - \frac{\sum (a x_i + b - y_i)^2}{\sum (y_i - \overline{y_i})^2} \tag{4.45}$$

and the *standard error of estimate*

$$s_{y,x}^2 = \frac{\sum (y_i - Y_i)^2}{N - m} , \tag{4.46}$$

where m is the number of unknowns in the fitting function. For example, for a linear fit, there are two unknowns, a and b, and, therefore, $m = 2$. When all differences vanish (*ideal fit*), $r^2 = 1$ and $s_{y,x}^2 = 0$.

Example 4.9 We are given 10 pairs of measured values $x_i, y_i, i = 1, 2, ..., 10$ of two properties that are related to each other, but we have no theoretical basis on which to derive their relationship. These measurements are shown in Fig. 4.10. Fit these data with different polynomial expressions and discuss the results.

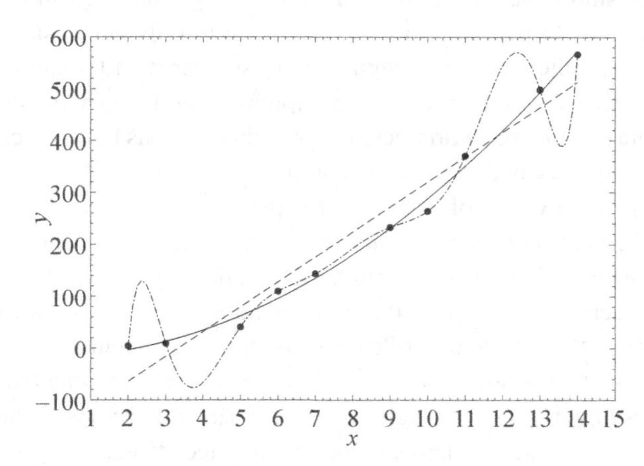

Figure 4.10 First- (dashed), second- (solid) and ninth-order (dot-dash) polynomials fitted to 10 measured pairs of values of related properties x and y.

Answer

First-, second- and ninth-order polynomials fitted to these data are shown in Fig. 4.10. The ninth-order polynomial is the solution of 10 linear algebraic equations with 10 unknowns (namely, the coefficients of the different terms in the polynomial) and so it fits perfectly to all measured points (x_i, y_i). Unsurprisingly, the coefficient of determination corresponds to the ideal fit value of $r^2 = 1$. Nevertheless, when plotted as a continuous function, this polynomial exhibits spurious oscillations between some of the data points, which cannot have a physical origin and, thus, demonstrate that this polynomial is a poor choice of a fit. The standard error of estimate cannot be determined for this fit, because $N = m = 10$. Visual inspection of the given data shows that there will likely be some scatter of values around a smooth fitted curve, but there is no apparent strong deviation from an overall trend. This directs us towards the use of a low-order polynomial, which would be more robust than high-order ones. A comparison between the first- and second-order polynomial fits in Fig. 4.10 shows that the latter follows better these data. This is confirmed by the fact that $s_{y,x} = 45.9$ for the first-order fit and $s_{y,x} = 16.2$ for the second-order fit. The second-order fit also seems to capture sufficiently any systematic relationship between these two variables, whereas the linear fit misses the apparent change of slope in the data. Third- or higher-order polynomials will certainly fit closer to the given 10 data points, but may also exhibit non-physical oscillations. All things considered, the second-order polynomial appears to be the best choice of fit for this set of data.

4.7 Pattern Detection and Trend Identification

4.7.1 Pattern Detection

An observation that a set of repeat measurements does not have a normal distribution is an indication that the measurement process may be influenced by systematic effects, which may manifest themselves by imposing a pattern in the data set. There are various statistical methods to detect and characterise patterns in measurements, but these are beyond the scope of the present discussion. It is, however, possible to identify some patterns simply by inspection of the measurement values plotted vs. the time they were obtained. Patterns to look for are visible step changes or monotonic upward and downward trends, even if these get somewhat obscured by scatter. If such patterns appear to be significant, one should search for their source. Sudden changes and drifts in experimental conditions and performance of instruments may introduce unacceptably large measurement errors, to the effect that such measurements should be discarded. On the other hand, it is also possible that a pattern in the data is associated with a physical phenomenon of interest and deserves close attention as part of the objectives of the experiment.

Example 4.10 Explore the presence of a pattern in the data presented in Example 4.5. Speculate on the source of such pattern, if it exists.

Answer

A plot of the data vs. the time they were obtained (Fig. 4.11) reveals an evident pattern: the first eight values seem to fluctuate about a lower average than the remaining ones. This observation indicates that, besides some random fluctuations, there is also a systematic sudden increase of the value at some instant during the experiment. Such a change could be the result of many causes, including a sudden change in temperature or pressure in the room, a change of power voltage that affects the speed of a motor, and drift or malfunction of an electronic component. Although one may be tempted to eliminate this pattern by offsetting the values of data in one or both groups, so that the averages of the two groups match, the proper approach would be to discard these values and repeat the experiment.

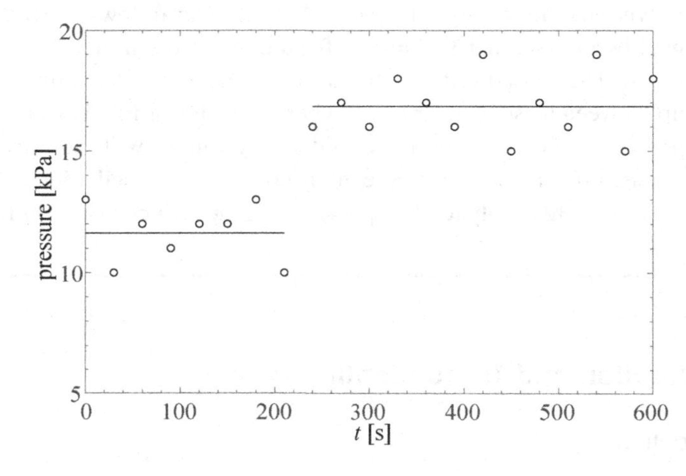

Figure 4.11 Plot of measured values in chronological order; the two solid lines are averages of values in two groups of data.

4.7.2 Trend Identification

A common type of pattern in experimental data is the *trend*, namely an overall monotonic increase or decrease in value, which usually coexists with random fluctuations (scatter) and which should be absent under conditions of statistical control. Trends may be revealed by inspection of the values plotted vs. time, but, especially when scatter is appreciable, it would be advisable to perform a quantitative statistical test to determine whether the trend is significant. The *reverse arrangement test* [2] is capable of detecting monotonic trends. Let $x_i, i = 1, 2, ..., N$ be a set of repeat measurements in chronological sequence. The *total number of reverse arrangements* R is defined as

$$R = \sum_{i=1}^{N-1} \sum_{j=i+1}^{N} R_{ij} , \tag{4.47}$$

where

$$R_{ij} = \begin{cases} 1 , & \text{if } x_i > x_j , \\ 0 , & \text{if } x_i \leq x_j . \end{cases} \tag{4.48}$$

If the process were under statistical control and N were sufficiently large, R would be a Gaussian random variable with a mean $\mu_R = N(N-1)/4$ and a variance $\sigma_R^2 = N(2N +$

5)$(N - 1)/72$. Thus, one may test the hypothesis that there is no trend in the data within a *level of significance* (the level of significance is 1 minus the level of confidence), by evaluating the probability that R would be within a certain interval of values. Trend removal should be done with caution to avoid eliminating true data features. If one wishes to correct a dataset for a linear trend, one may first fit a line to the original data, subtract this line from the data values and then offset the differences by adding the data average. The fitting of non-linear expressions to data for the purposes of non-linear trend removal should be avoided, unless there is sound theoretical reason to anticipate such trends.

Example 4.11 A set of repeat measurements consists of $N = 30$ values, which are known, although not listed here. Discuss how you would determine whether there is a trend in these data.

Answer

The mean and standard deviation of a Gaussian reverse arrangement set of 30 values are, respectively, $\mu_R = 217.5$ and $\sigma_R = 28.0$. The probability that $\mu_R - 2\sigma_R < R < \mu_R + 2\sigma_R$ would be approximately 0.95, while the probability that $\mu_R + 2\sigma_R < R$ would be 0.025, and the probability that $R < \mu_R - 2\sigma_R$ would also be 0.025. Let R be the number of reverse arrangements of the given dataset, calculated according to Eq. (4.47). If $161.45 < R < 273.55$, then one may say, at a level of significance of 5% (or a level of confidence of 95%) that there is no trend in the data. If $R < 161.45$, then one may say that the data have an upward trend, while, if $273.55 < R$, one may say that the data have a downward trend.

4.8 Outlier Identification

Occasionally, a set of measurements includes *spurious values*, namely values containing errors far greater than the measurement uncertainty. These may be due to gross human error (e.g., misreading of an instrument's output) or to a temporary or intermittent undesirable input (e.g., a power surge caused by the starting of a large electric motor). Such values are called *outliers* and are best to be discarded. Outliers may be identified by different statistical criteria.

4.8.1 Outlier Identification for Repeat Measurements

A commonly used method for identifying outliers in repeat measurements is *Chauvenet's criterion* [1, 4]. This criterion is based on the assumption that the repeat data belong to a Gaussian population. Moreover, this criterion is appropriate for sets of data that do not contain a distinct pattern. Chauvenet's criterion states that a value x_i is an outlier, if

$$\tau\sigma_{xN} \leq |x_i - \mu_{xN}| . \tag{4.49}$$

The parameter τ for a set of N repeat values is equal to the value z of a standardised normal random variable z, for which the Gaussian cumulative distribution function is

$$F_z(\tau) = P\{z \leq \tau\} = 1 - 1/(4N)\,. \tag{4.50}$$

Values of τ for different N are listed in Table 4.3.

Table 4.3 Parameter τ in Chauvenet's criterion for outlier identification

N	3	4	5	6	8	10	15	20	25	50	100
τ	1.38	1.53	1.64	1.73	1.86	1.96	2.13	2.24	2.33	2.57	2.81

It is clear that, as the number of measurements decreases, the effect of an outlier on the mean, the standard deviation and other statistical properties of the set of values would become stronger. For this reason, if outliers are detected in a small set, it would be advisable to repeat the experiment and collect additional data. Note that Chauvenet's criterion was developed for normal random variables and should not be applied indiscriminately to strongly non-normal random variables. Moreover, if a pattern is detected in the data, a different outlier identification method may be more appropriate.

Example 4.12 For the pressure measurements listed in the previous example, identify any possible outliers using Chauvenet's criterion. Discuss the relevance of this procedure to this set of values.

Answer

Using suitable software, we find that $\tau = 2.26$ for $N = 21$. It was found previously that $\mu_x = 14.9$ and $\sigma_x = 2.9$. According to Eq. (4.49), a data value would be an outlier, if $6.6 \leq |x_i - 14.9|$, namely, if the measured pressure is outside the interval 8.3 to 21.5 kPa. As none of the 21 given values lies outside this interval, we must conclude that this set of measurements contains no outliers. Nevertheless, the applicability of Chauvenet's criterion to the present set of values, which was found to be distinctly non-normal, is questionable. A search for possible patterns in the data and other statistical tests may provide clues for the presence of outliers, which were not detected by Chauvenet's criterion.

4.8.2 Outlier Identification for Correlated Data

Consider a set of simultaneous measurements of two measurands x_i and y_i. The correlation coefficient (see Section 4.2.4) between the two discrete random variables x_i and y_i is defined as

$$\rho_{x,y} = \frac{\sum_{i=1}^{N}(x_i - \mu_x)(y_i - \mu_y)}{\left[\sum_{i=1}^{N}(x_i - \mu_x)^2 \sum_{i=1}^{N}(y_i - \mu_y)^2\right]^{1/2}}\,. \tag{4.51}$$

Following common practice, we may assess, with 95% confidence, that the two variables are statistically correlated if

$$|\rho_{x,y}| > |\rho_{x,y}|_{min} = 2.04N^{-0.51} . \qquad (4.52)$$

This expression is valid for $N > 5$ and is a fit to values given in Ref. [6]. It shows that the threshold $|\rho_{x,y}|_{min}$ decreases with increasing N. Representatively, this threshold is equal to 0.90 for $N = 5$; it decreases to 0.28 for $N = 50$ and to a mere 0.09 for $N = 500$. To identify outliers among statistically correlated data, one may use the following procedure [7]:

- Determine a least-squares-fit function $Y(x)$ (see Section 4.6.2), which may be linear or non-linear.
- Calculate the *residuals* $e_i = y_i - Y(x_i),\ i = 1, 2, ..., N$.
- Identify the absolute maximum value $|e|_{max}$ of the residuals.
- Calculate the standard error of estimate $s_{y,x}^2$ using Eq. (4.46).
- Consider that the value (x_i, y_i) is an outlier, if $|e|_{max} > 2s_{y,x}$.

Note that only a single outlier can be identified each time this procedure is applied. If this outlier is removed, and provided that the remaining pairs of variables are still statistically correlated, the previous procedure may be repeated to identify additional outliers.

If two variables x_i, y_i are found to be statistically uncorrelated, that is, if $|\rho_{x,y}| \leq 2.04N^{-0.51}$, one may treat x_i and y_i as two separate sets of repeat measurements, and follow the procedure outlined in Section 4.8.1 to identify possible outliers.

4.9 Random Processes

4.9.1 Definitions

Consider an experiment, in which you measure one or more properties as functions of time. Assume that you repeat this experiment so that each repetition, to be referred to as a *realisation* ζ, is independent of any other. A *random process* or *stochastic process* is the process of assigning, according to a specified rule, a time-dependent function $\boldsymbol{x}(\zeta, t)$ to each realisation ζ. Therefore, a random process is a function of two variables, the particular realisation ζ and the time t. Associated with a particular realisation ζ_i is the time-function $\boldsymbol{x}(\zeta_i, t)$, which is called a *time series*. The values $\boldsymbol{x}(\zeta, t_o)$ of all realisations at a fixed time instance t_o constitute a random variable. The set of all time series is called an *ensemble*, while each time series is a *member* of the ensemble. A few realisations of a general random process are illustrated in Fig. 4.12a.

Example 4.13 Consider a large number of vertical tubes supplied with water by the same large head tank and assume that the flow in each tube does not affect the flow in any other. Let the velocity at the centre of the exit plane of each tube be continuously

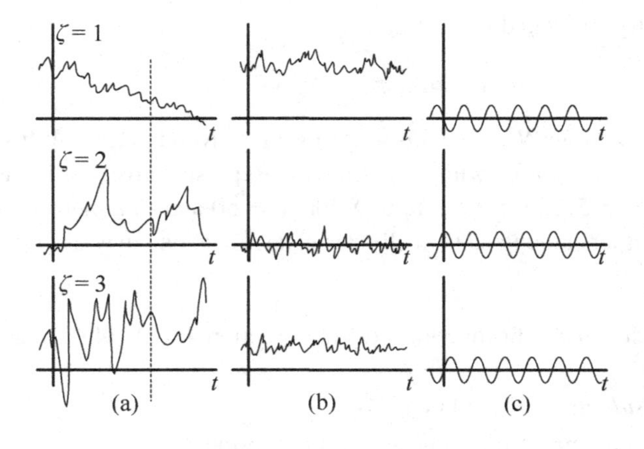

Figure 4.12 Three realisations of (a) an arbitrary (non-stationary) random process, (b) a stationary but non-ergodic random process and (c) the stationary and ergodic random process that is discussed in Example 4.17.

measured with a fast-response, accurate instrument. Discuss the random process that is associated with this experiment.

Answer

In this experiment, the random process is the magnitude of velocity in the centre of the exit plane of a tube; the velocity measurement in each particular tube is one realisation ζ_i of the experiment; the temporal record $V(\zeta_i, t)$ of the velocity in a particular tube is a time series; and the set of velocity values $V(\zeta_i, t_0), i = 1, 2, ...$ in all tubes at a given time instant t_0 constitutes a random variable.

So far, no assumption was made concerning the diameters and lengths of the tubes. A related but distinct random process can be constructed by considering a single tube connected to a large head tank. One realisation of the experiment is to open the valve connecting the tube to the tank and measure the velocity at one location as a function of time up to a certain time instance, when the valve is closed again. Additional independent realisations can be obtained by repeating this process after adding or removing fluid in the tank and allowing the fluid in the tank and tube to become still. Compared to the multiple-tube experiment, which, in principle may extend to infinite times, the duration of the single-tube experiment would be finite, a limitation that is common to all physical experiments. In addition, all realisations in the single-tube experiment would be constrained by the use of the same tube, while all realisations of the multiple-tube experiment would be constrained by the application of the same tank head at a given instant. Thus, some caution is required when using a single physical set-up as representative of a general random process.

4.9.2 Statistical Properties of Random Processes

Now, consider a random process $\boldsymbol{x}(\zeta, t)$ and the corresponding random variable at a fixed time instant t. Then, the *first order distribution function* of this random process at time instant t is defined as

$$F_x(x, t) = P\{\boldsymbol{x}(\zeta, t) \leq x\} , \tag{4.53}$$

and the *first-order probability density function* of the same random process is defined as

$$f_x(x,t) = \frac{\partial F_x(x,t)}{\partial x} \ . \tag{4.54}$$

Note that both of these functions depend on two variables, the real number x and time t. Now, consider the two random variables $\boldsymbol{x}(\zeta, t_1)$ and $\boldsymbol{x}(\zeta, t_2)$, corresponding to times t_1 and t_2. Then, one can define the *second-order distribution function* of the random process $\boldsymbol{x}(\zeta, t)$ as

$$F_x(x_1, x_2, t_1, t_2) = P\{\boldsymbol{x}(\zeta, t_1) \leq x_1 \text{ and } \boldsymbol{x}(\zeta, t_2) \leq x_2\} \ , \tag{4.55}$$

and the *second-order pdf* as

$$f_x(x_1, x_2, t_1, t_2) = \frac{\partial^2 F_x(x_1, x_2, t_1, t_2)}{\partial x_1 \partial x_2} \ . \tag{4.56}$$

In a similar manner, one can define the *nth-order distribution function* and the *nth-order pdf.* A random process is called *statistically determined*, if all its distribution functions are known.

The *mean* of a random process is a time function defined as

$$\langle \boldsymbol{x}(t) \rangle = \int_{-\infty}^{\infty} x f_x(x,t) \mathrm{d}x \ . \tag{4.57}$$

The *auto-correlation function* and the *auto-covariance function* of a random process are, respectively, defined as

$$R_x(t_1, t_2) = \int_{-\infty}^{\infty} \int_{-\infty}^{\infty} x_1 x_2 f_x(x_1, x_2, t_1, t_2) \mathrm{d}x_1 \mathrm{d}x_2 \tag{4.58}$$

and

$$C_x(t_1, t_2) = \int_{-\infty}^{\infty} \int_{-\infty}^{\infty} (x_1 - \langle \boldsymbol{x}(t_1) \rangle)(x_2 - \langle \boldsymbol{x}(t_2) \rangle) f_x(x_1, x_2, t_1, t_2) \mathrm{d}x_1 \mathrm{d}x_2 \ . \tag{4.59}$$

A *two-dimensional random process* consists of two random processes $\boldsymbol{x}(\zeta, t)$ and $\boldsymbol{y}(\zeta, t)$. One may easily extend the previous definitions to define the *second order joint distribution function* $F_{xy}(x_1, y_2, t_1, t_2)$, the *second order joint pdf* $f_{xy}(x_1, y_2, t_1, t_2)$ and joint functions of higher orders, as for example

$$F_{xy}(x_1, y_2, t_1, t_2) = P\{\boldsymbol{x}(\zeta, t_1) \leq x_1 \text{ and } \boldsymbol{y}(\zeta, t_2) \leq y_2\} \ . \tag{4.60}$$

Furthermore, one can also define the *cross-correlation function* and the *cross-covariance function*, respectively, of the two random processes as

$$R_{xy}(t_1, t_2) = \int_{-\infty}^{\infty} \int_{-\infty}^{\infty} x_1 y_2 f_{xy}(x_1, y_2, t_1, t_2) \mathrm{d}x_1 \mathrm{d}y_2 \tag{4.61}$$

and

$$C_{xy}(t_1, t_2) = \int_{-\infty}^{\infty} \int_{-\infty}^{\infty} (x_1 - \langle \boldsymbol{x}(t_1) \rangle)(y_2 - \langle \boldsymbol{y}(t_2) \rangle) f_{xy}(x_1, y_2, t_1, t_2) \mathrm{d}x_1 \mathrm{d}y_2 \ . \tag{4.62}$$

Two random processes are called *uncorrelated*, if $C_{xy}(t_1, t_2) = 0$ for any pair of times

(t_1, t_2). They are called *independent*, if, for any two sets of times $\{t_1, t_2, ...\}$ and $\{t'_1, t'_2, ...\}$, the set of random variables $\{\boldsymbol{x}(t_1), \boldsymbol{x}(t_2), ...\}$ is independent of the set $\{\boldsymbol{y}(t'_1), \boldsymbol{y}(t'_2), ...\}$.

Example 4.14 Identify two possible two-dimensional processes for the experiment described in Example 4.13.

Answer

One example is the pairs of values of the velocity and the temperature at the centres of the exit planes of tubes supplied with water by the tank. Another one is pairs of values of the velocity at two distinct locations in the same tube. One may expect that neither of these pairs of processes are independent of each other.

4.9.3 Stationary and Ergodic Random Processes

A random process $\boldsymbol{x}(\zeta, t)$ is called *stationary* (in the strict sense), if its statistical properties are not affected by a shift in the time origin, that is, if the statistical properties of $\boldsymbol{x}(\zeta, t)$ are the same as the corresponding properties of $\boldsymbol{x}(\zeta, t+\tau)$ for any time increment τ. An example of a stationary random process is shown in Fig. 4.12b.

Example 4.15 Discuss the types of conditions under which measurements taken in the multiple-tube system described in Example 4.13 can be considered as a stationary random process and a non-stationary random process.

Answer

A velocity measurement in the multiple-tube system would be a stationary random process, if the tank head were maintained constant and if sufficient time were allowed for the effects of valve opening to vanish; the latter requirement obviously limits the ranges of t and τ. A related non-stationary process would be the case of velocity measurement in the same multiple tubes, when supplied with fluid from a tank with a time-dependent head.

The following properties are direct consequences of the definition of a stationary random process:

- The first-order pdf of a stationary random process is independent of time
- The mean of a stationary random process is a constant.
- The second-order pdf of a stationary random process is a function of the time difference $\tau = t_2 - t_1$ only and not of the specific times t_1 and t_2.
- The autocorrelation function of a stationary random process is only a function of the time difference τ; thus, it may be denoted as $R_x(\tau)$.

The requirement of stationarity for a physical random process is quite severe and cannot be precisely verified experimentally. A more flexible requirement is that of *weak stationarity*. A random process is called weakly stationary, if its mean is a constant and its autocorrelation function is a function of the time difference τ only.

Two random processes, $\boldsymbol{x}(\zeta, t)$ and $\boldsymbol{y}(\zeta, t)$, are called *jointly stationary*, if their joint statistical properties at any time t are the same as the corresponding properties at $t + \tau$, for any time increment τ.

The auto- and cross-correlation functions of stationary random processes have the following properties [9]:

$$R_x(\tau) = R_x(-\tau) \ , \ \text{i.e., } R_x(\tau) \text{ is an even function} \ , \tag{4.63}$$

$$-R_x(0) \leq R_x(\tau) \leq R_x(0) \ , \tag{4.64}$$

$$R_{xy}^2(\tau) \leq R_x(0) R_y(0) \ , \tag{4.65}$$

$$2\left|R_{xy}(\tau)\right| \leq R_x(0) + R_y(0) \ . \tag{4.66}$$

As mentioned previously, the ensemble of realisations $\{\zeta_1, \zeta_2, ...\}$ of a repeated time-dependent experiment is represented by a set of functions $\{\boldsymbol{x}(\zeta_i, t), i = 1, 2, ...\}$ that constitute a random process, while the function $\boldsymbol{x}(\zeta_i, t)$ for each realisation constitutes a time series. Assume that the random process is stationary. Then, one may define a *time average* for each time series, $\boldsymbol{x}(\zeta_i, t)$ as

$$\overline{x}_i = \lim_{T\to\infty} \frac{1}{2T} \int_{-T}^{T} \boldsymbol{x}(\zeta_i, t) \mathrm{d}t \ . \tag{4.67}$$

Similarly, one may define time-averaged moments and central moments of different orders for each time series. In general, the time-averaged properties of each member of the ensemble would be different from those of any other member. To obtain the statistical properties of the random process, one must perform *ensemble averaging*, namely average across all (or, at least, a sufficiently large number of) members of the ensemble, which may prove to be a tedious procedure. In many experiments, however, there is sufficient control of the process such that time averages do not vary from one ensemble member to any other. Such random processes are called *ergodic*. Therefore, a stationary random process would be ergodic, if all its statistical properties could be determined from a single member of the ensemble. In other words, for stationary and ergodic random processes, ensemble averages would be equal to corresponding time averages. Notice that a stationary random process may or may not be ergodic, but stationarity is a prerequisite for ergodicity, otherwise, one would not be able to define time averages.

Example 4.16 Discuss the type of conditions under which the velocity measurement in the tube-tank system described in Example 4.13 would be an ergodic random process.

Answer

The velocity measurement would be a stationary and ergodic random process only if the tank head were maintained constant and if all tubes were identical. If the tubes were different from each other, then the process would be non-ergodic and statistical information from one tube would not represent statistics of the ensemble. One must be warned against taking ergodicity for granted, because many processes in the environment and in technology are non-ergodic, even if they can be considered as approximately stationary.

Following the previous definitions, the mean and the autocorrelation function of a stationary and ergodic random process can be computed from any single member ζ of the ensemble as, respectively,

$$\overline{x} = \lim_{T\to\infty} \frac{1}{2T} \int_{-T}^{T} \boldsymbol{x}(\zeta, t)\mathrm{d}t \tag{4.68}$$

and

$$R_x(\tau) = \lim_{T\to\infty} \frac{1}{2T} \int_{-T}^{T} \boldsymbol{x}(\zeta, t)\boldsymbol{x}(\zeta, t+\tau)\mathrm{d}t \,. \tag{4.69}$$

In a similar fashion, two stationary random processes $\boldsymbol{x}(\zeta, t)$ and $\boldsymbol{y}(\zeta, t)$ are called *jointly stationary and ergodic*, if all their joint statistical properties can be defined from any single pair of members of the ensemble. Then, their cross-correlation function can be computed as

$$R_{xy}(\tau) = \lim_{T\to\infty} \frac{1}{2T} \int_{-T}^{T} \boldsymbol{x}(\zeta, t)\boldsymbol{y}(\zeta, t+\tau)\mathrm{d}t \,. \tag{4.70}$$

Example 4.17 To assist with the understanding of the definitions and relationships among various statistical properties, consider the random process

$$\boldsymbol{x}(\boldsymbol{\theta}, t) = a\cos(2\pi f_o t + \boldsymbol{\theta}) \,,$$

where a is a positive constant number, f_o is a constant frequency and the phase $\boldsymbol{\theta}$ is a random variable, which is uniformly distributed in the interval $[-\pi, \pi]$, as discussed in Example 4.3. Compute the statistical properties of this random process.

Answer

Each realisation of this random process is a sinusoidal time series, corresponding to a fixed value of the phase $\boldsymbol{\theta}$ (see Fig. 4.12c). Ensemble averaging is obtained by fixing the value of time t and averaging over all realisations. It is evident that $f_x(x, t)$ should be independent of time, as, at any given time, the random variable $\boldsymbol{x}$ would take values corresponding to all possible values of the phase $\boldsymbol{\theta}$. Therefore, the process is stationary and, as a result, its statistics can be computed for an arbitrary choice of time origin. For convenience, we may select the time origin such that $t = 0$. Then, the pdf of the random process at any time would be the same as the pdf of the random variable $\boldsymbol{x}(\boldsymbol{\theta}) = a\cos\boldsymbol{\theta}$, as discussed in Example 4.3.

The analysis can be significantly simplified by noticing that this random process is also ergodic. This is evident, because each realisation is identical to any other with the exception of a phase shift, which does not affect time integration. Then, one may, for example, compute the autocorrelation function as

$$
\begin{aligned}
R_x(\tau) &= \lim_{T\to\infty} \frac{1}{2T} \int_{-T}^{T} x(\theta, t)x(\theta, t+\tau)\mathrm{d}t \\
&= a^2 \lim_{T\to\infty} \frac{1}{2T} \int_{-T}^{T} \cos(2\pi f_o t + \theta)\cos[2\pi f_o (t+\tau) + \theta]\,\mathrm{d}t \\
&= \frac{a^2}{2} \lim_{T\to\infty} \frac{1}{2T} \int_{-T}^{T} \{\cos(4\pi f_o t + 2\pi f_o \tau + 2\theta) + \cos(2\pi f_o \tau)\}\,\mathrm{d}t \\
&= \frac{a^2}{2} \cos(2\pi f_o \tau) \quad .
\end{aligned}
$$

4.9.4 Non-stationary Random Processes

Many measurement processes are non-stationary, even when considered within a relatively short time interval. A simple qualitative test that can be used to detect non-stationarity in a random signal or time series is to inspect its variation over the time of interest, trying to identify possible systematic upward or downward trends. A convenient procedure for long time series is as follows:

- Separate the time series into a number of sequential or overlapping blocks.
- Compute the means of each block.
- Plot these means vs. time and try to identify trends in the block means, rather than in the entire signal.

The absence of a trend in the mean does not necessarily preclude non-stationarity, as it is possible that, even if the mean remains constant within a time interval, other moments may be time-dependent. To increase the confidence in a qualitative assessment of stationarity, one may inspect the variation of additional block statistics, such as standard deviations and skewness and flatness factors, noting that the test sensitivity generally increases with the order of the moment. A more reliable approach would be to apply one of the available quantitative statistical non-stationarity tests to various block statistics. A relatively simple test is the reverse arrangement test (Section 4.7), which is capable of detecting monotonic trends. Assume that the time series has been separated into a number of blocks, and that statistical properties have been calculated and may be assumed to be independent of those in any other block. Then, one may apply this test to any such statistical block property, which itself is a random variable. In cases in which a monotonic trend has been detected for the mean, but not for higher statistical moments, and that trend can be plausibly attributed to spurious effects (e.g., instrumentation drift), one may replace the original time series by a corresponding *quasi-stationary* time series, determined by fitting a straight line to a plot of the block means and subtracting the values provided by this line at all times from all corresponding measured values. More sophisticated approaches have also been suggested. For example, a time series may be restored to a quasi-stationary form by the application of a *moving-average filter*, namely by subtracting from each value the average of a number of its neighbouring values, either equally weighted or weighted by a factor that decreases inversely to proximity. In such cases, one must be aware that, in addition to removing spurious effects that may introduce non-stationarity, it is possible to remove signal fluctuations, which are essential features of the physical process under study.

4.10 Conditional Sampling and Phase Averaging

Conditional sampling and phase averaging are techniques used in experimental fluid mechanics, as well as in computational fluid mechanics and many other fields, to enhance the detection of features of the measurand, which may be obscured by other physical phenomena or undesirable influences, such as noise. A primary application of such techniques is the detection of *coherent structures* in turbulent flows, which, in simple terms, are recurring, organised, vortical entities, coexisting with *non-coherent turbulence*. An in-depth coverage of this topic is beyond the present scope and has been the subject of numerous references [e.g., 8]. In this section, we will demonstrate the application of these techniques by simple examples.

First, consider that measurements are performed in a flow that is produced by a periodic mechanism, for example, by the motion of a piston driven at a given frequency. Time-averaging of measurands in this flow would smear out their phase dependence and so ensemble averaging is necessary, for which we produce *phase averages* at a sufficient number of phases during one period. To do so, one must synchronise the time origin of each cycle with a specific event of the flow-generating mechanism, for example, the time when the piston is at the start of its stroke. One can mark this time with the use of a relay, a photo-cell or other appropriate device and average only measurements taken after a fixed time interval following such a *triggering event*. This is called *conditional sampling*, as it only considers measured values that are obtained under a specified condition. Other types of conditions may be specified for different purposes. For example, triggering can be effected when the measurand or its derivative exceed a specified threshold in magnitude or when a measured property at some other part of the flow satisfies some given condition.

Example 4.18 Consider a sinusoidal signal that is contaminated by very strong, white (namely, having an infinite bandwidth), electronic noise, as shown in Fig. 4.13a. Perform phase averaging to reconstruct the sinusoidal signal.

Answer

In order to conduct the phase averaging, we must first determine the period of the signal. One method by which the period can be determined is through the autocorrelation function. As shown in Example 4.17, the autocorrelation function of a cosine is itself a cosine with the same period; the same holds for a sine wave. For this particular flow, the

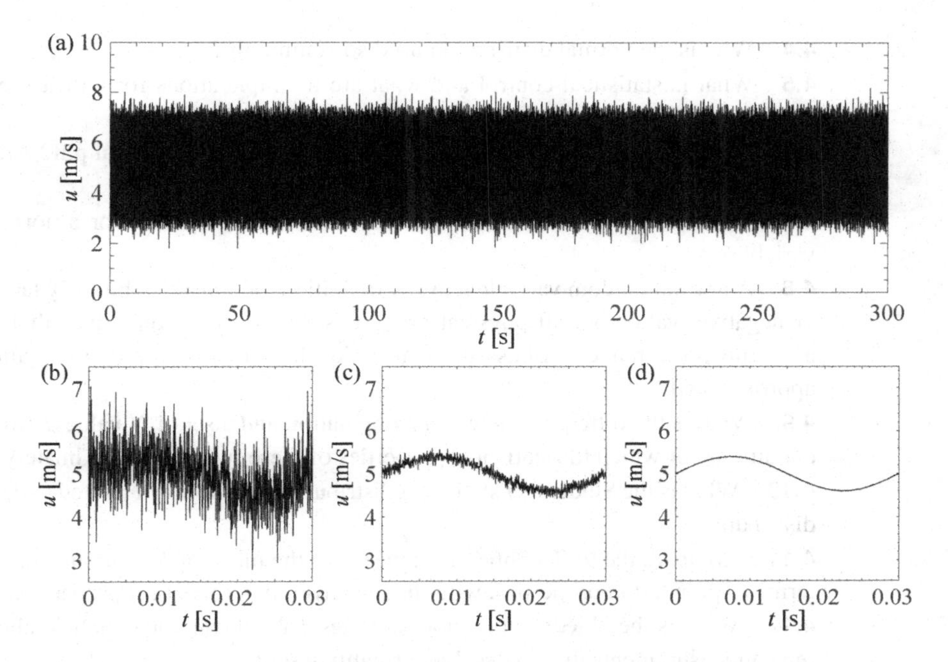

Figure 4.13 (a) A sinusoidal signal contaminated by white noise, (b) a period of the same signal, phase-averaged over 1 cycle, (c) a period of the same signal, phase-averaged over 50 cycles, and (d) a period of the same signal, phase-averaged over 10,000 cycles.

period is found to be $T = 0.03$ s. The time series can then be separated into segments extending over one period each, and phase averaging can be performed over them. As Fig. 4.13b, c and d illustrate, the reconstruction quality improves as the number of realisations is increased.

Phase averaging may also be used to eliminate the effect of a periodic undesirable input, when such input is independent of the measurand. For example, consider a signal that is contaminated not only by white noise, but also by periodic interference, which consists of many harmonics. One may reconstruct the waveform and phase of the interfering component by phase averaging over a time window that is inversely proportional to the fundamental frequency of the interference (e.g., the frequency of the electric power line). Then, one may subtract the reconstructed waveform from the recorded time series to produce an essentially interference-free signal.

Chapter Digest

4.1 What is probability?

4.2 What is the difference between a random property and a real random variable? Provide an example of a random property and a related real random variable.

4.3 Define the skewness and flatness factors. How is the flatness factor related to the kurtosis?

4.4 Why is the normal distribution used so commonly?

4.5 What is statistical control and what are its implications for a measurement process?

4.6 Which are the variables that completely characterise a normal pdf? Are the same variables also sufficient for characterising other types of pdf?

4.7 What are the values of the skewness and flatness factors for a normal random variable?

4.8 A normal random variable may, by definition, obtain any arbitrarily large, positive or negative, value, but all physical properties may only obtain values that are within a certain finite range. Discuss the convention by which this range of values can be approximated.

4.9 What is the difference between a population and a set of N values? Are there any conditions for which the statistical properties of these two are approximately the same?

4.10 What is the Student's t statistical distribution and how is it related to the Gaussian distribution?

4.11 Compare the 95% confidence interval of the mean of N random values, given in terms of population properties, with the one given in terms of set properties.

4.12 What is the objective of a normality test? What can you conclude about a set of repeat measurements that 'passes' a normality test?

4.13 What is an outlier? Provide two examples of outlier sources.

4.14 When should an outlier be discarded?

4.15 Discuss the benefits and shortcomings of linearisation.

4.16 How can you detect patterns in repeat measurements? What would you do, if you find some significant pattern?

4.17 What is a trend in a set of repeat measurements? How would you detect a significant trend?

4.18 What is the difference between a random variable and a random process?

4.19 Provide an example of a stationary and ergodic process, other than those mentioned in the book.

4.20 What is weak stationarity? Devise an example of a non-stationary random process that is weakly stationary.

4.21 What is phase averaging and when would it be useful in a fluid mechanics experiment?

Problems

4.1 Consider the repeated casting of a pair of dice, which may be assumed to be fair. Define a random variable as the sum of the values indicated by the two dice. Determine and plot the distribution function and the probability density function of this discrete random variable.

4.2 Plot isocontours of the joint pdf of two jointly normal random variables with correlation coefficients equal to $0, 0.5, -0.5$ and 1. Discuss the shapes of these contours.

4.3 Repeat measurements of the electric resistance of a resistor, measured with a digital multimeter, are as follows: 325, 315, 322, 309, 315, 318, 327, 319, 325, 313, 336,

310, 321, 316, 320, 319, 321, 302, 323 and 318 Ω. Perform normality tests of these data by histogram inspection and a chi-square test. Do you suspect that these data are subjected to undesirable influences? Identify and discard any possible outliers. Compute the effect of outliers, if any, on the mean and standard deviation of these values. Inspect the data and identify any trend or pattern that is apparent to you.

4.4 Consider the following sequence of repeat voltage measurements (in volts): 1.861, 1.005, 1.655, 1.642, 1.442, 1.776, 1.505, 1.498, 1.608, 1.296, 1.947, 1.282, 1.857, 1.008, 1.297, 1.302. Would you discard any of these measurements as spurious? If so, which ones and why?

4.5 According to Government of Canada records, the average annual temperature in Canada differed from a 'normal' value (the average of temperatures from 1961 to 1990) by an amount ΔT, as indicated in the following table.

Year	ΔT	Year	ΔT	Year	ΔT	Year	ΔT	Year	ΔT
1948	−0.2	1963	0.2	1978	−0.5	1993	0.4	2008	0.6
1949	−0.2	1964	−0.6	1979	−0.2	1994	0.5	2009	0.8
1950	−1.3	1965	−0.6	1980	0.4	1995	0.6	2010	3.0
1951	−0.6	1966	−0.3	1981	2.0	1996	0.0	2011	1.4
1952	0.8	1967	−0.4	1982	−1.0	1997	0.6	2012	1.8
1953	0.8	1968	0.2	1983	0.1	1998	2.3	2013	0.7
1954	0.0	1969	0.4	1984	0.2	1999	1.7	2014	0.6
1955	−0.2	1970	−0.2	1985	0.0	2000	0.9	2015	1.3
1956	−0.8	1971	0.0	1986	0.0	2001	1.8	2016	2.1
1957	−0.3	1972	−2.0	1987	1.5	2002	0.5	2017	1.5
1958	0.5	1973	0.6	1988	0.8	2003	1.1	2018	0.6
1959	−0.4	1974	−0.8	1989	−0.2	2004	0.0	2019	1.1
1960	0.4	1975	−0.1	1990	−0.1	2005	1.6	2020	1.1
1961	−0.2	1976	0.0	1991	0.4	2006	2.4	–	–
1962	0.0	1977	1.0	1992	−0.1	2007	0.8	–	–

(a) Assess the normality of these data by histogram inspection and with the use of the chi-square test. Discuss your observations. Can you reach any conclusion concerning temperature change, based on these observations?

(b) Inspect the data and identify any trend or pattern that is visually apparent to you. Perform a quantitative trend identification test. Can you reach any conclusion concerning temperature change, based on these observations?

(c) Considering the given temperature data as a set of repeat values, identify any possible outliers, noting the year each occurred, and determine the mean and standard deviation of the data, after the outliers have been removed. Discuss possible reasons for the presence of outliers in these data. Explain whether the assumptions for the criterion you used are satisfied and whether such a procedure would be reliable.

(d) Determine whether the temperature is statistically correlated with the year of observation. If so, derive a statistical relationship between the two. If indeed these properties are correlated, identify any possible outliers using the procedure for correlated data. Compare the results of the two outlier identification procedures and discuss your confidence in them.

4.6 We want to determine whether the flow temperature affects the velocity measurement in a calibration jet. Velocity data, collected for flow temperatures in the range $10°\text{C} \leq T \leq 55°\text{C}$, are shown in the following table.

(a) Determine whether the measurements are statistically correlated.

(b) Identify and remove possible outliers.

(c) Estimate the velocity at a temperature of 21°C.

T [°C]	10	15	20	25	30	35	40	45	50	55
U [m/s]	2.46	2.40	2.47	2.44	2.52	2.49	2.48	2.53	2.48	2.47

4.7 Additional data for the experiment described in the previous problem are collected for temperatures in the range $60°\text{C} \leq T \leq 105°\text{C}$ and shown in the following table.

(a) Determine whether the measurements are statistically correlated.

(b) Identify and remove possible outliers.

(c) Estimate the velocity at a temperature of 72°C.

T [°C]	60	65	70	75	80	85	90	95	100	105
U [m/s]	2.52	2.56	2.56	2.55	2.57	2.61	2.67	2.59	2.55	2.65

4.8 Consider both sets of data given in the previous two problems.

(a) Determine whether the measurements are statistically correlated.

(b) Identify and remove possible outliers.

(c) Estimate the velocity at temperatures of 21°C and 72°C. Compare these values to the ones you found in the previous two problems.

4.9 Derive a linearised expression for the speed of sound in air in terms of the absolute temperature, valid for temperatures near 300 K. Determine the relative error in the computation of speed of sound using this linearised expression, as a percent of the exact value, if the temperature is (i)) 305 K, (ii) 250 K and (iii) 450 K. If a relative error of ± 1% is acceptable in some engineering application, determine the range of temperature over which one may use the linearised expression.

4.10 A common expression for the response of a hot-wire anemometer is King's law, $V^2 = A + BU^n$, where V is the voltage output, U is the flow velocity and A, B and n are empirical constants determined by calibration. For a particular experiment, it was found that $A = 7.80$, $B = 3.12$ (both in SI units) and $n = 0.45$. Using Taylor's expansion, derive a linear relationship between E and U, valid in the vicinity of (i) $U = 10$ m/s and (ii) $U = 15$ m/s. Determine the maximum linearisation error for these cases, for velocity intervals of ±0.05 and ±0.50 m/s about the reference values. Then, assume that, when the anemometer is inserted in a turbulent flow, the measured standard deviation of the anemometer voltage output is 0.033 V. Estimate the standard deviation of the velocity for the two mean speeds mentioned previously.

4.11 Turbulence can be generated in a wind tunnel by passing a uniform stream with velocity U through a grid of parallel rods, spaced from axis to axis by the mesh size M. The turbulence intensity u'/U (where u' is the standard deviation of the velocity fluctuations) can be represented as a function of the downstream distance x from the grid by the empirical expression $u'^2/U^2 = a\,(x/M - x_0/M)^{-n}$, where the effective origin x_0 and the exponent n are to be determined by curve fitting. The usual ranges are $-10 < x_0/M < 10$ and $-1.4 < -n < -1.0$. During an experiment with $U = 10.0$ m/s and $M = 25.4$ mm, the following measurements have been taken using a hot wire anemometer:

x	[mm]	381	508	635	889	1,270
u'^2	$\left[\text{m}^2/\text{s}^2\right]$	0.205	0.135	0.098	0.062	0.039
x	[mm]	1,651	2,159	2,794	3,556	4,572
u'^2	$\left[\text{m}^2/\text{s}^2\right]$	0.028	0.020	0.015	0.011	0.008

Plot these results in log–log coordinates to determine the optimal values of the constants a, x_0/M and n. Hint: Plot the given data using different values of x_0/M and select the value that produces the best straight line.

References

[1] AGARD, Fluid Dynamics Panel Working Group 15. *Quality Assessment for Wind Tunnel Testing*. AGARD Advisory Report 304, North Atlantic Treaty Organization, Neuilly-sur-Seine, France, 1994.

[2] J.S. Bendat and A.G. Piersol. *Random Data Analysis and Measurement Procedures (4th Edition)*. Wiley, New York, 2010.

[3] W.H. Beyer. *CRC Handbook of Mathematical Sciences (6th Edition)*. CRC Press, Boca Raton, FL, 1987.

[4] H.W. Coleman and W.G. Steele. *Experimentation, Validation, and Uncertainty Analysis for Engineers (4th Edition)*. Wiley, Hoboken, NJ, 2018.

[5] W.S. Gosset (Student). The probable error of a mean. *Biometrika*, 6(1):1–25, 1908.

[6] R. Johnson and P. Kuby. *Elementary Statistics (11th Edition)*. Brooks/Cole Publishing Company, Boston, MA, 2011.

[7] D.C. Montgomery, G.C. Runger, and N.F. Hubele. *Engineering Statistics (5th Edition)*. Wiley, New York, 2010.

[8] H. Nobach et al. Conditional averages and stochastic estimation (section 22.5). In C. Tropea, A.L. Yarin, and J.F. Foss, editors, *Springer Handbook of Experimental Fluid Mechanics*. Springer, Berlin, 2007.

[9] A. Papoulis. *Probability, Random Variables and Stochastic Processes (3rd Edition)*. McGraw-Hill, New York, 1991.

5 Frequency Analysis of Signals

This chapter discusses two main methods that are used for the analysis of signals in the frequency domain, namely, spectral analysis and wavelet analysis. Additional mathematical methods that are used to decompose a time-dependent function into components and then to approximate this function by superimposing a relatively small number of these components are available, but their description is beyond the present scope. An example of such a method, which is widely used for turbulent flows, is the *proper orthogonal decomposition* (POD) [9, 15].

5.1 Fourier Transform

Measurements of a physical property are expressed either in the form of an analogue signal, namely, a continuous function of time, or as a time series, which is usually, although not universally, understood to be a discrete sequence of numbers. The analysis of such measurements using time as the independent variable is said to be conducted in the *time domain*. An alternative approach is to transform these properties, through a series expansion or an integral transform, into equivalent functions of frequency. Then, one may analyse these properties using frequency as the independent variable, in which case the analysis is said to be conducted in the *frequency domain*. In the following, we shall adopt the usual convention, that the term frequency denotes both the cyclic frequency f, measured in cycles/s (Hz), and the angular frequency $\omega = 2\pi f$, measured in rad/s (see also Section 2.4.2).

Fourier analysis, also called *harmonic analysis*, is the representation of a function or time series in terms of sinusoidal ('harmonic') functions [3, 4, 18]. The representation of a *periodic function* $x(t)$ as a Fourier series has been described in Section 2.4.2. As an example, Fig. 5.1 shows a square wave signal, approximated as the sum of 10 sinusoidal terms with decreasing amplitude and varying phase, in the time domain. Each of the 10 sinusoidal terms are also shown. The same approximation of the square wave signal in the frequency domain is represented as 10 delta functions, each with an amplitude that is proportional to the amplitude of the corresponding sinusoidal term in the Fourier series and at a frequency that is equal to the frequency of this term. A more general transformation from the time domain to the frequency domain, which applies both to periodic and non-periodic functions, is obtained with the use of the *Fourier transform*. The Fourier transform applies equally to real and complex functions of time, but, in view

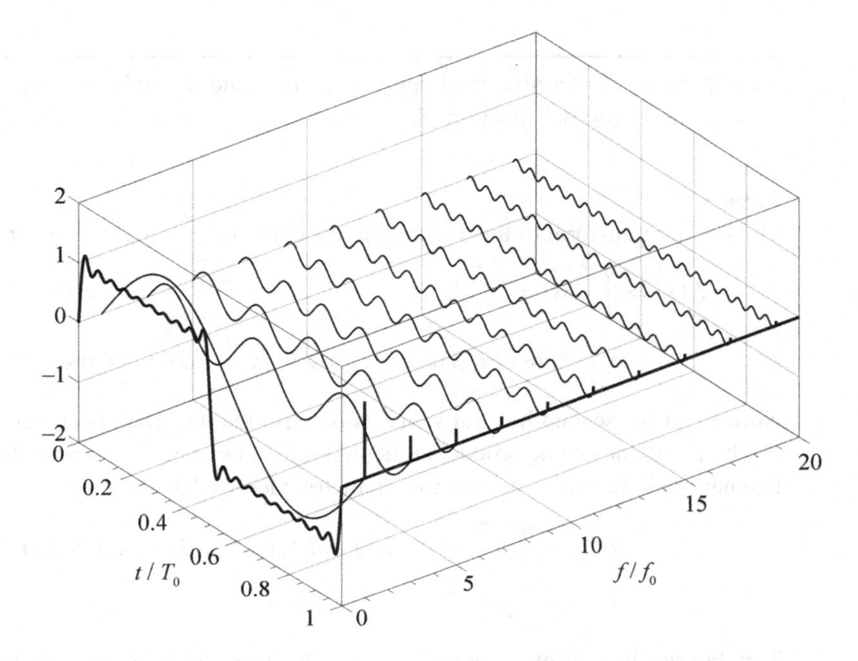

Figure 5.1 A square-wave signal with a period T_0 and a fundamental frequency $f_0 = 1/T_0$, approximated by a Fourier series with 10 terms, also shown in the figure.

of the fact that we are interested in measurement signals, we will present the analysis for real functions of time.

Consider any real function $x(t)$, which satisfies the condition $\int_{-\infty}^{\infty} |x(t)|\,\mathrm{d}t < \infty$ (this requires that $|x(t)| \to 0$, as $|t| \to \infty$). Then, one may define its *Fourier transform* as

$$F(f) = \int_{-\infty}^{\infty} x(t)\, e^{-j2\pi f t}\mathrm{d}t\,. \tag{5.1}$$

The Fourier transform is, in general, a complex function of the real variable ('frequency') f. However, if $x(t)$ is an even function, that is, $x(t) = x(-t)$, $F(f)$ is real. Given $F(f)$, one can compute the function $x(t)$ through the *inverse Fourier transform*

$$x(t) = \int_{-\infty}^{\infty} F(f)\, e^{j2\pi f t}\mathrm{d}f\,, \tag{5.2}$$

and so $x(t)$ and $F(f)$ are equivalent.

The Fourier transform can be defined only if $|x(t)| \to 0$, as $|t| \to \infty$. This condition is not satisfied by stationary functions (namely functions having statistical properties that do not change by a shift of the time origin), however, in practice, values of physical processes are only available within a finite time interval, that is, $[-T, T]$ and may be considered as vanishing outside this interval. So, for practical signals, one may use the *finite-interval Fourier transform*, defined as

$$F(f, T) = \int_{-T}^{T} x(t)\, e^{-j2\pi f t}\mathrm{d}t\,. \tag{5.3}$$

Example 5.1 Determine the Fourier transform and the finite-interval Fourier transform of the sinusoidal function $x(t) = a\cos(2\pi f_o t)$, where f_o is a constant frequency.

Answer

The Fourier transform of this function in the entire time domain ($-\infty < t < \infty$) is

$$F(f) = \int_{-\infty}^{\infty} x(t)e^{-j2\pi ft}\mathrm{d}t$$
$$= a\int_{-\infty}^{\infty} \cos(2\pi f_o t)\cos(2\pi ft)\,\mathrm{d}t - ja\int_{-\infty}^{\infty} \cos(2\pi f_o t)\sin(2\pi ft)\,\mathrm{d}t\,.$$

Notice that the second integral vanishes because the integrand is an odd function of t. Furthermore, the cosine product in the first integral can be transformed into the sum of two harmonic functions using a trigonometric identity. Then,

$$F(f) = \frac{a}{2}\int_{-\infty}^{\infty} \{\cos[2\pi(f - f_o)t] + \cos(2\pi(f + f_o)t)\}\,\mathrm{d}t$$
$$= \frac{a}{2}[\delta(f - f_o) + \delta(f + f_o)]\,.$$

This Fourier transform is shown in Fig. 5.2a. The easiest way to verify that this expression is correct is to use the inverse transformation (Eq. (5.2)), combined with the property of Dirac's delta function (Eq. (2.36)). The finite-interval Fourier transform of the sinusoidal function is

$$F(f,T) = \frac{a}{2}\int_{-T}^{T} \{\cos[2\pi(f - f_o)t] + \cos(2\pi(f + f_o)t)\}\,\mathrm{d}t$$
$$= \frac{a}{2\pi}\left\{\frac{\sin[2\pi(f - f_o)T]}{f - f_o} + \frac{\sin[2\pi(f + f_o)T]}{f + f_o}\right\}\,.$$

This expression has been plotted, in normalised form, in Fig. 5.2b. As can be seen by comparing the two previous expressions, the infinite-interval Fourier transform of a sinusoidal function has non-zero values only at the two frequencies f_o and $-f_o$, whereas

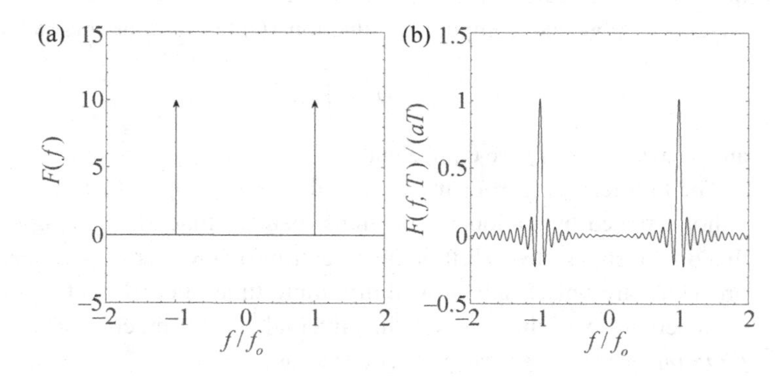

Figure 5.2 (a) Fourier transform of the function $x(t) = a\cos(2\pi f_o t)$; (b) detail of the finite-interval Fourier transform of the same function for $T = 10/f_o$.

the finite-interval Fourier transforms has values expanding into the entire frequency domain. The latter is of oscillatory nature with positive and negative peaks, called *sidelobes*, the amplitudes of which diminish as $|f|$ increases.

Example 5.2 Determine the Fourier transform of the *rectangular* or *boxcar function*

$$w(t) = \begin{cases} 1\,, & -T \le t \le T\,, \\ 0\,, & T < |t|\,. \end{cases}$$

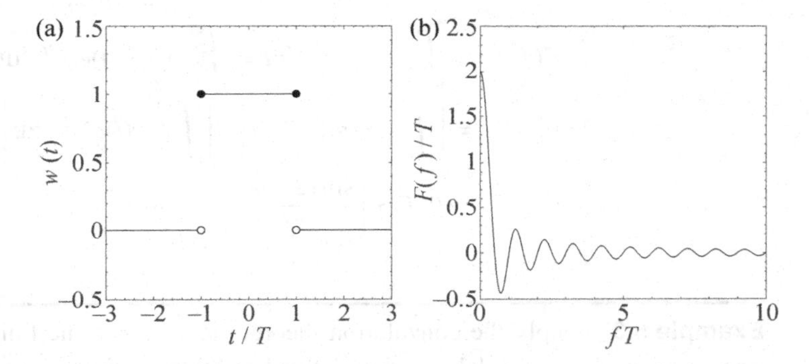

Figure 5.3 (a) The boxcar function and (b) its Fourier transform for a range of positive frequencies.

Answer

The boxcar function has been plotted in Fig. 5.3a. The Fourier transform of this function, which is equal to a finite-interval Fourier transform, can be found as

$$\begin{aligned} F(f) &= \int_{-T}^{T} e^{-j2\pi ft} dt = \int_{-T}^{T} [\cos(2\pi ft) - j\sin(2\pi ft)]dt \\ &= \int_{-T}^{T} \cos(2\pi ft)dt = \frac{1}{2\pi f}[\sin(2\pi ft)]\Big|_{-T}^{T} \\ &= \frac{\sin(2\pi fT)}{\pi f}\,. \end{aligned}$$

As can be seen in Fig. 5.3b, $F(f)$ has strong sidelobes.

5.2 Convolution Theorem

Consider two functions $x_1(t)$ and $x_2(t)$ of the real variable t. Their *convolution* is defined as

$$x_1(t) * x_2(t) = \int_{-\infty}^{\infty} x_1(t')x_2(t-t')\,\mathrm{d}t' = \int_{-\infty}^{\infty} x_1(t-t')x_2(t')\,\mathrm{d}t' \,. \tag{5.4}$$

The *convolution theorem* states that the Fourier transform of the convolution of two functions is equal to the product of the Fourier transforms of these functions. An alternative form of the same theorem states that the Fourier transform of the product of two functions is equal to the convolution of the Fourier transforms of these functions. This theorem can be applied to the computation of finite-interval Fourier transforms from corresponding infinite-interval ones, by noticing that the finite-interval Fourier transform of a function $x(t)$ can be considered as the finite-interval Fourier transform of the product of the function $x(t)$ and the rectangular function $w(t)$, as

$$\begin{aligned} F(f,T) &= \int_{-T}^{T} x(t)e^{-j2\pi ft}\mathrm{d}t = \int_{-\infty}^{\infty} x(t)w(t)e^{-j2\pi ft}\mathrm{d}t \\ &= \left[\int_{-\infty}^{\infty} x(t)e^{-j2\pi ft}\mathrm{d}t\right] * \left[\int_{-\infty}^{\infty} w(t)e^{-j2\pi ft}\mathrm{d}t\right] \\ &= F(f) * \left[\frac{\sin(2\pi fT)}{\pi f}\right] \,. \end{aligned} \tag{5.5}$$

Example 5.3 Apply the convolution theorem to compute the finite-interval Fourier transform of the sinusoidal function defined in Example 5.1 over one period on each side of the time axis.

Answer

The finite-interval Fourier transform of this function can be computed as

$$\begin{aligned} F(f,T) &= F(f) * \left[\frac{\sin(2\pi fT)}{\pi f}\right] \\ &= \int_{-\infty}^{\infty} \frac{a}{2}\left[\delta(f-f_o-f') + \delta(f+f_o-f')\right] \frac{\sin(2\pi f'T)}{\pi f}\mathrm{d}f' \\ &= \frac{a}{2\pi}\left\{\frac{\sin[2\pi(f-f_o)T]}{f-f_o} + \frac{\sin[2\pi(f+f_o)T]}{f+f_o}\right\} \,. \end{aligned}$$

5.3 Frequency Spectra

In this section, we shall consider exclusively real, time-dependent functions $x(t)$ (namely, signals), which are realisations of a real, stationary, random process (see Section 4.9).

We shall further assume that the time average has been computed and subtracted from all values.

5.3.1 Autocorrelation Function

As discussed in Section 4.9, the autocorrelation function of $x(t)$ is defined as (see also Eq. (4.69))

$$R_x(\tau) = \lim_{T\to\infty} \frac{1}{2T} \int_{-T}^{T} x(t)x(t+\tau)\,\mathrm{d}t\,. \tag{5.6}$$

The autocorrelation function value at $\tau = 0$ is equal to the variance of the signal, that is, $R_x(0) = \sigma_x^2$, and this value is always a maximum, that is, $|R_x(\tau)| \le R_x(0)$. Moreover, the autocorrelation function is an even function, that is, $R_x(\tau) = R_x(-\tau)$. Normalising the autocorrelation function by the variance of the signal results in the *autocorrelation coefficient*

$$r_x(\tau) = \frac{R_x(\tau)}{R_x(0)} \equiv \frac{R_x(\tau)}{\sigma_x^2}\,,$$

which also has a maximum $r_x(0) = 1$ at the origin. The integral of the autocorrelation coefficient,

$$\mathcal{T} = \int_0^\infty r_x(\tau)\,\mathrm{d}\tau\,, \tag{5.7}$$

is called the *integral time scale*. $\mathcal{T}$ quantifies the 'memory' of the process, namely, it is a measure of how far back current values are correlated with past values. $\mathcal{T}$ is a useful time scale for random processes, whereas periodic processes are characterised by their period.

5.3.2 Power Spectrum

The *frequency spectrum*, also called the *power spectrum* or *power spectral density function* of a mean-free signal $x(t)$ is defined as the Fourier transform of its autocorrelation function, that is, as

$$S_x(f) = \int_{-\infty}^{\infty} R_x(\tau)\, e^{-j2\pi f\tau}\mathrm{d}\tau\,, \tag{5.8}$$

where $j = \sqrt{-1}$ and the variable f has dimensions of frequency (inverse time). Because $R_x(\tau)$ is an even function, the power spectrum is always a real, non-negative, even function (i.e., $S_x(f) = S_x(-f) \ge 0$) and can be computed as

$$S_x(f) = 2\int_0^\infty R_x(\tau)\cos(2\pi f\tau)\,\mathrm{d}\tau\,. \tag{5.9}$$

It is interesting to note that

$$S_x(0) = 2\int_0^\infty R_x(\tau)\,\mathrm{d}\tau = 2\sigma_x^2 \int_0^\infty r_x(\tau)\,\mathrm{d}\tau = 2\sigma_x^2\mathcal{T}\,, \tag{5.10}$$

and hence the integral time scale can also be found by extrapolating the power spectrum to zero frequency.

Inversely, the autocorrelation function can be computed from the corresponding spectrum by using the inverse Fourier transform, as

$$R_x(\tau) = \int_{-\infty}^{\infty} S_x(f)\, e^{j2\pi f\tau} \mathrm{d}f = 2\int_{0}^{\infty} S_x(f) \cos(2\pi f\tau)\, \mathrm{d}f\,. \tag{5.11}$$

Letting $\tau = 0$, one gets

$$R_x(0)\ (= \sigma_x^2) = \int_{-\infty}^{\infty} S_x(f)\, \mathrm{d}f\,. \tag{5.12}$$

This expression demonstrates that the area under the spectrum in the positive frequency axis is equal to half the signal variance, that is,

$$\int_{0}^{\infty} S_x(f)\, \mathrm{d}f = \frac{1}{2}\sigma_x^2\,, \tag{5.13}$$

thus being a measure of the total fluctuation energy of the signal.

The reader is alerted to the fact that some sources define the power spectrum as *twice* the Fourier transform of the autocorrelation function, which makes the integral of the spectrum over the positive frequency half axis equal to the variance itself and not its half. This approach is followed in the analysis of turbulent flows [19].

Example 5.4 Compute the power spectrum of a sinusoidal signal $x(t) = a\cos(2\pi f_o t)$.

Answer

As shown in Example 4.17, the autocorrelation function of this signal is

$$R_x(\tau) = \frac{a^2}{2}\cos\left(2\pi f_o \tau\right)\,.$$

Taking the Fourier transform of this function, one gets the power spectrum as

$$S_x(f) = \frac{a^2}{4}\left[\delta\left(f - f_o\right) + \delta\left(f + f_o\right)\right]\,.$$

Example 5.5 Compute the autocorrelation functions of *white noise* and band-limited white noise.

Answer

White noise may be considered as a random signal $n(t)$, which has a uniform power spectrum over the entire frequency axis, namely,

$$S_n(f) = a > 0 \ \text{ for } \ -\infty < f < \infty\,. \tag{5.14}$$

Its autocorrelation function can be computed as

$$R_n(\tau) = a\delta(\tau) \ , \tag{5.15}$$

which indicates that the value of white noise at any instant is uncorrelated with its value at any other instant. The total energy (i.e., variance) of ideal white noise would be infinite, which is unrealistic for physical systems. Random noise values are often assumed to have a normal pdf, but this is of no consequence for the autocorrelation function or power spectrum and so white noise may or may not be Gaussian.

A more realistic random signal is *band-limited white noise*, the power spectrum of which is, by definition,

$$S_n(f) = \begin{cases} a > 0 \ , & -f_n \le f \le f_n \ , \\ 0 \ , & |f_n| < f \ . \end{cases} \tag{5.16}$$

The autocorrelation function of this signal can be computed as

$$R_n(\tau) = \frac{a \sin(2\pi f_n \tau)}{\pi\tau} \ , \tag{5.17}$$

which is an oscillatory function with diminishing amplitude. Its maximum value is $2af_n$ and occurs at $\tau = 0$.

In general, the area under the spectrum within a range of frequencies from f_1 to f_2 is a measure of the energy of fluctuations having frequencies within the specified range. Because, however, in many applications the bandwidth of plotted spectra extends over more than one decades, and even the magnitude of the power spectrum may span a wide range of values, it is customary to plot spectra in logarithmic coordinates, which makes it difficult to visualise the contribution of different frequency bands to the total fluctuation energy of the signal. An equivalent approach is to plot the *pre-multiplied power spectrum* $fS_x(f)$ vs. $\ln f$. Noting that

$$\int_{f_1}^{f_2} fS_x(f)\,\mathrm{d}\ln f = \int_{f_1}^{f_2} fS_x(f)\frac{1}{f}\,\mathrm{d}f = \int_{f_1}^{f_2} S_x(f)\,\mathrm{d}f \ , \tag{5.18}$$

it is evident that the area under the pre-multiplied spectrum within a frequency band would be proportional to the energy contained in the signal fluctuations within that band.

Example 5.6 Consider the frequency spectrum of a turbulent velocity, measured with a hot-wire anemometer and plotted in logarithmic coordinates in Fig. 5.4a. This spectrum exhibits an unusual bump. Similar bumps were observed during calibration of the hot-wire in velocity spectra that were measured with the same measurement system and in the same facility, but in steady laminar flow. This observation makes it plausible that the bump is due to some, yet undetermined, quasi-periodic interference, which is independent of the turbulent field. Quantify the relative strength of this interference by comparison to the energy of the turbulent velocity.

Answer

The log–log plot of the spectrum in Fig. 5.4a shows that the interference is present over

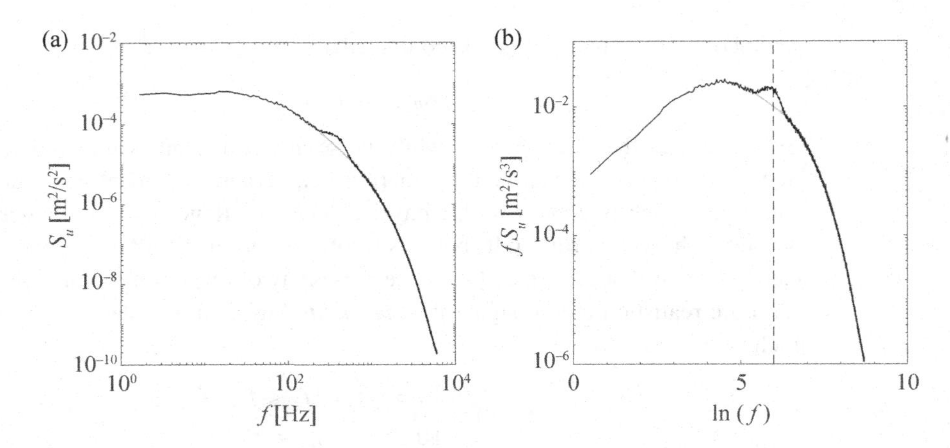

Figure 5.4 (a) The frequency spectrum of a turbulent velocity, which possibly contains the contribution of some unknown interference; the dashed line shows the smoothly interpolated spectrum under the bump; (b) the same spectrum, plotted in pre-multiplied form.

a range of frequencies; however, the absence of a clear peak in it makes it difficult to determine the central frequency of this interference. Based on established theoretical treatment of turbulence spectra, we expect that in the intermediate range of frequencies, where the bump is located, the velocity spectrum would approximately follow a power-law relationship of the form $S_u = Af^B$, where the exponent B would be in the vicinity of the Kolmogorov inertial-range prediction $-5/3$. This supports our conjecture to separate the interference effect by interpolating the spectral values under the bump by a smooth spline. We can see from Fig. 5.4a that the interference extends approximately over the range of frequencies from 200 to 600 Hz. This plot, however, cannot quantify the percentage of the measured signal energy that is due to interference. This is achieved with the use of the pre-multiplied spectrum, shown in Fig. 5.4b. In this plot, we can observe a distinct peak at 400 Hz, which provides us with a clue for our search for the source of this interference. Moreover, we may determine that the bump area, namely the area between the measured and the interpolated spectra, is about 6% of the area under the measured spectrum. This means that the fluctuation energy that is due to interference is roughly 6% of the turbulence energy. In this particular example, we have sufficient confidence in the statistical independence of the interference from the turbulence to apply a correction to the measured spectrum and the turbulence energy. This may not apply to cases with coupled energies, for example, grid-generated turbulence with superimposed periodic vortices.

5.3.3 Cross-Spectrum

The cross-correlation function of two mean-free signals $x(t), y(t)$, which are realisations of two jointly stationary random processes, is defined as (see also Eq. (4.70))

$$R_{xy}(\tau) = \lim_{T\to\infty} \frac{1}{2T} \int_{-T}^{T} x(t)y(t+\tau)\,\mathrm{d}t\,. \tag{5.19}$$

The *cross-spectrum* or *cross-spectral density function* of two such signals is defined as the Fourier transform of the cross-correlation function $R_{xy}(\tau)$, namely, as

$$S_{xy}(f) = \int_{-\infty}^{\infty} R_{xy}(\tau)\, e^{-j2\pi f \tau} \mathrm{d}\tau \,. \tag{5.20}$$

Inversely,

$$R_{xy}(\tau) = \int_{-\infty}^{\infty} S_{xy}(f)\, e^{j2\pi f \tau} \mathrm{d}f \,. \tag{5.21}$$

The cross-spectrum is generally a complex function, consisting of a real part, called the *coincident* or *co-spectral density function*, and an imaginary part, called the *quadrature* or *quad-spectral density function.* A real-valued quantity, which is related to the cross-spectrum, is the *coherence function* $\gamma_{xy}(f)$, defined as

$$\gamma_{xy}^2(f) = \frac{\left|S_{xy}(f)\right|^2}{S_x(f) S_y(f)} \,. \tag{5.22}$$

The coherence function is often used in fluid mechanics as a means of quantifying the correlation between periodic components of two random properties. As an example, consider the quasi-periodic meandering of a heated wake, which would generate quasi-periodic fluctuations of both the velocity and the temperature at a fixed point in space. One would intuitively expect that the average periods of the quasi-periodic fluctuations of the two properties would be equal, although their amplitudes would be different. Consequently, one would expect to see a strong coherence at the frequency of meandering (namely, the inverse of its period) and a weak, if any, coherence at other frequencies.

5.3.4 Finite-Interval Spectra and Spectral Windows

An alternative way of computing the finite-interval power spectrum, which does not require the calculation of the autocorrelation function, is directly from the signal $x(t)$ as

$$S_x(f,T) = \frac{1}{2T} \left| \int_{-T}^{T} x(t)\, e^{-j2\pi f t} \mathrm{d}t \right|^2 \equiv \frac{1}{2T} \left|F(f,T)\right|^2 \,. \tag{5.23}$$

An assumption for computing the finite-interval Fourier transform in this way is that the signal is periodic with a period $2T$. For the vast majority of signals, however, this would not be the case, so that these signals would appear to have discontinuities at the boundary points. Consequently, spectral estimates with the use of Eq. (5.23) would be distorted, when compared to those computed via the general definition (Eq. (5.1)).

Consider the case of a the sinusoidal function from Example 5.1. As mentioned previously, the finite-interval Fourier transform is equal to the convolution of the infinite-interval Fourier transform and the Fourier transform of the rectangular function. Thus, the finite-interval power spectrum of the sinusoidal function, found in Example 5.1, is

$$\begin{aligned} S(f,T) &= \frac{1}{2T} \left|F(f,T)\right|^2 \\ &= \frac{a^2}{2T} \left\{ \frac{\sin\left[2\pi (f - f_o) T\right]}{2\pi (f - f_o)} + \frac{\sin\left[2\pi (f + f_o) T\right]}{2\pi (f + f_o)} \right\}^2 , \end{aligned} \tag{5.24}$$

rather than the delta-function-like $S(f)$. Although both $S(f)$ and $S(f,T)$ have peaks at $f = f_o$ and $f = -f_o$, the former has no energy, except at these two peaks, while the latter has its energy distributed over the entire frequency range and, in fact, it has a sequence of secondary peaks (sidelobes) with slowly diminishing amplitudes. Thus, it appears that energy of the peaks at $\pm f_o$ 'leaks' to other frequencies, distorting the spectrum. This distortion is a consequence of the abrupt termination of the signal at the boundaries of the interval $[-T, T]$ and can be reduced by introducing a smooth transition to zero near these boundaries. This is implemented by multiplying the signal by a smooth *window function*, instead of the rectangular function $w(t)$.

A number of suitable window functions have been suggested, which have sidelobes with much lower amplitudes than those of the rectangular window. Because the use of windows generally changes the total energy of the spectrum, a correction needs to be applied. This is readily achieved by dividing the power spectrum by the mean of the squared values of the applied window function.

In general, window functions reduce smoothly the signal values near the two ends of the finite interval. A commonly used window function that partially corrects the spectral estimates for finite-interval effects is the *cosine-tapered window* (or *Tukey window*). This window multiplies the signal near the ends of the time interval by a half-cycle cosine function, which takes values from 1, towards the middle of the signal time interval, to 0, towards the two ends of this interval. A cosine-shaped taper from 0 to 1 is typically applied to the first 10% of the signal, and a similar taper from 1 to 0 is applied to the last 10% of the signal. When a (full-cycle) cosine function is used, the cosine-tapered window becomes the *Hann window* (often incorrectly referred to as a *Hanning window*), which, for a finite interval signal, is defined as (see also Eq. (3.37))

$$h(t) = \frac{1}{2}\left[1 + \cos\left(\frac{\pi t}{T}\right)\right], \text{ for } -T \leq t \leq T \,. \tag{5.25}$$

The Fourier transform of the Hann window has much lower sidelobe amplitudes than those of the rectangular window and is suitable for broadband signals, which have their energy spread across a wide range of frequencies. It is also advisable to use a Hann window for signals the nature of which is not known beforehand. The Hann window and other spectral window functions are used for different purposes, one of which is to reduce 'spectral leakage', which broadens narrow frequency peaks in the spectrum and makes it difficult to distinguish a spectral peak at one frequency from that at another. An assortment of spectral windows can be easily applied to time series with the use of functions available in scientific software packages.

5.4 Discrete Fourier Transform and Fast Fourier Transform

Besides continuous functions, the finite-interval Fourier transform can be also applied, in the form of the *discrete Fourier transform* (*DFT*), to a real or complex time series. In the present discussion, we will consider the DFT of a real time series $x_i, i = 0, 1, 2, ..., N-1$, sampled at a rate $f_s = 1/\Delta t$ over a time interval $T = N\Delta t$. It is emphasised that the time interval for this time series is defined as $[0, T]$, which is different from the time interval $[-T, T]$ that was used so far for the finite-interval Fourier transform. Accordingly, some definitions (e.g., the definition of the Hann window function) would need to be adjusted for the different significance of the symbol T. Moreover, to prevent possible misunderstandings, we note that the presently defined T is the length of a block of data from which the DFT is computed, and not the total length of a data record, which may consist of several data blocks, the spectral values of which may be ensemble-averaged.

The discrete Fourier transform of a time series is computed as

$$F_k = \frac{T}{N} \sum_{i=0}^{N-1} x_i e^{-2\pi j \frac{ik}{N}} \ , \ k = 0, 1, 2, ..., N-1 \ . \tag{5.26}$$

F_k has N discrete complex values at frequencies $f_k = k/T$, but only the first $N/2$ terms (i.e., those with $k = 0, 1, 2, ..., N/2 - 1$) are independent, while the others may be found by symmetry conditions. The frequency increment among consecutive terms of the DFT is $\Delta f = 1/T$. To compute all terms in a DFT according to Eq. (5.26), one would have to perform N^2 complex multiplications and additions, which would require an excessive computer time for large values of N. Fortunately, efficient algorithms have been developed for the computation of DFT, requiring $2N \log_2 N$ complex operations, which are significantly less than N^2. Such algorithms are known as *fast Fourier transforms* (*FFT*). Since the original invention of the first FFT by Cooley and Tukey in 1965 [5], a variety of FFT have been devised and become a standard feature of scientific software packages. Their common requirement is that the number of discrete values in the time series must be equal to a power of 2, because they all involve a number of intermediate computation steps arranging the data in groups of numbers of data that are powers of 2. If the number of values in a time series is not equal to a power of 2, one must simply add zeros to increase the number of values up to the closest power of 2. As an example of the efficiency of FFT, assume that $N = 2^{13} = 8,192$; then, computation of the DFT by application of the definition (Eq. (5.26)) would require $N^2 = 67,108,864$ complex operations, whereas that by FFT would require $2N \log_2 N = 212,992$ operations, which is a reduction by a factor of 315.

Equation (5.23) is very useful in signal analysis, because it permits the computation of spectra and autocorrelation functions with the use of FFT algorithms, which greatly reduce the required computational time. The discrete power spectrum of a time series $x_i, i = 0, 1, 2, ..., N-1$ can be found as

$$S_k = \frac{1}{N} \sum_{i=0}^{N-1} \left| x_i e^{-2\pi j \frac{ik}{N}} \right|^2 \ , \ k = 0, 1, 2, ..., N/2 - 1 \ . \tag{5.27}$$

Note that only values at the lower $N/2 - 1$ frequencies need to be considered, as the remainder are related to them through symmetry.

Besides its use for spectral estimates, FFT can be used for the efficient determination of auto- and cross-correlation functions. For this, one needs to apply the FFT once to compute the spectrum, from which the corresponding correlation function can be found by applying an inverse FFT.

5.5 Sampling Considerations

It is evident that the highest observable frequency in a power spectrum is related to the sampling rate as $\omega_{max} \equiv 2\pi f_{max} = \omega_s/2 \equiv \pi f_s$. The lowest frequency in the power spectrum is related to the data block length as $\omega_{min} \equiv 2\pi f_{min} = 2\pi/T$. In accordance with Section 3.3.4, the length of each block of data from which the DFT is computed should, in general, include at least 100 cycles at the lowest frequency f_0 of interest in a signal (if values $T \geq 100/f_0$ are not achievable for some reason, one may compromise somewhat the low-frequency measurement precision by choosing a lower value, down to a minimum of $T = 20/f_0$). For example, for a purely periodic signal with a period $T_0 \equiv 1/f_0 \equiv 2\pi/\omega_0$, the block length should be chosen as $T \geq 100T_0 \equiv 100(2\pi/\omega_0)$, and, hence, $\omega_{min} \leq \omega_0/100$. The use of a larger block length would have the additional benefit of reducing spectral leakage effects. For example, consider once more the finite-interval power spectrum of a sinusoidal function (Eq. (5.24)), which has a period T_0. The appearance of this spectrum depends on the value of T/T_0. As $T/T_0 \rightarrow 1$, the magnitudes of the sidelobes increase and the two peaks merge, thus obscuring the identification of the correct value of f_0. Inversely, as $T/T_0 \rightarrow \infty$, the sidelobes diminish and the peaks become easier to locate.

Both ω_{max} and ω_{min} need to be considered when designing an experiment to ensure that the full frequency range of interest is resolved. Spectra obtained from a single block of data tend to display considerable scatter. Ensemble averaging over several data blocks is a very effective method of reducing this scatter. One should strive to take as many data blocks as possible, with a ceiling imposed by the practicality of the data record length (namely the cumulative length of all blocks). For example, consider an experiment, for which it was found that $T = 12$ min. Taking 100 blocks would take 20 hrs, which may not be possible for different reasons, and so fewer blocks must be taken by necessity.

Example 5.7 Consider a signal, consisting of the sum of a sinusoidal signal, having an amplitude of 1 V and a cyclic frequency $f_o = 100$ Hz, and strong white noise, that is,

$$x(t) = \cos(2\pi f_o t) + n(t) .$$

Specify the sampling requirements for measuring the power spectrum and the autocorrelation function of this signal.

Answer

This analogue signal is the sum of a periodic component, having a period equal to $1/f_o = 0.01$ s, and a random component, which has energy over an unlimited frequency band. First, we discretised this signal, selecting a sampling rate $\omega_s = 100(2\pi f_o)$, which is fast enough to capture well the periodic component. This corresponds to a sampling time of $\Delta t = 1/(100 f_o) = 0.1$ ms. If we do not apply any low-pass analogue filtering before discretisation, the high-frequency energy of noise will result in aliasing of the low-frequency spectrum. According to the sampling theorem, we need to filter out the energy of noise at frequencies $\omega > \omega_s/2 = 50(2\pi f_o)$. To do so, we used a fourth-order, low-pass Butterworth filter with a cut-off frequency $\omega_c = 40(2\pi f_o)$. We discretised this signal with a 12-bit analogue-to-digital converter, having a range from −5 V to 5 V and, thus, a resolution of roughly 2 mV, which is sufficient for the present purposes. The next step was to choose a record length. As we know that the lowest frequency of interest is $1/f_o$, we would like to capture at least 100 cycles of the sinusoidal component in each block and so we recorded data blocks with a time duration $T = 100/f_o = 1$ s each. Considering that T is relatively short, to implement ensemble averaging and improve our statistics, we recorded 100 such blocks of data, each extending over 1 s, so that the overall record length was 100 s.

Example 5.8 Compute and plot the autocorrelation function and the power spectrum of the random signal discussed in Example 5.7.

Answer

To reduce the finite-interval effect, we applied a cosine-tapered window, by which the first 10% and the last 10% of the values of each data block were attenuated by a half-cycle cosine function. For efficiency of calculations, we first computed the power spectrum with an FFT-based algorithm and then the autocorrelation function with an FFT-based inverse Fourier transform. In addition, we performed ensemble averaging over 100 independent blocks. It is noted that ensemble averaging is a very effective method of reducing scatter in statistical properties of random signals. In the present example, one would get more precise estimates of the power spectrum and other statistical properties by averaging values obtained from M blocks, each extending over a time $100/f_o$, than from a single record, extending over $100M/f_o$, even though both approaches would use exactly the same data.

Figure 5.5 shows the given signal, its power spectrum and its autocorrelation function, the latter in the form of the *autocorrelation coefficient* $R_x(\tau)/R_x(0)$ (one is reminded that $R_x(0) = \sigma_x^2$). The sinusoidal signal seems to be largely obscured by noise. As expected, the spectrum has a sharp spike (an approximation of a delta function) at f/f_o and a roughly uniform value at other frequencies. The autocorrelation function has a sharp spike at the origin, which is the result of noise, and a sinusoidal appearance at other times. As noise is an undesirable distortion of the signal, one may produce a more accurate measurement of the autocorrelation function of the signal of interest (in this case, the sinusoidal part) by removing the spike, extrapolating the remainder to $\tau = 0$, and renormalising the values, so that the value at $\tau = 0$ is maintained at 1.

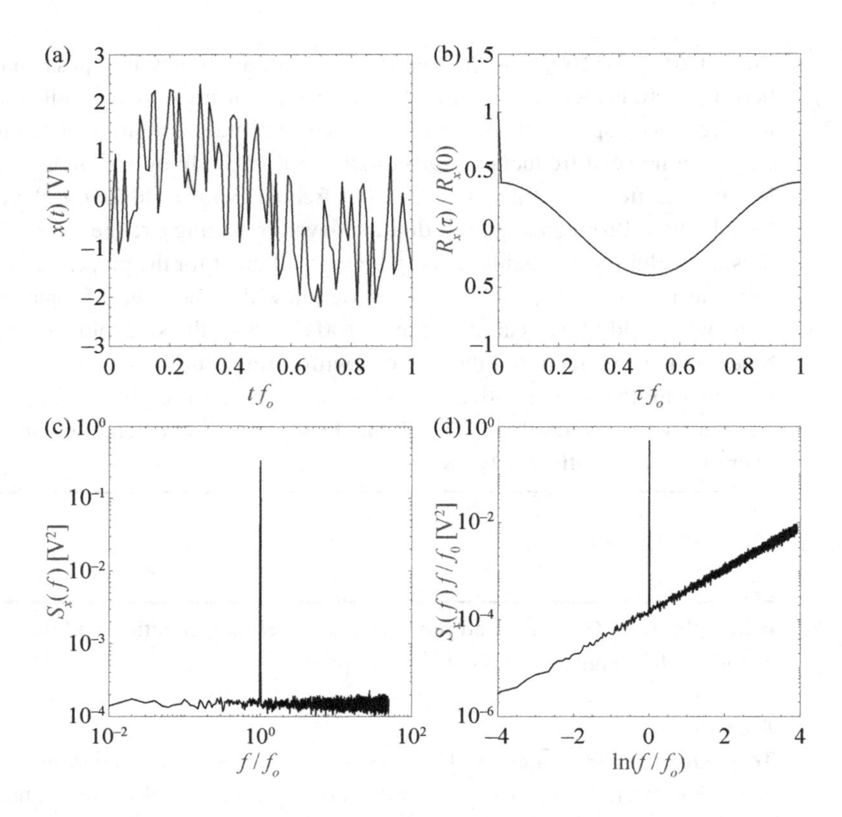

Figure 5.5 (a) One period of a noisy sinusoidal signal; (b) the autocorrelation coefficient of this signal; (c) the power spectrum of this signal; and (d) the pre-multiplied power spectrum of this signal

---

Example 5.9 In an experimental study of grid-generated turbulence [14], the local flow velocity was measured with a hot-wire anemometer. Based on theoretical considerations about the frequency content of turbulent signals, we low-pass filtered the signal with a fourth-order Butterworth filter, having a cut-off frequency $f_c = 14$ kHz. To conform with the sampling theorem requirement, we discretised the signal at a sampling rate $f_s = \omega_s/(2\pi) = 35$ kHz, so that $f_s/f_c > 2$. The measurements were recorded over a record length of 10 min. This data record was divided into 200 blocks, each extending over a block length $T = 3$ s, to allow ensemble averaging. Every other block was discarded, so that the remaining 100 blocks that were processed were essentially independent from each other. Assuming that you have access to this signal, compute and plot the autocorrelation function and the power spectrum of the turbulent velocity.

Answer

The velocity fluctuations during the first block of the signal (i.e., a single time interval with $T = 3$ s) and the corresponding autocorrelation coefficient are, respectively, shown in Fig. 5.6a and Fig. 5.6b. The signal clearly displays an apparent randomness, whereas the autocorrelation coefficient decreases from the value 1 at $\tau = 0$, crosses zero and

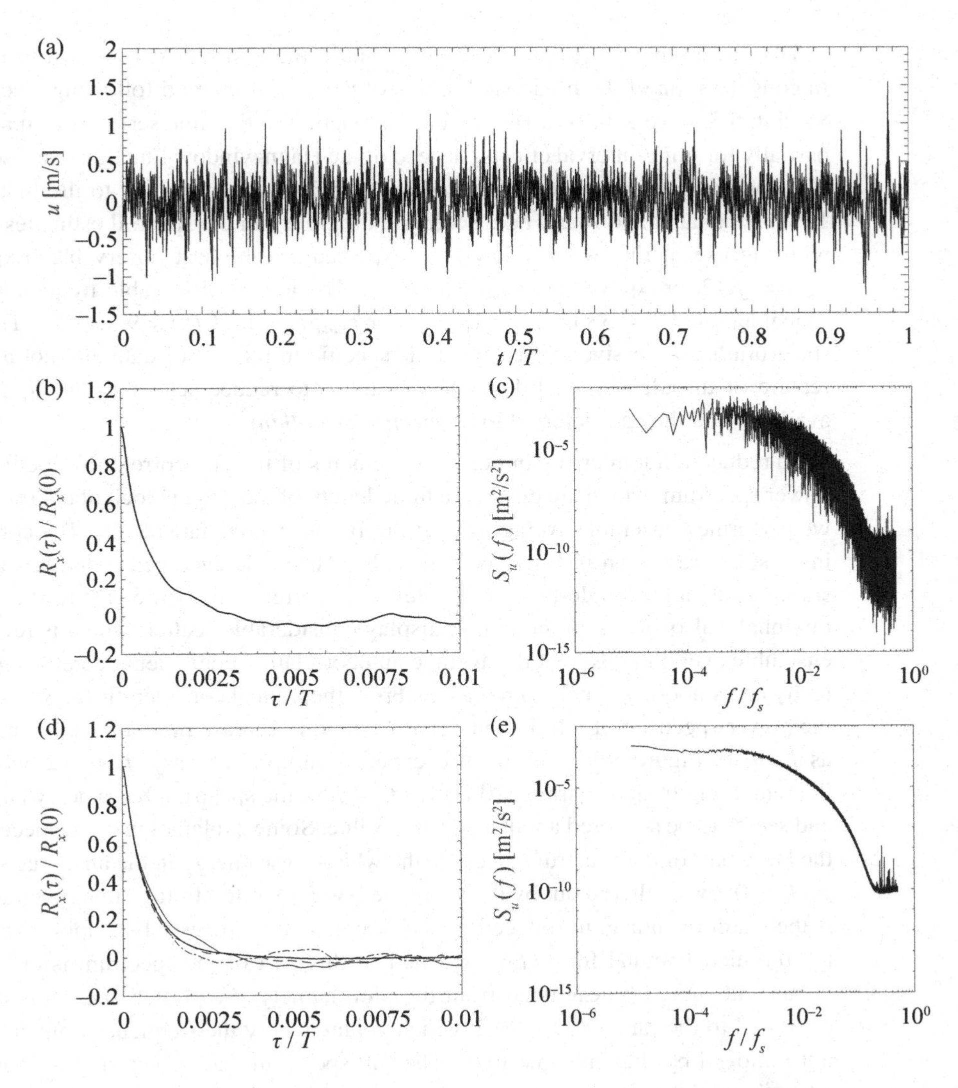

Figure 5.6 (a) The signal representing turbulent velocity fluctuations (i.e., with the mean velocity removed) for the first block of data; (b) the initial part of the autocorrelation coefficient for the first block of data; (c) the power spectrum for the first block of data; (d) the ensemble-averaged autocorrelation coefficient (thick solid line) and three curves obtained from three different blocks; and (e) the ensemble-averaged power spectrum.

then wanders about, while maintaining values not exceeding the range $(-0.1, 0.1)$. A low value of the correlation coefficient at large τ indicates that local velocity fluctuations separated by sufficiently large time differences do not likely have the same source. Assuming that the autocorrelation coefficient for $\tau/T > 0.01$ (not shown in the figure) is negligible, we may infer that the longest period of physical phenomena that generate significant turbulent fluctuations is in the vicinity of $T_t = 0.03$ s, which justifies our choice of $T = 3$ s as block length, because it is equal to $100T_t$ (see Section 3.3.4).

The power spectrum, computed from data in the first block, is shown in Fig. 5.6c. In consideration of the broadband nature of this spectrum, and following discussion in Section 5.3.4, we applied a Hann window function to the time series, which accounted partially for finite-interval effects. Instead of the Hann window function, we also applied a Tukey window function to the time-series, with a taper applied to the first and last 10% of the data, and found no discernible difference in the spectral estimates obtained with the use of the two windows. As expected, the highest observable frequency is $f_{max} = f_s/2$, or equivalently, $f_{max}/f_s = 0.5$. The lowest observable frequency for this time interval of $T = 3$ s is $f_{min} = 1/(T f_s)$, or $f_{min}/f_s = 1/(T f_s^2) \approx 9.5\times10^{-6}$. This power spectrum has excessive scatter, which is specific to this set of data and not a physical feature of turbulence. Well-developed methods to reduce scatter in the spectrum are available and they are referred to as *spectral smoothing*.

To reduce random errors in our measurements of the autocorrelation coefficient and power spectrum, which are due to the finite length of the single block of data used so far, we performed ensemble averaging over the 100 measured data blocks. The corresponding results are shown in Fig. 5.6c and d. The ensemble-averaged properties are much smoother than the single-block estimates. Autocorrelation coefficients for three blocks of signals, also shown in the figure, display considerable scatter, which is removed by ensemble averaging, so that the average autocorrelation coefficient decreases monotonically and smoothly to near-zero values. From theoretical considerations, we expect that the power spectrum of a turbulent velocity should decrease monotonically and sharply as $f \to \infty$. Figure 5.6e confirms this expectation up to about $f/f_s \approx 0.2$, whereas, in the remaining frequency range ($0.2 < f/f_s \leq 0.5$), the spectrum has a noisy appearance and seems to be scattered about a constant value. Some explanation seems necessary for the latter part of the spectrum, because the white-noise energy in the analogue signal for $f_c/f_s > 0.4$ was filtered out by the analogue low-pass filter. In fact, this spectral activity is the result of 'noise' introduced by the discretisation process. This noise overwhelms the turbulence signal for $f/f_s > 0.2$ and so this part of the spectrum is clearly erroneous and should be excluded from the plot, namely, the plotted spectrum should be restricted to the range $f/f_s \leq 0.2$, which contains only the frequency content that was not removed by the analogue filter. The full spectrum was shown in this example for educational purposes.

Example 5.10 Extend the analysis of the data described in Example 5.9 to illustrate the effect of aliasing on the power spectrum of a turbulent flow. To do so, decimate the recorded time series by retaining one every 128th data point, namely, reduce the effective sampling frequency to $f_s/128 \approx 273$ Hz.

Answer

The power spectra, before and after decimation, are shown in Fig. 5.7. Within its entire frequency range, the decimated spectrum is larger. This is a direct result of aliasing and illustrates that the energy content at frequencies exceeding the Nyquist frequency was *folded over* to lower frequencies. Note that the mean and the variance of this random signal were unaffected by aliasing.

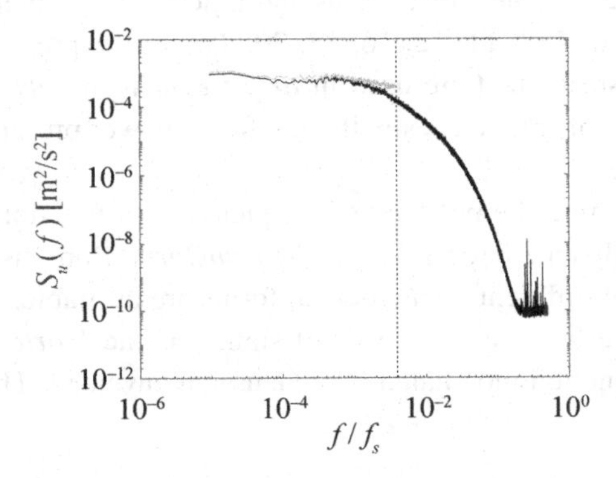

Figure 5.7 Power spectra obtained, respectively, using sampling frequencies 35 kHz (black curve) and 273 Hz (grey curve); the former spectrum is not aliased, but the latter one is.

5.6 Spectral Analysis of Randomly Spaced Time Series

Direct spectral analysis of discrete time series through FFT requires that the samples be evenly spaced. Certain measuring systems, however, with the notable example of laser Doppler velocimeters, provide discrete data at randomly spaced intervals. Two general approaches have been applied successfully to the spectral analysis of such signals [1, 2, 10]. The first one, known as the *slot correlation method*, separates all possible pairs of sample points into bins, according to the difference in their times of arrival. If the width of each bin is relatively narrow, the average product of all sample pairs within the bin would be a good estimate of the covariance of the signal for a time lag equal to the mid-time of the bin. This way, one could estimate the autocorrelation of the time series, from which one could estimate its spectrum. This method does not introduce additional distortion to the spectrum, but requires a large number of samples, which may necessitate impractically long measurement times. A more practical approach is *resampling at a constant rate*. This means that the randomly spaced time history is replaced by a continuous function, which is subsequently resampled at a fixed rate and to which FFT can be applied. Among the various types of interpolation between samples that can be used to construct the continuous function, most widely used is the simple method of holding the value of each sample until the next sample becomes available. This procedure distorts the spectral estimates in two ways. First, it acts as a first-order low-pass filter with a cut-off frequency equal to $N/(2\pi T)$, where N is the number of original samples over the sampling time T. And second, it introduces white noise, due to the step-like character of the resampling process; the variance of this noise decreases as $(N/T)^{-3}$. Thus, for data rates that are significantly higher than the frequencies of interest, the resampling method would produce acceptable spectral estimates.

5.7 Wavelet Analysis

Wavelets have been used widely for a variety of purposes, including data compression and detection of periodicity. Their mathematical basis and practical aspects have been documented in several sources [6, 7, 8, 11, 12, 16, 17, 20, 21]. In the present discussion, we shall focus on those aspects that are used in *time-frequency analysis* of signals, namely in the identification of frequencies with significant power present in non-stationary signals at a given time.

Wavelets are capable of extracting the peak signal frequency as a function of time. In order to do so, one has to apply an appropriate *wavelet transform*. Both discrete and continuous wavelet transforms of different mathematical forms are available. The most commonly used wavelet for time-frequency analysis of signals is the *Morlet wavelet*, which is a complex harmonic function modulated by a Gaussian envelope. The Morlet wavelet [13] can be defined as

$$\psi(\tau) = \pi^{-1/4} e^{j2\pi\gamma\tau} e^{-\tau^2/2} , \tag{5.28}$$

where $j = \sqrt{-1}$, τ is a dimensionless variable (e.g., a normalised time) and γ is the *dimensionless centre frequency* of the wavelet. The real and imaginary parts of the Morlet wavelet are shown in Eq. (5.28). Rigorous definition of the Morlet wavelet requires the addition of a correction term to the right-hand side of Eq. (5.28). However, in practice, the value of γ is taken to be sufficiently large (typically, larger than 1) for this term to be negligible. The value of γ also determines the amplitudes of the successive peaks of the harmonic function relative to that of the central peak at $\tau = 0$. In his original work, Morlet selected γ such that the second peak of the real part of Eq. (5.28) would be equal to half the value of central peak. For simplicity, we shall choose the value $\gamma = 1$, which gives second, third and fourth positive peak amplitudes equal to 0.614, 0.142, and 0.012 times the central peak amplitude, respectively. As illustrated in Fig. 5.8, this wavelet essentially vanishes after three oscillations. The Morlet wavelet is capable of identifying the presence of a wavetrain (i.e., a set of periodic cycles) in a random signal. Due to the decaying exponential in Eq. (5.28), the Morlet wavelet has a localised character, because its magnitude decreases rapidly with increasing $|\tau|$. Other popular wavelets include the *Morse*, *bump* and *Ricker* ('Mexican hat') wavelets.

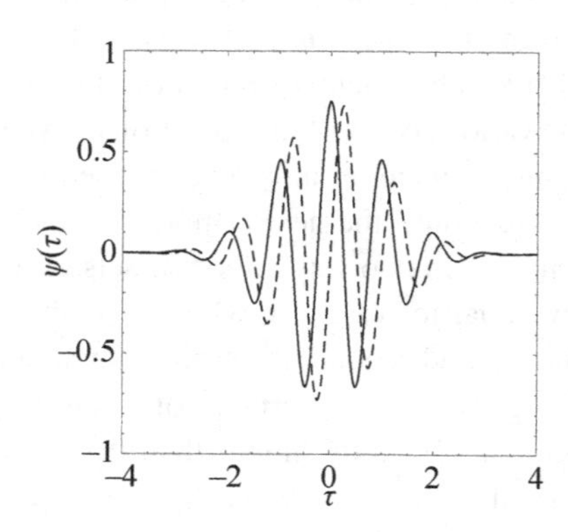

Figure 5.8 Real part (solid line) and imaginary part (dashed line) of the Morlet wavelet.

The *continuous wavelet transform* of a function $x(t)$ consists of its convolution by a wavelet function as

$$W(a,b) = \int_{-\infty}^{\infty} \frac{1}{\sqrt{a}} x(t)\, \psi^*(\tau) \mathrm{d}t \,, \tag{5.29}$$

where $\tau = (t-b)/a$, the asterisk indicates complex conjugate, a is the *scale* or *width* of the wavelet and b is the *temporal translation* of the wavelet. The function $W(a,b)$ is commonly called the *wavelet coefficient* and is generally complex and expressed by its amplitude $|W(a,b)|$ and phase $\varphi(a,b)$. The variable $|W(a,b)|^2$ is referred to as the *wavelet power spectrum*. Variation of the scale adjusts the width of the wavelet so that its frequency matches as best as possible the local frequency of the fluctuating signal. For a given b and wavelet function, the function $W(a,b)$ would depend on the scale a alone. The common way of presenting wavelet transform results is the *wavelet map*, in which contours of constant amplitude $|W(a,b)|$ are plotted vs. time and frequency axes. If a nearly periodic wavetrain is present in the signal, then $|W(a,b)|$ would present a maximum at a certain value of a, from which the signal peak frequency can be calculated.

For a Morlet wavelet with $\gamma \geq 1$, the signal frequency corresponding to a particular wavelet scale a can be computed as

$$f = \frac{\gamma}{a} \,. \tag{5.30}$$

The range of scales a applied to a wavelet transform is bounded by the sampling rate f_s of the time series and the chosen frequency resolution f_r of the wavelet map; for a Morlet transform, $\gamma/f_s \leq a \leq \gamma/f_r$. Once the map is produced, a peak in the wavelet amplitude $|W(a,b)|$ at a particular time would indicate the presence of dominant fluctuations at the corresponding frequency.

Example 5.11 Consider the signals

$$x(t) = \begin{cases} A_1 \sin(2\pi f_1 t)\,, & t < t_o\,, \\ A_2 \sin(2\pi f_2 t)\,, & t \geq t_o\,, \end{cases}$$

and

$$x'(t) = A_1 \sin(2\pi f_1 t) + A_2 \sin(2\pi f_2 t)\,, \ -\infty < t < \infty\,.$$

Plot these signals for $f_1 = 2$ Hz, $f_2 = 1$ Hz, $A_1 = 1$, $A_2 = 2$, and $t_o = 20$ s. Plot the power spectra and the Morlet wavelet map $|W(a,b)|$ of these two signals. Discuss the capabilities of the wavelet maps by comparison to the corresponding power spectra.

Answer

The signals and the corresponding power spectra and wavelet maps are shown in Fig. 5.9. The power spectra of both signals have two peaks at the dominant frequencies f_1 and f_2. A comparison between the areas under the two peaks in each spectrum shows that the energy of fluctuations with frequency f_1 is lower than that with f_2. A comparison of the total areas under the two spectra further shows that the variance of $x(t)$ is smaller

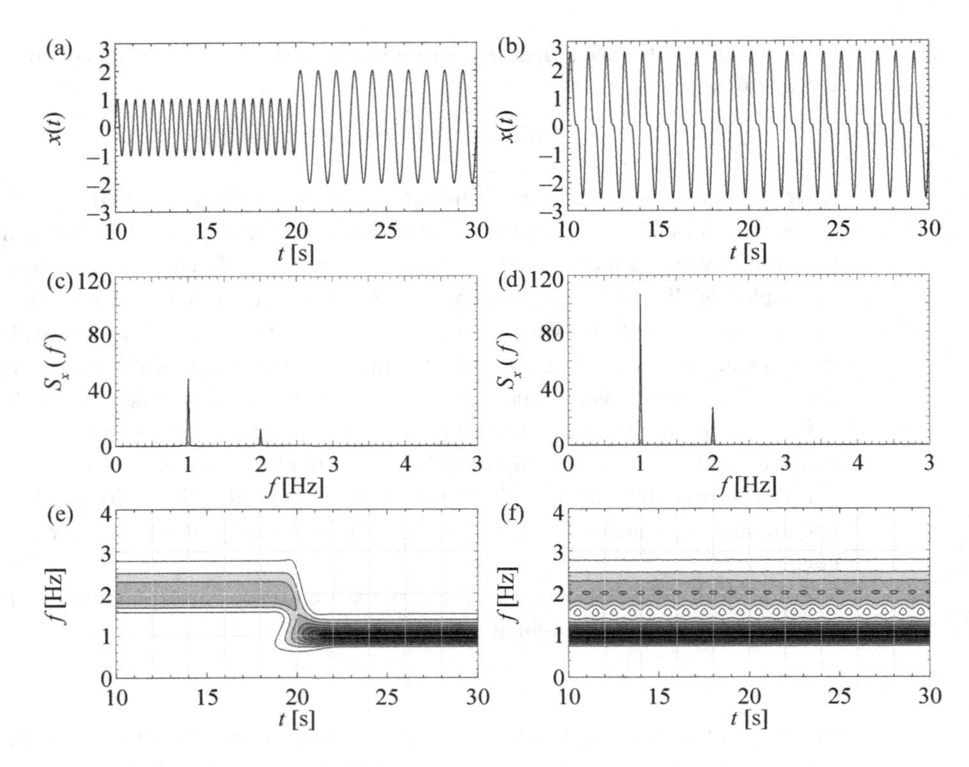

Figure 5.9 A sinusoidal signal with a step change in frequency and amplitude (a), its power spectrum (c) and its Morlet wavelet map (e). The sum of two sinusoidal functions with different frequencies and amplitudes (b), its power spectrum (d) and its Morlet wavelet map (f). Darker grey tones indicate larger values of the wavelet amplitude.

than the variance of $x'(t)$. Nevertheless, the two spectra look alike and cannot discriminate between the signal that has an abrupt frequency change and the signal that has two frequencies at all times. Like the power spectra, the wavelet maps can identify the two dominant frequencies. Moreover, Fig. 5.9e reveals that $x(t)$ has an abrupt change in frequency at $t = 20$ s, whereas Fig. 5.9f shows that $x'(t)$ has both frequencies at all times. The wavelet maps also indicate that the amplitude of fluctuations having a frequency f_1 is larger than the one having a frequency f_2. In summary, the wavelet map contains more information about the signal than the power spectrum.

5.8 Spatial vs. Temporal Analysis

All definitions and properties concerning time-dependent signals and time series can be extended to the one-dimensional space domain by replacing time t as the independent variable by the length r along a straight axis. Then, one would consider the random process $\mathbf{x}(\zeta, r)$ and should appropriately modify all definitions to reflect this change. The equivalent of a stationary process in the space domain would be a *homogeneous random process*, namely a process whose statistical properties are independent of coordinate origin translations, such that the properties of $\mathbf{x}(\zeta, r)$ are the same as those of $\mathbf{x}(\zeta, r+\Delta r)$, where Δr is the *separation distance*. In the absence of a different term, one may also use the term *ergodic* to specify a homogeneous random process whose statistical properties can be computed by appropriate space averaging in one member of the ensemble. The autocorrelation function $R_x(\Delta r)$ of a homogeneous random process would be a function of the separation distance and its power spectrum $S_x(\kappa)$ would be defined in terms of the spatial Fourier transform as

$$S_x(\kappa) = \int_{-\infty}^{\infty} R_x(\Delta r)\, e^{-j\kappa\Delta r} \mathrm{d}r\,. \tag{5.31}$$

The parameter κ, which has a dimension of inverse length, is called the *wavenumber* and is analogous to the frequency. Thus, spectral and wavelet analyses can be performed in the one-dimensional space domain by direct analogy to the time domain. When, however, one considers two- or three- dimensional space, the direct analogy ceases to exist, because the position vector $\vec{r}$ is of higher dimension than time t. Even so, one-dimensional space analysis can be extended to higher dimensions by modifying definitions and relationships. In this case, properties such as the autocorrelation function and power spectrum would have to be replaced by tensors, whose components would in general depend on orientation. The properties of two- and three-dimensional Fourier and wavelet transforms are beyond the scope of the present text.

Chapter Digest

5.1 Discuss the common features and differences between Fourier series and Fourier transforms.

5.2 Discuss the common features and differences between Laplace transforms and Fourier transforms.

5.3 What is the use of the convolution theorem?

5.4 Why is the power spectrum always real, although it is defined by a complex operation?

5.5 What is the coherence function between two identical copies of the same signal? How would a phase shift of one copy affect the result?

5.6 What is white noise? Is there any physical system that generates white noise?

5.7 How are the sampling rate and the sampling time related to the frequency range in the power spectrum?

5.8 What is the benefit of using ensemble averaging when computing the autocorrelation function and the power spectrum?

5.9 How would the Hann window be defined, if the time interval were $0 \leq t \leq T$ instead of $-T \leq t \leq T$?

5.10 What is the advantage of a Fast Fourier transform and how is it achieved?

5.11 What can wavelet analysis show that cannot be shown by spectral analysis?

Problems

5.1 A sensor is placed below a rotating propeller and produces a voltage output of 1 V, while a blade passes above the sensor, and an output of 0 V, while the sensor is under the space between two blades. For a two-bladed propeller rotating at 5000 RPM (revolutions per minute), sketch the voltage output of the sensor vs. time, marking the scales of the axes. Sketch the corresponding frequency spectrum, and indicate how the RPM of the propeller would be determined from the spectrum.

5.2 We wish to measure the rotational speed N_p in RPM of a three-bladed propeller, using the sensor mentioned in the previous problem. The expected operating range is $5000 \leq N_p \leq 15000$. Which is the range of frequencies that you expect to have the most prominent peak in the spectrum? Which sampling frequency and sampling time should you use for discretising the sensor signal in this experiment? Which sampling frequency and sampling time would you use, if a four-bladed propeller were used instead?

5.3 Verify that the power spectra in the following list are related to the corresponding autocorrelation functions and plot both sets of functions.

(a)

$$R(\tau) = e^{-\lambda|\tau|}\,, \; S(f) = \frac{2\lambda}{\lambda^2 + (2\pi f)^2}\,.$$

(b)

$$R(\tau) = e^{-\lambda\tau^2}\,, \; S(f) = \sqrt{\frac{\pi}{\lambda}}\, e^{-\frac{(2\pi f)^2}{4\lambda}}\,.$$

(c)

$$R(\tau) = \begin{cases} 1 - \frac{|\tau|}{T}\,, & |\tau| \leq T \\ 0\,, & T < |\tau| \end{cases}\,, \; S(f) = \frac{4\sin^2\frac{2\pi f T}{2}}{(2\pi f)^2\, T}\,.$$

5.4 Plot the finite-interval power spectrum of the function $x(t) = \sin(20\pi t/T)$ in the interval $[-T, T]$. Compute and plot the finite-interval spectrum of this function after the Hann window has been applied to it. Compare the results.

5.5 Plot the finite-interval power spectrum of the periodic function $x(t) = \sin(20\pi t/T) + \sin(25\pi t/T)$ in the interval $[-T, T]$. Compute and plot the finite-interval spectrum of this function after the Tukey window with different taper lengths has been applied to it. Compare the results.

5.6 Determine the coherence function of the two random processes

$$\mathbf{x}(t) = a\cos(2\pi f_o t) + \mathbf{n}(\zeta, t)\,,$$
$$\mathbf{y}(t) = b\sin(2\pi f_o t) + \mathbf{n}(\zeta, t)\,.$$

How would the coherence function change if the fundamental frequencies of the two random processes were different?

5.7 Determine the integral time scale for a sinusoidal function with superimposed white noise from the power spectrum. Compare this estimate with the value obtained directly from the autocorrelation coefficient. How would you calculate the integral time scale if the white noise were removed?

5.8 Derive the spectral corrections for a Hann window function and a Tukey window function with various taper lengths. Which window function has a greater influence on the energy of the power spectrum?

5.9 The following table contains measurements of the power spectrum S_{Uhw} of atmospheric flow velocity with a hot-wire anemometer (deemed to be accurate at all reported frequencies) and measurements of the same spectrum S_{Upa} with a propeller anemometer (deemed to be accurate only at the lowest frequencies); both spectra have been normalised by their values at the lowest frequency. Plot the two spectra in the same graph, first in logarithmic coordinates and, then, in pre-multiplied form, in semi-logarithmic coordinates. Discuss the appearance of the plots. Determine the percentage of the energy of the velocity fluctuations that is resolved by the propeller anemometer.

f [Hz]	1	3.5	10	35	100	350	1,000	3,500	10,000
$\frac{S_{Uhw}(f)}{S_{Uhw}(0)}$	1.0	1.0	1.0	0.70	0.45	0.15	0.040	0.003	0.0001
$\frac{S_{Upa}(f)}{S_{Upa}(0)}$	1.0	1.0	0.85	0.40	0.090	0.003	0.0001	–	–

5.10 Sound is a wave that propagates through a medium. Such waves are audible (namely, detected by the ears of humans), only within a range of frequencies and a range of amplitudes, which vary from one individual to another and generally become narrower with age. The perceived sound intensity of a standardised human ear in air is characterised by the *sound pressure level* (SPL), measured in dB and defined as $\text{SPL} = 20\log_{10}(p'/(20 \times 10^{-6}\ \text{Pa})$, where p' is the standard deviation of air pressure fluctuations, expressed in Pa. The standardised threshold SPL_{th} of sound audibility for selected frequencies is given in the following table. SPL is used in Physics and Engineering applications, but, in audiology, a different property is used, called the *hearing level* (HL). The standardised hearing level HL_{th} for audibility can be found from SPL_{th} by adding the corresponding values shown in the table. Setting HL_{th} as reference (thus, plotting it at 0 dB at all frequencies in the range of interest), audiologists express the hearing level HL_p of a particular individual ('patient') as the value in dB, exceeding HL_{th}, of the weakest sound that can be recognised by this individual. An example of HL_p for one ear of an elderly patient is also given in the table.

(a) In the same graph, plot vs. frequency the standard deviation of pressure fluctuations corresponding to SPL_{th}, HL_{th} and HL_p.

(b) Consider a synthesiser–loudspeaker system, which produces air pressure fluctuations at narrow frequency bands at all frequencies that are indicated in the table, such that all bands have equal powers p'. For a selected value of p', determine the frequencies that would be audible to a standard ear and to the elderly patient we considered previously. Repeat this for a few values of p', covering the range that shows some interesting facts. Discuss the effect of only hearing some of these frequencies on symphonic music devotees.

f	[kHz]	0.125	0.25	0.5	1	1.5
SPL_{th}	[dB]	20	6	4	2	3
$HL_{th}-SPL_{th}$	[dB]	−45	−25.5	−11.5	−7	−6.5
HL_p	[dB]	20	20	20	25	25

f	[kHz]	2	3	4	6	8
SPL_{th}	[dB]	−2	−6	−5	5	12
$HL_{th}-SPL_{th}$	[dB]	−9	−10	−9.5	−15.5	−13
HL_p	[dB]	35	55	60	60	50

5.11 The Ricker wavelet is defined as $\psi(\tau) = 2\pi^{-1/4}3^{-1/2}\left(1-\tau^2\right)e^{-\tau^2/2}$. Plot the wavelet as a function of τ. Write a code to obtain the wavelet map for the two signals described in Example 5.11. Which information does the Ricker wavelet provide about the signal that the Morlet wavelet does not? Note that the signal frequency for a Ricker wavelet is related to the scale factor as $f = 1/(5\gamma a)$.

References

[1] R.J. Adrian and C.S. Yao. Power spectra of fluid velocities measured by laser doppler velocimetry. *Exp. Fluids*, 5:17–28, 1987.

[2] L.H. Benedict, H. Nobach, and C. Tropea. Estimation of turbulent velocity spectra from laser Doppler data. *Meas. Sci. Technol.*, 11:1089–1104, 2000.

[3] P. Bloomfield. *Fourier Analysis of Time Series: An Introduction*. Wiley, New York, 1976.

[4] H.S. Carlslaw. *An Introduction to the Theory of Fourier's Series and Integrals (3rd Edition)*. Dover Publications, New York, 1950.

[5] J.W. Cooley and J.W. Tukey. An algorithm for the machine calculation of complex Fourier series. *Mathematics Comput.*, 19:297–301, 1965.

[6] I. Daubechies. *Ten Lectures on Wavelets*. SIAM, Philadelphia, PA, 1992.

[7] M. Farge. Wavelet transforms and their application to turbulence. *Ann. Rev. Fluid Mech.*, 24:395–457, 1992.

[8] S.V. Gordeyev and F.O. Thomas. Temporal subharmonic amplitude and phase behaviour in a jet shear layer: Wavelet analysis and hamiltonian formulation. *J. Fluid Mech.*, 394:205–240, 1999.

[9] P. Holmes, J.L. Lumley, G. Berkooz, and C. Rowley. *Turbulence, Coherent Structures, Dynamical Systems and Symmetry*. Cambridge University Press, Cambridge, UK, 2012.

[10] A. Host-Madsen and C. Caspersen. Spectral estimation for random sampling using interpolation. *Signal Process.*, 46:297–313, 1995.

[11] D. Jordan and R.W. Miksad. Implementation of the continuous wavelet transform for digital time series analysis. *Rev. Sci. Instrum.*, 68 (3):1484–1494, 1997.

[12] J. Lewalle. Three lectures on the application of wavelets to experimental data analysis, 1998. Von Karman Institute Lecture Series on Advanced Measurement Techniques, April 6–9, 1998 VKI LS 1998-06, F.A.E. Breugelmans, editor.

[13] J. Morlet. Sampling theory and wave propagation. In *Proc. 51st Annu. Meet. Soc. Explor. Geophys.*, Los Angeles, 1981.

[14] J. Nedić and S. Tavoularis. A case study of multi-structure turbulence: Uniformly sheared flow distorted by a grid. *Int. J. Heat Fluid Fl.*, 72:233–2429, 2018.

[15] H. Nobach et al. Proper orthogonal decomposition: POD (section 22.4). In C. Tropea, A.L. Yarin, and J.F. Foss, editors, *Springer Handbook of Experimental Fluid Mechanics*. Springer, Berlin, 2007.

[16] H. Nobach et al. Wavelet transforms (section 22.6). In C. Tropea, A.L. Yarin, and J.F. Foss, editors, *Springer Handbook of Experimental Fluid Mechanics*. Springer, Berlin, 2007.

[17] D.B. Percival and A.T. Walden. *Wavelet Methods for Time Series Analysis*. Cambridge University Press, Cambridge, UK, 2000.

[18] L.A. Pipes and L.R. Harvill. *Applied Mathematics for Engineers and Physicists*. McGraw-Hill, New York, 1970.

[19] S.B. Pope. *Turbulent Flows*. Cambridge University Press, Cambridge, UK, 2000.

[20] A. Teolis. *Computational Signal Processing with Wavelets*. Birkhäuser, Boston, MA, 1998.

[21] C. Torrence and G.P. Compo. A practical guide to wavelet analysis. *Bull. Amer. Meteor. Soc.*, 79:61–78, 1998.

6 Measurement Uncertainty

This chapter outlines definitions, sources and estimation methods of measurement uncertainty, as well as conventional practices for the presentation of measurements. Information on measurement uncertainty can be found in many elementary engineering textbooks, but some caution is necessary when using older sources, because they may contain definitions and procedures, which do not follow current international conventions. The present discussion is based on a simplified version of the current standards set by the International Organization for Standardization (ISO) [5], which is deemed to be suitable for engineering experimentation [1, 3].

6.1 Measurement Accuracy and Measurement Errors

6.1.1 Error and Accuracy

Measurement error: A measurand can be assumed to have a *true value* x_{true}, which ideally matches the value of the physical property of interest and the determination of which is the objective of the measurement. The true value cannot be known; if it were known, there would not be a need for measurement. The *measured value* x is an estimate of the true value provided by the measuring system; it is generally different from the true value, because no measuring system is perfect. The difference ε_x between the measured value and the true value, expressed in the same units as the measured property, is called the *measurement error* and is defined as

$$\varepsilon_x = x - x_{true} \,. \tag{6.1}$$

The ratio of the measurement error and the true value, expressed as a percentage, parts per thousand, or a fraction, is called the *relative measurement error*. Like the true value, the measurement error cannot be determined exactly. It is the width of the range of probable error values that is estimated and specified in terms of the *measurement uncertainty*.

Although, in general, one cannot know the exact true value and the exact measurement error, one may, under specific conditions, obtain close estimates of them. One such case is when a reproducible physical phenomenon is used to define the true value, as, for example, the value of 100°C for the boiling temperature of pure water under atmospheric pressure. Another case is the estimation of a true value from the reading of a different measuring system, which is deemed to be very accurate for the purposes of a particular experiment. Then, the measurement error of a less accurate system can be estimated as the difference between the readings of this system and the more accurate one, which may thus serve as a *calibration standard*. If one has a reliable error estimate, one should apply an appropriate correction to the measurement, rather than including the error in the reported value.

Accuracy: *Accuracy* of a measurement is an indication of how close the measured value is to the true value. *Inaccuracy* is an indication of how imperfect the measurement is. The term accuracy may be defined as the inverse of the relative error, expressed as a fraction of the measured property. Nevertheless, it is more common to use this term qualitatively and, instead, to quantify inaccuracy by providing values for the measurement uncertainty. Inaccuracy in a measurement may be caused by errors during manufacturing of measuring system components, calibration, data acquisition and data reduction. A component may also malfunction during operation. In addition to errors due to imperfections of the measuring system and procedures, human errors may also be present in measurements, for example those caused by incorrect or inconsistent readings of analogue instruments or by inaccurate eye averaging of fluctuating digital metres, or by misplacing the decimal point when recording a digital output. Such errors are classified as *gross error*. Although gross error may contaminate a measurement, even to the point of rendering it useless, it does not, in general, follow any pattern and cannot be accounted for, either on a case-by-case basis or statistically. Thus, gross errors will not be considered further, under the assumption that the experimenter will expend every effort to avoid such errors.

Systematic errors and random errors: Measurements generally contain two types of error that can be dealt with by statistical means: a *systematic* (also referred to as *bias* or *fixed*) error ε_s and a *random* (or *precision*) error ε_r. The *total measurement error* is the sum of the two, namely,

$$\varepsilon_x \equiv x - x_{true} = \varepsilon_s + \varepsilon_r \,. \tag{6.2}$$

The systematic error remains fixed for all repeat measurements, whereas the random error is different for each repeat measurement.

Accurate and inaccurate measurements: A measurement is called *accurate*, if both its systematic and random errors are small, that is, when

$$|\varepsilon_s|\,, |\varepsilon_r| \ll |x_{true}| \,. \tag{6.3}$$

Otherwise, the measurement is considered to be *inaccurate*. An inaccurate measurement may be classified in one of the following three categories:

- **Precise but biased:** This would be the case, if $|\varepsilon_r| \ll |x_{true}|$ but $|\varepsilon_s|$ is not small compared to $|x_{true}|$.
- **Unbiased but imprecise:** This would be the case, if $|\varepsilon_s| \ll |x_{true}|$ but $|\varepsilon_r|$ is not small compared to $|x_{true}|$.
- **Biased and imprecise:** This would be the case, if neither $|\varepsilon_r|$ nor $|\varepsilon_s|$ are small compared to $|x_{true}|$.

To illustrate the concepts of accurate and inaccurate measurements, consider the

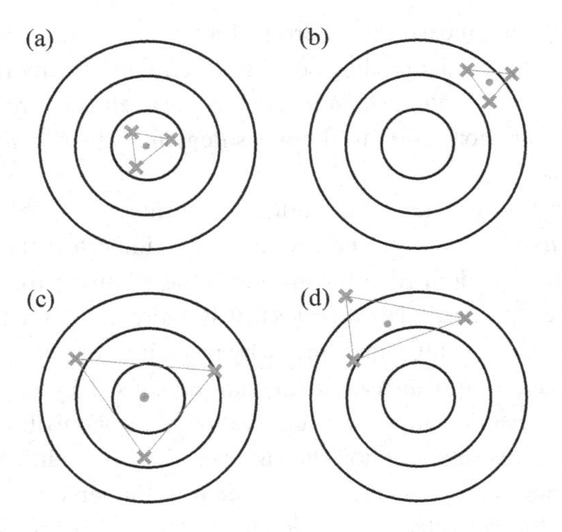

Figure 6.1 An illustration of (a) accurate, (b) precise but biased, (c) unbiased but imprecise and d) biased and imprecise measurements by their analogy to different positions of darts thrown on a dartboard. The filled circles mark the centroid of the triangle formed by the three darts.

throwing of three darts on a dartboard (see Fig. 6.1). The position of each of the three darts may be considered as a repeat value of a measurement. The measurement error of each dart throw is represented by the distance of the dart from the centre of the dartboard. All errors would vanish in the ideal case, for which all three darts hit the dartboard centre. The three-dart throw ('measurement') may be considered to be accurate, if the distances of all darts from the centre of the dartboard were very small by comparison to the outer diameter of the dartboard; in practice, we may declare that the throw was accurate, if all three darts were inside the central circle ('bull's eye' – Fig. 6.1a). Now, consider the average position of the darts, defined as the centroid of a triangle that connects the three darts. The bias error may be represented by the distance between the average position and the dartboard centre, whereas the precision error may be represented by the distance between an individual dart and the average position. Figure 6.1b–d illustrate the three types of measurement inaccuracy.

Example 6.1 Consider Example 4.4. Which of the five groups of 20 amplifiers would you say contains amplifiers with the most accurate gain, if, for educational purposes, we approximate the true value of the gain by the mean gain of the set of all 100 amplifiers?

Answer

We may evaluate this accuracy by visual inspection of Fig. 4.7 as well as by comparing the values listed in the legend of this figure. The least biased among the groups is the one that has a mean that is the closest to the mean of the full set. Therefore, the least biased is the third group ($|\mu_{gn} - \mu_{gN}| = 0.02$) and the most biased is the fourth group ($|\mu_{gn} - \mu_{gN}| = 0.25$). On the other hand, the most precise of the groups is the one that has the smallest spread, namely the smallest standard deviation of the corresponding 20 values. Therefore, the most precise is the second group ($\sigma_{gn=0.49}$) and the least precise is the third group ($\sigma_{gn=0.99}$). These observations show that the third group is both the least biased and the least precise among the five groups, which prevents us from concluding that this group is more or less accurate than the others. This example illustrates that accuracy is a relative concept, which depends on the criteria that are used to evaluate it.

Repeatability and reproducibility: A situation that is sometimes interpreted erroneously to indicate measurement accuracy is the occurrence of a good *repeatability*. Repeatability is the closeness of measurements of the same property repeated under the same conditions, with the same measuring system and during a relatively short time interval. Good repeatability may be considered as a positive indication, albeit not conclusive evidence, that random errors are small. On the other hand, good repeatability cannot exclude the presence of bias, which could be responsible for a large error. The *reproducibility* (or *replicability*) of a measurement is the closeness of measurements of a property collected by different experimenters in different laboratories and using different measuring systems. Good reproducibility is evidence that both the systematic and the random errors are small and so it is a stronger indicator of accuracy than repeatability.

6.1.2 Systematic Errors

It has already been mentioned that the systematic error remains the same for all repeat measurements and could be positive or negative. Following ISO standards [5], the systematic error is defined as '*the mean that would result from an infinite number of measurements of the same measurand carried out under repeatability conditions minus a true value of the measurand*', namely, as

$$\varepsilon_s = \mu_x - x_{true} \; . \tag{6.4}$$

In general, the systematic error is unknown, as it depends on the unknown true value, and so it contributes statistically to the uncertainty of the measurement. When, however, the true value can be estimated by a different, very accurate, measuring system or by theoretical analysis, one is able to determine the systematic error. In such cases, if this systematic error is found to be non-negligible, it should be subtracted from the measured value, which would then contain only a random error. The following example discusses a case in which the systematic error can be determined and, therefore, removed from the measured value.

Example 6.2 We wish to construct an inverting amplifier by connecting two resistors R_1 and R_f to an operational amplifier, as shown in Fig. 6.2a. As discussed previously, the amplifier gain would be $g = -R_f/R_1$. Our electronics shop has a large assortment of mass-produced, inexpensive, carbon-film resistors. The nominal resistance of such commercial resistors is indicated, to two significant digits, by three colour bands on the resistor, whereas a fourth band indicates the manufacturer's tolerance, as a percentage of the nominal resistance value (Fig. 6.2b). We select as R_1 a resistor with brown, black, orange and silver bands, which indicate a nominal resistance $R_{1n} = 10\ \text{k}\Omega$ within a tolerance of $\pm 10\%$, and as R_f a resistor with brown, black, yellow and silver bands, which indicate a nominal resistance $R_{fn} = 100\ \text{k}\Omega$ within a tolerance of $\pm 10\%$. Before connecting the resistors to the circuit, we measure their resistances with a multimeter that provides readings that are considered to be accurate within four significant digits. We find the corresponding readings to be $R_1 = 9.521\ \text{k}\Omega$ and $R_f = 106.3\ \text{k}\Omega$. Determine the gain of this amplifier, as well as the gain error that would have occurred, if we had relied on the nominal, rather than the measured, values of the two resistances.

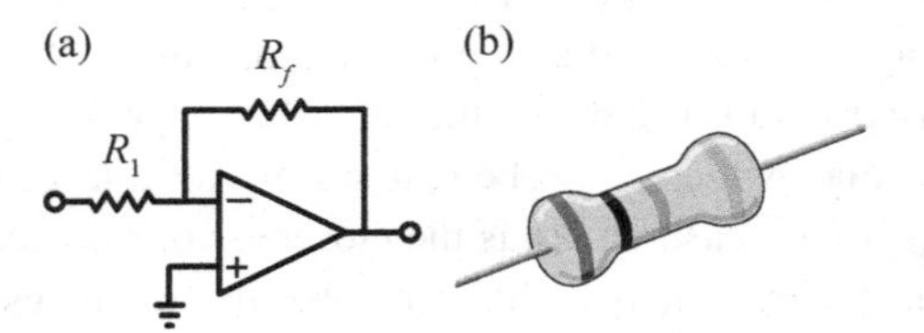

Figure 6.2 (a) An inverting amplifier and (b) a carbon film resistor with its image converted to greyscale.

Answer

We may assume that the multimeter readings are very close to the true resistance values and so we may determine the actual gain, rounded to four significant digits for consistency with the resistance readings, as $g = -R_f/R_1 = -106.3/9.521 = -11.16$. The use of nominal values instead of the measured ones would have resulted in a nominal gain $g_n = -R_{fn}/R_{1n} = -100/10 = -10$. Therefore the gain error would have been

$$\bar{\varepsilon} = g_n - g = -10 - (-11.16) = 1.16\ .$$

This error would remain unchanged during the operation of the amplifier circuit and so it would be a systematic error. Assuming that the multimeter has a negligible random error, one may conclude that the random gain error would also be negligible.

Calibration errors: One of the most common sources of systematic errors are those introduced through the calibration of a measurement device, whereby the user determines a functional relationship between the input and the output. Although, in the discussion of measuring systems, it was the output that was expressed as a function of the input, we presently express the calibration function in a manner that it provides the value of the input to the measuring system (measurand, e.g., a pressure difference) in terms of the value of the device output (e.g., a voltage). For a wide range of measurement devices, including commercial pressure transducers and load cells, this functional relationship is often assumed to be linear, that is, of the form $x = ay + b$, where x is the measurand, y is the measured output and a, b are empirical coefficients, determined by linear least square fitting to the calibration data. For such devices, the following four distinct systematic errors may arise:

- **Linearity error:** This error is defined as the difference between the output of the measurement device and the linear fit. In practice, a linearity error is not specified for each calibration value, but for the device as a whole, which means that it is the *maximum* linearity error that is specified, as shown in Fig. 6.3a.

- **Sensitivity error:** This is the error in estimating the value of the inverse sensitivity a, because the fitted line may have a slope that is different from the slope of a line fitted to a very large number of perfectly accurate calibration values, as shown in Fig. 6.3b. A direct consequence of this error is that the operating span of the device may be different from the span of the calibration data; this type of error is also referred to as *span error*.
- **Zero-offset error:** This is the error in estimating the intersection b of the linear fit to calibration data with the output axis. As the name suggests, this error offsets (i.e., shifts) the calibration line either up or down, as shown in Fig. 6.3c. Most devices have a functionality that will allow the user to offset this error, such that a zero output is read when the device is unloaded.
- **Hysteresis error:** This error is defined as the difference between the upward and downward branches of the hysteresis loop; as with the linearity error, one typically states the maximum difference, rather than values corresponding to specific inputs. This error is shown in Fig. 6.3d.

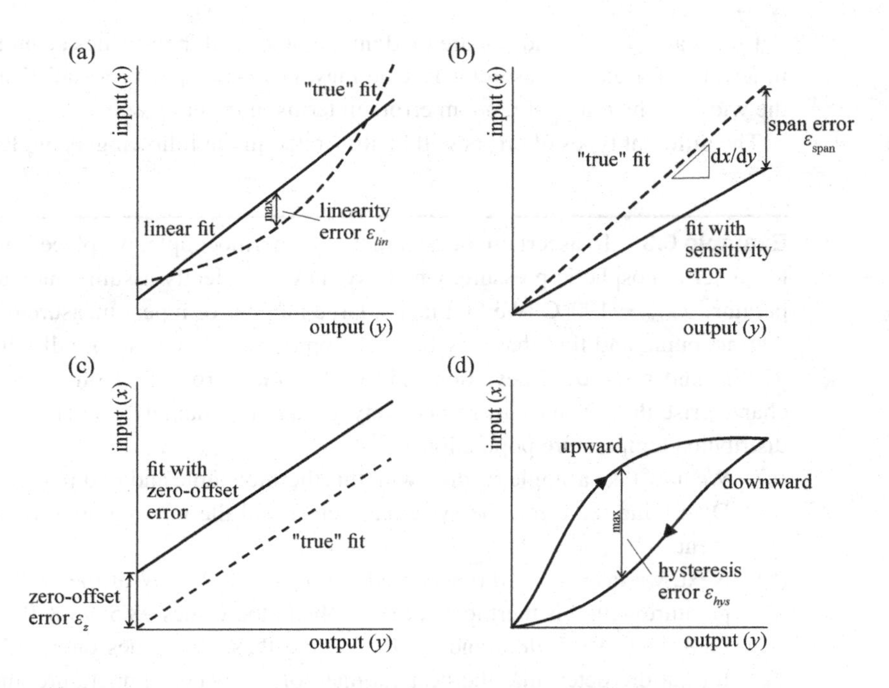

Figure 6.3 Systematic (bias) errors associated with the calibration process: (a) linearisation error, (b) sensitivity error, (c) zero-offset error and (d) hysteresis error.

6.1.3 Random Errors

In accordance with the previous discussion, the random error of a measured value x can be defined as

$$\varepsilon_r = x - x_{true} - \varepsilon_s \ . \tag{6.5}$$

Considering the definition of the systematic error (Eq. (6.4)) and assuming that the measured value belongs to a population obtained under conditions of statistical control that has a normal distribution with a mean μ_x, one may determine the random error of a single measurement as

$$\varepsilon_r = x - \mu_x \ . \tag{6.6}$$

If, instead of using a single measured value, one uses the mean μ_{xN} of N values, the random error of the measurement will be

$$\varepsilon_{rN} = \mu_{xN} - x_{true} - \varepsilon_s = \mu_{xN} - \mu_x \ . \tag{6.7}$$

As discussed in Section 4.4, μ_{xN} is a random variable. Let σ_x be the standard deviation of the population and σ_{xN} be the standard deviation of the set of N values. The property $(\mu_{xN} - \mu_x)/(\sigma_x/\sqrt{N})$ would be a standardised normal random variable with a pdf given by Eq. (4.24), whereas the property $(\mu_{xN} - \mu_x)/(\sigma_{xN}/\sqrt{N})$ would have a Student's t distribution with a pdf given by Eq. (4.29); the latter tends to the former, as $N \to \infty$.

Like x and μ_{xN}, ε_r and ε_{rN} are random variables and their values cannot be specified in advance for each measurement. One may, however, specify a statistical measure of the width of the range of random errors in terms of σ_x or σ_{xN}.

The different types of errors will be illustrated in the following example.

Example 6.3 To ascertain the accuracy of a thermocouple, we place it in boiling water under atmospheric pressure, which we may confidently assume that has a true temperature $T_{true} = 100°C$. We first take a large number of repeat measurements with this thermocouple and find that they have an approximately Gaussian pdf with an average 97.8°C and a standard deviation 0.55°C. For the purposes of this exercise, we may characterise these values as, respectively, μ_T and σ_T, namely properties of a normally distributed temperature population.

(a) We now take a single reading with this thermocouple and find it to be $T_t = 99.0°C$. Determine the error, the systematic error and the random error of this measurement.

(b) Next, we take $N = 10$ repeat readings $T_i, i = 1, 2, ..., N$ of the boiling water temperature with this thermocouple and obtain the values 98.5, 98.8, 98.1, 98.4, 98.9, 98.8, 99.0, 98.7, 98.0 and 98.9°C. Using these 10 values only and no other information, determine the best estimate of the water temperature and compare it to the true value and to the estimate that is based on the large number of repeat measurements. Then, using the 10 values only, determine the systematic error and the random error of the value $T_7 = 99.0°C$. Finally, determine the random measurement error of the mean of the 10 values.

(c) Plot and discuss the data and your findings.

Answer

(a) The error of this measurement is $T_t - T_{true} = -1.0°C$. The systematic error of the temperature population is $\varepsilon_s = \mu_T - T_{true} = -2.2°C$. T_t belongs to this population and, like all other values, it has the same systematic error, namely $-2.2°C$. The random error of this reading is $\varepsilon_r = T_t - T_{true} - \varepsilon_s = 1.2°C$.

(b) The given set of 10 values has a mean $\mu_{TN} = 98.6°C$ and a standard deviation $\sigma_{TN} = 0.35°C$. The best estimate of temperature, based only on the 10 readings, is their mean μ_{TN}, which is 1.4°C lower than the true value T_{true} and 0.8°C higher than the estimate μ_T of the population mean. If we only took into account the information contained in the 10 readings, we would estimate the systematic error of the T_7 reading as $\mu_{NT} - T_{true} = -1.4°C$ and its random error as $T_t - T_{true} - (-1.4°C) = 0.4°C$. Both values are different from the corresponding values that were based on the large set of readings, which are, therefore, more accurate. Finally, the random error of the mean μ_{TN} of the 10 measurements would be $\varepsilon_{rN} = \mu_{TN} - \mu_T = 0.8°C$.

(c) The 10 given values are plotted in Fig. 6.4. It is evident that a systematic error (bias) is present in the measurements, because all repeat values, as well as their average, are measurably lower than the true temperature $T_{true} = 100°C$. This figure also shows the systematic error of all measurements, based on the 10 values, and the one based on the large set of values, as well as the random error of the seventh value, based on the small and large sets of repeat measurements, and the random error of the mean of the 10 measurements.

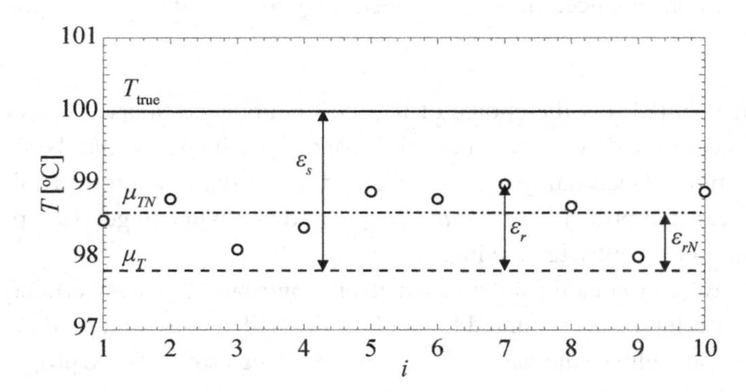

Figure 6.4 Repeat temperature readings of a thermocouple placed in boiling water, the corresponding average μ_T (dashed line), the systematic error ε_s, a representative random error ε_r, and the random error ε_{rN} of the mean of the 10 measurements; the true temperature is marked by a solid line.

6.2 Measurement Uncertainty

6.2.1 Uncertainty

Error vs. uncertainty: Both ε_s and ε_r generally contain multiple errors, which are due to different sources, may be positive or negative and may be called *error components*. The total error would obviously be the algebraic sum of all error components. It is most improbable that all error components for a certain measurement would have the same sign and would achieve their highest magnitude simultaneously, in which case the total error would have its highest possible magnitude. It is recognised that error components of opposite sign would, at least partially, cancel each other, thus reducing the total error magnitude to a value that is lower than its upper limit. Because we do not know the values of error components, we cannot compute the total error. We can, however, specify measures of the widths of the probable ranges of the systematic, random and total errors, which are always positive and are referred to as *uncertainties*. Although errors are additive, summing uncertainty components would imply the simultaneous occurrence of large error components and would overestimate the combined uncertainty. A more appropriate estimate of a combined uncertainty is as the square root of the sum of the squares of uncertainty components [3].

Definition of uncertainty: The *uncertainty* of a measurement specifies the width of the range of values, which the true value of a measurand can take with a certain probability. Let x be a measured value of a property. Following international convention [1], the uncertainty u of this measurement is defined such that the experimenter is 95% confident that the true value of this property lies in the interval $[x - u, x + u]$, namely, that

$$x - u \leq x_{true} \leq x + u \,. \tag{6.8}$$

For convenience, the previous inequality may be alternatively presented as

$$x_{true} = x \pm u \,. \tag{6.9}$$

It is noted that the choice of the 95% confidence interval is a convention adopted by international standards, but other confidence intervals may be used for specific applications. Uncertainty can be specified as *absolute*, expressed in the same units as the measured property, or *relative*, expressed as a percentage, parts per thousand, or a fraction of the measured value.

Both systematic and random errors contribute to the uncertainty and the contribution of each type is accounted for statistically with the use of a property that is representative of the width of the range of the corresponding errors. These properties, are, respectively, called the *bias limit* (or *systematic uncertainty*) b and the *precision limit* (or *random uncertainty*) p. Both b and p are always positive, in contrast to systematic and random errors, which may be positive or negative. The measurement uncertainty is then determined from these limits as

$$u = \sqrt{b^2 + p^2} \,. \tag{6.10}$$

The bias and precision limits will be defined in the following sections, where we will also present statistical methods by which they can be estimated, if suitable data are available. Our objective is to estimate the uncertainty of a measurement that belongs to a very large (ideally, infinite) population, which has a normal pdf with a mean μ_x and a standard deviation σ_x. In practice, however, the population properties may be unknown and we may be forced to estimate this uncertainty from a relatively small set of N values that belong to this population and have a mean μ_{xN} and a standard deviation σ_{xN}. It is evident that, as $N \to \infty$, $\mu_{xN} \to \mu_x$ and $\sigma_{xN} \to \sigma_x$, but, for finite N, the corresponding properties are different. The statistical distribution of the difference $\mu_{xN} - \mu_x$ was discussed in Section 4.4.

6.2.2 Bias Limit

The bias limit, or systematic uncertainty, b signifies that the experimenter is 95% confident that

$$|\varepsilon_s| < b \,, \tag{6.11}$$

where ε_s is the (unknown) systematic error, defined in Eq. (6.4). Equivalently, we may say that b is such that the experimenter is 95% confident that the true value would be within the interval

$$\mu_x - b < x_{\text{true}} < \mu_x + b \,. \tag{6.12}$$

Bias may be introduced to individual pieces that belong to mass-produced batches by tolerances in the materials and manufacturing procedures followed by a manufacturer, who may provide a nominal value for the batch rather than values of individual pieces. Given the opportunity and sufficient data, a user may be able to estimate directly the bias limit of such products, as will be shown in the following examples. Unfortunately, there are also cases for which there is little basis from which to estimate the bias limit and specifying a non-zero value for it may simply serve as a form of a safety factor.

Example 6.4 Consider, as in previous examples, temperature measurements by different thermocouples of the same type. We have $N = 100$ thermocouples, we dip each of them in boiling water and average many readings of each thermocouple. The resulting N mean readings are found to have an approximately Gaussian pdf with a mean $\mu_{TN} = 100.7°\text{C}$ and a standard deviation $\sigma_{TN} = 0.9°\text{C}$. Now, we pick randomly one of these thermocouples and, without knowing its previous readings, use it to measure the temperature of the surrounding air. If possible, find the systematic error (bias) and the bias limit of this thermocouple. Should we apply any correction to the reading of this thermocouple?

Answer

The fact that we average many readings of each thermocouple implies that the random error of the resulting mean would be negligible. Also, the fact that the pdf of the averages of the 100 thermocouples is approximately Gaussian implies that 95% of their values would be in the range $\mu_{TN} \pm 2\sigma_{TN}$.

The set of these 100 thermocouples has an average systematic error $\varepsilon_s = \mu_{TN} - T_{true} = (100.7 - 100.0)°C = 0.7°C$. This, however, is not the systematic error of any individual thermocouple, which can only be found from its own average reading in boiling water. What we can find is the bias limit of this set of 100 thermocouples, which would be approximately $2\sigma_{TN} = 1.8°C$; the same bias limit applies to each individual thermocouple.

Although we cannot correct the reading of an individual thermocouple for its bias, we would gain a statistical advantage, if we corrected all readings by subtracting the average systematic error (i.e., 0.7°C) of the 100 thermocouples.

Example 6.5 Let us revisit the amplifier circuits that were discussed in Example 4.4. Assume that all statistical information obtained by the previous class is available, but the circuits were stored without any individual identification and each student picks randomly a single amplifier from the set of 100. What value of the gain should the student use? What value of the bias limit should be used in the uncertainty analysis?

Answer

In this particular problem, we are given no information about the statistical distribution and properties of the population of amplifier gains. We can, however, perform normality tests on the set of 100 gain values. As Fig. 6.5 shows, the departure of the histogram of these gains from the Gaussian pdf seems to be relatively small. We further found that the gain distribution is slightly skewed, with a skewness factor of 0.23, and slightly flatter than the Gaussian pdf, with a flatness factor of 2.67 (the Gaussian values are, respectively, 0 and 3). For the present purposes, we may consider this distribution to be approximately Gaussian.

The best estimate of the gain, which should be used for all 100 amplifiers, is the set average $\mu_{gN} = 10.23$. As we consider the gain distribution to be Gaussian, we may approximate the bias limit of the set of gains as $b \approx 2\sigma_{gN} \approx 0.16$. An implicit assumption made for this calculation is that random errors in the determination of the gain of each

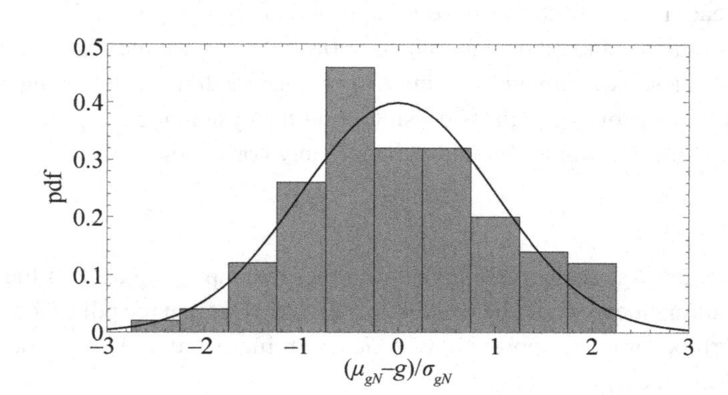

Figure 6.5 Histogram of the standardised amplifier gain values; the solid line shows the Gaussian pdf.

amplifier were negligible. Consequently, as already discussed in Example 4.4, all students should be 95% confident that their gains will be in the range from $\mu_{gN} - 2\sigma_{gN}$ to $\mu_{gN} + 2\sigma_{gN}$, namely, from 10.07 to 10.39.

When biases are introduced by K statistically independent components or steps of the measuring process (e.g., by an amplifier, a filter, a calibration procedure and a data acquisition system), the total bias limit is determined as

$$b = \sqrt{\sum_{k=1}^{K} b_k^2}\,, \tag{6.13}$$

where b_k are the bias limits of the components.

Bias limits may be difficult to estimate, but this does not mean that they should be disregarded. In some cases, one may estimate a bias limit from the manufacturer's specifications, theoretical considerations using material properties or personal experience with similar measuring systems.

Bias limit from manufacturer's specifications: First, let us assume that we wish to estimate the bias limit of a device based entirely on manufacturer's specifications about the device. Such specifications may be described by the terms error, uncertainty or accuracy, which may not necessarily follow the definitions and standards presented in this chapter. Moreover, one should take note of an important conceptual difference between the presently defined bias limit and usual manufacturer's specifications: we defined the bias limit for a particular input value of the measuring system, but manufacturers specifications apply to the entire range of operation of a device. Manufacturers do not usually state explicitly the bias limit of a device, but may typically specify some or all of the following values:

- The maximum linearity error, either as a percentage of full scale (FS) or as a percentage of the reading (RDG).
- The sensitivity error, as a percentage of RDG.
- The maximum hysteresis error, as a percentage of FS or RDG.

One may interpret these values as rough estimates of the corresponding bias limits, under the assumption that the positive and negative values of these errors bound roughly 95% of the possible cases [1, 6]. The total bias limit can then be determined with the use of Eq. (6.13).

Example 6.6 A thermometer manufacturer states that the accuracy due to linearity is 0.13% FS and that the accuracy due to zero offset is 0.02% FS. Find the inverse sensitivity and the bias limit for a thermometer with a full span of 250°C and an output range from 4 to 20 mA.

Answer

From the manufacturer's reference to a linearity error, we may assume that the fitted calibration function is of the form $T = aE + b$, where T is the temperature and E is the read voltage. We can then calculate the inverse sensitivity as

$$a = [250/(20-4)]°\mathrm{C/mA} = 15.6°\mathrm{C/mA}\,.$$

Assuming that the two specified accuracies are fair estimates of the corresponding bias limits, we may compute the total bias limit as

$$b = \sqrt{(0.0013 \times 250)^2 + (0.0002 \times 250)^2} = 0.3°\mathrm{C} = 0.12\%\ \mathrm{FS}\,.$$

Although it is possible that additional sources of bias error are present, we have no means of estimating the corresponding bias limits.

Bias limit from in-house calibration: Now, let us assume that we do not have sufficient manufacturer specifications concerning the bias of a device, or that we are uncertain whether such specifications are still valid. One way to estimate the bias limit is to perform our own calibration and use the calibration data for this purpose. In doing so, we make the assumption that there are no random errors during the calibration process. In particular, we assume that the measurement devices being used as standards for the calibration process are significantly more accurate and more precise than the device being calibrated. Having detailed calibration data allows us also to compare the simplified approach by which manufacturers present their bias specifications and the statistical approach that was discussed in this chapter.

Example 6.7 A differential pressure transducer is calibrated in the laboratory against a very accurate standard and has provided precisely repeatable values for different inputs, which indicates that random errors during calibration were negligible. The operating range of the transducer is from 0 to 1,000 Pa, hence the operating span is P_s = 1,000 Pa. The calibration data, consisting of pairs of values of the output voltage E and the actual ('true') input pressure difference Δp_a, are shown in Fig. 6.6a. Before conducting the calibration, the zero offset bias was removed, through manual tuning of a knob on the device, to ensure that a zero voltage reading was observed when no pressure was applied to the transducer. Although this transducer has a measurable non-linearity, which may be effectively represented by a polynomial fit, it is convenient to use a linear relationship, in which case, it is necessary to quantify the systematic error that would be introduced as the transducer operates within a range of input values. Determine the bias limit due to non-linearity, using the simplified and statistical approaches.

Answer

We fit the calibration data by a linear fit of the form $\Delta p_l = a(E - E_0)$, where E_0 is the voltage output at zero pressure difference input. Given that the zero offset was removed

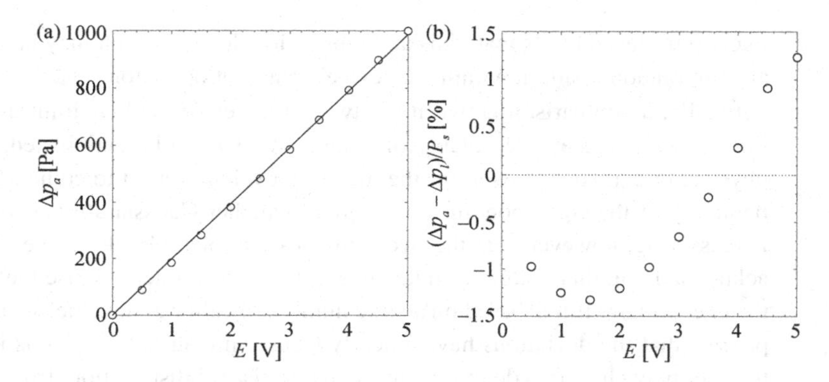

Figure 6.6 (a) Calibration data showing the voltage output E of a pressure transducer subjected to different input pressure values Δp_a; the solid line shows the linear fit aE; (b) differences between input pressure values and values from the linear fit, given as a percentage of full scale.

prior to the start of the calibration, we set $E_0 = 0$. It is important to note, therefore, that the present analysis is only valid for measurements for which the zero offset is removed.

Using least square regression techniques, as outlined in Section 4.6.2, we compute the transducer's average inverse sensitivity as $a = 197.5$ Pa/V. The difference between the applied pressure and the pressure calculated from the linear fit, normalised by the full scale, is shown in Fig. 6.6b, which clearly illustrates the systematic non-linearity of the calibration data. This normalised difference takes values in the range $-1.33\% \leq 100(\Delta p_a - \Delta p_l)/P_s \leq +1.23\%$. It is obvious that, if one were to use the linearised response equation for all readings, each reading would have a different bias error.

First, we estimate the bias limit due to non-linearity b_{lin} using the simplified approach. This approach approximates b_{lin} by the linearity of the device, defined as the *maximum* absolute deviation $d_{max} = |\Delta p_a - \Delta p_l|$ from the calibration line (see Section 2.3.2). For the present calibration data, we find that $b_{lin} = 13.3$ Pa.

Next, we compute the bias limit using the statistical approach. The standard deviation of the differences from the linear fit is found to be $\sigma_d = 8.96$ Pa. We used $N = 11$ calibration values and we have a single empirical parameter (i.e., the sensitivity) to estimate, so that $m = 1$ and the number of degrees of freedom for the t-score is $\nu = N - m = 10$. The t-score at 95% confidence interval for $\nu = 10$ (Table 4.1) is $t_\nu = 2.228$ (see Eq. (4.30)), which is only slightly larger than the Gaussian value of 2.0. Therefore, the bias limit due to non-linearity of the response is estimated to be $b_l = t_\nu \sigma_d = 20.0$ Pa. It is important to emphasise, however, that this statistical approach is based on the assumption that the distribution of the differences between actual calibration values and the linear fit is approximately Gaussian. For many devices, this assumption may be weakly satisfied, if at all. We calculated the skewness and flatness factors of the present calibration data to be, respectively, 0.62 and 2.04, which are significantly different from the Gaussian values 0.00 and 3.00. Increasing N would not necessarily improve Gaussianity, because the non-linearity in the device response is likely introduced by its materials and fabrication details, which may have a systematic, and not random, pattern within the device's range of operation.

Finally, a comparison between the two estimates of the bias limit due to linearisation shows that they are notably different. This was to be anticipated, as there is no physical or statistical reason for the maximum value of a set to coincide with the 95% boundary of the corresponding distribution, whether Gaussianity is assumed or not. It is reassuring, however, that the two estimates are tolerably close to each other. While acknowledging that neither estimate is exact, we recommend to use the statistical estimate, especially if it is based on a large number of calibration values N, again under the proviso that the deviations have a nearly Gaussian distribution; if this is not the case, the user may choose to determine the actual statistical distribution, from which the corresponding 95% confidence limits can be found accurately. If, on the other hand, only a maximum-error estimate is available, one should use this as a bias limit; such choice would definitely be better than neglecting bias altogether.

The bias limit of sensitivity errors can be similarly determined from the calibration data, provided that they are fitted by a linear expression of the form $x = ay + b$, which is simplified to $x = ay$, if the zero offset is removed. For education purposes, we shall focus on the specific case of $x = ay + b$, although one should take advantage of scientific software, either open source or under licence, which have built-in capabilities providing the user with measures of the uncertainty of the 'best' empirical coefficients for many types of fitted functions, including polynomials of different order, power functions, exponential functions and others. The reader is strongly encouraged to explore the capabilities of their chosen software.

Under the assumption that the differences between the calibration data and the fitted line have ideally a Gaussian distribution [3, 4], several scientific software provide the uncertainties of the regression coefficients a and b. The uncertainty on the regression coefficient a can be considered to be an estimate of the bias limit of the sensitivity error, within a 95% confidence interval. This uncertainty, and based on N calibration points, is given as

$$b_a = \frac{s_{y,x} t_{N-2}}{\sqrt{\sum (y_i - \bar{y})^2}}, \tag{6.14}$$

where $s_{y,x}$ is the standard error of the estimate (Eq. (4.46)), y_i is the output of the measuring system, and $\bar{y}$ is the mean of the N measured outputs, obtained during the calibration. Note that the number of degrees of freedom for the t-score is $\nu = N - m$, where m is the number of empirical coefficients in the calibration function; hence, for a linear fit, $m = 2$, or, if the line is forced to pass through the origin, $m = 1$. Similarly, one can determine the uncertainty in the regression coefficient b as

$$b_b = s_{y,x} t_{N-2} \sqrt{\frac{\sum y_i^2}{N \sum (y_i - \bar{y})^2}}. \tag{6.15}$$

It is interesting to note that both of these uncertainties are reduced, as the calibration range is widened and as N is increased. One should not misinterpret the uncertainty in the intercept b_b as a bias limit due to zero-offset.

We have so far been concerned about estimating the bias limit of a measurand. Additionally, we also often need to determine the bias limit of a statistical property of the measurand. This topic is associated with the uncertainty of derived properties (Section 6.2.4), which is often referred to as propagation of uncertainty. At this point, we will only consider a simple case, to be illustrated by the following example.

Example 6.8 Assume that the transducer that was discussed in Example 6.7 is used to measure the standard deviation σ_P of pressure fluctuations, which is much smaller than the transducer span. Determine the bias limit of σ_P, which corresponds to errors caused by the difference between the actual transducer sensitivity for each pressure value and the average sensitivity, which is used for all pressures within the transducer range.

Answer

In Example 6.7, we distinguished between the actual pressure difference Δp_a and the value Δp_l provided by the linear fit to the calibration data. This difference is measured by the bias limit of the pressure difference. The standard deviation of the pressure is not subject to this bias, because the (possibly biased) mean pressure is subtracted from the fluctuating measurements before the standard deviation is computed. The standard deviation, however, is subject to a different bias, which is due to the difference between the *actual sensitivity* of the transducer at the measured mean pressure, and the sensitivity value that is determined from the calibration fit, in this case a linear one. To determine the actual sensitivity of the transducer for each pressure value, we fit the calibration data by a function of the form $\Delta p_a(E) = a_0 E^{a_1}$, where $a_0 = 185.1$ PaV$^{-1.05}$ and $a_1 = 1.05$. We assume that this fit is very accurate, at least by comparison to the linear fit, and sufficiently smooth to provide accurate estimates of its local derivative $\mathrm{d}\Delta p_a(E)/\mathrm{d}E$. Then, the actual standard deviation of pressure fluctuations would be

$$\sigma_{P_a} \approx \frac{\mathrm{d}\Delta p_a(E)}{\mathrm{d}E}\sigma_E ,$$

where σ_E is the standard deviation of the output voltage. On the other hand, the standard deviation of pressure fluctuations computed from the linearised expression would be

$$\sigma_{Pl} \approx \frac{\mathrm{d}\Delta p_l(E)}{\mathrm{d}E}\sigma_E = a\sigma_E .$$

Now, the normalised bias error in σ_P for each value of the calibration data is $\varepsilon_{b,\sigma_p} = (\sigma_{Pl}-\sigma_{Pa})/\sigma_E = a - \frac{\mathrm{d}\Delta p_a(E)}{\mathrm{d}E}$. The 95%-confidence bias limit for σ_P/σ_E can be found by taking the standard deviation of ε_{b,σ_p} for the $N = 11$ calibration values, which is 61.35 Pa/V, and multiplying it with the corresponding t-score of 2.228, to give 136.7 Pa/V. The bias limit on the pressure fluctuations is, therefore, a function of the measured standard deviation of voltage σ_E. Of course, if the calibration function was originally chosen to be of the form $\Delta p_a(E) = a_0 E^{a_1}$, the bias limit on the fluctuations would effectively be removed. In closing, we may mention that the variance, the skewness factor, spectra and other calculated properties of the measurand are also subject to different sensitivity biases, which can be determined by similar analyses.

6.2.3 Precision Limit

Definition: The precision limit p, or precision uncertainty, or random uncertainty, of a measured value x signifies that the experimenter is 95% confident that the (unknown) random error ε_r, defined by Eq. (6.6), satisfies the inequality

$$|\varepsilon_r| < p\,. \tag{6.16}$$

Equivalently, we may say that p is such that the experimenter is 95% confident that the population mean would be within the interval

$$x - p \leq \mu_x \leq x + p\,. \tag{6.17}$$

Precision limit for multiple sources of random error: As for the bias limit, when random errors are introduced by K different components or steps of the measuring process (e.g., by an amplifier, a filter, a calibration procedure and a data acquisition system), the total precision limit is determined as

$$p = \sqrt{\sum_{k=1}^{K} p_k^2}\,, \tag{6.18}$$

where p_k are the precision limits of the components.

Precision limit from $\sigma_{\mathbf{x}}$: The usual practice in measurements is to collect N repeat values, average them to obtain their mean μ_{xN} and report this mean as the measurement. In the present discussion, we will assume that the population of the random variable $\mathbf{x}$ has a normal distribution with a standard deviation σ_x, which may be known or unknown. If σ_x is known, or can be estimated accurately from a large set of repeat measurements, the precision limit of μ_{xN} will be

$$p = \frac{2\sigma_x}{\sqrt{N}}\,. \tag{6.19}$$

It is obvious that the precision limit is reduced as N increases. Equation (6.19) is valid for all values of N, including $N = 1$ (single measured value), in which case,

$$p = 2\sigma_x\,. \tag{6.20}$$

Precision limit from $\sigma_{\mathbf{xN}}$: If σ_x is unknown, but the set standard deviation σ_{xN} is known, one may follow the discussion in Section 4.4 to determine the precision limit of μ_{xN} as

$$p = \frac{t_\nu \sigma_{xN}}{\sqrt{N}}\,, \tag{6.21}$$

where the values of the 95% confidence t-score for different N are listed in Table 4.1.

Example 6.9 Consider Example 6.3, which describes temperature measurements by a thermocouple in boiling water. Determine the precision limits of single readings and of the mean of 10 readings, if (a) the population standard deviation is given or estimated from a large set of repeat measurements as $\sigma_T = 0.55°\text{C}$ and (b) if the standard deviation $\sigma_{TN} = 0.45°\text{C}$ of the particular set of 10 values has been determined, σ_T is unknown.

Answer

(a) The precision limit of a single measurement would be

$$p = 2\sigma_T = 2 \times 0.55°\text{C} = 1.10°\text{C}$$

and the precision limit of the average of *any* 10 values would be

$$p = \frac{2\sigma_T}{\sqrt{N}} = \frac{2 \times 0.55°\text{C}}{\sqrt{10}} = 0.35°\text{C}\,.$$

(b) Without knowing σ_T, we cannot determine the precision limit of a single measurement. For $N = 10$, hence $\nu = 9$, we find from Table 4.1 that $t_\nu = 2.262$ at the 95% confidence level. Then, the precision limit of the average of the 10 values would be

$$p = \frac{t_\nu \sigma_{xN}}{\sqrt{N}} = \frac{2.262 \times 0.45°\text{C}}{\sqrt{10}} = 0.32°\text{C}\,.$$

The latter limit would, in general, be different for different sets of 10 values.

Calibration-based precision limit: In the previous section, we illustrated how one may estimate a bias limit from manufacturer's specifications and from in-house calibration data. Of course, that approach has some limitations, because the sources and levels of bias errors during manufacturer's and in-house calibrations may not coincide with those during measurement. Such limitations are much more prominent when it comes to estimating the precision limit. Some random errors, for example those due to thermal noise (see Section 6.5.1) in some component of the measuring system, may have the same statistical properties, irrespectively of the use of this system (even thermal noise, however, depends on the temperature of the component, which may be different in different settings and may also change from one instant to another). More significantly, random errors due to external sources, such as interference from nearby electromagnetic sources or from occasional air drafts, would be specific to each setting and, possibly, to the time of use; therefore any calibration-based estimate of the precision limit may not apply at all to measurements at arbitrary conditions. Although it is straightforward to adapt the methodology developed for the bias limit, so that it can be used for estimating the precision limit, we will not pursue this here, as we discourage such use. Instead, in Section 6.2.5, we are going to adapt this methodology for estimating the total uncertainty, in full awareness of the possible limitations of this approach.

6.2.4 Uncertainty of Derived Properties

The discussion on uncertainty in the previous subsection applies to properties that are measured directly. Most commonly, however, a *derived property* y is computed from direct measurements of one or more *subsidiary properties*, $x_m, m = 1, 2, ..., M$, with the use of an analytical expression, namely, as

$$y = y(x_1, x_2, ..., x_M) . \tag{6.22}$$

The uncertainty of any measured property generally contributes to the uncertainty of any other property that depends on this measurement. This is commonly referred to as *propagation of uncertainty*. In some cases, the value of a subsidiary property is determined from a graph or table as a function of one or more subsidiary properties; for example, the viscosity of water for a given temperature may be determined by interpolation from a table. Graphs and tables are equivalent to analytical expressions, to which they can be converted by curve fitting. We shall further assume that $y(x_1, x_2, ..., x_M)$ is continuous in all variables, so that partial derivatives $\partial y/\partial x_m$ can be defined. Each of these derivatives represents the *sensitivity* of y to changes in the corresponding subsidiary property x_m. When a graph or table is involved in the determination of y, partial derivatives may be computed from the fitted expression or estimated as finite differences between neighbouring values.

Derived uncertainty from total subsidiary uncertainties: First, let us consider a case in which the total uncertainties $u_{x_m}, m = 1, 2, ..., M$ of all subsidiary properties are specified, without a breakdown to bias and precision uncertainties. Further assume that all subsidiary properties are measured or estimated independently, so that there is no indication that the biases of some of them are correlated. Then, one may calculate the uncertainty of the derived property as

$$u_y = \sqrt{\sum_{m=1}^{M} \left(\frac{\partial y}{\partial x_m} u_{x_m} \right)^2} . \tag{6.23}$$

This expression shows that u_y depends not only on all u_{x_m}, but also on all sensitivities $\partial y/\partial x_m$. Thus, subsidiary properties with relatively large sensitivities tend to contribute more to u_y.

A convenient form of Eq. (6.23), which applies only to derived properties that can be expressed as products of powers of the subsidiary properties, namely, as

$$y = (\text{const.})\, x_1^{n_1}\, x_2^{n_2} \cdots x_M^{n_M} , \tag{6.24}$$

is

$$\frac{u_y}{|y|} = \sqrt{\sum_{m=1}^{M} \left(\frac{n_m u_{x_m}}{x_m} \right)^2} . \tag{6.25}$$

Example 6.10 Determine the uncertainty in the gain of the amplifier described in Example 6.2, based on the tolerances of the resistors, as indicated by their fourth colour bands.

Answer

Manufacturer catalogues specify that the resistor tolerance is the maximum (positive or negative) difference between the nominal and actual resistor values. This implies that the actual amplifier gain would be bounded by values computed from the corresponding combinations of the most extreme possible values of the two resistors, namely, $g_{min} = -R_{fmax}/R_{1min} = -(100 \times 1.1)/(10 \times 0.9) = -12.2$ and $g_{max} = -R_{fmin}/R_{1max} = -(100 \times 0.9)/(10 \times 1.1) = -8.2$. Thus, the amplifier gain should be within the range

$$-12.2 < g < -8.2 .$$

The relationship, if any, between the resistor tolerance, as specified by the manufacturer, and its uncertainty, as defined presently, is unclear. Let us assume that the uncertainty of a resistor, within an unspecified confidence interval, is equal to the absolute value of its tolerance. This implies that the uncertainty of a silver-band resistor is 10% of its nominal value. Because the actual amplifier gain would usually be unknown, we will approximate it by the nominal gain $g_n = -R_{fn}/R_{1n} = -10$. Under these assumptions, the uncertainty of the gain would be

$$u_{g_n} = |g_n| \sqrt{\left(\frac{-u_{R_{1n}}}{R_{1n}}\right)^2 + \left(\frac{u_{R_{fn}}}{R_{fn}}\right)^2} = 10\sqrt{0.1^2 + 0.1^2} \approx 1.4 .$$

In general, the gain uncertainty of an amplifier that uses carbon film resistors with a silver-coloured fourth-band would be 14% of the nominal gain. This approach implies that, within the confidence interval corresponding to this uncertainty, the amplifiers made with the type of resistors discussed previously would have a gain within the range

$$-11.4 < g < -8.6 .$$

We note that the two previously found gain ranges are somewhat different, as the result of the difference between the ways they are computed. The tolerance approach literally infers that 100% of the gains will be within the former range. On the other hand, if one assumes that the tolerance specified by the manufacturer follows current international practice and is equal to the uncertainty of the resistor within a 95% confidence interval, then one may state that 95% of the amplifiers would have a gain that is within the latter range.

Example 6.11 Given the uncertainties u_{x_1}, u_{x_2} of the subsidiary properties x_1, x_2, determine the uncertainty of the derived property

$$y = \frac{ax_1^{3/4}}{\sqrt{x_2}}, \quad a = \text{const.}\,.$$

Answer

We may rewrite the previous expression as

$$y = ax_1^{3/4}x_2^{-1/2},$$

which is a product of powers. Consequently, we may use Eq. (6.25) to find

$$u_y = |y|\sqrt{\left(\frac{3u_{x_1}}{4x_1}\right)^2 + \left(\frac{u_{x_2}}{2x_2}\right)^2}\,.$$

Example 6.12 Given the uncertainties u_{x_1}, u_{x_2} of the subsidiary properties x_1, x_2, determine the uncertainty of the derived property

$$y = ax_1^{3/4}e^{x_2}, \quad a = \text{const.}\,.$$

Answer

We note that this expression is not a product of powers of the subsidiary properties and so we cannot use Eq. (6.25). Instead, we use the general expression, Eq. (6.23), which gives

$$\begin{aligned} u_y &= \sqrt{\left(\frac{\partial y}{\partial x_1}u_{x_1}\right)^2 + \left(\frac{\partial y}{\partial x_2}u_{x_2}\right)^2} \\ &= \sqrt{\left(\frac{3a}{4}x_1^{-1/4}e^{x_2}u_{x_1}\right)^2 + \left(ax_1^{3/4}e^{x_2}u_{x_2}\right)^2} \\ &= |y|\sqrt{\left(\frac{3u_{x_1}}{4x_1}\right)^2 + u_{x_2}{}^2}\,. \end{aligned}$$

Derived uncertainty from subsidiary systematic and random uncertainties: An approach that, in some cases, may provide a more accurate value of u_y is to compute it from the corresponding bias limit b_y and precision limit p_y as

$$u_y = \sqrt{b_y^2 + p_y^2}\,. \tag{6.26}$$

These bias and precision limits are estimated from the corresponding bias limits b_{x_m} and precision limits p_{x_m} of the subsidiary properties as follows.

If the value of each subsidiary property is determined independently of all others, so that all b_{x_m} would be independent of each other, the bias limit b_y can be computed as

$$b_y = \sqrt{\sum_{m=1}^{M}\left(\frac{\partial y}{\partial x_m}b_{x_m}\right)^2}\,. \tag{6.27}$$

It may happen, however, that the biases of two or more subsidiary properties arise from the same source, for example, the same biased instrument or the same biased experimenter. Equation (6.27) shows that the contributions of all biases are cumulative, but, depending on the form of $y(x_1, x_2, ..., x_M)$, it is possible that some biases may partially or totally cancel each other, so that b_y would be lower than predicted by Eq. (6.27).

First, consider the relatively simple case for which, among the M subsidiary properties $x_m, m = 1, 2, ..., M$, there are only two, let's say x_1 and x_2, that have identical biases ($b_{x_1} = b_{x_2}$). If the two sensitivities $\partial y/\partial x_1$ and $\partial y/\partial x_2$ have the same sign, both b_{x_1} and b_{x_2} contribute to b_y, which may be determined from Eq. (6.27). If, however, $\partial y/\partial x_1$ and $\partial y/\partial x_2$ have opposite signs, the two biases would cancel each other and b_y should be computed from

$$b_y = \sqrt{\sum_{m=1}^{M}\left(\frac{\partial y}{\partial x_m}b_{x_m}\right)^2 + 2\frac{\partial y}{\partial x_1}\frac{\partial y}{\partial x_2}b_{x_1}b_{x_2}}\,. \tag{6.28}$$

When the biases of two subsidiary properties are due to more than one sources, it is possible that only portions of the total biases are correlated. In this case, the correction term in Eq. (6.28) should only include the correlated portion of the subsidiary biases, which means that the second term in Eq. (6.28) should be multiplied by a weight with a value between 0 and 1.

Example 6.13 Let T_1, T_2 be, respectively, the temperature values at the inlet and outlet of a pipe flow, measured with liquid-in-glass thermometers having equal bias limits $b_1 = b_2 = b_T$. Compute the bias limits of derived properties for the following cases:

(a) Bias limit of the average temperature $T_{av} = (T_1 + T_2)/2$.

(b) Bias limit of the temperature difference $\Delta T = T_1 - T_2$, when different thermometers are used.

(c) Bias limit of the temperature difference $\Delta T = T_1 - T_2$, when the same thermometer is used for both temperatures.

(d) Bias limit of the temperature difference $\Delta T = T_1 - T_2$, when the same thermometer is used, but two different persons read it at the inlet and the outlet.

Answer

(a) In this case, both sensitivities are positive and so, irrespectively of whether T_1, T_2 are measured by the same or different thermometers, one can use Eq. (6.27) to compute

$b_{T_{av}}$ as

$$b_{T_{av}} = \sqrt{\left(\frac{\partial T_{av}}{\partial T_1} b_1\right)^2 + \left(\frac{\partial T_{av}}{\partial T_2} b_2\right)^2} = \sqrt{\left(\frac{1}{2} b_1\right)^2 + \left(\frac{1}{2} b_2\right)^2} = \frac{\sqrt{2}}{2} b_T .$$

(b) In this case, the two biases are uncorrelated and one can use Eq. (6.27) to compute $b_{\Delta T}$ as

$$b_{\Delta T} = \sqrt{\left(\frac{\partial \Delta T}{\partial T_1} b_1\right)^2 + \left(\frac{\partial \Delta T}{\partial T_2} b_2\right)^2} = \sqrt{(b_1)^2 + (-b_2)^2} = \sqrt{2} b_T .$$

(c) In this case, the two biases are perfectly correlated and, moreover, the two sensitivities have opposite signs. Then, one must use Eq. (6.28), which gives $b_{\Delta T}$ as

$$\begin{aligned} b_{\Delta T} &= \sqrt{\left(\frac{\partial \Delta T}{\partial T_1} b_1\right)^2 + \left(\frac{\partial \Delta T}{\partial T_2} b_1\right)^2 + 2 \frac{\partial \Delta T}{\partial T_1} \frac{\partial \Delta T}{\partial T_2} b_1^2} \\ &= \sqrt{(b_1)^2 + (-b_1)^2 + 2(1)(-1) b_1^2} = \sqrt{b_T^2 + b_T^2 - 2 b_T^2} = 0 . \end{aligned}$$

(d) It has been observed that each person introduces a bias in temperature measurement by tending to offset readings on the glass scale upwards or downwards. Thus, there are two sources of bias, the thermometer and the reader, but only the thermometer biases are correlated in the present experiment. The bias limit b_T of each temperature measurement can be computed from the thermometer bias limit b_{th} and the reader bias limit b_r as

$$b_T = \sqrt{b_{th}^2 + b_r^2} .$$

Then, Eq. (6.28) gives $b_{\Delta T}$ as

$$\begin{aligned} b_{\Delta T} &= \sqrt{\left(\frac{\partial \Delta T}{\partial T_1} b_{th}\right)^2 + \left(\frac{\partial \Delta T}{\partial T_1} b_r\right)^2 + \left(\frac{\partial \Delta T}{\partial T_2} b_{th}\right)^2 + \left(\frac{\partial \Delta T}{\partial T_2} b_r\right)^2 + 2 \frac{\partial \Delta T}{\partial T_1} \frac{\partial \Delta T}{\partial T_2} b_{th}^2} \\ &= \sqrt{(b_{th})^2 + (b_r)^2 + (-b_{th})^2 + (-b_r)^2 + 2(1)(-1) b_{th}^2} = \sqrt{2} b_r . \end{aligned}$$

Now, let us discuss how to determine the precision limit of a derived property. When a derived property is estimated from a set of single subsidiary measurements x_m, its precision limit can be estimated from the precision limits of x_m as

$$p_y = \sqrt{\sum_{m=1}^{M} \left(\frac{\partial y}{\partial x_m} p_{x_m}\right)^2} . \tag{6.29}$$

As in the case of directly measured properties, however, the precision uncertainty of a derived property y can be reduced substantially by averaging a number of estimates of y based on repeat subsidiary measurements. Let us assume that $N > 30$ sets of subsidiary measurements $(x_1, x_2, ..., x_M)_i, i = 1, 2, ..., N$ are available, providing N estimates of y

as $y_1, y_2, ..., y_N$. Then, one may compute the average and variance of the sample as, respectively,

$$\mu_{yN} = \frac{1}{N}\sum_{i=1}^{N} y_i \tag{6.30}$$

and

$$\sigma_{yN}^2 = \frac{1}{N-1}\sum_{i=1}^{N}\left(y_i - \mu_{yN}\right)^2 . \tag{6.31}$$

If one uses μ_{yN} as the estimate of y, rather than a single value y_i, the precision limit should be estimated as

$$p_y = \frac{t_\nu \sigma_{yN}}{\sqrt{N}} . \tag{6.32}$$

Subsidiary and derived random variables: In general, all measured and derived properties are random variables. The uncertainty of any random variable is defined statistically, usually under the assumption that the random variable has a normal distribution. The statistical distribution of a derived property is related to the distributions of the subsidiary properties, in a manner discussed in Section 4.2.3, where an example was given. As a simpler example, consider a derived property, which is equal to the square of a normally distributed subsidiary property. The derived property would would not have a normal distribution, but a chi-square one with one degree of freedom. It is evident that, even if all subsidiary properties had a normal distribution, the derived property would, in general, have a different one, except under conditions for which the central limit theorem applies. Consequently, the confidence limits of derived uncertainties would not, in general, coincide with the confidence limits of subsidiary uncertainties. In practice, however, this distinction is disregarded and one considers that derived uncertainties are at the same confidence level as subsidiary ones. For example, if all subsidiary uncertainties are at the 95% confidence level, one typically considers that the derived uncertainty is also at the 95% confidence level. In some cases, this presumption may have adverse consequences, but these are beyond the present scope.

6.2.5 Estimating Uncertainty from Calibration Data

In general, any measurement, including calibration measurements, would have both systematic and random errors. When the experiment and calibration are performed in the same apparatus (e.g., a hot-wire calibration in the same wind tunnel that is used for the main experiments), it is more likely that the sources and level of random errors would be the same, than when the calibration is performed in a dedicated apparatus under well controlled conditions. In any case, determining the overall measurement uncertainty during an in-house calibration is a useful enterprise, because, in the absence of better approaches, it can provide an estimate of the overall measurement uncertainty during an experiment.

For simplicity, we will assume, once more, that the calibration data are fitted by the line $x = ay + b$. The uncertainty in x depends on the uncertainties in the regression coefficients a and b, which makes x a derived property and a, b subsidiary properties. Nevertheless, we cannot use Eq. (6.29), which is based on the assumption that all subsidiary properties are statistically independent, because the coefficients in the regression fit depend on each other's value. Instead, we may determine the uncertainty in the derived property x for a measurement y within the calibration range as [4]

$$u_x = s_{y,x} t_{N-2} \sqrt{\frac{N+1}{N} + \frac{(y_m - \bar{y})^2}{\sum y_i^2 - N\bar{y}^2}} \,. \tag{6.33}$$

Similar approaches can be used to determine the uncertainty from other types of fitted functions, obtained by least square regression [3, 4]. In practice, there is no need to perform such analyses, because many scientific software readily provide these uncertainties to the user. Note that the presented uncertainty in the calibrated value does not consider the uncertainty in the measurement y itself, which would have to be determined separately and added to the uncertainty due to calibration.

Example 6.14 Using the calibration data shown in Example 6.7, determine the uncertainty of the estimated pressure due to the calibration, within a 95% confidence interval, for a voltage reading of 2.3 V. Assume a linear calibration fit of the form $\Delta p_l = aE + b$.

Answer

Using linear least square fits, we obtain a calibration function of $\Delta p_l = 201.36E - 13.41$; hence the expected pressure for a read voltage of $E_i = 2.3$ V is 449.73 Pa. For a linear fit with two unknowns ($m = 2$) and $N = 11$ calibration points, the t-score is 2.262. The standard error of the estimate (Eq. (4.46)) is $s_{y,x} = 6.66$ Pa. Finally, $\bar{E} = 2.5$ V, and $\sum E_i^2 = 96.25V^2$. Inserting these values into Eq. (6.33) gives an uncertainty of $u_{\Delta p} = 15.73$ Pa. For a voltage reading of 2.3 V, the corresponding pressure value, with its uncertainty (based on the calibration data), would be 450 ± 16 Pa, which have been rounded according to the guidelines presented in the next section.

6.3 Rounding of Reported Values

6.3.1 Significant Digits

Readings of digital instruments, outputs of calculators and computers, results of mathematical operations and values in one system of units obtained via conversion of values in another may appear in an arbitrary number of digits, but, in general, this is not the way these values should be reported. When presenting numerical results of an experiment, whether direct measurements or calculated values, one must report all values rounded to the appropriate number of *significant digits*, which are also referred to as *significant figures*.

The number of significant digits may be different from the actual number of digits appearing in the value, depending on whether the value contains zeros. To determine the number of significant digits of a value, one may use the following conventions:

- All non-zero digits are always significant. For example, 234 has three significant digits.
- Zero is always significant, when it is between two non-zero digits. For example, 2,305 has four significant digits.
- Zero is usually not significant, when it appears before all other digits in a decimal number that has an absolute value less than 1. However, the same zero may be significant, if the number in which it appears is part of a sequence in which other numbers have non-zero digits of the same decimal rank as the zero of interest. For example, 0.0123 may have three, four or ever more significant digits. It has three significant digits, if it is part of the sequence 0.0645, 0.0034, 0.0456, and four significant digits, if it is part of the sequence 0.02345, 0.01234, 0.00456, in which case, it is preferable to write this number as 0.01230. Zero is usually significant, when it appears as the last digit after the decimal point. For example, 2.0 has two significant digits and 2.00 has three significant digits.
- When it appears at the end of a number, zero may or may not be significant, depending on the sequence of numbers it is part of. To indicate unambiguously the number of significant digits, one should present this number in engineering notation, namely as a number with the proper number of digits times a power of 10. For example, a measured reading of 53,400 may have three, four or five significant digits. To declare that it has three significant digits, one should write this number as 53.4×10^3.

6.3.2 Rounding Conventions

A number is said to be rounded to N significant digits, if all digits to the right of the Nth digit are replaced by zeros or discarded. For example, 123,456 is rounded to three significant digits, if it is reported as 123,000 (or, better, as 1.23×10^5), and to two significant digits, if it is reported as 120,000 (or, better, as 1.2×10^5). The rounding of experimental values should be done according to the following convention:

- If the discarded $(N + 1)$th digit is between 0 and 4 inclusive, the Nth digit should be left as is.
- If the discarded $(N + 1)$th digit is between 6 and 9 inclusive, or it is 5 followed by at least one non-zero digit, the Nth digit should be incremented by one.
- If the discarded $(N + 1)$th digit is 5 and it is not followed by other digits, or followed only by zeros, the Nth digit should be left as is, if even, and incremented by one, if odd.

Example 6.15 Measurements from a particular experiment produced the following readings: 0.2452, 3.672, 0.366, 0.3250 and 0.435. Round the values to two decimal digits.

Answer

Following the previous conventions, 0.2452 should be rounded to 0.25; 3.672 should be rounded to 3.67; 0.366 should be rounded to 0.37; 0.3250 should be rounded to 0.32; and 0.435 should be rounded to 0.44.

An overview of literature indicates that there are no universally enforced rules for determining the number of significant digits that appear in reported experimental or analytical results. Conventions vary with area of application and it appears that number rounding is often left to the discretion of the experimenter. Nevertheless, it is self-evident that a reported value of a directly measured property should be rounded to a number of digits that is consistent with the accuracy of measurement, as expressed by its uncertainty. In the following, we will present some recommendations for rounding experimental results, which conform with good experimental practices.

Rounding of uncertainty: An estimated uncertainty value is necessarily approximate and should preferably be presented in one significant digit, for example, as 20, 0.3 and 0.006 (units have been omitted). In some cases, especially if the first significant digit is 1, one may report uncertainty in two significant digits, as for example, 17 and 0.025. It is the decimal rank of the last significant digit of the uncertainty, which sets the way experimental values should be rounded. For example, this rank is tens for the number 30 and hundredths for the number 0.06.

Rounding of measured values: As a rule, a directly measured value should be reported such that its last significant digit is of the same rank as the last significant digit of its uncertainty. Examples: 256 ± 3 and 0.234 ± 0.005. When, however, a directly measured property is used to compute derived properties through mathematical operations, it is recommended to retain one or two additional significant digits, in order to avoid excessive rounding off of the derived property.

Rounding of derived values: The accuracy of a derived property depends on the accuracies of the subsidiary properties, which are used for the determination of this derived property, as well as on the corresponding sensitivities. Therefore, the way the derived property value is reported must be consistent with the accuracies and the sensitivities of the subsidiary properties. One may intuitively expect that the accuracy of the derived property is set by the least accurate subsidiary property (the 'weak link'), but this may not be the case when the corresponding sensitivity is very small. Thus, one must examine the relationship between derived and subsidiary values. We will do so for two common types of relationships: summation and multiplication.

Rounding of sums: For the purpose of adding or subtracting decimal numerals, the *most specific* among them is not the one with the most significant digits, but the one for which the rightmost significant digit corresponds to the lowest power of 10 (note that we use the word specific, rather than accurate or precise, to avoid confusion with other sections of this book). For example, among the three numbers 33.5, 78,974 and 0.335, the most specific, for the purposes of addition, is 0.335 (having a rightmost significant digit that corresponds to 10^{-3}) and the least specific is 78,974 (having a rightmost significant digit that corresponds to 10^{0}), despite the fact that the latter has more significant digits than the two other numbers. As a rule, the sum of different numbers should be rounded so that its rightmost significant digit corresponds to the same power of 10 as that of the least specific number. One may wonder whether values should be rounded or not before addition. Some literature recommends to round the more specific numbers to one rightmost significant digit that corresponds to one power of 10 lower that that of the least specific number. For the previous example, this means that, to add these three numbers, one must first round 33.5 to 33.5 (no change) and 0.335 to 0.3, then calculate the sum as 79,007.8 and round it to 79,008 for final presentation. A different approach is to add the numbers as they are given and then round the sum. For our example, the sum would be 79,007.835, which should also be rounded to 79,008. Although it is possible that the two methods give slightly different results, especially if a large number of values are added, it seems preferable to adopt the second approach, which does not require intermediate operations.

Rounding of products: For the purposes of multiplication and division of different numbers, the least specific number would be the one with the smallest number of significant digits. For example, the most specific among 33.5, 78,974 and 0.335 is 78,974, which has five significant digits, and the least specific are both 33.5 and 0.335, which have three significant digits. As a rule, the product or quotient of different numbers should be rounded to the same number of significant digits as that of the least specific among the multiplied or divided numbers. Once more, there are two approaches for performing intermediate operations. In one approach, the more specific numbers should be first rounded to one significant digit more than the number of significant digits of the least specific number, whereas, in another approach, the numbers should be multiplied or divided in the same form as they are given. For the previous example, the first approach requires that, before multiplying these three numbers, one should first round 78,974 to 78,970, calculate the product as 886,240.825 and round it as 886,000, or, preferably, present it as 0.886×10^6. For the second approach, we compute the product as 886,285.715 and round it again as 0.886×10^6. The two approaches may give the same or slightly different final result, but the second approach seems to be preferable for the sake of economy.

6.4 Application

In closing the discussion on uncertainty and rounding the reported values, we will present an example of the application of the presented principles and conventions to a basic experiment in fluid mechanics.

Example 6.16 An experiment consists of measurements of water flow in a smooth circular tube. The following measurements were made:

- The tube diameter, measured with a caliper that has an uncertainty $u_D = 0.005$ in, was $D = 2.571$ in.

- The volume flow rate, measured with a calibrated ultrasonic flow meter that has a total uncertainty $u_Q = 0.01$ l/s, was $Q = 0.143$ l/s.
- The water temperature, measured with a digital thermometer that has an uncertainty $u_T = 1$°C, was $T = 61.23$°C.
- The kinematic viscosity ν of water for different temperatures is found in the following table.

T [°C]	0	5	10	20	30	40
ν [m^2/s $\times 10^{-6}$]	1.787	1.519	1.307	1.004	0.801	0.658
T [°C]	50	60	70	80	90	100
ν [m^2/s $\times 10^{-6}$]	0.553	0.475	0.413	0.365	0.326	0.294

Determine the Reynolds number and its uncertainty. While performing your calculations, respect the conventions for rounding numbers, as described in Section 6.3.

Answer

The Reynolds number, expressed in terms of directly measurable or estimable properties, is

$$\mathrm{Re} = \frac{UD}{\nu} = \frac{4Q}{\pi D \nu(T)} ,$$

where $U = Q/(\pi D^2/4)$ is the bulk velocity. We first find the appropriate values of the subsidiary properties and their uncertainties in SI units as follows:

- The measured value of D is given as 2.571 in, the last digit of which is of the same rank as its given uncertainty of 0.005 in. When converted to SI units and rounded to a single digit, the uncertainty of D is $u_D = 0.127$ mm ≈ 0.1 mm $= 1 \times 10^{-4}$ m. Conversion of D to SI units and rounding of it in a manner consistent with its uncertainty gives $D = 2.571 \times 0.0254$ m $= 0.0653034$ m ≈ 0.0653 m $= 65.3 \times 10^{-3}$ m.
- The measured volume flow rate was given as $Q = 0.143$ l/s. The rank of the last decimal digit of this value (namely, the number 0.003) is lower than the rank of the uncertainty $u_Q = 0.01$ l/s and so, in the presentation of the results, Q must be rounded to 0.14 l/s. Conversion to SI units gives $Q = 0.14 \times 10^{-3}$ m^3/s and $u_Q = 0.01 \times 10^{-3}$ m^3/s.
- The kinematic viscosity ν is not measured directly, but its value can be estimated by interpolation of values in the given table at the measured temperature T. An empirical function $\nu(T)$ can be found by fitting a polynomial to the data in the table. One should not select a polynomial of very high order, because it may provide unrealistic interpolated values and particularly erroneous values of the derivative, which is needed in this problem. A fifth-order polynomial, fitted with MATLAB, appears to represent well the data in the table (Fig. 6.7). Another point of caution is that some other popular software, which may be used for fitting polynomials to data, report coefficients of high order terms with a single significant digit, which produces an error, particularly for large values of the independent variable. Evaluated at the measured temperature, the polynomial provides the value $\nu = 0.47 \times 10^{-6}$ m^2/s. This value

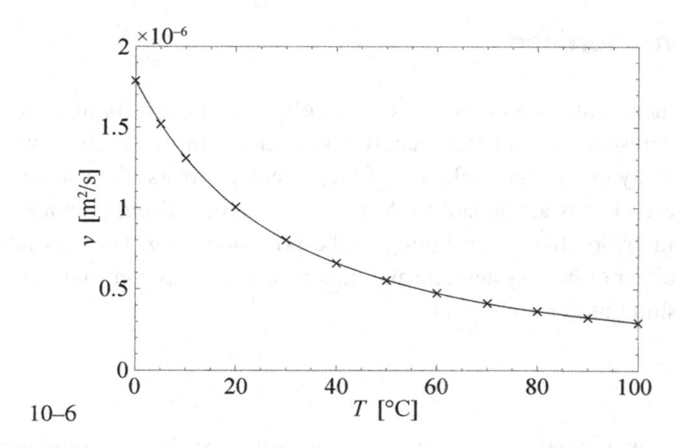

Figure 6.7 Fifth-order polynomial (solid line) fitted to values of the kinematic viscosity of water (×) at different temperatures.

was rounded retroactively to be consistent with its uncertainty, which was calculated as

$$u_\nu = \left| \frac{\partial \nu}{\partial T} u_T \right| = | - 0.0070 \times 1 \times 10^{-6} | \text{ m}^2/\text{s} \approx 0.01 \times 10^{-6} \text{ m}^2/\text{s} .$$

Now, we proceed to compute the Reynolds number and its uncertainty, considering that this number is a derived property found by multiplying or dividing subsidiary properties. Thus, its relative uncertainty can be found from Eq. (6.25) as

$$\frac{u_{\text{Re}}}{\text{Re}} = \sqrt{\left(\frac{u_Q}{Q}\right)^2 + \left(\frac{u_D}{D}\right)^2 + \left(\frac{u_\nu}{\nu}\right)^2} = \sqrt{\left(\frac{0.01}{0.14}\right)^2 + \left(\frac{0.1}{65.3}\right)^2 + \left(\frac{0.01}{0.47}\right)^2} = 0.0745 .$$

We may follow two approaches for intermediate calculations, either to use all subsidiary values as given and round the result at the end, or to round these values consistently first and further round the final result.

If we do not perform any preliminary rounding of subsidiary values and substitute these values as presented previously, we find Re = 5,808 and u_{Re} = 433. With the uncertainty rounded to one significant digit (i.e., to 400) and Re rounded for consistency with its uncertainty, the result should be presented as

$$\text{Re} = 5{,}800 \pm 400 .$$

To perform a preliminary rounding of subsidiary values, we note that, among the values of Q, D and ν, the ones of Q and ν are the least specific for the purposes of multiplication and division, because they have a lower number of significant digits. The convention is to round the more specific numbers to one significant digit more than that of the least specific number. In this particular case, however, this condition is already satisfied and so no preliminary rounding is required. Therefore, both approaches give the same results.

6.5 Noise and Interference

Measurement uncertainty is a measure of the likely magnitude of both systematic and random errors, irrespectively of their nature and sources. In this section, we will discuss two commonly encountered classes of measurement errors, the sources of which are known. These classes are broadly referred to as *noise* and *interference*. Noise introduces random errors, thus contributing to the precision limit, whereas interference may introduce either or both systematic and random errors, thus contributing to the bias limit, the precision limit or both.

6.5.1 Noise

The electric voltage (signal) at the output of a measuring system or component would generally fluctuate in time. Part of the fluctuations may be inherent to the measurand (e.g., the velocity in an unsteady or turbulent flow), but another part, which is broadly termed as *noise*, is due to various, internal or external, undesirable influences on the measuring device. The noise is generally random and cannot be known precisely, so that its effects are analysed statistically. The *signal-to-noise ratio* S/N may be defined as the ratio of the average value of the signal and the standard deviation of the noise [8]. It may also be expressed in dB (Section 3.2.2), in which case it is defined as

$$\left.\frac{S}{N}\right|_{dB} = 20\log_{10}\frac{\text{signal amplitude}}{\text{noise amplitude}} = 10\log_{10}\frac{\text{signal power}}{\text{noise power}} . \tag{6.34}$$

A related parameter is the *noise figure*, defined as the signal-to-noise ratio referred to the input, normalised by the signal-to-noise ratio referred to the output,

$$F = \frac{S/N_{input}}{S/N_{output}} . \tag{6.35}$$

Noise is best considered in the frequency domain. There are many sources of noise, depending on the type of measured property, the measuring system design and the measuring conditions [7, 8]. Some of the most common types of noise will be discussed briefly in the following.

Thermal noise: A type of noise that is ubiquitous in electric/electronic components, is *thermal* or *Johnson noise*, which is due to fluctuations of the electron gas in a conductor caused by random thermal motions. This type of noise has contributions from fluctuations over a wide range of frequencies, including very high frequencies. A common assumption is to consider that thermal noise is *white*, namely having a uniform energy distribution over the entire frequency spectrum (Section 5.3), which is a non-physical condition, as it implies that white noise has an infinite energy. In practice, however, the bandwidth of any circuit would be finite. Thus, actual circuits generate *band-limited white noise*, which has a finite energy, which is measured by the variance of the corresponding noise signal. The variance of noise in the voltage across a resistor R at a temperature T within a bandwidth Δf has been estimated as

$$\sigma_V^2 = 4kTR\Delta f , \tag{6.36}$$

where $k = 1.38 \times 10^{-23}$ J/K is the *Boltzmann constant*. Thermal noise cannot be eliminated and the only way that it can be reduced is by filtering the signal, usually with the use of a low-pass filter. It is obvious that, the narrower the frequency band of the signal is, the lower the noise energy would be. In practice, the optimal filter choice would be such as to produce a signal that contains as much as possible of the fluctuations of the measurand and as little noise as possible. The choice of filter settings may require a detailed analysis of the measuring system, followed by trial-and-error-type experimentation. White noise has a normal pdf and is uncorrelated not only with the signal, but also with its own value at a different time (Example 5.5). This allows one to apply statistical corrections to noisy signals, if it is possible to characterise statistically the noise of a circuit in the absence of the signal.

Example 6.17 Consider a signal $x(t)$, which consists of mean-free fluctuations $u(t)$ that correspond to a turbulent flow velocity and band-limited, Gaussian white noise $n(t)$. With all properties converted to equivalent velocities, the instantaneous values are related as

$$x(t) = u(t) + n(t) .$$

Assume that all random processes are stationary and ergodic, so that ensemble averages can be computed as time averages. Further assume that the noise has been characterised statistically by performing measurements in a steady laminar flow that has the same mean but no velocity fluctuations, with the use of the same measuring system and the same experimental settings (e.g., low-pass analogue filter cut-off frequency and sampling rate). Derive corrections for the flow velocity variance and skewness factor.

Answer

For simplicity, we will present the analysis as if it were performed on continuous functions, although it is actually performed on time series. As all functions are mean-free, statistical moments coincide with corresponding central moments. All averages, to be denoted by overbars, will be computed by integrating corresponding functions over a sufficiently long time interval T, as, for example,

$$\overline{u^m n^k} \approx \frac{1}{T} \int_0^T u^m(t) n^k(t)\, dt .$$

We may confidently assume that the noise is statistically independent of the velocity fluctuations, which implies that covariances (e.g., $\overline{un}$) between noise and velocity fluctuations vanish.

To compute the velocity variance, we decompose the signal variance as

$$\overline{x^2} = \overline{(u+n)^2} = \overline{u^2} + \overline{n^2} + 2\overline{un} = \overline{u^2} + \overline{n^2} .$$

With $\overline{n^2}$ known from the measurements in laminar flow, we may thus correct the signal variance to estimate the velocity variance as

$$\overline{u^2} = \overline{x^2} - \overline{n^2} .$$

We note that, in all cases, $\overline{u^2} < \overline{x^2}$.

To compute the skewness factor, we first decompose the signal third moment as

$$\overline{x^3} = \overline{(u+n)^3} = \overline{u^3} + \overline{n^3} + 3\overline{u^2 n} + 3\overline{un^2} \, .$$

Because of the statistical independence of noise from velocity fluctuations, $\overline{u^2 n} = \overline{un^2} = 0$. Moreover, for Gaussian noise, $\overline{n^3} = 0$. Then,

$$\overline{u^3} = \overline{x^3}$$

and the velocity skewness factor can be found as

$$S_u \equiv \frac{\overline{u^3}}{\overline{u^2}^{3/2}} = \frac{\overline{x^3}}{\left(\overline{x^2} - \overline{n^2}\right)^{3/2}} \, .$$

This analysis demonstrates that noise affects different statistical properties in different ways and that noise effect may not be cancelled by normalisation. In the present case, for example, the velocity skewness factor is always larger, in absolute value, than the signal skewness factor, namely, $|S_u| > |S_z|$.

Shot noise: *Shot noise* occurs when the measuring system contains devices that collect electrons at an electrode or divert them over a barrier. Such devices include photomultiplier tubes, junction diodes and junction transistors (Section 7.5.2). Like thermal noise, shot noise may be termed as white. The collected or diverted electrons form a current with a mean i and a random component that, when limited to the bandwidth Δf, has a variance

$$\sigma_i^2 = 2 q_o i \Delta f \, , \tag{6.37}$$

where $q_o = 1.6 \times 10^{-19}$ C is the elementary electric charge. Once more, low-pass filtering is the usual means of reducing the effect of shot noise.

$1/f$ **noise:** Besides white noise, measuring systems are sometimes subjected to noise that is concentrated in particular frequency ranges. A type of noise that is not well understood, but is present in semiconductors, is $1/f$ *noise*, sometimes also referred to as *pink noise*, which has equal energy within each frequency octave. In electronic devices, $1/f$ noise is generated by slow fluctuations in material properties, a condition referred to as *flicker noise*, which seems to decrease with time, as component design and manufacturing keep improving. The variance of $1/f$ noise may be estimated from *Hooge's law*

$$\sigma_V^2 = \frac{\alpha}{N_f} \frac{\Delta f}{f} \, , \tag{6.38}$$

where $\alpha = 2 \times 10^{-9}$ W/m^3 is an empirical constant and N_f is the concentration of free charges (electrons or holes) in the semiconductor. It is clear that $1/f$ noise increases with decreasing frequency, and, although it can be reduced by filtering, its contamination of slowly-varying signals cannot be eliminated.

6.5.2 Interference

A source of distortion of electric signals is various interfering inputs, which could be internal in the circuitry or external [7, 9]. Some common types of interference are discussed in the following.

Thermoelectric voltages: A problem that is especially important in low-voltage applications is *thermoelectric voltages*, created when junctions of different materials (e.g., wire junctions, mating connectors and even solder used in circuitry) are exposed to different temperatures. This problem can be reduced by a careful selection of materials and by avoiding thermal gradients in the equipment.

Ground loops: A very common problem in electronic circuitry is the generation of *grounding currents*. Such currents flow between two *grounded* points in the circuit, which ideally should be at the same electric potential, namely the *ground*, but, in practice, have a potential difference, because they are connected to the external ground through connectors with a finite resistance. This creates *ground loops*, namely, multiple paths towards the external ground, with associated undesirable currents. The proper *grounding* of circuits often proves to be a tedious and unpredictable ordeal.

Electromagnetic waves: Electric circuits may be coupled to electromagnetic fields, generated by nearby transformers, electric motors, and even transmitters of radio, television and other types of electromagnetic waves. This coupling becomes intensified by the presence of stray capacitances and inductances in the circuit, the source of which may be traced to loose connections or looped wires. Sensitivity to electromagnetic waves would create undesirable fluctuating voltages that interfere with the signals of interest. The most common interference is from electric power lines, at a frequency of 60 Hz in North America or 50 Hz in Europe and elsewhere. When a signal is transmitted in the form of a voltage difference through a pair of conductors, both the positive and the negative side may be subjected to the same interference, called *common mode*. Thus, it may happen that a very weak signal is superimposed on a very strong common mode. Ideally, a differential amplifier should remove the common mode, but, in practice, due to input asymmetry and offsets, part of the common mode may be amplified and appear at the output. The term *common-mode error* [9] signifies the residual common-mode voltage at the output of the amplifier, divided by the amplifier's gain, so that it is presented in the form of noise superimposed to the input voltage.

The elimination of interference could be a very difficult problem; although suggestions for proper circuit design and other rules of thumb, including shielding, guarding and grounding [2, 7, 9]) are available, this process usually requires trial-and-error-type troubleshooting and adjustment. Even the most experienced experimenter must be prepared to spend many hours, or days, in tracing and minimising interference effects.

Chapter Digest

6.1 What is the true value? Can it be known?

6.2 What is error? Can it be known?

6.3 What is the difference between inaccuracy and uncertainty?

6.4 The objective of an experiment is to measure the drag coefficient on a circular cylinder inserted in a wind tunnel as a function of the flow Reynolds number. You purchase a cylinder from a supplier, which specifies the *nominal* diameter of the cylinder as 50 mm. Discuss how a possible difference between the true diameter of the cylinder and the nominal diameter will introduce a systematic error in this experiment.

6.5 Provide an example of a biased but precise measurement process and an example of an unbiased but imprecise one.

6.6 What is bias error? Provide an example.

6.7 What is precision error? Provide an example.

6.8 Explain the difference between bias and precision errors.

6.9 What happens to the bias and precision errors when you estimate a property from the average of repeat measurements, rather than a single measurement?

6.10 When would you consider a measurement to be accurate?

6.11 What is repeatability? How is it related to accuracy?

6.12 What is reproducibility? How is it related to accuracy?

6.13 Define uncertainty.

6.14 Define bias limit and precision limit.

6.15 How can you calculate the uncertainty of a derived property from the uncertainties of its subsidiary properties?

6.16 How can you account for subsidiary properties that have the same bias?

6.17 What is propagation of uncertainty? What does it depend on?

6.18 How would you round the sum of two numbers with different uncertainties?

6.19 How would you round the product of two numbers with different uncertainties?

6.20 Is it possible to have an entirely noise-free electronic instrument? Explain your answer.

6.21 Assume that you have a noisy random signal. Would its skewness and flatness factors be insensitive to the noise level, because they are normalised and both their numerators and denominators would contain some correction?

6.22 What is the difference between noise and interference?

Problems

6.1 The manufacturer of a load cell (i.e., a force sensor) has provided the following specifications:

Span: 5 kN

Linearity: ±0.1% FS

Sensitivity: 2.25 N/mV ±0.23% RDG

Hysteresis: ±0.01% FS.

Estimate the bias limit, both in value and in full scale, for a voltage reading of 1.03 V.

6.2 You calibrate a load cell that you find in the lab by loading it with known, accurate weights F, and measure the voltages V shown in the following table.

F [N]	0.0	2.5	5.0	7.5	10.0	12.5	15.0	20.0
V [V]	0.01	0.07	0.27	0.61	1.08	1.67	2.34	4.11

Estimate the bias limits due to the calibration fit, if the calibration data have been fitted by the following relationships:

(a) A linear relationship of the form $F = aV + b$.

(b) A power law relationship of the form $F = aV^b + c$.

(c) A quadratic relationship of the form $F = aV^2 + bV + c$.

Based on your results, which calibration relationship do you recommend?

6.3 For the load-cell calibration discussed in Problem 6.2, determine the uncertainty associated with each of the calibration functions for a read voltage of 3.42 V.

6.4 To determine the hysteresis characteristics of a load cell, we first apply loads F in an increasing order and measure the output voltages V shown in the table that is included in Problem 6.2. In addition, we also apply loads in a decreasing order and obtain the results shown in the following table.

F [N]	20	15	12.5	10	7.5	5	2.5	0
V [V]	4.11	3.08	2.55	2.03	1.52	1.03	0.52	0.01

Based on both data sets, determine the linearity, sensitivity and hysteresis errors, as well as the corresponding bias limits.

6.5 Assume that the response relationship of a measuring system is actually $y = 20.25x^{1.1}$; however, you are not aware of this and perform a calibration to determine such a relationship. You collect N equally spaced data points in the range $0 \leq x \leq 5$ and fit a line of the form $y_l = cx$ to them.

(a) How does the linearity error change, as the number of calibration points increases?

(b) Plot the pdf of the difference between the actual and linearised responses. Comment on the shape of this pdf and compare it to the Gaussian pdf.

(c) Determine a suitable bias limit for the linearity, assuming a 95% confidence level.

6.6 King's law, $E^2 = A + BU^n$, is a semi-theoretical relationship between the voltage E of a hot-wire probe and the flow velocity U. You calibrate the probe by immersing it in flows with a known velocity and record the voltage for each velocity; the calibration data are shown in the following table.

U [m/s]	1.60	1.97	2.37	2.70	3.08	4.06	5.11	6.29	7.49	9.25
E [V]	1.50	1.53	1.56	1.58	1.61	1.66	1.71	1.75	1.79	1.84

To determine the flow velocity during an experiment, you need an expression that provides U as a function of E. King's law can be easily solved for U to provide such a relationship. An alternative approach is to fit the calibration data with the fourth-order polynomial $U = aE^4 + bE^3 + cE^2 + dE + e$, where the empirical coefficients $a, ..., e$ are determined by linear regression.

(a) Based on the given calibration data, which calibration function has the lowest uncertainty?

(b) Which calibration function imposes the lowest uncertainty on the fluctuations? Consider the more accurate fit in (a) to be the true fit to the data.

6.7 In a simple experiment, you wish to measure the Joule power consumed by an electrically heated resistor. You have available three instruments, a voltmeter, an ammeter and an ohmmeter, all of which have the same measuring uncertainty, expressed as a percent of the measured value. Would the power uncertainty depend on which instruments you use? If so, which instruments do you propose to use?

6.8 Consider a $3^1/_2$ digit digital voltmeter with a full scale of 10.00 V (this means that this voltmeter reads values at increments of 0.01 V). The manufacturer has specified that the bias limit for this instrument is 0.3% of full scale. Repeat readings of a voltage using this voltmeter are as follows: 9.52, 9.65, 7.39, 7.00, 7.84, 8.59, 7.03, 9.60, 9.00, 8.74 V.

(a) Determine the uncertainty of this measurement, if (i) a single reading is used (for this case, make a drastic approximation, explaining your logic) and (ii) the average of all previous repeat readings is used. Present the uncertainty values rounded appropriately. Round the first value of the measured voltages (namely, 9.52 V) to the appropriate number of significant digits.

(b) What can you say about the error of this measurement? Can you determine the value of the error? Can you determine the accuracy of this voltmeter?

(c) In your opinion, are these measurements significantly biased? Are they precise? The answer to this question is subjective and would depend on the requirements for the accuracy of these measurements. Discuss the context of your answer.

6.9 A supplier's catalogue lists a mass-produced resistor as having a nominal resistance of 2×10^3 Ω and a 10% uncertainty (assume that this includes both bias and precision uncertainties). A voltmeter with an uncertainty of 1 mV is used to measure the voltage across the resistor, from which the electric current is estimated using Ohm's law. The reading of the voltmeter is 3.650 V.

(a) Which should be the number of significant digits of this resistance and how should the resistance value be written in an unambiguous way using engineering notation?

(b) Estimate the current through the resistor, rounded to the appropriate number of significant digits.

(c) Apply propagation of uncertainties analysis to estimate the uncertainty of electric current, rounded to the appropriate number of significant digits.

6.10 We have a set of 100 thermocouples, each connected to an electronic circuit with a digital readout showing a number in three decimal digits (e.g., 12.345), that ideally corresponds to a temperature in degrees Celsius. We also have an insulated flask that contains a well-mixed bath of water and ice.

(a) We pick randomly one of these thermocouples and dip it into the bath. We take a single reading of the readout, which is 0.553. Determine the error of this reading. Can you estimate the systematic and random errors of this reading?

(b) We record a large number of readings with the same thermocouple and find that the set of these readings has an approximately Gaussian pdf with a mean 0.364 and a standard deviation 0.421. Determine the systematic and random errors of the previous reading (i.e., 0.553).

(c) Determine the precision limit of this particular thermocouple system, if (i) a single reading is used and (ii) the average of 10 readings is used.

(d) Now, we take, in turn, each of the 100 thermocouples, dip it into the bath and average a large number of readings of this particular unit. The mean readings of all units are found to have an approximately Gaussian pdf with a mean −0.359 and a standard deviation 0.603. Determine the systematic error of this set. How can you use the systematic error to improve statistically the accuracy of any reading by any thermocouple? Next, determine the bias limit of this set, when each thermocouple is used independently from the others.

(e) Assume that any single reading of any single thermocouple in any environment has a systematic error and a bias limit as those found in (d) and a precision limit as that found in (c). Determine the uncertainty of such readings, rounded to the appropriate number of significant digits. Rationalise your choice.

(f) We take a randomly picked thermocouple, place it in air flow and take a single reading, which is 34.563. Correct this reading, if necessary, and round it appropriately. How would you report this reading in an engineering report?

6.11 To measure its resistance, you connect a resistor to a DC power supply and measure the voltage across it and the current through it. A voltmeter with an uncertainty of 0.01 V gives you a reading of 10.435 V and an ammeter with an uncertainty of 0.01 mA gives a reading of 0.845 mA.

(a) If you were to report the voltage and the current in an engineering report, how would you report them?

(b) Determine the resistance value and its uncertainty, both rounded to the appropriate number of significant digits.

6.12 The discharge coefficient of a concentric orifice plate meter with corner pressure taps for large Reynolds number flows is given by the empirical relationship

$$C = 0.5959 + 0.0312\beta^{2.1} - 0.1840\beta^{8.0} ,$$

obtained by curve fitting to a large number of measurement data. The orifice diameter ratio is defined as $\beta = d_t/d$, where d_t is the throat diameter and d is the pipe diameter. It is given that $d_t = 50.00 \pm 0.50$mm and $d = 65.00 \pm 0.50$ mm (95%-confidence uncertainties). Determine the absolute and relative uncertainties of the estimated discharge coefficient for the following two cases: (i) the given diameter uncertainties are independent and include both bias and precision errors, and (ii) the given diameter values have been measured with the same instrument, whose total uncertainty is 0.50 mm and whose bias limit is 0.25 mm. For this analysis, it may be assumed that the coefficients in the previous expression have negligible uncertainty. Discuss the validity of this assumption and suggest a procedure by which one may include the uncertainty of the curve-fitting process into the overall uncertainty of the discharge coefficient.

6.13 Convective heat transfer in fully developed, turbulent flow through a smooth circular tube can be estimated by the so-called Dittus–Boelter empirical expression

$$\mathrm{Nu} = a\mathrm{Re}^b\mathrm{Pr}^c ,$$

in which the Nusselt, Reynolds and Prandtl numbers are defined, respectively, as $\mathrm{Nu} = hd/k$, $\mathrm{Re} = Ud/\nu$ and $\mathrm{Pr} = \nu/\gamma$; h is the heat transfer coefficient; d is the diameter of the tube; k is the thermal conductivity; U is the bulk velocity; ν is the kinematic viscosity;

and γ is the thermal diffusivity (all properties are evaluated at the bulk temperature T); the coefficients a, b, c, determined by curve fitting to highly scattered experimental data plotted in logarithmic coordinates, are specified as $a = 0.023 \pm 0.001, b = 0.80 \pm 0.01, c = 0.40 \pm 0.01$. For the purposes of uncertainty analysis, we may treat these coefficients as subsidiary properties. Consider air flow in a tube for which the following results have been obtained by direct measurement: $U = 40.3 \pm 0.1$ m/s, $d = 105 \pm 1$ mm and $T = 593 \pm 2$ K. In all expressions, ± values indicate 95%-confidence uncertainties. Further assume that the values of k, μ, γ have a relative uncertainty of 1%. Estimate h and its uncertainty, presented with the appropriate number of significant digits.

References

[1] AGARD, Fluid Dynamics Panel Working Group 15. *Quality Assessment for Wind Tunnel Testing*. AGARD Advisory Report 304, North Atlantic Treaty Organization, Neuilly-sur-Seine, France, 1994.

[2] T.G. Beckwith, R.D. Marangoni, and J.H. Lienhard. *Mechanical Measurements (6th Edition)*. Pearson, Upper Saddle River, NJ, 2007.

[3] H.W. Coleman and W.G. Steele. *Experimentation, Validation, and Uncertainty Analysis for Engineers (4th Edition)*. Wiley, Hoboken, NJ, 2018.

[4] A. Hayter. *Probability and Statistics for Engineers and Scientists (4th Edition)*. Cengage Learning, Boston, MA, 2012.

[5] ISO. *Uncertainty of Measurement – Part 3: Guide to the Expression of Uncertainty in Measurement (GUM:1995) ISO/ICE Guide 98-3:2008*. International Organization for Standardization, Geneva, Switzerland, 2008.

[6] NIST/SEMATECH. e-handbook of statistical methods.

[7] H.W. Ott. *Noise Reduction Techniques in Electronic Systems*. Wiley, New York, 1976.

[8] M. Sayer and A. Mansingh. *Measurement, Instrumentation and Experiment Design in Physics and Engineering*. Prentice-Hall of India, New Delhi, 2000.

[9] D.H. Sheingold, editor. *Transducer Interfacing Handbook*. Analog Devices, Inc., Norwood, MA, 1980.

7 Background for Optical Experimentation

Visual and optical techniques occupy a prominent role in experimental fluid mechanics. Their understanding, even at an elementary level, requires some familiarity with concepts of light propagation, which, in a typical undergraduate engineering curriculum, are taught as part of a general physics course, without focus on issues of present concern. This chapter briefly reviews some definitions and other background material that is necessary for explaining the visual and optical methods to be introduced in following chapters. While the various phenomena associated with light propagation have, over the centuries, been the subject of several theories of increasing complexity, some important aspects remain to be explained in an entirely satisfactory manner. Consequently, the present discussion will be of limited scope and focused on the needs of our subject. First, we shall review the fundamental properties of light and the principles of light emission and propagation through media; next, we shall present some common instrumentation used for the generation, conditioning, detection and recording of light; and, finally, we shall discuss the optical and dynamic characteristics of particles that are inserted in fluids to serve as flow markers.

7.1 The Nature of Light

7.1.1 Light as Waves

According to *classical electromagnetic theory*, *light* is considered to be radiation, which propagates through vacuum in free space in the form of electromagnetic waves, oscillating both transversely to the direction of wave propagation and normal to each other, as illustrated in Fig. 7.1 [24, 52]. The intensities of the electric and magnetic fields E_y and B_z oscillate harmonically both in time t and along their direction of propagation x, as, respectively,

$$E_y(x,t) = E_{yo} \sin 2\pi \left(\frac{x}{\lambda} - \frac{t}{T} \right) , \tag{7.1}$$

$$B_z(x,t) = B_{zo} \sin 2\pi \left(\frac{x}{\lambda} - \frac{t}{T} \right) , \tag{7.2}$$

where λ is the *wavelength* and T is the *period* of oscillation. The reciprocal $\nu = 1/T$ of the period is called the *frequency* of oscillation, while the reciprocal $\kappa = 1/\lambda$ of the wavelength is called the *wavenumber*. Under certain conditions, the same model may be used to describe light propagation through various media. The speed of propagation

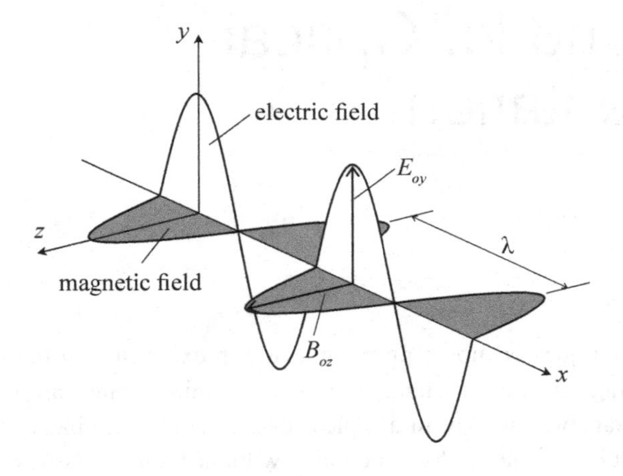

Figure 7.1 Sketch of light viewed as a travelling electromagnetic wave.

v of any point in the wave that maintains a constant phase difference from a reference point in the cycle (e.g., a crest or a zero) is called the *phase speed* and can be found as

$$v = \frac{\lambda}{T} . \tag{7.3}$$

Thus, frequency and wavelength are related through the wave speed as

$$v = \nu\lambda . \tag{7.4}$$

The speed of light propagation in free space (vacuum) is equal to $c = 2.998 \times 10^8$ m/s, or roughly 300,000 km/s. This value is considered to be the highest possible speed that can be achieved under any conditions.

Now, consider light propagating along different paths in three-dimensional space. The locus of all points along the different paths that have the same phase is a surface called a *wavefront*. If all wavefronts are plane, then the light is considered to be a *plane wave*. Besides plane waves, light may propagate in the form of *spherical* or *cylindrical waves*.

As mentioned previously, light propagation is associated with electric and magnetic fields. The two fields are in phase and have amplitudes which are related as

$$E_{yo} = cB_{zo} . \tag{7.5}$$

It is usually sufficient to analyse electromagnetic waves by considering only the electric field. The term *polarisation* is associated with the orientation of the plane of oscillation of the electric field. If the oscillating electric field lies on a single plane at all times, then the light is called *plane-* or *linearly polarised*. When two light waves with the same frequency, but out-of-phase, travel along the same path and are both linearly polarised, but on two mutually perpendicular planes, the resulting light wave is called *elliptically polarised*. If the two waves have the same amplitudes but a phase difference of $\pi/2$, the resulting wave is called *circularly polarised*. If the amplitudes are equal but the phase changes randomly with time, the wave is called *unpolarised* or, more appropriately, *randomly polarised*. Natural light is essentially randomly polarised.

7.1.2 The Colours

Visible light consists of radiation with wavelengths in the range 380 to 750 nm (1 nm = 10^{-9}m), which corresponds to the frequency range between 4.0×10^{15} and 7.9×10^{15} Hz. The colours, as perceived by a 'standard' human eye, are customarily defined as radiation with wavelengths in the ranges specified in Table 7.1 [27]. For comparison, the typical ranges of other types of electromagnetic radiation, in order-of-magnitude, are summarised in Table 7.2. Light that contains components with all visible wavelengths at equal intensities is called *white light*. Sunlight is white and so is, approximately, light that is produced by certain light sources, such as some types of light-emitting diodes.

Table 7.1 Summary of wavelength ranges of light for different colours; the boundaries of the different ranges reported by different sources vary somewhat

Colour	Wavelength range [nm]
Ultraviolet	$0.85 < \lambda < 380$
Violet	$380 < \lambda < 424$
Blue	$424 < \lambda < 491$
Green	$491 < \lambda < 575$
Yellow	$575 < \lambda < 585$
Orange	$585 < \lambda < 647$
Red	$647 < \lambda < 750$
Infrared	$750 < \lambda < 1{,}000$

Table 7.2 Wavelength ranges of different types of radiation

Radiation type	Wavelength range [nm]
Cosmic rays	$\lambda < 10^{-4}$
Gamma rays	$10^{-4} < \lambda < 10^{-1}$
X-rays	$10^{-2} < \lambda < 10^{2}$
Disinfecting radiation	$10 < \lambda < 380$
Visible light	$380 < \lambda < 750$
Space heating	$750 < \lambda < 10^{7}$
Microwaves	$10^{6} < \lambda < 10^{9}$
Radar	$10^{7} < \lambda < 10^{9}$
Radio and television	$10^{8} < \lambda < 10^{13}$
Electrical power waves	$10^{14} < \lambda < 10^{17}$

7.1.3 Light as Photons

Now, let us go one step back and discuss the mechanism of light emission from different materials [19, 24]. First consider an *isolated atom*, which is unaffected by other particles and influences, as in the case of gases at very low pressure. According to a simplified model, this atom is said to be at its *ground state*, when its nucleus is surrounded by its electrons at their lowest energy levels. The ground state is the normal, stable state of an atom and would last indefinitely, unless the atom is disturbed by the 'pumping' of energy to it, in the form of a collision with another atom, approach of an electron, or absorption of radiation energy. In general, the electrons of atoms may only exist at distinct and well-defined energy levels. When an atom absorbs energy, one or more electrons may move to energy states above the ground state, called the *excited states*. An electron may undertake a *quantum jump* from the ground state to an excited state, following absorption of a fixed amount of energy, equal to the difference in energy levels of the two states. The excited states are unstable and tend to undergo *transition* to a lower energy state extremely rapidly, typically within 10^{-9} to 10^{-8} s. One type of transition is the return to the ground state or any other lower energy state by emission of a *quantum* of radiant energy, called a *photon*. For atoms having many electrons, it is only the outermost electrons that emit photons.

The motion of a photon may be reconciled with the propagation of an electromagnetic pulse by viewing each photon as a *wavetrain* of extremely short duration, whose energy is proportional to the frequency of oscillation, as

$$\mathcal{E} = h\nu \,, \tag{7.6}$$

where $h = 6.624 \times 10^{-34}$ Js is the *Planck constant*. Thus, an isolated atom would ideally absorb and emit radiation at distinct frequencies only, called *resonant frequencies*. In practice, these phenomena occur within narrow bandwidths in the *atomic spectrum*, rather than at discrete frequencies. This may be due to various causes, for example, *Doppler broadening* due to thermal motion of the atom. Isolated atoms may only be encountered in rarefied gases. In liquids and solids, atoms are relatively close to each other, affecting each other's energy states and further broadening the frequency bands. Thus, emitted and absorbed radiation would extend over essentially continuous, relatively broad, frequency ranges.

7.1.4 Geometrical Optics

Electromagnetic wave theory and quantum electrodynamic theory provide explanations for different phenomena associated with light propagation through media [24]. For the sake of simplicity, however, it is also possible to analyse a number of light-related phenomena by assuming that light propagates through a medium in the form of *rays*, which are lines normal at all their points to the wavefronts. This approach is known as *geometrical optics* and describes the phenomena of light transmission, refraction, reflection and dispersion at a macroscopic level.

7.2 Light Propagation Through Media

This section will apply mainly simple geometrical optics concepts to describe light propagation through media and at interfaces between different media. A deeper understanding of these phenomena may be based on light scattering considerations, to be briefly discussed in Section 7.4.

Table 7.3 Refractive indices of some common liquids and solids at a temperature of 293 K for light at a wavelength of 589 nm; (A) indicates an aqueous solution or a mixture with water

Liquid	n	Solid	n
Liquid helium	1.025	Fluorinated ethylene propylene (Teflon FEP)	1.344
Water	1.333	Fused quartz	1.46
Ethyl alcohol	1.361	Pyrex glass	1.47
Turpentine	1.472	Crown glass	1.52
Benzene	1.501	Flint glass	1.57–1.89
Glycerol (A)	1.33–1.47	Acrylic	1.51
Olive oil	1.47	Polycarbonate resin	1.58
Sodium iodide (A, 60%)	1.50	Polystyrene	1.59
Ammonium Thiocyanate (A)	1.33–1.50	Sapphire	1.77
Sodium thiocyanate (A)	1.33–1.48	Zirconium silicate	1.92
Silicone oil	1.47	Diamond	2.42

7.2.1 Refractive Index

The *index of refraction* or *refractive index n* of a medium is defined as

$$n = \frac{c}{v} \,, \tag{7.7}$$

where c is the speed of light in free space and v is the speed of light in the medium. Because c is the highest possible speed, it follows that $n \geq 1$. Considering that $v = \nu\lambda$, and that the frequency ν of light is not affected by the propagation medium, one may conclude that

$$n = \frac{\lambda_o}{\lambda} \,, \tag{7.8}$$

where λ_o is the wavelength of monochromatic light in free space and λ is the wavelength of the same light propagating through a medium.

For air and other gases, the refractive index is only slightly greater than unity, whereas, for liquids and solids, it usually varies in the range 1.3 to 3.5. Typical values of n for some common materials are listed in Tables 7.3 and 7.4 [24, 36, 52].

The speed of propagation of monochromatic light in a medium generally decreases slightly with a decrease in its wavelength. This phenomenon, is known as *dispersion of light* and is significant in liquids and solids. Therefore, if a light beam that contains components with different wavelengths propagates through a liquid or solid medium, each component will propagate with a different speed and the index of refraction of the material would generally be slightly larger for a component with a lower wavelength. The composition of a substance and its density are also factors that affect the speed of light in a medium. A general relationship between the refractive index of transparent materials and their density is the *Lorentz–Lorenz* (or *Clausius–Mosotti*) expression

$$\frac{1}{\rho}\frac{n^2 - 1}{n^2 + 2} = K_L \,, \tag{7.9}$$

in which the constant K_L depends on material properties and the wavelength of light.

Table 7.4 Values of the refractive index and the Gladstone–Dale constant for common gases at a temperature of 273 K and a standard atmospheric pressure for a light wavelength of 589 nm

Gas	n	K [m³/kg]
Air	1.00029	0.226×10^{-3}
O_2	1.00027	0.190×10^{-3}
N_2	1.00030	0.238×10^{-3}
H_2	1.00013	n.a.
He	1.000036	0.196×10^{-3}
CO_2	1.00045	0.229×10^{-3}

For a given material and wavelength, this expression gives

$$n = \sqrt{\frac{1 + 2K_L\rho}{1 - K_L\rho}} . \tag{7.10}$$

For gases, the *refractivity* $n - 1$ is extremely small and the previous expression can be simplified to the *Gladstone–Dale formula*

$$n - 1 \approx K\rho , \tag{7.11}$$

in which K is the *Gladstone–Dale constant* . Typical values of K for gases consisting of neutral (i.e., non-ionised) molecules are given in Table 7.4 [20]. Note that the value of K of an ionised gas is different from that of the same gas in a neutral state. For gas mixtures, the constant K may be found as the sum of the constants of each component, weighted by the corresponding mass fraction.

Variations in the refractive index within a fluid volume would be generated when part of the fluid is heated or cooled and when streams of the same fluid having different temperatures are mixed, as this would generate density variations. The dependence of the refractive index of water upon temperature T_c, given in °C and valid in the range of 20 to 34°C, for $\lambda = 632.8$ nm is [18, 21]

$$(n - 1.332156) \times 10^5 = -8.376(T - 20) - 0.2644(T - 20)^2 + 0.00479(T - 20)^3 . \tag{7.12}$$

For common gases under constant pressure, density changes $\delta\rho$ are related to temperature changes δT through the ideal gas law, as

$$\frac{\delta\rho}{\rho} \approx -\frac{\delta T}{T} . \tag{7.13}$$

The mixing of two different fluids would also result in a fluctuating refractive index field. For compressible flows, density variations would occur any time other properties change. For example, for an isentropic flow of a ideal gas, it is easy to show that

$$\frac{\delta\rho}{\rho} \approx \frac{1}{\gamma}\frac{\delta p}{p} , \tag{7.14}$$

where p is the pressure and γ is the ratio of specific heats.

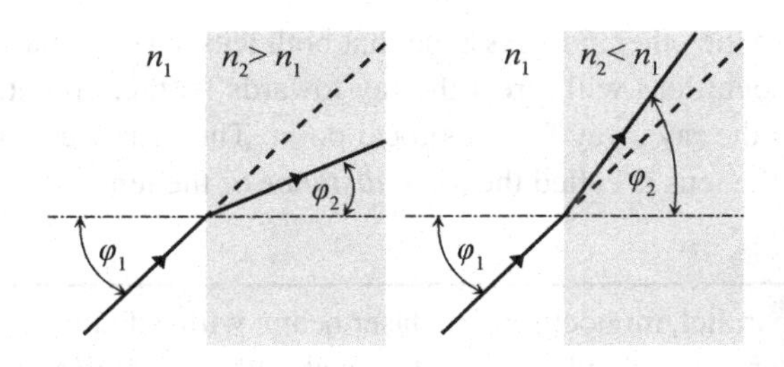

Figure 7.2 Refraction of a light ray crossing the interface between two media.

7.2.2 Refraction

When light propagates through a homogeneous medium, its path would be straight. In contrast, if the medium is non-homogeneous or if the light crosses from one medium to another, its path may change direction gradually or abruptly. This change of direction of propagation of light is called *refraction*. The refraction of light crossing the interface of two media with refractive indices n_1 and n_2, respectively, obeys the *law of refraction*, or *Snell's law* (Fig. 7.2),

$$\frac{\sin \varphi_1}{\sin \varphi_2} = \frac{n_2}{n_1}, \tag{7.15}$$

where φ_1 and φ_2 are, respectively, the angles between the directions of the ray in the two media and the normal to the tangent plane on the interface. Thus, a ray will be refracted towards the normal ($\varphi_2 < \varphi_1$), if it enters an *optically denser* medium ($n_1 < n_2$), and away from the normal, otherwise. The refractive index of a fluid depends on its density, which in turn depends on its composition and temperature. As a consequence, a ray of light propagating in a medium with a non-uniform density will follow a curved path. The actual path of a ray between two points is determined by *Fermat's principle*, which states that the path will be such that the time it takes for the light to traverse this path will be minimum. Refraction may distort the appearance of objects immersed partially in each of two fluids in contact or in non-homogeneous fluids; examples include the perceived 'bending' of a spoon in a glass of water or of an oar in the sea, the 'dancing' of shadows over a radiation heater or over a car hood exposed to the sun, and 'mirages' appearing in deserts or over a hot asphalt road.

As a simple application of refraction, consider the deflection of rays of light by convergent and divergent glass lenses (Fig. 7.3). Assume that a thin ray of light, which is propagating in air, enters one side of the lens while being parallel to the lens' axis and

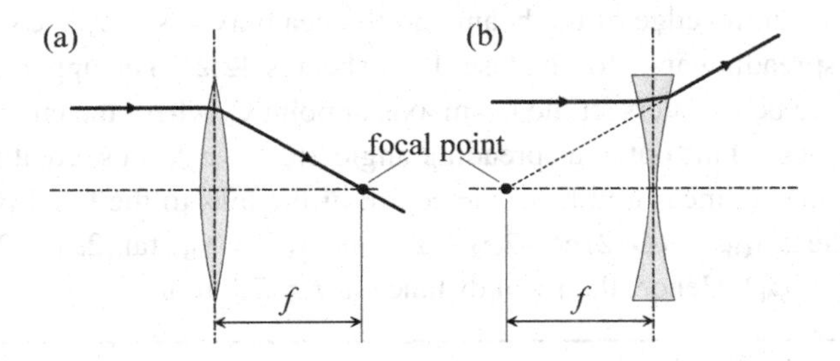

Figure 7.3 Light refraction through (a) a convergent lens and (b) a divergent lens.

then exits back to air from the other side. Assume that both lens surfaces have spherical shapes. Then, the convergent lens will direct the ray towards its focal point, while the divergent lens will direct the ray away from its focal point. The distance f of the focal point from the centre of the lens is called the *focal distance* of the lens.

Example 7.1 A thin, parallel, monochromatic laser beam, with a diameter $d_b = 5$ mm and travelling in air, enters a solid cylindrical glass rod with a diameter $D = 10$ mm. Determine the location of the beam focus and the spreading angle of the subsequent light sheet.

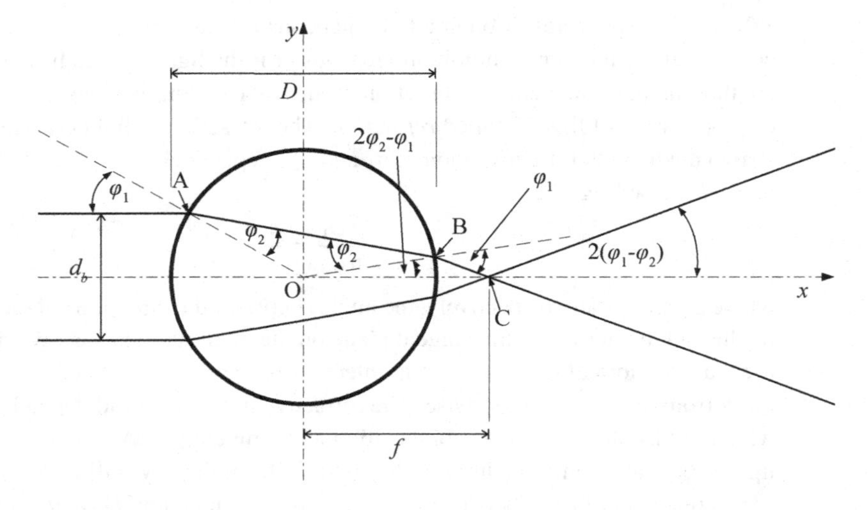

Figure 7.4 Propagation of a light beam through a glass rod in air.

Answer

We take the refractive index of air to be $n_1 \approx 1.0$ and that of glass to be $n_2 \approx 1.5$. A sketch of the paths of the edges of the light beam through the glass rod is shown in Fig. 7.4. The angle between the upper edge of the beam and the normal to the surface at location A, where it strikes the glass rod, is $\varphi_1 = \arcsin[(d_b/2)/(D/2)] = 30°$. From Eq. (7.15), the internal angle of the edge of the beam is $\varphi_2 = \arcsin[\sin\varphi_1(n_1/n_2)] \approx 19.5°$. The upper edge of the beam will exit the glass rod at point B, at an angle φ_1 to the radial line OB. The angle between OB and the beam axis direction is $2\varphi_2 - \varphi_1 \approx 8.9°$, thus, the angle between the edge of the beam and the beam axis is $2(\varphi_1 - \varphi_2) \approx 21.1°$. Consequently, the spreading angle of the laser light sheet is 42.2°. The upper, as well as the lower, edge of the beam intersect the beam axis at point C, where the entire beam is focused. Then, the beam fans out at a spreading angle $2(\varphi_1 - \varphi_2)$, thus, creating a light sheet. The focal distance, measured from the centre of the lens to the focal point C, is $f = x_{OB} + x_{BC}$, where $x_{OB} = (D/2)\cos(2\varphi_2 - \varphi_1)$ and $x_{BC} = y_B/\tan(2\varphi_1 - 2\varphi_2)$, with $y_B = (D/2)\sin(2\varphi_2 - \varphi_1)$. Hence, the focal distance is $f \approx 7.0$ mm.

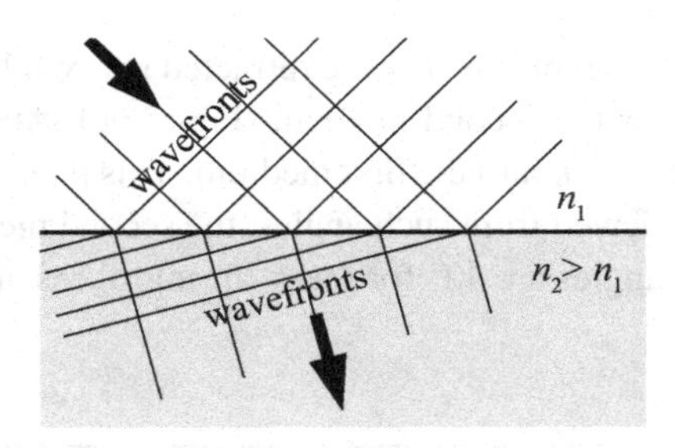

Figure 7.5 Interpretation of light refraction at the interface between two media as the rotation of wavefronts.

A physical interpretation of refraction can be made by considering the wave-like nature of light. A light ray represents waves propagating in a direction normal to plane wavefronts, the distance between which is equal to the wavelength of the wave. When the ray enters an optically denser medium, that is, a medium with a larger refractive index, its speed would decrease and the distance between wavefronts would decrease as well. Thus, as shown in Fig. 7.5, the ray would appear to get deflected towards the normal.

Now, consider that a beam of white light propagating in air enters into a glass prism at an inclination with respect to the prism surface. Dispersion of white light, namely the dependence of the refractive index upon wavelength, would cause its separation into different colours, each of which exits the prism at a different location. This phenomenon is known as *Newton's experiment.*

7.2.3 Reflection

In general, when a beam of light that propagates through a medium reaches a smooth interface with a second medium, part of its energy will be transmitted through the second medium according to the law of refraction. Another part, however, will be turned back into the first medium. This phenomenon is called *reflection* and follows the *law of reflection* (Fig. 7.6a)

$$\varphi_2' = -\varphi_1 . \tag{7.16}$$

When a ray travelling in a medium having a refractive index n_1 reaches an interface with a second medium having a refractive index $n_2 < n_1$ at the *critical angle*

$$\varphi_1 = \varphi_c = \arcsin(n_2/n_1) , \tag{7.17}$$

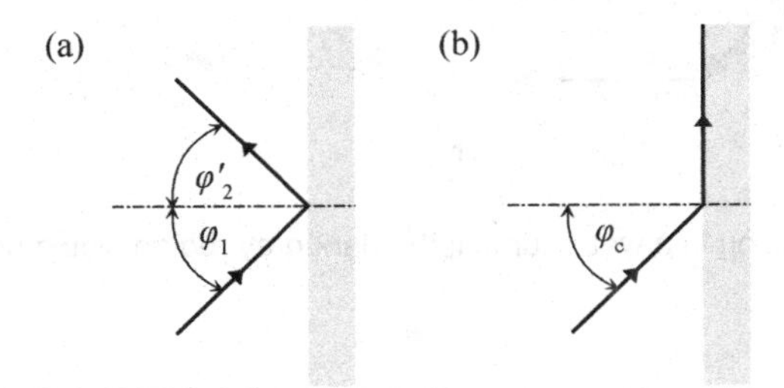

Figure 7.6 (a) Reflection and (b) total internal reflection of a light ray at the interface between two media.

the law of refraction gives $\varphi_2 = \pi/2$, which means that the refracted ray will be parallel to the interface, so that it will not enter the second medium (Fig. 7.6c). At incidence angles $\varphi_1 > \varphi_c$, the ray will be reflected back into the first medium. This phenomenon is called *total internal reflection*. When viewed from such angles, the second medium will be invisible. Typical values of critical angles are 42° for glass–air interfaces and 62° for glass–water interfaces.

Example 7.2 We wish to visualise flow in a water tunnel by illuminating the water with a light source and observing the flow, either by eye or with the use of a camera, outside the tunnel. Assume that the refractive indices of air, glass and water are $n_a \approx 1.00, n_g \approx 1.52$ and $n_w \approx 1.33$. Determine the critical angle and discuss the effect of total reflection on the visibility of flow inside the water tunnel test section.

Answer

We first note that total reflection cannot occur as light leaves water and enters glass, because $n_w < n_g$. However, total reflection may occur as light leaves glass and enters air, because $n_g > n_a$. The smallest angle φ_1, for which total internal reflection would occur, is $\varphi_1 = 90°$. Application of Snell's law relates the smallest angles $\varphi_1, \varphi_2, \varphi_3$ for which total reflection would occur, as

$$\frac{\sin \varphi_3}{\sin \varphi_1} = \frac{\sin \varphi_3}{\sin \varphi_2} \frac{\sin \varphi_2}{\sin \varphi_1} = \frac{n_g}{n_w} \frac{n_a}{n_g} = \frac{n_a}{n_w} .$$

Then, φ_3 can be found as

$$\varphi_3 = \arcsin[(n_a/n_w) \sin \varphi_1] = \arcsin(1/1.33) = 48.6° .$$

This means that an observer will be unable to see a certain part of the flow inside the water tunnel, as shown in Fig. 7.7.

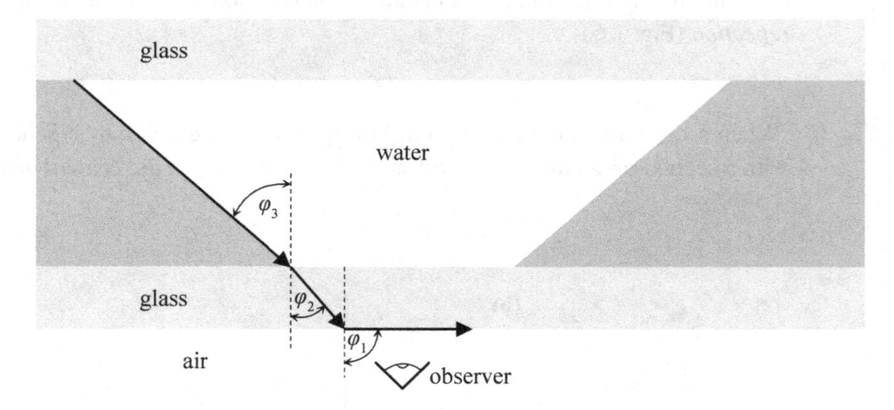

Figure 7.7 Total internal reflection in a water tunnel; the dark grey region would not be visible to the observer.

Total internal reflection causes the familiar phenomenon of the *rainbow*. Sunlight is refracted upon entering into the nearly spherical rain droplets suspended in the atmosphere and undergoes total internal reflection on the far side of each droplet, thus exiting the droplet from its near side. While propagating inside the droplet, the white sunlight is dispersed into coloured beams, which propagate along separate paths and form the rainbow.

7.2.4 Refractive Index Matching

In view of the previous discussion, serious image distortion may occur when light propagates in non-homogeneous materials or encounters interfaces between different materials. A common situation in experimental fluid mechanics is the use of transparent walls or windows, made of glass or plastics, for viewing a contained gas or liquid flow. In some cases, models and internal sections of the apparatus are also made of transparent solids to allow optical access. When fluids are viewed through curved walls, their images get particularly distorted. In some cases, this may result in entire flow regions becoming invisible or multiple images of the same region appearing simultaneously. These distortions can be reduced by using walls as thin as possible and avoiding curved sections. Corrections for the refraction of laser beams and other collimated light beams through curved walls are also available [7, 40]; such corrections are difficult to apply to broad images.

An effective method to reduce or eliminate optical distortion is to *match the refractive indices* of the contained fluid and the transparent wall. As shown in Table 7.4, the refractive indices of air and other gases are much smaller than those of glass and other transparent wall materials, so there is no possibility of matching the refractive index of a gas and a solid wall. The refractive index of water is also substantially smaller than those of glass, acrylics and other common transparent wall materials, and so significant refraction would occur through most water-containing vessels. There is one commercially available material, fluorinated ethylene propylene (FEP), the refractive index of which nearly matches that of water, but this material is opaque and light may penetrate it only when it is in the form of a thin film or machined to a thickness that is less than 1 mm. On the other hand, a number of liquids and solutions are available with refractive indices in the same range as those of common wall materials. Thus, it is possible to precisely match the refractive indices of glass and acrylic materials by either mixing different liquids, for example glycerol and water or various natural and mineral oils (e.g., silicon oil, xylene, naphtha and turpentine), or by dissolving various salts (e.g., ammonium thiocyanate at near-saturation concentrations) in water or other solvents [10]. This procedure is by no means routine, as most liquids with a relatively large refractive index happen to be flammable, volatile, toxic, corrosive, foul-smelling or unsafe. Besides a careful consideration of potentially hazardous effects of the selected liquid on the experimenter and destructive effects on the apparatus and the laboratory, consideration must be given to their clarity, density, viscosity, sensitivity to temperature, environmental impact and price. Because the refractive index of materials is sensitive to many factors, it may also be necessary to conduct on-site measurements of the refractive index of both the wall material and the liquid. This can be achieved with the use of a *refractometer*.

Even when a contained liquid is matched optically with the surrounding walls, optical distortion will be inevitable, if light propagating through air encounters a curved wall (e.g., a circular tube) before it enters the liquid. In such cases, it is advisable to create a plane wall–air interface, either by machining the transparent wall, if this is possible, or by immersing the viewing section into a rectangular *viewing tank*, filled with the same liquid, but still and not communicating with the liquid inside the test section. In the latter case, one must take care to eliminate possible temperature differences between the still and flowing liquids. If such differences persist, they may introduce optical distortion due to refractive index variations; in addition, temperature-related density variations may generate convection currents, which could cause flow distortions, especially in very low Reynolds number flows.

When dealing with two-phase or variable density flows, differences in refractive index are encountered not only between fluids and surrounding walls, but also within the fluid itself. In most cases, these differences are impossible to eliminate and one has to work within the limitations of the adopted visual or optical technique. Under certain conditions, however, it is possible to match refractive indices by selecting specific materials. For example, dense solid suspensions in liquids have been simulated by the use of liquid/particle combinations consisting of chemicals carefully mixed such as to match their refractive indices as well as having a desired density ratio, including unity [16]. The composition of fluid becomes more complicated, when one wishes not only to match the refractive index of a solid wall, but also the viscosity of a particular fluid. This may be achieved by mixing several fluids. An example of such an application is a mixture of water, glycerol and sodium iodide, which matches the refractive index of silicone and the viscosity of blood [23].

7.2.5 Absorption

When light or other radiation is transmitted through a material along a path with length l, it will be absorbed by the molecules of this material according to *Beer's law*

$$I = I_o e^{-\alpha l}, \tag{7.18}$$

where I is the radiant intensity (see Section 7.3) of the passing light, I_o is the radiant intensity of the incident light, and α is the *absorption* or *attenuation coefficient*, which depends on the material and the wavelength of radiation [24]. The length $1/\alpha$, called the *skin* or *penetration depth*, represents the thickness of this material that will absorb 63% of the incident light energy. Opaque materials have an extremely large value of α, while transparent materials have a relatively small value. Metals, in general, have a very small penetration depth and reflect most of the incident light. For example, copper has a penetration depth that varies between 0.6 nm for λ = 100 nm (ultraviolet) and 6 nm for λ = 10,000 nm (infrared); thus, a sheet of copper with a thickness of 2 nm will act as a high-pass light filter.

7.2.6 Birefringence

Birefringence, also known as *double refraction*, is the separation of light into two linearly polarised components, an 'ordinary' ray and an 'extraordinary' one, which have polarisation planes normal to each other and travel through the medium at different speeds, thus having different indices of refraction [24]. It is exhibited by some crystalline solids as well as some liquid polymers and colloidal solutions. Thus, when illuminated with linearly polarised light of the appropriate frequency, such materials could be opaque in one polarisation direction, while being transparent in another. Birefringent solids have uniaxial crystalline structure, with hexagonal, tetragonal, and trigonal crystals; the most commonly used such material is calcite. Materials with biaxial crystals, including orthorhombic, monoclinic and triclinic ones, exhibit three indices of refraction, thus being *trirefringent*. Combinations of calcite prisms joined at different angles have been used as *polarisers*, removing one ray by total reflection (*Nicol prism*, *Glan–Foucault prism*), and as *beam splitters*, splitting non-polarised light into two diverging, linearly polarised rays (*Wollaston prism*).

7.3 Flow Illumination

Visual and optical measurement methods usually employ specialised light sources and illumination techniques. In simple terms, the most important characteristics of a light source are its brightness, the duration of light it produces and the distribution of power of light over its wavelength bandwidth (*light spectrum*). Most light sources produce light in the visible range, which may be directly observed by the human eye as well as recorded by photodetectors and cameras. There are occasions, however, on which infrared or ultraviolet radiation may be preferable; such radiation, although invisible to human eye, can be detected and recorded by certain types of photodetectors and cameras. It is, therefore, necessary to distinguish the characteristics of light as recognised by the human eye, which are subjective, from its objective characteristics, which quantify light as electromagnetic radiation. Accordingly, there are two ways to characterise light: an objective one, called *radiometry*, and a subjective one, called *photometry* [20, 52]. Although both radiometry and photometry deal with the same physical properties, they use different names, symbols, definitions and units.

Light sources include *thermal sources*, *light-emitting diodes* (*LEDs*) and *lasers*. Light sources are further classified into *continuous-wave* (CW) sources, which produce radiation continuously, and *pulsed* sources, which produce single or repetitive radiation pulses of relatively short duration.

Besides using a suitable light source, visual and optical measurement methods require an optimised illumination arrangement, which is appropriate for observation and/or recording under the specific conditions of each experiment. The choice of illumination arrangement should be guided by theory and previous experience, but it is always worthwhile to optimise the setup by trial-and-error adjustments.

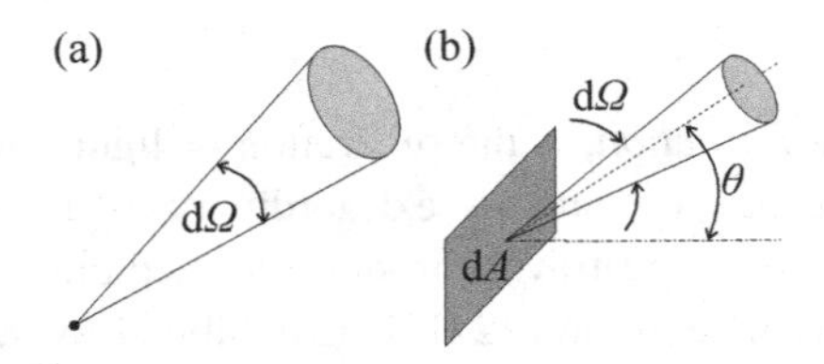

Figure 7.8 Sketches of (a) a point source of radiation and (b) a plane source of radiation.

7.3.1 Radiometric and Photometric Definitions

Radiometric properties: A *point source* of light is an idealised source of electromagnetic radiation, which is concentrated at a point in space and radiates uniformly in all directions. Its *radiation power* Φ_e is defined as the total emitted radiation energy per unit time. The *radiant intensity* I_e of this source, which in following sections is also denoted as I, is defined as the radiation power per unit solid angle Ω (Fig. 7.8a), that is, as

$$I_e = \frac{\mathrm{d}\Phi_e}{\mathrm{d}\Omega} \,. \tag{7.19}$$

A *plane source* of light emits energy uniformly from all points on a plane surface. An elementary plane source with area dA is characterised by its *radiance* L_e, defined as

$$L_e = \frac{\mathrm{d}^2\Phi_e}{\mathrm{d}\Omega \mathrm{d}A \cos\theta} \,, \tag{7.20}$$

where dΩ is an elementary solid angle centred at the centre of the source and θ is the angle between the axis of this solid angle and the direction normal to the source plane (Fig. 7.8b). The *spectral radiance* $L_{e\lambda}$ of this source is defined as the radiance per unit wavelength of the emitted radiation, that is, as

$$L_{e\lambda} = \frac{\mathrm{d}L_e}{\mathrm{d}\lambda} \,. \tag{7.21}$$

Now, consider a plane surface element with area dA, receiving an amount of radiation power dΦ_e. The *irradiance* E_e of this element is defined as the irradiation power per unit area, that is, as

$$E_e = \frac{\mathrm{d}\Phi_e}{\mathrm{d}A} \,. \tag{7.22}$$

Notice that the irradiance is independent of the orientation of the surface with respect to the direction of oncoming radiation. All previous properties have conventional dimensions and are measured in conventional units, called *radiometric units*. For example, the radiation power is measured in W, while the radiant intensity is measured in W/sr.

Photometric properties: When dealing with visible radiation it is common practice to use *photometric properties*, defined in terms of the response of a 'standard' human eye. The power of visible radiation sensed by the standard human eye is called *luminous power* or *luminous flux* Φ_v and it is measured in lumens (lm). Luminous properties are defined in a manner which is analogous to the definition of the corresponding radiation properties. The *luminous intensity* is defined as

$$I_v = \frac{d\Phi_v}{d\Omega} \tag{7.23}$$

and is measured in candelas (cd; 1 cd = 1 lm/sr). The *luminance*, commonly referred to as *brightness*, is defined as

$$L_v = \frac{d^2\Phi_v}{d\Omega dA \cos\theta}\,. \tag{7.24}$$

The *spectral luminance* is defined as

$$L_{v\lambda} = \frac{dL_v}{d\lambda} \tag{7.25}$$

and, finally, the *illuminance* is defined as

$$E_v = \frac{d\Phi_v}{dA} \tag{7.26}$$

and measured in luxes (1 lux = 1 lm/m^2). When the oncoming radiation has power over multiple wavelengths, the *total illuminance* is found by integrating the narrow-band illuminance over the entire visible spectrum.

7.3.2 The Human Eye

Before presenting the correspondence between radiometric and photometric units, it is necessary to discuss briefly the human eye, sketched in Fig. 7.9. The eye is enclosed within three membranes: the outer membrane is called the *cornea/sclera*, the middle one is called the *choroid* and the inner one is called the *retina*. A *lens*, which is flexible and adjusts its curvature for focusing purposes, images the received radiation onto the retina, which is lined with a large number of receptors sensitive to light. There are two types of such receptors: the *cones* (roughly 7 million), which respond only to bright light and are sensitive to colour, and the *rods* (of the order of 100 million), which are sensitive to

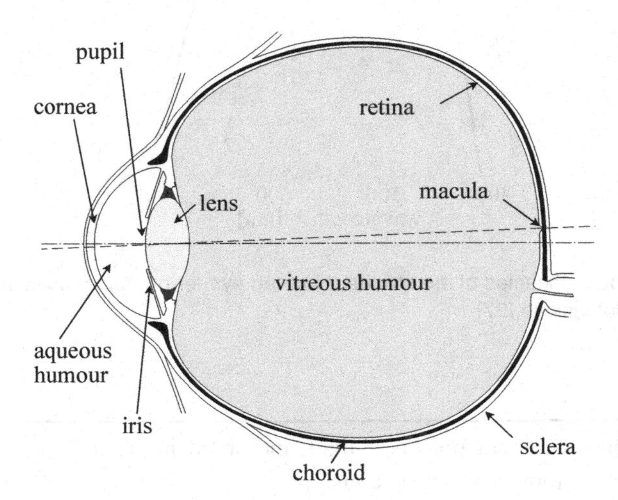

Figure 7.9 Sketch of the human eye.

dim light but cannot separate the different colours. The central part of the retina, called the *macula*, contains a particularly large concentration of cones and, therefore, has the highest resolution of colour; it is on this part that images are focused when highest clarity of vision is achieved.

The sensitivity of the eye to different colours depends on the brightness of light. When the light is bright, the cones are activated, resulting in *photopic* or *bright-adapted* vision, which resolves colours. In contrast, when the light is dim, such as during twilight, the main contributors to vision are the rods, which do not distinguish colours. This is called *scotopic* or *dark-adapted* vision. Consequently, the ratio K of luminous power to radiant power, called *luminous efficacy* and measured in lm/W, is different for photopic and scotopic visions. When normalised by the corresponding maximum value, the luminous efficacy is called *spectral sensitivity* or *luminous efficiency*. The luminous efficacies of the two types of vision are plotted in Fig. 7.10 [27]. It may be seen that the maximum eye sensitivity to bright light is 673 lm/W at about $\lambda = 555$ nm (green colour), while the maximum sensitivity to dim light is 1725 lm/W at about $\lambda = 510$ nm (blue-green colour). Notice also that scotopic vision is insensitive to yellow, orange and red colours, which explains why colourful objects appear to be blueish under dim illumination.

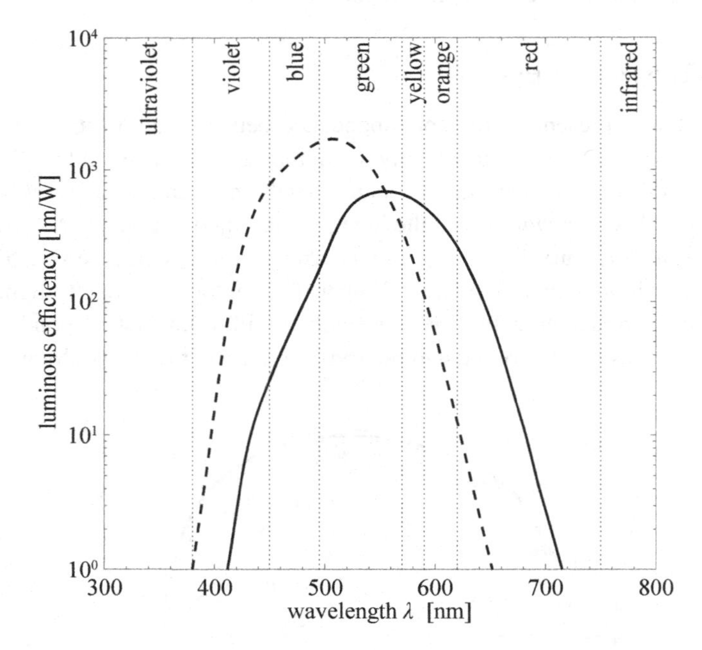

Figure 7.10 Luminous efficacies of the standard human eye for photopic (solid line) and scotopic (dashed line) visions [27].

Example 7.3 Convert radiant power of light, measured in W, to luminous power, measured in lm, for photopic and scotopic visions.

Answer

For this conversion, one needs to multiply the radiant power by the luminous efficacy K at a particular wavelength, as provided by the corresponding curve in Fig. 7.10. First, assume that the oncoming light has a wavelength $\lambda = 555$ nm (green colour), which is at the peak of the photopic curve, which also happens to be near the intersection of the photopic and scotopic curves. Then, $K \approx 673$ for both types of vision, which means that the two types of vision would have equal strength. At other wavelengths, 1 W of radiation power will produce less than 673 lm of luminous power for photopic vision, but more or less power for scotopic vision, depending on the wavelength. At the peak of scotopic luminous efficiency ($\lambda \approx 520$ nm – blue-green colour), scotopic vision will be five times stronger than photopic one, whereas, for orange light ($\lambda \approx 615$ nm), it is photopic vision that would be five times stronger. 1 W of radiation power at $\lambda = 650$ nm (red) will produce roughly 65 lm of luminous power for photopic vision and a mere 4 lm for scotopic vision.

Each light source emits light that has a specific spectral radiance at each wavelength. The perceived brightness of the source is characterised by the *overall luminous efficacy*, which can be used to convert radiation power (in W) to luminous power (in lm). Typical ranges of this parameter for some common light sources are given in Table 7.5.

Table 7.5 Typical ranges of the overall luminous efficacy

Light source type	Overall luminous efficacy [lm/W]
Sunlight	70–130
Incandescent (tungsten)	10–15
Incandescent (halogen)	17–25
LED	75–170
Fluorescent	45–100

The human eye can distinguish variations in brightness over an enormous range (10 orders of magnitude) of luminous powers, bounded upwards by the *glare limit* and downwards by the *scotopic threshold* [22]. The *subjective brightness* of the collected light is a logarithmic function of the luminous power, following two different curves, one for photopic and one for scotopic vision. The range of brightness differences that can be recognised simultaneously is much narrower, as it takes time for the eye to adapt to each average brightness level. The range of recognisable colours also depends on the brightness and the adaptation time. Another important parameter that concerns the recognition of visual patterns is the sensitivity of the eye to changes in contrast. Typically, the minimum detectable difference between the brightness of a spot and a uniform background brightness is about 2% for a wide range of brightness levels. For varying background brightness, this limit could be significantly higher [22].

7.3.3 Colour-Related Terminology

The acquisition, processing, display and interpretation of images are common activities in experimental fluid mechanics, and so it seems advisable to clarify some relevant concepts [45]. Colour-related metrics and terminology are quite complex and have been the object of standardisation by the International Commission on Standardization (CIE – Commission Internationale de l'Eclairage) [12]. CIE defines *colour* as an attribute of visual perception consisting of any combination of *chromatic* and *achromatic* content. The former includes common colour names, such as yellow, red etc., while the latter includes names, such as white, black and gray, and qualifications, such as bright or dim and light or dark. Colour is subjective and does not exist independently of the observer. *Related colours* are those perceived to belong to an area of an object seen in relation to other colours, while *unrelated colours* are those perceived to belong to an area of an object seen in isolation from other colours. Brown and gray can only be perceived as related colours and do not exist in isolation. *Hue* indicates whether an area appears to be similar to one of the perceived colours: red, yellow, green and blue, or to a combination of two of them. *Brightness* signifies whether an area appears to emit more or less light. *Lightness* is the brightness of an area judged relative to the brightness of a similarly illuminated area that appears to be white or highly transmitting. *Colourfulness* indicates whether the perceived colour of an area appears to be more or less chromatic, while *saturation* is the ratio of the colourfulness and the brightness of an area. Thus, the vividness of a colour is indicated by its saturation, while the intensity of a colour is indicated by its lightness. Hue, saturation and lightness have been identified as the parameters that determine colour perception, according to the popular *HSL model*. When displaying or reproducing colour images, one needs to simulate human perception of colour. Computer and television monitors display colours as mixtures, in different proportions, of the three primary colours red, green and blue (*RGB model*). Mixing equal amounts of all three produces gray, maximum amounts of all produces white and zero amounts of all produces black. More complex models are also available, such as the *CMYK* (cyan, magenta, yellow, black) *model*, employed by some printers.

7.3.4 Thermal Radiation

An idealised source of thermal radiation is the *black body*, which radiates at all wavelengths. The spectral radiance of the black body is given by *Planck's radiation law*

$$L_{e\lambda} = \frac{2\pi hc^2}{\lambda^5} \frac{1}{e^{hc/(\lambda k_B T)} - 1}, \tag{7.27}$$

where h is the *Planck constant* and $k_B = 1.38042 \times 10^{-23}$ J/K is the *Boltzmann constant*. For the infrared–visible–ultraviolet wavelength range and temperatures in the range below 10^4 K, the unit may be neglected compared to the exponential term in the denominator of Eq. (7.27) and Planck's radiation law may be simplified to *Wien's radiation law*

$$L_{e\lambda} = \frac{2\pi hc^2}{\lambda^5} e^{-hc/(\lambda k_B T)}. \tag{7.28}$$

Differentiation of this equation with respect to λ provides a maximum spectral radiance at

$$\lambda_{\max} = \frac{0.002898}{T} , \tag{7.29}$$

where $\lambda_{\max}$ is given in m and T in K. This relationship, called *Wien's displacement law*, shows that the peak of the radiation spectrum shifts towards lower wavenumbers as the temperature increases, as shown in Fig. 7.11. Approximate integration of Planck's radiation law over the entire wavelength range gives the total radiation power emitted by a black body as

$$\Phi_e = \sigma A T^4 , \tag{7.30}$$

where $\sigma = 5.67033 \times 10^{-8}$ W/m^{2}K is the *Stefan–Boltzmann constant*, A is the radiating area and T is the absolute temperature. This expression is called the *Stefan–Boltzmann law*. Although some opaque objects and dense gases may be treated approximately as black bodies, most objects may not. However, the radiation emitted by many objects may be referred to the black-body radiation, by introducing a correction factor ε, called the *total emissivity*, as

$$\Phi_e = \varepsilon \sigma A T^4 . \tag{7.31}$$

The emissivity of shiny metallic surfaces could be as low as 0.02 to 0.03, while that of black, flat surfaces may exceed 0.95, approaching the black-body emissivity of 1. In addition to the total emissivity, one may also define the *monochromatic emissivity* ε_λ as the ratio of the spectral radiance of the body and the spectral radiance of a black body at the same wavelength and temperature. In general, ε_λ is a function of λ, T, and surface conditions. Objects whose monochromatic emissivity is independent of wavelength, $\varepsilon_\lambda = \varepsilon$, are called *grey bodies*.

7.3.5 Thermal Light Sources

Thermal light sources emit electromagnetic radiation as a result of being heated to a high temperature [20, 50]. The important characteristics of thermal light sources are their radiation power and spectral radiance. A variety of thermal light sources with spectra in the visible, ultraviolet and infrared ranges are available. They are divided into *line sources*, which produce radiation at one or more narrow spectral bands, and *continuum* sources, producing wideband radiation.

Among the continuous-wave thermal light sources, the most commonly used are the *incandescent lamps* and the *electric discharge lamps*.

Incandescent lamps: These contain an electrically heated tungsten filament in an evacuated glass container. They have a smooth continuous spectrum across the visible range, following Planck's law, Eq. (7.27), with a peak at 900 nm at a temperature of 2854 K. *Halogen lamps* are filled with a halogen (iodine or bromine) compound, which prolongs the life of the tungsten filament and allows it to operate at higher temperatures.

Electric discharge lamps: Common type *fluorescent lamps* are filled with mercury vapour at low pressure and utilise an electric discharge through it to produce light in the

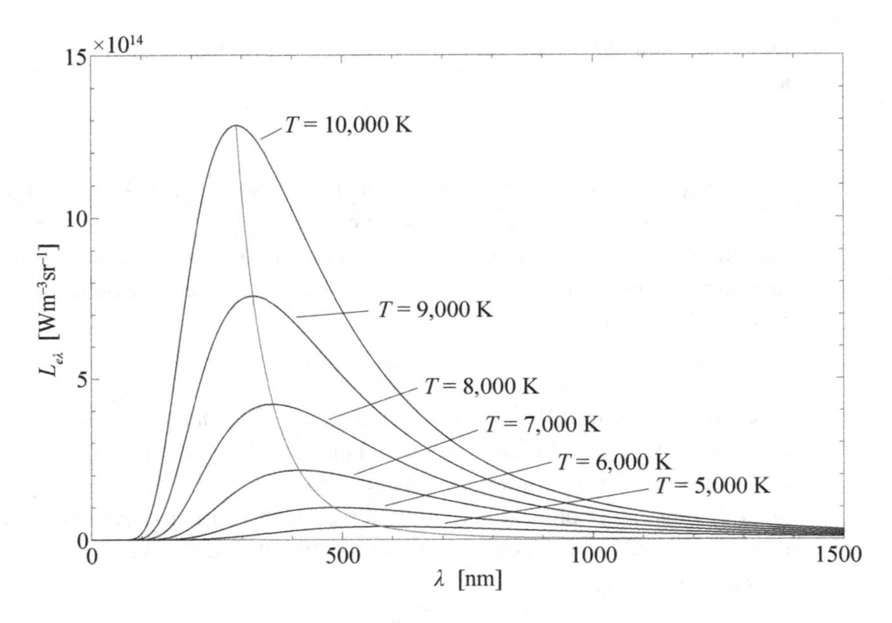

Figure 7.11 Spectral radiance for a range of temperatures according to Planck's radiation law. Dotted line indicates Wien's displacement law.

ultraviolet range, which, through fluorescence, is converted to the visible range. They have a higher efficiency, compared to incandescent lamps and have a spectrum that is comparable to that of natural light. *High-pressure arc lamps* consist of a quartz container filled with a gas, usually mercury, xenon, or a mixture of the two, under high pressure, and produce light at the arc between two tungsten electrodes. This light has a wideband continuous spectrum having superimposed spectral lines. Mercury lamps can be excited to emit at one or only a few wavelengths. A commonly used, nearly monochromatic, electric discharge lamp is the *low-pressure sodium lamp*, which emits in the yellow ($\lambda = 589$ and 589.6 nm) range and also has the highest efficiency of operation.

A variety of thermal light sources that provide radiation pulses of short duration are available, including the following:

Flash lamps: These are tubes containing a noble gas, usually xenon, krypton or argon. For their operation, high voltage stored in a capacitor is discharged through the gas producing a highly luminous *corona discharge*. Single-flash and stroboscopic devices are available. The duration of each light pulse varies typically between 1 μs and 1 ms.

Sparks: These are produced by the electric breakdown of a gas (helium, neon, argon or air) during an electric discharge between two electrodes. The choice of different electrodes produces sparks of different shapes.

Exploding wires: These are metallic (e.g., Cu-Mg) wires, which evaporate explosively when an extremely high current (of order 10^5 to 10^6 A/mm^2) causes them to heat rapidly to temperatures of tens of thousands K [6].

Explosive flashes: These consist of small amounts of explosive materials in a noble-

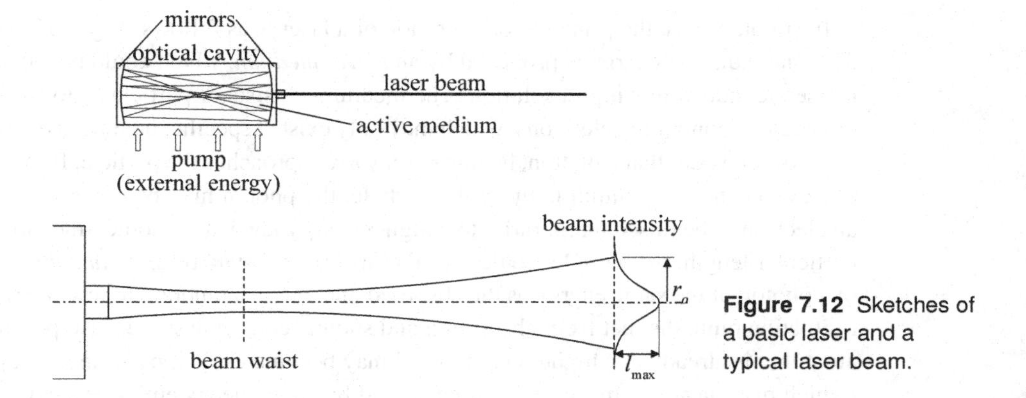

Figure 7.12 Sketches of a basic laser and a typical laser beam.

gas environment, detonated to produce shock waves, which are accompanied by short-duration (10 to 100 μs), high luminosity.

Plasmafocus discharges: These techniques utilise high temperature plasmas to produce short pulses in the ultraviolet range.

7.3.6 Light-Emitting Diodes

Light-emitting diodes (*LEDs*) are semiconductors that emit light when current flows through them, a phenomenon called *electroluminescence*. LEDs producing visible light of different colours, as well as infrared and ultraviolet radiation, are used for many different purposes in instrumentation and a host of other applications, but of main interest in the present context is the use of white LEDs, which generate bright white light that is suitable for flow illumination. LEDs have several advantages over incandescent light sources: they are smaller, they consume much lower electric power for the same luminous power and they have significant spectral luminance values over the entire visible spectrum. Another desirable feature for flow illumination is that LEDs generate very little heat, thus avoiding a change in temperature-dependent fluid properties and a possible deformation of or damage to apparatus walls.

7.3.7 Lasers

The word laser is an acronym for 'light amplification by stimulated emission of radiation'. Since the invention of the ruby laser in 1958, followed by the He-Ne laser in 1960, continuous development has produced a great variety of lasers and rendered them indispensable to many different applications. Compared to thermal light sources, lasers have several important advantages: laser beams are highly *coherent* (with all light wavefronts in phase), *collimated* and *concentrated* (essentially parallel, with a small cross-sectional area) and *monochromatic* (with spectral energy concentrated in one or more extremely narrow bands). In experimental fluid mechanics, lasers are used routinely as light sources for flow visualisation, but also in various techniques for measuring velocity, pressure, temperature and composition, in both liquids and gases.

In simple terms, the principle of operation of a laser is as follows (Fig. 7.12) [44, 46, 52]. The radiation energy is produced by an *active medium*, which could be a gas, crystal, semiconductor or liquid solution. The medium consists of particles (atoms, ions or molecules), containing electrons, which may only exist at specific, quantised energy levels. Now consider that a photon, having energy $h\nu$, approaches the particle. If the photon energy matches a quantum jump of the particle, the photon may be *absorbed*, causing an electron to be raised temporarily to a higher energy level; after some time, not of any particular length, the particle would return to its original state by *spontaneous emission* of a photon of the same energy as the absorbed one, but in a random direction. However, a situation quite distinct from absorption and spontaneous emission is also possible: an atom that is already at a higher energy level may become excited by an incident photon (which may have, for instance, been generated by spontaneous emission) and, without absorbing it, it may be stimulated to return to a lower energy level by *coherent emission* of another photon, namely one that is identical in energy (frequency), phase and direction with the incident photon. This process, called *stimulated emission*, results in amplifying the original photon energy, while maintaining its frequency and phase. For the process to be effective, the population of atoms at the higher energy state must be larger than that at the lower energy state, a situation referred to as *population inversion*, because it contradicts the normal state of equilibrium of matter, in which lower energy states occur with a higher probability than higher ones. Obviously, external power is required to maintain a sufficient population of atoms at the higher energy state. This is supplied by the *energy pump source*, in the form of electromagnetic or chemical energy provided by an electric discharge, a flash lamp, or another laser. The so achieved light amplifier is turned into an oscillator by being placed in an *optical cavity*, namely a tube that has two plane or concave mirrors at its two ends. The distance between the mirrors is adjusted precisely to an integral multiple of the light wavelength, which produces standing light waves in the cavity. One of the mirrors is slightly transparent, so that some of the light escapes in the form of a laser beam.

A performance characteristic of the optical cavity is the *quality factor*, or Q, which characterises energy decay within the cavity (high Q means better quality). A common method to produce extremely large light pulses is *Q-switching*. This method consists of a temporary disruption of stimulated emission within the cavity by various means (e.g., by rotating or blocking one mirror), which results in a large increase of inverted population of atoms; then, the cavity is suddenly restored and a massive emission of coherent photons is produced.

The intensity of light in the laser beam is not uniform and it depends on the number and types of *modes* of standing waves present in the optical cavity. There are two types of modes: *longitudinal modes*, forming along the length of the cavity, and *transverse modes*, normal to it. For a longitudinal mode with wavelength λ to be present, the distance between mirrors must be an integral multiple of $\lambda/2$. Thus, mode selection can be made by adjusting the mirrors. The most commonly used mode is the *fundamental mode*, in which the light intensity amplitude I_a in the beam decreases monotonically with radial distance r from the axis, following approximately a Gaussian-type shape, as

$$I_a(r) = I_{a\,\max} e^{-2(2r/d_e)^2}, \tag{7.32}$$

where the laser beam diameter d_e is defined as the diameter of the circle on whose perimeter the intensity amplitude is $e^{-2} I_{a\,\max}$, namely 13.5% of the maximum intensity on the axis. Other modes have intensity profiles with peaks off the axis; for example, the first-order transverse mode can be made to have an approximately annular intensity distribution (*doughnut mode*). One should further note that the laser beam is not cylindrical but it first converges to a section of minimum diameter (*beam waist*), where the average intensity is maximum, and then diverges (Fig. 7.12), with the exact shape depending on the arrangement of the mirrors in the optical cavity. Although multi-mode operation of the laser would provide higher total power than the Gaussian mode alone would, the latter is generally preferable, as it is spatially coherent, results in the smallest possible beam diameter, can be focused to the smallest possible spot and has the lowest beam divergence; a typical divergence angle for the fundamental mode is of the order of 1 mrad or less. Special lens systems are available to modify the Gaussian intensity profile to an approximately uniform one or other shapes, as required in specific applications.

One disadvantage of laser light is its *diffraction* of dust particles inside the optical cavity, resulting in a beam distorted by fringes and patterns, called *speckles*. Such diffraction patterns may be partly removed by passing the beam through a *pinhole*, which acts as a spatial filter.

There is a great variety of lasers using many different media. The ones most commonly used in experimental fluid mechanics and combustion research are the following.

Helium-neon (He-Ne) lasers: These are of the CW type, producing powers between 0.3 and 15 mW, at $\lambda = 633$ nm (red). Due to their easy operation and relatively low cost, they are frequently used for flow visualisation purposes and in a variety of other applications. The active medium is a mixture of helium and neon atoms, while the energy pump source is a high voltage (~ 2,000 V) electric field. Free electrons detached from a cathode are accelerated by the electric field and occasionally collide with helium atoms, which are excited by the electron's kinetic energy to a higher energy state. This state is referred to as *metastable*, because it is relatively long-lasting, although it eventually returns to the ground state by spontaneous emission. Some of these excited helium atoms collide with neon atoms at their ground state and excite them to their higher energy state; this process works because the excited energy states of the helium and neon atoms are essentially identical. Thus, a population inversion of more neon atoms at the excited state than at lower states is generated. When a photon of a particular energy (corresponding to $\lambda = 633$ nm) strikes the excited neon atom, it triggers the stimulated emission of an identical photon, resulting in reducing the energy state of the atom to a lower level, which very quickly returns to the ground state. Doppler spreading, due to the random motion of neon atoms, causes some widening of the emitted spectrum around the $\lambda = 633$ nm line.

Argon-ion (Ar-Ion) lasers: These are also CW, producing powers between 100 mW and 10 W or more, at seven wavelengths, with the strongest peaks at $\lambda = 488.0$ nm (blue) and 514.5 nm (green). They have an extremely low power efficiency, and, while some air cooled laser at 100 mW power is available, the higher power ones require continuous water cooling. These are the most commonly used lasers in laser-Doppler velocimetry and are also used for flow visualisation and phase-Doppler particle analysis (PDA), as well as for particle image velocimetry (PIV) in low-speed flows. The active medium is argon atoms maintained at the ion state through collisions with electrons accelerated by an electric field, all contained within a *plasma tube*. Some of these ions are further excited to a higher energy state by collisions with electrons; stimulated emission is triggered by interactions of the excited ions with photons at the appropriate frequency. **Nd:YAG lasers:** These are solid-state lasers, containing the rare earth neodymium (Nd^{+3}) as the active medium, incorporated, as an impurity, into a crystal of yttrium aluminum garnet (YAG), which serves as a host. Energy pumping is usually produced optically by a flash lamp. In a single-pulse mode, these lasers produce pulses with energy between 100 and 400 mJ and duration of the order of 100 ps to 10 ns. They generally require a Q switch to generate a short intense laser pulse. Their main emission is at λ = 1064 nm (infrared), but a frequency-doubled output at 532 nm (green), or a frequency-tripled output in the ultraviolet range may be produced with the use of a special crystal. Dual configurations are available for PIV, in which two lasers are fired between 0 and 10 ms apart, at repetition rates of up to 30 Hz. With pumping provided by a CW diode laser (see below), Nd:YAG lasers can give at train of pulses at repetition rates exceeding 1 kHz, although the energy of each pulse would diminish with increasing repetition rate.

Nd:YLF lasers: These lasers use neodymium as the active medium and yttrium lithium fluoride (YLF) as the host. Their operation is similar to that of Nd:YAG lasers, with the difference that energy pumping is achieved with a diode instead of a flash lamp. This form of pumping yields a longer fluorescence lifespan, which allows for higher repetition rates, in some cases up to 20 kHz. The pulse widths of Nd:YLF lasers are typically up to 100–200 ns, thus longer than those of Nd:YAG lasers. To increase the repetition rate, one must necessarily reduce the energy level per pulse. Nd:YLF lasers produce a maximum energy of around 35 mJ per pulse at a repetition rate of up to about 1 kHz. These lasers are better suited to time-resolved particle image velocimetry measurements in both liquids and gases, especially in turbulent flows, for which spectral measurements are required.

Copper-vapour (Cu) lasers: These lasers typically produce repetitive pulses of duration 15 to 60 ns, energy of 10 mJ per pulse and rate of 5 to 15 kHz, mainly at λ = 510.6 nm (green) and 578.2 nm (yellow). Unlike most other lasers, instead of requiring cooling, they are insulated to achieve a high operating temperature. These lasers are particularly suitable for particle tracking, flow visualisation and particle image velocimetry.

Dye lasers: These lasers have an active medium that consists of complex, multiatomic, organic molecules, with essentially continuous emission spectra over relatively wide wavelength bands in the range from 200 to 1,500 nm; their name derives from the fact that such substances are used for dying fabrics. They are pumped by flash lamps or other types of lasers. When pumped by a Nd:YAG laser, in combination with frequency filters contained in the optical cavity, they can be ‘tuned’ to emit radiation at narrow bands matching the resonant frequencies of various molecules contained in the fluid of interest. They are used for the measurement of species concentration in gas mixtures and combustion products.

Excimer lasers: These lasers have an active medium with diatomic molecules, whose atoms are bound attractively only if one of them is at an excited electronic state (*exciplexes*), while they dissociate if it is at the ground state; examples include KrF and XeCl. Excimer lasers emit in the ultraviolet range and provide high energy pulses at relatively high repetition rate. They are used in specific measurement techniques in combustion research.

Carbon dioxide lasers: These lasers oscillate at $\lambda = 10.6$ μm (infrared), either continuously or in pulsed mode. They are used for heating materials, cutting and welding. The active medium is carbon dioxide molecules at their ground electronic state. Population inversion of CO_2 molecules to higher vibrational energy states is caused by collisions with nitrogen molecules at a metastable, excited vibrational mode.

Laser diodes: These are semiconductor devices which emit coherent radiation in the visible or infrared ranges when current passes through them. They are used extensively in optical-fiber communication systems, compact disc players, laser printers, remote controls, and intrusion detection systems. They are much smaller than conventional lasers, and they have a much lower power requirement. High-power laser diodes have been used in fluid mechanics research for flow visualisation, laser-Doppler velocimetry and particle image velocimetry, particularly for experiments in water.

7.3.8 Diffraction

Diffraction is the transverse spreading of a light beam when it passes through a small opening or near the edge of an object [14, 24, 29]. This phenomenon contradicts geometrical optics concepts, but can be explained by considering the wave-like character of light. Consider a beam of radiation produced by a light source and passing through a small *aperture*, such as a slit or a hole on an opaque surface. If the radiation is then projected onto a screen, it will form a *diffraction pattern*. Depending on the arrangement of the basic components, a diffraction process can be classified as *Fraunhofer* or *Fresnel* diffraction. Fraunhofer or far-field diffraction occurs when both the light source and the screen are at essentially infinite distances from the aperture. In practice, this process can be produced by collimating the light of the source with a collimator lens and by projecting the light passing through the aperture to the screen through a collecting lens focused on the screen. Fresnel, or near-field diffraction, occurs when either the source or the screen, or both, are at finite distances from the aperture. The description of Fresnel diffraction is more complex than the description of Fraunhofer diffraction. In the following, we shall deal exclusively with the latter.

First consider a parallel, monochromatic, coherent light beam with an intensity I_o and a wavelength λ, approaching an opaque plane at an angle θ_i and passing through a single slit with a width w (Fig. 7.13a). The intensity of the light that is diffracted by the slit would vary with the diffraction angle θ_d according to the relationship

$$I = I_o \left(\frac{\sin\beta}{\beta} \right)^2 \quad , \; \beta = \frac{\pi w}{\lambda} (\sin\theta_i + \sin\theta_d) \; . \tag{7.33}$$

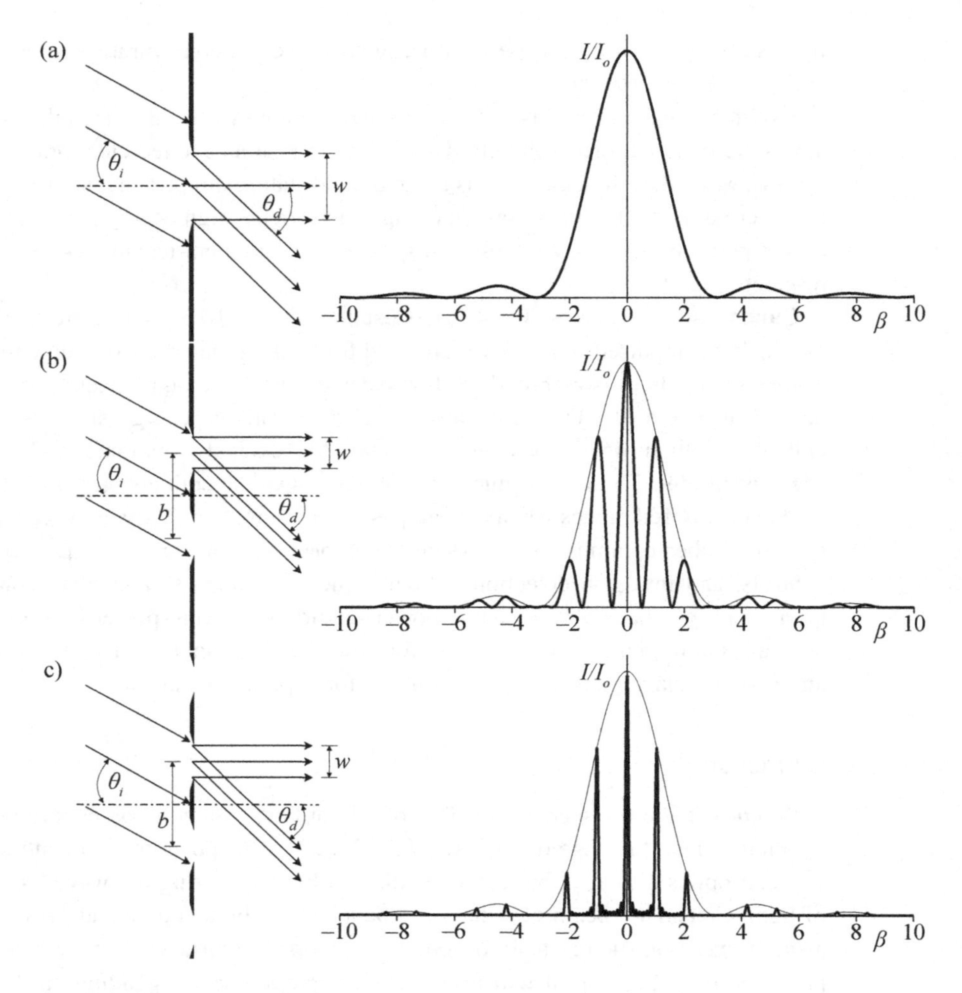

Figure 7.13 Diffraction of light by (a) a single slit, (b) a double slit and (c) a diffraction grating. The plots show the variation of the relative intensity I/I_o for normal incidence ($\theta_i = 0$) through (a) a single slit, (b) a double slit with $b/w = 3$ and (c) a diffraction grating with $b/w = 3$ and $N = 8$. Envelopes in plots (b) and (c) indicate the corresponding single-slit relative intensity variation, in case (c) multiplied by N^2.

Thus, the intensity of the diffracted light would alternate between relative maxima, at angles having $\tan\beta = \beta$, and relative minima equal to zero, at angles having $\beta = \pm k\pi, k = 1, 2, 3, ...$ The absolute maximum $I = I_o$ (*principal maximum*) would occur at $\beta = 0$, that is, when $\theta_d = -\theta_i$. In the case of normal incidence, for which $\theta_i = 0$, the absolute maximum would be at $\theta_d = 0$ and minima would occur at $\theta_d = \pm \sin^{-1}(k\lambda/w)$. The light pattern on the far screen would be a set of parallel bright fringes, corresponding to intensity maxima, and dark fringes corresponding to intensity minima, with the principal maximum being much brighter than any other. A simple explanation of this phenomenon can be based on *Huygens–Fresnel principle* [24], which states that '*every unobstructed point of a propagating wavefront at a given instant serves as the source of spherical secondary wavelets with the same frequency as that of the primary wave; the amplitude of the optical field at any point beyond is the superposition of all these wavelets considering their amplitudes and relative phases*'. At normal incidence, all light would reach the slit plane in-phase. Each point across the slit can be viewed as the source of spherical wavelets, which propagate in all directions and interfere with wavelets produced by other points. The light wave at a location after the slit would be the result of superposition of all wavelets emitted by points between the two edges of the slit, which reach that location with different phases. Integration of these wavelets across the slit leads to Eq. (7.33). One may interpret the parameter β at some location as one-half the phase difference between the two wavelets that reach this location having been emitted at the same time by the two edges of the slit.

A similar situation arises when monochromatic coherent light passes through a circular aperture. If this happens at normal incidence, and the aperture diameter is d, the intensity of the diffracted light would be

$$I = I_o \left[\frac{2J_1(\beta)}{\beta} \right]^2 \ , \ \beta = \frac{\pi d}{\lambda} \sin \theta_d \ , \tag{7.34}$$

where J_1 is the Bessel function of first order. The pattern on the far screen would be a bright central spot, called *Airy's disk*, surrounded by dark and bright rings. The first dark ring would occur at $\beta = 1.22\pi$.

Next, consider coherent light passing through two slits of equal widths w and separated by a distance b centre-to-centre (Fig. 7.13b). The diffracted light intensity beyond the slit plane would be

$$I = I_o \left(\frac{\sin \beta}{\beta} \right)^2 \cos^2 \gamma \ , \ \beta = \frac{\pi w}{\lambda} (\sin \theta_i + \sin \theta_d) \ , \ \gamma = \frac{\pi b}{\lambda} (\sin \theta_i + \sin \theta_d) \ , \tag{7.35}$$

where the reference for measuring angles would be the axis of symmetry of the double slit. Thus, the single slit angular distribution will be modulated by the $\cos^2 \gamma$ factor, so that the principal fringe in the single-slit configuration would now be replaced by a number of fringes, with maxima at $\gamma = k\pi, k = 0, 1, 2, ...$

A device consisting of a number of parallel, equidistant slits, with equal widths is called a *diffraction grating* (Fig. 7.13c). The intensity of monochromatic coherent light passing through a diffraction grating with N slits with widths w and separation distances b would be

$$I = I_o \left(\frac{\sin \beta}{\beta} \right)^2 \frac{\sin^2 N\gamma}{\sin^2 \gamma} \ , \ \beta = \frac{\pi w}{\lambda} (\sin \theta_i + \sin \theta_d) , \gamma = \frac{\pi b}{\lambda} (\sin \theta_i + \sin \theta_d) \ , \tag{7.36}$$

The maxima of the modulating function $\sin^2 N\gamma / \sin^2 \gamma$ are equal to N^2 and occur when $\gamma = k\pi, k = 0, 1, 2....$ This function represents the interference among wavelets produced by the N slits. Although the locations of these peaks are independent of N and the same as for the double-slit configuration, their strengths increase with increasing number of slits and their widths decrease accordingly. The diffraction angles θ_d at which the peaks occur would be such that

$$\sin\theta_i + \sin\theta_d = k\frac{\lambda}{b}, k = 0, 1, 2, \ldots, \quad (7.37)$$

namely they depend on the wavelength of light. Thus, it is possible to determine the wavelength of incident light by measuring the distances between the principal fringes on the screen. For this reason, diffraction gratings are used as *spectrometers*, a term denoting devices that measure radiation wavelength.

The slit-array is only one of several kinds of diffraction gratings. Owing to its function, it is known as *transmission amplitude grating*. A related device is the grating produced by scratching parallel grooves on a clear glass plate. Thickness variations would create phase differences across a light beam transmitted through the plate, and so the spherical wavelets emitted from the surface of the plate would interfere creating bright and dark fringes. This is known as *transmission phase grating*. The latter device may also serve as a *phase reflection grating*, if light is reflected on it, rather than being transmitted through it. Reflection gratings can be produced on opaque, reflective materials, such as aluminium. Most modern diffraction gratings are of the *blazed reflection phase grating* type, with a sawtooth-like reflecting surface.

7.3.9 Illumination Techniques

Sufficient illumination may often be achieved by simple means, but it is worthwhile to spend some time refining the selected technique. Thermal and LED light sources are usually operated in a flood-light arrangement. Even so, the position of the light source with respect to the visualised part of the flow and the observation/recording plane may be optimised by trial and error in order to obtain the highest clarity and contrast possible. For example, when markers are introduced into a flow, it is best to view it along a line perpendicular to its direction, while illuminating it along an axis inclined by 120°, either from the front or the back of the apparatus (Fig. 7.14a).

Collimators (Fig. 7.14b) are combinations of lenses, such as the lens of a projector, used to produce a cylindrical, or slightly diverging, light beam, with high intensity, possible to direct towards specific areas to reduce reflections. Combinations of collimators and slits, or cylindrical lenses, may be used to produce a 'sheet of white light' with a thickness of a few mm or cm. Further reduction of light reflections and increased contrast may be achieved by painting illuminated solid surfaces white or black and draping parts of the apparatus.

A laser beam provides high intensity illumination, albeit only of a narrow volume of fluid, with a typical cross-sectional diameter of the order of a mm. Whether the beam path in the fluid and intersected solids will be straight or not depends on whether the refractive index of materials along this path is uniform or not, and on the angle of intersections with interfaces. A number of optical arrangements have been used to increase the volume of laser-light illuminated fluid. These can be divided into *plane-illumination* and *volume-illumination* techniques. **Laser-sheet:** A commonly used and very effective method of illumination is the *laser-sheet* technique [35]. In its simplest form, a laser beam is passed through a cylindrical lens (e.g., a glass rod), which fans the beam into a diverging sheet of light of a thickness of a few mm (Fig. 7.15a). Further use of a converging lens may reduce the sheet thickness to typically 1 mm or less, while passing the sheet through a plano-convex lens will eliminate its lateral divergence and turn a triangular sheet into a rectangular one. Specially shaped lenses (*Powell lenses*) are also available to compensate for the variation of the ideal light intensity across the beam, and produce a light sheet of nominally uniform intensity. Even so, imperfections inside the laser and on the various lenses create non-uniformities in the intensity of the sheet, usually in the form of darker and brighter stripes. The use of an inexpensive, rotating, cylindrical rod has been suggested as a means to reduce the effect of stripes [47]. Rapid oscillatory scanning of the laser sheet across the illuminated plane, achieved with the use of commercial scanners or rotating-mirror galvanometers, has also been utilised to eliminate such non-uniformities. An example of the use of a rotating polygonal mirror to generate a light sheet by sweeping a laser beam is shown in Fig. 7.15b. Another advantage of the latter method, compared to the conventional light sheet method, is that the laser beam maintains its full intensity at any instant, rather than being spread over the entire sheet. An alternative approach to correct for intensity non-uniformity is to calibrate the optical system in situ. As an example, consider the laser sheet illumination of dye patterns in a turbulent flow in a water tunnel, for which one wants, in addition to qualitative visualisation, also to measure the local concentration of dye. Non-uniformity of the light sheet will produce erroneous values of dye concentration. This error can be corrected by applying a correction coefficient obtained by calibrating the output of the optical system using a container filled with a well-mixed dye solution of known concentration, inserted in the water tunnel at the location of interest [31]. Such calibration

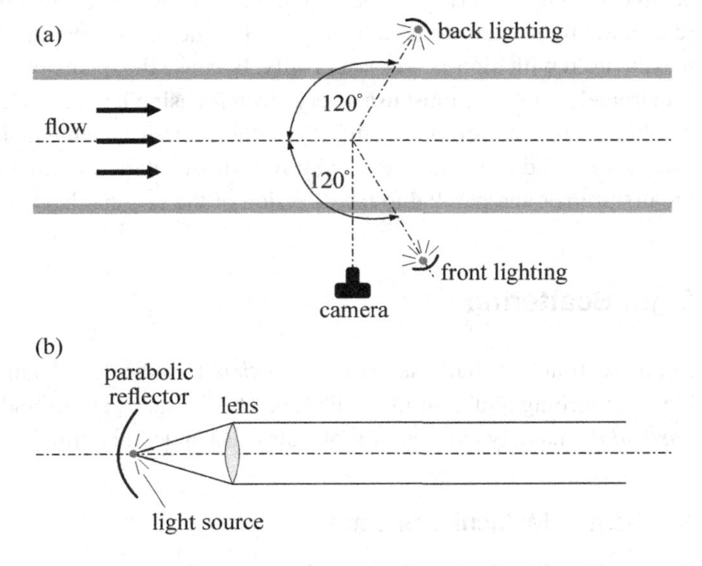

Figure 7.14 Sketches showing (a) optimal illumination arrangements and (b) an elementary light collimator.

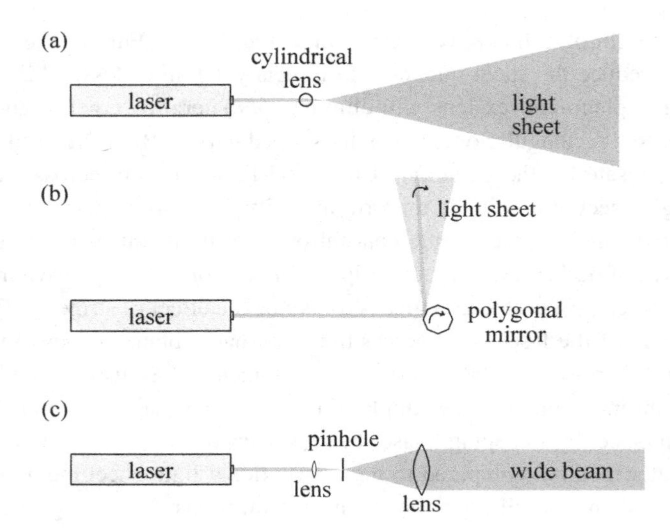

Figure 7.15 Conversion of a laser beam into (a) a light sheet with the use of a lens, (b) a light sheet with the use of a polygonal mirror and (c) a wide beam.

can also compensate for other imperfections in the optical components. The application of the light-sheet to the flow could be continuous or interrupted, with the latter achieved either by using a pulsed laser or by interrupting the beam of a CW laser by a shutter, rotating perforated disk or other means.

Volume illumination: Volume illumination can be achieved either by expanding a laser beam or by sweeping a beam or a light-sheet. A lens arrangement, consisting of a converging lens to focus the beam, a pinhole on the focal plane to remove unwanted light, and another confocal, converging lens, can be used to expand the original laser beam to a *wide beam* (Fig. 7.15c), having a diameter of several cm. Obviously, this will reduce the light intensity by a factor equal to the beam area ratio, so that this approach may be used with high-power lasers only. Because the intensity of a focused laser beam is extremely high, one must use a very clean focusing lens. To sweep a light sheet across a volume, one may simply modify the light-sheet arrangement shown in Fig. 7.15b by adding a second rotating mirror, with its axis of rotation normal to the rotation axis of the first mirror and parallel to the direction of the original laser beam [9, 48].

7.4 Light Scattering

In this section, we shall use the term *particle* to indicate a small parcel of mass, capable of absorbing and emitting radiation. This includes atoms and molecules as well as *particulate*, namely clusters of molecules suspended in a fluid.

7.4.1 Atomic and Molecular Spectra

As mentioned in Section 7.1, atomic spectra contain peaks at characteristic resonant frequencies, flanked by narrow spectral bands, which are created by thermal motion, atomic collisions and other effects. Molecular spectra are more complex. Molecules have *electronic energy*, associated with the states of electrons in each atom, but also *vibrational* and *rotational energies*, corresponding to vibrations and rotations of the molecule nucleus. All energies are *quantised*, which means that each molecule is stable at specific energy levels only, which distinguishes it from other types of molecules. Each electronic state has a number of vibrational and rotational energy levels associated with it. A representative sketch of the possible energy levels for a diatomic molecule, plotted vs. the separation distance between the nuclei of the two atoms, is shown in Fig. 7.16. Molecular energy transitions involve changes of any or all of these energies. Matters are further complicated by the fact that the three energy modes are coupled to one another: the vibrational energy levels depend on the electronic energy state and the admissible rotational energy levels depend on the vibrational energy of the molecule. Thus, molecular spectra have more peaks than atomic ones, broadened by thermal motion, intramolecular interactions and other effects.

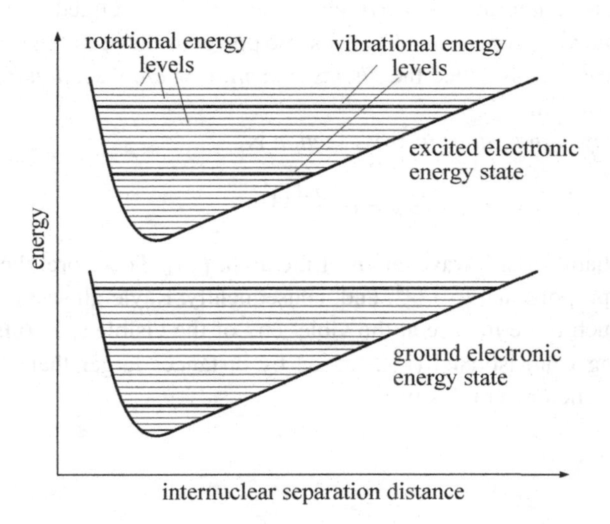

Figure 7.16 Schematic representation of the ground electronic energy state and an excited electronic energy state of a diatomic molecule, together with corresponding vibrational and rotational energy levels.

7.4.2 Elastic and Inelastic Scattering

Consider that an atom is approached by a photon, whose energy is associated with a frequency ν, according to Eq. (7.6). The photon will be *absorbed* by the atom, if its frequency matches one of the *resonant frequencies* of the atom. Normally, in solids and liquids as well as gases at ordinary pressures, the absorbed energy of the photon will be converted to *thermal energy*, that is, random motion, without emission of another photon. This process is called *dissipative absorption*. However, if the frequency of the photon is significantly lower than all resonant frequencies of the atom, the absorbed energy will momentarily cause an electron to oscillate, while the atom remains in its ground state. Thus, the oscillating electron and the positive nucleus become an *electric dipole* (i.e., a pair of oscillating electric charges), radiating an electromagnetic wave at the same frequency as the incident one. This process is called *non-resonant scatter-*

ing or *elastic scattering* and takes place essentially instantly, within less than 10^{-15} s. In general, the photons are emitted in random directions, so that, when continuously irradiated, the atom effectively becomes a source of spherical electromagnetic waves. Besides elastic scattering, molecules are also capable of scattering light *inelastically*, namely at frequencies that are higher or lower than the incident frequency. When a molecule is excited by absorbing a photon, it is possible that it will emit a photon with energy that is different from the energy of the absorbed photon, with the (positive or negative) balance being converted to vibrational and/or rotational energy.

7.4.3 Rayleigh Scattering

Light scattering from particles that are smaller than about $\lambda/15$ (λ is the incident light wavelength) is called *Rayleigh scattering* [24]. The efficiency of light scattering from a particle is expressed in terms of its *scattering cross section*, defined as the equivalent area of the incident wavefront which has the same power as that emitted by the particle. For a single electron, this is called the *Thomson scattering cross section* σ_T, which has a value of $6.65 \times 10^{-29} \mathrm{m}^2$.

The classical *Rayleigh scattering cross section* is

$$\sigma_R = \sigma_T \left(\frac{\lambda_o}{\lambda} \right)^4 , \tag{7.38}$$

where λ_o is the characteristic wavelength of the atom [14]. Therefore, the intensity of scattered light is proportional to $1/\lambda^4$, and, consequently, Rayleigh scattering of white light would be much more intense at the violet end of the visible spectrum than at the red end. Scattering from isolated (i.e., spaced by distances larger than λ) atoms and ordinary molecules belongs to this type.

7.4.4 Mie Scattering

When many atoms and molecules are close to each other, their electromagnetic waves interfere with each other and lateral scattering diminishes, while becoming direction-dependent. The intensity of light scattered by spherical particles of arbitrary size can be calculated based on a theory referred to as *Mie scattering theory* [8, 17, 20, 30, 39]. Rayleigh scattering is the limit of Mie scattering as the particle diameter diminishes towards zero. Extensions of Mie's theory include predictions of light scattering from coated and optically inhomogeneous spheres and from cylinders [30]. In practice, the term Mie scattering is often used to describe light scattering not only from spherical particles but generally from aggregates of molecules of arbitrary shape. Like Rayleigh scattering, Mie scattering is an elastic process.

Consider a spherical particle with a diameter d_P and a refractive index n_P, immersed in a fluid with a refractive index n_F, and exposed to a collimated beam of randomly polarised, monochromatic light of wavelength λ and intensity (radiant power per unit cross-sectional area) I_o, as shown in Fig. 7.17. According to Mie scattering theory, the radiant intensity (power per unit solid angle) of light scattered by the particle towards a

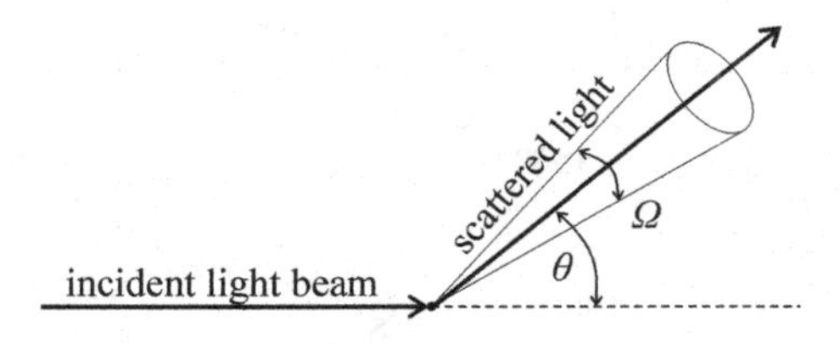

Figure 7.17 Sketch showing light scattered by a spherical particle towards a collecting lens.

particular direction forming an angle θ with the incident light direction would be equal to $I_o\sigma_\lambda$, where the *monochromatic angular scattering cross section* σ_λ is given by

$$\sigma_\lambda = \frac{\lambda_F^2}{8\pi^2}\left[i_1\left(x, n_P/n_F, \theta\right) + i_2\left(x, n_P/n_F, \theta\right)\right] . \tag{7.39}$$

In this expression, $x = \pi d_P/\lambda_F$, $\lambda_F = \lambda/n_F$, and the *intensity* or *phase functions* i_1 and i_2, in units of inverse steradians, correspond to the two components of the scattered light, both of which are plane-polarised, with the corresponding waves oscillating on planes perpendicular and parallel to, respectively, the plane formed by the incident and the scattered beams. The intensity functions for a particle with a specified size and refractive index can be computed using analytical expressions provided by Mie theory. If the incident light is linearly polarised, the scattered light will be linearly polarised as well, but on a plane perpendicular to the polarisation plane of the incident light; its monochromatic angular scattering cross section will be

$$\sigma_\lambda = \frac{\lambda_F^2}{4\pi^2} i_2\left(x, n_P/n_F, \theta\right) . \tag{7.40}$$

In contrast to the Rayleigh scattering cross section, which is independent of orientation, the Mie scattering cross section is a strong function of the scattering angle θ. The maximum amount of scattered light always corresponds to *forward scattering* ($\theta = 0°$), while a local maximum or minimum also occurs for *backscattering* ($\theta = 180°$). Local maxima and minima may also occur at intermediate angles, the number of which increases with increasing x. To illustrate the dependence of σ_λ on the scattering angle, we present three representative cases in Fig. 7.18. This figure shows that the directional asymmetry of scattered light increases drastically as the particle diameter increases, while, as the diameter decreases, light scattering becomes less dependent on the scattering angle, thus approaching a Rayleigh scattering behaviour.

The total power of scattered light that reaches a certain aperture within a solid angle Ω (Fig. 7.17) would be

$$\Phi_\Omega = I_o \int_\Omega \sigma_\lambda \mathrm{d}\Omega . \tag{7.41}$$

It is evident that, the larger this aperture is, the smoother the variation of the collected light power with scattering angle θ would be.

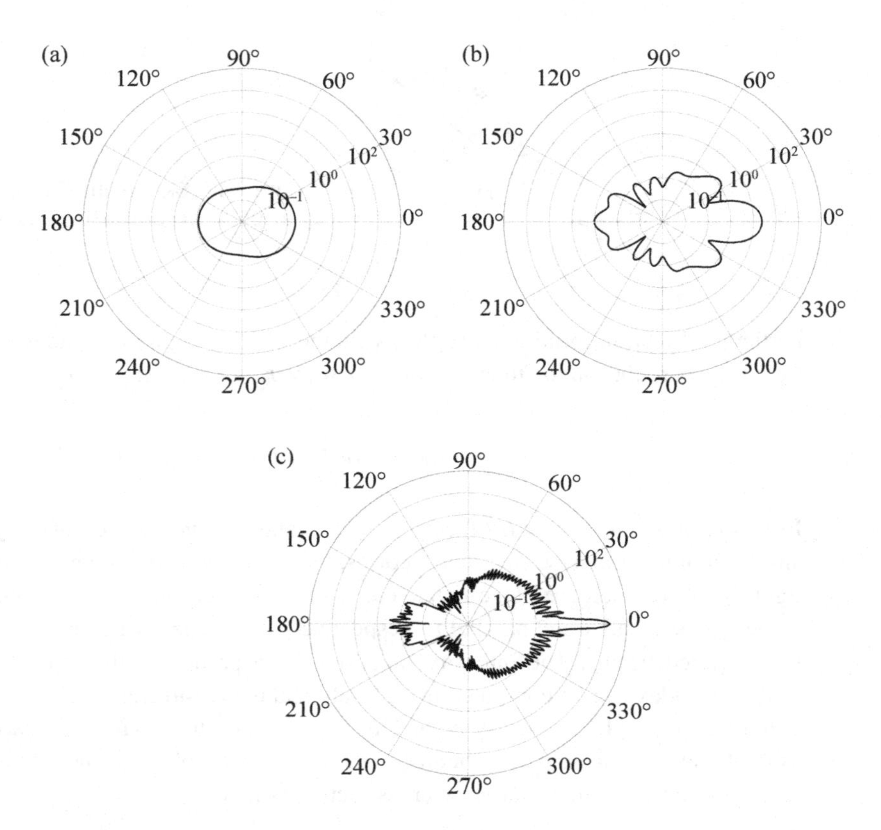

Figure 7.18 Monochromatic ($\lambda = 0.488$ μm) scattering cross sections of glass spheres in air with (a) $d_p = 0.1$ μm, (b) $d_p = 1$ μm and (c) $d_p = 10$ μm, according to Mie scattering theory. Note that the radial axis is logarithmic, with values changing by an order of magnitude from one circle to the next. The angle of 0° corresponds to forward scattering.

7.4.5 Raman Scattering

The phenomenon of inelastic light scattering from molecules is called the *spontaneous Raman effect*. This phenomenon occurs quite rarely, typically with an occurrence frequency that is 10^{-5} to 10^{-2} times lower than that for Rayleigh scattering. Thus, the *Raman scattering cross section* σ_{Rm} for each species and a specific energy level is several orders of magnitude smaller than the corresponding Rayleigh scattering cross section. If the energy of the emitted photon is higher than that of the absorbed photon, the process is called *Stokes transition*. In addition to Stokes transition, it is possible for a molecule to emit a photon of energy lower than that of the absorbed photon; this process is called *anti-Stokes transition* [24]. The time between photon absorption and emission for Raman scattering is of the order of 10^{-14} s, which is negligible for most practical purposes. Table 7.6 summarises the Raman-scattered spectral lines of common molecules in air at room temperature and pressure [51]. It can be seen that both Stokes and anti-Stokes

Table 7.6 Wavelengths of Raman-scattered radiation for some common molecules in air at standard atmospheric pressure and a temperature of 295 K; excitation was provided by a ruby laser

Molecule	Anti-Stokes line [nm]	Stokes line [nm]
Incident light	694.30 (Rayleigh line)	
CO_2	638.23	761.17
CO_2	637.92	762.23
CO_2	633.25	768.37
CO_2	632.42	769.61
O_2	626.57	778.44
O_2^+	615.50	796.23
NO	614.26	798.31
CO	604.34	815.73
N_2^+	603.19	817.82
N_2	597.57	828.39
CH_4	577.45	870.462
H_2	538.67	976.408

lines are present, but it must also be noted that the anti-Stokes lines have significantly higher energy at relatively low temperatures.

7.4.6 Fluorescence and Phosphorescence

The Rayleigh and Raman transitions occur essentially instantaneously, thus not allowing other energy conversion phenomena to occur. Certain molecules, however, are capable of emitting radiation after a certain delay time following the absorption of a photon. When such emission takes place relatively rapidly, typically within the range between 10^{-10} and 10^{-5} s, the phenomenon is called *fluorescence* [19]. When it is relatively slow, following time delays between 10^{-4} s and up to hours, it is called *phosphorescence*. Fluorescence allows sufficient time for collisions of molecules to take place and photon energy to be converted to chemical reaction, dissociation and ionisation energies, before emission takes place. These phenomena, referred to as *quenching*, interfere with the emission process and create measurement uncertainty in fluorescence-based diagnostic and measurement methods. Besides photon absorption, fluorescence may be caused by electron bombardment, heating or chemical reaction (*chemiluminescence*). Although *resonant fluorescence* (i.e., emission at the same wavelength as that of the absorbed light) is possible, it is *non-resonant fluorescence* (i.e., emission at a wavelength that is longer than the incident one) which is utilised for flow visualisation and measurement, as it permits the separation of fluorescence radiation from the incident light and Mie scattering. This happens when the molecule absorbs a photon that causes transition from the ground energy state to a higher vibrational level of an excited electronic state, but then returns to the ground state from a lower vibrational level, so that the energy of the emitted radiation is lower than the absorbed one. Fluorescence of several species is quite vigorous, exhibiting cross sections that are several orders of magnitude larger than the corresponding Raman scattering cross sections. The intensity of fluorescence

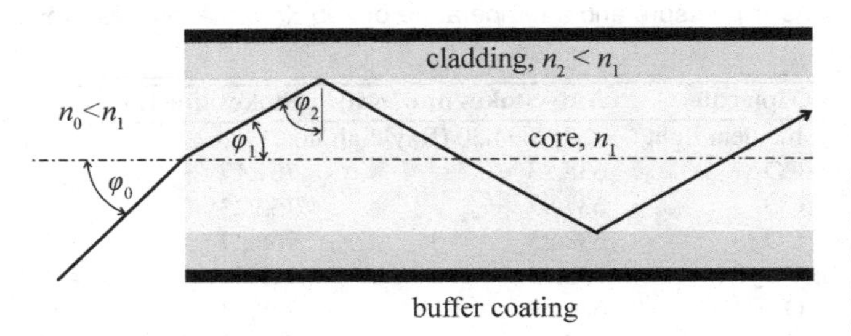

Figure 7.19 Schematic diagram of a step-index optical fibre and the propagation of a light ray along it.

is independent of orientation and the fluorescent radiation is randomly polarised, even though the incident radiation may be linearly polarised.

7.5 Light Transmission, Sensing and Recording

Observation of flow patterns by eye may be sufficient for some simple visual and optical experiments. In general, however, it would be necessary to transmit light to a convenient location and then to measure and record its intensity, wavelength and other properties, which may be further processed later. The present section summarises commonly used instrumentation and techniques for the transmission, sensing and recording of light and images.

7.5.1 Fibre Optics

Optical fibres, or *optical waveguides*, are thin cables capable of transmitting internally light over considerable distances with high efficiency [4, 5, 24, 44]. They consist of a *core*, which is a fine fibre of glass, quartz or plastic (commonly methyl methacrylate), a *cladding*, which is a thin shroud of glass or plastic surrounding the core, and a *buffer coating*, which protects the fibre (Fig. 7.19).

The refractive index of the fibre usually changes abruptly from a value n_1 in the core to a smaller value n_2 in the cladding, although fibres with a gradual decrease of n from the core towards the cladding are also available. Transmission of light along the core is effected through reflections (see Section 7.2.3) at the interface with the cladding (Fig. 7.19). To ensure that all light is reflected at this interface and none penetrates through the cladding, the incidence angle φ_2 must remain larger than the critical angle $\varphi_c = \arcsin(n_2/n_1)$ for total internal reflection. Application of Snell's law gives the *maximum acceptance angle* for light entering the fibre to be transmitted along the core as

$$\varphi_o = \varphi_{\max} = \arcsin \frac{\sqrt{n_1^2 - n_2^2}}{n_o}, \tag{7.42}$$

where n_o is the refractive index of the surrounding medium. For typical values of $n_o \approx$

1.00 (air), $n_1 \approx 1.62$ and $n_2 \approx 1.52$, one may calculate $\varphi_{\max} \approx 34°$. The parameter

$$\mathrm{NA} = n_o \sin \varphi_{\max} = \sqrt{n_1^2 - n_2^2} \tag{7.43}$$

is called the *numerical aperture* of the fibre. Commercial fibres are available with NA in the range between 0.2 and 1.

An important distinction among optical fibres is with respect to the *modes of propagation* of light along them. This arises from the fact that a ray propagating along the axis of the fibre will traverse the fibre faster than a ray that undergoes successive reflections at the cladding. The number of reflections, and thus the delay in propagation, increases with increasing entry angle φ_o. This phenomenon is called *modal dispersion*, which is distinct from *chromatic dispersion*, due to variation of refractive index with light wavelength. Thus, waves propagating along the fibre accumulate phase shifts; when different waves are in-phase, they interfere constructively adding their energies, otherwise they interfere destructively and eventually fade away. As a result, only certain modes propagate effectively through the fibre, appearing as distinct light patterns. A parameter that describes the number of these modes is the *V-number*

$$V = \frac{2\pi}{\lambda} r_1 \mathrm{NA}, \tag{7.44}$$

where r_1 is the radius of the core and λ is the wavelength of propagating light. When $V < 2.405$, only a single mode, called the *fundamental mode*, will exit the fibre, whereas, when $V > 2.405$, multiple modes will do so. Thus, by maintaining the core diameter sufficiently small, typically less than 10 μm, it is possible to create *single-mode fibres*, which are preferable in most applications in fluid mechanics.

7.5.2 Light Sensing

Light may be converted to an electric current with the use of devices called *photodetectors* or *photosensors* [28]. In the following, we will restrict the discussion to phenomena and devices that are of main interest in fluid mechanics research. Moreover, we will focus on *quantum sensing*, which refers to light sensing following absorption of one or more photons. *Thermal sensing*, namely, sensing of light by the increase of temperature of an irradiated surface, will be discussed in Chapter 13.

Photosensing: Electric conduction in solid materials is generally associated with the motion of charges (electrons or holes) in two bands: an inner *valence band* and an outer *conduction band*. In metals, this distinction is of little consequence, but, in semiconductors, these bands are separated by a gap, which cannot be occupied by electrons and so an electron from the non-conducting valence band requires a minimum amount of energy to transition into the conduction band. The *photoelectric* or *photoemissive* effect is the phenomenon of absorption of a photon by a photodetector and the subsequent emission of a free electron. This happens when the photon has sufficiently large energy (or, equivalently, sufficiently large frequency or sufficiently short wavelength) for an electron that occupies the conduction band of a metallic detector to transition outwards to a free state within a gas or in vacuum. The motion of these electrons is an electric current, which is proportional to the number of received photons. A phenomenon that is related to, but is distinct from, the photoelectric effect is the *photovoltaic* effect, which occurs in semi-conductors. This phenomenon refers to the absorption of a photon with sufficient energy to excite an electron occupying the valence band so that it transitions to the conduction band, thus generating a measurable current.

Quantum efficiency and responsivity: The *quantum efficiency* of a photodetector is defined as

$$\eta_q = \frac{N_e}{N_p} , \tag{7.45}$$

where N_p is the number of photons that are incident on the photodetector during time δt and N_e is the number of electric charges induced by the photodetector during the same time. The quantum efficiency depends on the frequency ν, or, equivalently, the wavelength λ, of incident radiation. The radiation power collected by the photodetector is $\Phi_e = N_p h\nu/\delta t$, where h is the Planck constant. The electric current produced by the photodetector would be

$$i = \frac{N_e q_o}{\delta t} = \eta_q \frac{\Phi_e q_o}{h\nu} , \tag{7.46}$$

where q_o is the electron charge. Another commonly used property of photodetectors is the *responsivity*

$$R \equiv \frac{i}{\Phi_e} = \eta_q \frac{q_o}{h\nu} . \tag{7.47}$$

Considering the incident radiation as the input and the current as the output of the photodetector, one may view the responsivity as the gain of the light detection system. The responsivity is inversely proportional to the light frequency and so the gain of detectors of infrared radiation would generally be higher than that of visible-light detectors.

Photomultiplier tubes and photodiodes: *Photomultiplier tubes* (PMT) collect light on a semi-transparent *photocathode*, which absorbs photons and emits electrons according to the photoelectric effect. The electrons are subjected to electric fields between successive pairs of *grids-dynodes* and increase in numbers until they are captured by an *anode grid*. PMT have a low quantum efficiency, but provide a strong electric output due to internal amplification. *Photodiodes* (*PD*) or *photoelectric cells* are p-n junctions of semiconductors, commonly silicon–silicon type. They have a high quantum efficiency but no internal amplification and so require an external amplifier in order to produce a current at a usable level. By comparison to PMT, photodiodes are much less expensive and less bulky, but they generally provide a lower signal-to-noise ratio, except at very high light levels. A variance of the photodiode is the *avalanche photodiode*, which has some internal amplification and so has characteristics intermediate between those of PMT and common photodiodes.

Digital cameras: Having undergone spectacular technological progress in recent decades, digital cameras have entirely replaced conventional cameras and they are still progressing in quality and speed of imaging. Some current digital cameras are based on *charge-coupled device* (CCD) technology. A CCD contains a photosensitive sensor, referred to as the *image sensor*, which consists of an array of coupled metal-oxide semiconductor (MOS) capacitors. Each capacitor is capable of transferring its charge to neighbouring ones across the array, whereby generating a detectable current. CCD image sensors are being replaced by complementary metal oxide semiconductor (CMOS) image sensors, which provide higher speed of imaging, up to 100,000 frames per second (fps) in some cases, compared to the typical 15 fps speed of CCD cameras. All digital cameras provide images consisting of discrete elements, called *pixels* (pixel is a composite of parts the words 'picture' and 'element'), which are arranged in a grid-type manner. The information in each of these pixels is produced by a corresponding element on the image sensor, which is also referred to as a pixel. The input to the measuring system, which provides a pixel of an image as output, is light, which is sensed and converted to a current by a *photodiode* and then it is conditioned and discretised by different electronic circuits. In CMOS cameras, much of this circuitry has been integrated with the sensor to produce an *active pixel sensor* (APS). Typically, images obtained with CCD cameras tend to have more pixels than those from CMOS cameras; for example, images obtained with cameras in current mobile telephone devices have up to 16 MP (megapixels), whilst high-grade scientific CMOS cameras have typically 4 MP. Scientific CMOS (sCMOS) cameras were developed recently to capitalise on the benefits of both CMOS and CCD technologies. Such cameras have a higher speed of imaging than CCD cameras, currently around 30 Hz, whilst maintaining a relatively large number of pixels.

Noise of photodetectors: As mentioned in Section 6.5, photodetectors are subjected to two types of noise: (a) *shot noise* (or photon noise), due to random fluctuations of the rate of photon collection and to background illumination, and (b) *thermal noise*, caused by amplification of the current inside the photodetector and, if present, by an external amplifier. Even in the absence of a desirable source of light, photodetectors produce a current, called *dark current*. Dark current decreases with decreasing photodetector temperature. This observation has led to the development of cooled low-noise cameras.

7.5.3 Image Recording

Basic camera parts and terminology: A basic camera consists of a *camera lens*, which is actually a system of lenses and an *aperture stop*, and a photosensitive surface, referred to as the *image sensor* (Fig. 7.20). Scientific cameras normally have a *main body*, onto which the user can attach different camera lenses, including *macro* and *telescopic* lenses, as well as other accessories, such as optical filters. The camera lens collects light and focuses images on the sensor, while the aperture stop controls the amount of light that enters the camera. Camera lenses have either a fixed or a variable focal length. Fixed lenses provide a superior image quality, because light through them is transmitted through fewer components. The camera sensor only receives light that originates from part of the space and passes through the lens. In general photography sources, the imaged part of space is called the *field of view* w, which is the size of the in-focus, planar area that is imaged on the sensor of the camera. The size of the field of view depends on the working length, that is, the distance x from the *optical centre* (also referred to as the *nodal point*) of the camera lens to the in-focus plane, the *focal length of the camera lens*

z_0 and the size s of the sensor. The camera lens focal length z_0 is defined as the distance between the image sensor and the optical centre, which is the location on the camera lens axis where the entering light rays converge. Therefore, one would naturally expect that a camera lens with a longer focal length would be larger than a lens with a shorter focal length; for example, a 150 mm lens is larger than a 28 mm lens. For a rectangular sensor, having a height s_h and a width s_w, the field of view would have a height w_h and a width w_w, which are given by the simple expressions

$$w_h = s_h x / z_0 \quad \text{and} \quad w_w = s_w x / z_0 \, . \tag{7.48}$$

It is common for camera manufacturers to specify an *angular field of view* α, from which one can obtain the size of the field of view as, that is, $w = 2x \tan(\alpha/2)$. It can be shown, therefore, that *wide-angle lenses* correspond to small camera lens focal lengths z_0; for example, a 28 mm camera lens is considered to be a wide-angle lens. Fixed focal length camera lenses typically have a limited working length, allowing for minor adjustments to ensure that the image is in focus; hence the user would have to move the camera closer to or further from the object, if the desired field of view, and hence focus, is not attained. Variable focal length camera lenses, or *zoom lenses*, allow the user more flexibility in terms of focusing an image for a given working length; however, this comes at the expense of a change in the size of the field of view.

The *magnification factor* M is defined as the ratio of the corresponding dimensions of the sensor and the field of view, namely, as

$$M = s_h / w_h = s_w / w_w \, , \tag{7.49}$$

and is typically expressed either as a ratio, for example, 1:2, or as a multiplier, for example, 0.5X, which in this case indicates that the field of view is twice the size of the image sensor. If the magnification factor is relatively small, say $M < 0.1$, or the working length is relatively large, an approximation to the magnification factor can be obtained using the focal length of the camera lens and the working length as $M \approx z_0/x$. Regular camera lenses have magnification factors that are smaller than 1, whereas *macro lenses*

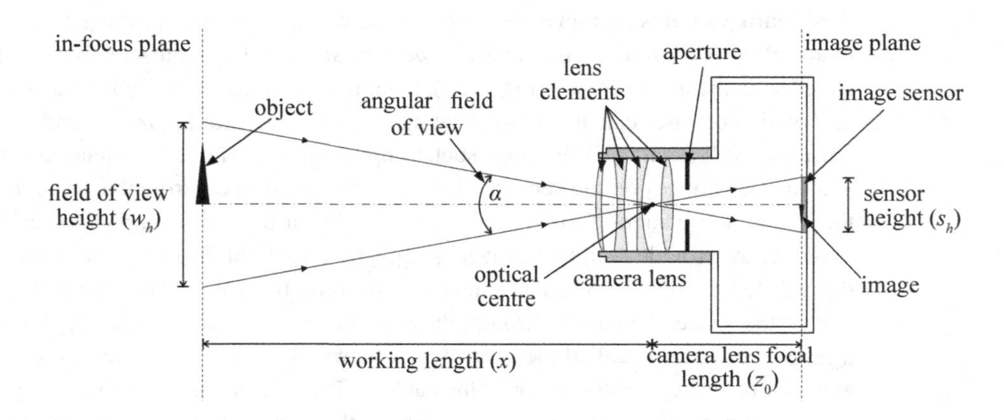

Figure 7.20 Simplified sketch of a basic digital camera with a camera lens system.

have magnification factors of 1:1 or greater, that is, they magnify the object onto the image plane (image sensor).

The *clear aperture* is the unobstructed area of the lens that is exposed to light. Its size is set by the aperture stop, which is either the rim of the lens or a diaphragm with an adjustable opening, called the *iris*. For digital cameras, the photosensitive surface is an array of photodetectors. The time interval during which the photosensitive surface is exposed to light is controlled by the *shutter*, which could be mechanical (e.g., a metal plate covering the lens) or electronic, which controls the time images are recorded.

Camera characteristics: The main performance characteristics of a camera, as they pertain to measurements in fluid mechanics, are its *resolution* and its *dynamic range*. The resolution of a digital camera specifies the number of pixels that are used to discretise an image; for example, a 1 MP camera would discretise an image into a grid of 1024×1024 pixels, for a total of 1,048,576 pixels. Most scientific cameras are monochromatic, because this ensures a relatively high sensitivity. The dynamic range of such cameras specifies the number of shades of grey that the image is displayed in and is given in terms of bits, for example, 1-bit, 2-bit etc. A 1-bit camera will only display black and white colours, whilst a 4-bit camera will also display 14 shades of grey between black and white, for a total of $2^4 = 16$ shades. Both of these characteristics are illustrated in Fig. 7.21.

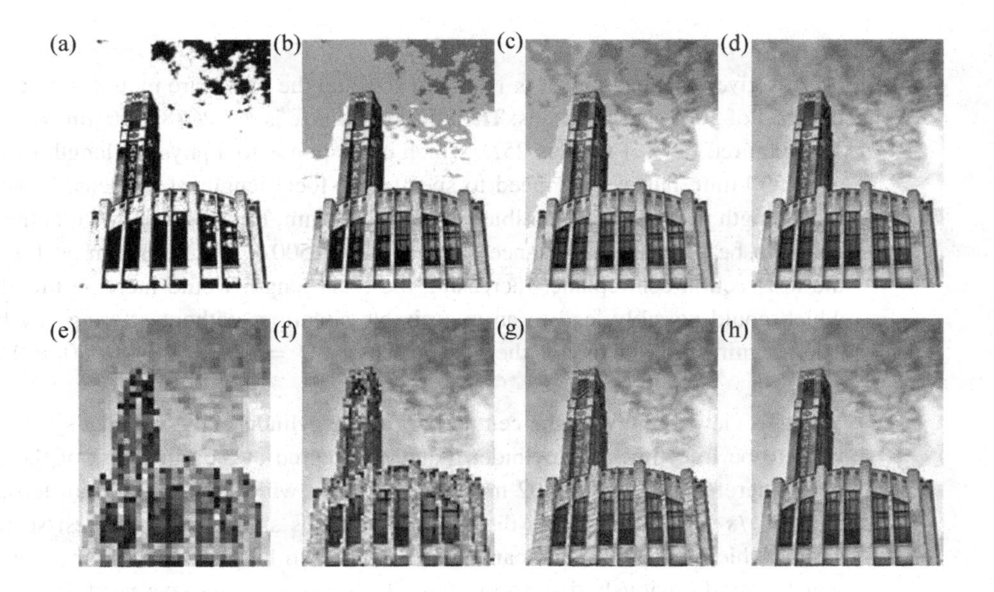

Figure 7.21 Images with (a) 1-bit, (b) 2-bit, (c) 3-bit and (d) 4-bit dynamic ranges, all with a resolution of 3024×4032 pixels. Images with a fixed dynamic range, but with spatial resolutions of (e) 32×42 pixels, (f) 64×85 pixels, (g) 256×341 pixels and (h) 3024×4032 pixels.

Example 7.4 A 50–110 mm zoom lens is attached to a 2 MP (2048 × 2048 pixels) digital camera, having a pixel size of 10 μm. This camera is used to view the flow around a cylinder with a diameter $D = 0.02$ m, inserted in a wind tunnel with a test section that is 0.5 m wide, 0.5 m high and 1 m long. The cylinder is placed in the middle of the test section, spanning its full width. A sketch of the experimental set-up is shown in Fig. 7.22. (a) If we wish to view the flow $12.5D$ upstream and downstream of the cylinder, how far away should we place the camera? Determine the corresponding magnification factor. (b) What should we do, if we want to view the flow near the central part of the cylinder only? Suggest how this can be accomplished. Also determine the corresponding magnification factor.

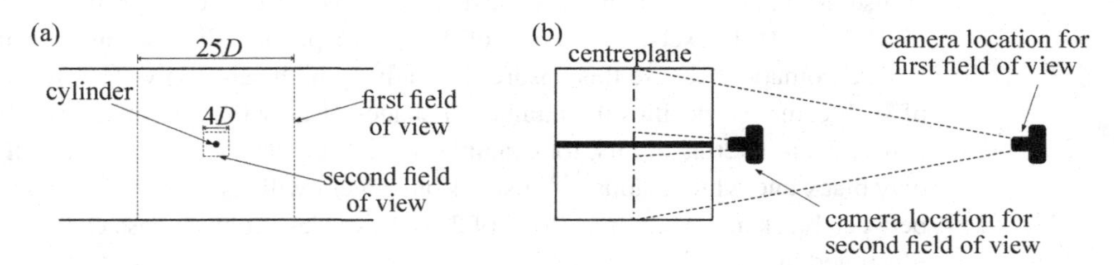

Figure 7.22 (a) Side view and (b) front view of the experimental set-up, showing the location of the cylinder and the two fields of view with the corresponding camera positions.

Answer

(a) It is given that each pixel is 10 μm in size and the pixels are mounted on the sensor in a grid of 2048 × 2048 pixels. Then the sensor size is $s = 2048 \times 10\ \mu\text{m} = 20.48$ mm. The required field of view is $25D$, which corresponds to a physical length of $w = 0.5$ m = 500 mm, but we also need to specify the focal length of the lens. If we set the focal length to the lowest possible value $z_0 = 50$ mm, Eq. (7.48) shows that the camera needs to be placed at a distance $x = wz_0/s = 500 \times 50/20.48$ mm ≈ 1.2 m from the test section centreplane. Increasing the focal length would increase this distance, which could possibly lead to an undesirable situation, without offering any benefits. The magnification factor for the 50 mm lens is $M = s/w = 20.48/500 \approx 0.04$X (or 1:25).

(b) To view the flow in the central part of the cylinder only, we may set the field of view to be four times the cylinder diameter, centred at the mid-point of the cylinder axis. Therefore, $w = 4 \times 0.02$ m $= 0.08$ m, from which, for the 50 mm lens, we get $x = wz_0/s \approx 195$ mm. This distance, however, is smaller than the test section half width, which means that the camera would have to be placed inside the test section, which would obviously disturb the flow. To avoid this issue, we need to increase the focal distance of the lens. To set the minimum image distance to just over half the width of the tunnel, say at $x = 0.3$ m, we should select a focal length $z_0 = sx/w = 76.8$ mm. The magnification factor for this setting would be $M = s/w = 0.26$X (or 1:3.19).

Camera settings: The basic settings of a camera are the *f-number*, the *shutter speed* and the *ISO speed*.

The f-number, denoted as $f^{\#}$, of the lens, also called *f-stop* or *focal ratio*, is defined as the ratio of the focal distance of the lens and its clear aperture diameter. Camera lenses indicate the ratio of the lens focal distance to the f-number, for example, $f/2.8$, $f/4$, $f/5.6$, $f/8$, $f/11$, $f/16$, in which case $f^{\#} = 2.8, 4, 5.6, 8, 11$, and 16, respectively. The larger $f^{\#}$ is, the smaller the aperture opening will be. This will result in less light being collected by the lens, which in turn would result in a deeper field of view. Conversely, a lower $f^{\#}$ would be preferable when one wishes to have a shallower *depth of field*, thus allowing one to focus on a particular plane with the background blurred. This is particularly important for the particle image velocimetry technique. To determine the depth of field for general photography purposes, one is referred to photography references and manuals; a measure of the depth of field, as it pertains to experimental fluid mechanics, is presented in Section 7.6.3.

The shutter speed, also referred to as *exposure time*, is indicated in seconds, as, for example, ..., 1/1,000, 1/500, ..., 2, 4, ..., and indicates the length of time during which the camera sensor is exposed to light. The primary function of this setting is to keep image blur as low as possible, while at the same time producing a sufficiently bright image. If the object moves while the shutter is open, this motion will result in some image blur, because successive object positions during the exposure time are superimposed on each other. For example, the image of a waterfall taken with a relatively long exposure time would have a blurred, but smooth, sheet-like appearance.

The ISO speed, which is a combination of the ASA and DIN specifications for conventional photographic films, allows recording of images in progressively darker environments. The ISO speed is specified in numbers that are doubled (e.g., ISO 100, ISO 200, ISO 400, ISO 800), which indicates doubling of the image brightness. Increasing the ISO speed allows one to photograph a darker subject, but also creates more 'grainy' images. An improvement of the quality of digital images and extraction of their features can be made by post-processing them with a variety of image processing software, which are available both in the public domain and under licence.

7.6 Characteristics of Seeding Particles

Visual and optical methods provide visible images of fluids in motion and measure different flow properties either from measured variations of the refractive index of the fluid itself or from the measured motion of visible *markers*, consisting of foreign material that is immersed in the fluid and transported by it. For a correct interpretation of marker images and their relationship to flow characteristics, it is necessary to understand the following conditions:

- The reason for the visibility of a pattern in marker location or motion, according to the principles of optics.
- The physical process of generation or injection of markers.

- The space-time relationship between marker motion and fluid motion.
- The physical significance of the observed images *vis-à-vis* streamlines, pathlines, streaklines and timelines (see Chapter 1).

7.6.1 Flow Seeding

Untreated air and water flows normally contain a significant concentration of natural foreign particles, such as dust, lint, microorganisms etc.These may actually be utilised as flow markers by certain visual and optical methods. Because, however, there is no control of the size, density, shape and optical properties of natural markers, it is often preferable to filter them, either entirely or down to a tolerable maximum size, and then to introduce deliberately particles whose characteristics are more suitable for the purposes of the study [3, 20, 34]. This process is called *seeding*. Some popular seeding materials and their refractive indices are listed in Table 7.7.

Table 7.7 Common seeding materials with their typical diameter d_P, refractive index n_P and density ρ_P

Seed material	d_P [μm]	n_P	ρ_P [kg/m^3]	Fluid
Water	1–10	1.33	1000	Gas
NaCl crystals	$\gtrsim 1$	1.54	2,160	Gas
Diethylhexylsebacat (DEHS)	1–3	1.45	910	Gas
TiO_2	0.01–0.25	2.5	4,230	Gas
Oil mists	1–10	1.4–1.5	900–1,000	Gas
Helium-filled soap bubbles	300	1.2–1.4	≈ 1.2	Gas
Polystyrene	0.05–100	1.59	1,005	Gas/liquid
Polyamid	20–100	1.5	1,030	Gas/liquid
Al_2O_3	2–5	1.76	3,960	Gas/liquid
SiC	1–75	2.6	3,200	Gas/liquid
Hollow glass beads	10	1.52	1,100	Liquid

Many visual and optical methods work best when all particles have a uniform size. An indicator of size uniformity is the *monodispersity*

$$\sigma_g = \sqrt{d_2/d_1}\,, \tag{7.50}$$

where the diameters d_1 and d_2 are defined such that approximately 15.9% of the particles have diameters smaller than d_1 and 15.9% of the particles have diameters larger than d_2. This means that, for a Gaussian particle distribution, d_2 and d_1 are, respectively, larger and smaller than the mean diameter by one standard deviation. For a perfect monodispersity, $\sigma_g = 1$. If $\sigma_g \approx 1$, a particle distribution is called *monodisperse*, otherwise it is called *polydisperse*. Particles may be generated and/or distributed in a flow by several methods, using either home-made or commercial devices. For liquid flows, a usual approach is to mix a powder (e.g., SiC or Al_2O_3) with the same liquid to a suitable concentration of particles, taking care to break down agglomerates, and then to release the suspension into the flow. Particle seeding of gas flows requires more care in order to achieve a relatively uniform particle concentration. Among the methods commonly used, one could mention the following:

Dispersion of powders: This may be achieved by using a fluidised bed, namely by passing an air stream through a porous screen on which there is a supply of the powder.

Atomisation: This consists of creating an aerosol of liquid droplets suspended in a gas stream by feeding the gas into a liquid through special nozzles (*Laskin nozzles*). A variation of the method is to create an aerosol from an aqueous solution of sugar, salt or other solvable substances, or from a very dilute suspension of solid particles; evaporation of the water in the gas stream leaves a suspension of solid crystals or other solid particles, which act as tracers. Atomisation usually produces polydisperse seeds.

Evaporation and condensation: An oil, either natural or mineral, is heated to evaporation and then the vapour is cooled and allowed to condense into droplets (*oil mist*). With proper control, this process produces monodisperse droplets. Water-based fluids can also be used in commercially available *fog generators* to produce non-hazardous *fogs* in small or large quantities.

7.6.2 Dynamic Response of Seeding Particles

The motion of particles immersed in a fluid is an extremely difficult problem, for which there is no general analytical solution and of which even the mathematical formulation remains a topic of research. The formulation and solution of governing equations for particle motion can be treated with confidence only for some idealised cases, which will be discussed in the following.

A general dynamic model of particle motion: Consider an isolated, rigid particle with density ρ_P and mass m_P, moving at a speed $\overrightarrow{u_P}$ while immersed in an unbounded fluid with density ρ_F and viscosity μ_F. We further assume that the fluid velocity at the location of the particle, but in the absence of the particle, would be $\overrightarrow{u_F}$. In general, the particle and fluid velocities would differ by an amount

$$\overrightarrow{u_s} = \overrightarrow{u_F} - \overrightarrow{u_P} \,, \tag{7.51}$$

which is called the *slip velocity*. Newton's second law for the particle motion is written as

$$\overrightarrow{F} = m_P \frac{\mathrm{d}\overrightarrow{u_P}}{\mathrm{d}t} \,, \tag{7.52}$$

where $\overrightarrow{F}$ is the resultant of all external forces acting on the particle. The most common external forces are the following:

- the *gravitational* force, namely, the *weight* of the particle;
- the *buoyancy* force, exerted by the fluid on the immersed particle in a direction opposite to that of the weight;
- the *drag* force, due to friction between the particle and the fluid, when the particle is in relative motion with respect to the fluid;
- the *added-mass* force, accounting for the change of momentum of a fluid mass in the proximity of the particle;
- the *pressure-gradient* force, when a pressure gradient is present in the undisturbed flow; and

- the history force, which is a cumulative effect of the unsteadiness of the flow.

For liquid particles and bubbles, the analysis would be further complicated by internal motions, which tend to reduce the drag, and possible deformation of the particle, which alters the near field. The treatment of such effects is beyond our scope and so we shall restrict the discussion to rigid particles.

An important parameter affecting the dynamic response of particles is the density ratio $\gamma = \rho_P/\rho_F$. It is intuitive to expect that the larger the deviation of this parameter from 1 is, the more the particle motion would deviate from the fluid motion. In fact, particles with $\gamma = 1$, which are referred to as *neutrally buoyant* particles, may, for many purposes, be dynamically indistinguishable from the fluid. Several neutrally buoyant solid particles are available for liquid flows. For gas flows, the only nearly neutrally buoyant particles that have been developed are helium-filled soap bubbles. Another important parameter that affects the dynamic response of particles is the relative, or slip, Reynolds number, defined as

$$\mathrm{Re}_s = \frac{\rho_F \left|\overrightarrow{u_s}\right| d_P}{\mu_F} , \tag{7.53}$$

where d_P is the particle diameter, if it is spherical, or some other characteristic length otherwise.

Dynamics of heavy particles: A particularly simple dynamic model of particle motion can be derived when the particle is spherical and much more dense than the fluid ($\gamma \gg 1$), as in the case of solid or liquid particles in a gas, and inertia effects are negligible ($\mathrm{Re}_s \ll 1$), as in the case of very small particles or very viscous fluids. We also disregard gravitational forces for the time being. Then, the only significant external force would be the drag force, which is estimated as the *Stokes drag* [43]

$$\overrightarrow{F}_D = 3\pi\mu_F d_P \overrightarrow{u_s} . \tag{7.54}$$

To avoid complications, let us consider that the fluid and the particle move in the same direction. Then, the dynamic equation of the particle motion (Eq. (7.52)) becomes

$$\frac{\rho_P d_P^2}{18\mu_F}\frac{\mathrm{d}u_P}{\mathrm{d}t} + u_P = u_F . \tag{7.55}$$

Equation (7.55) shows that this particle may be viewed as a first-order system, in which the undisturbed fluid velocity is the input and the particle velocity is the output. As the static sensitivity is 1, the dynamic response of this particle depends on a single parameter, the time constant

$$\tau_P = \frac{\rho_P d_P^2}{18\mu_F} , \tag{7.56}$$

which depends only on the size and density of the particle and the viscosity of the fluid. Thus, heavy particles act as first-order low-pass filters of fluid motion with a -3 dB cut-off frequency equal to $\omega_c = 1/\tau_P$. Once τ_P is found, then the response of the particle to different inputs and for different initial conditions can be easily calculated.

Equation (7.55) shows that the time constant is proportional to the cross section and density of the particle and inversely proportional to the viscosity of the fluid. It is evident that, if the fluid and particle material properties are specified, the only way to improve the dynamic response of the particles is to decrease their size.

One is reminded that the Stokes drag approximation would be inaccurate for $\mathrm{Re}_P \gtrsim 1$, as may be the case for large particles or relatively large slip velocities. Fortunately, estimates of particle response based on the Stokes approximation would be conservative, because the actual drag would be larger than the Stokes estimate. If a more accurate estimate is required, one may estimate the drag in the equation of motion from one of several available empirical expressions for the drag coefficient of incompressible flow on rigid spheres, as, for example, [13]

$$C_D = \frac{F_D}{\frac{1}{2}\rho_F \left|\vec{u_s}\right|^2 \pi d_P^2/4} = \frac{24}{\mathrm{Re}_P}\left(1 + 0.15\mathrm{Re}_P^{2/3}\right), \tag{7.57}$$

and solve this equation numerically.

The previous analysis would describe fairly well the motion of sparsely distributed liquid or solid particles suspended in gases. To avoid particle interference with each other, it is advisable to maintain the distance between particles greater than $1000d_P$. The motion of non-spherical particles operating in the Stokes-flow regime would also be described by expressions applicable to spheres, as long as the particle dimensions do not vary much in different directions. Differences would become stronger with increasing Re_P and increasing particle aspect ratio [13].

In the majority of conventional flow seeding applications, the fluid may be treated as a continuum, which allows one to use the no-slip condition on the particle surface, which is an assumption in the derivation of the Stokes drag expression. In nanofluid studies, however, the particle size may be small by comparison to the mean free path λ between collisions of molecules in the containing gas. In such cases, there is slip of gas molecules on the particle surface and the resulting drag force on the particle turns out to be smaller than the Stokes drag. A common approach for such cases is to divide the Stokes expression by the *Cunningham slip correction factor* C, which, for air at standard sea-level pressure and temperature, is given as

$$C = 1 + \mathrm{Kn}\left[1.257 + 0.400\, e^{-1.1/\mathrm{Kn}}\right], \tag{7.58}$$

where $\mathrm{Kn} = 2\lambda/d_P$ is the Knudsen number. The value of λ can be calculated from the kinetic theory of gases.

Dynamics of non-heavy and light particles: As the density ratio γ decreases, the previous analysis would become increasingly inaccurate. The particle relative motion sets in motion part of the surrounding fluid, whose inertia must be taken into consideration. In simple terms, in addition to the particle acceleration, one must also consider the acceleration of a mass of fluid, called the *added mass*, which, for spherical particles, is equal to half the particle mass. The unsteady motion of an isolated, rigid, spherical particle in an infinite, incompressible, uniform flow stream, for which the Stokes approximation for the relative velocity applies, has been described by the *Basset–Boussinesq–Oseen* (*BBO*) equation. In addition to the terms contained in Eq. (7.55), the BBO equation contains terms representing the added mass and a time integral representing additional resistance due to the unsteadiness of the flow. A number of solutions of the BBO equation can be found in the literature [11, 20, 26, 34]. This type of equation would be more appropriate than Eq. (7.55) for solid particles suspended in a liquid medium, for which γ is usually of order 1. A consequence of the added mass and unsteady drag effects is that a gravitating particle would reach its terminal velocity at a slower rate than predicted by a first-order model. For the case of bubbles rising in a liquid ($\gamma \ll 1$), it has been calculated that the time required for a bubble to reach 90% of its terminal velocity would be many orders of magnitude smaller than τ_P [13, 15]. A characteristic time of the particle–fluid interaction for an arbitrary value of γ, which takes into account the added mass and is applicable to spherical, non-deformable particles in relative Stokes flow is [37]

$$\tau_{Pa} = \tau_P \left(1 + \frac{1}{2\gamma}\right) . \tag{7.59}$$

In the case of bubbles suspended in a liquid, this characteristic time becomes $\tau_P/2\gamma \gg \tau_P$.

Terminal velocity of particles in a gravitational field: For simplicity, let us assume that the fluid is either still or moves vertically with a constant speed u_F and that the particle is spherical, isolated from other particles, far from flow boundaries and non-deformable. Under the effect of gravity, the particle is subject to its weight $\rho_P \pi d_P^3 g/6$ and the buoyancy force $\rho_F \pi d_P^3 g/6$, which are both vertical. We may further assume that the relative velocity of the particle is small enough for the drag to be approximated by the Stokes expression, Eq. (7.54). The drag always opposes the relative velocity and tends to reduce it from its initial value until the relative acceleration vanishes and the relative velocity reaches an asymptotic value $(u_P - u_F)_\infty$, called the *terminal velocity*. When this happens, the momentum equation takes the steady form

$$3\pi\mu_F d_P (u_P - u_F)_\infty + \rho_P \pi d_P^3 g/6 - \rho_F \pi d_P^3 g/6 = 0 , \tag{7.60}$$

from which the terminal velocity can be found as

$$(u_P - u_F)_\infty = \left(\tfrac{1}{\gamma} - 1\right) g \tau_P . \tag{7.61}$$

If $\gamma > 1$, the particle sinks relative to the fluid and, if $\gamma < 1$, the particle rises. In flows in which the local velocity changes with time or position, and especially in turbulent flows, the relative vertical velocity of particles will also be variable and may never approach its terminal value. In such cases, the terminal velocity may be considered as an upper bound of the difference between the vertical velocities of the particle and the fluid.

Frequency response of particles: An issue of great significance in many flow measurement methods is the response of suspended particles of arbitrary density in unsteady fluid flows. Solutions of the BBO equation for a few types of unsteady motion are available in the literature. More elaborate momentum equations for rigid sphere motion in non-uniform flow have also been formulated [32, 38] and solved analytically by various authors for a few sets of simple conditions. As representative, we shall present some results based on a formulation and solution that applies not only to Stokes flow but also to flows with significantly higher Re_P [33].

The objective is to find the steady-state response of a rigid, spherical particle suspended in a gravity-free flow field, which oscillates sinusoidally with a frequency ω_F and an amplitude A. It may be assumed that, in steady-state, the particle would also oscillate (at least approximately) sinusoidally with an amplitude $B = \eta A$ and a phase shift φ relative to the fluid oscillation. We mainly need to estimate the dependence of the amplitude ratio η upon the density ratio γ and the product $\omega_F \tau_P$, which represents the ratio of the characteristic times of the particle and the flow. The latter ratio is referred to as the *Stokes number* and can be expressed as [33]

$$S = \sqrt{\frac{\omega_F d_P^2}{8 \nu_F}} . \tag{7.62}$$

One is warned, however, that different authors use the term Stokes number to describe parameters different from the one defined by Eq. (7.62). The following analytical solution for the amplitude ratio η in terms of S and γ has been derived by asymptotic analysis [33]:

$$\eta = \left\{ \frac{(1+S)^2 + \left(S + \frac{2}{3}S^2\right)^2}{(1+S)^2 + \left[S + \frac{2}{3}S^2 + \frac{4}{9}(\gamma - 1)S^2\right]^2} \right\}^{1/2} . \tag{7.63}$$

It is interesting that this solution is accurate in two distinct asymptotic limits:

- for Stokes flow ($\mathrm{Re}_P \to 0$) and any Stokes number S; and
- for large Stokes numbers ($S \to \infty$) and Re_P not restricted to small values, but significantly exceeding 1.

For intermediate combinations of Re_P and S values, this solution may not be accurate; however, this is not a serious limitation for the purposes of particle selection, for which we are mainly interested in ensuring that the amplitude ratio remains close to 1. Equation (7.63) for representative values of γ has been plotted in Fig. 7.23, from which one can make the following practical observations:

- As $S \to 0$ (steady flow), $\eta \to 1$ in all cases. This means that all particles would have a perfect dynamic response at sufficiently low frequencies.
- Neutrally buoyant particles ($\gamma = 1$) would have a perfect steady-state frequency response and their steady-state motion would be indistinguishable from that of the surrounding fluid. Even particles with density ratios in the range $0.56 \le \gamma \le 1.62$ would be within the ± 3 dB tolerance limit for the amplitude ratio at all frequencies and for $0 \le \mathrm{Re}_P \le 20$. This confirms the earlier observation that, from the viewpoint of dynamic characteristics, it is preferable to use particles whose density is close to the fluid density.
- In general, when $\gamma > 1$, $\eta < 1$, and when $\gamma < 1$, $\eta > 1$. This means that particles that are heavier than the surrounding fluid would oscillate with smaller amplitude than the fluid, whereas particles that are lighter than the fluid would oscillate with larger amplitude than the fluid.

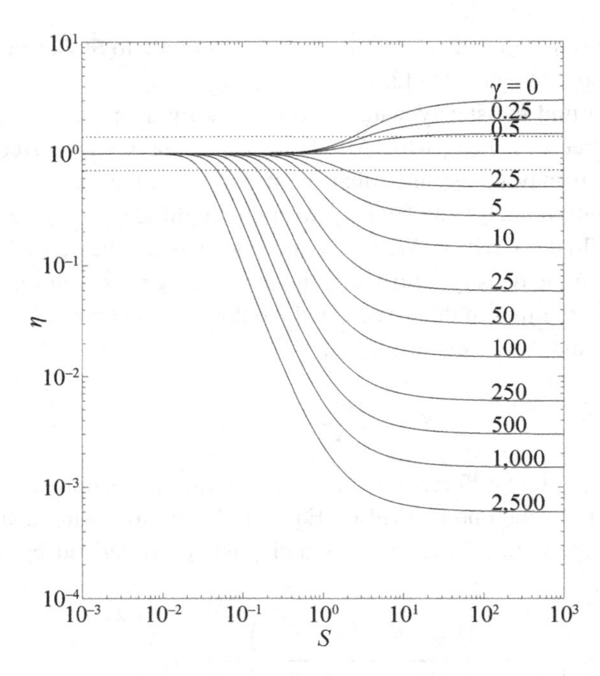

Figure 7.23 Amplitude ratio of rigid spherical particles immersed in a fluid; the intersections of the curves with the two dashed lines mark the corresponding ±3 dB cut-off frequencies.

- In the high-frequency limit of $S \to \infty$, $\eta \to (3/2)\,/\,|\gamma - 1|$. For very heavy particles ($\gamma \gg 1$), this implies an entirely negligible response ($\eta \ll 1$), whereas, for very light particles ($\gamma \approx 0$), $\eta \approx 3$, which indicates a motion amplification by a factor of 3.
- The *critical Stokes numbers* S_c, corresponding to the ±3 dB cut-off frequencies of particles with different density ratios, can be found from the intersections of the dashed lines in Fig. 7.23 with the corresponding curves. For additional accuracy, one may use the following interpolation relationships:

$$S_c \approx \begin{cases} \left[2.380^{0.93} + \left(\frac{0.659}{0.561-\gamma} - 1.175\right)^{0.93}\right]^{\frac{1}{0.93}}, & \text{for } 0 < \gamma < 0.561\,, \\ \left[\left(\frac{3}{2\sqrt{\gamma}}\right)^{1.05} + \left(\frac{0.932}{\gamma-1.621}\right)^{1.05}\right]^{\frac{1}{1.05}}, & \text{for } \gamma > 1.621\,. \end{cases} \tag{7.64}$$

Given the particle diameter d_P, one can thus estimate the cut-off frequency (in Hz) as

$$f_c \approx \frac{4\nu_F}{\pi}\frac{S_c^2}{d_P^2}\,. \tag{7.65}$$

If, instead, the cut-off frequency is specified, the maximum particle diameter can be calculated as

$$d_{P\,\max} \approx 2S_c\sqrt{\frac{\nu_F}{\pi f_c}}\,. \tag{7.66}$$

The cut-off frequencies for a few representative particle/fluid combinations have been plotted vs. particle diameter in Fig. 7.24.

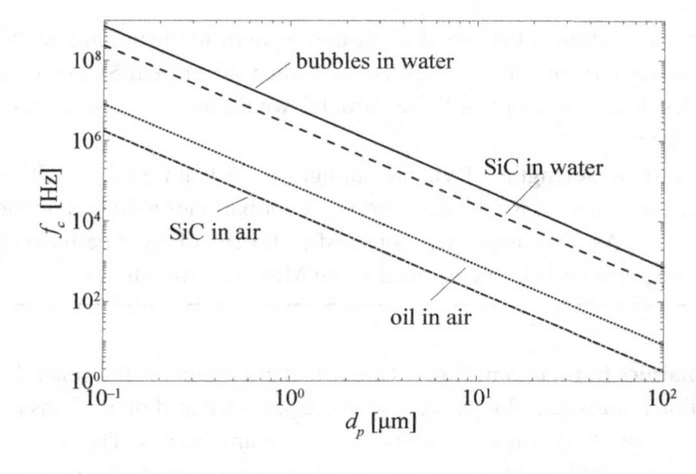

Figure 7.24 3 dB cut-off frequencies for representative particle/fluid combinations.

Example 7.5 Consider a sinusoidal, horizontal, flow of water at 20°C, seeded with spherical particles having materials and diameters shown in the following table. Neglect gravitational effects and take the kinematic viscosity of water to be 10^{-6} m^2s^{-1}.

Material	d_P [μm]
Hydrogen bubbles	50
Polystyrene beads	20
Al_2O_3	1

(a) Determine the 3 dB cut-off frequencies (in Hz) of the motions of these particles.

(b) Assuming that the measured period and amplitude of hydrogen bubble oscillations are 0.1 s and 5 mm, respectively, estimate the water motion amplitude and the amplitudes of oscillation of the two other types of particles.

(c) Discuss whether sunlight scattered by these particles would be best described by Rayleigh or Mie scattering theory.

Answer

(a) The 3 dB cut-off frequencies are obtained using Eq. (7.65) and Eq. (7.64), which require values of $\gamma = \rho_P/\rho_F$. For the hydrogen bubbles, we get $\gamma_{\text{bub}} = \rho_{H_2}/\rho_{water} \approx$ 1.2 kg m^{-3}/1(,)000 kg m^{-3} ≈ 0.0012. The values of γ for the polystyrene beads and Al_2O_3 particles, with the use of Table 7.7, are $\gamma_{\text{poly}} \approx 1$ and $\gamma_{\text{Al}_2\text{O}_3} = 3.96$ respectively. Given that $\gamma_{\text{poly}} \approx 1.0$, the polystyrene beads will not have a 3 dB cut-off frequency and will in fact closely follow flow oscillations of any frequency (see Fig. 7.23). The hydrogen bubbles and Al_2O_3, however, will have 3 dB cut-off frequencies. Their respective critical Stokes numbers are $Sc_{\text{bub}} = 2.38$ and $Sc_{\text{Al}_2\text{O}_3} = 1.12$, which gives the corresponding 3 dB cut-off frequencies as $f_{c\text{bub}} \approx 2.9$ kHz and $f_{c\text{Al}_2\text{O}_3} \approx 1.6$ MHz.

(b) If the period of the oscillation is 0.1 s, the corresponding frequency of the oscillation is 10 Hz, or equivalently $\omega_F = 2\pi f_F = 20\pi$ rad/s. The Stokes number for this motion, using Eq. (7.62), is $S = 0.14$, which is below the critical Stokes number for the hydrogen bubbles, found previously to be $Sc_{\text{bub}} = 2.38$. Hence, the observed amplitude of oscillation of the hydrogen bubbles directly corresponds to the amplitude of the water. Indeed, a Stokes number of $S = 0.14$ is lower than the critical Stokes number for all three particles, hence, any one of these particles would be a suitable choice for this particular experiment.

(c) The range of wavelengths of visible sunlight is $380 \text{ nm} \leq \lambda \leq 750 \text{ nm}$ (Table 7.1), and the condition for Rayleigh scattering is for particles to have a diameter that is smaller than $\lambda/15$. As this condition is not satisfied for any of these particles, the light scattered by these particles is best described using Mie scattering theory.

Particle dynamics in turbulent flows: Of particular difficulty is the general problem of particle motion in turbulent flows, because of the presence of fluid motions ('eddies') with a wide range of characteristic lengths, times and amplitudes. The particle would respond differently to different eddies and its final motion would be the result of a non-linear superposition of the different contributions. One would anticipate that a heavy particle would generally act as a non-linear low-pass filter of fluid motion and it would not adequately respond to motions the characteristic frequency of which is higher than a certain threshold. Many attempts have been made to predict particle motion in turbulent flows, without yet reaching a solution of general validity. Considering the available results, it is safer to conclude that the various recommended expressions might be adequate for selecting particles that would be suitable for a particular application (e.g., for selecting a maximum particle size for a given turbulent flow), but rather uncertain as a means of correcting experimental results. A conservative condition for the particle cut-off frequency to ensure that it would follow all motions present in a turbulent flow is

$$f_c > (1/2\pi)(\nu_F/\varepsilon)^{-1/2}\,, \tag{7.67}$$

where ε is the turbulent kinetic energy dissipation rate [25, 41, 49]. In addition, the size of the particle should be small compared to the size of the smallest turbulence eddies, which is represented by the Kolmogorov microscale $\eta = (\nu_F^3/\varepsilon)^{1/4}$. Such requirements would be excessive, if one is only interested in large-scale turbulent motions and not the fine structure.

Additional effects on particle dynamics: Besides gravitational forces, a particle may also be subjected to other types of body forces, including electric forces for electrically charged particles in an electric field, magnetic forces for ferromagnetic particles in a magnetic field, centrifugal forces in swirling flows, Coriolis forces in rotating flows, and aerodynamic lift for rotating or non-spherical (e.g., flakes) particles or due to shearing (*Saffman force*). Lift due to shearing would be particularly significant near solid walls, tending to move particles away from the wall. Irrespectively of shearing effects, particle concentration near walls would tend to diminish due to deposition, either through gravity or through adherence. In vortical flows, particles heavier than the fluid would tend to centrifuge away from the axis of rotation, whereas lighter particles would tend to concentrate in the core. For this reason, gas bubbles are suitable for flow visualisation of vortices in liquids. Finally, in flows with a temperature gradient, collisions with fluid molecules in Brownian motion would tend to move small particles from warmer towards cooler regions.

7.6.3 The Visibility of Seeding Particles

Particle illumination: When comparing the visibility of two different types of particles, one should not only consider their actual sizes and refractive indices, but also a number of other factors. An important parameter is the scattering cross section σ_λ. In general, the larger σ_λ is, the more visible the particle would be. For a given particle size, incident radiation intensity and scattering angle, the radiation scattered by the particle may vary by one or more orders of magnitude, depending on its own refractive index and the refractive index of the surrounding fluid. For example, a particle with $n_p > 1.33$ will appear to be larger in air ($n \simeq 1.00$) than in water ($n \simeq 1.33$). Furthermore, as illustrated in Fig. 7.18, in the Mie scattering regime, σ_λ is a complicated function of orientation and very sensitive to the value of the scattering angle. Forward scattering produces the *maximum maximorum* σ_λ and so, in principle, it should be the preferable mode of particle observation. However, because of considerations of depth-of-field and convenience in position of transmitting and receiving optical components with respect to the observed flow, backscattering and sidescattering are more common arrangements. Backscattering also permits the use of the same lenses and other components for both transmitting and receiving purposes. The apparent sensitivity of scattered light to the scattering angle is smoothed out by two effects: the averaging effect of the finite aperture of the collecting lens (see Eq. (7.41)) and multiple scattering, namely, the successive scattering of light by more than one particles. When the light is polychromatic, the overall scattering cross section would be a weighted average of all applicable monochromatic scattering cross sections, a process which also tends to smooth out intensity non-uniformity. In the Mie scattering regime ($d_P \gg \lambda$), the average energy scattered by a particle over a solid angle collected by a lens would approximately increase as $(d_P/\lambda)^2$, whereas, in the Rayleigh scattering regime ($d_P \ll \lambda$), the same energy would increase as $(d_P/\lambda)^4$ [2].

Particle visibility enhancement: Various techniques have been developed to enhance the visibility of particles. One approach is to coat solid particles with a highly reflective material, such as silver, to increase their refractive index. Another is to either coat them with a fluorescent dye or embed a fluorescent substance in their material. A flow laden with fluorescent particles is illuminated by a laser sheet with a spectral peak near the absorption wavelength of the fluorescent substance and the flow is observed through an optical band-pass filter, which removes all radiation except that in the narrow emission band of the substance. This way, only particles are visible, while all incident illumination, reflections from the apparatus walls and other interfering radiation are removed (see also Section 15.5).

Particle imaging: When viewing or recording particle images through a collecting lens, one must also consider the effect of diffraction on particle visibility [1, 2]. Ideally, if the lens magnification factor (namely, the ratio of the sensor size and the physical size of the field of view) is M, the diameter of the particle image would be Md_P. This is actually the case for relatively large particles, however, the image diameter of small particles would be larger than Md_P, due to diffraction on the lens aperture. A length that characterises the image of a spot viewed through a diffraction-limited lens with an f-number ($f^\#$) is the diameter of the first dark ring of the Airy disk light intensity distribution (see Section 7.3.8).

$$d_s = 2.44\,(M + 1)\,f^\#\lambda\,. \tag{7.68}$$

The image diameter of sufficiently small particles would be approximately equal to d_s and independent of d_P. More generally, one may interpolate between the two extreme cases to express the diameter of the particle image as

$$d_e \approx \sqrt{(Md_P)^2 + d_s^2}\,. \tag{7.69}$$

A consequence of diffraction is that the images of closely spaced small particles may overlap to the point that they cannot be distinguished from each other. An additional blur (more than 20%) of particle images would occur, when the particles are outside the depth of field. For the present purposes, the depth of field (DOF) can be defined as the depth of the field of view within which the minimum diffraction-limited particle size is visible, which can be calculated as [42]

$$\delta_z = 4.88\left(1 + \frac{1}{M}\right)^2 f^{\#2}\lambda\,. \tag{7.70}$$

It is therefore only a function of the camera settings and the wavelength of the light source used to illuminate the particles, with the depth decreasing as the $f^\#$ decreases for a given magnification factor. Finally, when a particle image is recorded, it may be further enlarged as a result of inadequate resolution of the photodetector array. It is evident that no image can be smaller than the size of a pixel.

Example 7.6 We wish to use 1 μm Al_2O_3 particles to study the flow of fluid due to natural heat convection in a small container. The central plane of the container is illuminated with a green light sheet produced by a Nd:YAG laser (λ = 532 nm). We view the seeded flow with the same camera as in Example 7.4, but use a fixed 50 mm camera lens with $f^\#$ = 1.8, 2.8, 4, 5.6, 8, 11, 16 and 22. The camera is placed 300 mm away from the flow. Identify any possible experimental problems that you may encounter, when viewing these particles with this set-up.

Answer

In Example 7.4, we found that the sensor width is s = 20.48 mm and that the pixel size is 10 μm. Equation (7.48) provides the field of view width for the proposed set-up as w = 122.88 mm. Hence, the magnification factor is $M = s/w = 20.48/122.88 = 0.17$X (or 1:6). The normalised image diameter $d_e/(Md_P)$ of the particles, calculated with the use of Eqs. (7.68) and (7.69), is plotted as a function of $f^\#$ in Fig. 7.25. It is evident that the particle images on the camera sensor would be considerably larger than their undistorted images for all $f^\#$, when viewed by the camera. The particle images range from just over 16 times the ideal image for $f^\#$ = 1.8, to almost 200 times the ideal image for $f^{\#} = 22$. The equivalent pixel size of the camera sensor is shown in Fig. 7.25 as a dashed line; any visible particle with a normalised image diameter that is larger than this value (in this case, all particles) will occupy more than one pixel on the sensor.

In order to distinguish the image of one particle from another, we would need to ensure that the distance between the particles is larger than $200 d_P M \approx 33$ μm. Let us assume that the particle concentration is sufficiently high for many particles to be closer than this distance. If we set the camera at $f^{\#} = 22$, then the particle images will overlap and, thus, we would observe clusters of particles in the flow, rather than individual particles. By reducing the $f^{\#}$, this issue would be mitigated, which further highlights the benefit of using low $f^{\#}$ settings. One is reminded that a low $f^{\#}$ setting results in a shallow field of view (see Section 7.5.3).

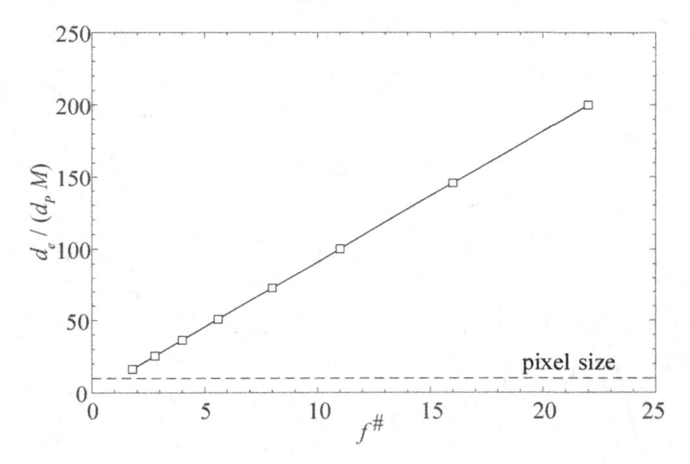

Figure 7.25 Effective particle size as a function of *f*-number for a $d_P = 1$ μm particle, illuminated by 532 nm green light. The dashed line indicates the size of pixels on the camera sensor.

Example 7.7 Air bubbles are released from a small nozzle near the bottom of a vertical 10 mm × 10 mm square channel, having thin glass walls and containing still water with a depth of 1 m. From a previous study, we know that the actual diameters of the bubbles are normally distributed with a mean $\mu_{d_P} = 25$ μm and a standard deviation $\sigma_{d_P} = 5$ μm and we may assume that all released bubbles have diameters within the range $\mu_{d_P} \pm 3\sigma_{d_P}$. The bubbles are illuminated with a red laser, having a wavelength $\lambda = 670$ nm, and their images are recorded with a camera lens with a 1:40 magnification factor, and with $f^{\#} = 1.8, 2.8, 4, 5.6, 8, 11, 16$ and 22. The camera is located 1 m away from the channel wall. We may assume that the bubbles rise with their terminal velocity, while retaining a spherical shape with an approximately constant diameter.

(a) Compute the terminal velocities of the bubbles and plot them vs. bubble diameter.

(b) Assuming that the camera settings are such that the images of still bubbles have accurate diameters, determine the maximum exposure time, so that the blur of any bubble image due to its rising motion does not exceed 5% of its diameter.

(c) Neglecting blur due to bubble rising, determine the pdf of the bubble images for different f-number settings of the camera. Discuss the accuracy of these pdf.

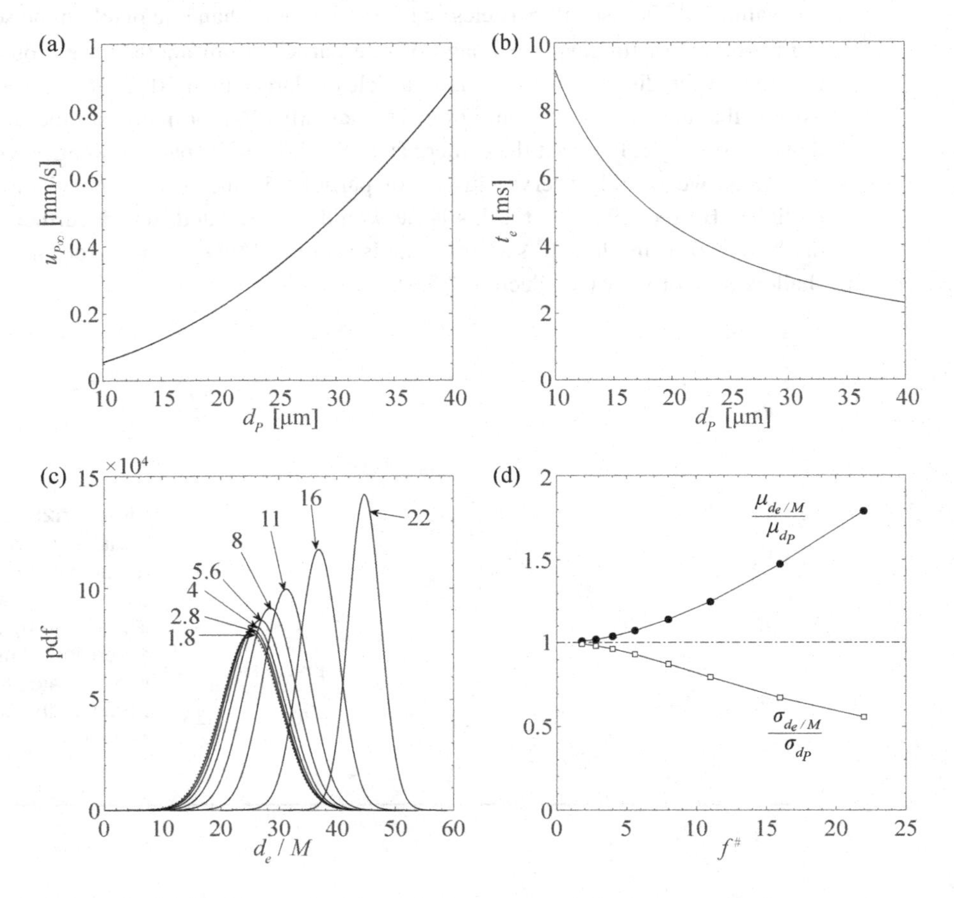

Figure 7.26 (a) Bubble terminal rise velocity vs. the bubble diameter. (b) Exposure time vs. the bubble diameter. (c) Probability density functions of air bubble diameter images, recorded with different f-numbers (solid lines) and the pdf of the ideal images (filled symbols). (d) The ratios of the mean and standard deviation of bubble image diameters and the corresponding undistorted values vs. f-number.

Answer

(a) The terminal rise velocity of the bubbles, calculated using Eq. (7.61) for a density ratio of $\gamma = 0.0012$ and the time constant obtained from Eq. (7.56), is shown as a function of bubble diameter in Fig. 7.26a. The terminal velocity increases rapidly as the bubble diameter increases, from 0.05 mm/s for $d_P = 10$ μm to 0.87 mm/s for $d_P = 40$ μm .

(b) If the exposure time (shutter speed) is t_e and the bubble speed is $u_{P\infty}$, the bubble would rise by a distance $u_{P\infty}t_e$, while its image is being recorded, and the recorded image would be blurred in the vertical direction, having a length equal to $d_P + u_{P\infty}t_e$, rather than d_P. We want the blur not to exceed $0.05d_P$, from which we get the maximum exposure time as $t_{emax} = 0.05d_P/u_{P\infty}$. This expression does not indicate which bubbles would have the shortest t_{emax}, because, as d_P increases, $u_{P\infty}$ increases as well. The exposure times for the range of interest in this experiment, that is, 10 μm $\leq d_P \leq$ 40 μm, are shown in Fig. 7.26b. The shortest maximum exposure time is 2.3 ms and corresponds to bubbles with the largest diameter (d_P = 40 μm). The standard shutter speed of digital cameras that is closest to this exposure time is 1/500 s. Setting the camera to this or any smaller shutter speed would ensure that all bubble images would have a blur that does not exceed 5% of their respective diameters.

(c) Considering that the bubble diameter has a normal pdf, we expect that 99% of the bubbles would have diameters within five standard deviations from the mean, in this case, in the range 0 μm $\leq d_P \leq$ 50 μm. In the previous example, we showed that the recorded particle image size can greatly exceed its ideal size, when the particle is very small. Through similar calculations, we find that, for the largest f-number of $f^{\#} = 22$, the 'smallest' bubbles, let's say those having a diameter d_P = 1 μm, would appear to be roughly 39 times larger. On the other hand, the 'largest' bubbles, having a diameter d_P = 50 μm, would appear to be only roughly 25% larger. Therefore, the pdf of the bubble size distribution would be distorted more on its left tail than on its 1 m tall right tail and the pdf would become skewed, thus, non-Gaussian. Moreover, as all apparent bubble diameters would be larger than the ideal ones, the mean of the pdf would be increased and the entire pdf would move to the right. The previous expectations are fully confirmed by Fig. 7.26c, which shows the pdf of the bubble images obtained with different f-numbers. For a lower f-number, the effects are less pronounced, with the mean and standard deviation of the images tending to match the undistorted values, as shown in Fig. 7.26d. This example clearly demonstrates that camera settings may introduce a significant bias in the results.

Chapter Digest

7.1 Which are the three different ways to describe light propagation?

7.2 What is a wavefront? What shapes can a wavefront take?

7.3 How is the energy of a photon related to the frequency of light? How is the wavelength related to the frequency?

7.4 What is the refractive index and on which material property does it depend?

7.5 Describe how sunlight gets refracted as it propagates towards the bottom of the sea. Which properties of the seawater would affect this refraction? Would refraction be stronger in a lake or in an ocean?

7.6 Explain how one-way mirrors work.

7.7 Explain the reasons for refractive index matching.

7.8 Explain how a tinted windshield/windscreen works.

7.9 What is the difference between radiometry and photometry?

7.10 Why do things look bluish under the moonlight?

7.11 What is a black body and what is a grey body?

7.12 Describe in plain language the operation of a laser.

7.13 How does light diffraction affect the appearance of small particles? What is Airy's disk?

7.14 Would backscatter or forward scatter be stronger?

7.15 Explain the difference between elastic and inelastic scattering.

7.16 Explain the common features and the differences between Rayleigh scattering and Mie scattering.

7.17 Why is the sky blue and the sunset red?

7.18 Explain fluorescence and phosphorescence.

7.19 How does the diameter of an optical fibre affect transmission of light?

7.20 Describe the advantages and disadvantages of photomultiplier tubes by comparison to photodiodes.

7.21 Which two properties quantify the efficiency of photodetectors?

7.22 Which are the sources of noise of photodetectors?

7.23 Why are LEDs more economical than incandescent lights?

7.24 Why is it preferable to use monodisperse than polydisperse seeding particles?

7.25 When do you consider a particle to be heavy? Why is the dynamic analysis of heavy particles easier than that of non-heavy ones?

7.26 Given the size and material properties of a heavy particle and the surrounding fluid, how would you determine its dynamic response?

7.27 Would bubbles rising in a glass of beer accelerate, decelerate or move at a constant speed? Would they change size and shape and how? Is mass conserved within each of these bubbles? Is there a difference, other than composition and size, between bubbles in a glass of beer and hydrogen bubbles generated by electrolysis?

7.28 What is the Stokes flow approximation and when can it be made?

7.29 Which properties affect the visibility of a particle?

7.30 If we have particles of different sizes, is it possible for all of their images obtained by a camera to have the same size?

Problems

7.1 Compute the Gladstone–Dale constant for monochromatic light with a wavelength of 580 nm passing through the atmosphere of a strange planet, which consists of a homogeneous mixture of the following neutral molecules, with relative number densities given in parentheses: oxygen (2%), nitrogen (10%), carbon dioxide (88%).

7.2 Consider water in a rectangular tank made of clear acrylic material. The beam of a He-Ne laser approaches the wall of the tank at an angle of 30° with respect to the normal. Determine the direction of the laser beam in the water, if its temperature is (i) 20°C and (ii) 34°C.

7.3 Water flows in a water tunnel having a rectangular cross section with a width of 0.500 m and a height of 1.000 m, which is surrounded by glass walls with a thickness of 10 mm. A laser beam that lies on a vertical plane enters the front vertical wall at its centre-height, while forming an angle of 20° with respect to the normal. Sketch accurately the path of this beam, indicating its inclinations with respect to the direction normal to the vertical walls, as it crosses the different interfaces between air, glass and water and exits from the other side of the water tunnel. Determine the vertical distance between the entry and exit points of the beam. Furthermore, determine the location of the intersection of the laser beam with the room wall, which is located 3 m away from the water tunnel back wall. Finally, determine whether it is possible to have a total reflection of this beam and, if so, find the conditions for which this happens.

7.4 Consider a thin-walled glass tube containing water and surrounded by air. Two parallel beams of a He-Ne laser located on a plane normal to the tube axis approach the tube symmetrically about a plane containing the tube axis. Explain whether it would be possible for the beams to intersect inside the tube. If not, orient and position the beams so that they would intersect at a position on the symmetry plane, half a radius away from the distant end of the tube.

7.5 Consider water flowing in a cylindrical glass tube with an inner diameter of 100 mm and a wall thickness of 10 mm, surrounded by air. A laser beam is directed parallel to the horizontal plane of symmetry of the tube and at a distance of 25 mm from this plane. Plot the path of the beam through the tube, indicating the values of the different angles. Now consider that the glass tube is immersed in a viewing tank filled with water. Plot again the path of the beam and compare it with the one in the previous case. Finally, consider that the viewing tank and the tube contain a fluid the refractive index of which matches the refractive index of glass. Compare the path of the same beam in the latter case with those in the previous two cases.

7.6 Plot the spectral radiance of a black body vs. temperature in the range between 0 and 10,000 K, for light wavelengths $\lambda = 300, 500, 700$ and 900 nm. Also plot the spectral radiance of a black body vs. light wavelength in the ultraviolet, visible and infrared ranges, for temperatures $T = 100, 1{,}000$ and 10,000 K.

7.7 Consider light generated by a low-pressure sodium lamp passing through isolated circular apertures with diameters equal to 0.1, 1 and 10 μm. Plot the light intensity exiting the aperture, per unit incident light, vs. the diffraction angle.

7.8 A laser beam with diameter d_e is passed through a material with an absorption coefficient α. If the intensity of the incident beam has a Gaussian-shaped variation, determine the thickness variation of the material such that the exiting beam would have a uniform intensity over a cylindrical core with a diameter d_e. In your analysis, allow the minimum possible light absorption.

7.9 Two identical laser beams are positioned such that their axes intersect perpendicular to each other. For this problem disregard the wave-like character of light and assume that, at each location within a beam, the light intensity is equal to its amplitude I_o. Determine the intensity of light at different positions within the intersection volume. Plot contours of constant light intensity on the plane of symmetry and on other planes parallel to it. Describe the shapes of surfaces of constant intensity.

7.10 Using an online, interactive, Mie scattering calculator (https://omlc.org/calc/mie_calc.html), plot, in logarithmic polar coordinates, the monochromatic scattering cross sections of crown glass spheres, immersed in water, illuminated with light having a wavelength $\lambda = 0.488$ μm and having diameters equal to 0.1, 1.0, 10.0 and 100 μm.

7.11 Consider that glass beads with a refractive index of 1.55 and a diameter of 1 μm are injected in air flow and are illuminated by a He-Ne laser. Plot the monochromatic angular scattering cross section of the beads vs. the angle of observation.

7.12 Consider spherical particles with diameters $d_P = 0.5$, 1, 5 and 50 μm, suspended in air. The particle images are recorded with a digital camera equipped with macro lenses having magnification factors $M = 1$, 10 and 100 and f-numbers in the range from 1 to 90, while being illuminated by a He-Ne laser, an Ar-Ion laser, a frequency-doubled Nd:YAG laser or a low-pressure sodium lamp.

(a) Assuming that the correct image diameter would be equal to Md_P, derive an expression of the relative error in the recorded particle image diameter d_e in terms of the various parameters of the problem. Explain how this error depends on each parameter and suggest which values of each parameter, among the available choices, would be preferable for keeping this error as small as possible, if your objective is to measure accurately the diameters of a polydisperse aerosol containing particles with diameters in the range 0.5–50 μm.

(b) Explain the dependence of the depth of field upon the various parameters of the problem and the importance of the depth of field in imaging a polydisperse aerosol (hint: consider the ratio of the depth of field and the particle diameter for different particles).

(c) Determine the relative error in image diameters and the depth of field for a few representative combinations of the problem parameters, including some extreme cases. Based on these results, discuss whether it is possible to obtain accurate images of all droplets with the type of equipment that is available.

7.13 Consider the following experiment taking place in the International Space Station, while it is in orbit around the Earth. A container filled with air is put on a shaker and is oscillated sinusoidally with a frequency of 5 Hz and an amplitude of 100 mm. Determine the amplitude of oscillation of suspended droplets of some medication, which have a density approximately equal to that of water and a diameter of 100 μm. Also determine the amplitude of oscillation of suspended, neutrally buoyant helium-filled soap bubbles with the same diameter.

7.14 A wind tunnel has a horizontal test section, which discharges into the laboratory and is 5.00 m long, 100 mm wide and 100 mm high. The flow velocity in the wind tunnel is laminar, uniform and equal to 5 m/s. A nozzle produces an aerosol, consisting of water droplets having a Gaussian size distribution with an average diameter of 85 μm and a standard deviation of 20 μm. Assume that there are no droplets with diameters larger or smaller than the average by more than three standard deviations. The aerosol is injected isokinetically (i.e., at a horizontal speed equal to that of the flow) through a fine tube at the centre of the inlet of the test section.

(a) Compute the monodispersity of this aerosol.

(b) Determine the terminal sink velocities of the smallest, the average and the largest droplets, assuming Stokes drag. Discuss the validity of the Stokes drag assumption. Without computation, but based on qualitative arguments, discuss whether the actual terminal velocities would be higher or lower than the previously determined values.

(c) Neglecting the effects of boundary layers along the wind tunnel walls and assuming that the test section were infinitely long and that the Stokes drag law were valid, determine the downstream distances from the point of injection at which the smallest, the average and the largest droplets would contact the bottom of the wind tunnel. Do not assume that the vertical droplet speeds are equal to the corresponding terminal velocities and determine the vertical speeds of these droplets when they reach the bottom. Now, considering that the test section is 5.00 m long, determine the percentage of the droplets in this aerosol that would sink to the bottom and, thus, would not be discharged with the air flow.

7.15 A sealed tank containing air at 20°C is mounted on a support oscillating horizontally and sinusoidally with an amplitude of 40 mm and a frequency of 2.0 Hz. A glass bead with a diameter of 50 μm is released isokinetically at the top of the tank while the tank is at the midpoint of its horizontal stroke. Assume that the tank height is sufficiently large for the bead to reach a steady-state response before it touches the bottom. Derive expressions describing the vertical and horizontal velocities of the bead, assuming that the two motions are independent of each other. Describe how you would calculate the position of the bead at different times. Determine the steady-state amplitude of oscillation of the bead.

7.16 A diver breathes through a tube 20 mm in diameter and 300 mm long. The capacity of the diver's lungs is 3 l and the diver's breathing rate is 20 times per min. Assume that the flow in the tube is uniform and varies sinusoidally with time. Liquid droplets with density of water are released in the air near the inlet of the tube.

(a) Neglecting gravity and diffusion, determine the maximum size of the droplets that will reach the diver. State clearly all assumptions and approximations that you make.

(b) Compute the length of the previously mentioned tube that is required in order for the diver to avoid breathing droplets larger than 10 μm in diameter. Do you foresee any other problems with the use of this tube?

References

[1] R. Adrian. Particle-imaging techniques for experimental fluid mechanics. *Annu. Rev. Fluid Mech.*, 23:261–304, 1991.

[2] R.J. Adrian and C.-S. Yao. Pulsed laser techniques application to liquid and gaseous flows and the scattering power of seed materials. *Appl. Opt.*, 24:44–52, 1985.

[3] H.-E. Albrecht, M. Borys, N. Damaschke, and C. Tropea. *Laser Doppler and Phase Doppler Measurement Techniques.* Springer-Verlag, Berlin, 2003.

[4] J.E. Anderson. *Fiber Optics: Multi-Mode Transmission.* Technical Memorandum 200, Burle Technologies, Inc., Lancaster, PA, 2001.

[5] Anonymous. *Fiber Optics: Theory and Applications.* Technical Memorandum 100, Burle Technologies, Inc., Lancaster, PA, 2001.

[6] F.D. Bennett. Exploding wires. *Sci. Amer.*, 206:103–112, 1962.

[7] A.F. Bicen. Refraction correction for LDA measurements in flows with curved optical boundaries. *TSI Quarterly*, 8, Issue 2 (April–June 1982):10–12, 1982.

[8] C.F. Bohren and D.R. Huffman. *Absorption and Scattering of Light by Small Particles*. Wiley, New York, 1983.

[9] C. Bruecker. Digital-particle-image-velocimetry (DPIV) in a scanning light-sheet: 3d starting flow around a short cylinder. *Exp. Fluids*, 19:339–349, 1995.

[10] R. Budwig. Refractive index matching methods for liquid flow investigations. *Exp. Fluids*, 17:350–355, 1994.

[11] B.T. Chao. Turbulent transport behaviour of small particles in a turbulent fluid. *Oesterreichisches Ingenieur-Archiv*, 18:7, 1964.

[12] CIE. *International Lighting Vocabulary*. Number CIE Publ. No. 17.4. CIE Publ. No. 17.4, Vienna, Austria, 1987.

[13] R. Clift, J.R. Grace, and M.E. Weber. *Bubbles, Drops and Particles*. Academic Press, New York, 1978.

[14] E.R. Cohen, D.R. Lide, and G.L. Trigg. *AIP Physics Desk Reference (3rd Edition)*. Springer, New York, 2003.

[15] C.F.M. Coimbra and R.H. Rangel. General solution of the particle momentum equation in unsteady stokes flows. *J. Fluid Mech.*, 370:53–72, 1998.

[16] M.M. Cui and R.J. Adrian. Refractive index matching and marking methods for highly concentrated solid-liquid flows. *Exp. Fluids*, 22:261–264, 1997.

[17] H.C. Van de Hulst. *Light Scattering by Small Particles*. Wiley, New York, 1957.

[18] H.M. Dobbins and E.R. Peck. Change of refractive index of water as a function of temperature. *J. Opt. Soc. Amer.*, 63:318–320, 1973.

[19] A.C. Eckbreth. *Laser Diagnostics for Combustion Temperature and Species (2nd Edition)*. Gordon and Breach Publishers, Philadelphia, PA, 1996.

[20] R.J. Emrich, editor. *Methods of Experimental Physics, Vol. 18A and B: Fluid Dynamics*. Academic Press, New York, 1981.

[21] H. Fiedler, K. Nottmeyer, P.P. Wegener, and S. Raghu. Schlieren photography of water flow. *Exp. Fluids*, 3:145–151, 1985.

[22] R.C. Gonzalez and P. Wintz. *Digital Image Processing (2nd Edition)*. Addison–Wesley, Reading, MA, 1987.

[23] L. Haya and S. Tavoularis. Effect of bileaflet mechanical heart valve orientation on fluid stresses and coronary flow. *J. Fluid Mech.*, 806:129–164, 2016.

[24] E. Hecht. *Optics (4th Edition)*. Addison–Wesley, Reading, MA, 2002.

[25] J.O. Hinze. *Turbulence (2nd Edition)*. McGraw-Hill, New York, 1975.

[26] A.T. Hjemfelt and L.F. Mockros. Motion of discrete particles in a turbulent fluid. *Appl. Sci. Res.*, 16:149–161, 1966.

[27] H.L. Anderson (Editor in Chief). *A Physicist's Desk Reference (2nd Edition)*. American Institute of Physics, New York, 1989.

[28] B. Jähne. Data acquisition by imaging detectors (chapter 24). In C. Tropea, A.L. Yarin, and J.F. Foss, editors, *Springer Handbook of Experimental Fluid Mechanics*. Springer, Berlin, 2007.

[29] F.A. Jenkins and H.E. White. *Fundamentals of Optics (3rd Edition)*. McGraw Hill, New York, 1957.

[30] M. Kerker. *The Scattering of Light and Other Electromagnetic Radiation*. Academic Press, New York, 1969.

[31] M.M. Koochesfahani and P.E. Dimotakis. Mixing and chemical reactions in a turbulent liquid mixing layer. *J. Fluid Mech.*, 170:83–112, 1986.

[32] M.R. Maxey and J.J. Riley. Equations of motion for a small rigid sphere in a nonuniform flow. *Phys. Fluids*, 26:883–889, 1983.

[33] R. Mei. Velocity fidelity of flow tracer particles. *Exp. Fluids*, 22:1–13, 1996.

[34] A. Melling. Tracer particles and seeding for particle image velocimetry. *Meas. Sci. Technol.*, 8:1406–1416, 1997.

[35] W. Merzkirch. *Flow Visualization (2nd Edition)*. Academic Press, New York, 1987.

[36] W. Merzkirch. Flow visualization (chapter 11). In C. Tropea, A.L. Yarin, and J.F. Foss, editors, *Springer Handbook of Experimental Fluid Mechanics*. Springer, Berlin, 2007.

[37] E.E. Michaelides. Hydrodynamic force and heat/mass transfer from particles, bubbles and drops – The Freeman Scholar Lecture. *J.Fluids Eng.*, 125:209–238, 2003.

[38] E.E. Michaelides and Z.-G. Feng. The equation of motion of a small viscous sphere in an unsteady flow with interface slip. *Int. J. Multiphase Flow*, 21:315–321, 1995.

[39] G. Mie. Beitraege zur optik trueber medien, speziell kolloidaler metalloesungen. *Ann. Physik*, 25:377–452, 1908.

[40] A.J. Parry, M.J. Lalor, Y.D. Tridimas, and N.H Woolley. Refraction corrections for laser-doppler anemometry in a pipe bend. *Dantec Information*, No. 09 (September 1990):4–6, 1990.

[41] S.B. Pope. *Turbulent Flows*. Cambridge University Press, Cambridge, UK, 2000.

[42] M. Raffel, C.E. Willert, F. Scarano, C.J. Kähler, S.T. Wereley, and J. Kompenhans. *Particle Image Velocimetry: A Practical Guide (3rd Edition)*. Springer, 2018.

[43] L. Rosenhead. *Laminar Boundary Layers*. Oxford University Press, Oxford, 1963.

[44] M. Sayer and A. Mansingh. *Measurement, Instrumentation and Experiment Design in Physics and Engineering*. Prentice-Hall of India, New Delhi, 2000.

[45] G. Sharma, editor. *Digital Color Imaging Handbook*. CRC Press, Boca Raton, FL, 2003.

[46] K. Shimoda. *Introduction to Laser Physics*. Springer, Berlin, 1984.

[47] D.J. Shlien. Inexpensive method of generation of a good quality laser light sheet for flow visualization. *Exp. Fluids*, 5:356–358, 1987.

[48] A.J. Smits and T.T. Lim, editors. *Flow Visualization Techniques and Examples*. Imperial College Press, London, 2000.

[49] S. Tavoularis. Turbulent flow (chapter 31). In J. Saleh, editor, *Fluid Flow Handbook*, pages 31.1–31.33. McGraw-Hill, New York, 2002.

[50] T. Vo-Dinh. Basic instrumentation in photonics (chapter 6). In T. Vo-Dinh, editor, *Biomedical Photonics Handbook*, pages 6.1–6.30. CRC Press, Boca Raton, FL, 2003.

[51] G.F. Widhoff and S. Lederman. Specie concentration measurements utilizing raman scattering of a laser beam. *AIAA J.*, 9(2):309–316, 1971.

[52] M. Young. *Optics and Lasers*. Springer-Verlag, Berlin, 1984.

8 Fluid Mechanical Apparatus and Experimental Practices

Measurement in fluid mechanics encompasses two types of physical flow systems: (a) those in which the flow is set by factors beyond the control of the experimenter, for example environmental flows and flows in existing industrial facilities; and (b) those which are designed and built in the laboratory and over which the experimenter has full, or at least limited, control. Our interest focuses on the second type, namely, laboratory apparatus, and particularly on apparatus intended to generate, as much as possible, idealised flows (e.g., a uniform stream or a fully developed axisymmetric jet), and, to a lesser degree, those intended to reproduce a complex technological flow. The importance of using properly designed and operating flow apparatus cannot be overemphasised. It is obvious that experimental skill and good instrumentation cannot compensate for flaws in the flow itself. Most commonly, experiments in a fluid mechanics laboratory are performed using available apparatus. Even then, understanding of the function of various components would be very useful, in case performance improvements or modifications to suit specific needs become necessary. When considering building new apparatus, awareness of various options available and careful comparison of the corresponding advantages and disadvantages become absolutely essential. In the first part of this chapter, we will summarise the basic elements of flow apparatus, directing the reader to selected references for design details. As a second part, we will outline some established good practices in conducting experimental research, particularly in a fluid mechanics laboratory.

8.1 Fluid Mechanical Apparatus

8.1.1 Producing the Desired Flow

The centrepiece of all fluid mechanical apparatus is the *test section*, or *working section*, in which measurements are taken. All other parts of the apparatus serve the purpose of producing the desired flow in the test section. The design and construction of such apparatus usually requires considerable skill, effort and expense. A common type of fluid mechanical apparatus is the *general purpose wind tunnel*, which, ideally, has a test section with a uniform, disturbance-free stream. One type of such a wind tunnel is shown in Fig. 8.1. Other types of apparatus may be designed to produce a certain type of turbulent flow or flows in which variations of other properties, such as temperature and concentration, are present. In general, beside the test section, fluid mechanical apparatus includes

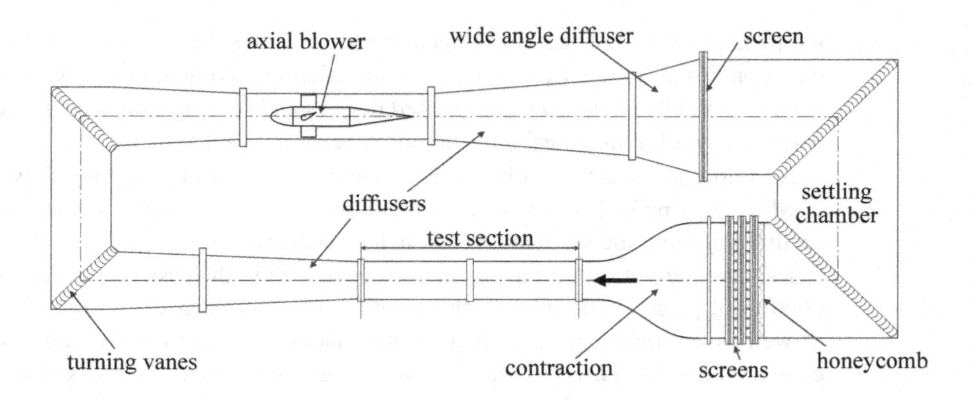

Figure 8.1 A general-purpose, closed-circuit, medium-size and speed, wind tunnel, designed and built at Imperial College London, London; this facility has a square test section with a side of 0.91 m and a length of 5.5 m.

devices and sections that are designed to provide *flow generation* or *flow management.* Flow generation is usually accomplished with the use of a *fluid mover,* such as an axial or mixed-flow fan or a pump. When placed upstream of the test section, a fluid mover would generally create significant flow disturbances, which one must remove before the flow enters the test section. An alternative approach of producing a flow is to store a substantial amount of fluid in a *container under pressure* and then release it towards the test section. For water and other liquids this container would be a head-tank, whereas for gases it would be a sealed pressure vessel.

A common section of fluid mechanical apparatus is the *settling chamber*, or *plenum*, which is a duct located upstream of the test section with a cross-sectional area that is much larger than that of the test section itself. As its name indicates, the settling chamber is a relatively low-speed region, in which the fluid is allowed to relax, at least partially, from upstream disturbances. Flow management, consisting of a reduction of speed non-uniformity and undesired turbulence, is most effectively accomplished inside the plenum by passing the flow through a series of distributed obstructions mounted across the stream. These include *flow straighteners* (such as honeycombs), *screens* of various sorts and *perforated plates*. The same type of devices can be used to generate flows with specified non-uniform characteristics and additional flow management devices can be placed in other sections of the flow facility.

Downstream of the low-speed plenum, the flow must be accelerated to the test section speed. Sudden area expansions and reductions are generally avoided in fluid mechanical apparatus, because they cause energy losses, as well as complications in the flow pattern, including flow separation, recirculation and unsteadiness. Instead, area changes are accomplished gradually, with the use of specially designed transition sections. Additional transition pieces may also be required to change the shape of the cross section or the direction of flow within the apparatus. A *contraction* is a smooth transition duct, which is designed to reduce the cross-sectional area of the apparatus, for example, from the plenum to the test section. A section that increases the cross-sectional area in the direction of the flow, for example, a section coupling the flow mover with the settling chamber, is called a *diffuser*. Specialised devices and sections may also be required for particular types of facilities, for example, supersonic wind tunnels.

An important decision to be made in the early stages of designing flow apparatus is whether to make it open- or closed-circuit. Open-circuit facilities release the test-section fluid into the surroundings, which, in a broad sense, constitutes the return. At times, open-circuit operation becomes necessary, as in the case of gas flow loops supplied from pressurised tanks. Such operation has the drawback of wasting fluid mass as well as kinetic energy, which could be a factor if kinetic energy loss is significant compared to other energy losses. In some cases, open-circuit facilities may be preferable as occupying less laboratory space and generally subjected to lower disturbance levels, compared with equivalent closed-circuit facilities. When the flow is supplied with heat or chemical contaminants or contains phase changes or chemical reactions, open-circuit operation removes the burden of restoring the fluid to its original state. On the other hand, closed-circuit facilities have the advantage of containing and controlling the entire fluid volume, thus allowing operation of the apparatus independently of its surroundings. Closed-circuit flow facilities require changes in the flow direction, which produce additional energy losses and may generate secondary flows, flow separation and other undesired effects. The management of such effects may necessitate the use of additional devices, for example, turning vanes and flow straighteners.

8.1.2 Contractions

A *subsonic contraction* (Fig. 8.2a) decreases the flow cross-sectional area from A_{in} at its entrance to A_{out} at its exit, thus accelerating the flow entering the test section. The main parameter of a contraction is the *contraction ratio* $c = A_{in}/A_{out} > 1$. Moreover, contractions are characterised as *axisymmetric*, *square* or *rectangular*, depending on their cross-sectional shape.

A contraction has the following three important effects on the flow:

- It accelerates the flow to a speed $U_1(L)$ at its exit, which is larger than the speed $U_1(0)$ at its entrance.
- It tends to make the velocity profile more uniform.
- It reduces the streamwise turbulence intensity $u_1'/\overline{U}_1$, which is defined as the ratio of the standard deviation u_1' of the streamwise velocity fluctuations and the local mean velocity $\overline{U}_1$.

These effects are quantified by a simple inviscid flow analysis, originally performed by Prandtl [11], as follows. Assume that the entrance and exit streams are parallel to the contraction axis, so that the entrance and exit static pressures are uniform, and neglect friction, gravity and flow mixing. Consider that the entrance flow velocity is equal to $U_1(0)$, except over some part in which it is disturbed to the velocity $U_1(0) + \delta U_1(0)$, where $|\delta U_1(0)| \ll U_1(0)$. The exit stream will have a uniform velocity $U_1(L)$, except in a

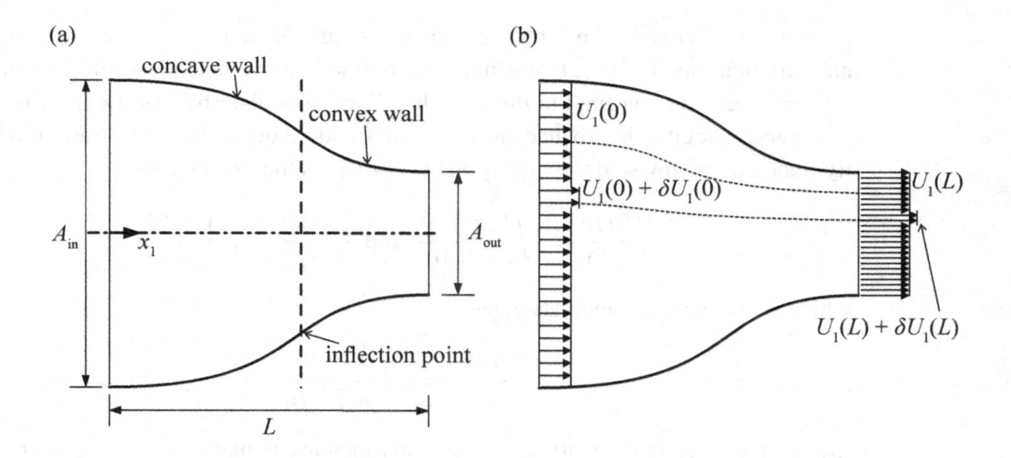

Figure 8.2 Sketches of (a) a subsonic contraction and (b) the same contraction with two streamlines connecting the entrance and exit undisturbed and disturbed parts of the corresponding velocity profiles.

corresponding disturbed part in which the velocity will be $U_1(L) + \delta U_1(L)$. Application of the continuity equation gives

$$U_1(L) \approx cU_1(0)\,. \tag{8.1}$$

Applying Bernoulli's equation in the two streams (Fig. 8.2b), neglecting viscous shear forces at the boundary between the two streams and linearising the corresponding relationships, one may derive the expression

$$\frac{\delta U_1(L)}{U_1(L)} \approx \frac{1}{c^2}\frac{\delta U_1(0)}{U_1(0)}\,, \tag{8.2}$$

which shows that streamwise disturbances, normalised by the local speed, would be reduced by a factor inversely proportional to the square of the contraction ratio. For example, for the moderate contraction ratio of $c = 10$, the relative non-uniformity at the exit would be reduced to a mere 1% of the entrance level. Equation (8.2) would also apply approximately to the streamwise turbulence intensity. Thus, the entrance and exit streamwise turbulence intensities would be approximately related as

$$\frac{\overline{u_1'(L)}}{\overline{U_1}(L)} \approx \frac{1}{c^2}\frac{\overline{u_1'(0)}}{\overline{U_1}(0)}\,. \tag{8.3}$$

An idealised viscous flow analysis [3] connects the entrance and exit streamwise turbulence intensities by the expression

$$\frac{\overline{u_1'(L)}}{\overline{U_1}(L)} \approx \frac{1}{c^2}\sqrt{\frac{3}{4}\left(\log 4c^3 - 1\right)}\,\frac{\overline{u_1'(0)}}{\overline{U_1}(0)}\,. \tag{8.4}$$

Equation (8.4) gives larger values of the intensity ratio than those given by Eq. (8.3), for example, by about 30% for $c = 10$ and by 54% for $c = 20$. Even so, both expressions predict that a contraction would drastically reduce streamwise velocity disturbances and turbulent fluctuations. Unfortunately, contractions are much less effective in reducing transverse (i.e., normal to the flow direction) flow disturbances. Let δU_2 be a local transverse velocity disturbance and u_2' be the standard deviation of the transverse velocity fluctuations. Inviscid flow analysis [11] provides the expressions

$$\frac{\delta U_2(L)}{U_1(L)} \approx \sqrt{\frac{1}{c}}\frac{\delta U_2(0)}{U_1(0)} \quad \text{and} \quad \frac{u_2'(L)}{\overline{U_1}(L)} \approx \sqrt{\frac{1}{c}}\frac{u_2'(0)}{\overline{U_1}(0)}, \tag{8.5}$$

whereas viscous flow analysis gives

$$\frac{u_2'(L)}{\overline{U_1}(L)} \approx \sqrt{\frac{3}{4c}}\frac{u_2'(0)}{\overline{U_1}(0)}. \tag{8.6}$$

Transverse velocity fluctuations will actually increase in magnitude along a contraction, while the transverse turbulence intensity will decrease only by a moderate amount, for example, by a factor of roughly 3.5 for $c = 10$ and roughly 5 for $c = 20$.

Contractions are not always beneficial, as they may actually generate or amplify velocity disturbances, when the flow density at the entrance is not uniform. Density non-uniformity could be generated by temperature non-uniformity, as would, for example, be the case of a flow inside a facility that is warmer or cooler than the surroundings and develops thermal boundary layers near walls. For a rough estimate, assume that the entrance flow has a uniform velocity $U_1(0)$ and consists of a main part with a uniform density ρ and a disturbed part with a density $\rho+\delta\rho$. Making simplifying approximations as previously, it is possible to calculate that the exit flow will contain a region with a velocity $U_1(L) \approx cU_1(0)$ and a small region with a disturbed velocity $U_1(L) + \delta U_1(L)$, where [11]

$$\frac{\delta U_1(L)}{U_1(L)} \approx -\frac{1}{2}\frac{\delta\rho}{\rho}\left(1 - \frac{1}{c^2}\right). \tag{8.7}$$

For example, in air flow, a positive temperature disturbance of 5 K at the entrance would generate a positive velocity disturbance of the order of 1% at the exit. A source of density disturbances in wind tunnel flows is heating of boundary layers due to friction between the fluid and the wall.

Contraction design is an art rather than a science and, whenever possible, it is advisable to copy designs which are known to perform well under comparable conditions. The contraction ratio c should be selected as large as possible, within the constraints of available laboratory space, fabrication capabilities and budget. A value of $c \geq 16$ would generally be sufficient, although values near 10 are also commonplace. The cross-sectional shape is usually dictated by the shape of the test section or other practical needs and it is not based on fluid mechanical considerations. Axisymmetric shapes are the choice for contractions leading to circular test sections, whereas rectangular shapes are easier to fabricate and should be preferred when the test section is rectangular. Note that the aspect ratio (height to width ratio) of rectangular contractions may be varied along the flow direction to match the dimensions of adjacent sections.

The most important factor for contraction performance is the shape of its walls, as it dictates the pressure gradient along the flow and, thus, the state of boundary layers. All contraction profiles have an inflection point, which separates a concave section (towards the entrance) from a convex section (towards the exit). A favourable pressure gradient (i.e., a monotonically decreasing wall pressure) in the entire contraction may only be achieved if the contraction length becomes infinite. Finite-length contractions are subjected to local adverse pressure gradients in both the concave and convex sections, particularly in the concave one; these correspond to the well-known phenomena of *overshoot* and *undershoot* of the wall velocity in potential flow through the contraction. Contraction profiles, optimised to eliminate boundary layer separation as predicted by the Stratford separation criterion [67], have been presented for both axisymmetric [45] and rectangular [14] contractions, while shapes optimised to produce the shortest possible contraction that meets the separation criterion have also been proposed [43]. These profiles usually consist of two or more circular, elliptical, cubic or other polynomial-type curves, matched tangentially to each other and to the axial direction at the entrance and exit. As practical guidelines, one is advised to maintain the concave section considerably longer and more gradual than the convex one and to select the contraction length L approximately 50% larger than the average entrance dimension. In fact, the exact shape of the profile is not particularly important and the experienced designer may exercise judiciousness in respecting these rules while making a totally empirical choice. To be on the safe side, however, when using novel designs, it seems worthwhile to compute the wall pressure distribution using potential flow simulation and then verify that a boundary layer separation criterion is respected.

Example 8.1 Design a contraction connecting the plenum of a wind tunnel, which has a square cross section with a side of $W_p = 2$ m, and the test section, which has a square cross section with a side of $W_t = 0.5$ m.

Answer

The contraction ratio is $c = W_p^2/W_t^2 = 16$. We further select the length of the contraction to be $L = 1.5W_p = 3$ m. These two parameters are sufficient for designing the contraction following the procedure described by Downie, Jordinson and Barnes [14], which produces the shapes of two elliptic arcs, one for the concave section and another for the convex section, which are tangent to each other at the inflection point. Note that this work is expressed in terms of the *two-dimensional contraction ratio r*, which, for the given contraction, is found as $r = W_p/W_t = 4$. The profile of the contraction is calculated in terms of the normalised coordinates $x_1/L, x_2/W_1$ and with the origin set on the contraction axis at the exit.

The design requires specification of the location of, and slope at, the inflection point.

These are calculated from given default expressions as

$$x_{1i}/L = 1/(r+1) = 0.2\,,$$
$$x_{2i}/W_p = 1/(r+1) = 0.2\,,$$
$$\left.\frac{\mathrm{d}(x_2/W_p)}{\mathrm{d}(x_1/L)}\right|_i = \frac{2(r-1)}{r} = 1.5\,.$$

Then, the equation for the convex section ($0 \le x_1/L \le 0.2$) is given by

$$\frac{x_2}{W_p} = a - \left(a - \frac{1}{2r}\right)\left[1 - \left(\frac{x_1}{L}\right)^2 \frac{\left.\frac{\mathrm{d}(x_2/W_p)}{\mathrm{d}(x_1/L)}\right|_i \left(a - \frac{x_{2i}}{W_p}\right)}{\frac{x_{1i}}{L}\left(a - \frac{1}{2r}\right)^2}\right]^{1/2},$$

where

$$a = \frac{\left(\frac{x_{2i}}{W_p}\right)^2 - \frac{x_{1i}}{L}\frac{x_{2i}}{W_p}\left.\frac{\mathrm{d}(x_2/W_p)}{\mathrm{d}(x_1/L)}\right|_i - \frac{1}{4r^2}}{2\frac{x_{2i}}{W_p} - \frac{x_{1i}}{L}\left.\frac{\mathrm{d}(x_2/W_p)}{\mathrm{d}(x_1/L)}\right|_i - \frac{1}{r}}\,.$$

The equation for the concave section ($0.2 \le x_1/L \le 1.0$) is given by

$$\frac{x_2}{W_p} = d + \left(\frac{1}{2} - d\right)\left[1 - \left(\frac{x_1}{L} - 1\right)^2 \frac{\left.\frac{\mathrm{d}(x_2/W_p)}{\mathrm{d}(x_1/L)}\right|_i \left(\frac{x_{2i}}{W_p} - d\right)}{\left(1 - \frac{x_{2i}}{L}\right)\left(\frac{1}{2} - d\right)^2}\right]^{1/2},$$

where

$$d = \frac{\left(\frac{x_{2i}}{W_p}\right)^2 + \left(1 - \frac{x_{1i}}{L}\right)\frac{x_{2i}}{W_p}\left.\frac{\mathrm{d}(x_2/W_p)}{\mathrm{d}(x_1/L)}\right|_i - \frac{1}{4}}{2\frac{x_{2i}}{W_p} + \left(1 - \frac{x_{1i}}{L}\right)\left.\frac{\mathrm{d}(x_2/W_p)}{\mathrm{d}(x_1/L)}\right|_i - 1}\,.$$

Substituting the given values, we get $a = 0.2375$ and $d = 0.0500$, from which we can get the equations for the convex and concave sections as, respectively,

$$\frac{x_2}{W_p} = 0.2375 - 0.1125\left[1 - 22.22\left(\frac{x_1}{L}\right)^2\right]^{1/2}, \quad 0 \le x_1/L \le 0.2$$

and

$$\frac{x_2}{W_p} = 0.0500 + 0.4500\left[1 - 1.3889\left(\frac{x_1}{L} - 1\right)^2\right]^{1/2}, \quad 0.2 \le x_1/L \le 1.0\,.$$

The computed profile is shown in Fig. 8.3a. Let's assume that the contraction will be built by welding together four metallic sheets, rolled gradually to the shape of the profile. Then, we need to design the pattern of these four walls on a flat sheet before they are rolled. Obviously, the length L_p of each pattern would be equal to the length of the rectified profile curve, which is larger than L. From elementary calculus, one gets

$$L_p = \int_0^L \left[1 + (\mathrm{d}x_2/\mathrm{d}x_1)^2\right]^{1/2} \mathrm{d}x\,.$$

The streamwise coordinate of the pattern can be similarly calculated as

$$s(x_1) = \int_0^{x_1} \left[1 + (\mathrm{d}x_2/\mathrm{d}x_1)^2\right]^{1/2} \mathrm{d}x_1' ,$$

whereas the width of the pattern is simply $x_2(x_1)$. This pattern has been illustrated in Fig. 8.3b. Note that, because this is a square contraction, all four walls are identical. If this were a rectangular contraction instead, the top/bottom walls would be different from the side walls, but the design procedure would be essentially the same, with the difference that the local width of the pattern for the top/bottom walls would be equal to the value of x_2 that was found for the side-wall profile and vice versa.

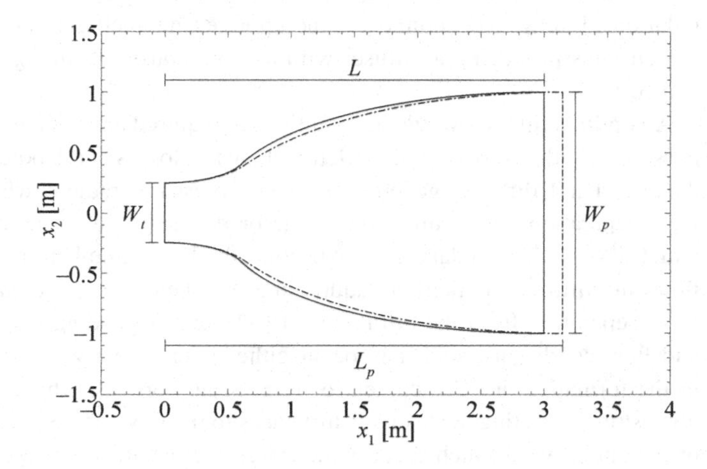

Figure 8.3 Computed profile (solid line) of a square contraction with a contraction ratio of 16, a length L and an entrance (plenum) width W_p. Pattern of a wall, with a length L_p, before it is rolled (dot-dash line) to the previous profile.

8.1.3 Diffusers

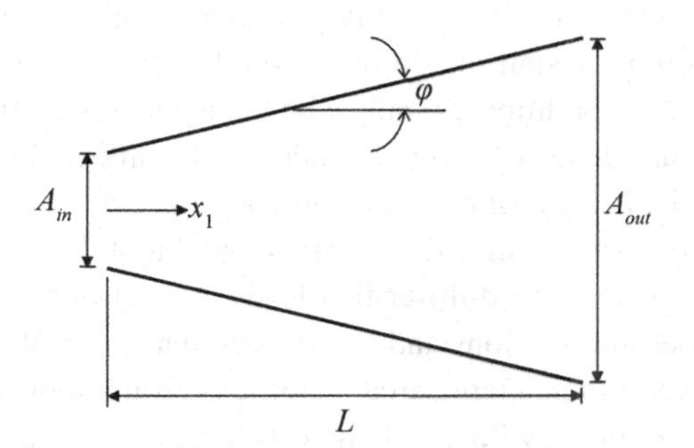

Figure 8.4 Sketch of a subsonic diffuser.

A *subsonic diffuser* (Fig. 8.4) is a duct with a continuously increasing cross-sectional area, which results in a gradual reduction of the flow velocity and an ensuing static pressure recovery. Wind tunnels and other flow facilities may contain more than one diffuser. Diffusers are commonly placed between the flow mover and the settling chamber, but also following the test section. In open-circuit facilities, a diffuser placed at the test section exit increases the test section speed, acting like a suction. In blower-type wind tunnels, an exit diffuser also decreases kinetic energy losses and laboratory drafts.

Although shape optimisation would generally lead to a non-linear diffuser wall profile, convenience in fabrication suggests a straight wall and, thus, diffusers commonly have the shape of a truncated cone or a truncated, right, square or rectangular pyramid. However, in some cases, for example, when connecting a circular section to a rectangular one, it may be economical to incorporate a geometrical transition into the diffuser design, thus producing a diffuser with a continuously changing cross-sectional shape and size.

An optimal diffuser would accomplish the required area change by using the shortest possible length, as to minimise frictional energy losses and laboratory space usage. On the other hand, diffusers generate an adverse pressure gradient, which induces boundary layer separation when sufficiently large or if acting over a sufficiently long distance, which allows the boundary layer to become thicker. A number of studies have dealt with flows in diffusers, particularly addressing the phenomenon of *stall*, namely, boundary layer separation from the diffuser wall [32, 62]. Depending on the diffuser geometry and flow conditions, stall may occur either intermittently (*transitory stall*), resulting in the formation and detachment of large-scale vortices, which would be transported downstream creating unsteady disturbances to the flow, or steadily (*full stall*), in part of or the entire wall, which reduces the effective flow area and results in a strongly non-uniform velocity profile. To prevent stall of low-speed diffusers, the diffuser half angle φ (see Fig. 8.4) must be kept relatively small. For two-dimensional diffusers, operating under typical wind-tunnel conditions, full stall would occur, essentially irrespectively of the diffuser length, when $\varphi \gtrsim 20°$ [32]. For smaller angles, transitory stall may occur, depending on the diffuser length. Two-dimensional diffusers would stall for $\varphi \gtrsim 9°$, when the length-to-inlet-height ratio is equal to 2, but at much smaller angles ($\varphi \gtrsim 1°$), when this ratio is increased to 15. Diffusers with different cross-sectional shapes would perform similarly, although their stall angles may be different. Another factor that affects diffuser performance is the state of the boundary layer at its inlet. The thicker this boundary layer is, the more prone to stall the diffuser would be. One is also reminded that, under the same free-stream conditions, laminar boundary layers would be much more likely to separate than turbulent ones. Diffuser stall is also affected by the shape of the incoming velocity profile. In view of all these complications, one should limit the area increase of an unobstructed, small-angle diffuser to about 3 to 4. In cases of highly non-uniform or turbulent flow entering a diffuser that leads to a settling chamber, it is preferable to divide the diffuser into sections and insert flow management devices (see Section 8.1.4) between sections. When a large area increase is required or the available length is limited, a *wide-angle diffuser*, with a half angle as large as 70°, may be used. A diffuser with such a large half angle will definitely separate, unless fitted with a number of pressure-reducing screens or perforated plates, which would generate a pressure drop that exceeds the ideal pressure rise across the unobstructed diffuser [61]. In extreme cases, the flow discharges directly into the settling chamber, in the form of a *sudden expansion*; this situation is to be avoided, if possible, as it leads to total loss of kinetic energy and to a jet-like flow, which would require substantial effort for its management before entering the test section.

Wind tunnel designers often meet specific diffuser requirements of their facilities using imaginative approaches. For example, the designer of the wind tunnel shown in Fig. 8.1 connected the fan section with the settling chamber using two diffusers *in tandem*: first, a diffuser with an area ratio of 2.76 and a half angle of 4.6°, followed by a wide-angle diffuser with an area ratio of 2.28 and a half angle of 14.2°. Compared to a single diffuser with a half angle of 4.6° and the same total area ratio of 6.29, the two-part diffuser is shorter by 5.65 m, which may have been necessary to fit this wind tunnel in the laboratory room or avoid crowding the open space. Moreover, this may have also reduced the cost of the facility, as a result of eliminating materials and labour required to extend both the main section and the return section of this closed circuit wind tunnel. The same rationale brings naturally a second question: why did the designer not further shorten the wind tunnel (by up to 3.06 m) by replacing part or all of the upstream diffuser by an extension of the wide-angle diffuser? The answer may have been that lengthening the wide-angle diffuser would have increased its chance of stalling and may have required the installation of several screens, which were avoided in the present design. In fact, there is a screen (having a pressure loss coefficient of 2.5) installed at the exit of the wide-angle diffuser, but this screen would not have reduced the possibility of diffuser stall, although having beneficial effects on the quality of flow in the settling chamber and the ensuing test section.

Diffuser analysis: A rough prediction of diffuser stall is possible with the use of simple methods: potential flow analysis to determine the pressure field and an integral method to predict boundary layer growth and possible separation. For a more accurate analysis, one could solve numerically the 3-D viscous flow equations (namely, the continuity and momentum equations) in the diffuser with the use of a computational fluid dynamics (CFD) code, taking care to model turbulence adequately. The accuracy of any analytical study would be affected adversely by the anticipated large uncertainty of inlet conditions, particularly if a blower is located close to the diffuser inlet. A more reliable, and likely more costly and time consuming, approach for optimising diffuser design would be to experiment with physical diffuser models. In the remainder of this section, we will present a simplified analysis, with the objective to identify the existing phenomena and trends.

Potential flow analysis: The idealised pressure field in a diffuser, in the absence of stall, can be easily calculated by solving the continuity and Bernoulli equations, under the assumption that the velocity in each cross section is uniform (1-D analysis). Let us assume that we have a conical diffuser with an inlet diameter $d(0)$, an outlet diameter $d(L)$ and a length L. Then, the diffuser half angle would be

$$\varphi = \arctan\{[d(L) - d(0)]/(2L)\}\ . \tag{8.8}$$

The diffuser diameter at a distance x_1 from the inlet will be

$$d(x_1) = d(0) + 2x_1 \tan\varphi \,, \tag{8.9}$$

which gives the corresponding cross-sectional area as

$$A(x_1) = \pi[d(x_1)]^2/4 = \pi[d(0) + 2x_1 \tan\varphi]^2/4 \,. \tag{8.10}$$

Using the continuity and Bernoulli's equations along the diffuser, one can easily derive the *pressure recovery coefficient* as

$$C_p(x_1) \equiv \frac{p(x_1) - p(0)}{{}^1\!/\!{}_2\rho[U_1(0)]^2} = 1 - \left[\frac{d(0)}{d(x_1)}\right]^4 , \tag{8.11}$$

where $U_1(0)$ is the inlet velocity. The gradient of this coefficient would be

$$\frac{\mathrm{d}C_p(x_1)}{\mathrm{d}x} = \frac{4[d(0)]^4}{[d(x_1)]^5}\frac{\mathrm{d}d(x_1)}{\mathrm{d}x_1} = \frac{8[d(0)]^4 \tan\varphi}{[d(x_1)]^5} = \frac{8[d(0)]^4 \tan\varphi}{[d(0) + 2x_1 \tan\varphi]^5} \,. \tag{8.12}$$

The last equation shows clearly that the pressure gradient in the diffuser decreases with increasing distance x from the inlet. This adverse pressure gradient would be strongest at the inlet and would vanish far downstream from it. Does this mean that the diffuser would be most likely to separate very close to its inlet? Possibly, but not necessarily, because flow separation is also more likely to occur to thicker boundary layers and boundary layer thickness would be smallest at the inlet and increase continuously with increasing x. To illustrate the combined effect of these two contradicting trends with x, we may apply the classical *Stratford separation criterion* [34, 65].

The Stratford separation criterion: To apply the Stratford criterion, we need to know the conditions of the boundary layer at the diffuser inlet. Such information is, in general, hard to find, but, for the sake of illustrating the approach, we can make some drastic simplifications. If there is a blower upstream of the diffuser, we may plausibly assume that the boundary layer is turbulent from the start, as a result of strong free-stream disturbances. We may further assume that the upstream boundary layer develops along a cylindrical section that has a length x_{1u} and perform the analysis for a representative inlet flow velocity $U_1(0)$. Thus, the boundary layer Reynolds number at the diffuser inlet would be $Re_x \approx U_1(0)x_{1u}/\nu$. Finally, we note that, for flow in a diffuser, $\mathrm{d}^2p/\mathrm{d}x_1^2 > 0$. The *Stratford turbulent separation parameter* is defined as

$$C_{ST} = \frac{C_p(x_1)\left[(x_1 + x_{1u})\frac{\mathrm{d}C_p}{\mathrm{d}x_1}\right]^{0.5}}{(Re_x/10^6)^{0.1}} \,, \tag{8.13}$$

and the Stratford criterion specifies that separation will occur as soon as C_{ST} exceeds 0.39. The behaviour of C_{ST} along the diffuser clarifies some aspects of diffuser stall. First, $C_{ST} = 0$ at the inlet, because $C_p(0) = 0$. It is also easy to show that, as $x_1 \to \infty$, $C_{ST} \to 0$. Therefore, C_{ST} will increase with increasing x_1 up to a maximum at $x_1 = x_{1max}$ and then it will decrease monotonically. If $C_{ST}(x_{1max}) < 0.39$, the Stratford criterion predicts that the diffuser will not stall, irrespectively of its length. If this is not the case, let x_{1sep} be the smaller of the two distances for which $C_{ST} = 0.39$. A diffuser that is short enough for $L < x_{1sep}$ will not stall, according to this criterion, but a longer one will.

Effect of screens: An effective way to prevent diffuser stall is to segment the diffuser into sections and install screens between sections. The effect of a screen on a complex flow is hard to assess accurately, but we point out that a screen would tend to mix the near-wall flow, thus reducing the boundary layer thickness. Let us assume that we have an unobstructed diffuser that stalls at $x_1 = x_{1sep}$. Installing a suitable screen at $x_{1scr1} < x_{1sep}$ would prevent early stall and reduce the value of C_{ST}. As x_1 increases, however, C_{ST} may exceed 0.39 again, if the diffuser is long enough. Thus, a second screen may need to be installed at x_{1scr2}, and more screens may be required until the entire diffuser is stall-free. Considering the large uncertainty of stall prediction, it seems advisable to overdesign diffusers.

8.1.4 Flow Management Devices

A variety of flow management devices have been used to reduce or eliminate undesired disturbances and other effects from the flow, before it enters the test section. Such devices usually have the form of a planar, periodic array of elements (e.g., cylinders, square rods, aerofoils, orifices with different shapes, disks etc.) that is fastened on the facility walls normal to the stream. The main characteristics of flow management devices are the following:

- The centre-to-centre spacing between elements. For square elements, this spacing is called the *mesh size* M and, for simplicity, we will use this term to characterise all shapes of elements. The mesh size is normally much smaller than the cross-sectional dimensions of the facility section in which the device is positioned.
- The frontal flow blockage by the device. This is quantified by the *solidity* σ, defined as the ratio of the projected blocked area to the total projected area of the device. An alternative term is the *porosity*, which is equal to $1 - \sigma$.
- The thickness of the device, measured in the flow direction.
- The shape of the elements.
- The material properties of the device, particularly properties pertaining to its deformation and its possible degradation by the flowing fluid or impurities carried with the stream.
- The most important property of the device, which depends on the previously listed device characteristics, is the mechanical energy loss it produces. This is expressed as the *permanent pressure loss*

$$\Delta p = K \frac{1}{2} \rho U_1^2 , \tag{8.14}$$

where K is the *pressure loss coefficient* of the device and U_1 is the velocity of the undisturbed stream. The higher the value K is, the stronger the influence of the device on the flow, and, thus, its capacity to reduce undesired effects, would likely be. On the other hand, an increase of Δp would reduce the energy efficiency of the apparatus and, when excessive, it may generate flow instability, unsteadiness and velocity non-uniformity. To combine relatively large values of K and relatively small values of Δp, one is advised to position flow management devices in the settling chamber or other low-speed sections of the apparatus. It should be mentioned that, for a given device, the pressure loss coefficient may be sensitive to the device Reynolds number, which is based on the element characteristic length and would thus be much smaller than the Reynolds number of the apparatus. In compressible flows, K would also depend on the Mach number.

Besides causing pressure drop, a flow management device can mix the fluid stream as it passes through it, tending to make the flow velocity more uniform. The mixing effect may be particularly significant in the boundary layers, where high-momentum fluid from the free stream is injected towards the wall region. This decreases the boundary layer thickness and stabilises the boundary layer, so that it can resist separation in the presence of an adverse pressure gradient.

8.1.5 Flow Straighteners

Turning vanes: When a wind tunnel or other flow apparatus has corners or other sections in which the flow direction changes, it is a good practice to guide the flow to the new direction with the use of *turning vanes* [4, 24, 42]. Properly designed vanes eliminate flow separation and secondary flows due to streamline curvature, at the expense of some pressure loss. Rolled or bent plates, with ends aligned with the desired flow directions, are the simplest types of turning vanes and have been used extensively in closed-circuit wind tunnels. When it is necessary to keep pressure loss as low as possible, one may use aerofoils with an optimised design [59].

Honeycombs: The most common device for straightening a non-parallel flow within a straight duct is the *honeycomb*, which consists of an array of cellular channels, usually of hexagonal cross-sectional shape (other shapes are also available), through which the flow is forced to pass. Honeycomb materials include aluminium and various alloys as well as paper and plastics. Cell widths range typically between 0.5 and 20 mm and lengths vary from a few mm to hundreds of mm [27]. A cellular length-to-width ratio in the range 6 – 10 is recommended for optimal performance [37, 42]. The wall material is quite thin, resulting in a low solidity, in the range between 1 and 5%. For small facilities, flow straighteners can be made by packing plastic drinking straws or other thin-walled tubes. In general, honeycombs pose relatively small pressure losses to the flow. This pressure loss increases linearly with increasing flow speed [60], which is consistent with the behaviour of pressure loss in laminar pipe flows. As a representative example, the pressure loss coefficient for an aluminium honeycomb with a cell size of 6.4 mm, a length of 25.4 mm and a solidity of 4% is $K \approx 0.4$, whereas even smaller values have been reported for different types of honeycombs installed in wind tunnels [2, 60]. The main function of flow straighteners is to obstruct transverse velocity components, including swirl, thus generating a flow that is nearly parallel to their walls. In suction wind tunnels, honeycombs are usually positioned near the flow inlet, whereas, in blower tunnels, they are usually positioned at the inlet of the plenum. Because of their small K-value, honeycombs are ineffective in removing streamwise velocity non-uniformity present in the stream. Honeycombs will suppress upstream turbulent motions with a scale that is larger than the cell width, but also introduce their own turbulence, which, however, is of small scale and will decay at a relatively short downstream distance. Honeycomb-produced turbulence can be reduced drastically by a fine-mesh screen positioned closely downstream of the honeycomb. In general, honeycombs are followed by a number of screens, which have much higher loss coefficients. For additional discussion on honeycombs, one may consult the available literature [19, 36, 37, 38, 39, 42].

8.1.6 Screens, Grids and Perforated Plates

A *screen*, per se, is a thin, fine mesh, made of metal, plastic or other material and fabricated by weaving, welding, etching or other method; such a screen is also referred to as a *gauze*. More generally, the term screen may also be used to describe *grids* of parallel rods, *perforated plates*, honeycombs, foam and fibre sheets, loosely woven fabrics and other relatively thin flow obstructions mounted across the flow stream and introducing a pressure loss that is characterised by the pressure loss coefficient K. Usually, screens are planar and inserted normal to the flow direction, but they may also be used in inclined or curved configurations. The present section is concerned with the use of screens to improve flow uniformity and to reduce turbulence, while some other uses will be described in Section 8.1.9. For more details, the reader is advised to consult the extensive literature which is available on screens and turbulence management [11, 35, 37, 66].

The pressure loss coefficient K depends on the geometrical solidity σ of the screen, the Reynolds number (with a diminishing effect as this number increases) and the shape of the screen's elements. For example, a grid of parallel square rods would have a K larger than that of a grid of circular rods with the same solidity, because the wakes of the square rods are wider than those of circular ones. Thus, an *effective solidity*, taking into account the geometrical shape and even material of the screen, seems to be more appropriate to characterise its dynamic performance. Among the various relationships that have been suggested for square-mesh, woven gauzes, one may use as representative the following [33]:

$$K = \left(0.52 + \frac{17}{\mathrm{Re}_d}\right)\frac{\sigma(2-\sigma)}{(1-\sigma)^2}\,, \tag{8.15}$$

where $\mathrm{Re}_d = Ud/\nu$ is the Reynolds number, based on the diameter d of the wire.

Consider a uniform screen normal to a flow having a spatially varying velocity with a scale of non-uniformity that is larger than the mesh size. The drag force on one particular mesh of the screen will be

$$F_{DM} = \frac{1}{2}\rho K M^2 U_1^2\,, \tag{8.16}$$

where U_1 is the local flow velocity. Thus, the drag force on a mesh that is in a high-velocity region will be larger than that on a mesh that is in a low-velocity region and, in relative terms, high-velocity regions will lose more momentum than low-velocity regions. The net effect of the spatial variation of momentum loss is that the flow downstream of the screen tends to be more uniform that it was upstream. Because continuity requires that the velocity averaged over the entire screen remains constant, the low-speed regions of the flow will actually be accelerated across the screen. The permanent pressure loss across the screen adds to the power requirement of the flow mover and reduces the energy efficiency of the facility. As an example, the mean velocity gradient of incompressible flow was found to decrease to about 60% of its upstream value, when crossing a fine screen with a solidity $\sigma = 0.3$, and to about 40% of its upstream value, when the solidity was increased to $\sigma = 0.4$ [33]. One must also be aware that two or more screens would have a stronger reduction of non-uniformity than a single screen with a pressure loss coefficient equal to the sum of those of the individual screens used separately. Thus, multiple screens of relatively low solidity would be preferable to a smaller number of high-solidity screens. At any rate, it is recommended that the solidity should not exceed the value of about 0.40 to 0.45, because, at high solidities, screens tend to produce large-scale flow non-uniformity [11].

A more thorough understanding of screen–flow interaction is required in order to use screens effectively for free-stream turbulence reduction. The flow closely downstream of screens consists of a cluster of jets and wakes, which, within the Reynolds number range of most fluid mechanics experiments, are unstable and break down to turbulence that tends to be nearly homogeneous and isotropic. Thus, a screen necessarily generates turbulence; however, this turbulence receives no additional energy away from the screen and decays with a distance from the screen. The screen-generated turbulence kinetic energy k, normalised by the mean velocity $\overline{U}_1$, decays following the empirical law

$$\frac{k}{\overline{U}_1^2} \sim a\left(\frac{x_1 - x_{10}}{M}\right)^n \quad , \quad \frac{x_1}{M} > 5 \, , \tag{8.17}$$

where the coefficient a depends on the geometry of the screen, x_1 is the distance from the screen and x_{10} is an empirical effective origin. Sufficiently far from the grid, namely, for $x_1 \geq 30M$, the exponent n is in the range -1.2 to -1.4 and x_{10} is typically equal to $3M$ [11, 57]. At intermediate distances from the grid, namely for $5M \leq x_1 \leq 30M$, $x_{10} \approx -3M$ and $-2.3 \leq n \leq -2.9$ [30, 52, 69]. It is generally accepted that, for $x_1/M > 500$, screen-generated turbulence is negligible. A more tolerant limit, for flow facilities that have significant disturbances from other sources, would be $x_1/M = 100$.

Why then are screens used to reduce turbulence? Screens are usually positioned in parts of the apparatus in which the turbulence of oncoming flow has dominant eddies with relatively large size, comparable, for example, to the dimensions of the blades of the fluid mover. As these eddies cross a screen with a substantially smaller mesh size, they are broken down to smaller sizes. The smaller the eddies, the more they are subjected to viscous dissipation (i.e., reduction of local velocity differences by friction) and the faster they decay. The optimal range of mesh sizes of screens that are capable of effective eddy breakdown extends to about one or two orders of magnitude less than the scale of oncoming turbulence, while much finer screens of the same solidity would have a lesser effect. Thus, a strategy emerges on how to reduce flow turbulence with the use of screens. To reduce pressure losses, screens must be positioned in low-speed sections of the apparatus, preferably the settling chamber or the downstream half of a diffuser. Typically four or more screens should be selected, arranged in order of diminishing mesh size and with sufficient distance from each other for the turbulence generated by each screen to decay before the next screen is approached. To avoid generating flow non-uniformity, the screens must be of good quality and free of defects, kinks and bulges, which could have been caused, for example, by rolling the screen for shipping. They must be maintained very taut, with uniform tension applied on all sides. One must also ensure that the flow approaches each screen in an approximately normal direction over its entire surface. Finally, during operation of the apparatus, the screens must be maintained clean and rust free; this may require periodic inspection, cleaning and maintenance.

Example 8.2 Design a set of flow management devices for a small, blower-type wind tunnel, having a test section entrance with a height of 12″ (304.8 mm) and a width of 18″ (457.2 mm) and supplied with air by an axial blower with a duct diameter of 30″ (762 mm). Dimensions are given in feet (1′ = 12″ = 304.8 mm) and inches (1″ = 25.4 mm) to conform with material specifications in North America.

Answer

There is no unique solution to this problem. The following suggestions are representative among several alternatives and they are merely meant to illustrate some good design practices. The proposed design may actually be improved with the use of parametric studies, computational fluid dynamics analysis and experimental tests with small-scale models and samples. Additional considerations must be given to availability and cost of materials. The design choices and the related rationale are as follows:

(a) A contraction with a contraction ratio $c = 16$ will be used. The two-dimensional contraction ratio will be kept at $r = 4$ for both pairs of opposing contraction walls, so that the entrance of the contraction will have a height of 4′ and a width of 6′. The length of the contraction will be taken as 1.5 times the width, which is 9′. The wall profiles of the contraction will then be determined as in Example 8.1. This contraction will be made from rolled and welded aluminium or steel sheets.

(b) The plenum will necessarily have a height of 4′ and a width of 6′. Its length will be determined after the flow management devices have been chosen.

(c) A transition piece will be used to connect the circular blower exit to the square entrance of the diffuser, which will have a side that is equal to the blower duct diameter, namely, 30″. This transition piece can be made from a formed metal sheet, molded fibreglass or wood. Its length can be also set at 30″.

(d) We choose a diffuser length of 15′, which gives an angle of 6.7° for the side walls and an angle of 2.9° for the top and bottom walls. Because one of these angles exceeds somewhat the recommended limit, we are going to fit the diffuser with a couple of screens, which are also expected to improve the flow quality.

(e) To straighten the flow and remove blower-induced swirl, we will use two honeycombs, one between the transition piece and the diffuser and another between the diffuser and the plenum. Among available products, we choose a honeycomb made of aluminium and having hexagonal cells with a width of 1/4″ and a length of 2″. We will condition the flow exiting each honeycomb with the use of a perforated plate, positioned 1.4″ (namely, 5.6 cell widths [37]) downstream of the honeycomb. We choose perforated plates made of stainless steel and having a thickness of 1/4″ and staggered round perforations with a diameter of 1/4″ and a centre-to-centre spacing of 5/16″. The solidity of this plate is 0.42, which is marginal for flow stability purposes, but it is noted that low-solidity, small-mesh perforated plates could not be identified in a search. A particularly cautious designer may use a screen instead.

(f) We will use a number of screens kept taught across the diffuser and the plenum. We choose three sizes of screens, all made of woven galvanised steel wires and having a solidity of about 0.365 and mesh sizes equal to 1/4″, 1/8″ and1/16″. We will refer to these screens as 4M (i.e., four meshes per inch), 8M and 16M. The distance between screens will be maintained at the least equal to 100 times the mesh size of the upstream screen. We decided to use one 4M and one 8M in the diffuser and one 8M and two 16 M in the plenum. In consideration of this spacing, we set the length of the plenum to 7′.

A sketch of the presently designed wind-tunnel sections and flow management devices is shown in Fig. 8.5. A similar facility was designed and built at the University of Ottawa, Ottawa, Canada, serving mainly for the study of grid-generated turbulence and uniformly sheared turbulence. Its test section has a height of 0.305 m, an initial width of 0.457 m and a length of roughly 5 m. The vertical test section walls are made to diverge slightly in order to maintain an approximately constant pressure within the entire test section. The unobstructed flow at the test section inlet has a turbulence intensity of roughly 0.15%.

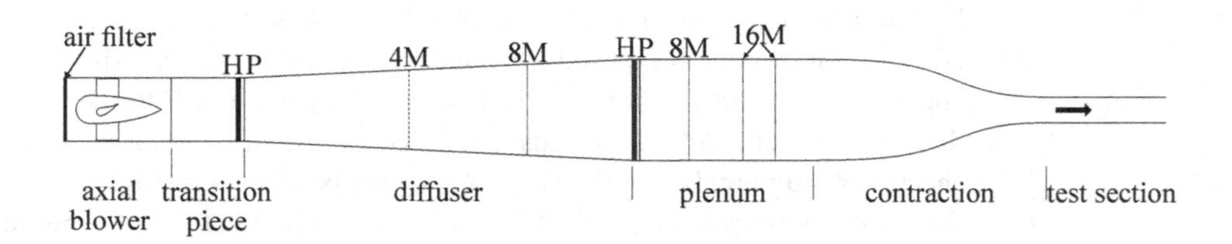

Figure 8.5 Sketch of a small, low-speed, blower-type wind tunnel (not to scale); H indicates a honeycomb, P indicates a perforated plate and 4M, 8M and 16M indicate screens.

Example 8.3 Design a conical diffuser that will connect a blower with a duct diameter $d_b = 0.500$ m to a cylindrical plenum with a diameter $d_p = 1.500$ m, taking into consideration that the available laboratory space is very limited.

Answer

The plenum-to-blower diameter ratio is $d_p/d_b = 3$ and the diffuser area ratio is $c = (\pi d_p^2/4)/(\pi d_b^2/4) = 9$. A conical diffuser with an angle $\varphi' = 6°$, which is within the likely stall-free range, would have a length $L' = (d_p - d_b)/(2\tan 6°) = 4.757$ m. Assume, however, that the laboratory space can fit at most a diffuser with a length $L = 2$ m. Such a diffuser would have an angle $\varphi = \arctan[(d_p - d_b)/(2L)] \approx 14°$, which is certainly within the stall range. For this reason, this diffuser needs to be fitted with screens. Once more, there is no unique solution to this problem and the choice would depend, among other factors, on availability and cost of materials and ease of fabrication. In the following, we will apply the simplified diffuser analysis that was outlined previously. In consideration of the rough nature of this analysis, it is wise to overdesign somewhat this facility.

Let us assume that the boundary layer entering the diffuser develops along a cylindrical upstream section with a length $L_d = d_b = 0.500$ m (see Fig. 8.6a) and that it is turbulent from the start. We will perform the analysis for the representative flow velocity $U_1 = 5$ m/s at the diffuser inlet, which corresponds to a boundary layer Reynolds number at the diffuser inlet $Re_x = U_1 L_d/\nu \approx 0.17 \times 10^6$.

The pressure recovery coefficient, its gradient and the Stratford parameter are, respectively, plotted against the normalised distance $x - 1/L$ from the inlet in Fig. 8.6b, c and d. It is evident that the pressure gradient is strongest just downstream of the inlet and tends to vanish near the outlet. $C_{ST} > 0.39$ in most of the diffuser, with the first crossing at $x_{1sep} \approx 0.08$ m, or $x_1/L \approx 0.04$, namely, very close to the diffuser inlet. All indications point to the expectation that this diffuser would stall close to its inlet. To suppress stall, we will fit the diffuser with screens.

Next, we proceed to select screens. These screens should have a large enough solidity σ to produce a substantial pressure drop, but not larger than about 0.4, as to avoid the possibility of flow instability. The choice of screens with $\sigma = 0.365$, like the ones used in Example 8.2, seems to be appropriate. Neglecting the Reynolds number correction in Eq. (8.15), we get the conservative value $K \approx 0.77$. It would be desirable for the same screens to also serve for some flow management, besides introducing pressure losses, or at least not to introduce much turbulence of their own; however, such considerations introduce conflicting criteria for the mesh size M of the screen. We will adopt the smallest of the screens used in Example 8.2, namely screen M16, which has $M \approx 1.6$ mm and is sturdy enough to be kept taut across the diffuser cross section.

We finally need to determine how many screens to use and where to insert them. To be conservative, we should insert the first screen upstream of the estimated separation location $x_{1sep} \approx 80$ mm, but measurably downstream of the diffuser inlet. Let us then put the first screen at $x_{1scr1} = 40$ mm $= 0.020L$. It is hard to determine exactly how the screen will affect the oncoming stream, but it will almost certainly mix the flow sufficiently to reduce the boundary layer thickness, thus, making the flow less vulnerable

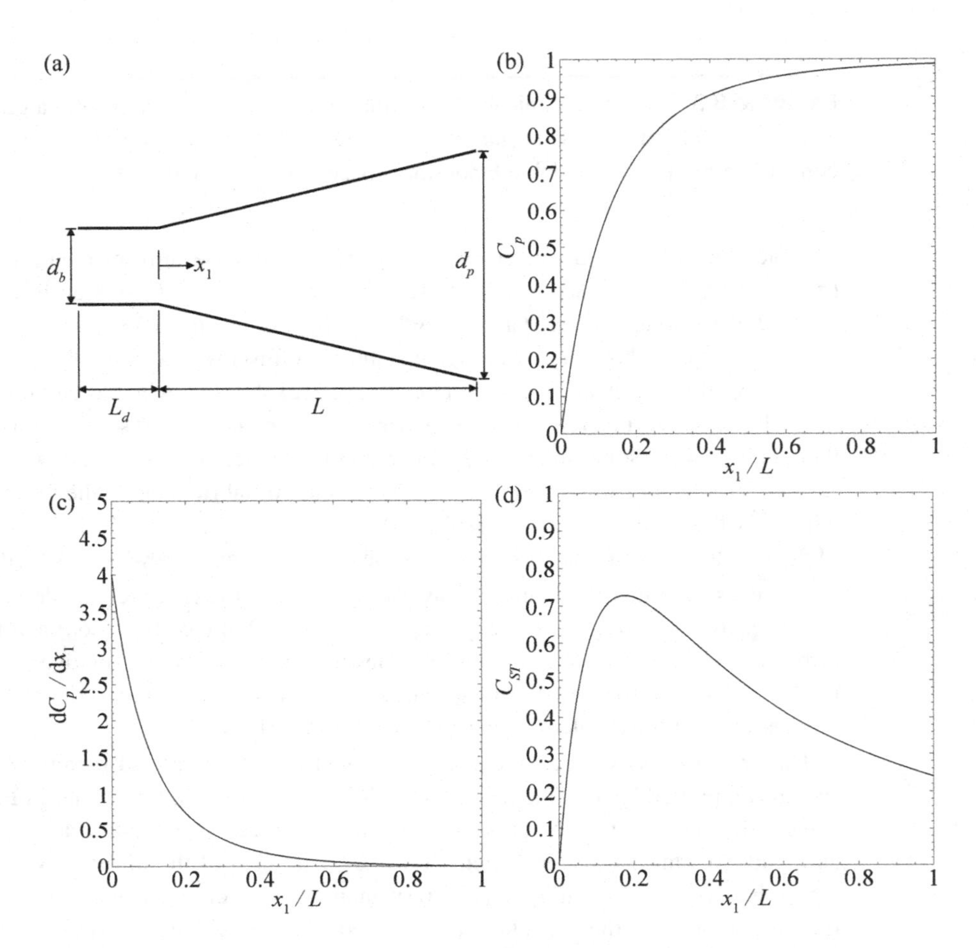

Figure 8.6 (a) Sketch of a wide-angle conical diffuser; (b) pressure recovery coefficient vs. the normalised distance from the inlet; (c) the normalised pressure gradient; and (d) the Stratford turbulent separation parameter.

to flow separation. Intuitively, a good strategy seems to be to locate the second and subsequent screens at a spacing that increases downstream. It is also difficult to predict how many screens to use, but we may postulate that the total pressure drop introduced by the screens should be at least equal to the pressure rise in the unobstructed diffuser. Screen i, positioned at a location x_{1scri}, where the diameter is d_{scri}, will introduce a normalised pressure loss

$$\frac{\Delta p_i}{p(L) - p(0)} = K_i \frac{(d_b/d_i)^4}{1 - (d_b/d_p)^4},$$

where the pressure recovery $p(L) - p(0)$ in the absence of screens is estimated using inviscid flow analysis (namely, the continuity and Bernoulli's equations). Our target is to use N screens, which will generate a total pressure loss $\sum_1^N \Delta p_i$ that exceeds $p(L)-p(0)$. The first screen will produce a pressure drop $\Delta p_1/[p(L) - p(0)] \approx 0.67$. Therefore, it

Table 8.1 Pressure drop caused by screens

i	$\frac{x_{1scri}}{L}$	$\frac{d_{scri}}{d_b}$	$\frac{U_1(x_{1scri})}{U_1(0)}$	$\frac{\Delta p_j}{p(L)-p(0)}$	$\frac{\sum_1^i \Delta p_i}{p(L)-p(0)}$
1	0.020	1.04	0.924	0.667	0.67
2	0.075	1.14	0.770	0.462	1.13
3	0.200	1.35	0.549	0.235	1.36
4	0.500	1.75	0.327	0.084	1.45

seems that two screens would satisfy the pressure drop postulate. Nevertheless, our rough analysis cannot guarantee that two screens would be sufficient to prevent stall under all operating conditions of the diffuser. To be on the safe side, we suggest to use three screens, in addition to the first one, locating them judiciously at, respectively, $x_{1scr2} = 150$ mm, $x_{1scr3} = 400$ mm and $x_{1scr4} = 1{,}000$ mm. Table 8.1 clearly shows that the pressure drop caused by these screens decreases rapidly with increasing distance from the inlet. Additional screens would have increasingly weaker effects.

8.1.7 Wind Tunnels

A wind tunnel is a duct containing flow of air or, rather rarely, another gas. Wind tunnels are commonly used for the study of air flow past models of aircraft, vehicles and structures as well as for fundamental fluid mechanical experiments, such as studies of vortices, the structure of turbulent flows and turbulent diffusion. There is a great number and variety of wind tunnels around the world, including many large-scale and specialised national facilities [2, 48, 49, 50, 55, 56]. The present section will be concerned with a few general-purpose wind tunnel designs, of the kinds that can be designed and built with a moderate amount of effort and resources [4, 5, 26, 42] and likely to be found in an academic institution or a small industry. The vast majority of such facilities are classified as *low-speed wind tunnels*, which means that their maximum speed would be lower (or, in many cases, considerably lower) than about 100 m/s, so that compressibility effects would be negligible or very small. Such facilities can also be classified as small- or medium-size, having test section cross-sectional areas in the range between a fraction of and a few square metres. Representative designs of such wind tunnels are the following.

Suction wind tunnels: These consist of a straight, open-circuit arrangement, in which air is drawn from the surroundings into the test section through suction generated by a fan located downstream of the test section (Fig. 8.7). The flow usually passes through an air filter, a honeycomb and screens before entering the contraction which leads to the test section. The latter is connected to the blower section through a diffuser, in order to reduce kinetic energy losses at the tunnel exit. Compared to other types, suction tunnels are easier to design, less expensive to built and occupy less laboratory space. Although free of disturbances caused by an upstream fan, they are vulnerable to disturbances due to external obstructions and large-scale recirculation in the room.

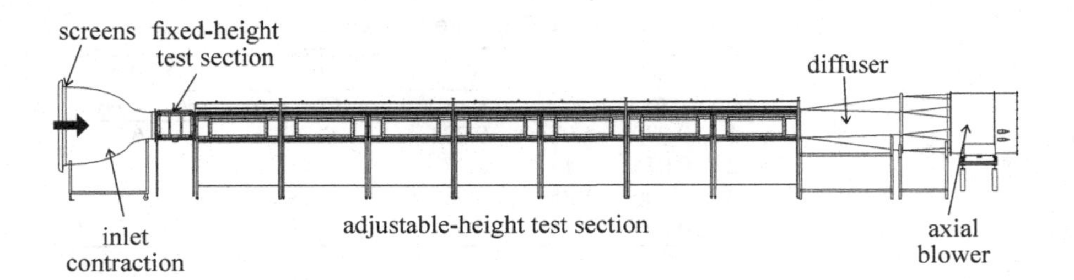

Figure 8.7 The suction-type boundary layer wind tunnel at the University of Ottawa, having a test section with a fixed width of 0.91 m, an adjustable height starting at 0.61 m and a length of approximately 16 m.

Blower wind tunnels: Like suction tunnels, blower tunnels are of the open-circuit type. They are also arranged in a straight configuration, with the difference that a blower is positioned upstream of the test section (Fig. 8.5). Flow disturbances introduced by the blower are reduced with the use of a settling chamber and numerous flow management devices, but the energy efficiency of the facility is reduced by the need to balance the pressure losses that occur in such devices.

Because open-circuit wind tunnels freely communicate with their surrounding space, whether it is the open atmosphere or an enclosed laboratory, they are subjected to contamination by dust and other impurities. For the same reason, they are generally unsuitable for studies requiring injection of smoke, seeding particles or other contaminants. On the positive side, the use of a large volume of air as a return nearly eliminates the temperature increase due to accumulation of self-heating effects, thus eliminating the need for a cooling system.

Closed-circuit wind tunnels: Compared to open-circuit tunnels, closed-circuit tunnels feature higher power efficiency, reduced noise level, flow containment and better flow management. For these reasons, most large wind tunnels are of the closed-circuit type (Fig. 8.1). Many variants of this design are available, some suitable mainly for conventional aerodynamic studies, others offering special capabilities. Many contain a refrigeration unit to remove the self-heating load, which would otherwise introduce temperature differences and possible flow non-uniformity.

Specialised facilities: A variety of specialised wind tunnels and related facilities have been constructed in many academic, government and industrial institutions around the world. Representative examples include:

- *Open test section wind tunnels*, also called *Eiffel-type* wind tunnels. The test section in these tunnels is not bounded by solid walls but is in the form of a free jet, in which the models are inserted. Compared to wind tunnels with enclosed test section, they offer the advantage of being free of wall effects (see Section 8.2). Open test sections can be used in closed-circuit configuration by being surrounded by a much larger enclosure leading to the return. Such designs, with acoustic padding and microphones placed outside the flow, are suitable for far-field aero-acoustic studies, for example, [8, 9, 51]. To eliminate the need for separate blowers, modern large-scale wind tunnels of this type are usually combined with conventional types by employing flow gates and diverters. A wind tunnel that is a hybrid of solid-wall and open test section designs is the *slotted-wall test section* wind tunnel. A slotted wall poses a smaller distortion to the flow that a solid wall, while being free of the unsteadiness that may be observed in free-jet test sections due to instabilities that develop in the shear layer along the perimeter of the free jet.

- *High-speed wind tunnels*, providing compressible air flow; these include transonic, supersonic, and hypersonic tunnels and require the addition of a converging–diverging nozzle to produce the desired Mach number in the test section.
- *Pressurised wind tunnels*, which maintain the test section pressure at levels substantially higher than the atmospheric pressure, by either releasing the exhaust air through a throttling valve or by being entirely enclosed within a pressurised vessel; the advantage of pressurisation is an increase in the flow Reynolds number, following an increase in air density.
- *Cryogenic wind tunnels*, which maintain a gas stream at an extremely low temperature by injection of the same gas in liquid form; this also results in increased Reynolds number due to the reduced viscosity, while pressurisation keeps the density at high values.
- *Environmental wind tunnels*, which simulate the characteristics of the lower atmospheric boundary layer and are suitable for wind engineering studies. Facilities capable of simulating tornadoes and strong wind gusts, such as the WindEEE Dome at Western University, London, Canada, are included in this category. Wind tunnels operating above the free surface of a water tank have also been constructed for the study of wave motion and air-sea interaction.
- *Low-temperature wind tunnels*, specialised for the study of low-temperature flow phenomena, such as icing on aircraft wing and power cables.
- *Combustion wind tunnels*, with wall materials suitable for the study of flames and other high-temperature or corrosive phenomena.
- *Vertical* or *short take-off and landing (V/STOL) wind tunnels*, which require a particularly large test section and are usually combined with conventional wind tunnels.
- *Shock tubes*, for studies of compressible flows and the effects of shock waves on various flow phenomena [6].
- A variety of devices in which relative motion is produced by moving the model and instrumentation through still air, rather than placing them in an air stream; these include *whirling arms*, *towing systems* mounted on rails, *rotating frames* and *swinging arms* (pendulum-like devices).
- *Free jets*, issuing from nozzles and supplied with air from a fan or a compressed air tank; these are particularly suitable as calibration rigs for velocity and temperature transducers.

8.1.8 Water Tunnels and Towing Tanks

These are facilities that use water as a medium, although other liquids have also been used for specific purposes.

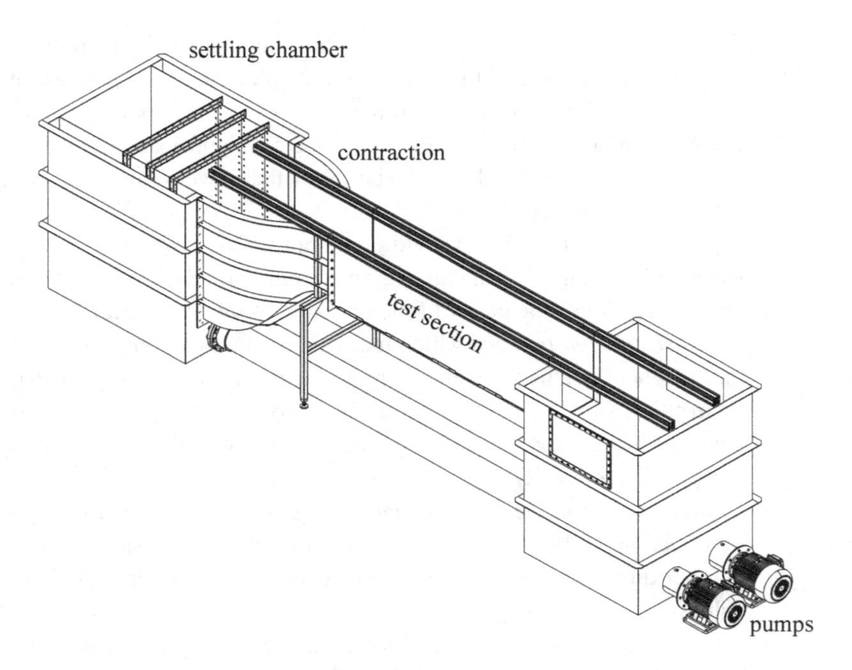

Figure 8.8 The recirculating-flow water tunnel at McGill University, designed by *Hambleton Instruments*.

Water tunnels: With the possible exception of some very small flow rigs, water tunnels are normally closed-circuit (see example in Fig. 8.8). They can be subdivided into entirely enclosed facilities, which can be properly called water tunnels, and facilities with a free surface, often called *water channels* or *flumes*. Water tunnels consist, in principle, of the same types of components as wind tunnels, namely a flow mover (pump), a settling tank, a contraction, diffusers and flow management devices, adapted to the use of water rather than air [18, 26, 55, 70]. General-purpose water tunnels are used widely for aerodynamic/hydrodynamic and fundamental fluid mechanical studies.

In addition, specialised facilities have also been developed, a few types of which are as follows:

- *Stratified flow channels*, usually with stable density stratification, that is, having heavier fluid layers underneath lighter layers. Density stratification is produced by maintaining salinity or temperature differences between adjacent layers. The closed-loop operation of such facilities requires prevention of vertical mixing, which excludes the use of a single pump. An example of a suitable design recirculates the different layers without vertical mixing with the the use a version of a *Tesla pump*, consisting of two sets of horizontal, meshing, rotating disks [53]. In a different design, different horizontal layers are skimmed separately and recirculated through separate pumps [64].
- *Low Reynolds number facilities*, containing a liquid with very high viscosity, such as glycerol or silicon oil. In view of flow heating caused by strong viscous dissipation of

kinetic energy and the strong temperature sensitivity of viscosity, such facilities may require a cooling system.

- *Refractive-index-matching facilities*, containing a liquid with a refractive index that matches that of transparent walls and/or immersed objects, thus enabling undistorted optical access to the flow.
- *Cavitation tunnels*, which include a mechanism capable of reducing the static pressure in the enclosed test section below atmospheric level to facilitate cavitation.

Towing tanks: Compared to water tunnels of the same size and operating at the same relative velocity between fluid and towed object, towing tanks [21, 55] have the advantages of lower power consumption, lower free-stream disturbances and the absence of boundary layer effects. Due to their low background turbulence, towing tanks are particularly suitable for studies of hydrodynamic stability and transition to turbulence. Moreover, stratified and multilayered inhomogeneous liquids can be most conveniently produced when still. In combination with two- or three-dimensional traversing and/or rotation of towed objects, towing tanks offer the possibility of simulating complex object paths and unsteady flows, as it is much easier to control the motion of towed objects and instrumentation than that of a flowing fluid. The main disadvantage of towing tanks is their finite length, which limits the testing time for each experimental run. They may also impose a significant 'dead time' between runs, in order to allow the liquid to acquiesce. Inevitably, towing also generates mixing, which is a limiting factor for studies in inhomogeneous and stratified fluids. The carriage could be simply made to slide on a smooth track, driven by a falling weight through a pulley. Most commonly, however, towing-tank carriages employ the use of an electric motor, preferably one with accurate speed control, which turns a rubbing wheel or a pinion gear. A point of concern in many experiments is to minimise disturbances of the motion due to carriage vibration; the use of continuous lubrication and backlash-free gears would certainly help in this respect. *Ripple tanks* are shallow tanks dedicated to the study of surface waves, as models of other types of wave motion.

Surface waves: A common problem with all facilities that have a liquid free surface is the generation of travelling, *shallow-water surface waves*, which get reflected at the end of the facility and travel back and forth along the test section, thus creating a periodic disturbance to the flow. The travelling speed of surface waves is [54]

$$c = f\lambda = \sqrt{\frac{g\lambda}{2\pi} \tanh \frac{2\pi h}{\lambda}}\,, \tag{8.18}$$

where f is their frequency, λ is their wavelength and h is the depth of the channel, assumed to be uniform. For relatively shallow channels ($h < 0.07\lambda$), one gets a simplified expression for the wave speed, as

$$c \approx \sqrt{gh}\,, \tag{8.19}$$

which is independent of the wavelength. Given an average depth and the length l of the facility, one may estimate the period of disturbance as

$$T \approx \frac{2l}{\sqrt{gh}}\,. \tag{8.20}$$

Shallow-water surface waves are difficult to eliminate. One may neglect their effects only if their amplitude is very small and their period is much larger than the time scales of interest in a particular experiment. Water tunnels that are entirely enclosed by walls and have no free surface are free of such waves.

8.1.9 Turbulence and Shear Generation

Devices similar to those serving for flow management, as described in previous sections, may be also used to generate flows with a desired mean velocity profile and/or desired turbulence characteristics. In this section, we will discuss a few relatively simple examples that illustrate the means by which some idealised types of flow can be approximately realised in the laboratory.

Nearly isotropic turbulence: The statistically simplest type of turbulence is homogeneous and isotropic, namely a flow whose statistical moments and other averaged properties are independent of location and orientation. This is a flow which has been studied extensively on its own merit, but also a good environment for testing the effect of turbulence on a great variety of flow-related phenomena, for example on separation from a wing or on combustion efficiency [10, 57]. One can generate a flow that is nearly homogeneous and isotropic in a wind tunnel or a water tunnel by passing a stream through a normal array of periodically spaced obstructions, such as a grid, perforated plate or honeycomb, with the most commonly used devices being grids of parallel rods, or biplanar, square-mesh grids. The two most important macroscopic properties of isotropic turbulence are its kinetic energy per unit mass k (or, equivalently, the *turbulence intensity*, usually defined as the standard deviation of the streamwise velocity fluctuations normalised by the mean velocity) and its integral length scale (i.e., the size of the most energetic motions). The kinetic energy of grid-generated turbulence decays according to Eq. (8.17), where the coefficient a increases with increasing solidity and also depends on the shape of the grid [57]. A preferable solidity range is 0.30 to 0.38, which should not be exceeded substantially to avoid flow unsteadiness and large-scale non-uniformity. The mesh size should be maintained sufficiently small for wall effects to be negligible in the core of the stream. Thus, the mesh size should be smaller than about one tenth of the smallest transverse dimension of the test section, considering that the integral length scale of isotropic turbulence starts from a value comparable in order of magnitude to the mesh size and grows downstream following a power law like Eq. (8.17), but with an exponent between 0.4 and 0.5. *Active grids*, consisting of arrays of randomly rotating elements, are becoming increasingly popular for the generation of turbulence with an intensity that is much larger than the one produced by passive grids [40, 46, 47].

Uniformly sheared flow: A rectilinear flow with a constant transverse velocity gradient (*shear*) is the simplest type of shear flow and is often used as an idealised environment to study the effects of shearing on a variety of phenomena. Also known as *Couette flow*, it is the paradigm used to illustrate the concept of viscosity, representing the laminar flow in the narrow gap between two infinite parallel walls, one of which is fixed and the other moves parallel to itself with a constant velocity [20]. By definition, the mean velocity of such a flow in a channel with a rectangular cross section having a height h would be (Fig. 8.9)

$$U_1 = U_{1c} + \frac{\mathrm{d}U_1}{\mathrm{d}x_2}\left(x_2 - \frac{h}{2}\right), \tag{8.21}$$

where $\mathrm{d}U_1/\mathrm{d}x_2 = \text{const.}$ and U_{1c} is the centreline speed. The strength of shearing is measured by the value of the *shear parameter*

$$\beta = \frac{h}{U_{1c}}\frac{\mathrm{d}U_1}{\mathrm{d}x_2}, \tag{8.22}$$

which, to avoid flow reversal, is limited to the range between 0 and 2.

The generation of Couette flow in the laboratory by moving plane walls in the form of belts is, in principle, possible, but rather impractical, while the use of coaxial cylinders, at least one of which rotates (*circular Couette flow*), is mostly utilised for the measurement of viscosity or the study of rotation effects, rather than studies of turbulent shear flows. The most popular way of generating shear is by passing a parallel stream through a non-uniform transverse obstruction, which is referred to as the *shear generator* (Fig. 8.9a). A variety of such obstructions that have been used in different laboratories, including planar grids, screens, honeycombs, perforated plates and aerofoil cascades with a varying solidity, produced by variable spacing of identical elements, variable thickness of equally space elements or variable mesh size [33]. Such elements would introduce significant turbulence, which would be sustained and even amplified by production due to shear. Mesh size non-uniformity also results in length-scale non-uniformity, which is an undesirable feature if the apparatus is meant to generate homogeneous turbulence. For this reason, it is advisable to use a *flow separator*, placed immediately following the shear generator and consisting of a set of evenly spaced plates parallel to the flow, thus separating the cross section into parallel channels. The flow separator helps straighten the flow, which otherwise may start with curved streamlines. Moreover, it imposes its own spacing as an initial scale for the energy containing eddies of device-generated turbulence. The use of a flow separator also introduces the possibility of generating mean shear by supplying each channel with fluid from a separate fluid mover or head tank. If variable obstruction is achieved by varying the solidity of a perforated plate, one may achieve an approximately uniform shear starting with a linearly varying solidity and then improve the shear uniformity by trial-and-error minor adjustments.

As mentioned previously, a shear generator/flow separator type of device would introduce substantial turbulence, much stronger than typical free-stream levels in wind tunnels and unacceptably high for certain types of experiments in which turbulence might obscure shearing effects. A means of producing low turbulence shear, typically at intensities about 0.5 to 1%, is the use of curved screens (Fig. 8.9b). The determination of the shape of a uniform fine screen that would produce uniformly sheared flow with a desired value of shear parameter β can be achieved using idealised theory [15, 35, 41]. According to this analysis, the screen is modelled as a discontinuity in the tangential velocity component, which generates circulation about the screen elements (wires), and

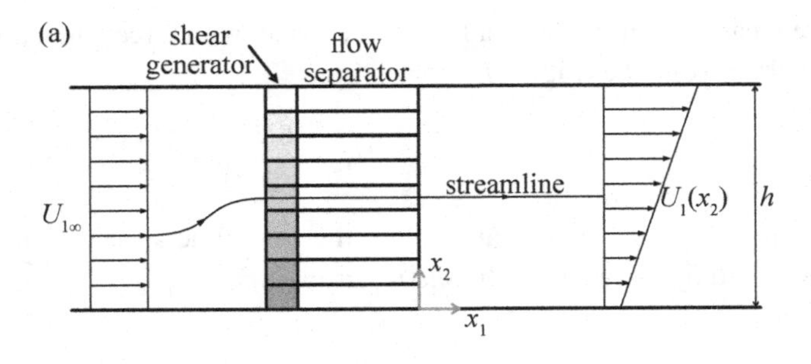

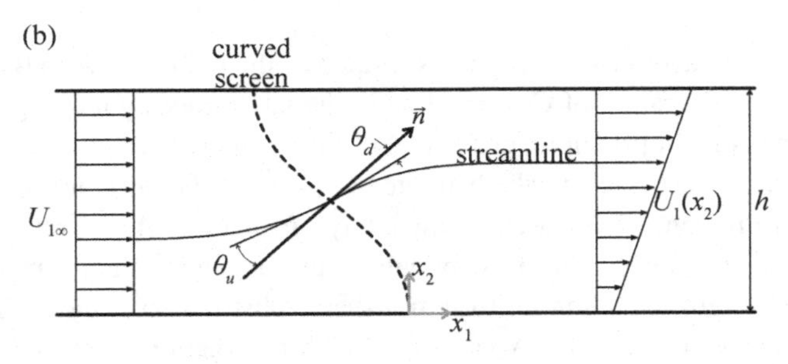

Figure 8.9 Sketches of shear generation by (a) a shear generator/flow separator combination and (b) a curved screen; boundary layers have been disregarded, for simplicity.

thus lift. At the same time, the screen also produces a pressure drop, corresponding to drag and characterised by the pressure loss coefficient K, referenced to the normal velocity V_n, which, by continuity, remains unchanged across the screen. Thus, the screen effectively deflects the velocity vector towards its normal. This is indicated in Fig. 8.9b, by the reduction of the upstream incidence angle θ_u to the lower downstream value θ_d. For small incidences, the ratio of these angles reaches a finite limit [35, 68]

$$\alpha = \lim_{\theta_u \to 0} \frac{\theta_d}{\theta_u}, \quad 0 \leq \alpha \leq 1 , \tag{8.23}$$

which depends on K. Among the available empirical expressions that describe the relationship between α and K, one may use the following [25]:

$$\alpha = \left[\left(\frac{K}{4} \right)^2 + 1 \right]^{1/2} - \frac{K}{4} , \quad 0.7 < K < 5.2 . \tag{8.24}$$

Let V_{tu} and V_{td} be the velocity components tangential to the screen just upstream and just downstream of it. Then, one may define the *deflection coefficient*

$$B = 1 - \frac{V_{td}}{V_{tu}} , \tag{8.25}$$

which would normally be in the range between 0 and 1. The following expression was obtained by polynomial least squares fitting to a theoretical estimate, also including an empirical correction [41], of the shape of a screen that would produce uniform shear with a shear parameter β:

$$\frac{KB}{(2+K-B)\beta}\frac{x_1}{h} = -0.738\left(\frac{x_2}{h}\right)^6 + 2.812\left(\frac{x_2}{h}\right)^5 - 3.839\left(\frac{x_2}{h}\right)^4 + 2.687\left(\frac{x_2}{h}\right)^3 - 1.224\left(\frac{x_2}{h}\right)^2 - 0.0054\frac{x_2}{h}, \tag{8.26}$$

where the deflection coefficient was estimated as

$$B = 1 - \frac{1}{\sqrt{1+\sqrt{K}}}. \tag{8.27}$$

The value of K in Eq. (8.27) could be based on direct measurement or estimated from Eq. (8.15). This procedure has been applied successfully by several investigators for the generation of shear in wind and water tunnels, with typical values of β about 0.4 or lower.

Turbulent boundary layers: Experimental realisations of two-dimensional turbulent boundary layers in wind and water tunnels are very useful configurations, both for fundamental turbulence studies, flow separation studies, as well as for simulations of flows in the lower atmosphere in a variety of applications. In such studies, it is desirable for the turbulent region to be as thick as possible, to reduce the spatial resolution requirements of measuring sensors and to permit the use of relatively large models. The generation of boundary layers that grow gradually without external intervention requires a long test section. An increase in the boundary layer thickness at an early stage of development has been achieved with the use of large roughness elements. Turbulent boundary layers, roughly simulating wind in the lower atmosphere, have also been generated in short test sections with the use of large *spires* or *strakes*, in combination with distributed wall roughness elements [7, 12, 13, 29, 63]. A growing boundary layer in a duct with a fixed cross-sectional area would result in free-stream acceleration and a favourable streamwise pressure gradient. To maintain the pressure constant along the test section, it is necessary to increase the cross-sectional area gradually or to bleed part of the free stream. A common practice in wind-tunnel design is to use *corner fillets* to reduce secondary flows. An increasing cross-sectional area can be achieved by tapering the fillets or by using slightly divergent walls. Boundary layers with favourable or adverse pressure gradients can be generated by similar means, most commonly by adjusting the shape of a flexible or articulated wall opposing the one on which the boundary layer grows.

Other shear flows: Variable-solidity shear generators and curved screens can be also used to generate flows with non-uniform mean shear, making it possible, in principle, to simulate any desired two-dimensional velocity profile. A splitter plate separating two streams with different speeds is normally employed to generate mixing layers, whereas obstructions of various shapes are used to generate wake flows. A *swirl generator*, is used in certain applications to create a circumferential velocity component in the test section. Considering the large variety of possible configurations, the reader is advised to consult the relevant experimental literature before attempting to generate a complex flow field.

8.2 Model testing

8.2.1 Model Design

Model sizing: The laboratory study of flows through and past various objects often employs the use of geometrically scaled models. In most cases, the model is scaled down, compared to the object of interest (e.g., models of aircraft and hydraulic turbines), but full-size or scaled-up models (e.g., models of insects and micromachines) are also used to permit convenient flow visualisation and measurement. Scaling normally preserves geometric similarity, although the use of different scales in different directions can be employed in specific studies, as, for example, in hydraulic models of large terrains, in which the vertical system-to-model scale ratio is taken to be larger than the horizontal one.

Reynolds number matching: For an exact correspondence between model studies and actual system properties, it is necessary to respect *dynamic similarity*, which requires that the values of corresponding relevant dimensionless parameters (see Section 1.6.2) should be identical in the model and actual systems. For incompressible, single phase flows of homogeneous fluids, it is normally the Reynolds number value that has to be matched. In the majority of cases, the fluid in the model studies is, by necessity, air or water, namely, the same fluid as in most actual systems. As a result, the fluid properties in the model studies have nearly the same values as those in the actual system. Then, the use of a small-scale model would necessitate a higher-than-actual speed for the tests, in order to maintain Reynolds number similarity. The generation of adequately high speeds may not be feasible in the available facilities and, even when it is feasible, it may introduce undesirable compressibility effects. In many internal and external flow configurations, Reynolds number effects on properly non-dimensionalised forces, pressures etc., as well as on flow patterns (e.g., flow separation from sharp edges, vortex shedding from bluff objects), are known to be weak for large Reynolds numbers, and so it may be acceptable to conduct model tests at lower Reynolds numbers. For improved estimates of actual system properties, it may be possible to apply empirical corrections (e.g., corrections for hydraulic turbines and pumps, based on size or Reynolds number) or extrapolate trends to higher Reynolds numbers.

Matching of other dimensionless parameters: When, in addition to viscous effects, other physical phenomena are important in a flow, one has also to match additional dimensionless parameters. For example, in studies involving a free surface of a liquid, the Froude number must be matched, in addition to the Reynolds number; in high-speed gas flows, it is also the Mach number that has to be matched; when cavitation, surface tension, buoyancy, rotation or other effects are significant, additional parameters (see Section 1.6.2) have to be also matched. The matching of more than one parameters may not be possible for scaled models, especially if additional constraints, such as use of the same fluid, are imposed. In such cases, one is forced to compromise similarity to a weak or incomplete form and it is best to follow established practices in the particular field of application.

Flow blockage: An important effect that distorts the correspondence between model measurements in wind and water tunnels and properties in actual systems is the confinement of the laboratory flow by solid walls. In this discussion, we exclude the case of open-test-section wind tunnels, for which information may be found in the cited references. When a solid model is inserted in a duct flow, it blocks part of the stream, which is diverted around the model. To maintain the conservation of fluid mass crossing each cross section, the flow speed at cross sections intersecting the model or near it would be higher than the *undisturbed velocity* V_∞ in the duct, defined as the flow velocity (assumed to be uniform) in the absence of the model. This effect is referred to as *solid blockage*. Defining an average velocity increase as ΔV_∞, one can also deduce that the various dimensionless parameters (e.g., normalised local velocities, pressure differences, forces and moments) should be evaluated using $V_\infty + \Delta V_\infty$ as a velocity scale, rather than V_∞. If, instead, such properties were normalised by V_∞, they would have erroneously large values. Blockage does not cease at the downstream end of the model, as its wake (which is a low-speed region) also produces a velocity increase in the free stream, an effect known as *wake blockage*. Associated with the velocity increase in the free stream is a static pressure drop across the model, which increases the drag force by an amount called *longitudinal buoyancy drag* or simply *buoyancy drag*. This additional drag occurs in all confined flows, irrespectively of their orientation with respect to the gravitational direction. A similar drag is produced by flow acceleration due to boundary layer growth on the duct walls; this drag can be removed by equalising the pressure through deliberately introduced test section wall divergence. Several types of corrections for blockage effects are available, based on theoretical, numerical and experimental findings for flows around both streamlined and bluff objects, which usually provide estimates of the velocity ratio $\Delta V_\infty / V_\infty$ [2, 16, 17, 23, 44]. For incompressible, single-phase flows, it is generally accepted that such effects may be neglected when the projected frontal area of the model is less than 5% of the test section area, and tolerable when this ratio is lower than 10% [2].

Other confinement effects: Confinement also affects flow motion around models, as it tends to restrict streamline displacement away from the model walls, as well as to direct streamlines towards directions parallel to the wall planes. Moreover, it may alter the nature of the flow in ways that would not occur in the absence of walls. For example, the increased effective velocity and resulting increased effective Reynolds number may dramatically affect the flow separation pattern from bluff objects with rounded edges operating under critical or near-critical conditions, at which the drag coefficient presents a well-known sudden drop [20, 71]. On the other hand, separation from bluff objects with sharp edges is known to be insensitive to Reynolds number changes, as such objects have no critical regime. Pressure drop due to blockage in liquid streams may introduce or affect cavitation, while, in high-speed gas flows, it may affect shock wave patterns. An aspect that is particularly relevant to measurements of forces and moments on models in a facility is the need to isolate the model from the surfaces in the test section; in other words, one needs to ensure that the measured forces are those due to the interaction between the stream and the model, and not due to supports that hold the model or from interactions with the boundary layers on the test section walls and flow restriction by such walls.

8.2.2 Measurement of Forces and Moments on Models

The measurement of forces and moments on a stationary model immersed in a moving fluid, or on a model towed through a stationary fluid, is a common task in a fluid mechanics laboratory. Practical aspects of model testing, including the design of different mounting mechanisms and the types of devices that measure the required forces and moments, are discussed in several sources [2, 28]. Although each experiment has its own unique design aspects and associated challenges, there are certain common aspects that apply to all model testing.

Coordinate system: It is important to note that commercial F/T sensors are calibrated to measure forces and moments about a particular reference point, usually on the mounting surface of the sensor. When conducting an experiment, however, we may require the forces and moments in a different reference frame; in such cases, we need to perform a coordinate transformation. Established standards specify the reference frame that is appropriate for a specific application: ANSI/AIAA R-004-1992 standards [1] apply to aerospace applications; J1594-201007 [58] standards apply to automotive applications; and European aerospace standards follow ISO 1151 [31].

Balances: The measurement of forces and moments on models can either be achieved via the use of *internal balances*, located within the model, or *external balances*. The objective of model testing is to determine the forces and moments on the model itself, ideally without any interference from other systems. In practice, however, one must mount the model into a wind or water tunnel, in order to take the measurements, and the model mounting mechanism may introduce undesirable forces and moments, thus, a systematic error, which one should strive to eliminate or, at least, reduce to a tolerable level. Wherever possible, it is preferable for the model and its mounting mechanism not to be in physical contact with the frame of the wind or water tunnel, in order to avoid mechanical coupling effects.

An internal balance, fitted to a model as shown in Fig. 8.10a, does not introduce additional aerodynamic forces and moments, because it is not exposed to the flow. The internal balance is connected to a streamlined external strut via a *sting*, which is typically a cylindrical rod. The combined internal balance/sting mechanism is sometimes referred to as a *sting balance*. To minimise any possible interference of the sting with the flow around the model, it is customary to have the sting come out of the rear of the model, in the downstream direction, as shown in Fig. 8.10a. The sting should be as long as is structurally feasible, so that the strut onto which it is mounted poses minimal disturbance on the upstream flow.

Supports and mounting considerations: A model, fitted or not with a sting, is mounted to an external balance via a support, called a *strut* (see Fig. 8.10b). In addition to performing a coordinate transformation, one must also account for the forces/moments on the strut in order to get accurate measurements. The most rudimentary technique is to perform a *tare test*, whereby one measures the forces and moments on the strut, in the absence of the model, under all all required flow velocities and strut orientations, and then subtracts the corresponding values from the measurements in the presence of the model. The simplest type of strut is a cylindrical rod, but struts are usually designed to

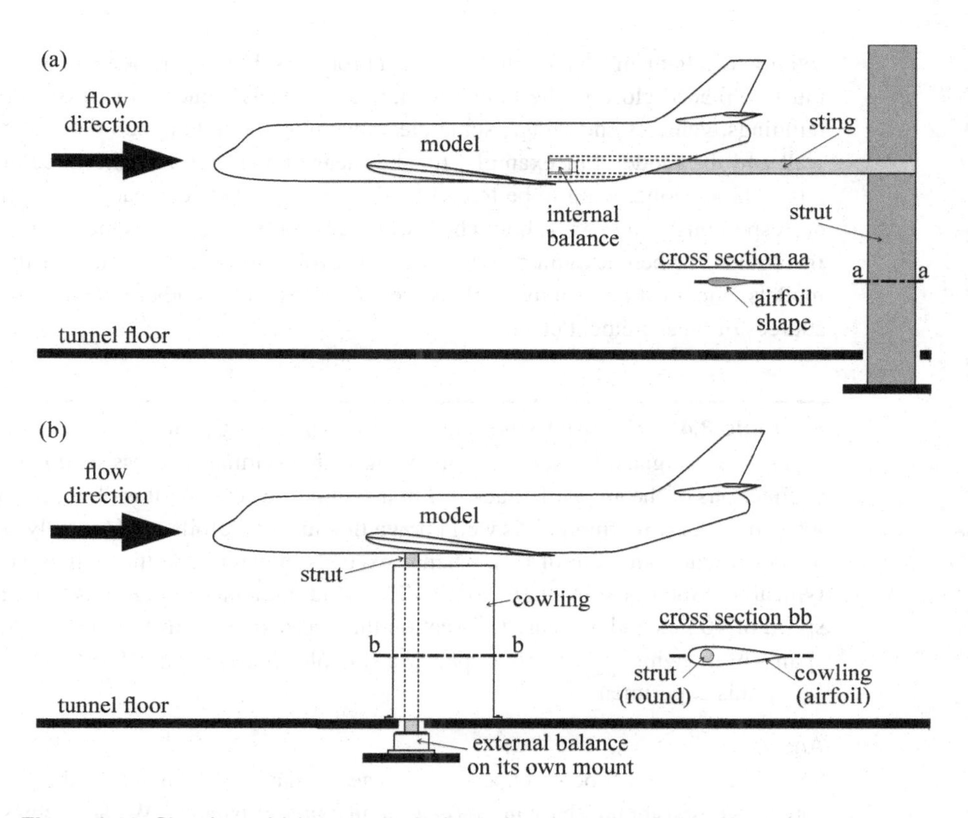

Figure 8.10 Sketches of (a) an internal sting balance, mounted through a sting on a streamlined external strut; (b) an external balance with a cylindrical strut and a streamlined cowling.

have a symmetric, low drag profile, which will not introduce oscillatory forces due to vortex shedding. If, either for structural rigidity or ease in fabrication, a circular rod or other bluff support is used as a strut, it would be preferable to encase it within a *cowling* (or *shields*), which has a streamlined profile, as shown in Fig. 8.10b. It is important for the cowling not to be in contact with either the model or the strut. The use of a cowling drastically reduces aerodynamic forces/moments on the strut, which, in its largest part, is shielded from the flow. Any residual force/moment on the unshielded part of the strut may be corrected for by a tare test.

For tests of models in different orientations, such as an aeroplane model at different pitch, yaw and roll angles, one must mount the model rotation mechanism on the same frame as the balance itself. For example, if a second strut was attached to the rear of the model in Fig. 8.10b to allow for a change in pitch of the model, the other end must be attached to the external balance (not the tunnel floor), so that no reaction forces/moments are introduced to the measurement. Moreover, any changes in the orientation of the model by rotating the supports would bring the model close to, and eventually in contact with, the tunnel walls; aerodynamicists have crafted special designs of rotation mechanisms that circumvent such problems. For experiments in which a model must be placed close to the tunnel wall, such as measurements of forces/moments on buildings, vehicles and wings, some clearance must be maintained between the tunnel wall and the model. For example, for the measurement error in the force/moment on a two-dimensional wing to be tolerable, the clearance between each model tip and the corresponding tunnel wall should be lower than $0.005b$, where b is the span of wing [2]. In summary, when designing an experiment involving force/moment measurements on models, one must keep in mind blockage effects, as well as other general model testing sources of measurement error.

Example 8.4 We wish to measure the lift coefficient C_L and the drag coefficient C_D of a newly designed, two-dimensional wing with a symmetric cross section ('aerofoil'), as functions of the angle of attack α. For a symmetric aerofoil, it is obvious that $C_L = 0$, when $\alpha = 0$. Furthermore, it is well known that most aerofoils *stall* (namely, experience a catastrophic reduction of C_L), when α exceeds in magnitude the stall angle, which is typically in the range from 10° to 20°. The wind tunnel available to us has a maximum speed of 40 m/s and a square test section that has a side width $W = 0.5$ m. Specify the span b of the wing model. Also, specify a suitable chord length c. Explain how you will set up this experiment.

Answer

This is a common type of experiment in aerodynamics. To measure the lift and drag forces, we will mount the wing model on an external balance. We are interested in the performance of a two-dimensional wing, which must be free, as much as possible, from finite wing effects. To avoid such effects, the wing model must, ideally, span the entire width of the test section. On the other hand, contact of the wing tips with the walls would generate interfering forces and impede easy changes in the angle of attack. Therefore, we need a small clearance δ between each wing tip and the corresponding side wall of the test section. Literature recommends a maximum recommended clearance $\delta_{max} = 0.005b$, which implies that the minimum wing span would be $b_{min} = W - 2\delta_{max} = W - 0.01b_{min}$, from which we find $b_{min} = W/1.01 \approx 495$ mm.

Next, we need to determine the chord length of the wing. A requirement for this experiment is to ensure that the blockage ratio is lower than 5% for the entire range of angles of attack considered. We will assume that measurements will be taken in the range $-25° \leq \alpha \leq 25°$, in order to verify the symmetry of the results and to ensure the detection of stall. Approximating the frontal area of the inclined wing as $bc \tan \alpha$, we find the maximum model blockage, which would occur when $\alpha_{max} = \pm 25°$, as $bc \tan \alpha / W^2$. To maintain the blockage below 5%, we need to select a chord $c < 54$ mm. The aspect ratio of a wing with this chord is $b/c \approx 9.2$. A remaining issue is whether the wing performance may depend on the side-wall boundary layer characteristics. In the absence of further details about the wind tunnel and the test conditions, and the fact that the aspect ratio is relatively large, we may plausibly assume that boundary layer effects would not be very significant.

8.3 Towards a Sound Experiment

The success of an experiment relies on the skill, experience and background preparation of the experimenter, as much as it depends on the suitability, quality and condition of the apparatus and instrumentation. Personal enthusiasm and interest are propitious, but not sufficient, qualities of the good fluid mechanics experimenter. In addition, consideration must be given to time-tested sound practices, which should be followed while preparing for and performing an experiment. This section summarises such considerations.

8.3.1 Planning the Experiment

When it comes to fluid mechanics experiments, there is no substitute for hard and arduous work. Nevertheless, the task can be lightened significantly by proper preparation and planning. A non-exclusive list of preliminary actions and considerations is as follows:

- Understand the physical problem at hand and study its theoretical aspects.
- Establish the need for experimental work and identify the information that would be desirable to obtain experimentally. Try to be realistic in your expectations.
- Clearly define the objectives and the scope of the experiment.
- Identify the ideal apparatus and equipment that would provide the desired information; list and compare alternatives.
- Identify all resources available for your experiment and those that can be used by arrangement with other projects or laboratories. Besides hardware, this list could include space, utilities (e.g., compressed air, distilled water etc.) technical services, computational resources and financial support. In cases where there is more than one suitable alternative (e.g., availability of multiple wind tunnels), compare them and rank them in terms of desirability. As a rule, use the simplest, least expensive and least time-consuming approach that would produce the desired results within an acceptable uncertainty.
- From the previous list, choose the most suitable items and identify what is missing for carrying out your experiment.
- Explore the possibilities of borrowing, leasing, purchasing or constructing needed apparatus, equipment and software. Identify and compare suppliers and sources, obtain quotations and cost estimates and establish delivery and construction times. Prepare a budget for the project. Consider the need for writing proposals or requests and obtaining approvals.
- Before using any piece of equipment for actual measurement, become familiar with its principle of operation, specifications, range and general capabilities and limitations.

8.3.2 Qualitative Assessment

As emphasised in Section 8.3.1, a principle of the experimentalist should be that all required information ought to be collected by the simplest and most economical means available. In order to illustrate this principle, we shall consider a few simple and inexpensive experimental approaches, which should always precede measurement in any experiment. More often than not, preliminary qualitative assessment has revealed unexpected problems and led the experimenter to radically rethink the experiment, in some cases identifying flaws in the objectives and critical limitations of the earlier planning process.

Benefits from a qualitative assessment: The general objective of a qualitative assessment is to provide some preliminary insight into the overall appearance and characteristics of a flow, before engaging into possibly cumbersome, expensive and lengthy experimentation. It provides an opportunity to reassess the need for detailed measurement, to evaluate the appropriateness of the setup and to select the most suitable measuring instrumentation and procedures. Among the many flow features that may be conveniently detected by qualitative assessment, one may identify the following:

- possible defects (e.g., leakage) and undesirable influences (e.g., motor-induced vibrations) in the operation of the flow apparatus (Fig. 8.11a, b);
- the flow direction, particularly the presence of separated and recirculating flow regions (Fig. 8.11c, d);
- flow boundaries and material interfaces;
- the degree of mixedness of different streams (Fig. 8.11e);
- unsteadiness and instability (Fig. 8.11f, g);
- the presence of impurities or a second phase (Fig. 8.11h);
- regions with chemical reactions or combustion (Fig. 8.11e);
- differences in density or temperature, density stratification and natural convection (Fig. 8.11i);
- turbulence and typical eddy structure; interface between turbulent and non-turbulent flow regions; orders-of-magnitude of the turbulence intensity and the macroscopic length scale (Fig. 8.11j);
- special flow patterns and coherent structures (Fig. 8.11k);
- shock waves, expansion waves and compression waves (Fig. 8.11l);
- interfacial and internal waves.

Qualitative assessment vs. measurement: A qualitative assessment technique is usually a simple version of a more sophisticated qualitative or quantitative method used in scientific studies. However, at the preliminary stage, one should resist the temptation to refine the technique at the expense of time, effort and cost. For the present purposes, an experimental procedure may be considered as qualitative assessment of a flow, if it meets the following conditions:

- It is relatively easy and fast to implement.
- It utilises already available or inexpensive equipment and materials and does not require the design or purchase of sophisticated instrumentation.
- It is non-destructive and non-contaminating, such that the facility could be easily returned to its original condition at the end of the assessment.
- Its results are relatively easy and clear to interpret.

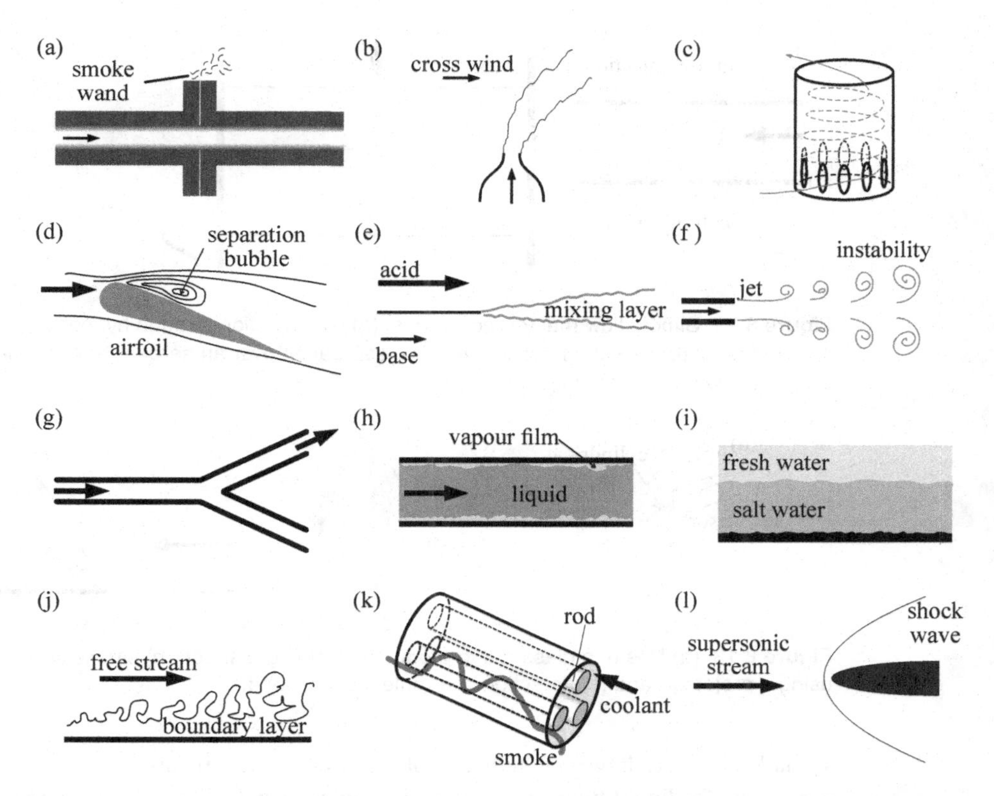

Figure 8.11 Examples of application of qualitative assessment techniques: (a) leakage detection, (b) detection of a draft or cross stream, (c) swirling flow pattern in a Diesel engine, (d) separation bubble over an aerofoil, (e) mixing layer of reacting streams resulting in color change, (f) jet instability and break-up, (g) flow direction in a fluidic switch, (h) steam-water flow in a heated tube, (i) density stratified flow, (j) turbulent boundary layer, (k) flow pulsations and coherent structures in gap regions of rod bundles and (l) shock wave in supersonic flow over a bluff object.

Visual assessment: The vast majority of qualitative assessment techniques are visual, because of the relative ease by which many flow phenomena can be made visible. Some flows have inherently visible characteristics, which can be observed without any special effort, as, for example, liquid jets in air, bubbly flows of liquids, flames, and flows carrying impurities. However, in most situations, fluids are optically homogeneous and isotropic and, thus, their internal motions are invisible to the naked eye. In such cases, flow visualisation may be achieved by the addition of *flow markers* or the use of suitable illumination and recording methods.

Easily visible flow markers [2] are relatively large, typically with dimensions between 50 and 300 μm. These can be released either from an orifice on a surface or from a hypodermic tube in the stream (Fig. 8.12a). Among the common flow markers, which are suitable for a qualitative assessment of flows, are the following:

- natural markers, such as dirt, lint or dust;

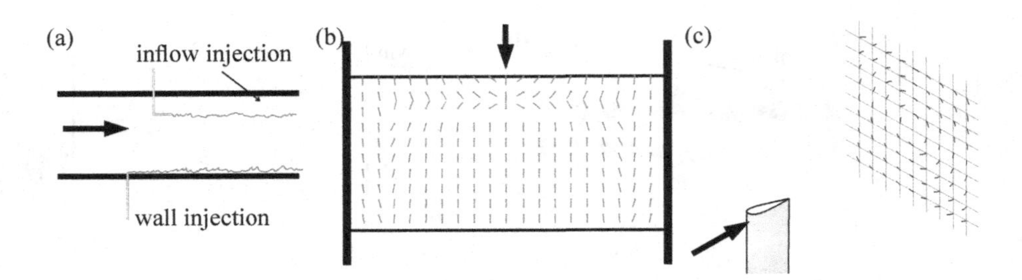

Figure 8.12 Simple flow marker techniques: (a) dye injection through hypodermic tubes and wall taps; (b) use of tufts to identify flow separation over an aerofoil; and (c) use of a tuft screen to visualise a wing-tip vortex.

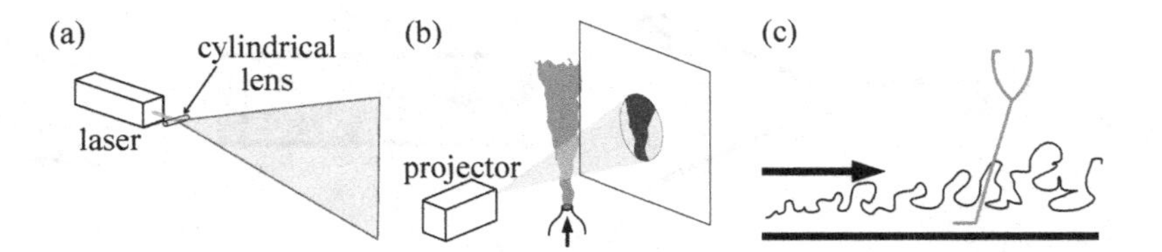

Figure 8.13 (a) Use of a glass rod to produce a laser light sheet; (b) simple shadowgraph using a projector; and (c) detection of turbulence by sound.

- markers for gas flows, including smoke, powders and aerosols;
- markers for liquid flows, including air bubbles, powders, metal filings, plastic beads and dyes;
- tufts, attached to a surface (Fig. 8.12b) or a thin-wire screen across a stream (Fig. 8.12c).

Illumination can be conveniently and easily achieved by room light, or other flood-type light, optimised in position and intensity by trial and error. For better results, one may utilise the nearly collimated white light produced by a projector, a stroboscope or a camera flash. One may also consider the use of an inexpensive, low-power laser, which could be of the continuous-wave type or pulsed; laser beams may be turned into a light sheet with the use of a cylindrical glass or acrylic rod (Fig. 8.13a). In flows with density variations, a simple version of a shadowgraph, consisting of a collimated light beam and a projection screen, may be appropriate (Fig. 8.13b). More details on natural markers and illumination methods can be found in Chapter 7. Although the human eye has a fairly good resolution and a wide range of operation, image recording and post-processing may reveal features that cannot be seen by naked eye.

Non-visual assessment: In addition to visual methods, qualitative assessment of flows may be achieved by the use of any other sense. By touching the apparatus or feeling the flow, one may identify vibrations, flow direction, temperature differences or leakage. By listening to a flow through a small tube connected to a stethoscope, one may detect transition and turbulence (Fig. 8.13c). Sounds in a flow may reveal cavitation, flow-induced vibration or other phenomena that are accompanied by characteristic noise. Finally, in some cases, leaks and chemical reactions can be detected by smell.

Interpretation of preliminary observations: The interpretation of the results of a qualitative assessment method has to be made with caution, particularly because such methods are, by definition, relatively crude and of moderate sensitivity. For example, separation on the surface of a wing may not be detectable by heavy or stiff tufts, although the same separation may be clearly indicated by finer ones. Injection of dye or smoke in excessive amounts or non-isokinetically (namely at speeds substantially different from the local flow speed) may generate patterns, unrelated to the flow of interest. Under certain conditions, non-isokinetic marker injection or the presence of the injection tube may result in prolonged memory effects, trigger flow instability and premature transition to turbulence or change the turbulence structure. Finally, one is warned to avoid misinterpretation of visual patterns, for example, pathlines, streaklines and streamlines, as described in Chapter 1. In unsteady flows, even carefully conducted flow visualisation may sometimes lead to erroneous or misleading results [22].

8.3.3 Record Keeping

When performing any experiment, it is necessary to be able to retrace one's steps in order to document the experimental conditions, explain unexpected results and even reconstruct, if necessary, the experiment under conditions identical to the previous ones. For these reasons, it is a sound practice to maintain a detailed, chronologically arranged, scientific journal of all conditions, actions and results obtained.

The logbook: The traditional format of a scientific journal is a *hardcover logbook*, of the type available in University bookstores, with pages lined and numbered and preferably containing some pages with graph grids. It is advisable to leave the first few pages blank, to be filled as the experiment progresses in the format of a table of contents. The logbook should be updated on a continuing basis, with the date and time of day of all actions clearly marked.

The following information should be typically recorded in the logbook:

- A cover page, containing the author's name and contact information and the date when the logbook was started.
- Sequential numbers for all pages. Pages may be crossed out or be left blank, but should not be removed.
- A table of contents, listing each experiment recorded, the date(s) that it was conducted and the corresponding page numbers. Sufficient blank space should be allowed for the table of contents to be expanded by new entries as they get recorded in the logbook.
- Starting with a new page for each experiment, the full title of the experiment and a short title for easy reference.
- The objective of the experiment.
- The location of the apparatus used.
- The date and time of each entry.
- References (e.g., textbooks, manuals, articles etc.), personal notes or other information used in preparation for the experiment.

- A list of all equipment used with brief technical details (e.g., dimensions, output units, resolution of readings), sketches of connections between different pieces of equipment and availability of technical manuals or instructions for use.
- A list of all acquired measurements that were recorded during each experiment, including the name, symbol and units of each recorded property. For each set of measurements, record the experimental settings and relevant reference values (e.g., room temperature, barometric pressure and humidity). The actual values of the measurands can be recorded on the logbook, when convenient. If large data records have been recorded electronically or otherwise, list the filename and location of each record.
- Personal observations concerning the experiment, including possible difficulties, malfunctions of equipment, notes for the interpretation of the results etc.
- Copies of photographs or printouts acquired during the experiment.
- All sources of measurement uncertainty and any information that would be relevant to the uncertainty analysis. Some preliminary uncertainty analysis, to ensure that all necessary information has been recorded.
- Some post-processing of results. This includes samples of calculations, but only to the extent necessary for verifying that these are possible based on the recorded information and that the results do not contain evident inconsistencies or gross inaccuracies. Possibly include also some preliminary plots of various parameters to verify that the experiment is progressing as planned and to detect spurious phenomena that may affect the results.

All information should be recorded neatly and coherently, so that it can be reviewed easily not only by the author but also by other members of the group and the supervisor. If any additional material is attached to a physical logbook, it must be pasted firmly on blank pages, should not be oversize and should not interrupt the page numbering or the continuity of the logbook.

Electronic records: As an alternative to a physical logbook, one may maintain an integrated, single-file, *electronic logbook* for each experiment. As much as possible electronic material should be stored in a folder on a shared online platform that can be accessed by all members of the team, as well as future users. The following is a non-exhaustive list of files to be placed in this folder, meant to identify ways by which the knowledge gained from the experiment can be easily passed on to the next user:

- The electronic logbook, which is a document describing briefly the equipment used, the objectives and methodology of the experiment, links/hyperlinks to related technical specifications and manuals and any other information that is necessary for a user to make proper use of the data collected during the experiment. An essential task is to explain the structure and format of data files, that is, the physical property and units corresponding to the values in the file, clearly identified by row and column. For example, the author of the document informs the user that the first column in the data file is the x-position in mm, the second column is the y-position in mm, the third column is the streamwise velocity in m/s, the fourth column is the transverse velocity in m/s, the fifth column is free-stream temperature in Celsius etc.

- Depending on the size of files of data collected during the experiment, such data files could be located in the common folder or be stored in a separate mass storage device, the location and access to which must be identified in electronic logbook.
- Photographic images of the experimental set-up, which may later be used in scientific reports.
- When appropriate, computer-aided design (CAD) files of the experimental set-up, especially if custom-made items were used for the experiment.
- When appropriate, a video recording of the set-up and connections, with instructions how to operate the equipment and conduct the experiment.
- Computer scripts/codes of the data processing used for the experimental data, which in itself should be sufficiently commented so that a new user can follow the data processing steps and analysis.

8.3.4 Laboratory Safety

Any laboratory user must, at all times, safeguard without compromise their personal health and safety, as well as the health and safety of other persons working in the laboratory and the public at large. Laboratory usage is subjected to safety legislation and regulations at national, provincial/state, municipal and institutional levels, and many organisations require safety training before granting permission to work in a laboratory. Besides being familiar with regulations and policies, one must also exercise common sense to prevent accidents and mishaps. Common rules for laboratory users include the following:

- Ideally, a laboratory user should be in good physical condition and free of stress, and should not act under extreme time pressure.
- If at all possible, avoid working in the laboratory alone, especially during late hours or holiday periods.
- Identify alternative nearby locations of room exits, fire extinguishers, fire alarms and first-aid kits.
- Avoid wearing loose clothing and footwear and unrestricted long hair, which can easily be entangled in moving components or suction ports.
- Secure gas cylinders and other unstable objects.
- Particular attention should be given to rotating and reciprocating components, high electric voltages, ungrounded equipment, electric cables, high pressure releases from valves, hot surfaces and fluids, unpleasant odours, toxic and combustible substances and sharp or pointed objects.
- When dealing with lasers or other sources of radiation, follow meticulously all regulations and instructions of use. Wear protective glasses and avoid exposure of skin to radiation, even at low levels. Besides immediate hazards, such as damage to the cornea by exposure to a laser beam, there are long-term effects of low-level radiation, which may not yet be understood.
- Release of harmful substances to the environment is strictly prohibited and their disposal should be made according to the existing regulations.

- Acids, flammable substances and other hazardous materials must be stored in specially designated cabinets. Use ventilation fans and fume hoods, when necessary.
- Be aware that older or specialised apparatus and instruments may contain currently regulated substances such as mercury, chlorofluorocarbons, banned refrigerants or asbestos.
- Report immediately any hazardous spill, accident, dangerous event or safety concern.

8.3.5 Scientific Ethics

While performing experiments and when reporting results one must comply with appropriate scientific and professional ethics and practices and exercise honesty and collegiality. Particular attention should be given to the following issues.

Co-authorship: A contentious issue in scientific research is co-authorship in joint publications and public presentations. This refers to both the names to be included as co-authors and the order of names. Practices vary widely among different fields and one should not venture to refer to general rules. It is generally accepted, however, that a co-author is a person who has made a substantial intellectual contribution to the project and/or the report. When one perceives that the co-authorship is not entirely clear, it would be desirable to obtain the consensus of all persons concerned before the completion, or, preferably, the start, of a project.

Acknowledgement of contributions: All ideas that contributed measurably to the project and any non-trivial assistance provided during the various stages of the work (e.g., technical services, help with collecting and analysing the measurements and preparation of graphics or a manuscript) should be duly acknowledged.

Documentation: An important aspect of all scientific reports and publications is *documentation*. This means that all presented material or ideas that have been borrowed from other sources must be accompanied by *citation* of the proper references. Thus, it is understood that any undocumented material contained in a report is either common knowledge (e.g., the First Law of Thermodynamics) or an original creation of the author.

Copyright: The duplication of copyrighted material in a thesis or publication requires written permission, usually granted by the publisher of the journal or book in which the material first appeared, or, for unpublished material, by the author. Even material that is not specifically identified as copyrighted is protected. This includes laboratory manuals and operating procedures, course assignments and handouts, and examination questionnaires and solutions. Copying or scanning such documents and disseminating them by posting them on a website is therefore a violation of copyright and is illegal. In addition, making photographic, audio or video recordings of presentations requires the explicit permission of the author. Copyright, trademarks and intellectual property in general are protected by legislation and specific regulations and guidelines. Issues concerning intellectual property can be very complex, as evidenced by the large number of legal contests and the on-going updating of laws and regulations. When uncertain how to approach such an issue, it is best to consult with an experienced person or an authority on the subject.

Plagiarism: Verbatim copying of material from unacknowledged sources or intentional presentation of previously published material as the author's creation constitutes *plagiarism* and is punishable at different levels. Submitting the same piece of work for two different purposes, for which only original material can be submitted, constitutes *self-plagiarism*.

Academic fraud: The fabrication, falsification or 'doctoring' of experimental results constitutes *academic fraud*. Both are intolerable in the scientific community and also punishable at different levels. A researcher should feel compelled to preserve all data and experimental records for several years, in case a question of their authenticity arises.

Chapter Digest

8.1 Identify common flow generating and flow management devices for fluid mechanical apparatus.

8.2 Why do we use a diffuser, a plenum and a contraction in wind tunnels?

8.3 Identify the advantages and disadvantages of open-circuit wind tunnels by comparison to closed-circuit ones.

8.4 Derive Eq. (8.2) and Eq. (8.7).

8.5 Which are the functions of a honeycomb and a screen? Are these devices interchangeable?

8.6 Discuss the advantages and disadvantages of using screens in wind tunnels.

8.7 Why do we use wide-angle diffusers? Why do we position screens inside a wide-angle diffuser? Should we space them equally along the entire length of the diffuser or make some more careful considerations?

8.8 Identify the advantages and disadvantages of suction wind-tunnels by comparison to blower types.

8.9 Identify the advantages and disadvantages of entirely enclosed water tunnels by comparison to water tunnels that have a free surface.

8.10 Identify differences in the conditions around an aircraft in free flight and tests on a small-scale model of this aircraft in a wind tunnel.

8.11 Identify the largest wind tunnel in the world. Summarise its main features and uses.

8.12 Provide a possible scenario for each of the following cases: (a) a contribution by another person that does not need to be acknowledged formally; (b) a contribution by another person that should be acknowledged in a note; and (c) a contribution by another person that qualifies this person as a co-author of a publication.

8.13 In general, you need to cite all sources from which you borrowed material. Is there an exception to this rule? Provide an example.

Problems

8.1 Consider the contraction discussed in Example 8.1.

(a) Compute the length L_p of the plane pattern.

(b) Scale down the dimensions so that the pattern of each of the four walls can fit on a 8.5 in × 11 in or A4 sheet of paper. Then print and cut four patterns, roll them and paste them together along the edges to build a 3-D small-scale model of this contraction. Take a photograph of this contraction and attach it to your answer.

8.2 Consider a wind-tunnel contraction with a contraction ratio c. Two parallel streams of air enter the contraction, the first one with speed U_1 and density ρ and the second one with speed $U_1 + \delta U_1$ and density $\rho + \delta\rho$, where $|\delta U_1| \ll U_1$. Determine the density difference $\delta\rho$ required for the flow at the exit of the contraction to have uniform velocity. Neglect friction, heat transfer and mixing.

8.3 Flow in a wind tunnel is produced by a mixed-flow fan, the pressure-flow-rate characteristic of which at a rotational speed of 1,400 RPM is shown in Fig. 8.14. The wind tunnel has a square test section with a height $h_t = 0.50$ m and a square plenum with a height $h_p = 1.50$ m. When no screens are installed in the settling chamber, and while the fan is running at 1,400 RPM, the test section speed is 10.0 m/s.

(a) Define a plenum equivalent system characteristic as $\Delta p = 0.5K\rho(Q/h_p^2)^2$, where Δp is the total pressure loss, $\rho \approx 1.2$ kg/m^3 is the air density and Q is the volume flow rate. Determine the value of the equivalent pressure loss coefficient K and the operating point of the fan in the absence of screens (to assist you, the plot already shows the equivalent system characteristic, but you still need to determine K).

(b) To improve the flow quality, you consider installing a number of screens in the settling chamber, which have a total pressure loss coefficient $K' = 5$. Determine the new system characteristic, the new operating point of the fan and the corresponding test section speed.

(c) Determine the rotational speed of the fan for the test section speed, after the screens are installed, to be 10 m/s.

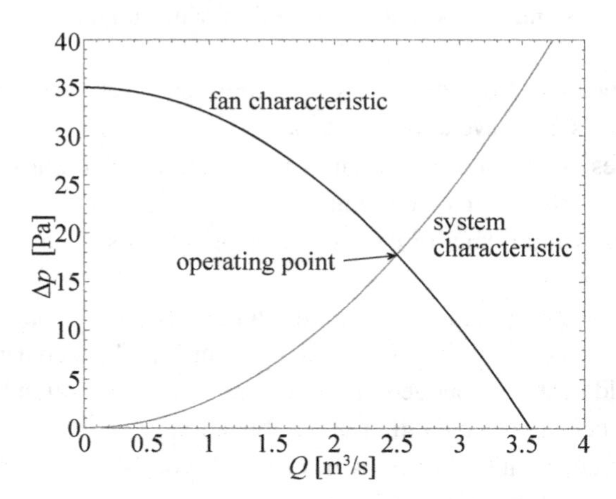

Figure 8.14 Pressure-flow-rate characteristic of a mixed-flow fan, operating at a rotational speed of 1,400 RPM.

8.4 Turbulence generated by a particular grid in a wind tunnel is weakly anisotropic, with the ratio of the standard deviations of streamwise and transverse velocity fluctuations being $u_1'/u_2' \approx 1.3$. To improve the isotropy level, namely to achieve $u_1'/u_2' \approx 1.0$, we consider passing this flow through a secondary contraction, before entering the test section. Determine the required contraction ratio using both inviscid and viscous flow analyses.

8.5 Consider the right-angle bend in the return duct of a closed-circuit wind tunnel. To prevent secondary flows, consider stretching inclined plane screens at appropriate positions. The air speed in the duct is 10 m/s and the duct cross section is rectangular. Explain the rationale for this design. Give an example of a proper screen arrangement, specifying the mesh sizes, solidities, locations and orientations of the screens. Explain a procedure by which you would optimise this arrangement so that the combined pressure loss would be minimised.

8.6 Apply the Stratford separation criterion to determine the sensitivity of diffuser stall onset upon various flow parameters. In particular, consider the conical diffuser discussed in Example 8.3, keeping the same values of all but one geometrical and kinematic parameters. Examine the following cases: (a) the inlet speed is increased to 30 m/s; (b) the upstream section length is doubled to 1.00 m; and (c) the diffuser length is increased to 5.00 m. Compare your results with the ones in Example 8.3 and explain whether you should reconsider the screen placement for each of these cases.

8.7 Consider flow in a water tunnel with a water depth of 1 m and a desired centreline speed of 0.2 m/s. Determine the shape of a curved, woven-wire screen with a solidity of 0.40 and a mesh size of 2.5 mm that would produce a uniformly sheared flow with a velocity gradient of 0.06 s^{-1}.

8.8 A hydraulics laboratory has just completed the construction of a scale model of a river and its surrounding flood planes. It is desired to assess, quickly and inexpensively, the flow quality in the model for different water levels, and particularly to determine whether there is flow separation at the bends, whether the flow is laminar or turbulent and whether there is a hydraulic jump. Describe the type of tests that you recommend and list the instrumentation and materials required. Estimate the time and cost of these tests, if they have to be repeated for three flow rates, including the maximum flow rate that can be provided by the available pumping system.

8.9 A student team is modifying the exterior shape of a classic car in the hope of achieving a better aerodynamic performance. They have no access to wind-tunnel facilities and their budget is very limited. Describe the type of tests that you would recommend, including materials and a cost estimate.

8.10 A pressure tube is used to measure flow properties in a transonic stream. Discuss whether there is a simple type of test that can be used to determine whether there are shock waves forming around the tube.

References

[1] American National Standards Institute. *Recommended Practice for Atmosphere and Space Flight Vehicle Coordinate Systems, ANSI-AIAA R-0004-1992*. American National Standards Institute, 1992.

[2] J.B. Barlow, W.H. Rae, and A. Pope. *Low-Speed Wind Tunnel Testing (3rd Edition)*. Wiley, New York, 1999.

[3] G.K. Batchelor. *The Theory of Homogeneous Turbulence*. Cambridge University Press, Cambridge, UK, 1953.

[4] P. Bradshaw. *Experimental Fluid Mechanics*. Pergamon Press, Oxford, 1970.

[5] P. Bradshaw and R.C. Pankhurst. The design of low-speed wind tunnels (chapter 1). In D. Kuechemann and L.H.G. Sterne, editors, *Progress in Aeronautical Science*, volume 5, pages 1–69. Pergamon Press, Oxford, 1964.

[6] G. Briassulis, J.H. Agui, J. Andreopoulos, and C.B. Watkins. A shock tube research facility for high-resolution measurements of compressible turbulence. *Exp. Thermal Fluid Sci.*, 13:430–446, 1996.

[7] J.E. Cermak. Wind tunnel design for modeling of atmospheric boundary layers. *J. Eng. Mech. Div. ASCE*, 107:623–642, 1981.

[8] T.P. Chong, P.F. Joseph, and P.O.A.L. Davies. Design and performance of an open jet wind tunnel for aero-acoustic measurement. *Appl. Acoust.*, 70(4):605–614, 2009.

[9] D. Chow. Design and characterization of the UTIAS anechoic wind tunnel. Master's thesis, University of Toronto, 2016. https://hdl.handle.net/1807/71668.

[10] G. Comte-Bellot and S. Corrsin. The use of a contraction to improve the isotropy of grid-generated turbulence. *J. Fluid Mech.*, 25:657–682, 1966.

[11] S. Corrsin. Turbulence: Experimental methods. In S. Fluegge and C. Truesdell, editors, *Handbuch der Physik – Encyclopedia of Physics*, volume 8 (2), pages 524–590. Springer, Berlin, 1963.

[12] J. Counihan. An improved method of simulating an atmospheric boundary layer in a wind tunnel. *Atmosph. Environ.*, 3:197–214, 1969.

[13] J. Counihan. Simulation of an adiabatic urban boundary layer in a wind tunnel. *Atmosph. Environ.*, 7:673–689, 1973.

[14] J.H. Downie, R. Jordinson, and F.H. Barnes. On the design of three-dimensional wind tunnel contractions. *Aeron. J.*, 88:287–295, 1984.

[15] J.W. Elder. Steady flow through non-uniform gauzes of arbitrary shape. *J. Fluid Mech.*, 5:355–368, 1959.

[16] Engineering Sciences Data Unit. Lift interference and blockage corrections for two-dimensional subsonic flow in ventilated and closed wind-tunnels. Technical Report Item Number 76028, Institution of Structural Engineers, London, 1976.

[17] Engineering Sciences Data Unit. Blockage corrections for bluff bodies in confined flows. Technical Report Item Number 80024, Institution of Structural Engineers, London, 1980.

[18] G.E. Erickson, D.J. Peake, J. Del Frate, A.M. Skow, and G.N. Malcolm. Water facilities in retrospect and prospect – an illuminating tool for vehicle design. Technical Report NASA Technical Memorandum 89409, NASA, November 1986.

[19] C. Farell and S. Youssef. Experiments on turbulence management using screens and honeycombs. *J. Fluids Eng.*, 118:26–32, 1996.

[20] R.W. Fox, A.T. McDonald, and J.W. Mitchell. *Fox and McDonald's Introduction to Fluid Mechanics (10th Edition)*. Wiley, New York, 2019.

[21] M. Gad-el-Hak. The water towing tank as an experimental facility. *Exp. Fluids*, 5:289–297, 1987.

[22] M. Gad-el-Hak. Splendor of fluids in motion. *Prog. Aerosp. Sci.*, 29:81–123, 1992.

[23] H.C. Garner, E.W.E. Rogers, W.E.A. Acum, and E.C. Maskell. Subsonic wind tunnel data corrections. Technical Report AGARDograph 109, Advisory Group for Aerospace Research and Development, NATO, Paris, October 1966.

[24] T.F. Gelder, R.D. Moore, J.M. Sanz, and E.R. McFarland. Wind tunnel turning vanes of modern design. Technical Report AIAA-86-0044, 1986.

[25] J.C. Gibbings. The pyramid gauze diffuser. *Ing. Arch.*, 42:225–233, 1973.

[26] R. Gordon and M.S. Imbabi. CFD simulation and experimental validation of a new closed circuit wind/water tunnel design. *J. Fluids Eng.*, 120:311–318, 1998.

[27] Hexcel. Honeycomb in air directionalizing applications. Technical Report TSB 102, Hexcel International, Arlington, TX, 1986.

[28] K. Hufnagel and G. Schewe. Force and moment measurement (chapter 8). In C. Tropea, A.L. Yarin, and J.F. Foss, editors, *Springer Handbook of Experimental Fluid Mechanics*. Springer, Berlin, 2007.

[29] H.P.A.H. Irwin. The design of spires for wind simulation. *J. Wind Eng. Industr. Aerodyn.*, 7:361–366, 1981.

[30] J.C. Isaza, R. Salazar, and Z. Warhaft. On grid-generated turbulence in the near- and far field regions. *Journal of Fluid Mechanics*, 753:402–426, 2014.

[31] ISO. *Flight Dynamics — Concepts, Quantities and Symbols, ISO 1151:1988*. International Organization for Standardization, Geneva, 2018.

[32] J.P. Johnston. Diffuser design and performance analysis by a unified integral method. *J. Fluids Eng.*, 120:6–18, 1998.

[33] U. Karnik and S. Tavoularis. Generation and manipulation of uniform shear with the use of screens. *Exp. Fluids*, 5:247–254, 1987.

[34] A.M. Kuethe and C.-Y. Chow. *Foundations of Aerodynamics: Bases of Aerodynamic Design (5th Edition)*. Wiley, New York, 1998.

[35] E.M. Laws and J.L. Livesey. Flow through screens. *Annu. Rev. Fluid Mech.*, 10:247–266, 1978.

[36] R.I. Loehrke and H.M. Nagib. Experiments on management of free-stream turbulence. Technical Report AGARD Report No. 598, AGARD, Neuilly-sur-Seine, France, 1972.

[37] R.I. Loehrke and H.M. Nagib. Control of free stream turbulence by means of honeycombs. *J. Fluids Eng.*, 98:342–353, 1976.

[38] J.L. Lumley. Passage of a turbulent stream through honeycomb of large length-to-diameter ratio. *J. Basic Eng.*, 86:218–220, 1964.

[39] J.L. Lumley and J.F. McMahon. Reducing water tunnel turbulence by means of a honeycomb. *J. Basic Eng.*, 89:764–770, 1967.

[40] H Makita and K Sassa. Active turbulence generation in a laboratory wind tunnel. In *Advances in Turbulence 3*, pages 497–505. Springer, 1991.

[41] D.J. Maull. The wake characteristics of a bluff body in a shear flow. In *The Aerodynamics of Atmospheric Shear Flow*, pages 16.1–16.13. AGARD C.P. 48, Paper 16, 1970.

[42] R.D. Mehta and P. Bradshaw. Design rules for small low speed wind tunnels. *Aeron. J.*, 73:443–449, 1979.

[43] M.N. Mikhail and W.J. Rainbird. Optimum design of wind tunnel contractions. Technical Report AIAA Paper 78-819, 1978.

[44] V.J. Modi and S. El-Sherbiny. Effect of wall confinement on aerodynamics of stationary circular cylinders. In *Proc. Third Intern. Conf. on Wind Effects on Buildings and Structures*, pages 365–375, Tokyo, Japan, 1971. Saikon Co Ltd.

[45] T. Morel. Comprehensive design of axisymmetric wind tunnel contractions. *J. Fluids Eng.*, 97:225–233, 1975.

[46] L. Mydlarski. A turbulent quarter century of active grids: from Makita (1991) to the present. *Fluid Dynamics Research*, 49(6):061401, 2017.

[47] L. Mydlarski and Z. Warhaft. On the onset of high-Reynolds-number grid-generated wind tunnel turbulence. *J Fluid Mech.*, 320:331–368, 1996.

[48] H. Nagib, M. Hites, J. Won, and S. Gravante. Flow quality documentation of the National Diagnostic Facility. In *AIAA 94-2499, Proc. 18th AIAA Aerospace Ground Testing Conference, June 20–23, 1994*, pages 1–13, Colorado Springs, CO, 1994.

[49] NASA Ames Research Center. Ames unitary plan wind tunnel, 2003. www.nasa.gov/centers/ames/orgs/aeronautics/windtunnels/index.html.

[50] National Research Council Canada. Aerospace Research Centre, 2021. https://nrc.canada.ca/node/453.

[51] J. Nedić, B. Ganapathisubramani, J.C. Vassilicos, J. Borée, L.E. Brizzi, and A. Spohn. Aeroacoustic performance of fractal spoilers. *AIAA J.*, 50:2695–2710, 2012.

[52] J. Nedić and S. Tavoularis. Measurements of passive scalar diffusion downstream of regular and fractal grids. *J. Fluid Mech.*, 800:358–386, 2016.

[53] G.M. Odell and L.S.G. Kovasznay. A new type of water channel with density stratification. *J. Fluid Mech.*, 50:535–543, 1971.

[54] R. L. Panton. *Incompressible Flow*. Wiley, New York, 1984.

[55] S.P. Parker. *Fluid Mechanics Source Book*. McGraw Hill, New York, 1987.

[56] F.E. Penaranda and M Shannon Freda. *Aeronautical Facilities Catalogue, Vol. 1: Wind Tunnels*. Research and Support Facilities (Air) NASA-RP-1132, National Aeronautics and Space Administration, Washington DC, 1985. https://ntrs.nasa.gov/citations/19850010682.

[57] P.E. Roach. The generation of nearly isotropic turbulence by means of grids. *J. Heat Fluid Flow*, 8:82–92, 1987.

[58] SAE International. *Vehicle Aerodynamics Terminology, J1594-201007*. SAE International, 2010. https://www.sae.org/standards/content/j1594_201007/.

[59] A. Sahlin and A.V. Johansson. Design of guide vanes for minimizing the pressure loss in sharp bends. *Phys. Fluids A*, 3:1934–1940, 1991.

[60] J. Scheiman and J.D. Brooks. Comparison of theoretical and experimental turbulence reduction from screens, honeycomb and honeycomb-screen combinations. *J. Aircraft*, 18:638–643, 1981.

[61] M.M. Seltsam. Experimental and theoretical study of wide-angle diffuser flow with screens. *AIAA J.*, 33:2092–2100, 1995.

[62] G. Sovran and E.D. Klomp. Experimentally determined optimum geometries for rectilinear diffusers with rectangular, conical or annular cross-section. In G. Sovran, editor, *Fluid Mechanics of Internal Flow*, pages 270–319. Elsevier Publishing Co., New York, 1967.

[63] N.M. Standen. A spire array for generating thick turbulent shear layers for natural wind simulation in wind tunnels. Technical Report Technical Report LTR-LA-94, National Aeronautical Establishment, Ottawa, Canada, 1972.

[64] D.C. Stillinger, M.J. Head, K.N. Helland, and C.W. Van Atta. A closed loop gravity driven water channel for density-stratified shear flows. *J. Fluid Mech.*, 131:73–90, 1983.

[65] B.S. Stratford. The prediction of separation of the turbulent boundary layer. *J. Fluid Mech.*, 5:1–16, 1959.

[66] J. Tan-Atichat, H.M. Nagib, and R.I. Loehrke. Interaction of free-stream turbulence with screens and grids: A balance between turbulence scales. *J. Fluid Mech.*, 114:501–528, 1982.

[67] S. Tavoularis. Flow past immersed objects (chapter 20). In J.M. Saleh, editor, *Fluid Flow Handbook*, pages 20.1–20.44. McGraw-Hill, New York, 2002.

[68] G.I. Taylor and G.K. Batchelor. The effect of a gauze on small disturbances in a uniform stream. *Quart. J. Mech. Appl. Math.*, 2:1–29, 1949.

[69] P.C. Valente and J.C. Vassilicos. The decay of turbulence generated by a class of multiscale grids. *J. Fluid Mech.*, 687:300–340, 2011.

[70] T.M. Ward. The hydrodynamics laboratory at the California Institute of Technology – 1976. *J. Fluids Eng.*, 98:740–748, 1976.

[71] F.M. White. *Fluid Mechanics (8th Edition)*. McGraw-Hill, New York, 2015.

Part II

Measurement Techniques

9 Measurement of Flow Pressure

The pressure in a fluid element is one of its most important properties and a primary measurand in experimental fluid mechanics. Measurements of pressure are used for the determination of the local flow velocity, the flow rate through a vessel, the fluid density, the wall shear stress, the forces and moments on immersed surfaces and the energy losses in ducts. Physically, pressure is a consequence of collisions between elementary particles, such as molecules, atoms and ions, but it is usually treated as a macroscopic property. Casual use of the term pressure may lead to confusion, which we will attempt to resolve in this chapter. Although some methods for measuring pressure are quite simple, the accurate determination of in-flow pressure is a most challenging problem, largely because of the necessarily intrusive nature of such methods. Pressure-measuring instrumentation and techniques have been described in numerous sources, both of general [4, 8, 10, 15, 19, 21, 31, 37, 43] and specific [22, 39] scope. The following sections represent a digest of material that is relevant to the measurement of pressure in the fluid dynamics laboratory.

9.1 What Exactly is Pressure?

Pressure as the average normal stress: The terms *pressure* and *static pressure* are often used interchangeably. In some sources, either term appears to describe casually the normal force per unit area, namely the normal stress. This definition is unambiguous, when one deals with fluid pressure on an immersed solid surface or in a static fluid; however, it is inappropriate for a moving fluid element, because normal stresses generally depend on direction, even though such dependence is often disregarded. The appropriate mechanical definition of pressure is as the *average normal stress* [7, 38]

$$p = -\frac{1}{3}\left(\sigma_{11} + \sigma_{22} + \sigma_{33}\right), \tag{9.1}$$

which is invariant under coordinate system rotations and reflections. The negative sign accounts for the fact that, by convention, pressure is positive when compressive, whereas a normal stress is considered positive when tensile. In static fluids, the normal stress is independent of orientation and, therefore, the pressure is equal in magnitude to any normal stress at a given position.

Hydrostatic pressure: In static fluids subjected to a gravitational field, the *hydrostatic pressure difference* between two locations A and B can be found as

$$p(B) - p(A) = -\int_A^B \rho g \, \mathrm{d}x_3 , \tag{9.2}$$

where x_3 is a vertical upwards direction, g is the gravitational acceleration and ρ is the fluid density; both g and ρ are generally functions of elevation, but such dependence is usually neglected for relatively small elevation differences.

Pressure in thermodynamics: Besides its mechanical definition, which relies on the continuum assumption, pressure has also been defined through thermodynamic concepts, which equate pressure and work per unit volume change [7, 38, 45, 50]. In static fluids, the thermodynamic and mechanical pressures coincide, while, in incompressible fluids, it is only the mechanical pressure that can be defined. To simplify matters, one may generally employ the *Stokes assumption*, by which the thermodynamic pressure may be taken as equal to the mechanical pressure. The Stokes assumption applies fairly well to Newtonian fluids, provided that their rate of expansion is not exceedingly high.

Absolute vs. gauge pressure: Pressure, as was defined here, is also referred to as the *absolute pressure*. Some pressure-measuring instruments may, however, indicate not the absolute pressure, but its difference from the reference pressure in the surrounding environment, usually the atmospheric pressure p_{atm}. The pressure difference $p_g = p - p_{atm}$ is called the *gauge pressure*. Note that atmospheric pressure varies with altitude and also depends on the wind speed and direction, the temperature and the humidity. The value of atmospheric pressure measured at a particular location and a particular instant is called the *barometric pressure*.

Dynamic and total pressures: Another parameter of importance in fluid dynamics is the *total pressure*, or *stagnation pressure*, p_o, defined as the static pressure p that a fluid particle would achieve, if it were decelerated isentropically (namely adiabatically and reversibly) to zero speed (stagnation). In gravity-free incompressible flows,

$$p_o = p + \frac{1}{2}\rho V^2 , \tag{9.3}$$

where $1/2\rho V^2$ is called the *dynamic pressure*. In compressible flows, however, static and total pressures are related as

$$p_o = p\left(1 + \frac{\gamma - 1}{2}\mathrm{M}^2\right)^{\frac{\gamma}{\gamma-1}} , \tag{9.4}$$

where γ is the ratio of specific heats and M is the Mach number (note that this expression applies to ideal gases with constant γ). Particular care must be taken when a shock wave is formed upstream of an instrument. Assuming that the shock wave is normal, one may compute the ratios of static and total pressures across the shock wave as, respectively [16],

$$\frac{p}{p'} = \left(\frac{2\gamma}{\gamma + 1}\mathrm{M}^2 - \frac{\gamma - 1}{\gamma + 1}\right)^{-1} \tag{9.5}$$

and

$$\frac{p_o}{p'_o} = \left(\frac{2\gamma}{\gamma+1}\mathrm{M}^2 - \frac{\gamma-1}{\gamma+1}\right)^{\frac{1}{\gamma-1}} \left[\frac{(\gamma-1)\,\mathrm{M}^2+2}{(\gamma+1)\,\mathrm{M}^2}\right]^{\frac{\gamma}{\gamma-1}}, \tag{9.6}$$

where non-primed and primed properties refer, respectively, to positions just upstream and just downstream of the shock wave. These expressions are called the *Rankine–Hugoniot conditions*.

9.2 Pressure-Measuring Instrumentation

9.2.1 Liquid-in-Glass Manometers

Static response of liquid manometers: Liquid-filled, U-shaped manometers are simple and effective devices, which can measure pressure differences that are constant or vary sufficiently slowly for the operation of the instrument to be considered as static. For the basic configuration, shown in Fig. 9.1, in which the manometer tubes are vertical, the difference between the pressures at positions *A* and *B* is given by

$$p_A - p_B = \rho_1 g\,(z_C - z_A) + \rho_2 g\,(z_D - z_C) + \rho_3 g\,(z_B - z_D)\ . \tag{9.7}$$

In this configuration, fluid 2 must be a liquid, while fluids 1 and 3 may be either gases or liquids that are immiscible with liquid 2. The input to this pressure-measuring system is the pressure difference $p_A - p_B$. The output consists of the values of z_C, z_D, but, if these values were known when $p_A - p_B = 0$, or any other given value, then one may consider the column height $z_D - z_C$ as the output, because all free-surface elevations can be calculated from this value.

The operation of this manometer is simplified, when fluids 1 and 3 are gases (in which case, $\rho_1, \rho_3 \ll \rho_2$) and the elevations of points *A* and *B* are not much higher than those of points *C* and *D*. Then, the static response equation of the U-tube manometer becomes

$$p_A - p_B \approx \rho_2 g\,(z_D - z_C)\ . \tag{9.8}$$

The static sensitivity of the U-tube manometer, when used in gases, is, therefore, $1/(\rho_2 g)$, clearly increasing with decreasing liquid density. On the other hand, for a given length

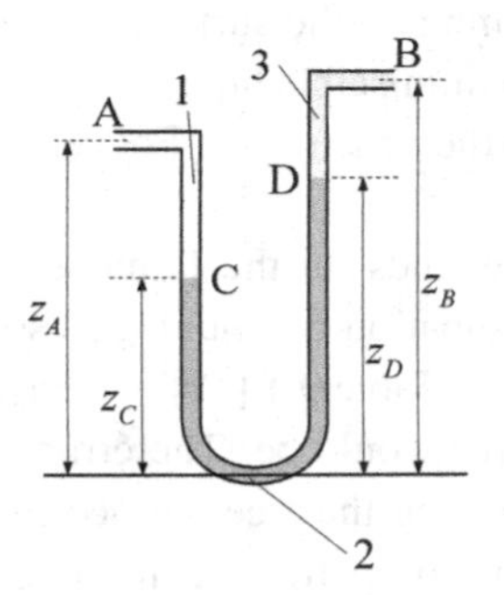

Figure 9.1 Sketch of liquid-in glass U-tube manometer.

of the U-tube, the range of this manometer would be proportional to $\rho_2 g$, thus, increasing with increasing liquid density. The choice of manometric fluid is done by compromise between the conflicting requirements of high sensitivity and wide range. With a density of about 13,550 kg/m^3 at room temperature, mercury is by far the liquid that would provide the widest range of a manometer, but this is no longer an option for open systems, because of the hazardous nature of mercury vapours. The heaviest allowable liquids are various types of oils, none of which, however, has a density exceeding about 3,000 kg/m^3. To widen the range of the manometer, one may use a longer tube (e.g., by mounting a transparent hose on the laboratory wall) or pressurise one side of it.

When a manometer is used to measure pressure differences in a liquid, its static sensitivity would be inversely proportional to the difference between the densities of that liquid and the manometric fluid. Therefore, to increase the sensitivity of this system, one needs to use a manometric fluid with a density that is close to that of the liquid of interest. Manometric fluids with densities that are slightly different from the density of water, as well as being immiscible with water, are available commercially.

Sources of error: The U-tube manometer response equations clearly indicate that errors in the measurement of pressure may be introduced by changes in the fluid densities (e.g., as a result of a temperature change) or the gravitational acceleration (e.g., as a result of a change in the altitude of the manometer); such errors are relatively easy to correct [10]. Other errors could be introduced by a misalignment of the manometer tubes and the vertical direction and by horizontal acceleration, which would generate a reading even if the pressure difference across the manometer is zero. When the manometer is used to measure gauge pressure, for example by opening side B in Fig. 9.1 to the atmosphere, an error in the value of the absolute pressure would be introduced by changes in the *barometric pressure*, as the result of changes in weather; to eliminate this error, one needs to measure the barometric pressure on the experimental site, for example, with a *barometer*.

A different source of error is *capillarity*, namely the rise or fall of a liquid free surface inside vertical or inclined tubes. Capillarity is due to adhesion forces between liquids and solids in contact. For the case of a vertical tube containing a liquid and a gas, as shown in Fig. 9.2, the capillarity error is the height of the liquid column

$$h_c = \frac{4\sigma \cos\theta}{\rho g d}, \tag{9.9}$$

determined by equating the weight of the column and the surface tension force on the meniscus perimeter [7]. If the tube contains two immiscible liquids, it seems appropriate to account for the buoyancy force by replacing the density ρ in Eq. (9.9) by the density difference.

In this expression, the surface tension σ depends on the liquid–gas combination, while the contact angle θ depends on the combination of liquid–gas–solid-wall materials. A few representative values are presented in Table 9.1 [10]. Clearly, the larger the tube diameter is, the smaller the capillarity error would be. The error can be positive (namely, the free surface in the tube is higher than the free surface in the surrounding tank) or negative, depending on the material properties of the fluids and the tube

Table 9.1 Surface tension values and contact angles for selected manometric material combinations

Material combination	σ [N/m]	θ [°]
Water–air–glass	72.8×10^{-3}	0
Mercury–air–glass	470×10^{-3}	140
Mercury–vacuum–glass	480×10^{-3}	140
Mercury–water–glass	380×10^{-3}	140

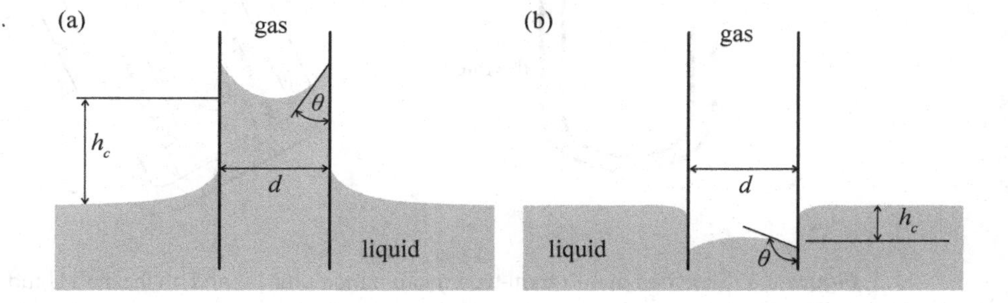

Figure 9.2 Schematic representation of (a) capillary rise ($h_c > 0$) and (b) capillary drop ($h_c < 0$).

wall. For example, the capillarity error is positive in a water–air–glass manometer and negative in a mercury–air–glass manometer.

Example 9.1 Determine the capillary rise inside a glass tube, which has a diameter $d = 10$ mm, is open at its top end and is dipped vertically at its bottom end into an open tank of clean water, as shown in Fig. 9.2. Repeat this calculation for $d = 2$ mm and for cases with mercury used instead of water.

Answer

Using Eq. (9.9) and values of σ, θ from Table 9.1, one can compute the capillary rise of water in the tube as $h_c \approx$ mm, when $d = 10$ mm, and $h_c \approx 15$ mm, when $d = 2$ mm. If mercury were used instead of water, similar analysis gives $h_c \approx -1.1$ mm, when $d = 10$ mm, and $h_c \approx -5.5$ mm, when $d = 2$ mm. Capillarity causes a drop of the mercury inside the tube below the undisturbed free surface of the surrounding tank.

Errors in manometer readings could also be caused by inconsistent readings of the meniscus (i.e., the free surface inside the tube) elevation or changes of the meniscus shape due to dirt, variations in the tube size and shape and inclination of the tube.

Improved manometers: Capillarity and meniscus-reading errors can be minimised by always positioning the meniscus at the same reference position, marked clearly by a line. This is the basis of the *Prandtl micromanometer*, which permits the vertical traversing of the liquid containing tank to restore the meniscus at the reference position

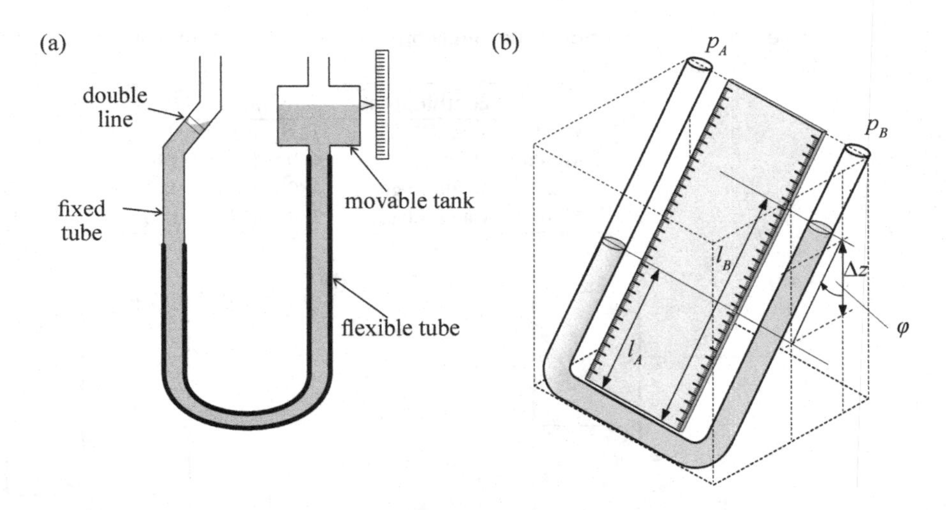

Figure 9.3 Sketches of a Prandtl-type micromanometer (a) and an inclined U-tube manometer (b).

(Fig. 9.3a). The precision of this instrument is increased with the use of a precision lead screw and a rotary scale [10].

A number of variants of the basic U-tube manometer have been proposed to enhance its operation for specific purposes. A useful configuration is the *inclined manometer* (Fig. 9.3b), the static sensitivity of which, when used in gases and if one considers the reading of the meniscus position along the inclined tube as the output, is $1/(\rho_2 g \cos\varphi)$; for an inclination of $\varphi = 80°$, the sensitivity of the inclined manometer would be nearly six times higher than that of a vertical manometer using the same liquid. It bears repeating that the measured pressure difference, for example, the value computed from Eq. (9.8), depends on the change in height in the direction of gravity, that is, Δz in Fig. 9.3b. Inclined manometers typically have graded indicators to display the change in height of the liquid column l, as shown in the side view of Fig. 9.3b. In such cases, Eq. (9.8) can be expressed as

$$p_A - p_B \approx \rho_2 g \cos\varphi \, (l_B - l_A) \ , \tag{9.10}$$

which reverts to Eq. (9.8) when $\varphi = 0°$.

Dynamic response of liquid-in glass manometers: Although liquid-in-glass manometers are not used to measure rapidly changing pressures, they do tend to exhibit oscillatory response, and so it is of interest to understand their dynamic characteristics and use them as a guide for selection or design of a manometer. The following is a summary of a simplified analysis [14], which has been roughly confirmed experimentally.

Consider a liquid-in-glass manometer, as sketched in Fig. 9.1, containing a liquid with a density ρ and connected on both sides to a gas. Let d be the diameter of the tube and L be the total length of the tube part that contains the liquid. The unsteady flow of the liquid inside the tube could be quite complex and different from fully-developed laminar or turbulent pipe flow. If, however, one assumes the flow to have a fully-developed, steady, laminar profile, and further neglects secondary effects such as surface tension and gas inertia, it is possible to show that the motion of the liquid in the manometer has the dynamic response of a second-order system with an undamped natural frequency and a damping ratio that are, respectively, equal to

$$\omega_n = 1.2\sqrt{\frac{g}{L}} \tag{9.11}$$

and

$$\zeta = \frac{9.8\nu}{d^2}\sqrt{\frac{L}{g}}, \tag{9.12}$$

where ν is the kinematic viscosity of the working fluid.

The consideration of turbulent flow leads to a non-linear system, which does not lend itself to an analytical solution. Further simplifying the problem by assuming that the liquid flow is oscillatory with a frequency ω and an amplitude Δz, but has the same wall shear stress as fully developed, stationary, turbulent pipe flow, allows one to determine a quasi-linear response with an undamped natural frequency

$$\omega_n = 1.4\sqrt{\frac{g}{L}}, \tag{9.13}$$

which is only 14% higher than the laminar flow value, thus creating confidence in the use of either Eq. (9.11) or Eq. (9.13) for rough predictions. An equivalent damping ratio can be also calculated as

$$\zeta_e = \left(\frac{9.8\nu}{d^2}\sqrt{\frac{L}{g}}\right)\left(\frac{1}{1{,}200}\frac{\omega\Delta z\, d}{\nu}\right)^{3/4}, \tag{9.14}$$

which is a function of the frequency $\omega \approx \omega_n$ and the amplitude Δz of oscillation, assuming that the system is underdamped. When a step-like pressure change Δp is applied to the manometer, one may compute an upper bound for the amplitude of oscillation as $\Delta z_{max} = \Delta p/(\rho g)$. The characteristic Reynolds number $\mathrm{Re}_\omega \equiv \omega \Delta z\, d/\nu$ of the oscillation should be greater than 1200, if the flow is to be turbulent over most of the cycle, and so $\zeta_e > \zeta$, which is consistent with the fact that frictional losses in a turbulent flow are larger than in a laminar flow with the same Reynolds number, if that can be sustained. Furthermore, this relationship shows that damping would decrease as the oscillations continue, which means that the amplitude of oscillation would diminish from one cycle to the next more slowly than that of a corresponding second-order system.

Example 9.2 A U-tube manometer has been constructed by bending semi-circularly a glass tube with an inner diameter $d = 6$ mm and a wall thickness $t = 1$ mm, such that the distance between the axes of the two vertical tube sections is $w = 50$ mm. The manometric fluid is water at 20°C. When both tubes are open to the atmosphere, the distance from the bottom of the manometer to the free surface is $h_{ref} = 750$ mm. Compute the natural frequency and the damping coefficient of this manometer.

Answer

The length of the bent tube section along its axis is $L_b = \pi w/2 = 78.5$ mm. The length of each vertical section of the tube that is occupied by water in its undisturbed state is $L_v = h_{ref} - w/2 - d/2 - t = 721$ mm. Thus, the total length of the tube, along its axis, that is occupied by water is $L = L_b + 2l_v = 1520.5$ mm. The kinematic viscosity of water is taken as $\nu = 10^{-6}$ m^2/s.

First, let us assume that the flow is laminar, in which case we can substitute the values of properties in Eq. (9.11) and Eq. (9.12) to determine the natural frequency and the damping coefficient of the manometer as, respectively, $\omega_n = 3.05$ rad/s and $\zeta = 0.107$. This indicates that the manometer would be strongly underdamped and oscillate with a large amplitude.

Next, let us assume that the flow is turbulent. The natural frequency of oscillation can be calculated from Eq. (9.13) as $\omega_n = 3.56$ rad/s. According to the previous discussion, for the water flow to be turbulent following a step-change in pressure, the pressure change must be $\Delta p > 1200\nu\rho g/(\omega_n d) \approx 550$ Pa. For the mild value $\Delta p = 1$ kPa, Eq. (9.14) gives $\zeta_e = 1.57\zeta = 0.168$, which also corresponds to a strongly underdamped system. Equation (9.14) predicts that increasing Δp will increase damping and the system will become overdamped, at least initially, when $\Delta p > 10.8$ kPa.

9.2.2 Mechanical Pressure Gauges

Deadweight gauges: *Deadweight gauges* are highly accurate devices, but fairly cumbersome in their use, so that they find application mainly as standards for the calibration of other pressure gauges, rather than for the measurement of pressure in the laboratory [10, 14]. An idealised deadweight gauge is shown in Fig. 9.4. Both hydraulic and pneumatic versions are available. Pressure is built-up inside the fluid chamber by the addition of weights on a platform on top of the plunger. If the plunger reaches the end of the cylinder, a pump is used to push it away. The plunger is typically spun before a measurement is taken to remove any friction effects between the plunger and cylinder. Then the gauge pressure inside the fluid chamber (omitting hydrostatic pressure) should ideally be equal to the weight (excluding the *tare*, namely the weight of the plunger and other accessories) divided by the plunger cross-sectional area. For a higher accuracy, and in order to account for the small clearance that must be provided between the plunger and the cylinder, an average area between the plunger and the cylinder cross-sectional areas should be used instead. Additional corrections can be made for buoyancy, altitude and temperature variation effects. Improved versions of deadweight gauges are available, using rotation or oscillation of the cylinder or the plunger to reduce friction and various compensation methods to account for leakage. Typical ranges of deadweight gauges are between 10^2 and 10^8 Pa within an uncertainty of 0.01 to 0.05% of the reading.

Elastic element gauges: These widely used, general-purpose pressure gauges contain an elastic element which deforms under pressure and creates a linear or angular displacement of a component that is either displayed on a dial via purely mechanical linkages or transformed to an electric signal that can be displayed or recorded at will.

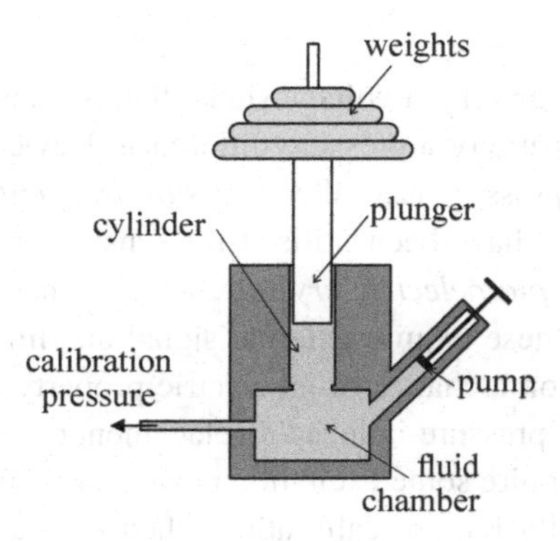

Figure 9.4 Sketch of an idealised deadweight gauge.

Although the operation of these devices can be explained from first principles, their output cannot be predicted accurately and so they require calibration. Manufacturers of precise commercial gauges supply calibrated devices and perform periodic calibration updates, which maintain accuracy following use. A common type of elastic element is the *Bourdon tube*, namely a sealed metallic tube with an oval or flattened cross-sectional shape (Fig. 9.5a) that tends to deform towards the circular shape, as the difference between the inner and outer pressures increases. Thus, curved tubes (Fig. 9.5a) tend to become straight, causing a linear displacement, and helical tubes (Fig. 9.5b) tend to unwind, causing an angular displacement. Other types of elastic elements include capsules, diaphragms and bellows (Fig. 9.5c–e). Elastic deformation gauges are available for both absolute and gauge pressures. As might be expected, they have a relatively slow response. Their accuracy is moderate, especially in the lower 10-20% of their ranges. They find extensive application in the monitoring of industrial flows, but not as much as laboratory instruments, other than for the purpose of monitoring supply pressures and the like.

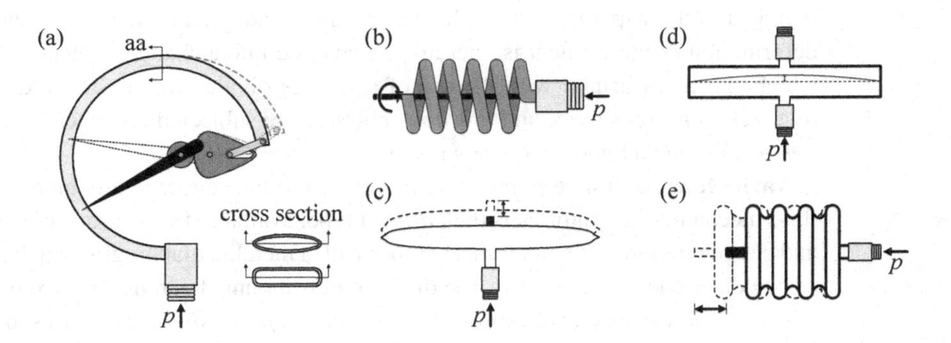

Figure 9.5 Sketches of elastic elements used in pressure gauges: (a) curved Bourdon tube; (b) helical Bourdon tube; (c) capsule; (d) diaphragm; and (e) bellows.

9.2.3 Electrical Pressure Transducers

Electrical pressure transducers provide an output voltage signal that is linearly or non-linearly dependent on the absolute pressure or a pressure difference. Devices that produce a current signal that is related to pressure are called *pressure transmitters*. There is a variety of physical phenomena that have been utilised for sensing pressure. The only truly passive pressure sensors are *piezoelectric crystals*, which generate an electric voltage when deformed, but even these require external signal amplification. The operation of all other devices is based on a change of an electric property (resistance, capacitance or inductance) as a result of pressure-induced displacement or deformation; such devices are active, namely, they require some excitation power for their operation.

Electrical pressure transducers usually require calibration when they are first employed and frequent calibration checks thereafter. They are generally susceptible to temperature and humidity effects, which could seriously contaminate the measurements. Temperature effects often manifest themselves in the form of a zero-drift, while they may not significantly influence the static sensitivity. In such cases, an adjustment of the zero or subtraction of the zero-offset value from the results may be sufficient to account for this effect. Humidity effects could be more unpredictable, particularly for variable capacitance transducers. An easy way to remove humidity from the chambers of pressure transducers used in air is to leave them connected overnight to a flask containing *hygroscopic crystals*. When using pressure transducers in liquids, care must be taken to remove gas pockets and bubbles from the transducer chambers and all connecting lines.

The great advantage of electrical transducers over liquid-in-glass manometers and mechanical gauges is their superior frequency response, which makes them suitable, under certain conditions, for the measurement of unsteady and turbulent pressures. In fact, several of the electrical pressure transducers (piezoelectric, variable capacitance and strain gauge types) originated as *microphones*, namely devices used for acoustical measurements. The disadvantage of some of these devices is that they do not measure the absolute or gauge pressure itself, but only pressure fluctuations.

Potentiometric transducers: These are simple devices, in which an elastic element is linked to the wiper of a rotary, linear or wirewound potentiometer. Pressure-induced deformation of the element is, therefore, converted into a change in electric resistance, which can be measured with the use of a voltage divider circuit. These devices have relatively high sensitivity and low cost, but they are subjected to hysteresis and have a low resolution and poor frequency response.

Variable capacitance transducers: A pressure transducer that is quite common in fluid mechanics laboratories is the *barocell* type, available for measuring both absolute and gauge pressures. It contains a chamber with a metallic diaphragm, which also serves as one electrode of a capacitor. The diaphragm is mounted parallel to a fixed backplate, which acts as the second electrode of a capacitor. A *polarising voltage* must be provided to the capacitor by a power supply. The pressure difference that is applied on the two sides of the diaphragm is measured via changes in capacitance between the diaphragm and the backplate, caused by diaphragm deformation.

The common-type *condenser microphones* or *capacitor microphones* [36] also operate on the same principle. The elastic diaphragm is usually made of nickel or plated mylar or glass. The microphones are also available in miniature forms, with diameters as small as a few millimetres, and have a wide frequency response, ranging from a few Hz to nearly 100 kHz, which makes them ideal as audible-sound transducers, although they cannot resolve the value of pressure. In addition to externally polarised microphones, a permanently polarised variation is available, called the *electret* microphone. Electret microphones contain a polymer diaphragm with embedded electric charges. Their advantages over capacitor microphones are lower cost and lower sensitivity to humidity.

Piezoelectric transducers: The most commonly used piezoelectric materials are lead zirconate titanate (PZT) and barium titanate, although several other crystalline materials have also been used as the passive elements [3, 30, 36]. A limitation of piezoelectric transducers is their relatively high sensitivity to vibrations and acceleration; to reduce such sensitivities, various compensation methods (e.g., the use of two crystals in tandem) have been proposed [19].

Strain gauge transducers: These transducers contain one or more strain gauges (i.e., components having an electric resistance that is proportional to the axial strain) attached to an elastically deformable element. These strain gauges are connected in a Wheatstone bridge and supplied with an excitation voltage to produce a pressure-dependent electric signal. Because their outputs require substantial amplification, they may be subjected to considerable drift and temperature sensitivity. An advantage of strain gauge transducers is that they can be made quite thin, with a thickness as low as about 1 mm, and so are suitable as wall-pressure sensors [8].

Variable reluctance transducers: *Reluctance* is the ratio of magnetic 'force' to magnetic flux in a magnetic circuit; its reciprocal is called *permeance* [20]. These transducers contain a magnetically permeable diaphragm, which is mounted between two symmetrically located magnetic coils. If the diaphragm is deflected towards one of the coils, as a result of applied pressure, its reluctance with respect to the magnetic field of one coil would increase and with respect to that of the other would decrease. Thus, the inductances of the two coils would be changed. The coils are connected in a bridge configuration such that the inductance ratio is measured as an electric voltage. Both DC and AC outputs can be provided.

Linear variable differential transformers (LVDT): In these transducers, pressure-induced deformation of an elastic element is transmitted to the core of a transformer consisting of a central primary coil, flanked by two secondary coils at either of its ends. When no pressure is applied, the core is in a symmetric position and the two secondary coils are in balance. When pressure is applied, imbalance in the circuitry produces an output voltage proportional to pressure.

Semiconductor and microelectromechanical pressure transducers: During the past few decades, semiconductor fabrication technology has been applied to the manufacturing of silicon-based pressure transducers, which can be made to sizes much smaller than the corresponding conventional transducers. Transducers with sizes of 1 mm or larger are usually referred to as *miniature-type*, while the term *microelectromechanical* (*MEMS*) applies to devices with sizes between 1 μm and 1 mm [1, 18, 25, 28].

Such transducers are usually manufactured by the use of photolithographic methods and are divided into the following three categories:

- **Piezoelectric transducers:** These have relatively low sensitivity and high noise level.
- **Piezoresistive transducers:** These have a relatively high sensitivity to pressure variations, but are also sensitive to temperature variations and stresses; they are usually connected in a Wheatstone bridge configuration and are compensated for temperature variation.
- **Capacitive transducers:** These operate similarly to capacitor microphones and require electronic preamplification; their advantage is a high sensitivity to pressure variations but low sensitivity to temperature variations; electret type MEMS transducers are also available.

MEMS transducers can be manufactured at low cost and can also be produced in the form of two-dimensional arrays, which makes them suitable for the simultaneous measurement of pressure fluctuations over a surface. Their frequency response extends to several tens of kHz, which is adequate for the measurement of pressure fluctuations in many turbulent flows.

9.2.4 Gauges for Extreme Pressures

The measurement of pressure in ranges that are several orders of magnitude higher or lower than the usual, near-atmospheric, laboratory range can only be made with special instrumentation.

Vacuum gauges: Liquid-in-glass micromanometers, including special designs that 'amplify' pressure, can be used effectively to measure pressures as low as about 0.1 Pa. *Vacuum*, namely the extremely low pressure range, can be measured with devices relating pressure to various other fluid properties, including viscosity, capacitance and thermal conductivity [5, 14, 22, 35, 44, 46]. Among the available instruments one may mention the ones mentioned in Table 9.2, keeping in mind that the indicated ranges may vary significantly from one model to another.

Table 9.2 Devices used for the measurement of low pressures

Device	**Pressure range** [Pa]
McLeod gauge	10^{-1} to 10
Pirani (resistance) gauges	10^{-3} to 10^{2}
Thermistor and thermocouple gauges	10^{-2} to 10^{2}
Philips cold-cathode gauges	10^{-3} to 1
Penning cold-cathode gauges	10^{-6} to 10^{-1}
Capacitance gauges	10^{-5} to 10^{-2}
Ionization gauges	10^{-11} to 10^{2}
Mass spectrometers	to less than 10^{-11}

High-pressure gauges: The high-pressure range may be conventionally defined as exceeding 10^{8}Pa. Strain gauge transducers and elastic element transducers can be made to measure pressures at most an order of magnitude larger than that value. For even higher pressures, up to a limit of about 10^{12} Pa, the only transducers available are of the variable resistance type [8, 14, 24, 39], but of a design quite different from that of strain gauge transducers. The high-pressure gauges contain a coiled wire of gold-chrome (2.1%) alloy or manganin, sealed inside metallic bellows filled with kerosene. The pressure is transmitted by the bellows to the kerosene, which compresses the wire. The gauge's output is proportional to the electric resistance R of the wire, which is related to pressure as [8]

$$\frac{\mathrm{d}R/R}{p} = \frac{2}{E} + \frac{\mathrm{d}\rho_e/\rho_e}{p} ,$$

where E and ρ_e are, respectively, Young's modulus of elasticity and the electric resistivity of the wire.

9.3 Local Wall Pressure Measurement

9.3.1 Static Pressure Taps

The simplest method of sensing steady or slowly varying static pressure at a solid wall is by machining a small orifice (*tap*) facing the flow and connecting it to a manometer or pressure transducer. This approach is used widely; however, it may introduce appreciable systematic errors, when not implemented carefully. To begin with, flow over a cavity generally induces motions of the fluid contained in it in the form of a sequence of counter-rotating vortices (Fig. 9.6a), each having a strength that weakens with increasing distance from the orifice exit. Such motions entrain high-speed fluid from regions at some distance from the wall, and tend to create a pressure inside the cavity, which is higher than the true wall static pressure. Thus, wall pressure will be measured accurately only by an infinitesimally small tap. The use of extremely small taps, however, is impractical for several reasons. Besides the technical difficulty of machining small holes that are clean and perpendicular to a surface, such holes would be amenable to blockage by flow impurities. Practical hole sizes typically range between 0.5 and 3 mm.

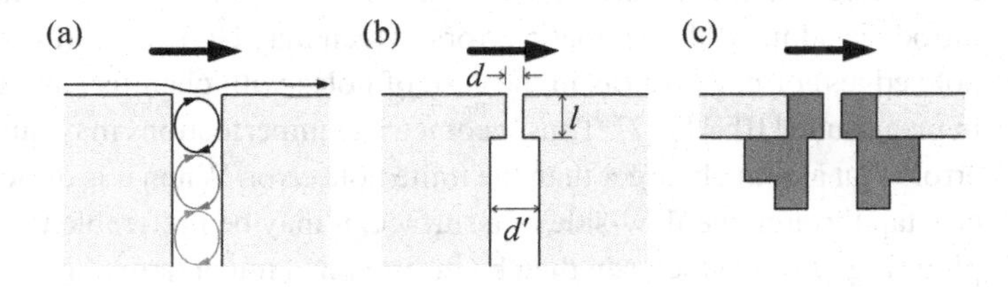

Figure 9.6 Sketches of wall pressure tap configurations.

9.3.2 Tap Pressure Measurement Error

The pressure p_m measured from a tap is subject to a systematic error $\Delta p = p_m - p$, even for the ideal case of a perfect hole that is normal to a smooth wall. This error, which is usually positive, depends on several geometric and dynamic parameters. In order to combine the benefits of a small tap diameter and a fairly fast response time, it is a common approach to connect the tap to a larger-diameter cavity (Fig. 9.6b), which may contain a pressure transducer or a tube leading to remotely located instrumentation. An important length scale characterising the dynamics of near-wall flow is the *viscous length* ν/u_τ, where ν is the kinematic viscosity of the fluid and $u_\tau = \sqrt{\tau_w/\rho}$ is the *friction velocity* (τ_w is the wall shear stress and ρ is the fluid density). Thus, an appropriate dimensionless tap diameter can be defined as $d^+ = d/(\nu/u_\tau)$ (d is the tap diameter), which may be also viewed as the *friction Reynolds number* of the tap. The error in pressure measurement can be normalised either by the local wall shear stress as $\Delta p/\tau_w$ or by the free stream dynamic pressure $q = \frac{1}{2}\rho V^2$ as $\Delta p/q$. Figure 9.7 summarises some of the available error measurements, plotted vs. d^+. Early measurements [47] have demonstrated that the error increases with increasing tap length-to-diameter ratio l/d, but becomes insensitive to this parameter for $l/d \geqslant 1.5$. It may be noted that, in a different set of experiments [27], this error was found to assume negative values for $l/d \leqslant 3$. In order to avoid a variable error, and considering that very large values of l/d would lead to slow response, it seems sensible to maintain l/d in the range 5 to 15. It has further been assessed [17] that, for sufficiently large l/d, the error $\Delta p/\tau_w$ increases with increasing d^+ to an asymptotic value around 3.5 for $d^+ > 2000$. In addition to these effects, however, it has been realised that the free stream Reynolds number also has an effect on the error [10]. Measurements in pressurised pipe flow [31, 32] have demonstrated that the asymptotic value of $\Delta p/\tau_w$ increases with increasing pipe Reynolds number $\mathrm{Re}_D = DU_b/\nu$ (D is the pipe diameter and U_b is the bulk velocity), reaching values near 7 for $d^+ > 2500$ and $\mathrm{Re}_D = 1.4 \times 10^7$ (Fig. 9.7). A power-law-type fit to the envelope of the overlapping regions of these measurements in the range $200 \leq d^+ \leq 2500$ gives

$$\frac{\Delta p}{\tau_w} = 4.623 d^{+0.1655} - 10.2\,. \tag{9.15}$$

The previous error estimates apply to smooth pipes and very precisely machined holes. Even slight imperfections in the shape or orientation of the hole are likely to introduce additional systematic errors, which may be positive (as in the case of rounded hole edges) or negative (as in the case of holes with chamfer) and typically up to $0.01q$ in magnitude [10, 41, 42]. Thus, geometrical imperfections may introduce unpredictable errors that are much larger than the finite hole error. When it is difficult to machine a precise tap through the flow-side of a surface, it may be preferable to do so on a removable plug (Fig. 9.6c), which can then be inserted in a much larger hole through the wall, thus reducing the effects of geometrical distortions. Tap errors become quite unpredictable when dealing with rough walls. Flow distortions due to the roughness elements tend to introduce additional errors, while roughness-induced turbulence tends to mix the flow

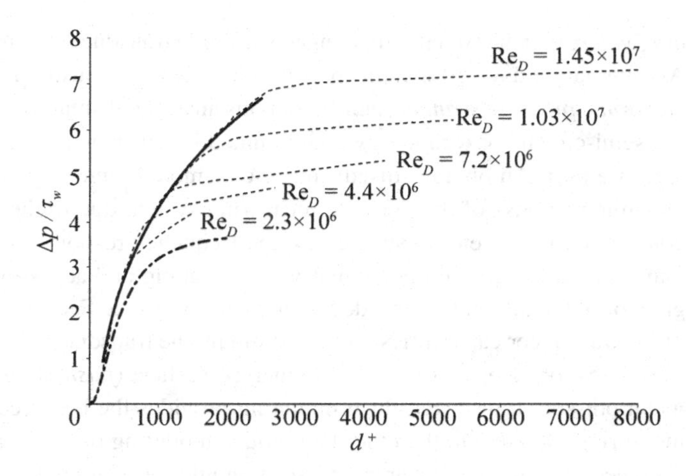

Figure 9.7 Influence of the tap diameter on the pressure measurement normalised error. The dot-dashed line corresponds to measurements on a flat plate at an unspecified Reynolds number [17]. The dashed lines represent measurements in pipe flows at the specified pipe Reynolds numbers [32] and the the solid line (Eq. (9.15)) is a fit to the overlapping ranges of these measurements.

near the tap, possibly reducing the error. When wall roughness is considerable, wall tap measurements may be totally unreliable.

Additional errors are due to compressibility [10, 31, 40]. In the case of transonic and supersonic flows, wall pressure measurement might be exceedingly complicated by the presence of unsteady shock waves and shock–boundary layer interactions. For subsonic flows, the error due to finite hole size tends to increase with increasing Mach number M. For example, the asymptotic ($d^+ \gg 1$) error $\Delta p/q$ increases by 13% for M = 0.4 and 45% for M = 0.8, with larger increases for smaller hole diameters [10, 41, 42].

Several authors have addressed the issue of turbulence effects on wall pressure measurement, without reaching a satisfactory conclusion. It appears that turbulence tends to decrease the finite hole error and so, in the absence of a better approach, turbulence effects may be disregarded [10, 11].

Alternative approaches are needed in experimental settings in which it is not possible or convenient to install pressure taps. An example is the surface-mounted disk probes, which, with proper corrections applied, may provide reasonably accurate wall static pressure measurements, while mountable externally on immersed surfaces [29].

9.3.3 Pressure Transducer Connection

Three common configurations of pressure transducers used in the measurement of wall pressure are illustrated schematically in Fig. 9.8. *Remote mounting* (Fig. 9.8a) through flexible or metallic tubing is the simplest way to connect any pressure transducer or manometer, irrespectively of size or operation principle. This approach can be applied to *multi-port measurement* through the use of a manual or automated *pressure scan-*

ning valve, which sequentially connects a single transducer to an array of pressure taps. An alternative device for multi-port measurement, also using remote mounting, is the *electronic pressure scanner*, which contains an array of small transducers, fabricated using semi-conductor technology and mounted together on a block, each with a separate pressure port. An obvious disadvantage of remote mounting is the deterioration of the dynamic response of the pressure-measuring system due to the interference of tubing, connectors, valves etc. In such cases, the frequency response of the measuring system may depend largely on the geometry of the tubing and accessories and to a lesser degree, or not at all, on the transducer characteristics (see Section 9.6). To take advantage of the transducer capabilities, one must mount the transducer close to (*cavity mounting*; Fig. 9.8b), or, if possible, on the immersed surface (*flush mounting*; Fig. 9.8c). For a given pressure transducer, flush mounting provides the best frequency response of the measuring system. On the other hand, flush mounting provides a pressure value that is averaged over the sensor area and so it should be restricted to miniature and MEMS types of transducers, unlike cavity mounting, which may be applied to a wide range of transducers, without strict requirements on the size or quality of the sensing surface. Another advantage of cavity mounting is its generally better spatial resolution, because the sensing area corresponds to the tap cross section, which can be made very small. When considering flush mounting, it is necessary to match closely the shape of transducer tip and the wall surface shape, in order to avoid local pressure distortion. Small transducers with flat tips are mainly of the piezoelectric and piezoresistive types. Such transducers may only be mounted on plane surfaces or on curved surfaced with a radius of curvature that is several orders of magnitude larger than the transducer diameter.

One important consideration in choosing a pressure-measuring configuration is cost. When temporal response requirements are not essential, it would be preferable to use a multi-port connection to a single, general-purpose transducer, or a small number of such transducers. Surface mounting of a large number of transducers may prove to be expensive, cumbersome to implement (need for room under the solid wall and between transducers) and possibly subjected to electronic interference and cross-talk among the transducers.

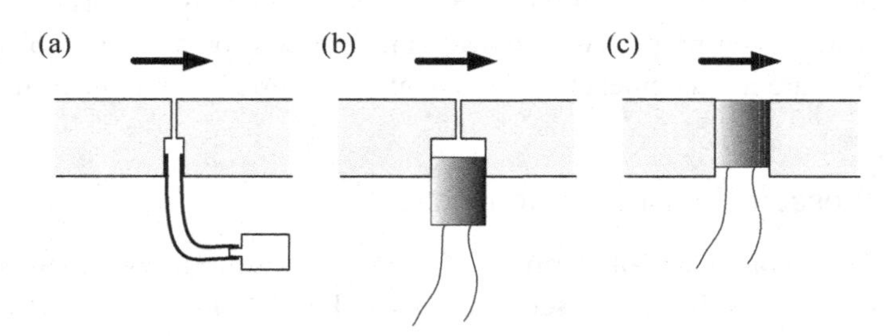

Figure 9.8 Various ways of connecting pressure transducers for the measurement of wall pressure: (a) remote connection; (b) cavity mounting; and (c) flush mounting.

9.4 Pressure Sensitive Paints

Surface coatings of *pressure sensitive paints* (*PSP*) are used for flow visualisation purposes (e.g., to indicate locations of flow separation or shock waves), as well as to map local surface pressure, which may then be integrated to provide forces and moments. This method has found application mostly in high-speed air flows ($M > 0.3$), but has also been used in low-speed air studies [2, 9, 26, 31, 33, 34, 49].

The paint consists of an active material (e.g., a ruthenium compound), the luminescence of which depends, in an inversely proportional fashion, on pressure, and a polymer binder (e.g., a silicone-based polymer). The principle of operation is a photophysical process, which may be summarised as follows. When the painted surface is illuminated by light with a suitable wavelength, photons are absorbed by the photosensitive molecules in the paint, which undergo transition to an unstable state. Some paints are excited by ultraviolet light, which is supplied by mercury-vapour or xenon-arc lamps or an excimer laser, whereas others require blue light, which is supplied by xenon-arc or halogen lamps, an argon-ion laser or a blue light-emitting diode. Some of the excited molecules return to their original state by emitting radiation at a higher wavelength (yellow or red), namely, by fluorescence. Other molecules convert the absorbed energy to vibrational energy, that is, heat, while others yet loose energy by collisions with oxygen molecules, a process called *oxygen quenching*. By *Henry's law*, the oxygen concentration in a material in contact with air is proportional to the partial pressure of oxygen in the air, which is proportional to the surface static pressure. Therefore, the coating at a location on the surface exposed to relatively high pressure will have a relatively high concentration of absorbed oxygen, which would result in more intense oxygen quenching, lower luminescence, and, therefore, a darker appearance, compared to locations of lower pressure. The image of the surface is monitored with a CCD camera, through an optical filter that removes the incident light and only allows the fluorescent radiation to pass.

In order to reconstruct the surface pressure field from PSP images, one needs to perform a calibration, for example, vs. readings from static pressure taps distributed on the surface; in this case, consideration must be given to maintaining the quality of the orifice geometry during PSP application, which is usually performed by spraying [31]. Calibration of a test model can be made in situ in a wind tunnel or a priori in a dedicated calibration chamber.

The luminescence I is related to the local pressure by the *Stern–Volmer equation*

$$\frac{I_{ref}}{I} = A(T) + B(T)\frac{p}{p_{ref}}\,, \tag{9.16}$$

where the subscript 'ref' refers to reference conditions, and the Stern-Volmer coefficients A and B depend on the local temperature. This relationship shows that the use of PSP maps for pressure measurement requires knowledge of the local temperature and that temperature variations would result in an increased pressure uncertainty. PSP calibration may be done by enclosing the model, or a sample coated with the same paint, in a sealed chamber, whose pressure and temperature can be controlled. Calibration for varying pressure must be repeated for different values of constant temperature, so that the coefficients $A(T)$ and $B(T)$ can be established.

The accuracy of this technique also depends on the exact knowledge of the intensity of incident light I_{ref}. In addition to difficulties in illuminating complex surfaces uniformly, light variations may be caused by a fluctuating power of the light source or deformation of the model when exposed to different air speeds. Correction methods for some of these effects have been devised. An alternative technique, which is insensitive to incident light variations, is the *lifetime mode* method. The fluorescent light intensity emitted by a surface decays exponentially with time as

$$I = I_o e^{-kt} , \tag{9.17}$$

where I_o is the initial intensity and the decay rate (or inverse time constant) is

$$k = k_R + k_{nR} + k_Q , \tag{9.18}$$

with k_R, k_{nR} and k_Q representing the decay rates due to radiation, non-radiative energy transfer and quenching. This technique consists of using two light detectors to integrate the fluorescent light intensity over two contiguous time intervals, both with a duration t_1, and then compute the ratio

$$\frac{\int_{t_1}^{2t_1} I\mathrm{d}t}{\int_0^{t_1} I\mathrm{d}t} = e^{-kt_1} , \tag{9.19}$$

which is sensitive to pressure and temperature, but not to the illumination intensity. The radiative decay rate k_R is relatively insensitive to temperature, while the non-radiative decay rate k_{nR} depends on the absolute temperature T as

$$k_{nR} \sim e^{-\frac{\Delta E}{RT}} , \tag{9.20}$$

where ΔE is the activation energy of the process and R is the molecular constant. Materials with relatively high ΔE would have a relatively low temperature sensitivity, which would make them suitable as PSP. Nevertheless, the dependence of the quenching rate k_Q on temperature as well as binder properties further complicates this method.

PSP luminophores and binders are quite complex, and, although possible to assemble from generic products, they would be best acquired in ready-to-use form from specialised suppliers.

9.5 In-Flow Pressure Measurement

The accurate measurement of the instantaneous, local, static pressure in a flowing fluid is extremely difficult, as there is no non-intrusive method for measuring static pressure directly. In-flow static pressure measurements using intrusive instruments, such as pressure tubes, may be sufficiently accurate in flows having a pressure that is nearly uniform in space and nearly constant in time. However, because such instruments have poor spatial and temporal resolutions, as well as distorting the pressure field, they cannot be used for measuring time-dependent local static pressure. Indirect methods for measuring in-flow, fluctuating, static pressure have been described by previous authors [53]. According to these methods, static pressure is determined as the difference between the total pressure, measured with a fast-response total-pressure probe, and the dynamic pressure, determined from time-resolved measurements of flow velocity slightly upstream of the total pressure probe tip and obtained by hot-wires or laser-Doppler velocimetry. More recently, non-intrusive techniques have been developed for determining the local static pressure from time-resolved velocity-field data, acquired using particle image velocimetry (Section 12.6).

9.5.1 Pressure Tubes

The in-stream pressure can be sensed with the use of thin tubes of various designs, collectively known as *pressure tubes* [6, 10, 12, 13, 37, 51]. Different types of pressure tubes are also used for measuring local flow velocity and these will be discussed in Chapter 12. Depending on their function, pressure tubes can be classified in the following categories:

- *Static-pressure tubes*, which sense the local static pressure.
- *Pitot tubes*, also known as *impact tubes* and *total pressure tubes*, which sense the local total pressure.
- *Pitot-static tubes*, which are combinations of the previous two types and provide the local dynamic pressure $q = \frac{1}{2}\rho V^2$, from which one can easily calculate the local flow velocity.
- *Multi-hole probes*, which consist of combinations of several tubes arranged such that they measure the local static and total pressures as well as the local velocity in both magnitude and direction.

Pressure tubes have relatively low spatial and temporal resolutions. Fast-response probes have been designed by embedding miniature and micro-pressure transducers within the probe body, in the vicinity of the pressure-sensing orifices [1, 25, 48].

Static-pressure tubes: Static-pressure tubes [6, 10, 12, 13, 31, 37] are thin hollow tubes, which are sealed at the tip facing the flow, while having holes or slits on their side. When a long static-pressure tube is inserted in uniform flow with its axis aligned with the flow direction, the pressure inside it, sensed through side holes, would be somewhat lower than the free-stream pressure, because the flow accelerates around the tube nose. Instead of straight tubes, however, a more practical configuration for measurements in pipes, ducts, wind tunnels etc. is the bent tube, in which the tip forms a right angle with the stem (Fig. 9.9a). For right-angle tubes, flow deceleration near the stem blockage would cause an increase in static pressure along the tube, which compensates, partially or totally, for the nose effect. This makes it possible, at least in principle, to design static tubes free of flow acceleration error by proper positioning of the side holes with respect to the nose and the stem. An optimal bent-tube design has been suggested to have a nose length that is equal to 14 tube diameters and holes that are located six diameters downstream of the nose [37]. The response of static pressure tubes depends on the shape of the nose, the number, size and location of holes and the location of the stem, but tube design is also dictated by ease and precision of construction and consideration of its sensitivity to misalignment with the flow direction, possible damage to the tube or the experimental apparatus during usage and possible hole blockage due to the presence of impurities in the stream.

A variety of static-pressure tube designs have been proposed and tested. Common designs available commercially have hemispherical or ellipsoidal noses and six or eight evenly spaced holes, typically located at the previously mentioned optimal position in the case of right-angle tubes. Some caution is required, as different suppliers may use the designation 'standard' to describe different designs. Moreover, it cannot be safely assumed that the nose and stem effects are perfectly balanced. Typically, even well-aligned, right-angle static pressure tubes could have an error (usually negative) in static pressure reading of up to $1\%q$. For higher accuracy, it would be advisable to calibrate each tube separately vs. the reading of a well-constructed static pressure tap. Static-pressure tubes are much more sensitive to flow direction than total pressure tubes are. As a result, even slight misalignment may introduce a significant error to the static pressure measurement. For example, the static pressure reading would typically drop by $1\%q$ for a yaw (i.e., rotation about the stem axis) angle of $5°$, $3\%q$ for $10°$ and $5\%q$ for $15°$. Therefore, a sizeable error might occur when exact alignment of the tube cannot be achieved or when the flow direction varies, as in the case of streamline divergence due to the presence of a model in a wind tunnel. In such cases, an on-site evaluation of the static tube response would be advisable. For wind tunnels and smooth pipes, one might also consider using wall pressure readings, rather than in-flow pressure measurements.

Other factors affecting the response of static tubes are flow turbulence, internal motions (see Section 9.3.1), cavitation, viscous effects and compressibility effects. Because of its dependence on tube geometry and operating conditions, it would be difficult to predict accurately the error of a specific tube when such effects are coupled. The literature on such topics (e.g., [13]) is contradictory and incomplete. For example, a reading of the few publications that have addressed the effects of turbulence would conclude that static pressure errors would be of the order of $1\%q$ (usually positive) or lower for turbulence intensities lower than 10%. However, correction methods for turbulence effects are not of general validity, as the error would depend not only on the turbulence intensity, but also on the length scale of turbulence, the turbulence structure and the presence of organised motions (coherent structures). For turbulence intensities greater than 15%, the errors are likely to be significant. Vibrations of the probe due to turbulence or vortex shedding from the stem would also tend to produce readings that are higher than the true static pressure.

Other static-pressure probes: Besides cylindrical tubes, in-flow static pressure can be measured or estimated using a variety of objects with different shapes. A useful type of devices for industrial flows containing significant amounts of suspended solids is the *disk static probes*, available in sharp-edge (Fig. 9.9b) and rounded-edge variations. Their main advantage over static tubes is that they can be produced with relatively large orifice openings (reportedly up to 50 mm), which would not be easily blocked by impurities. Such devices are sensitive to flow direction and may have systematic errors

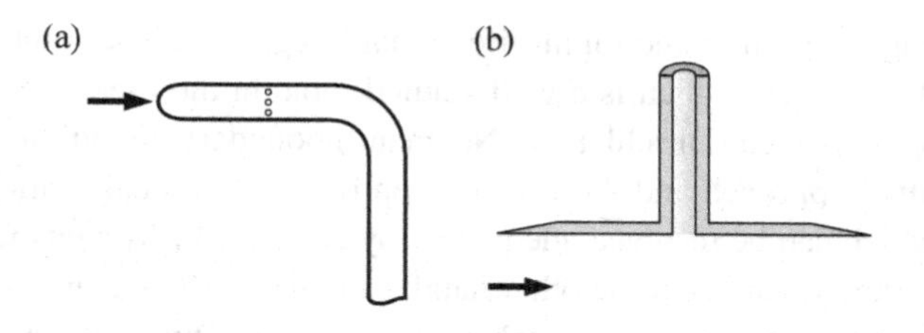

Figure 9.9 Sketches of (a) a static-pressure tube and (b) a sharp-edge static tube.

even when aligned, so some testing of their response is advisable before use. Finally, static pressure values can also be recovered from Pitot-static tubes and multi-hole probes of various shapes, which will be discussed in Section 12.1.

9.5.2 Pressure from the Velocity Field

The instantaneous local, in-flow, static pressure can be determined indirectly from instantaneous local velocity field measurements in incompressible flows, provided that analysis of these measurements can resolve the temporal and spatial derivatives of all velocity components. This method first computes the pressure gradient from the momentum (Navier–Stokes) equation as

$$\nabla p = -\rho \frac{\mathrm{D}\mathbf{u}}{\mathrm{D}t} + \mu \nabla^2 \mathbf{u} \,. \tag{9.21}$$

Integration of Eq. (9.21) can, in principle, provide the instantaneous pressure field. If the velocity field is also time-resolved, one may also use this approach to determine the time history of the pressure field.

A variation of this approach can provide the instantaneous pressure field, albeit not necessarily its time history, from spatially resolved velocity maps. Taking the divergence of both sides of Eq. (9.21) and using the incompressibility condition ($\nabla \cdot \mathbf{u} = 0$), one obtains the Poisson equation

$$\nabla^2 p = -\rho \nabla \cdot (\mathbf{u} \cdot \nabla)\mathbf{u} \,, \tag{9.22}$$

which can, in principle, be solved for the pressure, if all necessary spatial derivatives of the velocity are known.

Recent advances in particle image velocimetry (PIV), whose operating principle is discussed in Section 12.6, have enabled the measurement of velocity maps with sufficient spatial, and even temporal, resolutions to permit the solution of Eq. (9.21) or Eq. (9.22). The two approaches have different experimental requirements, follow different numerical approaches and introduce different errors. A detailed discussion of these methods can be found in the topical review by van Oudheusden [52]. A few pertinent points of this discussion will be summarised in the following.

When both temporal and spatial data are available for the velocity field, the use of Eq. (9.21) would, in principle, be preferable to that of Eq. (9.22), because it requires a single spatial integration for the pressure, rather than double integration, which is likely prone to larger error. For the solution of Eq. (9.21), two techniques have been used, space-marching integration and omni-directional integration. The solution of the Poisson equation (Eq. (9.22)), which is a well studied topic in the literature, typically requires specification of mixed (Dirichlet and Neumann) boundary conditions.

Both the momentum approach and the Poisson approach rely on calculations of the velocity gradient, which can be resolved adequately by current PIV systems for a large range of laboratory experiments. On the other hand, calculating the material derivative $\mathbf{Du}/Dt$ for the momentum equation approach requires at least two velocity fields to be taken at two closely spaced times. If one were to compute the material derivative directly using a finite difference method, then the temporal resolution of the experiment would play a pivotal role in its uncertainty. From practical and economical standpoints, determining the material derivative would be comparatively easier for experiments in flows with a relatively small convective speed, which include typical liquid flows in the laboratory and for which the required time difference between successive images is within reach of less expensive PIV equipment. An alternative approach for calculating the material derivative is from the tracked paths of particles contained in the flow. This approach also requires a sufficient temporal resolution of the measurements.

Current commercial PIV software allow the user to determine the pressure field from their PIV data, under the condition that these data are sufficiently accurate. This technique has typically been applied to unsteady flows and flows in which forces and moments acting on immersed bodies can be related to the pressure field. Although this method can, in principle, be applied to 3D fields, in practice, most applications are confined to flows which are predominantly 2D, primarily due to the relative ease in collecting such data experimentally and the comparatively simple numerical calculations. At the time of writing, effort of researchers and PIV system developers has focused on improving the accuracy and lowering the cost of time-resolved, volumetric PIV (tomographic PIV), as well as enhancing the spatio-temporal resolution of PIV data through the use of machine-learning algorithms. One may be hopeful that our capability to determine accurately instantaneous pressure fields will, in the near future, be extended to flows under more challenging conditions, notably turbulent flows at larger Reynolds numbers than presently possible.

9.6 Dynamic Response of Pressure-Measuring Systems

The dynamic response of a pressure-measuring system would generally be non-linear and difficult to analyse accurately. For instructional purposes, we will discuss a few idealised cases, which have been modelled as low-order linear dynamic systems. We will specifically consider a remotely mounted pressure transducer, such as the one shown in Fig. 9.8a, which is connected to a small pressure tap through a tube. The transducer is assumed to contain a deformable elastic membrane or diaphragm, mounted within an internal pressure chamber; the discussion also applies to transducers which are in contact with the fluid in an external cavity, like the one shown in Fig. 9.8b.

Effects that may influence the pressure-measuring system response include the following:

- **Transducer compliance:** Let V_{tr} be the volume of the pressure chamber or external cavity of the transducer. As the pressure in the chamber changes, the diaphragm will deform, causing a change in the fluid-filled volume. This is expressed in terms of the *compliance* of the transducer, defined as the change of volume per unit change of applied pressure, namely, as

$$C_{tr} = \mathrm{d}V_{tr}/\mathrm{d}p\,. \tag{9.23}$$

  The transducer compliance is sometimes provided by the transducer manufacturer and may also be estimated theoretically or experimentally. Assuming that the transducer can be exposed to water, we can determine experimentally its compliance by connecting it to a small sealed vessel filled entirely with water and then pressurising the system using a syringe, which also allows the measurement of volume change.
- **Tube compliance:** Volume changes due to expansion or contraction of the tube and the cavity walls may usually be neglected, with some exceptions, notably cases with long, easily deformable tubing subject to relatively large pressure changes. Assume that the tube has a diameter d, a length $l \gg d$, a wall thickness $t \ll d$, a Young's modulus of elasticity E and a Poisson's ratio ν. For a rough estimate of the *tube compliance*, we may use the following expression, which applies to elastic deformation of infinitely long, thin-wall, cylindrical vessels:

$$C_{tube} = \frac{\pi d^3 l(1-\nu)}{2Et}\,. \tag{9.24}$$

  C_{tube} should be included in the analysis, if found to be significant by comparison to C_{tr}. Flows in vessels with compliant walls are subject to fluid–structure interactions and their analysis would be quite involved, as well known from studies of pulsatile blood flow in arteries.
- **Frictional forces:** Friction impedes the motion of the fluid within any vessel, as well as the motion of solid components in contact with each other. Frictional forces are typically modelled as proportional to the relative velocity, or, in other words, the first derivative of displacement.
- **Acceleration:** The unsteady changes of momentum ('inertia forces') of the fluid and the diaphragm may or may not be significant, depending on the system design and the operating conditions.
- **Restoring forces:** A main external force to the fluid system is the restoring force of the diaphragm, following its elastic deformation. Restoring forces due to tube and cavity wall elasticity may usually be neglected.
- **Fluid compressibility:** The inclusion of fluid compressibility in the analysis would introduce several complications. First of all, a change of fluid density in the cavity and the tubing would create a flow into or out of the system, which would affect the frictional forces. This effect may be significant for pressure measurement in gases, but would be negligible in liquids. A different effect, which applies to both gases and liquids, is that a pressure change would propagate in the system with a finite speed (approximately equal to the speed of sound), which may cause a measurable delay in the pressure transmission through long tubing.

Depending on the transducer–tube configuration and the flow conditions, some or all of the previously mentioned factors would have to be considered [4, 14]. A few typical cases of dynamic analysis of transducer–tube combinations [14] are presented in the following.

First-order model for incompressible fluids: For the simplest dynamic model of such systems, one may assume that the fluid is incompressible. This assumption applies well to liquids, whereas gases may be also considered to be incompressible when pressure changes are very small. A second assumption that can be made is that inertia effects are negligible, which is satisfied when pressure changes take place slowly. Finally, one may assume that the instantaneous pressure difference across the connecting tube is related to the bulk velocity in the tube by the familiar relationship for steady, laminar, fully developed flow in tubes. Under these assumptions, the pressure p_m measured by the transducer is related to the pressure p at the inlet of the tube by a first-order differential equation with a time constant

$$\tau = \frac{128\mu l C}{\pi d^4} . \tag{9.25}$$

This expression clearly shows that the time constant is extremely sensitive to the tube diameter and that the use of small-diameter tubes would result in a very slow response of the pressure-measuring system. It also shows that the tube length should be kept as short as possible.

Example 9.3 Consider a pressure transducer connected through a tube to a tap in a pressure vessel containing water at 20°C. Estimate roughly the settling time of the pressure-measuring system as a function of the tube length and diameter. Assume that the transducer compliance was found to be $C \approx 10^{-10}$ m^3/Pa. Discuss the validity of your estimate.

Answer

We will assume that the pressure tap is sufficiently large not to affect the dynamic response of the pressure-measuring system and that the compliance of the tube is negligible. The viscosity of water at 20°C is approximately 10^{-3} Pa s. For specified values of l, d, we can calculate the time constant τ from Eq. (9.25) and then calculate the settling time as $T_s = 5\tau$. Representative values of T_s have been plotted in Fig. 9.10 vs. l for a few different values of d. Values of $T_s > 1$ s were computed only for tubes with $d < 4$ mm, whereas the most extreme case shown, the one for $d = 1$ mm and $l = 10$ m, had $T_s \approx 2$ min, which is too long for many practical purposes. Although the first-order model is based on several rough approximations, it may be usable for the purpose of checking whether the response of a transducer–tube system would be within acceptable limits. The assumption that the tube compliance is negligible is not a crucial one, as tube compliance would be larger for larger tubes, in which pressure losses would be smaller and so the two effects would partially compensate for each other.

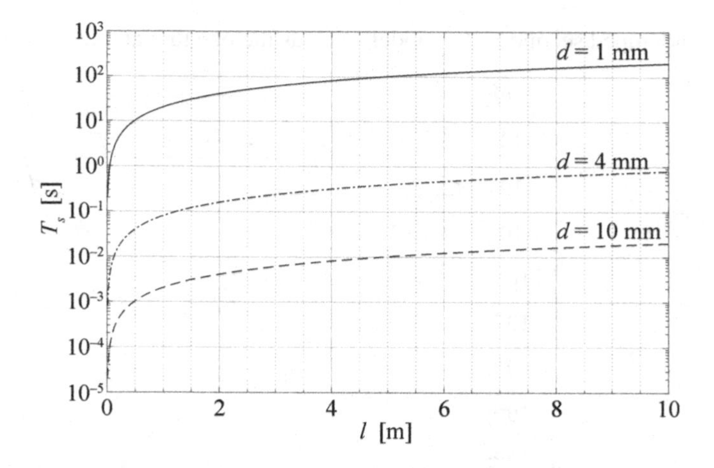

Figure 9.10 Settling time of a transducer–tube pressure-measuring system, following a first-order model.

Second-order model for incompressible fluids: An extension of the previous analysis to include inertia effects of the fluid motion, albeit also based on significant simplifications, results in a second-order system response with a natural frequency and a damping ratio, which are, respectively, equal to

$$\omega_n = \frac{0.767d}{\sqrt{\rho l C}} \quad \text{and} \quad \zeta = \frac{15.6\mu}{d^3}\sqrt{\frac{lC}{\rho}}. \tag{9.26}$$

Example 9.4 Consider the transducer–tube measuring system that was discussed in Example 9.3, but this time include fluid inertia effects, which lead to a second-order model. Determine the settling time of the pressure-measuring system as a function of the tube length l and diameter d, when measurements are conducted in water and in air. Compare these values with the ones obtained from the first-order model.

Answer

Substituting the values of the different properties in Eq. (9.26), we find that this pressure-measuring system with a $d = 1$ mm tube is heavily overdamped for $l \gtrsim 40$ mm, whilst the $d = 4$ and 10 mm tubes are underdamped. Consequently, we estimated T_s using Eq. (2.64) for the former case and Eq. (2.66) for the latter two cases. The values of T_s have been plotted vs. l in Fig. 9.11, on the same axes as in Fig. 9.10. It is interesting to note that the values of T_s from the first- and second-order models were very close to each other for the overdamped case, which means that a first-order model would be acceptable for such cases. In contrast, for the underdamped systems, values of T_s from the second-order model were orders of magnitude larger than those from the first-order model, and nearly constant for all tube lengths. Viewed qualitatively, these results are not surprising, because, in general, a first-order model behaves in a way that is similar to that of a heavily overdamped second-order model, but bares no similarity to a nearly undamped second-order model, such as the one that applies to the present two cases.

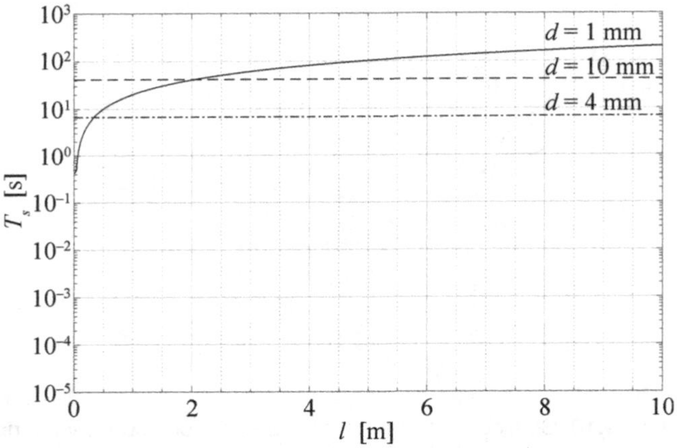

Figure 9.11 Settling time of a transducer–tube pressure-measuring system, following a second-order model.

A model for compressible fluids: The dynamic response of transducer–tube systems used in air and other gases could be affected by compressibility, which is a process that is analogous to the elastic deformation of a diaphragm. Accounting for gas compressibility effects introduces non-linearity, but the non-linear model can be linearised by assuming that pressure changes are small by comparison to the equilibrium value. Depending on the transducer connection type, the total volume occupied by the gas would be the sum of the volume V_{tr} of the transducer chamber (Fig. 9.8a) or the cavity (Fig. 9.8b) and the internal volume V_{tube} of the connecting tubing, which for cylindrical tubing would be $V_{tube} = \pi d^2 l/4$. This results in a second-order model [23] with a natural frequency and a damping ratio, which are, respectively, given by

$$\omega_n = \frac{c/l}{\sqrt{\frac{1}{2} + \frac{V_{tr}}{V_{tube}}}} \quad \text{and} \quad \zeta = \frac{16\nu l}{d^2 c}\sqrt{\frac{1}{2} + \frac{V_{tr}}{V_{tube}}}, \tag{9.27}$$

where $c = \sqrt{\gamma p/\rho}$ is the speed of sound in the gas at the reference pressure (γ is the specific heat ratio of the gas).

9.7 Dynamic Calibration of Pressure Transducers

In the majority of engineering applications, the dynamic responses of pressure transducers and pressure-measuring systems can be approximated by a second-order model, in which case it would be sufficient to determine the natural frequency and damping ratio of the system. These can be estimated most conveniently by monitoring the step response of the system exposed to a sudden change in pressure, as discussed in Section 2.4.8. A useful device for the generation of step-like changes in pressure is the *shock tube*, namely a tube containing a metallic or plastic diaphragm, which separates gas under sufficiently high pressure from low-pressure gas. At a certain time, the diaphragm is punctured and a shock wave propagates through the low-pressure side until it reaches the end of the tube where the pressure system under test is mounted. This method is good for testing systems with natural frequencies as high as 250 kHz, or even higher. Simpler pressure transducer testing devices, also consisting of two chambers containing gas under different pressures and separated by a diaphragm or a fast-opening valve are available commercially, and may also be fabricated relatively easily. Compared to shock tubes, such devices are suitable for much lower dynamic ranges, not exceeding a few kHz. For liquid systems, one may also utilise the water hammer pressure wave generated in a pipe by a fast-closing valve.

When the type of response of the pressure-measuring system is unknown, it is advisable to determine its frequency response rather than its step response (see also Section 2.4.8). To do so, it is necessary to generate a sinusoidal pressure variation with adjustable frequency and, if possible, amplitude. This can be achieved, within the frequency range of up to approximately 10 kHz, in a sealed chamber filled with a liquid or gas, by driving a diaphragm through a piston/connecting rod mechanism, a loudspeaker or an electrodynamic shaker. Another arrangement is to release the chamber pressure through an orifice, which is periodically aligned with each of multiple holes on a rotating disk. One may also use a variety of arrangements by which either the transducer is oscillated in front of a steady fluid stream or the stream is periodically interrupted. Each of these methods is subject to interference and must be applied with caution. In general, it is highly recommended to evaluate the performance of such apparatus by comparison to a reliable standard, normally another pressure transducer, which has a much higher accuracy and a much higher frequency response than the transducer under test.

9.8 The Accurate Measurement of Pressure Fluctuations

This section discusses conditions under which a pressure-measuring system (e.g., a transducer–tube combination) measures accurately pressure fluctuations within a given frequency bandwidth of interest. The same approach also applies to systems measuring other types of measurands. Let us consider a time dependent pressure field having fluctuations within the bandwidth from ω_l to ω_h ($\omega_h > \omega_l$). The lower boundary ω_l could be 0, for cases for which the mean pressure is to be measured as well, or some very low frequency, for cases for which only fluctuations are of interest. To measure accurately pressure fluctuations within this bandwidth, the pressure-measuring system must have a normalised amplitude ratio that is $|B/KA| \approx 1$. How close to 1 this value should be depends on the required measurement accuracy. To illustrate the concept, let us set the acceptable error to $\pm 5\%$, but other values (e.g., $\pm 2.5\%$ or $\pm 1\%$) may be more appropriate for some cases.

As discussed previously, a measuring system would typically have a cut-off frequency

ω_c, such that $|B/KA| < \sqrt{2}/2$, when $\omega > \omega_c$. It is evident that one must select a system with an ω_c that is substantially larger than ω_h, so that pressure fluctuations within the bandwidth of interest will not be attenuated by more than 5%. Moreover, one needs to ensure that the ±5% error bound will be observed for all $\omega < \omega_c$. For clarity, we will examine specifically the response of first- and second-order pressure-measuring systems, but the analysis can be extended to higher-order systems.

Let us assume that we have a first-order pressure-measuring system. The dynamic response of this system depends only on the time constant τ, which is assumed to have been determined by analytical means or by calibration. Because $|B/KA|$ for first-order systems decreases monotonically with increasing ω and is approaching 1 at very low frequencies, it is sufficient to ensure that $|B/KA| \geq 0.95$, when $\omega = \omega_c$. This will be the case, when $\tau \leq 0.329\omega_h$. In summary, a first-order pressure-measuring system with a time constant that is smaller than $0.329\omega_h$ will measure fluctuations with frequency in the range 0–ω_h with an amplitude error that is less than 5%.

Now, let us consider a second-order pressure-measuring system with a natural frequency ω_n and a damping ratio ζ. As was discussed in Chapter 2, the amplitude ratio of this system depends on both ω_n and ζ, and so one cannot use ω_n alone in any dynamic analysis. To observe the 5% error condition at $\omega = \omega_h$, we need to have (see Eq. (2.69))

$$\frac{B}{KA} = \frac{1}{\sqrt{[1-(\omega_h/\omega_n)^2]^2 + 4\zeta^2(\omega_h/\omega_n)^2}} \geq 0.95\,. \tag{9.28}$$

Equation (9.28) is a necessary condition for the pair of values ω_n and ζ. Unlike first-order systems, however, second-order systems with $0 < \zeta < \sqrt{2}/2$ will have a range of frequency values with $|B/KA| > 1$. One may also impose the 5% error bound on the amplitude overshoot, but, to be on the safe side, it is better to ensure that $\zeta \geq \sqrt{2}/2$, so that the amplitude ratio decreases monotonically with increasing frequency.

Chapter Digest

9.1 How is pressure related to stress?

9.2 When is pressure considered to be positive? When is normal stress considered to be positive?

9.3 What is the difference between static pressure and hydrostatic pressure?

9.4 How is mechanical pressure related to thermodynamic pressure?

9.5 What is gauge pressure and how is it related to the barometric pressure?

9.6 What is total pressure? How is it computed in incompressible and in compressible flows?

9.7 What is dynamic pressure?

9.8 Why is an inclined manometer more accurate than a vertical manometer?

9.9 Why is capillarity essential for trees?

9.10 Does capillarity always produce a higher-than-true reading of a manometer? If not, provide a counterexample and search for an explanation.

9.11 Under which conditions would a mercury manometer be preferable to an alcohol manometer?

9.12 Why do we not see any deadweight gauges in academic fluid mechanics laboratories?

9.13 When would a mechanical pressure transducer be preferable to an electrical one?

9.14 Why can we not use a liquid manometer to measure extremely high or extremely low pressures?

9.15 Which possible sources of error should one consider when installing pressure taps in pipes and which means should one use to keep such errors as low as possible?

9.16 When would be a pressure tap diameter be unacceptably small and unacceptably large?

9.17 Which possible sources of error should one consider when measuring pressure in pipe flow with the use of a bulky pressure transducer?

9.18 When would you consider using a pressure sensitive paint?

9.19 Which considerations should one make when selecting a flush-mounted pressure transducer for measuring wall pressure fluctuations in a turbulent boundary layer?

9.20 Why is it impossible to measure in-flow, turbulent, static pressure fluctuations with a probe, even if it has an extremely high frequency response?

9.21 Describe a non-intrusive in-flow pressure-measuring method. Is this a direct or indirect method?

9.22 What is the difference between a static-pressure tube and a Pitot tube? Can they be combined?

9.23 Can you use a static-pressure tube to measure pressure in flowing sewage? Justify your answer.

9.24 A pressure transducer which has been calibrated statically is used to measure pressure fluctuations in a turbulent flow. Discuss briefly the concerns that you have and any means to alleviate such concerns.

9.25 When can you use a first-order model for the response of a pressure-measuring system?

9.26 What is a shock tube and how can it be used in the fluid mechanics laboratory?

9.27 Consider that you measure wall pressure fluctuations in a turbulent pipe flow of air by connecting a pressure tap to an external fast-response pressure transducer with plastic tubing. Explain qualitatively how the accuracy of measurement will be affected, as the tubing length gets increased to large values.

Problems

9.1 A U-tube manometer filled with mercury has 1 mm markings. Determine the resolution of the pressure reading. How does the resolution change, as you incline the manometer to $\varphi = 10$, 20, and 45°? If, instead of mercury, water were used as the working fluid in the manometer, how would the resolution change?

9.2 Consider a U-tube manometer, which has two tubes with the same length, but different diameters, d_1 and $d_2 = 2d_1$. Express the pressure difference as a function of the free-surface elevation difference, assuming that both tubes are connected to the same gas and a liquid is the working fluid. Is this device more, less or equally sensitive than a U-tube manometer having tubes with the same diameter d_1?

9.3 Consider a U-tube manometer containing mercury and having scales marked every 1 mm. Assume that the uncertainty of the mercury density is ± 10 kgm^{-3} and the uncertainty in the manometer tilt angle is $\pm 1°$. The manometer reads a pressure difference of 1.0 kPa, while it is nominally vertical. Determine the absolute measurement uncertainty of this pressure. The same manometer is tilted by 45°, while connected to the same pressure difference. Determine the pressure uncertainty. Comment on your results.

9.4 The two tubes of the U-tube manometer shown in Fig. 9.1 are, respectively, connected to a pressure of 3 kPa at point A and a pressure of 1 kPa at point B. Assuming that fluids 1 and 3 are air with a density of 1.225 kgm^{-3} and fluid 2 is water with a density of 1,000 kgm^{-3}, determine the error introduced by using Eq. (9.8) to obtain the pressure difference instead of Eq. (9.7). You may assume that points A and B are at the same elevation, and that there are no capillarity errors. Comment on how large the error is and, hence, on the suitability of Eq. (9.8).

9.5 Consider the vertical manometer shown in Fig. 9.1. The performance of this manometer is dictated by the design parameters ρ_1, ρ_2, ρ_3, and the difference between the elevations of points A and B, that is, $\Delta z = z_A - z_B$. For most manometers, $\rho_1 = \rho_3$.

(a) Determine the bias error in the measurement of pressure difference that is based on Eq. (9.8).

(b) Suggest ways to reduce this measurement error by a judicious selection of the design parameters of the manometer.

(c) Show that the measurement error vanishes, when $z_A - z_B = -\Delta p_t/(g\rho_2)$, where Δp_t is the pressure difference obtained using Eq. (9.7).

(d) Derive an expression for the relative measurement error and show that it decreases as the pressure difference increases.

(e) If both openings of the manometer are at the same elevation, that is, $z_A = z_B$, estimate the relative measurement error of the manometer.

9.6 Consider a vertical U-tube manometer, as shown in Fig. 9.1. The manometer is made of a glass tube with an inner diameter $d = 1$ mm, fluid 1 is air, fluid 2 is mercury and fluid 3 is water, all at 20°C. Point A is open to a standard atmosphere and the elevations of different points are given as $z_A = z_B = 988$ mm, $z_C = 743$ mm and $z_D = 345$ mm.

(a) Neglecting capillarity errors, determine the absolute pressure $p_{B,unc}$ at point B.

(b) Determine the pressure $p_{B,cor}$ that is corrected for capillarity effects and the measurement error due to capillarity. How would you reduce the capillarity error?

9.7 Determine and plot the capillary rise or fall of manometric liquid in a glass tube with inner diameters in the range between 1 and 20 mm, for the liquid combinations contained in Table 9.1. Discuss your observations.

9.8 Consider a vertical U-tube manometer, as shown in Fig. 9.1, consisting of a glass tube with an inner diameter of 0.3 mm and connected to air on both sides. The manometric fluid occupies a length of 950 mm. Estimate the undamped natural frequency, damping ratio and damped natural frequency of this manometer, if the manometric fluid is (i) water, (ii) mercury and (iii) glycerol, all at 20°C. If a pressure difference of 100 Pa is suddenly applied on the manometer, plot the elevation difference of the manometer free surfaces vs. time. Which fluid would you recommend and why?

9.9 Consider fully developed, turbulent, air flow at 20°C, in a smooth pipe with an inner diameter $D = 500$ mm. Wall static pressure is measured using two pressure taps, an upstream one with a diameter $d_1 = 2$ mm and a second one with a diameter $d_3 = 4$ mm, located 1,000 mm downstream of the first one. The static pressure at the first tap is approximately 106.3 kPa, while the pressure difference between the two taps is 95 Pa.

(a) Estimate the wall shear stress assuming that there is no tap-size error.

(b) Estimate the bulk velocity of the flow in the pipe, assuming that there is no tap-size error. (Hint: Find the velocity by iteration using the Moody diagram to determine the friction coefficient.)

(c) Estimate corrections for the two tap readings.

(d) If you find these corrections to be significant, correct the estimated wall shear stress and bulk velocity; iterate until you are satisfied with the accuracy of your results. Determine the measurement error in bulk velocity and comment on how significant this is.

9.10 We wish to measure the wall pressure spectrum in a boundary layer, at least for frequencies up to 1 kHz. The free stream velocity is 10 m/s. The lab is equipped with a high-speed video camera, but no pressure transducers. Thus, you consider using the camera to record the change in elevation of the liquid free surface of a specially designed U-tube manometer and extracting the pressure time series from the images. Explain why you should not use Eq. (9.25) and Eq. (9.26) in the analysis of the measurements. Explain why you should design the manometer, such that its damping coefficient would be $\zeta = 1/\sqrt{2}$. If one opening of the U-tube manometer is mounted flush with the wall tap where the pressure is to be read, and the other opening is open to the atmosphere, determine the required length of the liquid-filled part of the tube, if the manometric fluid were water ($\nu = 1.00 \times 10^{-6}$ m^{2}s^{-1}), mercury ($\nu = 0.118 \times 10^{-6}$ m^{2}s^{-1}) or castor oil ($\nu = 292 \times 10^{-6}$ m^{2}s^{-1}). Approximately how long should the entire U-tube be, when containing each of these three liquids?

9.11 To revisit the previous problem, we consider buying a pressure transducer for the experiment. Design a set-up using compliant pressure devices (e.g., tubes and transducers) that would produce a bandwidth of 1 kHz with $\zeta = 1/\sqrt{2}$.

9.12 We wish to measure the fluctuating pressure due to vortex shedding from a hollow circular cylinder in a wind tunnel at a free stream velocity $U_\infty = 20$ m/s and temperature 20°C. A pressure tap with a diameter $d = 1$ mm is drilled into the base (downstream end) of the cylinder, which has an outer diameter $D = 50$ mm and a wall thickness of 5 mm. The cylinder can be split in half to allow an internal connection of a flexible tube with an inner diameter of 1 mm to the cylinder base tap, while the other end of the tube is guided through the cylinder to the pressure transducer, which is located outside the wind tunnel. The pressure transducer chamber has a volume of 600 mm^3. From literature, the non-dimensional frequency (Strouhal number) at which vortex shedding is expected to occur can be estimated as $St = f_{vs}D/U_\infty \approx 0.2$, and it is known that the frequency of pressure fluctuations at the cylinder base is $2f_{vs}$. Stating any assumptions and approximations that you make, investigate whether there is a value or range of values of the length of tubing between the tap and the transducer that would permit the accurate determination (within a 5% amplitude error) of the present base pressure fluctuations. (Hint: Use the dynamic response expressions for compressible fluids.)

9.13 Define the compliance of a diaphragm-type pressure transducer. Identify at least three properties that, in your judgement, would affect the value of the compliance. Would the compliance increase or decrease, as the value of each of these properties is increased? How would the response of the transducer be affected by an increase of its compliance? Justify your response by physical arguments and, as much as possible, by mathematical analysis.

References

[1] R.W. Ainsworth, R.J. Miller, R.W. Moss, and S.J. Thorpe. Unsteady pressure measurement. *Meas. Sci. Technol.*, 11:1055–1076, 2000.

[2] M. Anyoji, M. Numata, H. Nagai, and K. Asai. Pressure-sensitive paint technique for surface pressure measurements in a low-density wind tunnel. *J. Vis.*, 18:297–309, 2015.

[3] A. Arnau Vives, editor. *Piezoelectric Transducers and Applications*. Springer, Berlin, 2008.

[4] T. Arts et al. *Measurement Techniques in Fluid Dynamics (2nd Edition)*. von Kármán Institute for Fluid Dynamics, Rhode-Saint-Genese, Belgium, 2001.

[5] C.M. Van Atta. *Vacuum Science and Engineering*. McGraw-Hill, New York, 1965.

[6] J.B. Barlow, W.H. Rae, and A. Pope. *Low-Speed Wind Tunnel Testing (3rd Edition)*. Wiley, New York, 1999.

[7] G.K. Batchelor. *An Introduction to Fluid Dynamics*. Cambridge University Press, Cambridge, UK, 1970.

[8] T.G. Beckwith, R.D. Marangoni, and J.H. Lienhard. *Mechanical Measurements (6th Edition)*. Pearson, Upper Saddle River, NJ, 2007.

[9] J.H. Bell, E.T. Schairer, L.A. Hand, and R.D. Mehta. Surface pressure measurements using luminescent coatings. *Annu. Rev. Fluid Mech.*, 33:155–206, 2001.

[10] R.P. Benedict. *Fundamentals of Temperature, Pressure and Flow Measurements (2nd Edition)*. Wiley Interscience, New York, 1977.

[11] P. Bradshaw and D.G. Goodman. The effect of turbulence on static-pressure tubes. Technical Report Reports and Memoranda No. 3527, Aeronautical Research Council, 1966.

[12] D.W. Bryer and R.C. Pankhurst. *Pressure-Probe Methods for Determining Wind Speed and Flow Direction*. National Physical Laboratory, London, 1971.

[13] S.H. Chue. Pressure probes for fluid measurement. *Prog. Aerosp. Sci.*, 16(2):147–223, 1975.

[14] E.O. Doebelin and D.N. Manik. *Doebelin's Measurement Systems (SIE) (7th Edition)*. McGraw-Hill, New York, 2019.

[15] R.J. Emrich, editor. *Methods of Experimental Physics, Vol. 18A and B: Fluid Dynamics*. Academic Press, New York, 1981.

[16] R.W. Fox, A.T. McDonald, and J.W. Mitchell. *Fox and McDonald's Introduction to Fluid Mechanics (10th Edition)*. Wiley, New York, 2019.

[17] R.E. Franklin and J.M. Wallace. Absolute measurements of static-hole error using flush transducers. *J. Fluid Mech.*, 42:33–48, 1970.

[18] M. Gad-el-Hak, editor. *The MEMS Handbook*. CRC Press, Boca Raton, FL, 2002.

[19] R.J. Goldstein. *Fluid Mechanics Measurements (2nd Edition)*. Taylor & Francis, Washington DC, 1996.

[20] S. Handel. *A Dictionary of Electronics*. Penguin Books, Harmondsworth, Middlesex, 1962.

[21] P.W. Harland. *Pressure Gauge Handbook*. Marcel Dekker, New York, 1985.

[22] J.H. Henry. *Pressure Measurement in Vacuum Systems*. Chapman and Hall, London, 1964.

[23] J.O. Hougen, O.R. Martin, and R.A. Walsh. Dynamics of pneumatic transmission lines. *Control Engineering*, 10(9):114–117, 1963.

[24] W.H. Howe. What's available for high pressure measurement and control. *Control Engineering*, 2:53, 1955.

[25] P. Kupferschmied, P. Koeppel, W. Gizzi, C. Roduner, and G. Gyarmathy. Time-resolved flow measurements with fast-response aerodynamic probes in turbomachines. *Meas. Sci. Technol.*, 11:1036–1054, 2000.

[26] T. Liu, T. Campbell, S. Burns, and J. Sullivan. Temperature and pressure sensitive luminescent paints in aerodynamics. *App. Mech. Rev.*, 50:227–246, 1997.

[27] J.L. Livesey, J.D. Jackson, and C.J. Southern. The static hole error problem. *Aircraft Engineering*, 34:43–47, 1962.

[28] L. Loefdahl and M. Gad-el-Hak. MEMS-based pressure and shear stress sensors for turbulent flows. *Meas. Sci. Technol.*, 10:665–686, 1999.

[29] M. Mackay. Static pressure measurement with surface-mounted disk probes. *Exp. Fluids*, 9:105–107, 1990.

[30] W.P. Mason. *Physical Acoustics*. Academic Press, New York, 1964.

[31] B.J. McKeon and R.H. Engler. Pressure measurement systems (chapter 4). In C. Tropea, A.L. Yarin, and J.F. Foss, editors, *Springer Handbook of Experimental Fluid Mechanics*. Springer, Berlin, 2007.

[32] B.J. McKeon and A.J. Smits. Static pressure correction in high reynolds number fully developed turbulent pipe flow. *Meas. Sci. Technol.*, 13:1608–1614, 2002.

[33] B.M. McLachlan and J.H. Bell. Pressure-sensitive paints in aerodynamic testing. *Exp. Thermal Fluid Sci.*, 10:470–485, 1995.

[34] C. Mercer. *Optical Metrology for Fluids, Combustion and Solids*. Kluwer Academic Publishers, Dordrecht, 2003.

[35] J.F. O'Harlon. *A User's Guide to Vacuum Technology*. Wiley Interscience, New York, 1980.

[36] H.F. Olsen. *Elements of Acoustical Engineering*. Van Nostrand, Princeton, NJ, 1957.

[37] E. Ower and R.C. Pankhurst. *The Measurement of Air Flow*. Pergamon Press, Oxford, 1977.

[38] R.L. Panton. *Incompressible Flow*. Wiley, New York, 1984.

[39] G.N. Peggs, editor. *High Pressure Measurement Techniques*. Applied Science Publishers, London, 1983.

[40] W.J. Rainbird. Errors in measurements of mean static pressure of a moving fluid due to pressure holes. Technical Report Rep. DME/NAE, Quarterly Bulletin of the Division of Mechanical Engineering (3), National Aeronautical Establishment, National Research Council of Canada, Ottawa, 1967.

[41] R.E. Rayle. Influence of orifice geometry on static pressure measurements. New York. American Society of Mechanical Engineers. ASME Paper 59-A-234, 1959.

[42] R.E. Rayle. An investigation on the influence of orifice geometry on static pressure measurements. Master's thesis, Massachusetts Institute of Technology, Cambridge, MA, 1949.

[43] B.E. Richards, editor. *Measurement of Unsteady Fluid Dynamic Phenomena*. Hemisphere Publishing Corporation, Washington DC, 1977.

[44] A. Roth. *Vacuum Technology*. North-Holland Publishing Company, Amsterdam, 1982.

[45] R.H. Sabersky, A.J. Acosta, E.G. Hauptmann, and E.M. Gates. *Fluid Flow – A First Course in Fluid Mechanics (4th Edition)*. Pearson, Hoboken, NJ, 1998.

[46] M. Sayer and A. Mansingh. *Measurement, Instrumentation and Experiment Design in Physics and Engineering*. Prentice-Hall of India, New Delhi, 2000.

[47] R. Shaw. The influence of hole dimensions on static pressure measurements. *J. Fluid Mech.*, 7:550–564, 1960.

[48] T.E. Siddon. On the response of pressure measuring instrumentation in unsteady flow. Technical Report 136, University of Toronto Institute for Aerospace Studies, Toronto, 1969.

[49] A.J. Smits and T.T. Lim, editors. *Flow Visualization Techniques and Examples*. Imperial College Press, London, 2000.

[50] R.E. Sonntag, C. Borgnakke, and G.J. Van Wylen. *Fundamentals of Themodynamics (5th Edition)*. Wiley, New York, 1998.

[51] S. Tavoularis. Techniques for turbulence measurement (chapter 36). In N.P. Cheremisinoff, editor, *Encyclopedia of Fluid Mechanics, Vol. 1,*, pages 1207–1255. Gulf Publishing Co., Houston, TX, 1986.

[52] B.W. Van Oudheusden. PIV-based pressure measurement. *Meas. Sci. Technol*, 24:032001, 2013.

[53] W.W. Willmarth. Unsteady force and pressure measurements. *Annu. Rev. Fluid Mech.*, 3:147–170, 1971.

10 Measurement of Flow Rate

This chapter covers common instrumentation and techniques that are used to measure the *bulk flow rate*, namely the amount, per unit time, of a fluid that passes through a certain cross section of a pipe, duct, channel or other flow conduit. Bulk flow measurement is not concerned with local velocity variations across the cross section, nor with short-time (e.g., turbulent) fluctuations. The measured flow rate can be either the *mass flow rate* $\dot{m}$ or, when dealing with liquids or low-speed gases, the *volume flow rate* Q. Flow rate measurement is an essential activity in a variety of industries and utility services, but it is also performed regularly in the fluid mechanics laboratory, notably for its important role of monitoring and controlling the experimental conditions. The operation of flow rate measuring systems is based on diverse physical principles; with some exceptions, such systems require calibration or empirical corrections. The following presentation is mainly concerned with bulk flow measurement in 'simple' flows, which are single-phase and either steady or very slowly varying. Some of these methods can be extended to flows of multi-phase fluids, slurry's and granular materials, but the reader is advised to consult specialised instrumentation manufacturers manuals and other specific sources when dealing with such media. More details on general methods for the measurement of flow rate and specific instruments can be found in several monographs and books [3, 5, 8, 9, 10], handbooks [13, 15, 16, 17] and manufacturer catalogues (e.g., [12]).

10.1 Direct Methods

The simplest flow rate measurement methods are *direct*, which means that they measure a typical flow velocity or the amount of discharged fluid during a measured time interval. Such methods are more suitable for liquid than gas flows. For example, one may obtain a rough measurement of the bulk velocity of flows in water tunnels and open channels by timing the motion of suspended or floating objects. For flows of non-volatile liquids in an open-loop configuration, the volume flow rate can be measured by timing the filling of a container by the discharge of the apparatus; similarly, the mass flow rate can be measured by weighing the discharged fluid on a scale. In such cases, one must take care that discharging of fluid has no appreciable effect on the operation of the system, as for example would be the case if removal of liquid from the loop resulted in lowering the head of a feeding tank or shifting the operating point of a pump. Direct flow rate measurement methods, applicable to liquid and gas flows in both open- and closed-loop configurations, include the use of positive displacement flow meters, to be discussed in the next section.

10.2 Positive Displacement Flow Meters

Positive displacement (PD) flow meters are devices, which isolate a fixed volume of the fluid flowing into their inlet in sealed compartments and then discharge it to the outlet. Neglecting leakage and other possible deficiencies, one can easily compute the volume flow rate from the size and number of compartments in the device and the period of the cycle of their operation. In many cases, the same instrument can be configured to measure the total volume of fluid that passes through it over a time interval, and, for this reason, PD meters are commonly used to monitor the consumption of water, natural gas and hydrocarbon fuels. Some types of PD meters are operating passively by receiving power from the flowing fluid, whereas other types are driven by an external source to create the fluid motion, in which case they are referred to as *metering pumps*. There is a great variety of designs of PD meters, most of which can be classified as *rotary*, *reciprocating* or *nutating*. The important parameters for their operation are the *leakage* and the *pressure loss* across them. Leakage through narrow gaps between meshing parts of the device depends on the speed of operation, the viscosity of the fluid and the clearance between the moving components and the housing, which generally deteriorates with wear. To minimise leakage, the components are manufactured such as to have small tolerances and the clearances between meshing parts are kept small. Accurate operation of PD meters requires the use of clean fluids, and in most cases the fluid is passed through a strainer or a filter before entering the flow meter. PD meters are suitable for fluids having a wide range of viscosity. An increase in viscosity improves the sealing action, but also increases pressure loss, which has to be kept as low as possible in order to avoid significant loading of the physical system. Temperature variation affects the operation of PD meters in two ways: by affecting the viscosity of the fluid, with implications on leakage and pressure loss, and by affecting the density of the fluid, which is of concern when converting a measured volume flow rate to a mass flow rate. Manufacturers normally provide charts describing the operation characteristics of each model, such as the pressure loss for different flow rates and fluid viscosities, as well as correction factors and uncertainties for different speeds. Some representative designs of PD meters for liquid and gas flows are discussed in [2, 3, 4, 8, 13, 16].

Among the common PD flow meters that are used for measuring the flow rates of water and other liquids are the following:

- **Nutating disk meters:** The main element of these meters is a disk, which rotates in a nutating (precessing, wobbling) fashion, while both its sides are partially in contact with a dual conical housing (Fig. 10.1a). Fluid enters through the inlet port facing one side of the disk during half of the cycle, in isolation from the outlet port; it is then swept by the precessing motion of the disk to the outlet during the following half of the cycle, while being in isolation from the inlet port.

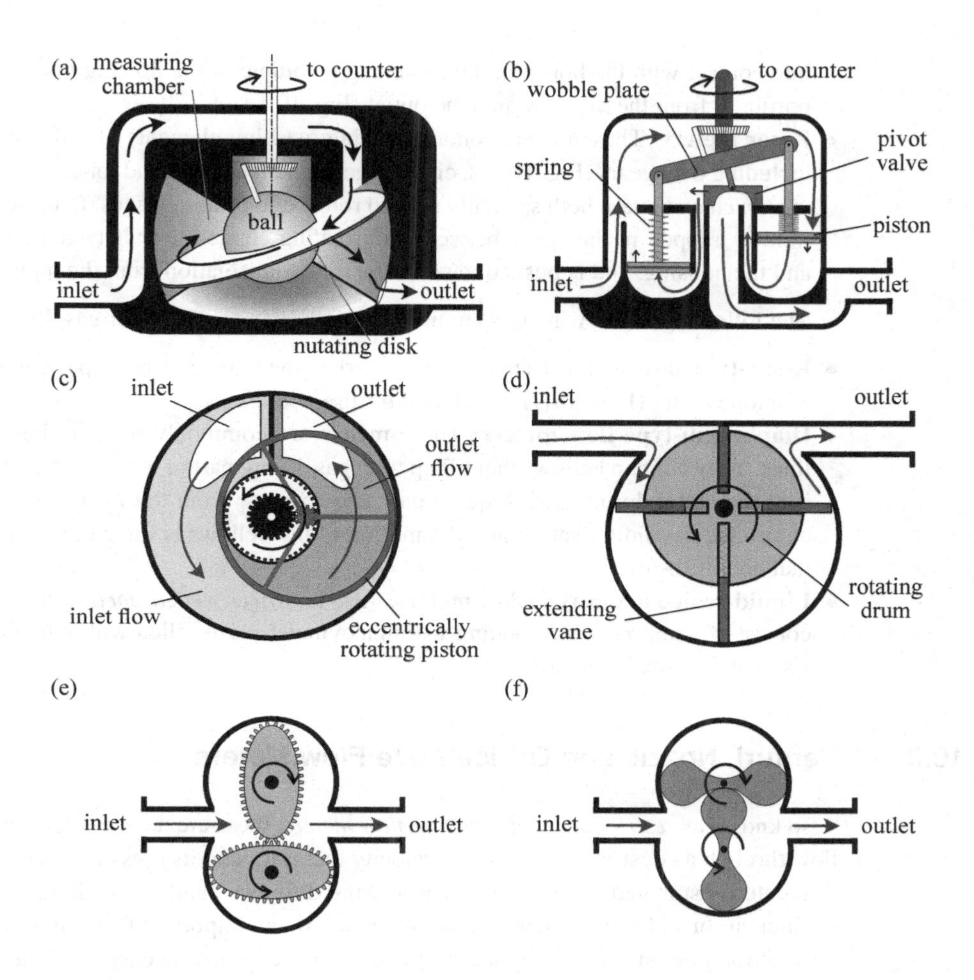

Figure 10.1 Sketches of representative positive displacement flow meters: (a) nutating disk meter, (b) reciprocating piston meter, (c) rotary piston meter, (d) rotary vane meter, (e) oval gear meter and (f) roots meter.

- **Reciprocating piston flow meters:** These meters contain a number of plungers or pistons, driven by a wobble plate and sweeping the volumes of corresponding cylinders, while at the same time opening and closing the input and output ports or valves (Fig. 10.1b).
- **Rotary piston flow meters:** These devices contain a cylindrical drum, which is mounted eccentrically inside a cylindrical housing and rotates with its outer surface in contact with the housing, while its inner surface maintains contact with an inner cylinder, which is coaxial with the housing (Fig. 10.1c).
- **Rotary vane flow meters:** Flat vanes are inserted into matching slots around the perimeter of a rotating cylindrical drum, located eccentrically within the housing. Centrifugal action or springs cause the vanes to slide out of the slots until they come

into contact with the housing, thus isolating a volume of the flowing fluid and transporting it from the inlet towards the outlet (Fig. 10.1d).

- **Rotor meters:** These meters contain rotating meshing elements of different shapes, including oval gears (Fig. 10.1e), circular gears, helical gears and lobes. *Rotary abutment meters* contain both specially shaped rotors and rotating vanes. In these devices, fluid is trapped in the space between the rotating elements, or between an element and the housing, and is pushed towards the outlet in isolation from the input.

The following PD flow meters are used commonly in air and other gas flows:

- **Roots-type flow meters:** This is a trademark name that describes a particular design of a lobe meter (Fig. 10.1f), developed for use with gases.
- **Diaphragm-type flow meters:** These meters are commonly used in domestic gas lines. They contain bellows that fill up with gas during part of the cycle and discharge it to the outlet during a subsequent part; the gas flow from the inlet to the outlet is controlled by sliding valves and the motion of the bellows is linked to a mechanism that counts the cycles.
- **Liquid-sealed drum-type flow meters:** Also known as *wet gas meters*, these devices consist of a hollow drum rotating within a cylinder partly filled with a liquid, which provides the sealing action.

10.3 Venturi, Nozzle and Orifice Plate Flow Meters

Also known as *restriction* or *obstruction flow meters*, these are devices that force a pipe flow through a constriction, where its velocity increases and its pressure decreases. The flow rate is estimated from a measured pressure difference and an empirical correction coefficient. In addition to their size, weight and cost, an important factor in selecting an obstruction flow meter is the amount of frictional energy loss occurring in them. Common low-loss restriction flow meters are the *Venturi tubes* (Fig. 10.2a,b) and the *Dall tubes* (Fig. 10.2c), whereas relatively high-loss meters include *flow nozzles* (Fig. 10.2d) and *orifice plates* (Fig. 10.2e). The designs of such devices are regulated by standards (e.g., ISO and ASME standards), so that they may be used interchangeably and without the need for individual calibration. A limitation of obstruction flow meters is their relatively narrow dynamic range, which is a result of the non-linearity of their response. The sensitivity of obstruction flow meters decreases rapidly, as the flow rate drops below about 25% of the full-scale value.

In order to determine the idealised response of these devices, consider steady, uniform, inviscid, incompressible flow, in the absence of body forces, flowing within a circular tube with a diameter D and guided to a restriction with a diameter d, as in the case of the classical Venturi tube, shown in Fig. 10.2a. Then, one may use the continuity equation and Bernoulli's equation to describe the ideal volume flow rate in terms of the pressure drop $\Delta p = p_1 - p_2$ between an upstream location and the throat as

$$Q_{id} = \frac{\pi D^2/4}{\sqrt{(D/d)^4 - 1}} \sqrt{\frac{2\Delta p}{\rho}} . \tag{10.1}$$

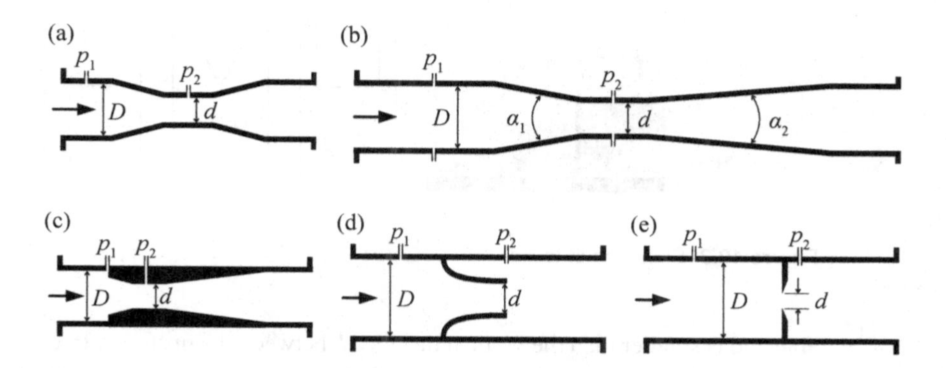

Figure 10.2 Sketches of obstruction flow meters: (a) classical Venturi tube, (b) ASME Venturi tube, (c) Dall tube, (d) flow nozzle and (e) orifice plate.

The actual flow rate through the device would be lower than Q_{id}, because of non-recoverable, frictional losses. Such losses are commonly accounted for with the use of an empirical discharge coefficient C_d, which provides the actual volume flow rate as

$$Q = C_d \frac{\pi D^2/4}{\sqrt{(D/d)^4 - 1}} \sqrt{\frac{2\Delta p}{\rho}} \,. \tag{10.2}$$

Well-designed Venturi tubes, operating at relatively large Reynolds numbers $4Q/(\pi D \nu)$, have very small losses, with typical values of C_d between 0.97 and 0.99; thus, one may approximate Eq. (10.2) by Eq. (10.1). However, Eq. (10.2) must be used for nozzle and orifice flow meters, for which C_d is significantly smaller than 1, with its value depending on the geometry of the device, the location of the pressure taps and the operating pipe Reynolds number. For Reynolds numbers that are sufficiently large for the flow to be in the fully turbulent regime, C_d becomes insensitive to Reynolds number and depends only on the shape of the device (e.g., the throat-to-pipe diameter ratio d/D). In order to keep energy losses low during operation, it is desirable to design the flow meter such that it has a large C_d. However, devices with a large C_d are typically longer, bulkier and more expensive than devices with lower C_d.

10.4 Open Channel Flow Measurement

The volume or mass flow rate of liquids in open channels or partially filled pipes and ducts is often measured with the use of direct methods (see Section 10.1). Another common approach is the use of flow restrictions, including *weirs* and *Venturi flumes* [1, 6, 8, 17, 18].

Weirs consist of obstructions positioned across the channel, so that they force the liquid to flow over a central opening. As the flow rushes over this opening, the free-surface elevation above the lowest point in the weir drops to a level H, which is easily measur-

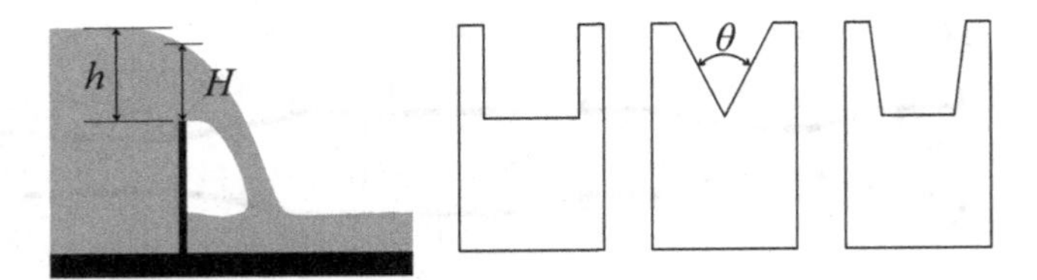

Figure 10.3 Sketches of weirs.

able and is smaller than the vertical distance h between the upstream free-surface elevation and the lowest point in the weir. There are several types of weirs, both sharp-crested and broad-crested, with rectangular, V-shaped or trapezoidal openings (Fig. 10.3). The most common type used in fluid mechanics and hydraulics laboratories is the sharp-crested, V-notch weir. An approximate expression for the flow rate over such weirs is [6]

$$Q \approx 2.5 \tan \frac{\theta}{2} H^{5/2} , \tag{10.3}$$

where θ is the full angle of the notch (usually equal to 90°).

Venturi flumes are converging–diverging channel constrictions, analogous to Venturi tubes used in pipes flows. They generally have low pressure losses and are available in several different designs, including flumes with rectangular, trapezoidal and U-shape cross sections; commonly used designs for sewage and irrigation flows are the *Parshall* and *Palmer–Bowlus flumes*. Their advantages over weirs is that they do not cause a water back-up and are less likely to be affected by deposited solids that may be transported by the water.

10.5 Averaging Pitot Tubes

These instruments consist of a tube spanning the cross section of the pipe or duct and having multiple frontal openings such that it measures a total pressure, which is roughly equal to the average total pressure over the cross section (Fig. 10.4a); they also have a second tube, facing backwards and monitoring the local static pressure. The volume flow rate is estimated from the pressure difference Δp as

$$Q = C_d \frac{\pi D^2}{4} \sqrt{\frac{2\Delta p}{\rho}} , \tag{10.4}$$

where C_d is an empirical correction coefficient accounting for deviations from the ideal response. Simplicity of operation and low cost are the main advantages of averaging Pitot tube. Their limitations include the need for clean fluid and a narrow dynamic range, extending to only about 30% of full scale.

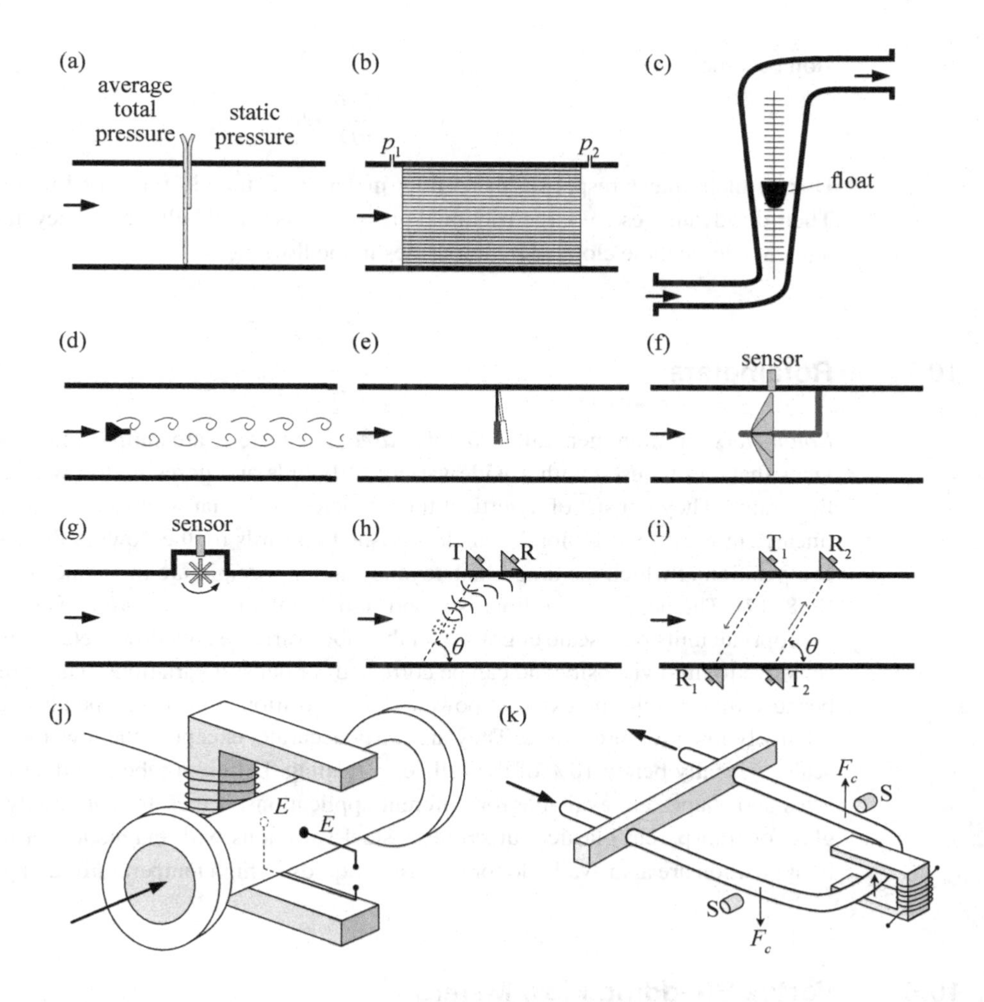

Figure 10.4 Sketches of common flow meters: (a) averaging Pitot tube, (b) laminar flow elements, (c) rotameter, (d) vortex shedding flow meter, (e) drag flow meter, (f) turbine flow meter, (g) paddlewheel flow meter, (h) Doppler ultrasonic flow meter, (i) time-of-flight ultrasonic flow meter, (j) electromagnetic flow meter and (k) Coriolis flow meter.

10.6 Laminar Flow Elements

Laminar flow elements comprise a section of a pipe or duct, which contains a thin-tube bundle or a long honeycomb (Fig. 10.4b). The fluid flow through this device is subdivided into flows through many elements, which are sufficiently narrow for the Reynolds number in each element to be lower than the transitional value, which is nominally about 2,300. Thus, the flow in each element is laminar and the pressure drop across the elements is related to the volume flow rate through the Hagen–Poiseuille expression, which is linear, in contrast with the quadratic expression for turbulent pipe flow. For fully developed laminar flow in a circular tube of length l and diameter D, this expression becomes

$$\Delta p = \frac{128\mu l}{\pi D} Q . \qquad (10.5)$$

Due to their linear response, these flow meters are suitable for very low flow rates. Their disadvantages are large frictional pressure losses and bulkiness. They also have a tendency to become clogged by impurities in the flow.

10.7 Rotameters

Rotameters, or, more generally, *variable area flow meters*, are simple and versatile devices that can be used with a wide variety of liquids and gases within wide ranges of flow rates. They consist of a vertical tube, tapered such that its cross section increases linearly upwards and a 'float', which is pushed upwards by the flowing fluid and stops at a position at which the drag, the buoyancy and the weight are in balance (Fig. 10.4c) [3, 8, 17]. The height of the float is proportional to the flow rate, which is displayed in appropriate units on a scale engraved on the tube. Variable area flow meters are not very sensitive to fluid viscosity and can be corrected for density variations. They are popular because they require no external power, can be positioned near pipe bends and present relatively low pressure losses. They are fairly accurate, except in the lower end of their scale, typically below 10% of the full-scale reading. Different tube and float materials, sizes and shapes are available for different applications. Tubes are commonly made of glass or transparent plastic, but stainless steel variations with magnetic sensing of the float position are also available for corrosive liquids or high temperatures and pressures.

10.8 Vortex Shedding Flow Meters

The main component of *vortex shedding flow meters* [3, 8, 13] is a bluff object immersed in the flowing fluid and spanning the pipe cross section (Fig. 10.4d). Their operation is based on the periodic shedding of vortices (*Kármán vortex street*) from the edges of the object; this occurs at a frequency f (in cycles per second), which is related to the frontal width h of the object and the flow velocity V. In dimensionless form, the shedding frequency is called the *Strouhal number*

$$\mathrm{S} = \frac{hf}{V} . \qquad (10.6)$$

For Reynolds numbers greater than a certain value (typically, about 5,000), the Strouhal number maintains an essentially constant value in the range 0.14 to 0.21, depending on the shape of the object and independent of V. The shedding frequency is detected by a variety of means, including piezoelectric pressure transducers, strain gauges, self-heated resistance elements and ultrasonic beams.

10.9 Drag Flow Meters

The operation of *drag flow meters*, which are also referred to as *target flow meters* [12, 16] (Fig. 10.4e) is based on the relationship between the drag force F_D on an immersed bluff object and the flow velocity, that is,

$$F_D = \frac{1}{2} C_D \rho A V^2 , \qquad (10.7)$$

where C_D is the drag coefficient and A is the frontal area of the object, namely, the area of its projection on a plane normal to the flow velocity. C_D is essentially constant for an object with sharp corners that is immersed in turbulent flow with a sufficiently large Reynolds number, typically greater than about 1,000. Thus, the volume flow rate through a pipe would be found as

$$Q = k \sqrt{F_D} , \qquad (10.8)$$

where k is a constant. In practice, the *target,* which is a disk-like object, is inserted in the pipe and mounted on a support instrumented with strain gauges or LVDTs (linear variable differential transformers), which measure the drag force through deflection. Such instruments are very sensitive and bidirectional and can be used at high pressures and with a variety of fluids. As the target is usually positioned in the centre of the pipe, they are not easily clogged by suspended impurities.

10.10 Turbine Flow Meters

Turbine flow meters measure the volume flow rate of fluids in pipes as proportional to the angular velocity of an immersed vaned rotor [8, 12, 17]. A very common type utilises an axial turbine with its axis aligned along the pipe centreline (Fig. 10.4f). The passage of each rotating blade is sensed electromagnetically by an externally mounted sensor and the flow rate is given by

$$Q = kn , \qquad (10.9)$$

where n is the number of pulses per unit time provided by the sensor and k is a constant, the value of which depends on the impeller design and size, the pipe diameter and the number of blades. Turbine flow meters introduce significant pressure losses and are prone to cavitation, when used with high-speed, low-pressure liquids. A low-cost version, called a *paddlewheel flow meter* (Fig. 10.4g), consists of a partially immersed rotor, with its axis normal to the flow direction. Besides flow rate, turbine flow meters may also provide the total fluid volume that passed through during a time interval. The common domestic water meters are of the turbine type.

10.11 Ultrasonic Flow Meters

Ultrasonic flow meters [3, 8, 13] utilise high-frequency (typically of the order of 10 MHz) pressure waves to compute the volume flow rate of liquids in pipes. There are two distinct types of such meters, *Doppler flow meters* and *time-of-flight flow meters*.

A representative Doppler flow meter consists of two piezoelectric crystals, a transmitter T, which transmits an ultrasonic wave through the pipe, and a receiver R, which receives the ultrasound reflected by solid particles or gas bubbles transported by the flowing fluid (Fig. 10.4h). The frequency f_r of the reflected sound is shifted from the frequency f_t of the transmitted sound by an amount Δf, called the *Doppler shift*, which is proportional to the velocity V of the reflector, as

$$\Delta f = f_t - f_r = \frac{2 f_t \cos\theta}{c} V \,, \qquad (10.10)$$

where c is the speed of sound in the fluid. Such devices are calibrated to provide an output that is equal to the average velocity of the fluid in the pipe, under the assumption that the flow is fully developed. They are non-invasive and can be hand-held or strapped to the exterior of the pipe.

A representative time-of-flight flow meter consists of two externally mounted pairs of piezoelectric transducers. Each transmitter emits sound waves towards the corresponding receiver, one of which is located downstream of its mate, while the other is upstream of it (Fig. 10.4i). Each transmitter emits a sound pulse each time the corresponding receiver receives the previous one. Because sound waves are transported by the flowing fluid, sound propagates faster downstream than upstream and the frequencies of pulsation of the two pairs differ by an amount

$$\Delta f = \frac{2\cos\theta}{l} V \,, \qquad (10.11)$$

where l is the distance between the transducers of each pair. This configuration makes the flow measurement independent of the speed of sound and, thus, flow temperature.

10.12 Electromagnetic Flow Meters

Electromagnetic flow meters provide the volume flow rate of electrically conducting liquids in pipes. Their operation is based on *Faraday's law of electromagnetic induction*, which states that, when a conductor with length l moves with speed V in a direction normal to the direction of a magnetic field with magnetic flux density B, an electric potential E is generated across it as

$$E = BlV \,. \qquad (10.12)$$

Practical electromagnetic flow meters [3, 8, 12] consist of an insulated pipe section of the same diameter D as the pipe of interest, surrounded by an alternating or pulsed magnetic field and having two surface electrodes embedded on the wall across a diameter that is normal to the magnetic field direction (Fig. 10.4j). The voltage difference between these electrodes is related to the volume flow rate as

$$E = \frac{4kB}{\pi D} Q \,, \tag{10.13}$$

where k is a numerical coefficient. Electromagnetic flow meters have an accuracy that exceeds 0.5% and are not overly sensitive to the velocity profile. On the other hand, they are bulky, heavy and relatively expensive.

10.13 Coriolis Flow Meters

Coriolis flow meters [3, 8, 14, 17] were developed relatively recently, but have become increasingly popular in a variety of industries and in many thermo-fluids laboratories, due to their versatility and their capacity to measure true mass flow rate, essentially independent of fluid properties and flow conditions. There are several different geometrical designs, all based on the Coriolis force principle, which may be explained as follows. Consider a fluid element having a mass δm and flowing with velocity V in a tube that rotates with angular velocity ω about an axis normal to its own axis; assume that the fluid element is at a radial distance r from the axis of rotation and, during time $\delta t = \delta r / V$, moves away from it to a radial distance $r + \delta r$. Then, the angular momentum $(\delta m)\, \omega r^2$ of this fluid element would increase to $(\delta m)\, \omega\, (r + \delta r)^2 \approx (\delta m)\, \omega \left(r^2 + 2r\delta r\right)$. This increase of angular momentum is attributed to a torque rF_c, where $F_c = 2(\delta m) V \omega$ is called the *Coriolis force*. The direction of the Coriolis force is circumferential and opposite in sense to the direction of rotation, for outward motion. In vector notation, the Coriolis force is written as $\vec{F}_c = 2(\delta m) \vec{V} \times \vec{\omega}$. It is also written as $\vec{F}_c = (\delta m) \vec{a}_c$, where $\vec{a}_c$ is called the *Coriolis acceleration*. The flowing fluid receives this force from the tube walls; by reaction, the fluid applies a force upon the containing tube, which is equal in magnitude and direction to the Coriolis force, thus affecting the tube motion.

In practical Coriolis flow meters, the tube is not rotated, but is immersed in an alternating magnetic field, which sets it in vibration at its natural frequency. As representative of this class of instruments, Fig. 10.4k shows a sketch of a *U-tube Coriolis flow meter*. The fluid is passed through a bent tube, whose ends are clamped, while its tip is set to vibration. The instantaneous angular velocity, and therefore the Coriolis force, increase towards the tip. The two legs of the tube receive forces in opposite directions and, thus, the tube is twisted in one sense during half of the cycle and in the opposite sense during the other half. The twist angle is measured by magnetic or optical position sensors, which sense the time delay Δt between the passage of the two legs through a transverse plane. This time delay is related to the mass flow rate as

$$\Delta t = \frac{8 r_t^2}{K_s} \dot{m} \,, \tag{10.14}$$

where r_t is the radius of the tube and K_s is a constant that, ideally, depends only on the tube material. Small deviations from this relationship may be caused by multi-phase effects and other variations in fluid properties. Even so, Coriolis flow meters are suitable for conventional as well as contaminated and non-Newtonian fluids. They can also be used in multi-phase flows and flows undergoing phase change.

10.14 Thermal Mass Flow Meters

Thermal mass flow meters are used to measure the mass flow rate of gases. They are not used for liquid flows due to the much higher power required to heat a liquid than a gas. For relatively small mass flow rates, the entire gas stream is passed through the meter, while, for higher flow rates, only part of the gas is heated inside a bypass tube. There are two types of such instruments, *heated tube flow meters* and *immersion probe flow meters* [12].

In heated tube flow meters, the flowing gas is passed through a piece of tube that is heated electrically and is instrumented with two temperature sensors, commonly thermocouples or resistance temperature detectors (RTD). The first sensor is located upstream of the heated section and the other one is downstream of it. The rate of heat transfer $\dot{H}$ to the fluid is

$$\dot{H} = \dot{m} c_p \Delta T \ , \tag{10.15}$$

where $\dot{m}$ is the mass flow rate of the gas, c_p is its specific heat under constant pressure and ΔT is the temperature difference across the heated section. Thus, the mass flow rate for a given gas can be measured from measurements of $\dot{H}$ and ΔT. Manufacturers supply instruments with an output that has been calibrated in air, nitrogen or some other gas. When used with different gases, this output has to be corrected by being multiplied by the ratio of specific heats of the two gases.

Immersion probe flow meters consist of a probe with two RTD connected in a Wheatstone bridge configuration. One RTD is used to measure the gas temperature, while the other is provided with a current so that it is heated to a temperature higher than the gas temperature by a fixed amount ΔT. The electric power required to heat the second sensor is related to the mass-weighted velocity ρV of the gas by a non-linear relationship, called *King's law*. Electronic circuitry is employed to linearise the output so that it is proportional to ρV for a given gas. To obtain the mass flow rate, one has to multiply this output by the pipe cross-sectional area. Corrections for use with different gases are also available.

10.15 Flow Meter Selection

Considering the diversity of designs and properties of flow meters and the wide ranges of flow conditions encountered in a fluid mechanics laboratories, it is advisable to compare carefully the different options that are available before purchasing a flow meter. Although it is possible that several devices may be equally suitable for a given application, it is also quite certain that several others would be unsuitable. To assist the selection

Table 10.1 Comparison of different flow meters

Flow meter type	Dirty fluid?[1]	Dyn-amic range	Pres-sure loss[2]	Uncer-tainty[3]	Upstream pipe[4]	Viscous effect[2]	Cost[2]
PD	N	10:1	H	±0.25 r	none	H	M
Venturi	Y	4:1	L	±1 fs	5–20	H	M
Nozzle	Y	4:1	M	± 1–2 fs	10–30	H	M
Orifice plate	Y	4:1	M	±2 – 4 fs	10–30	H	L
Weir (V-notch)	Y	100:1	VL	±2–5 fs	none	VL	M
Parshall flume	Y	50:1	VL	±2–5 fs	none	VL	M
Pitot	N	3:1	VL	±3–5 fs	20–30	L	L
Rotameter	Y	10:1	M	±0.5 r	none	M	L
Vortex	Y	10:1	M	±1 r	10–20	M	H
Drag	Y	10:1	M	±1-10 fs	10–30	M	M
Turbine	N	20:1	H	±1 r	5–10	H	H, M
Doppler	Y	10:1	none	±5 fs	5–30	none	H
Time-of-flight	N	20:1	L	±1–5 fs	5-30	none	H
Electromagnetic	Y	40:1	none	±0.5 r	5	none	H
Coriolis	Y	10:1	L	±0.4 fs	none	none	H
Thermal	Y	10:1	L	±1 fs	none	none	H

[1] Y: yes, N: no
[2] H: high, M: medium, L: low, VL: very low
[3] Percentage of full scale (fs) or reading (r)
[4] In diameters

of a suitable flow meter, we provide Table 10.1, which is based on information supplied by different manufacturers.

Chapter Digest

10.1 For which types of fluids and conditions would a volume flow rate measurement be equivalent to a mass flow rate measurement? Provide an example for which a volume flow rate measurement would not be appropriate, but a mass flow rate measurement would.

10.2 Explain the rationale behind the term 'positive displacement flow meters'. How do such meters differ from other types of flow meters?

10.3 Consider a pump that transfers water from a lake to an elevated tank. Explain how you would measure the flow rate through the pump in the most inexpensive manner.

10.4 Consider a reciprocating piston meter, as shown in Fig. 10.1b, with six cylinders having a diameter $D = 25.40$ mm and a piston stroke $\Delta x = 50.80$ mm. Derive an expression connecting the measured flow rate of water and the frequency f of rotation of the meter's shaft. Can you use this device as a metering pump?

10.5 Why do Venturi tubes have a long diffuser section? What would happen if this section were made much shorter?

10.6 Describe a method that can be used to measure the flow rate of water in a steel pipe without requiring any modification of the piping system.

10.7 Describe a method that can be used to measure the flow rate of a water–steam mixture flowing in a metallic tube.

10.8 Explain why most flow meters should be installed in a relatively long straight section of a pipe, rather than closely following a valve, a bend or other fitting.

10.9 Explain why a gel is used between the pipe wall and the transducers of ultrasonic flow meters.

10.10 Why do we use ultrasound, rather than sound in the audible range, for flow measurement?

10.11 Why is the direction of sound transmission in ultrasonic flow meters inclined with respect to the flow direction, rather than being normal to it?

10.12 Outline the main advantages and disadvantages of Coriolis flow meters by comparison to Venturi tubes.

10.13 Among all flow meters discussed in this chapter, identify those that can be used for measuring the flow rate of clean, dry air in a duct and those that cannot.

Problems

10.1 The discharge coefficient of orifice plate flow meters with a cylindrical orifice and pressure taps that are in close proximity of the two sides of the plate may be estimated from the empirical expression [7, 11]

$$C_d = 0.5959 + 0.0312\beta^{2.1} - 0.184\beta^8 + \frac{91.71\beta^{2.5}}{\mathrm{Re}_D^{0.75}} ,$$

where the diameter ratio $\beta = d/D$ is in the range from 0.2 to 0.75 and the pipe Reynolds number Re_D is in the range from 10^4 to 10^7. Plot this coefficient for several values of β within its specified range *vs.* Re_D within its specified range. Determine the minimum value of Re_D for which the Reynolds number effect is smaller than 1%, and plot it *vs.* β. Discuss your results and identify an obvious disadvantage of this flow meter, compared to a Venturi tube flow meter.

10.2 The discharge coefficient of ASME long-radius nozzle flow meters may be estimated from the empirical expression [7, 11]

$$C_d = 0.9975 - \frac{6.53\beta^{0.5}}{\mathrm{Re}_D^{0.5}} ,$$

where the diameter ratio $\beta = d/D$ is in the range from 0.25 to 0.75 and the pipe Reynolds number Re_D is in the range from 10^4 to 10^7. Plot this coefficient for several values of β within its specified range *vs.* Re_D within its specified range. Determine the minimum value of Re_D for which the Reynolds number effect is smaller than 1%, and plot it *vs.* β. Compare the losses of nozzle flow meters and orifice flow meters, as discussed in Problem 10.1.

10.3 A rotameter that has been calibrated for use in air is subsequently used for measuring the flow rate of helium. Discuss the sources of error in this measurement. Analyse the performance of this device in the two gases and develop a procedure for correcting the measurements in helium.

10.4 Design a laminar flow elements flow meter to measure the water flow rate in a pipe with a diameter $D = 100$ mm and expected bulk velocities in the range $0 < U_b \leq 20$ m/s. In your opinion, would that be a practical device? Which are the issues to be considered for this assessment?

10.5 Searching through manufacturers' catalogues, select specific flow meter models that would be suitable for measuring the flow rate in the following systems. Estimate the required range by appropriate arguments. In cases for which you cannot find appropriate products, describe briefly the design and specifications of a flow meter that would be suitable for the task.

(a) Natural gas through a 180 mm ID pipe.
(b) Blood transfused to a patient through a 3 mm ID tube.
(c) Helium through a 2 mm ID tube.
(d) Paper pulp through a 500 mm ID pipe.
(e) Water through a 2 m high, 3 m wide open channel.
(f) Ventilation air through a 150 mm × 300 mm duct.

References

[1] P. Ackers et al. *Weirs and Flumes for Flow Measurement.* Wiley, New York, 1978.

[2] Anonymous. 1969 guide to process instruments elements. *Chem. Eng.*, 76(12):137–164, 1069.

[3] R.C. Baker. *Flow Measurement Handbook.* Cambridge University Press, Cambridge, UK, 2000.

[4] R.C. Baker and M.V. Morris. Positive-displacement meters for liquids. *Trans. Inst. M.C.*, 7:209–220, 1986.

[5] R.P. Benedict. *Fundamentals of Temperature, Pressure and Flow Measurements (2nd Edition).* Wiley Interscience, New York, 1977.

[6] E.F. Brater and H.W. King. *Handbook of Hydraulics (6th Edition).* McGraw-Hill, New York, 1976.

[7] R.W. Fox, A.T. McDonald, and J.W. Mitchell. *Fox and McDonald's Introduction to Fluid Mechanics (10th Edition).* Wiley, New York, 2019.

[8] R.A. Furness. *Fluid Flow Measurement.* Longman, London, 1989.

[9] R.J. Goldstein. *Fluid Mechanics Measurements (2nd Edition).* Taylor & Francis, Washington DC, 1996.

[10] A.T.J. Hayward. *Flowmeters.* McMillan Press, London, 1979.

[11] R.W. Miller. *Flow Measurement Engineering Handbook (3rd Edition).* McGraw Hill, New York, 1996.

[12] Omega Engineering. *The Flow and Level Handbook.* Omega Engineering Inc., Stamford, CT, 2001.

[13] S.P. Parker. *Fluid Mechanics Source Book.* McGraw Hill, New York, 1987.

[14] K.O. Plache. Coriolis/gyroscopic flowmeters. *Mechanical Eng.*, 101(3):36–41, 1979.

[15] J. Saleh, editor. *Fluid Flow Handbook.* McGraw-Hill, New York, 2002.

[16] S. Tavoularis. Techniques for turbulence measurement. In N.P. Cheremisinoff, editor, *Encyclopedia of Fluid Mechanics*, volume 1, pages 1207–1255. Gulf Publishing Company, Houston, TX, 1986.
[17] J.G. Webster, editor. *The Measurement, Instrumentation and Sensors Handbook*. CRC Press, Boca Raton, FL, 1999.
[18] F.M. White. *Fluid Mechanics (8th Edition)*. McGraw-Hill, New York, 2015.

11 Flow Visualisation Techniques

Some of the most important discoveries in fluid mechanics were largely achieved by flow visualisation, using pioneering versions of modern techniques. As milestones in the development of visual and optical methods, one could mention Leonardo da Vinci's (1452–1519) detailed observations and sketches of river and channel flows, Antonie Van Leeuwenhoek's observation and timing of the motion of blood cells in order to measure blood flow velocity (1689), Ernst Mach's use of the reference beam interferometer and other optical methods to study gas dynamics phenomena (1878), Osborne Reynolds' experiments with dye in water, demonstrating the difference between laminar and turbulent pipe flows (1893), and Ludwig Prandtl's surface flow visualisation of flows around cylinders and aerofoils using aluminium filings (1930s). It is generally understood that flow visualisation is of qualitative nature and not primarily concerned with measurement. Nevertheless, many flow visualisation techniques have been refined and extended so that they can provide quantitative results, particularly measurement of flow velocity, temperature, pressure and composition. The term 'flow visualisation' has also recently been applied to the post-processing of computational fluid dynamics (CFD) results to enable the visual inspection of computed flow patterns, but computational techniques are outside the scope of this book. Interest on experimental flow visualisation techniques has been intense for a long time. Consequently, a great diversity of methods, each with many variants, are available to the fluid mechanics researcher. In the present chapter, we shall briefly outline some of the most popular methods only, referring the reader to the voluminous literature for additional methods and more details [14, 15, 28, 29, 30, 33, 44, 52].

11.1 Overview

Under certain conditions, some flow characteristics, such as fluid boundaries and even internal motion patterns, can be seen and recorded without any externally introduced disturbances or special instrumentation. Such flows may be called *naturally visible*. This may happen when the fluid contains surfaces across which there is an abrupt change of refractive index, for example, an interface between a liquid and a gas or two immiscible liquids. Such interfaces are encountered in liquids with a free surface and in fluid regions where condensation, boiling, cavitation, melting or solidification take place.

Flow visualisation techniques are available for nearly all sorts of fluids and for the entire speed range, from static fluids to hypersonic flows. The vast majority of flow visualisation techniques can be classified into two general classes, *marker techniques* and *optical techniques*. A distinct category of techniques, which, among diverse areas of science and medicine, have also found application in fluid mechanics, are the *tomographic techniques*.

Marker techniques identify patterns of fluid motion from the visible motion of foreign materials contained in the fluid. This class also includes surface marker techniques, which identify fluid velocity, temperature, wall shear stress and pressure patterns from visible changes of the properties of a coating that has been applied on a wall. Marker techniques are available for the majority of liquid and gas flows, excluding gases at very low densities.

Optical techniques identify patterns of fluid motion from variations of the refractive index within the fluid or from radiation emission from fluid atoms. Such techniques have been applied widely to gas flows and, to a lesser extent, to liquid flows. Because changes of the refractive index are associated with changes of density, optical techniques are particularly suitable for the study of compressible flows, natural convection, flames, combustion processes and, more generally, chemical reactions. Radiation emission techniques are applicable to very low density gas flows, for which no other flow visualisation technique has been successful.

As a rough selection guide, Table 11.1 lists some of the most common flow visualisation techniques for liquid and gas flows and the typical ranges of flow speeds to which they apply.

11.2 Marker Techniques

11.2.1 Tufts

Tufts are short pieces of yarn, string, or other flexible material, fastened at one end on an immersed wall or some thin support inside the flow [8]. They may be used in both gaseous and liquid flows. Their main utility is as flow direction indicators, but they may also serve to identify flow separation regions, flow instability and transition to turbulence. Under certain conditions, one may be even able to extract some quantitative results from tuft motion patterns, although with a very low resolution. The properties that affect the performance of tufts are their density, stiffness and length. Obviously, the dynamic response of tufts would improve as any of these properties diminishes.

Common materials for tufts are wool, various other natural and synthetic fibres, and even paper strips. For enhanced visibility, one may dip the tufts in a fluorescent dye, illuminate them with a monochromatic light and then record the images using an optical filter that removes the oncoming light, while permitting the fluorescent light to pass. One disadvantage of conventional tufts is that they are easily set in flapping motion when exposed to unsteady or highly turbulent flows. Small rigid *cones*, whose apex is attached to the surface through a thread, and *hinged tufts*, consisting of hinged sections of stiff materials, are reputed to present considerably less flapping. On the other ex-

Table 11.1 Common flow visualisation techniques; M1 denotes M = 1 etc.

Techniques for liquids	Speed range
Tufts: surface tufts, in-flow tufts, streamers, tuft screen	0.05–2 m/s
Surface marking:	
oil dots	0.5–4 m/s
oil film	0.1–25 m/s
electrolytic etching	0.01–0.1 m/s
Continuous dye injection (streakline marking):	0.5 mm/s–10 m/s
Particle tracing (pathline marking):	
suspended solid markers, droplets, bubbles	0.1 mm/s–30 m/s
floating solid markers	0.5 mm/s–5 m/s
Line marker generation (timeline marking):	
hydrogen bubbles	5 mm/s–10 m/s
thymol blue	
photochromic	
electrolytic precipitation	0.5 mm/s–0.1 m/s
Optical methods	
shadowgraph	
Schlieren	
interferometry	

Techniques for gases	Speed range
Tufts: surface tufts, in-flow tufts, streamers, tuft screen	0.1 m/s–M1
Surface marking:	
oil dots	20 m/s–M10
oil film	5 m/s–M6
sublimation	10 m/s–M2
soluble chemical film	0.01–4 m/s
temperature sensitive paint	150 m/s–M6
Continuous smoke injection (streakline marking):	0.1 m/s–M1
Particle tracing (pathline marking):	
suspended solid markers, droplets, bubbles	1–20 m/s
Line marker generation (timeline marking):	
smoke wire	0.3 m/s–8 m/s
sparks	2 m/s - M8
Optical methods	
shadowgraph	70 m/s–M4
Schlieren	2 m/s–M3
interferometry	70 m/s–M10

treme, relatively long tufts, called *streamers*, have also been used to indicate streamline patterns near and away from a surface.

In addition to being attached on a wall, tufts may be introduced in midstream using various devices, such as a *tuft screen*, which is a mesh of thin wires stretched across a stream that has tufts epoxied on its nodes; a *tuft wand*, which is a thin rod that may be easily traversed through a flow and has a tuft or a streamer attached to its tip; and *tuft pins*, which are tufts attached to thin needles protruding from a surface.

11.2.2 Surface Marking Methods

These techniques identify the direction and pattern of flow near an immersed solid surface, which become visible as a result of an effect that the flow has on an applied surface coating [28, 37]. Thus, surface flow visualisation is based on a relationship between near-wall flow and wall shear stress, pressure, temperature or mass transfer rate. Because in many cases such relationships are not known precisely and may be subjected to undesirable influences, the interpretation of surface maps is usually of qualitative nature. On the other hand, several surface visualisation methods have been refined or extended to become sufficiently accurate to provide not only surface maps, but also measurements of local wall properties.

Oil streak method: A relatively simple method that is commonly used in aerodynamics is the *oil streak* method, which is applicable to both gas and liquid flows. The surface of an immersed object is coated with a paint consisting of a pigment (e.g., TiO_2, china clay, lamp black, copier toner or fluorescent chrysene) suspended in some mineral oil (e.g., kerosene, diesel oil or light oil). When exposed to the flow, the pigment coagulates and is deposited on the surface in the form of small lumps, in the wake of which the oil is collected and forms short streaklines. At the end of the test, these patterns can be observed and photographed to provide information about flow direction, flow separation and reattachment and transition to turbulence (see, for example, Fig. 11.1a). Although an analysis of the relationship between oil patterns and flow properties is available [45], the oil streak method is generally used for qualitative purposes. It is also recognised that, in high-speed flows, the paint coating thickness may be comparable to the boundary layer thickness and thus the coating may distort the wall flow. A related method, called the *oil film* method, utilises interferometry (see Section 11.3.4) to measure wall shear stress. A variance of the oil streak method is the *oil dot* method, which consists of applying the paint in dots rather than coating the entire surface. An advantage of this method is that it may be applied to surfaces with complex shapes, such as aircraft models.

Mass transfer methods: A variety of techniques utilise change of colour of an immersed surface due to *mass transfer* towards or from it. Such techniques are mainly used to identify regions of turbulence, which enhances the mixing process. In liquid flows, this technique consists of coating the surface with a substance which reacts with a reagent dissolved in the flowing stream. A related technique exploits *electrochemical* deposition or release of a substance from a surface serving as the cathode or anode of an electrolytic configuration. In gas flows, mass transfer from a surface is utilised in the form of *sublimation* of a solid coating (e.g., hexachloroethane or naphthalene) or *evaporation* of a volatile liquid coating. In some cases, mass transfer is sufficiently substantial to leave *relief patterns* on the surface. This is the case of the *ablation* technique, used in high speed gas flows, and the *plaster of Paris* ($CaSO_4$) technique, used in water flows.

Pressure and temperature sensitive paints: Surface flow visualisation can be effected with the use of special surface coatings, the appearance of which is sensitive to pressure and/or temperature [22, 26, 27, 28, 44]. The same methods are also used

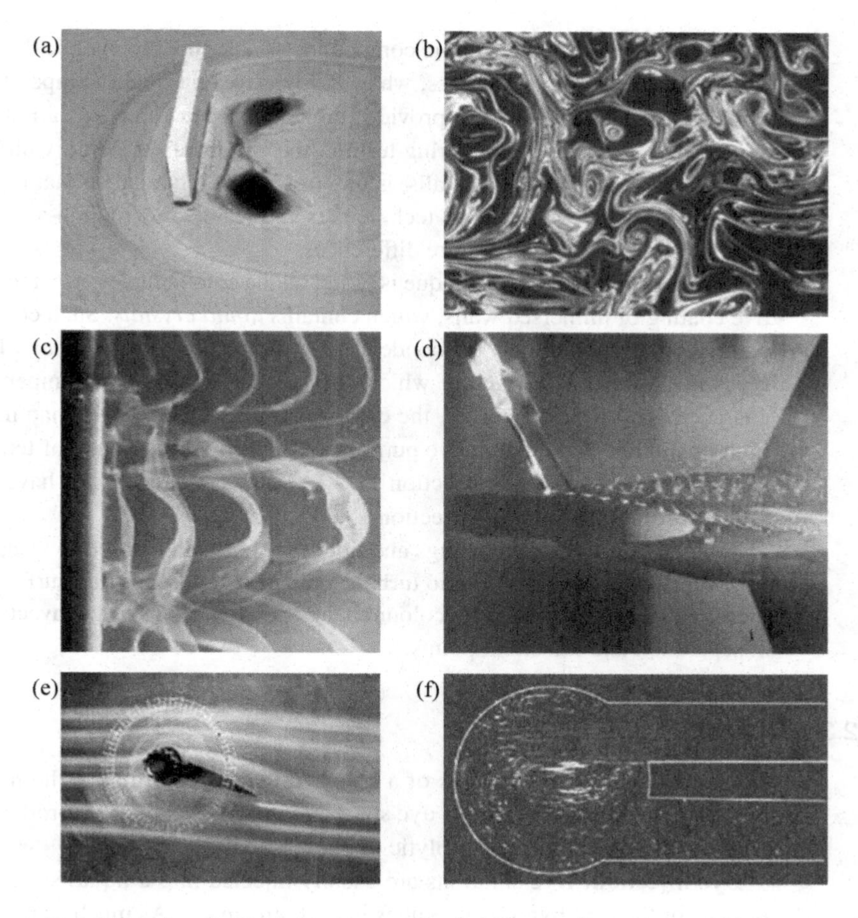

Figure 11.1 Flow visualisation with the use of relatively simple marker techniques: (a) oil streak visualisation of flow adjacent to a plane surface with a bluff protrusion; the mixture consists of a suspension of 3 ml of graphite powder in 70 ml of penetrating oil (photograph by Sean Bailey); (b) laser-sheet visualisation of fluorescent dye in highly turbulent shear flow (photograph by Sebastien Marineau-Mes); (c) visualisation of water flow past a cylinder with a step-change in diameter using electrolytic precipitation of lead (photograph by Warren Dunn); (d) hydrogen-bubble visualisation of the vortex produced by the leading-edge extension of a fighter aircraft model (photograph by Warren Dunn); (e) smoke-line (oil mist) visualisation of flow separation from an aerofoil (photograph by Warren Dunn); and (f) visualisation of pulsatile flow in a ventricular-assist-device model using laser-illuminated polystyrene particles impregnated with a fluorescent dye (photograph by Jean-Baptiste Vergniaud).

for the measurement of pressure and temperature, as discussed in Section 9.4 and Section 14.4.1. *Pressure sensitive paints* (*PSP*) contain fluorescent molecules, the luminescence of which is inversely proportional to the pressure, as well as sensitive to temperature. Similar materials, but with a relatively high temperature sensitivity, by comparison to their pressure sensitivity, are known as *temperature sensitive paints* (*TSP*). A different type of coatings contain metallic compounds which undergo chemical changes that are accompanied by colour changes, when exposed to a particular temperature over a certain time interval. Calibration provides the temperature change as a function of colour for a given exposure time. During testing, the colour of the paint, which has been applied on thermally insulated walls, is monitored and recorded so that high-temperature regions may be identified. This technique is applicable only to high-speed flows, which generate substantial temperature differences.

Liquid crystals: This technique is based on the colour change of a temperature sensitive coating of immersed walls, which contains *liquid crystals*. Such coatings are either painted on the wall or they are bonded to it in in the form of thin sheets. There are many types of liquid crystal, each of which changes colour when the temperature reaches a characteristic value. Therefore, the corresponding surface colour map may be used for both qualitative and quantitative purposes [18, 44] within a range of temperatures. Besides temperature mapping (Section 14.4.2), liquid crystal coatings have also been used to measure wall shear stress (Section 13.3.7).

Temperature sensitive coatings and liquid crystals can be used to visualise regions of flow separation and transition to turbulence. For this purpose, the surface is heated and these regions are identified by colour change due to a change in convective cooling rate at separation and transition points.

11.2.3 Dyes

In liquid flows, visible regions of a colour different from that of the main liquid may be created by either injecting a dye solution into the stream or by producing a coloured region in the fluid by an electrolytic or photolytic reaction.

Dye injection: Dye solutions are usually injected into a liquid stream through wall orifices or through hypodermic tubes in mid-stream [5]. As much as possible, the dyed stream should be injected at the local speed of the flow (*isokinetic injection*). Another requirement for the injected solution is to match, if possible, the density of the fluid (*neutrally buoyant dye*). Among commonly used dyes are milk, ink, food colourings, and various chemical dyes, such as Congo red, methylene blue and crystal violet. The choice of dye depends, among other factors, on its visibility to the particular light source that is used. To visualise complex motions and mixing, it is preferable to use two or more dyes with different colours. A special class of chemical dyes are the *fluorescent dyes*, notably rhodamine 6G, rhodamine B and fluorescein disodium; when illuminated by a laser or other monochromatic light source, and with the incident light removed by an optical filter, they remain visible through their fluorescence, while light reflections from immersed objects or the walls and windows of the apparatus become invisible (Fig. 11.1b). The *fluorescent dye-layer* method has been used in towing tanks to visualise near-wall and separated flow phenomena; in this technique, dye layers of different colours are applied on horizontal planes before the towing of a model starts [13]. *Reacting dyes* are particularly effective for visualising mixing streams; they consist of a pair of acidic and basic solutions, whose reaction produces a salt solution of a distinct colour. A disadvantage of dye injection is that, if continued over an extended period of time, it would saturate the liquid in the apparatus and the dye visibility would deteriorate. Certain dyes (e.g., milk) would also foul if they remain in the apparatus, while some chemical dyes may be toxic or corrosive and must be avoided. Dye injection is most effective in laminar flows, with speeds of the order of 0.01 m/s or less. Because the molecular diffusivity of most dyes is relatively large, they would diffuse rapidly in turbulent flows and would become indistinguishable fairly closely to their injection point. Dyes with a small diffusivity, such as fluorescent dyes, would be preferable for such cases. Special *shear-thickening dyes* have been developed, which produce streaks influenced mainly by large-scale, low-shear motions in the fluid and which are insensitive to small-scale motions [17].

Electrochemical methods: In electrochemical methods, a region of the fluid changes colour as a result of a chemical reaction, which is triggered by an electric current. In the *thymol blue technique* [1], the medium is a solution of thymol blue (pH indicator) dye in distilled water. In an acidic environment (pH < 8.0), this dye is orange-yellow but it changes colour to blue when the pH increases beyond 9.6. The water in the apparatus also contains an acid (HCl) and an electrolyte (NaOH), at proportions such that the pH is just below 8.0. A thin platinum wire (typically with a diameter of about 10 μm) is stretched across the flow and connected to the negative pole of a dc power supply, while the positive pole is connected to another electrode in contact with the water. When a current pulse is supplied to the wire, it acts as a cathode, releasing hydrogen ions in the solution, some of which combine to form molecular hydrogen. This leaves an excess of negative OH ions in the vicinity of the wire, which make the solution alkaline and cause the dye to turn blue within a thin cylindrical volume that surrounds the wire. Thus, this method creates a timeline, which is convected downstream with the flow. Because the marked fluid is locally neutrally buoyant, the thymol blue method has been used effectively in stratified and rotating flows, mainly related to atmospheric and oceanic flow phenomena. A similar method of producing timelines is the *tellurium wire* technique [11]. In this approach, a cathode, consisting of a thin tellurium wire, is supplied with current pulses. Tellurium ions are released into the flow and form a black colloidal suspension. The fluid is alkaline water ($9 < pH < 10$), also containing some H_2O_2, which releases O_2 for stabilising the suspension. A related approach is *electrolytic precipitation* [46]. Part of an immersed surface is coated with lead, tin, solder or copper and connected as an anode. Following a current pulse, a white crystalline salt is released in the form of a cloud of microscopic spherical particles. This method is particularly effective for the visualisation of separating boundary layers and vortex formation behind bluff objects Fig. 11.1c).

Photolytic methods: Finally, coloured fluid can be produced locally by *photocatalysis* [34, 35]. The working fluid contains a *photochromic* indicator, for example pyridines or spyrans, which is normally colourless. Typically, when a molecule of such substances absorbs a photon in the UV range, it temporarily changes its absorption properties, becoming capable of absorbing photons in the red-green range of the spectrum. When subsequently illuminated by white light, regions containing such molecules become visible, taking a dark blue appearance. Such substances can be dissolved in alcohol, kerosene and other solvents, but not in water, which limits the usability of this method to relatively small, contained facilities. The reaction is reversible, as the substance returns to its original state following a typical time of a few seconds. This property makes this method suitable for closed-circuit experiments, as it does not contaminate the fluid. A common application of the method uses a pulsed laser beam, which produces a time-line that can be followed in time to yield velocity profiles. It is also possible to mark small parcels of the fluid by focusing a pulsed light beam.

11.2.4 Hydrogen Bubble Method

Bubble production: When current flows between two electrodes separated by conductive water, it will cause its *electrolysis*, releasing hydrogen gas at the cathode and oxygen gas at the anode, according to the chemical reaction

$$H_2O \rightarrow H_2 + \frac{1}{2}O_2 \,. \tag{11.1}$$

The gas will be attached to the electrodes until buoyancy or drag from a flowing fluid detaches it in the form of bubbles. Although both gases can serve as flow markers, hydrogen is preferable, first because it is produced at twice the volume rate than oxygen, and second because oxygen would oxidise the material of many possible electrodes. Most commonly, a thin metallic wire, made of platinum, nickel or stainless steel with typical diameters in the range $d_w = 10{-}100$ μm, is used as a cathode, while a much thicker carbon rod is used as the anode. When a stream flows past a thin wire, connected as cathode, hydrogen bubbles would be released repeatedly, forming a sheet.

Bubble size: The bubble diameter d_h during its release depends on the cathode shape and quality, the cathode current, the surface tension and the flow speed. For a clean, thin, cylindrical cathode one may use the rough estimate $d_h \approx d_w/2$. The bubble diameter would change as the bubble moves to a location where the pressure is different from the pressure at the location of bubble production. Considering that the hydrogen mass inside each bubble remains fixed, while the bubble is transported, one may use the ideal gas law to relate the local bubble diameter to the local pressure. It is noted, however, that in many applications of interest, the relative change of pressure is quite small and the resulting relative change of bubble diameter would be even smaller, because the diameter is proportional to the third root of the bubble volume, which changes inversely proportionately to the pressure.

Bubble rate of production: The rate of mass production of hydrogen by a current I is, according to *Faraday's law of electrolysis*,

$$\dot{m} = kI \,, \tag{11.2}$$

where $k = 1.04 \times 10^{-8}$ kg/(sA) [47]. For simplicity, we will first assume that the entire length of the wire is immersed in water that has a uniform pressure p_w. Then, we may compute the volume flow rate of hydrogen as $Q = \dot{m}/\rho_h$, where ρ_h is the density of hydrogen inside the bubbles. This can be found from the ideal gas law and Eq. (1.12) as

$$\rho_h = \frac{p_h}{R_h T} = \frac{p_w + 4\sigma_h/d_h}{R_h T} \,, \tag{11.3}$$

where p_h is the pressure inside the bubbles, $R_h = 4124$ J/(kgK) is the hydrogen gas constant, T is the absolute temperature and the surface tension σ_h of the hydrogen–water interface may be taken as approximately equal to that of the air–water interface, namely, as $\sigma_h \approx 0.072$ N/m.

From this expression, one may calculate the number of bubbles $\dot{N}_h$ released from the wire per unit time as

$$\dot{N}_h = \frac{\dot{Q}}{\pi d_h^3/6} = \frac{6}{\pi} \frac{R_h T}{(p_w + 4\sigma_h/d_h)\, d_h^3} kI \, . \tag{11.4}$$

If we assume that hydrogen is released at a uniform rate along a wire that has a length l, then the number of hydrogen bubbles per unit time and per unit length of the wire would be $\dot{N}_h/l$.

If water pressure differences are significant along the wire, the rate of production of bubbles would vary from one point of the wire to another. If we assume that the number of released hydrogen molecules per unit wire length is uniform along the wire, then the rate of released hydrogen mass along an elementary wire length $\mathrm{d}s$ would be $kI\mathrm{d}s/l$ and the local rate of production of bubbles over a length $\mathrm{d}s$ at a distance s along the wire would be

$$\frac{\dot{N}_h(s)\mathrm{d}s}{l} = \frac{6}{\pi} \frac{R_h T}{(p_w + 4\sigma_h/d_h)\, d_h^3} \frac{kI\mathrm{d}s}{l} \, , \tag{11.5}$$

from which we can find the total rate of production of bubbles as

$$\dot{N}_{h,tot} = \frac{1}{l} \int_0^l \dot{N}_h(s)\mathrm{d}s \, . \tag{11.6}$$

Practical details: For a good visibility of the bubble sheet, one requires a sufficiently large bubble concentration and, thus, a sufficiently large current. With other conditions being fixed, it is clear that the required current should be increased proportionately to the flow speed. One must also consider that one can improve bubble visibility dramatically by optimising the illumination setting (see Section 7.3.9). Typical current values used in practice are between 0.02 and 1 A. The electrode voltage that must be applied to produce the desired current is equal to the current times the water path electric resistance. The latter depends on the distance between electrodes, the quality of surface of the electrodes, the water temperature and the presence of electrolytes. Untreated tap water usually contains a sufficient amount of salts to ensure good conductivity. To further reduce the water path resistance, it is customary to add a small amount of sodium sulfate (Na_2SO_4). Required voltages are in the range between a few volts and several hundred volts and one must take precautions against a possible electric hazard.

The use of a *straight wire* supplied with a dc current would produce a continuous sheet, which may be suitable for many flow visualisation purposes (Fig. 11.1d), but cannot differentiate between different velocities on the bubble plane. Supply of periodic short current pulses would essentially mark timelines, which may be followed in time to devise flow patterns. Another variant of this method is to release bubbles only at selected locations along the wire. To do so, one may insulate sections of the wire, while leaving short non-insulated sections in-between, but this approach will likely generate large bubbles at the boundaries of the insulated sections, which would disturb the local flow. A more successful approach is to use a *kinked wire*, fabricated, for example, by squeezing the wire between two fine gears; in this case, bubbles are generated on the entire wire length but released only at the apices, thus introducing streaklines with substantial bubble concentration. Another approach is to use a *ladder-type probe*, with the bubbles produced along the 'steps' of the ladder, which are aligned with the flow direction. Combination of current pulsation and pointwise bubble release produces a grid of small bubble markers, which can serve as combined *time-streaklines*, mapping the three-dimensional flow velocity field.

The hydrogen bubble technique has been used extensively in water flows and, to a lesser degree, in flows of aqueous solutions and mixtures. In particular, this technique has been successfully applied to the study of low Reynolds number phenomena, using water–glycerol mixtures, with glycerol concentrations exceeding 90% [39]. Although primarily used for qualitative purposes, hydrogen bubble images can be analysed to yield flow velocity fields and even some turbulence statistics [3, 9, 41]. Post-processing of simultaneous records of two perpendicular views allows the reconstruction of three-dimensional motions and velocity vectors [43]. To validate the accuracy of such calculations, one needs to estimate the relative velocity between a typical bubble and the surrounding fluid and, particularly, the rise speed of the bubbles due to buoyancy (see Section 7.6).

Example 11.1 Consider applying the hydrogen bubble method to visualise the vertical profile of horizontal flow in a water channel with a depth of 0.80 m and a temperature of 20°C, as shown schematically in Fig. 11.2. The bubble-producing wire has a diameter of 0.1 mm and has been kinked to form a sawtooth-like shape, in which each element is an equilateral triangle with a spacing of 4 mm along the wire axis. Assume that the wire is supplied with a direct current of 0.4 A. Compute the number of bubbles that will be released per unit time from each apex of the wire at different depths from the free surface. Assuming that the bubbles rise very slowly, estimate the sizes of bubbles released at different depths when they reach the free surface. Furthermore, making any simplifying assumptions that you find appropriate, estimate the vertical velocity of a bubble released at a certain depth, when it rises to a different depth (Hint: Assume that each bubble is always rising with its local terminal velocity.)

Answer

The number of apices acting as release points of hydrogen bubbles is 0.80/0.004 – 1 = 199. The vertical distance of apex i from the free surface is

$$h_i = 0.004(i - 0.5)\text{m} .$$

The diameters of all bubbles at their release time may be assumed to be equal to half the wire diameter, so that $d_h = 0.0005$ m, irrespectively of the release depth. Then, the

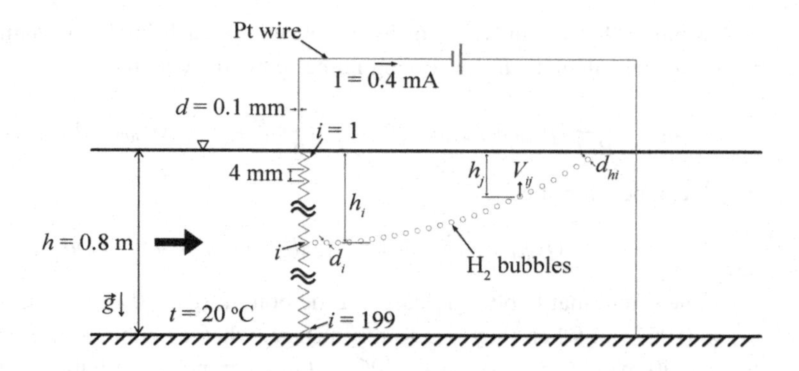

Figure 11.2 Sketch of water flow visualised with the use of the hydrogen bubble method (not to scale).

release pressure inside a bubble released at apex i would be

$$p_i = p_{atm} + \rho_w g h_i + 4\sigma_h / d_h \,,$$

where $p_{atm} = 101.3$ kPa is the atmospheric pressure, $\rho_w = 998$ kg/m^3 is the water density and $\sigma_h = 0.072$ N/m is the surface tension of gas–water interfaces. The total mass flow rate of hydrogen released from the entire length of the wire is $\dot{m} = kI$, where $k = 1.04 \times 10^{-8}$ kg/(sA). Assuming uniform release of electrons along the wire, the mass flow rate of hydrogen released at each apex would be $\dot{m}_i = \dot{m}/199 = kI/199$. Then, the number of bubbles released per unit time from apex i would be

$$\dot{N}_{hi} = \frac{6}{\pi} \frac{R_h T}{p_i d_h^3} \frac{kI}{199} \,.$$

As each bubble rises from its release depth to the surface, its diameter will increase from d_h to d_{hi}, which depends on the depth of the apex at which the bubble was released. The pressure inside a bubble when it has risen just below the free surface would be

$$p_s \approx p_{atm} + 4\sigma_h / d_{hi} \,.$$

As pressure decreases upwards, the size of rising bubbles would increase so that the density of hydrogen inside a bubble released at depth h_i would decrease from a value ρ_i at the time of release to a value ρ_{is}, when it rises just below the free surface. If the bubbles rise slowly enough for dynamic and thermodynamic non-equilibrium effects to be negligible, the two densities would be related by the ideal gas law

$$\frac{p_i}{\rho_i} = R_h T = \frac{p_s}{\rho_{is}} \,,$$

which gives

$$\rho_{is} = \rho_i \frac{p_{atm} + 4\sigma_h / d_{hi}}{p_{atm} + \rho_w g h_i + 4\sigma_h / d_h} \,.$$

Assuming that the mass m_h of hydrogen within each bubble remains constant as it rises, we can calculate d_{hi} by solving numerically the equation

$$m_h = \frac{1}{R_h T}(p_{atm} + \rho_w g h_i + 4\sigma_h/d_h)\frac{1}{6}\pi d_h^3 = \frac{1}{R_h T}(p_{atm} + 4\sigma_h/d_{hi})\frac{1}{6}\pi d_{hi}^3 ,$$

which can be simplified to

$$(p_{atm} + \rho_w g h_i + 4\sigma_h/d_h)d_h^3 = (p_{atm} + 4\sigma_h/d_{hi})d_{hi}^3 .$$

It can be found that bubbles released at the bottom ($h_i = 0.80$ m) will have a diameter $d_{hi} = 0.0513$ m when they rise to the surface, whereas bubbles released at mid-depth ($h_i = 0.40$ m) will have $d_{hi} = 0.0506$ m. One may note that bubble size changes only slightly. The reason for this is that the hydrostatic pressure change over the depth of this channel is much smaller than the atmospheric pressure.

The accurate calculation of the rising bubble speed is a difficult problem, as the bubble size changes with depth and even its shape could be affected by the motion. A very rough estimate would be to assume that, when a bubble that was released at a depth h_i rises to a depth h_{ij}, it would move with a terminal velocity V_{ij} corresponding to its local size. Under this assumption, the buoyancy, weight and drag force would be in balance. Assuming further that the bubbles are spherical and that Stokes' law is valid, we can derive the equation of motion as

$$(\rho_w - \rho_{ij})g\frac{1}{6}\pi d_{hj}^3 = 3\pi\mu d_{hj} V_{ij} .$$

The local bubble density and diameter can be found as previously, from which the bubble velocity can be found as

$$V_{ij} = \frac{(\rho_w - \rho_{ij})g d_{hj}^2}{18\mu} .$$

This problem can be further simplified, if we assume that the bubble density and diameter remain approximately constant as the bubble rises, in which case we can assume that $\rho_{ij} \approx (p_{atm} + 4\sigma_h/d_h)/(R_h T) = \text{kg/m}^3$ and $d_{hj} \approx d_h = 0.0005$ m.

11.2.5 Smoke, Mists and Fogs

In flow visualisation, *smoke* is a broadly understood term, which includes suspensions in air or other gases of solid products of combustion (which are properly referred to as smoke) as well as liquid droplets (*mists* or *fogs*) and visible vapours and tracer gases. The suspended particles are usually small enough (typically, smaller than 1 μm) to follow closely the fluid motion. Smoke visualisation has been largely used in wind-tunnel studies, but it can be effectively used in any apparatus that uses gas as a medium. A disadvantage of this method is the accumulation of smoke in the apparatus and the need for occasional cleaning of the windows and other exposed surfaces. The most common methods for generating smoke are described in the following.

Burning: Common combustible materials, such as incense, wood chips and paper, are suitable for many applications and a variety of *smoke generator* designs are available [31]. For improved performance, one may control the supply of oxygen in order to produce a denser smoke and pass the smoke through a filter to remove the larger particles before injection into the flow. Appropriate exhaust and ventilation are required, if considerable amounts of smoke are used.

Vaporisation and condensation: This is the most popular approach, usually involving mineral oils (e.g., kerosene). Smoke generators of this type are available at a relatively low cost. Oil flowing in a tube is heated by an electric heater and evaporates; then, it mixes with a cool air stream, in which it condenses, forming a mist of tiny droplets, in the sub-micron range (*oil mist*). Large supplies of smoke must be cooled before injection into the flow, in order to avoid buoyancy effects or damage to the test models or facility. For low-speed, small-scale air flows, one may use steam fog, dry ice (solid CO_2) or liquid nitrogen, which are non-polluting. Commercial fog generators are available to generate thick *fogs* of non-hazardous, water-based liquids in small or large quantities. A related method used exclusively in supersonic flows is the *vapour-screen method* [25]. For this method, moist air is supplied to the supersonic nozzle of the wind tunnel and, as it expands, it condensed into a fog. As the air flows around a model, the concentration of droplets becomes non-uniform and, thus, regions of separation, wakes etc.become visible.

Aerosol generation: A popular material is titanium tetrachloride ($TiCl_4$). In this method, the aerosol, created by passing compressed air through liquid $TiCl_4$ is passed through water, which reacts with it, producing titanium oxide (TiO) in the form of a dense white mist. The use of this material in uncontrolled environments should be avoided as it is toxic. Other organic or inorganic materials have also been used in a similar fashion.

Injection of smoke in bulk into a stream may be useful in identifying flow boundaries; however, the connection between visible smoke patterns and flow characteristics is quite uncertain. Smoke can be injected from an orifice at a wall or, through a thin tube, inside a stream, in the form of a streakline, called a *smoke line* (Fig. 11.1e). Injection from a rake of tubes in steady flow would mark a set of streamlines, which makes it a convenient method for visualising air flows past models in wind tunnels. A technique that introduces a very small amount of smoke in the wind tunnel is the *smoke wire method* [7, 51]. For this method, a thin wire is stretched vertically across the wind tunnel and a small quantity of oil from a reservoir is let to drip along it. Because of surface tension, the oil forms small droplets attached to the wire at even distances from each other. A pulse of electric current to the wire heats it and evaporates the oil, forming a set of streaklines. A camera, triggered in synchronisation with the electric pulse, is used to record the image. This method is suitable only for very small air speeds.

11.2.6 Solid Markers, Bubbles and Droplets

Besides hydrogen bubbles, dyes and smoke, a great variety of materials have been used as markers/tracers in flow visualisation studies [28] (Fig. 11.1f). One requirement for their selection is that they should follow the local fluid motion, at least to the degree that is resolved by the observation/recording system. In this respect, the requirement is not as severe as for velocity measurement, and markers which only follow the large-scale motions of the fluid may be acceptable. Ideally, the markers should be neutrally buoyant, a condition that is easily achievable for water and other liquid flows, but not for gas flows. Liquid free-surface motion can be visualised by sprinkling it with various powders, such as talcum, aluminium flakes, lycopodium and even paper shreds. Unless there is vigorous mixing, the solid particles would be suspended by surface tension and their motion would mark pathlines. The only neutrally buoyant markers used in gases are *helium-filled soap bubbles*, which may be produced by bubble generators to typical sizes in the range 0.3–1 mm [19]. The dynamic response of markers improves dramatically with decreasing size, although at the expense of visibility. Therefore, the optimal size of a marker is one that balances the requirements of dynamic response and visibility. The visibility of markers also depends on their reflectivity, which is proportional to the difference between their refractive index and that of the fluid. Thus, preferred marker materials would include aluminium, glass and droplets of liquids with a large refractive index. For enhanced visibility, solid markers may be coated with luminescent or fluorescent paints or even have fluorescent molecules substance incorporated in their chemical composition. Irrespectively of other requirements, markers which are hazardous to health or corrosive to the facility should be avoided.

Many types of flow markers (e.g., polystyrene or glass beads and liquid droplets) have approximately spherical shapes, as a result of their generation by methods that depend strongly on surface tension. Nevertheless, non-spherical markers have also been used in many applications. These include irregularly shaped solids, which are produced by crushing or filing a material, flakes and fibres. The latter two types have relatively small inertia, while being highly visible when viewed from certain angles. Flakes and fibres are aligned with surfaces of maximum shear, so they can also be used to indicate the direction of shear. A water-based suspension of microscopic crystalline platelets, known as *kalliroscope fluid*, is particularly effective for the viewing of complex vortical flows [23, 40, 48]. A limitation of this method and other methods that use dense suspensions of markers is that they render the fluid opaque beyond a short distance from the viewing window.

Example 11.2 Solid glass beads are used as markers to visualise a low Reynolds number, laminar flow of a mixture of 80% glycerol and 20% water at 20°C. The density and kinematic viscosity of the mixture can be found from tables as $\rho_F = 1{,}208$ kg/m^3 and $\nu_F = 49.6 \times 10^{-6}$ m^2/s, respectively. The glass beads have a diameter $d_g = 20$ μm and a density $\rho_g = 2{,}600$ kg/m^3, so that the density ratio is $\gamma = 2.15$. While the study is in progress, the supply of glass beads is exhausted and it becomes necessary to use different flow markers. It is required that such markers should have a dynamic response, which is equal to or higher than that of the glass beads.

(a) Compute the maximum diameter of aluminium oxide particles ($\gamma = 3.28$), polystyrene beads ($\gamma = 0.832$) and hydrogen bubbles ($\gamma = 69.5 \times 10^{-6}$), which are available for use as substitutes of the glass beads. State clearly and evaluate all assumptions that you make.

(b) Assuming that the maximum relative sinking or rising velocity of all particles should not exceed by more than 5% the true flow velocity, determine the minimum horizontal and vertical flow velocities of flows that can be visualised adequately by the glass beads and the particles found in (a).

(c) If the maximum horizontal flow velocity is 20 times larger than each of the values computed in (b), determine the maximum exposure time of the camera lens that would permit an approximately non-blurred recording of individual images of each of the previous particles, assuming that such recording requires that the particle should not move by more than 25% of its diameter during recording.

Answer

Table 11.2 lists values of the different properties of the available particles.

Table 11.2 Properties of different particles

Particle material	ρ_P [kg/m^3]	γ [–]	τ_P [μs]	τ_{Pa} [μs]	d_P [μm]	$\lvert u_{P\infty}\rvert$ [μm/s]	$\lvert u_{Fmin}\rvert$ [mm/s]	Δt [ms]
Glass	2,600	2.15	0.96	1.19	20.0	5.1	0.10	2.5
Al_2O_3	3,960	3.28	1.03	1.19	16.8	7.0	0.14	1.5
Polystyrene	1,005	0.832	0.74	1.19	28.2	1.5	0.029	0.12
Hydrogen	0.084	69.5 ×10^{-6}	0.16 ×10^{-3}	1.19	46.5	23.3	0.47	1.2

(a) Considering that the medium is liquid, the requirement of large density ratio ($\gamma \gg 1$) cannot be met by either the glass beads or any of the suggested particles. As a result, the heavy particle expression for the time constant would be inapplicable. A more appropriate characteristic time for spherical rigid particles in relative Stokes flow was given as (Eq. (7.59))

$$\tau_{Pa} = \tau_P\left(1 + \frac{1}{2\gamma}\right) = \frac{\rho_P d_P^2}{18\mu_F}\left(1 + \frac{\rho_F}{2\rho_p}\right).$$

The high viscosity of the medium hints to the likely validity of the Stokes expression. The only obvious deviation from these assumptions would be for the hydrogen bubbles, which are not rigid and may not be spherical, but, in the absence of any additional information, we may ignore these effects. So, two particles will have comparable dynamic responses, if they have equal values of τ_{Pa}. This implies that the diameter of alternative particles that would have the same characteristic time as the glass beads is

$$d_P = d_g\sqrt{\frac{2\rho_g + \rho_F}{2\rho_P + \rho_F}}.$$

The densities of the two solid markers are listed in Table 7.7 as $\rho_{AO} = 3{,}960$ kg/m^3 for aluminium oxide and $\rho_{Pol} = 1{,}005$ kg/m^3 for polystyrene. For the hydrogen bubbles, we may neglect the hydrostatic pressure change in the mixture and the surface tension term to assume that the pressure is approximately atmospheric, from which we estimate the density, using the ideal gas law, as $\rho_{H_2} \approx p_{atm}/(R_{H_2}T) \approx 0.084$ kg/m^3. Then, we find the equivalent diameters of the alternative particles as $d_{AO} = 16.8$ µm, $d_{Pol} = 28.2$ µm and $d_{H_2} = 46.1$ µm. We may iterate for the hydrogen bubbles by considering the surface tension effect, but note that the correction term in the first iteration is about 6,200 Pa $\approx 6\% p_{atm}$, which may be neglected for the present purposes.

(b) Obviously, an accurate estimate of the time-dependent relative velocity requires knowledge of a lot of information that is not available, and, even if more were known, this would be an extremely difficult problem to solve. So we need to make some drastic simplifications to get quick, but not very inaccurate, estimates.

First, let us assume that the undisturbed fluid velocity u_F is vertical and uniform. Then, the maximum relative velocity of the particle in the vertical direction will be its terminal relative velocity

$$|u_P - u_F|_\infty = |1 - 1/\gamma| g \tau_P \, .$$

For the previously determined particle diameters, the corresponding terminal velocities are found to be about 5 µm/s (sink) for glass beads, about 7 µm/s (sink) for aluminium oxide particles, about 1.5 µm/s (rise) for polystyrene and 23.3 µm/s (rise) for hydrogen bubbles. For vertical flow velocities, the particle terminal relative velocity would represent an estimate of a maximum error, so for this error to be no more than 5%, the vertical flow velocity (upwards or downwards) should be

$$|u_F| > 20|(u_F - u_P)_\infty| = 20|1 - 1/\gamma| g \tau_P \, .$$

Next, assume that the fluid moves horizontally with a uniform velocity u_F. After its transient response has vanished, the particle would asymptotically move horizontally with the same speed u_F as the fluid and vertically with its terminal velocity

$$|u_{P\infty}| = |1 - 1/\gamma| g \tau_P \, .$$

The particle velocity magnitude would be $\sqrt{u_F^2 + u_{P\infty}^2}$. In order for $|u_{P\infty}| < 0.05|u_F|$, we need

$$|u_F| > 20|u_{P\infty}| = 20|1 - 1/\gamma| g \tau_P \, ,$$

which is the same expression as for vertical flow. As the table shows, the minimum acceptable flow velocity for either vertical or horizontal flow is in all cases lower than 1 mm/s.

(c) Now, it is given that the fluid moves with a horizontal velocity u_F, which is 20 times larger than the corresponding value found previously. For such a large speed, the error due to the relative velocity of the particle may be neglected for all cases. During the exposure time Δt, the particle would move by a distance $\Delta x = u_F \Delta t$, which must be less than $0.25 d_P$. From this expression, it is easy to estimate the maximum exposure time as $\Delta t_{max} = 0.25 d_P / 20 |u_{Fmin}|$. These values are shown in the table.

11.3 Optical Techniques

In the following, we shall present the simplified principles of operation of some classical optical flow visualisation techniques. More details and descriptions of advanced methods can be found in specialised sources [24, 42].

11.3.1 Light Deflection in Variable Refractive Index Media

When a light ray propagates a medium that has a varying refractive index, it follow a path that will minimises the time it takes to pass through this medium, according to Fermat's principle (see Section 7.2.2). Use of variational calculus leads to a set of differential equations for the coordinates of this path, the solution of which provides the path of the ray as a function of the refractive index field [28, 50]. The following simplified analysis illustrates the procedure, while avoiding the use of advanced mathematical tools.

Consider that, as illustrated in Fig. 11.3, two closely spaced light rays first propagate along straight paths through medium 1, which occupies the space in the range $-\infty < x < x_1$ and has a uniform refractive index n_o. Then, the rays enter perpendicularly into medium 2, which extends in the range $x_1 < x < x_2$ and has a variable refractive index $n(x, y)$. Assume that $n(x, y)$ is only slightly larger than n_o and such that $\partial n/\partial y > 0$, which infers that the rays will follow paths curving upwards in medium 2. Finally, the rays enter into medium 3, which also has a uniform refractive index n_o, and so they resume straight paths. The distance between the two rays in medium 1 is δy and changes in medium 2 to δr, where r is the local radius of curvature of the ray. The symbols s and s' indicate the curvilinear distances along the two rays. The speed of light along the

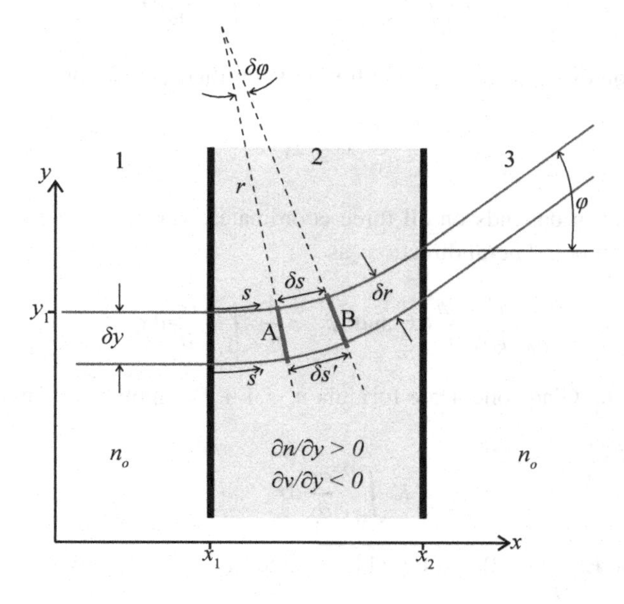

Figure 11.3 Sketch illustrating the passage of light rays through a medium with a variable refractive index; proportions have been distorted for clarity.

two rays would be $v(x, y)$ and $v(x, y) + \delta v(x, y)$, while it is reminded that $n = c/v$, where c is the speed of light in vacuum. We can assume that the deflection of the rays is very small, so that $\delta s \approx \delta x$ and $\delta r \approx -\delta y$. Now consider a wave front A inside medium 2; during a small time interval δt, this wave front will move to position B, rotating by an angle $\delta\varphi$, which can be expressed as

$$\delta\varphi = \frac{\delta s}{r} = \frac{v\delta t}{r} = \frac{\delta s'}{r + \delta r} = \frac{(v + \delta v)\,\delta t}{r + \delta r} = \frac{\delta v \delta t}{\delta r} . \tag{11.7}$$

This expression provides the radius of curvature of the top ray as

$$\frac{1}{r} = \frac{1}{v}\frac{\delta v}{\delta r} \approx -\frac{1}{v}\frac{\partial v}{\partial y} = -\frac{1}{c/n}\frac{\partial\,(c/n)}{\partial y} = \frac{1}{n}\frac{\partial n}{\partial y} . \tag{11.8}$$

Differential calculus provides another expression for the radius of curvature of the ray as

$$\frac{1}{r} = \frac{\mathrm{d}^2 y/\mathrm{d}x^2}{\left[1 + (\mathrm{d}y/\mathrm{d}x)^2\right]^{3/2}} \approx \frac{\mathrm{d}^2 y}{\mathrm{d}x^2} . \tag{11.9}$$

Equating Eq. (11.8) and Eq. (11.9), one gets a differential equation for the ray path as

$$\frac{\mathrm{d}^2 y}{\mathrm{d}x^2} \approx \frac{1}{n}\frac{\partial n}{\partial y} . \tag{11.10}$$

This equation, together with appropriate boundary conditions, can be solved for the path $y(x)$, $x_1 < x < x_2$. The angle $\delta\varphi$ also represents the rotation of the ray during time δt. The total deflection angle of the ray in medium 2 would approximately be

$$\varphi \approx \int_{x_1}^{x_2} \frac{\mathrm{d}x}{r} \approx \int_{x_1}^{x_2} \frac{1}{n}\frac{\partial n}{\partial y} dx \approx \frac{1}{n_o}\int_{x_1}^{x_2} \frac{\partial n}{\partial y}\mathrm{d}x . \tag{11.11}$$

If n is only a function of y, one may get the further simplified expression

$$\varphi \approx \frac{1}{n_o}\frac{\mathrm{d}n}{\mathrm{d}y}(x_2 - x_1) . \tag{11.12}$$

If, on the other hand, n depends on all three coordinates, x, y, z, one has to consider angular deflections on both lateral directions as

$$\varphi_y \approx \frac{1}{n_o}\int_{x_1}^{x_2} \frac{\partial n}{\partial y}\mathrm{d}x \text{ and } \varphi_z \approx \frac{1}{n_o}\int_{x_1}^{x_2} \frac{\partial n}{\partial z}\mathrm{d}x . \tag{11.13}$$

In gases for which the Gladstone–Dale formula $n = 1 + K\rho$ applies, one may rewrite Eq. (11.11) as

$$\varphi \approx K \int_{x_1}^{x_2} \frac{\partial \rho}{\partial y}\mathrm{d}x \tag{11.14}$$

and similarly rewrite Eq. (11.12) and Eq. (11.13) in terms of the density ρ.

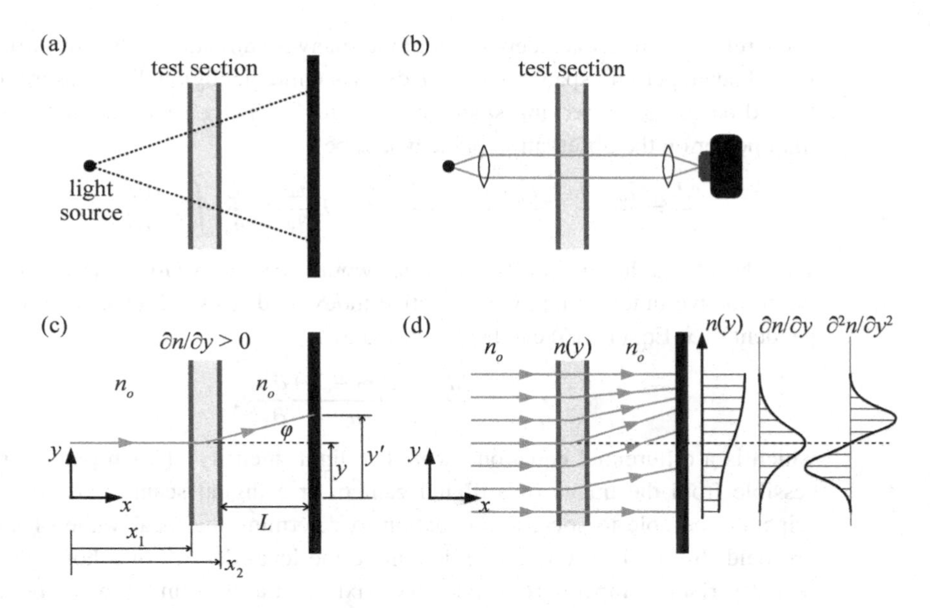

Figure 11.4 (a) Divergent-light shadowgraph; (b) collimated-light shadowgraph; (c) light beam deflection; and (d) light beam deflection through a mixing layer.

11.3.2 The Shadowgraph Method

This is a strictly qualitative flow visualisation method, possible to implement by relatively simple means. Like other optical methods, it is based on the variation of refractive index within a medium, and, thus, it is applicable to variable density flows. The shadowgraph operation may be explained by mainly geometrical optics concepts. A simple shadowgraph set-up consists of a concentrated light source, approximating a point source, and a projection plane (Fig. 11.4a). The diverging light from the source passes through the fluid and gets refracted by local refractive index variations, so that its projection appears to have variations of light intensity with darker ('shadows') and brighter regions. Because it would be difficult to record the projected image, due to its large size, a more usable shadowgraph uses one lens to collimate the light and a second lens to collect the light after it crosses the region of interest and focus it into the lens of a camera (Fig. 11.4b). To improve the contrast of shadowgrams, high-intensity point sources, such as sparks or pulsed lasers, would be preferable.

To understand the operation of the shadowgraph, consider the propagation of a light beam through a medium with $\partial n/\partial y > 0$, as shown in Fig. 11.4c. The beam will be deflected by a total angle φ and reach the observation plane at a position y', which has been shifted from the position y that would have been reached, if the refractive index were uniform along the entire beam path. If the optical disturbances are small, φ is given by Eq. (11.11) and one may calculate the beam deflection as

$$\Delta y = y' - y = L \tan \varphi \approx L\varphi \approx \frac{L}{n_o} \int_{x_1}^{x_2} \frac{\partial n}{\partial y} \mathrm{d}x\,. \tag{11.15}$$

If the refractive index is independent of the spanwise direction z, the power of the undisturbed beam per unit span would be $I_o \mathrm{d}y$. Assuming negligible light absorption, the deflected beam power per unit span would be $I\mathrm{d}y' = I_o\mathrm{d}y$. Then, the disturbance of the light power on the observation plane would be

$$\frac{\Delta I}{I_o} = \frac{I_o - I}{I_o} \approx \frac{I_o - I}{I} = \frac{\mathrm{d}y'}{\mathrm{d}y} - 1 \approx L\frac{\partial \varphi}{\partial y} \approx \frac{L}{n_o}\int_{x_1}^{x_2} \frac{\partial^2 n}{\partial y^2}\mathrm{d}x\,. \tag{11.16}$$

This shows that the shadowgram contrast would be sensitive to the second derivative of the refractive index. When the refractive index field is two-dimensional, that is, independent of x, Eq. (11.16) can be simplified to

$$\frac{\Delta I}{I_o} \approx \frac{L\,(x_2 - x_1)}{n_o}\frac{\partial^2 n}{\partial y^2}\,, \tag{11.17}$$

which is a differential equation for n. The light intensity of each pixel is readily accessible from the image of a digital camera or a digital scanner and one would, in principle, be able to solve this equation to determine the local value of n in the entire field, from which one could determine the local density of a homogeneous fluid with a variable temperature or a binary mixture at a uniform temperature. This is not done, however, because this process is distorted by several different effects. First of all, this approach applies to two-dimensional refractive index fields and so it would not be valid for three-dimensional fields and for strongly turbulent fields, even if the latter are two-dimensional on the mean. Second, the computation of n from Eq. (11.17) would require a double integration, which entails a drastic low-pass filtering action. Finally, because refractive index changes are very small, the shadowgraph contrast would need to be enhanced by different means, which is likely to introduce considerable noise. A more general expression for the light power disturbance caused by three-dimensional refractive index fields is

$$\frac{\Delta I}{I_o} \approx \frac{L}{n_o}\int_{x_1}^{x_2} \left(\frac{\partial^2 n}{\partial y^2} + \frac{\partial^2 n}{\partial z^2}\right)\mathrm{d}x\,. \tag{11.18}$$

As an illustration of shadowgraph application, consider a two-dimensional mixing layer of two fluid streams with different temperatures (or two different fluids), with the mean flow direction parallel to the z axis, as shown in Fig. 11.4d. The same figure shows the variation of the refractive index and its first two derivatives. It may be seen that the largest deviations from the average light intensity on the observation plane would occur at locations at which $\left|\partial^2 n/\partial y^2\right|$ is maximum, while, at locations at which $\left|\partial^2 n/\partial y^2\right|$ is small, the light intensity would be nearly equal to the undistorted intensity.

Shadowgrams are a valuable tool for the qualitative assessment of high-speed gas flows, flows involving heating or cooling of fluids and flows involving mixing of different fluids. They are particularly successful in visualising shock waves, expansion and compression waves, and the boundaries of high-speed or heated wakes and jets. The much cited shadowgram of a mixing layer, shown in Fig. 11.5, clearly illustrates the boundaries of the mixing layer and the formation of large-scale coherent structures, in this case, quasi-two-dimensional roller vortices. In interpreting shadowgrams of turbulent flows, one has to keep in mind that, because they are sensitive to second derivatives of the density, they tend to illustrate small-scale phenomena with much higher contrast than large-scale ones. The grainy appearance of shadowgrams is not necessarily an indication of isotropic turbulence structure, but an artefact of the method.

Example 11.3 Consider the shadowgram of a two-dimensional, helium–nitrogen mixing layer, shown in Fig. 11.5. Determine and plot the profiles of the refractive index of the mixture and the concentration of helium along a line normal to the mixing layer axis. Discuss the accuracy of this method.

Answer

This analysis requires information, which is only in part provided by the related references [2, 21, 36, 49] and was estimated roughly under some drastic simplifications and assumptions. To obtain this shadowgram, a photographic plate was positioned on the inner side of the back wall of the test section, which had a width (in the notation of Fig. 11.3) equal to $x_2 - x_1 = 101.6$ mm. The height of the test section was 50.8 mm, but the image shown in Fig. 11.5 has an estimated height of 36 mm. The available references do not specify the static pressure and the temperature T of the flow, but our best estimates were $p \approx 405$ kPa and $T \approx 293$ K. Using the ideal gas law, we then determined the densities of the two pure gases as $\rho_{He} \approx 0.66547$ kg/m^3 and $\rho_{N_2} \approx 4.65719$ kg/m^3 (retaining a large number of significant digits in these estimates is necessary for the subsequent analysis, because refractive index changes are very small). We further approximated the values of the Gladstone–Dale constants of the two pure gases by the values listed in Table 7.4, namely, $K_{He} = 0.196 \times 10^{-3}$ m^3/kg and $K_{N_2} = 0.238 \times 10^{-3}$ m^3/kg, from which we estimated the refractive indices of the two pure gases in the present mixing layer as, respectively, $n_{He} = 1 + K_{He}\rho_{He} \approx 1.00013$ and $n_{N_2} = 1 + K_{N_2}\rho_{N_2} \approx 1.00111$. Finally, we approximated the reference refractive index as $n_o \approx 1.00$.

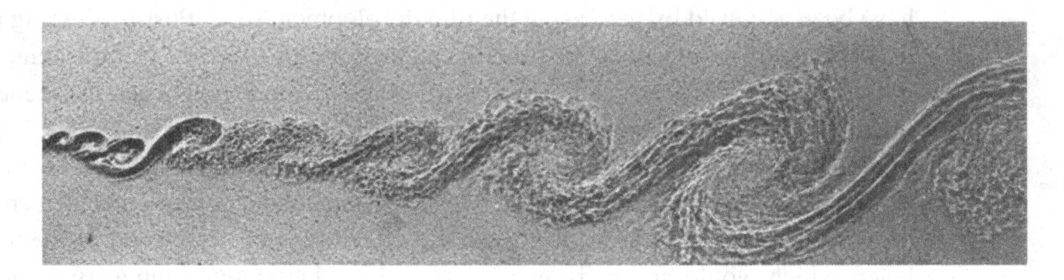

Figure 11.5 Shadowgram of a mixing layer between a helium stream (upper half of the image) at a higher speed and a nitrogen stream (lower half of the image) at a lower speed [49].

Let z, y be, respectively, the streamwise and transverse coordinate axes of the mixing layer, with the origin set at bottom left corner of the image, and let $I(z, y)$ be the light intensity, as it reached the photographic plate. Because the shadowgram was taken inside the test section, and not on a screen, as would be the case illustrated in Fig. 11.3, we cannot use Eq. (11.17) to relate $I(z, y)$ and the refractive index $n(z, y)$. We may, however, easily find a suitable relationship by considering Fig. 11.3. In this image, because of refraction, a light beam entering the test section with a height δy illuminates the opposite wall over a height $\delta y/\cos\varphi$, where the deflection angle φ is very small, so that $\cos\varphi \approx 1 - \varphi^2/2$. Therefore, the local light intensity on the back wall of the test section would be

$$I = I_o \cos\varphi \approx I_o \cos\left[\frac{1}{n_o}\frac{\mathrm{d}n}{\mathrm{d}y}(x_2 - x_1)\right] \approx I_o\left(1 - \frac{1}{2}\left[\frac{1}{n_o}\frac{\mathrm{d}n}{\mathrm{d}y}(x_2 - x_1)\right]^2\right).$$

It is noted that, irrespective of the sign of $\mathrm{d}n/\mathrm{d}y$, this expression gives $I \leq I_o$. Moreover, because φ is normally extremely small, $|I - I_o| \ll I_o$. This equation can be integrated with respect to y at each z-location, to provide the local refractive index as

$$n(z, y) = \frac{2n_o}{x_2 - x_1}\int_{-\infty}^{y}\left[\frac{I_o - I(z, y')}{I_o}\right]^{1/2}\mathrm{d}y' + n(z, -\infty).$$

Once the refractive index is found, one may also, in principle, determine the composition of the mixture, as follows. The refractive index and the density of the mixture are related by the Gladstone–Dale formula, in which both the constant K and the density ρ of the mixture are related to the corresponding properties of the constituent gases as

$$K = C_{He}K_{He} + (1 - C_{He})K_{N_2} \quad \text{and} \quad \rho = C_{He}\rho_{He} + (1 - C_{He})\rho_{N_2},$$

where C_{He} is the mass fraction of helium. Then, the Gladstone–Dale formula gives

$$n = 1 + [C_{He}K_{He} + (1 - C_{He})K_{N_2}][C_{He}\rho_{He} + (1 - C_{He})\rho_{N_2}],$$

which is a quadratic equation for C_{He} and can be easily solved.

In view of the fact that the deflection angle φ is extremely small, the ratio I/I_o would deviate only slightly from 1, which makes it difficult to resolve intensity differences on the image. To make intensity differences visible, the shadowgram contrast must have been boosted substantially, which, in the case of monochrome photographic films, could have been achieved by increasing the film development time. Post-processing of a digitised image by software can be used to enhance contrast further. Contrast enhancement is a non-linear process, which would distort the relationship between the recorded light intensity and the refractive index in an unknown way. Consequently, the scale of the light intensity in the available, post-processed shadowgram would be unknown and non-uniform. As a rough approximation, we may assume that contrast enhancement affected proportionately the difference between the recorded intensity $I^*(z, y)$ inside the mixing layer, which would appear in grey tone, and the background intensity I_o^* outside the mixing layer, which would ideally appear as pure white in the image. This assumption allows us to relate the recorded and incident light intensity differences as

$$\frac{I_o^* - I^*}{I_o^*} \approx \lambda\frac{I_o - I}{I_o},$$

where $\lambda \gg 1$ is an empirical constant. One may integrate the recorded light intensity to determine the local *apparent refractive index difference* as

$$\Delta n^*(z, y) = \frac{2n_o}{x_2 - x_1}\int_{-\infty}^{y}\left[\frac{I_o^* - I^*(z, y')}{I_o^*}\right]^{1/2}\mathrm{d}y',$$

from which one may estimate the contrast conversion coefficient as

$$\lambda = \left[\frac{\Delta n^*(z,\infty)}{n_{He} - n_{N_2}} \right]^2 .$$

Finally, the smoothed refractive index across the present mixing layer can be estimated as

$$n(z,y) = n_{He} + \frac{1}{\sqrt{\lambda}} \Delta n^*(z,y) .$$

To start the analysis, we digitised a section of the available image, so that it consists of 1,488 × 378 pixels, each containing, in an arbitrary scale, a value of the local light intensity $I^*(z,y)$ in grey tone. We applied the method along the line $z = 16.7$ mm, marked by a dashed line in Fig. 11.6a, which also shows the corresponding profile of the normalised light intensity differences. The background light intensity I_0^* was determined by fitting a straight line through the first, and last, 100 pixels in the image, which are outside the mixing layer. This normalised light intensity profile clearly identifies the limitations of the present approach. Our idealised refraction analysis applies to steady, 2-D refractive index fields. In the present, highly turbulent mixing layer, however, the refractive index has strong fluctuations and a three-dimensional structure. Moreover, the image is contaminated by strong noise, as manifested by the strong fluctuations of light intensity outside the mixing layer, where the refractive index should be uniform. One may expect that current shadowgraph techniques would provide much 'cleaner' images, which, however, would also have strong light intensity fluctuations within the mixing layer. Whichever the source and amplitude of fluctuations is, they end up giving non-physical results from this analysis, as the corresponding $\cos \varphi$ would exceed 1. Averaging of the corresponding values in a large number of images may alleviate somewhat this problem, but such images are not available. Our conclusion from the previous observation is that this method can at best predict approximately a steady, 2-D concentration field and by no means a noisy image of a turbulent flow like the present one. Although it seems reasonable to expect that the time-averaged light intensity across the mixing layer would have a near-Gaussian profile, instantaneous profiles, like the one in the present case, may be strongly non-Gaussian, due to large-scale transport of gases away from the corresponding free stream, even in the absence of noise. For educational purposes, we applied the analysis to a Gaussian-like light intensity profile, which was fitted by eye to the data so that its two tails approached the background light intensity. We calculated the refractive index scaling factor as $\lambda \approx 5.15 \times 10^3$, from which we estimated the average refractive index across the mixing layer, as plotted in Fig. 11.6b.

These values allowed us to determine the profile of the average mass fraction of helium, as also plotted in Fig. 11.6b. This profile can be compared to corresponding quantitative measurements, if such are obtained by traversing a concentration probe [2, 21].

This example illuminates the reasoning behind not using shadowgrams for quantitative purposes. Nevertheless, the present shadowgram contains invaluable qualitative information. It shows the boundaries of the mixing layer, from which one can determine its growth rate. Although light intensity fluctuations cannot be connected to non-

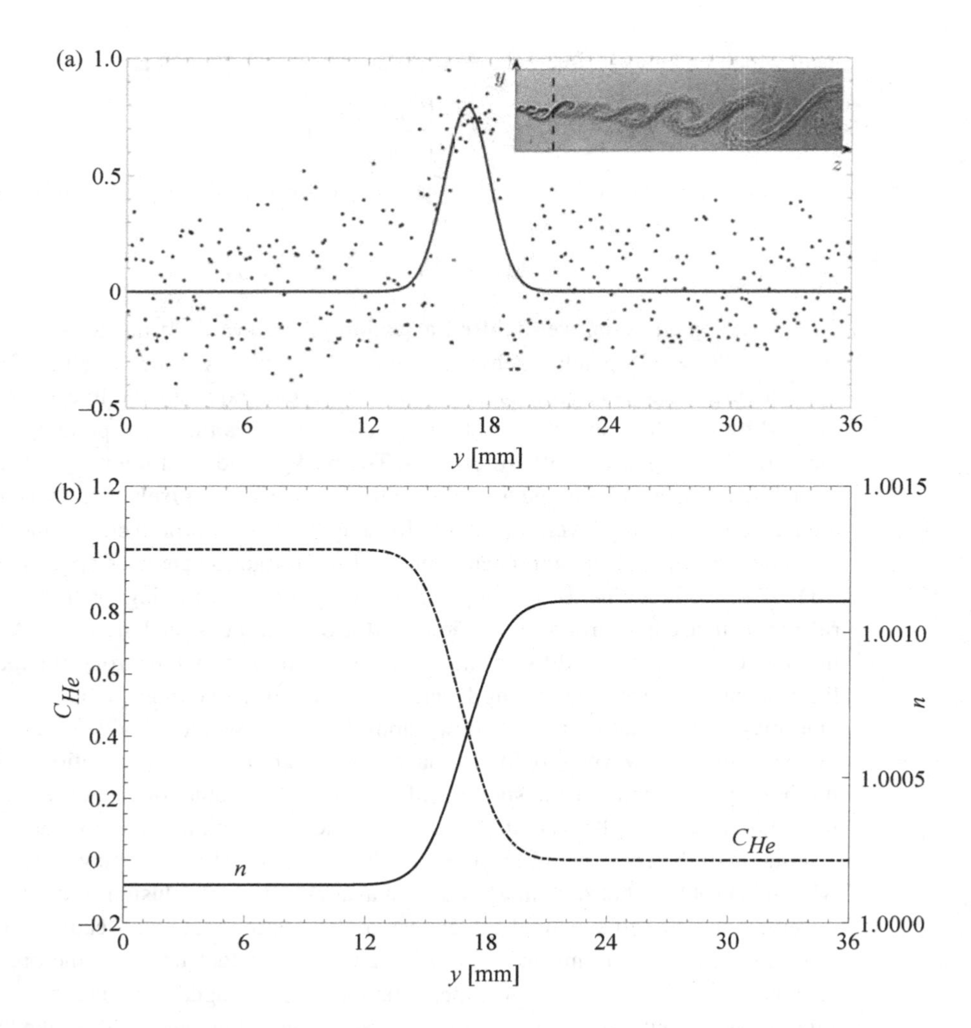

Figure 11.6 (a) The normalised recorded light intensity along the dashed line shown in the insert; (b) the computed refractive index profile in arbitrary scale and the estimated profile of the helium mass fraction.

coherent, three-dimensional, small-scale, turbulent motions, the presence of large-scale, quasi-2-D roller vortices and their interactions are captured unmistakably in the shadowgram. This and similar images played a historical role in the detection of coherent structures and the understanding of their significance in turbulence research.

11.3.3 The Schlieren Method

The Schlieren method is the most popular optical flow visualisation method, due to its good contrast and the relative ease by which it can be implemented. Its name is the plural of the German word *Schliere*, which indicates an optical inhomogeneity in glass.

As indicative of its many variations, we shall present the operation principle of the classical *Toepler method*. The optical system consists of a concentrated light source, a converging lens to collimate the light and a collecting lens, called the *Schlieren head*, which collects the light after it passes the region of interest. If a light ray passes through a fluid in which the refractive index varies as indicated in Fig. 11.7a, it will be deflected by an angle φ, as discussed earlier in this section. As a result of this deflection, the ray will cross the focal plane of the collecting lens at a distance $\Delta y \approx f_S \varphi$ from the crossing point of the undistorted ray, where f_S is the focal distance of this lens. Now consider that the light source has a rectangular shape, which, in the absence of any distortion, would be focused onto a rectangle with height h (shown by a dashed-line rectangle in Fig. 11.7b) on the focal plane of the Schlieren head. The distorted image of the source on the focal plane (shown by a solid-line rectangle in Fig. 11.7b) would be displaced by a distance Δy with respect to the undistorted image. The Schlieren method consists of using a sharp object, such as a knife edge or a razor blade, to partly block the focused image of the source, such that only a rectangle with height $h_S < h$, corresponding to light intensity I_o passes through. Due to light deflection, the distorted image will be blocked by a different proportion than the undistorted one. In the case of $\partial n/\partial y > 0$, as shown in Fig. 11.7, $\varphi > 0$ and the distorted source image would be deflected away from the edge, so that its light intensity I would be greater than I_o; in the case of $\partial n/\partial y < 0$, $\varphi > 0$ and the distorted image would be blocked more by the edge than the undistorted image, such that $I < I_o$. The relative difference in light intensity passing through the focal plane would be

$$\frac{\Delta I}{I_o} \approx \frac{\Delta y}{h_S} \approx \frac{f_S \varphi}{h_S} \approx \frac{f_S}{h_S} \frac{1}{n_o} \int_{x_1}^{x_2} \frac{\partial n}{\partial y} \mathrm{d}x \,. \tag{11.19}$$

If the refractive index field is two-dimensional, namely, if n is independent of x, this expression is simplified to

$$\frac{\Delta I}{I_o} \approx \frac{f_S \left(x_2 - x_1\right)}{h_S n_o} \frac{\mathrm{d}n}{\mathrm{d}y} \,. \tag{11.20}$$

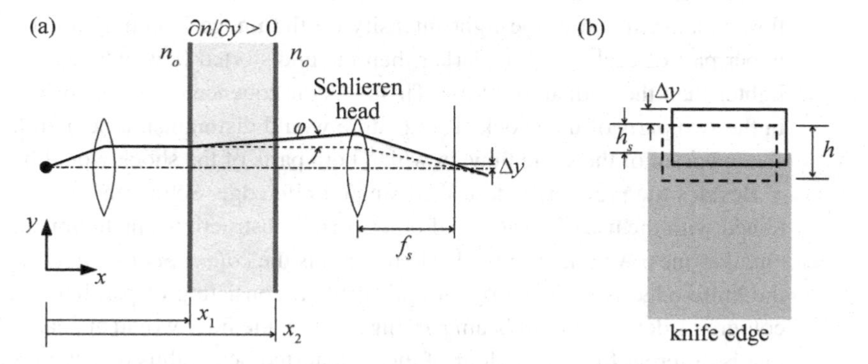

Figure 11.7 (a) Deflection of a light ray on the focal plane of a Schlieren head; (b) deflection of a rectangular light source image on the focal plane of a Schlieren head.

Figure 11.8 Shlieren image of a supersonic flow past a flat-ended cylinder, showing a bow shock upstream of the cylinder and an oblique shock further downstream [49].

The previous expressions indicate that the Schlieren image relative intensity change would be proportional to the refractive index gradient in the direction normal to the knife edge. Compared to shadowgraph images, Schlieren images are generally sharper and more likely to be usable for quantitative estimates of density, as they would require a single, rather than double, integration. Equation (11.19) shows that, for a given lens and test section width, the sensitivity of the method is inversely proportional to h_S, which means that, for improved contrast, one would have to block a larger proportion of the focused light source image. However, this has a limit, because, as h_S is decreased, there is an increasing risk that the distorted image would be either entirely unblocked, or entirely blocked by the knife edge, which would produce uniform illumination. An example of a Schlieren image is shown in Fig. 11.8. Note that this is an axisymmetric flow and so the change of light intensity on the image is not given by the 2-D Eq. (11.20), but by the 3-D Eq. (11.19). This results in integrating $\partial n/\partial y$ along the line of view and distorts the appearance of the flow features. For example, the shock waves in the image appear to have a significant thickness, because they are projections of axisymmetric shocks. If this flow were 2-D, the shocks in the Schlieren image would appear to be much thinner. Another observation that can be made in Fig. 11.8 is that, although the flow is axisymmetric, the light intensity on the image is anti-symmetric, namely, the upper part of each shock is darker than the undistorted flow, whereas the lower part is lighter than the undistorted flow. This is a consequence of the opposite signs of $\partial n/\partial y$ in the two parts of the shock. This feature would distinguish a Schlieren image from a shadowgram of the same flow, in which both parts of the shock would be dark.

Besides the previously described single knife edge, Schlieren images have been obtained with the use of a variety of other partial obstructions, including double-edge and circular ones. A variant of the basic method is the *colour Schlieren technique*, in which the knife edge is replaced by an optical filter, consisting of parallel strips of different colours. A deflected light beam passing through the filter would appear to have a colour that is different from the colour of the undistorted beam, thus revealing refractive index gradients. An advantage of the colour Schlieren technique over the knife-edge technique is that it can be applied to cases with relatively large light beam deflections.

11.3.4 Interferometry

The various available interferometric methods utilise the phase shift of light waves propagating through media of variable refractive index. Consider a light beam propagating through such a medium, as illustrated in Fig. 11.3. If the medium had a uniform refractive index n_o, the beam would cross the medium during a time

$$t_o = \int_{x_1}^{x_2} \frac{1}{\upsilon_o} \mathrm{d}x = \frac{1}{c} \int_{x_1}^{x_2} n_o \mathrm{d}x \,, \tag{11.21}$$

where υ_o is the (constant) light speed in the medium and c is the speed of light in vacuum. In the distorted medium, the corresponding light speed would be υ and the corresponding time would be

$$t = \int_{x_1}^{x_2} \frac{1}{\upsilon} \mathrm{d}x = \frac{1}{c} \int_{x_1}^{x_2} n \mathrm{d}x \,. \tag{11.22}$$

Two light waves entering in phase into two media with equal thicknesses and slightly different refractive indices equal to n_o and n, respectively, would exit these media with a phase difference

$$\Delta\theta \approx \frac{2\pi\upsilon_o}{\lambda}(t - t_o) \approx \frac{2\pi}{\lambda n_o} \int_{x_1}^{x_2} (n - n_o)\, \mathrm{d}x \,. \tag{11.23}$$

If n is independent of x, the phase shift would be simplified to

$$\Delta\theta \approx \frac{2\pi\,(x_2 - x_1)}{\lambda n_o} (n - n_o) \,. \tag{11.24}$$

Thus, the phase shift would be directly proportional to refractive index variations, rather than variations of its derivatives, as in the shadowgraph and Schlieren methods.

A direct application of the above analysis will be illustrated by the *Mach–Zehnder interferometry* method, which is the classical interferometric method in gas dynamics.

This method, illustrated in Fig. 11.9, consists of splitting a collimated light beam into two beams, one of which passes through the test section containing the flow of interest, while the second one (*reference beam*) passes through a pair of transparent

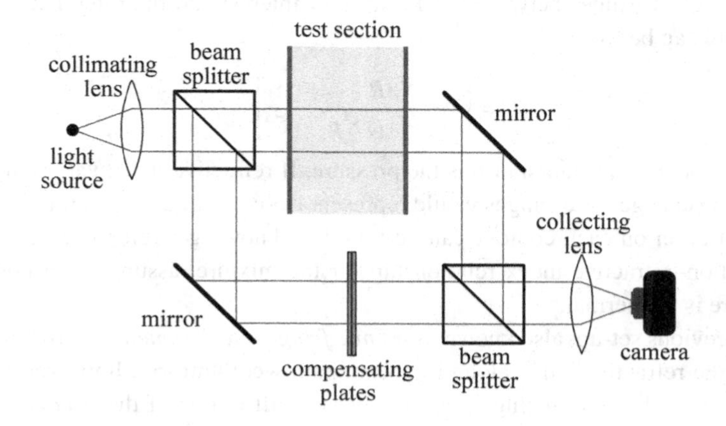

Figure 11.9 Schematic illustration of the Mach–Zehnder interferometer.

plates, identical to the walls of the test section. The two beams are made to *interfere* with each other on the observation plane or the recording film, where the disturbed beam would arrive with a phase lag given by Eq. (11.23), compared to the reference beam. The interfering planar light waves would combine to form a wave whose amplitude would be proportional to

$$\sin(2\pi\nu t) + \sin(2\pi\nu t - \Delta\theta) = 2\cos\left(\frac{\Delta\theta}{2}\right)\sin\left(2\pi\nu t - \frac{\Delta\theta}{2}\right). \tag{11.25}$$

The frequency response of a film, photodetector or the human eye would be insufficient to resolve the light frequency and so the captured power of the light on the recording plane would be proportional to $\cos^2(\Delta\theta/2)$. Therefore, the recorded light intensity would vary depending on the phase shift in the test section. When $\Delta\theta = (2i + 1)\pi, i = 0, \pm1, \pm2, ...,$ the light intensity would vanish, while, when $\Delta\theta = 2i\pi, i = 0, \pm1, \pm2, ...,$ the light intensity would be maximum. Thus, if the test section contains a fluid with varying refractive index, the interferogram would appear as a set of bright and dark *fringes*. Each fringe would be a contour of constant refractive index and the difference between the refractive indices of two adjacent dark (or bright) fringes would be

$$\Delta n = \frac{\lambda n_o}{x_2 - x_1}. \tag{11.26}$$

Therefore, the interferogram may also be used for quantitative purposes, because the fringes would be contours of constant n, namely of constant density. If, for example, the medium is a gas for which the Gladstone–Dale formula (see Section 7.2.1) applies, the density difference between two fringes of the same kind would be

$$\Delta\rho \approx \frac{\lambda}{(x_2 - x_1)K}, \tag{11.27}$$

where K is the Gladstone–Dale constant. If the medium is a perfect gas under constant pressure, then the fringes would also be isotherms (i.e., contours of constant temperature). To find the absolute temperature on a fringe, one must specify the temperature T_r on some reference fringe. Such reference temperatures can be determined by theoretical means or by independent measurement, for example using a thermocouple. If N is the number of fringes between the location of interest and the reference fringe, the temperature can be found as

$$T = \left[\frac{N\lambda R}{(x_2 - x_1)Kp} + \frac{1}{T_r}\right]^{-1}, \tag{11.28}$$

where R is the gas constant and p is the pressure. If refractive index variation is due to composition changes, the fringes would represent iso-concentration contours. The value of concentration on each contour can be found by knowing a reference value and the concentration–refractive index relationship for the mixture, assuming, of course, that the mixture is isothermal.

In the previous set-up, also known as *infinite fringe interferometry*, there would be no fringes if the refractive index along the beam path were uniform. Improved resolution can be achieved by introducing a variable phase shift to one of the beams by passing it through a prism. Superposition of the undistorted and distorted interferogram would produce a *Moiré pattern*, which may be analysed to resolve the refractive index variation; this technique is called *finite-fringe interferometry*. Mach–Zehnder interferometry requires careful alignment and high-precision optical components and test section windows. For these reasons, it has been largely superseded by holographic interferometry, to be discussed below.

11.3.5 Holography

Holography is an optical method that records and reconstructs three-dimensional images. It was invented in 1948 by D. Gabor [12], who was awarded a Nobel Prize for this discovery in 1971. The basic principle of holography is illustrated in Fig. 11.10. Consider a monochromatic point source A, producing a coherent spherical light wave, termed the *reference wave*. Now consider an object, for simplicity also assumed to be a point B, which is illuminated by the point source and, through reflection and scattering, also produces a coherent spherical light wave, called the *object wave*. As the reference and object waves propagate through the medium, they interfere with each other, producing standing waves whose amplitude depends on the phase relationship. Where the two waves are in phase, which will be along hyperbolic surfaces, maximum light intensity will be produced. If a photographic plate is exposed to this light, it will record a pattern of interference fringes. This is the *recording* step of holography. The developed photographic plate is called the *hologram*. During *reconstruction*, the same source is used to

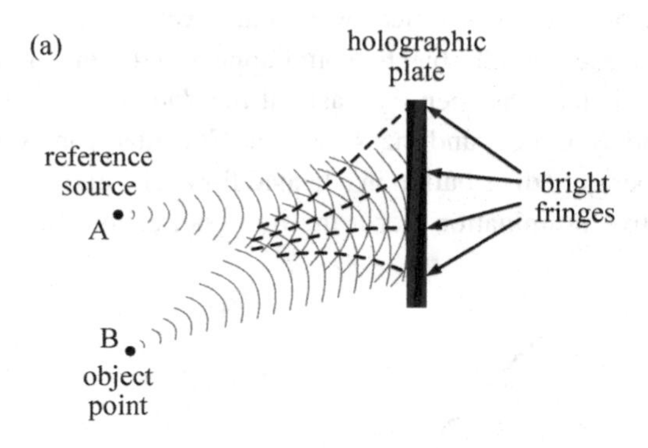

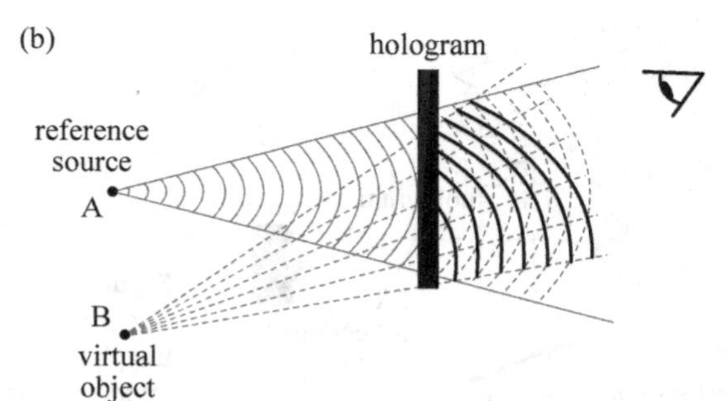

Figure 11.10 Principle of holography: (a) recording of image and (b) reconstruction of virtual object.

illuminate the hologram. The hologram will then act as a diffractive grating (see Section 7.3), not only allowing the reference wave to pass through, but also reconstructing the phase and amplitude characteristics of the object wave. Thus, an observer looking backwards through the hologram will see a *virtual object*, identical to the original one. This process applies not only to point objects, but also to any three-dimensional object, whose reconstruction allows its observation from different directions, like the original object. Although the concept of holography precedes the invention of lasers, its extensive application had to await the development of lasers, due to its dependence on a monochromatic, coherent light source. Holography is used in many applications and with many different variations [4, 6, 38], but our present interest is limited to configurations used in flow visualisation and measurement of fluid temperature, composition and velocity.

11.3.6 Holographic Interferometry

A typical optical arrangement for holographic interferometry is shown in Fig. 11.11. A single, monochromatic laser beam is split, expanded and collimated into an object beam and a reference beam, both of which are directed onto the photographic plate. The object wave crosses the test section and is phase shifted by the variable refractive index fluid, while the reference wave bypasses the test section and is undistorted. When the hologram is illuminated by the reference beam, the image of the test section would appear. Because the reconstruction of the object reverses all optical errors, there is no need for high-quality optical components, window materials or alignment. Each hologram may contain not only a single object wave but several, recorded at different instances. During reconstruction, all objects would appear and their images would interfere with each other. This fact has been the basis of the *double-exposure* interferometric technique, used widely in heat and mass transfer. Consider, for example, that one is interested in the temperature distribution in a heated flow. The photographic plate is exposed to two consecutive illumination steps: the first one, called the *comparison*

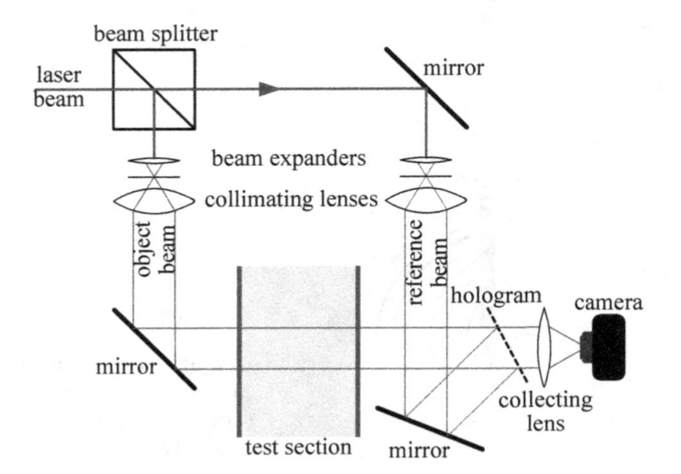

Figure 11.11 Schematic illustration of a typical holographic interferometry setup.

wave, is produced by the unheated flow, while the second one, called the *measuring wave*, is produced by the heated flow. Reconstruction produces an interference pattern due to the heat-related changes in refractive index in the test section. This technique eliminates the effects of light distortion due to curvature of the test section walls or other irregularities, but has the disadvantage that it can only be used to produce a single interferogram at a time, and so it is not suitable for transient flow phenomena. The *real-time* interferometric technique permits the continuous observation of the interference pattern. Like the double-exposure technique, it starts with exposing the photographic plate to the comparison wave, but then the plate is developed to produce a hologram. The hologram is re-positioned precisely in its original position and, when exposed to the measuring wave, it produces a time-continuous interference pattern, which can be recorded cinematographically or videoscopically.

11.3.7 Streaming Birefringence

This is a method of flow visualisation of certain liquids, which become *birefringent* (see Section 7.2.6) under the application of shear stresses [28, 52]. Such liquids include polymers and colloidal solutions of elongated crystals. When at rest, these liquids are optically isotropic, but, in shear flows, the long axis of the molecules or particles in the liquid tends to get oriented in a direction that is normal to the velocity gradient, thus introducing birefringence. A light beam passing through such a medium will be separated into two plane-polarised components, having polarisation planes that are normal to each other. Both components propagate in the same direction, but with different phase speeds, so that, upon exit from the medium, they have a phase shift with respect to each other. In other words, the refractive indices of the medium for the two components are different and superposition of the two exiting beams will introduce a fringe pattern, which may be correlated with flow characteristics. Methods introduced to detect the phase shift include the use of a polariscope (an optical instrument commonly used in photoelasticity experiments), Mach–Zehnder interferometry and a scattering-based technique. Milling yellow dye solutions have been used widely as a birefringent liquid, due to their high sensitivity.

11.3.8 The Wollaston-Prism technique

This is a hybrid method with elements of both the Schlieren method and interferometry [28, 42]. In a Schlieren optical setup, the knife edge is replaced by a *Wollaston prism*, which is a birefringent device (see Section 7.2.6) that splits the light beam through it into two linearly polarised components, propagating through the fluid with different phase speeds and at a very small angle to each other. When superimposed, the two components produce an interference pattern, which, similarly to the Schlieren method, is sensitive to the refractive index gradient, rather the refractive index value. Both finite-fringe and infinite-fringe patterns can be obtained, depending on whether the prism is located on the collected beam focal plane or not.

11.3.9 Radiation Emission Techniques

These are flow visualisation and measurement techniques suitable for low-density gases, in which they have superior sensitivity to other methods. Their operation is based on the emission of visible radiation by gas molecules following their excitation by an electric field. Radiation emission techniques have been used regularly for flow visualisation of shock waves and other flow patterns in supersonic and hypersonic wind tunnels or shock tubes, as well as for the measurement of temperature, pressure, density and composition variations. Such techniques include the following [28, 52]:

Spark tracing: This technique is suitable for absolute gas pressures between about 100 kPa and 1 Pa. Depending on the shape of the electrodes, the spark could form along a line, a plane or other surface. The basic configuration consists of two electrodes pointed at each other across the flow. A high-voltage (in the tens of kV), short duration (0.1 to 1 ms), electric pulse applied across the electrodes causes a spark to form along the shortest path between them, ionising the air along its path. Subsequent electric pulses generate sparks that travel along the original ionised path, which in the meantime has been convected downstream, thus marking timelines in the gas. The electrodes are usually needles or thin wires, between 8 and 130 mm apart. A thin wire stretched across the electrodes helps to create a straight initial timeline. The ionisation persists for about 1 ms, therefore, the frequency of electric pulses must be at least 1 kHz for a single timeline and higher for multiple timelines. This method is suitable for gas speeds between 1 m/s and several hundreds of m/s. As with any other timeline method, spark tracing cannot resolve the motion of individual fluid particles and, therefore, cannot identify the flow direction. To remedy this, a method has been suggested, in which the flow is seeded with AlN particles, which burn along the spark's path, marking short bright pathlines.

Glow discharge method: This is suitable for pressures between 0.01 Pa and 10 Pa. A strong electric field applied to the flow accelerates free electrons and ions which are contained in the gas. These particles collide randomly with neutral molecules, which are excited and emit radiation, which is proportional to the gas density. Thus, density fields can be mapped from measured variations of light intensity or changes in colour.

Electron beam method: This method is the only means of visualising gases in the extremely low pressure range below 0.01 Pa. The flow is bombarded with high-speed electrons from an *electron gun*, which excite the gas molecules to emit radiation, whose intensity is proportional to the local density [10, 16, 32, 53].

11.4 Tomographic Methods

The term *tomography* encompasses a host of sensing techniques and mathematical algorithms, which reconstruct three-dimensional images and fields of various properties of an object by processing multiple one-dimensional or two-dimensional scans of the object [24]. They are used extensively in medicine, process engineering and many other areas of science and technology. A great variety of sensing techniques have been employed, based on absorption, refraction, scattering and emission of various forms of radiation and sound waves. Among the many commonly used tomographic methods are *positron emission tomography* (*PET*), *nuclear magnetic resonance* (*NMR*), *magnetic resonance imaging* (*MRI*), *computerised axial tomography* (*CAT*) and *electric capacitance* (*ECT*), *resistance* (*ERT*) and *inductance* (*EMT*) *tomography* methods. Typically, a sensor records the cumulative effect of property change along a linear path across the object, a number of such records along intersecting paths on the same plane are obtained simultaneously or sequentially, and the multiple records are processed by a mathematical algorithm capable of reconstructing the local values of the desired property over a planar section of the object. The three-dimensional field of the property can be also reconstructed by processing multiple planar sections. In fluid mechanics research, tomographic methods are used for flow visualisation of three-dimensional patterns and for construction of three-dimensional maps of velocity, temperature and concentration. The resolution and accuracy of such methods vary significantly from one method to another, with some methods being capable of mapping transient and unsteady phenomena. Optical tomographic methods for three-dimensional flow visualisation employ refractive-index-based sensing, including interferometric and Schlieren techniques.

11.5 Enhancement of Flow Visualisation Records

Computers and software are indispensable components of most flow visualisation methods. They can be used to improve the quality of flow images and to assist the observer in the presentation and interpretation of visual patterns [52]. In the present section, we are mainly concerned with qualitative flow visualisation techniques, or those with a relatively low resolution. The use of computers and related tools to extract quantitative information from images, as for example in particle image velocimetry, will be considered in the appropriate sections in later chapters. In addition to optical images, similar techniques can be used to enhance and process two- and three-dimensional image-like plots, reconstructed from transducer measurements (e.g., multiple hot-wire signals) or from results of analytical calculations and numerical simulations.

Image processing is performed on digital images, which may be grey-tone or colour ones. A grey-tone image is an array of $\mathrm{M} \times \mathrm{N}$ pixels, each containing the corresponding light intensity, approximated by one of 2^n (from $2^1 = 2$ to $2^8 = 256$) levels. A colour image typically consists of three superimposed arrays, each containing the intensity of one of the three primary colours, red, green and blue (see Section 7.3).

One may filter a digital image, in the same way as a discrete time series, in order to remove certain features or accentuate others. Low-pass filtering replaces the original intensity of each pixel by an average over neighbouring pixels. Boxcar (i.e., unweighted) averaging or more sophisticated algorithms, using variable weights, may be used for this purpose. Low-pass filtering reduces optical noise as well as softens the appearance of small-scale features of the image, while preserving the large-scale features. High-pass filtering can be achieved by subtracting the low-pass filtered intensities from the original ones; this sharpens the appearance of the image, accentuating the boundaries between flow patterns and small-scale features; it also removes large-scale, systematic variations of intensity due to non-uniform illumination.

In some cases, it is desirable to enhance the contrast of an image by replacing the intensity at each pixel by one of two (black and white) or more (black, white and grey) levels. This is achieved by setting appropriate intensity thresholds and replacing values of the original intensity within a certain range by the limiting value. This method can be used to replace grey images by pseudocolour images, by selecting a colour for each range between two thresholds.

A common task of flow visualisation is to identify the boundaries between flow regions with different features, a process called *edge detection*. The most precise definition of a boundary is achieved when it has a thickness of one pixel, which is rarely, if ever, the case in unprocessed images. Optimised thresholding and low-pass filtering can be used for this purpose. The same methods can be used to identify lines, such as timelines, in the flow (*line reduction*). Other methods that accentuate changes of intensity are *gradient methods*, namely the computation and display of the intensity gradient, and *differencing methods*, by which 'bas-relief' illusions are produced by inverting the image to create a negative, shifting the negative slightly in space and superimposing it on the positive.

Image reconstruction includes methods that combine information from different images in an attempt to visualise features, which are not visible in any single image. When periodic or quasi-periodic phenomena occur in a flow, as for example in vortex shedding from bluff objects, and phase information is recoverable from each image, one may *phase-average* several images to enhance the appearance of common patterns, which in individual frames may be heavily distorted by random effects. One may also superimpose overlapping images of different flow regions to produce images of larger regions. If simultaneous images of the same flow patterns are taken from different orientations, they could be statistically correlated so that three-dimensional images may be reconstructed, which could be subsequently viewed from any orientation. Similarly, three-dimensional images can be reconstructed by correlating simultaneous images on parallel planes (tomography, see Section 11.4) or, in a convected frame, by correlating images on a fixed plane but at different time instances.

Chapter Digest

11.1 Discuss the simplest and least expensive flow visualisation technique that would be suitable for locating regions of flow separation from a racing car model tested in (i) a wind tunnel and (ii) a water tunnel. Identify all required materials and instrumentation.

11.2 Discuss the simplest and least expensive flow visualisation technique that would be suitable for locating the positions of shock waves generated by a Pitot tube in a supersonic wind tunnel. Identify all required materials and instrumentation.

11.3 Describe the applicability of the hydrogen bubble method in a channel filled with silicone oil.

11.4 Describe the applicability of the photocatalysis method in a water tunnel.

11.5 Describe the applicability of the tufts method as a means of roughly estimating the turbulence intensity in a wind tunnel.

11.6 State a general disadvantage of optical flow visualisation methods.

11.7 How would you distinguish a shadowgram of supersonic flow past a wedge from a Schlieren image of the same flow?

11.8 Under which conditions would a radiation emission technique be more suitable than others for flow visualisation?

11.9 Which is the main advantage of holographic flow visualisation methods?

11.10 In your opinion, are medical imaging techniques likely to find extensive application in laboratory-based fluid mechanics research? Explain your rationale.

Problems

11.1 Timelines, as shown in Fig. 11.12, have been marked in water flowing steadily through a two-dimensional contracting channel by supplying a thin vertical wire with brief electric pulses, which produce hydrogen bubbles. Assume that the channel inlet height is h = 200 mm and the electric pulse rate is 1 Hz. Neglecting the boundary layer at the channel bottom and the small velocity drop near the free surface, determine quantitatively and sketch a few vertical profiles of the flow velocity. Then, sketch some streamlines, considering that a streamline cannot be crossed by the fluid. Explain in detail your procedures and state all approximations and assumptions you make.

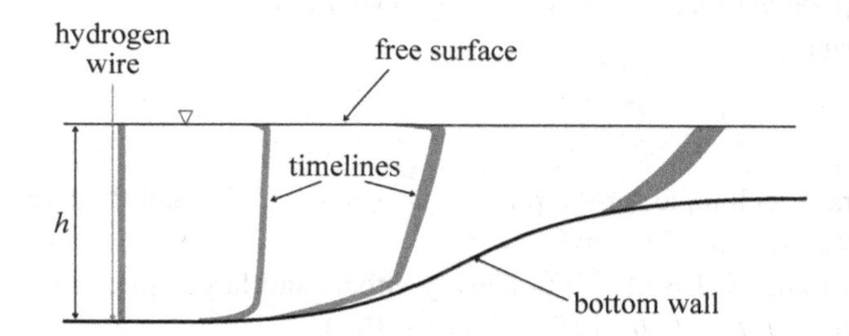

Figure 11.12 Timelines visualised by hydrogen bubbles (adapted from Ref. [20]; the leftmost timeline nearly coincides with the bubble generating wire; for clarity, it is the negative of the photograph that is shown).

11.2 A laser beam containing light with two wavelengths, λ_1 = 509.7 nm and λ_2 = 912.5 nm, passes through the square test section of a wind tunnel containing vertically stratified air flow, the density of which can be approximated by the function

$$\rho = \rho_o\left(1 - 0.2\frac{y}{h}\right),$$

where ρ_o = 1.2 kg/m^3, y is the distance from the bottom wall, and h = 0.300 mm is the test section height. The beam is observed on a plane wall located 2.5 m away from the test section. The Gladstone–Dale constants for air for the two wavelengths are K_1 = 0.2274 m^3/kg and K_2 = 0.2239 m^3/kg.

(a) Explain what will happen to the laser beam as it passes through the test section.

(b) Determine the distance between the positions of the two beam components on the observation plane. Would this distance depend on the vertical position of entry to the test section? Then, determine the phase shift between the two light components on the observation plane.

11.3 Consider horizontal, stratified, laminar air flow in a wind tunnel having glass walls. The tunnel's width is 0.5 m and its height is $h = 1$ m. The air density, measured with a sampling probe and fitted with a polynomial equation as a function of the vertical distance y from the bottom of the tunnel, is approximately

$$\rho = \rho_o \left\{1 - 0.05 \left[3\,(y/h)^3 + 4(y/h)^2 + 2y/h\right]\right\},$$

where $\rho_o = 1.25$ kg/m^3. Illumination is provided by a 'wide' laser beam with a wavelength of 589 nm. Determine and plot the variation of light intensity that would be obtained with (i) a shadowgraph, (ii) a Töppler Schlieren system and (iii) a Mach–Zehnder interferometer. Present the plots in dimensionless form and make simplifying assumptions, when required. Now assume that the density variation was unknown. Would you be able to estimate it, based on the visual patterns provided by each of these three methods? Discuss any possible difficulties.

11.4 Take a photograph of an object that has some sharp edges, such as a table. Open the image file with a scientific software package that gives you light intensity values at each pixel. Apply low-pass filtering by replacing the value at each pixel by the average of values in a number of surrounding pixels and replot the image. Repeat this for different numbers of pixels that you average. Compare and discuss the different images you obtained.

11.5 Apply high-pass filtering to the image you analysed in the previous problem by subtracting the pixel values obtained following low-pass filtering. Compare and discuss the resulting images.

References

[1] D.J. Baker. A technique for the precise measurement of small fluid velocities. *J. Fluid Mech.*, 26:573–575, 1966.

[2] G.L. Brown and A. Roshko. On density effects and large structure in turbulent mixing layers. *J. Fluid Mech.*, 64:775–816, 1974.

[3] S.D. Bruneau and W.R. Pauley. Measuring unsteady velocity profiles and integral parameters using digital image processing of hydrogen bubble timelines. *J. Fluids Eng.*, 117:331–340, 1995.

[4] H.J. Caulfield and Sun Lu. *The Applications of Holography*. Wiley Interscience, New York, 1970.

[5] B.R. Clayton and B.S. Massey. Flow visualization in water: A review of techniques. *J. Sci. Instrum.*, 44:2–11, 1967.

[6] C.B. Collier, C.D. Burckhardt, and L.H. Lin. *Optical Holography*. Academic Press, New York, 1971.

[7] T. Corke, D. Koga, R. Drubka, and H. Nagib. A new technique for introducing controlled sheets of smoke streaklines in wind tunnels. In *7th International Congress on Instrumentation in Aerospace Simulation Facilities, Shrivenham, England, September 6-8, 1977, Record (A79-15651 04-35)*, 1977.

[8] J.P. Crowder. Tufts. In W.-J. Yang, editor, *Handbook of Flow Visualization*, pages 125–175. Taylor & Francis, 1989.

[9] W. Davis and R.W. Fox. An evaluation of the hydrogen bubble technique for the quantitative determination of fluid velocities within clear tubes. *J. Basic Eng.*, 89:771–781, 1967.

[10] B. Diop, J. Bonnet, T. Schmid, and A. Mohamed. Compact electron gun based on secondary emission through ionic bombardment. *Sensors*, 11:5202–5214, 2011.

[11] R. Eichhorn. Flow visualization and velocity measurement in natural convection with the tellurium dye method. *J. Heat Transfer*, 83:379–381, 1961.

[12] D. Gabor. A new microscopic principle. *Nature*, 161:777–778, 1948.

[13] M. Gad-el-Hak. The use of the dye-layer technique for unsteady flow visualization. *J. Fluids Eng.*, 108:34–38, 1986.

[14] M. Gad-el-Hak. Visualization techniques for unsteady flows: An overview. *J. Fluids Eng.*, 110:231–243, 1988.

[15] M. Gad-el-Hak. Splendor of fluids in motion. *Prog. Aerosp. Sci.*, 29:81–123, 1992.

[16] L.A. Gochberg. Electron beam fluorescence methods in hypersonic aerothermodynamics. *Prog. Aerospace Sci.*, 33:431–480, 1997.

[17] J.W. Hoyt and R.H.J. Sellin. A turbulent-flow dye-streak technique. *Exp. Fluids*, 20:38–41, 1995.

[18] N. Kasagi, R.J. Moffat, and M. Hirata. Liquid crystals. In W.-J. Yang, editor, *Handbook of Flow Visualization*, pages 105–116. Taylor & Francis, 1989.

[19] J.C. Kent and A.R. Eaton. Stereophotography of neutral He-filled bubbles for 3-D fluid motion studies in an engine cylinder. *Appl. Opt.*, 21:904–912, 1982.

[20] S.J. Kline. Film notes for flow visualisation. In A.H. Shapiro, editor, *Illustrated Experiments in Fluid Mechanics: The NCFMF Book*. National Committee for Fluid Mechanics Films, Cambridge, MA, 1972.

[21] J.H. Konrad. *An experimental investigation of mixing in two-dimensional turbulent shear flows with applications to diffusion-limited chemical reactions*. PhD thesis, California Institute of Technology, Pasadena, CA, 1977.

[22] T. Liu, T. Campbell, S. Burns, and J. Sullivan. Temperature and pressure sensitive luminescent paints in aerodynamics. *App. Mech. Rev.*, 50:227–246, 1997.

[23] P. Matisse and M. Gorman. Neutrally buoyant anisotropic particles for flow visualization. *Phys. Fluids*, 27:759–760, 1984.

[24] F. Mayinger and O. Feldmann, editors. *Optical Measurements (2nd Edition)*. Springer, Berlin, 2001.

[25] I. McGregor. The vapour-screen method of flow visualization. *J. Fluid Mech.*, 11:481–511, 1961.

[26] B.M. McLachlan and J.H. Bell. Pressure-sensitive paints in aerodynamic testing. *Exp. Thermal Fluid Sci.*, 10:470–485, 1995.

[27] C. Mercer. *Optical Metrology for Fluids, Combustion and Solids*. Kluwer Academic Publishers, Dordrecht, 2003.

[28] W. Merzkirch. *Flow Visualization (2nd Edition)*. Academic Press, New York, 1987.

[29] W. Merzkirch. Density-based techniques (chapter 6). In C. Tropea, A.L. Yarin, and J.F. Foss, editors, *Springer Handbook of Experimental Fluid Mechanics*. Springer, Berlin, 2007.

[30] W. Merzkirch. Flow visualization (chapter 11). In C. Tropea, A.L. Yarin, and J.F. Foss, editors, *Springer Handbook of Experimental Fluid Mechanics*. Springer, Berlin, 2007.

[31] T.J. Mueller. Flow visualization by direct injection. In R.J. Goldstein, editor, *Fluid Mechanics Measurements (2nd Edition)*, pages 367–450. Hemisphere, Washington DC, 1996.

[32] E.P. Muntz. The electron beam fluorescence technique. Technical Report in AGARDograph 132, Open Library OL13957449M, Neuilly-sur-Seine, France, 1968.

[33] Japan Society of Mechanical Engineers. *Visualized Flow*. Pergamon Press, Oxford, 1988.

[34] M. Ojha, R.L. Hummel, R.S.C. Cobbold, and K.W. Johnston. Development and evaluation of a high resolution photochromic dye method for pulsatile flow studies. *J. Phys. E: Sci. Instrum.*, 21:998–1004, 1988.

[35] A.T. Popovich and R.L. Hummel. A new method for non-disturbing turbulent flow measurements very close to a wall. *Chem. Eng. Sci.*, 22:21–25, 1967.

[36] M.R. Rebollo. *Analytical and experimental investigation of a turbulent mixing layer of different gases in a pressure gradient*. PhD thesis, California Institute of Technology, Pasadena, CA, 1973.

[37] R. Reznicek. Surface tracing methods. In W.-J. Yang, editor, *Handbook of Flow Visualization*, pages 91–103. Taylor & Francis, 1989.

[38] E.R. Robertson and J.M. Harvey, editors. *The Engineering Uses of Holography*. Cambridge University Press, Cambridge, UK, 1970.

[39] F.W. Roos and W.W. Willmarth. Hydrogen bubble flow visualization at low Reynolds numbers. *AIAA J.*, 7:1635–1637, 1969.

[40] O Savas. On flow visualization using reflective flakes. *J. Fluid Mech.*, 152:235–248, 1985.

[41] F.A. Schraub, S.J. Kline, J. Henry, P.W. JR. Runstadler, and A. Littell. Use of hydrogen bubbles for quantitative determination of time-dependent velocity fields in low-speed water flows. *J. Basic Eng.*, 87:429–444, 1965.

[42] G.S. Settles. *Schlieren and Shadowgraph Techniques*. Springer, Berlin, 2001.

[43] C.R. Smith and R.D. Paxson. A technique for evaluation of three-dimensional behavior in turbulent boundary layers using computer augmented hydrogen bubble-wire flow visualization. *Exp. Fluids*, 1:43–49, 1983.

[44] A.J. Smits and T.T. Lim, editors. *Flow Visualization Techniques and Examples*. Imperial College Press, London, 2000.

[45] L.C. Squire. The surface oil flow technique: The motion of a thin oil sheet under the boundary layer of a body. Technical Report AGARDograph, AGARD-AG-70/1, pages 7-28, Neuilly-sur-Seine, France, 1962.

[46] S. Taneda, H. Honji, and M. Tatsuno. The electrolytic precipitation method of flow visualization. In A. Asanuma, editor, *Flow Visualization*, pages 209–214. Hemisphere, Washington DC, 1979.

[47] D.H. Thompson. Flow visualization using the hydrogen bubble technique. Technical Report Aerodynamics Note 338, Aeronautical Research Laboratories, Australian Defence Scientific Service, Melbourne, Australia, 1973.

[48] S.T. Thoroddsen and J.M. Bauer. Qualitative flow visualization using colored lights and reflective flakes. *Phys. Fluids*, 11:1702–1704, 1999.

[49] M. Van Dyke. *An Album of Fluid Motion*. Parabolic Press, Stanford, CA, 1982.

[50] F.J. Weyl. Analysis of optical methods. In R.W. Ladenburg, editor, *Physical Measurements in Gas Dynamics and Combustion*, pages 3–25. Princeton University Press, Princeton, NJ, 1954.

[51] H. Yamada. Use of smoke wire technique in measuring velocity profiles of oscillating laminar air flows. In T. Anasuma, editor, *Flow Visualization*, pages 265–270. Hemisphere, Washington DC, 1979.

[52] W.-J. Yang, editor. *Handbook of Flow Visualization, 2nd edition*. Taylor & Francis, 2001.

[53] A.S. Yaskin, V.V. Kalyada, A.E. Zarvin, and S.T. Chinenov. A high-efficiency method for scanning supersonic jets of rarefied gases. *Instrum. Exp. Techn.*, 63:430–434, 2020.

12 Measurement of Local Flow Velocity

The measurement of local flow velocity is a main objective of experimental fluid mechanics. It may be accomplished by direct measurement of the speed of flow markers, or by a variety of other methods, which utilise the relationship between flow velocity and other properties of the flow. In general, one may classify most flow velocity measurement methods in one of the following categories:

- *Pressure difference* methods. These utilise analytical relationships between the local velocity and the static and total pressures. The *Pitot-static tube* is the most common representative of instruments that measure local flow velocity from a pressure difference.
- *Thermal* methods. These compute flow velocity from its relationship to the convective heat transfer from heated elements. Common instruments of this type are the *hot-wire* and *hot-film anemometers*.
- *Frequency-shift* methods. These are based on the Doppler phenomenon, namely the shifting of the frequency of waves scattered by moving particles. The main instruments in this category are the *laser Doppler velocimeter* and the *ultrasonic Doppler velocimeter*, utilising light and sound waves, respectively.
- *Marker tracing* methods. These trace the motion of suitable flow markers, optically or by other means. Common optical methods include *chronophotography* and *particle image velocimetry*, while *pulsed-wire anemometry* is a method that traces heat emitted from local sources.
- *Mechanical* methods. These take advantage of the forces and moments that a moving stream applies on immersed objects. Examples of instruments of this type are the *vane*, *cup* and *propeller anemometers*.

12.1 Pressure Difference Methods

12.1.1 Pitot and Pitot-Static Tubes

A *Pitot tube* is an open-ended, hollow, cylindrical tube facing the flow (Fig. 12.1a), thus forming a stagnation region on its upstream tip, where the pressure would be equal to the total pressure p_o. The ratio d_i/d_o of the inner and outer diameters of the tube, is typically 0.6, although it can be as small as 0.15 and as large as 0.9. For relatively large Reynolds numbers and relatively small Mach numbers, the flow upstream of the tube

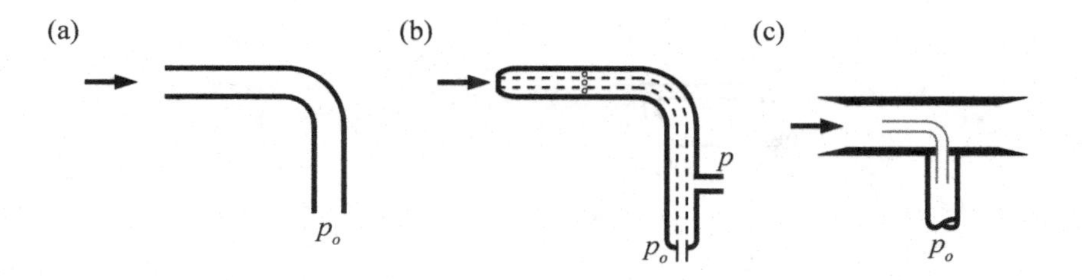

Figure 12.1 Sketches of (a) a Pitot tube, (b) a Pitot-static tube and (c) a Kiel probe.

may be accurately described by Bernoulli's equation, from which one may determine the ideal pressure coefficient as

$$C_P = \frac{p_o - p}{\frac{1}{2}\rho V^2} = 1\,. \tag{12.1}$$

Then, one may use the measured p_o, together with a local measurement of the static pressure p, to compute the flow velocity as

$$V = \sqrt{\frac{2\,(p_o - p)}{\rho}}\,. \tag{12.2}$$

Arrays of parallel Pitot tubes (*Pitot rakes*) are used to measure velocity profiles in wakes and other two-dimensional flows. *Pitot-static tubes* consist of an external static-pressure tube and a coaxial internal total-pressure tube (Fig. 12.1b). They are routinely used to measure the local flow velocity from the directly indicated pressure difference $p_o - p$.

Pitot tubes are simple, versatile instruments with the additional advantage of not requiring calibration, as their response is based on a basic principle. Nevertheless, their applicability is limited by a number of factors, the most common of which are the following:

Misalignment effect: Pitot tubes do not have to be precisely aligned with the flow direction, as the indicated total pressure reading is insensitive to orientation for small misalignment angles. However, as soon as such angle exceeds a critical value, the indicated C_P would decrease rapidly from the ideal value. For thin-walled, square-ended cylindrical tubes, the critical angle is about 20°, whereas, for similar tubes with the commonly used inner-to-outer diameter ratio $d_i/d_o = 0.6$, it decreases to about 12°. Tubes with rounded or conical faces are significantly more sensitive than square-ended tubes and would require more careful alignment. Specially designed probes consisting of tubes shielded by a surrounding cylinder, sometimes referred to as *Kiel probes* (Fig. 12.1c), are particularly insensitive to flow direction, with typical critical angles of about 45° [126]. As mentioned earlier, static pressure tubes are quite sensitive to misalignment, much more than total pressure tubes are. For this reason, in three-dimensional flows, it would be preferable to measure static pressure separately and not with the use of a Pitot-static tube.

Shear effects: When inserted in a shear flow with a mean velocity gradient $\mathrm{d}V\,/\,\mathrm{d}y$ normal to its axis, a Pitot tube will introduce an obstruction, displacing the stream-

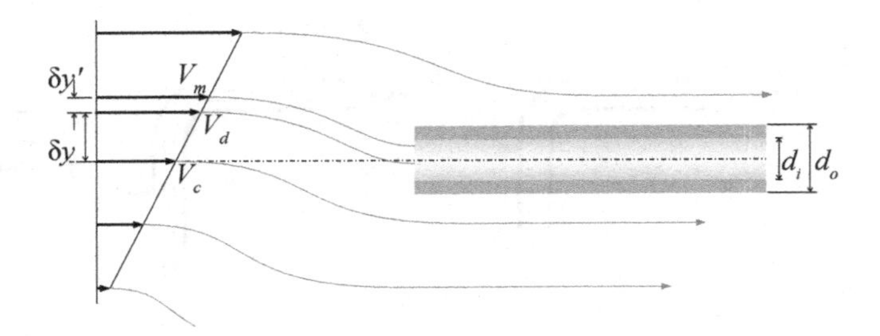

Figure 12.2 Streamline displacement in shear flow around a Pitot tube.

lines towards lower velocities, as shown in Fig. 12.2. Let δy be the displacement of the streamline that passes through the tube axis. Then, the total pressure at the centre of the tube orifice would not correspond to the free stream velocity V_c at the extension of its axis, that is, the *true* velocity, but to a higher value

$$V_d = V_c + \frac{\mathrm{d}V}{\mathrm{d}y}\delta y\,. \tag{12.3}$$

This error is known as *displacement effect* and is particularly significant in the inner boundary layer at large Reynolds numbers, or other flows where the shear is very large. It is reasonable to assume that the error in velocity due to the displacement effect would depend on the shear and the tube size and shape and would vanish in the absence of shear. Defining the shear parameter as

$$\alpha = \frac{d_o}{2V_c}\frac{\mathrm{d}V}{\mathrm{d}y}\,, \tag{12.4}$$

one may estimate the displacement δy as

$$\frac{\delta y}{d_o} = 0.15\tanh\left(4\sqrt{\alpha}\right). \tag{12.5}$$

This expression has been fitted to measurements obtained with Pitot tubes with a diameter ratio of 0.6 and outer diameters in the range 0.30–1.83 mm in pipe flows with pipe Reynolds numbers in the range 31×10^3–146×10^6 [81]. Equation (12.5) is reduced to $\delta y/d_o \approx 0.15$ for $\alpha \gtrsim 0.5$.

To reduce the displacement effect, one may use thinner tubes, thin-wall tubes, and tubes with conically shaped tips. There is a limit, however, on tube diameter, as a decrease in cross-sectional area would slow down the tube's response. For this reason, in highly sheared flows, tubes with nearly rectangular cross sections, obtained by pressing the tips of circular tubes, would be more usable than circular ones.

A secondary source of error is the averaging of pressure over the tube's orifice. This effect, which is due to the non-linearity of the relationship between total pressure and flow velocity, results in the tube reading an average total pressure, which is always larger than the total pressure on the extension of the tube's centreline. Thus, this error can be expressed as an additional displacement $\delta y'$ of the central streamline, which can

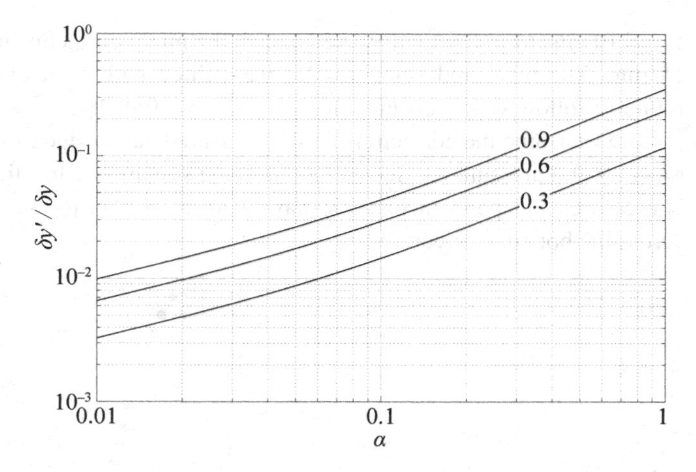

Figure 12.3 Ratio $\delta y'/\delta y$ of displacements due to obstruction and pressure averaging vs. the shear parameter α for various tube ratios d_i/d_o.

be estimated as [82]

$$\frac{\delta y'}{d_o} = \frac{d_i/d_o}{2\alpha}\left[\left(1 + \frac{\alpha^2}{4}\right)^{1/2} - 1\right] . \tag{12.6}$$

Combining the two streamline displacement errors, one may estimate the velocity measured by the tube as

$$V_m = V_c + \frac{\mathrm{d}V}{\mathrm{d}y}(\delta y + \delta y') . \tag{12.7}$$

Figure 12.3 shows the ratio $\delta y'/\delta y$ of displacements due to obstruction and pressure averaging vs. the shear parameter α for representative values of the diameter ratio d_i/d_o. For $\alpha \leq 0.1$, and for $d_i < 0.9d_o$, the displacement due to pressure averaging is less than 5% of the displacement due to the obstruction.

To estimate the previous displacements and the corresponding velocity corrections, one would need values of α, which in turn requires values of V_c and $\mathrm{d}V / \mathrm{d}y$. These values are not available in advance, because Pitot tube measurements would generally provide only a set of measured values V_{mi} at discrete y_i locations. Therefore, it is necessary to follow an approximate procedure, like the following one. First, fit the measurements of V_{mi} by a smooth function $V_m(y)$, which is, as much as possible, representative of the expected velocity profile and does not have kinks and other sudden changes of slope. Then, estimate α by substituting $V_c \approx V_m(y_c)$ and $\mathrm{d}V / \mathrm{d}y \approx \mathrm{d}V_m / \mathrm{d}y|_{y_c}$ in Eq. (12.4), where y_c is the location of the tube axis. Once the velocity corrections are determined, one may construct the corrected velocity profile.

Example 12.1 A Pitot tube with an inner diameter of 0.18 mm and an outer diameter of 0.30 mm is used to take measurements of the streamwise velocity profile in air flow across a boundary layer with a thickness $\delta = 10$ mm and a free-stream velocity of

$V_\infty = 10$ m/s. Measurements were taken at 0.5 mm increments in the range 1 mm $\leq y \leq$ 15 mm. The measured velocities V_m were fitted with a quadratic function and were found to follow well the function $V_m(y)/V_\infty = -0.982(y/\delta)^2 + 1.986(y/\delta) - 0.004$ for $y \leq \delta$. Determine the corrections in the measured values due to shear effects. The static pressure at a pressure tap, located at the same streamwise location as the Pitot tube tip, is measured independently and it may be assumed that the static pressure is constant across the boundary layer.

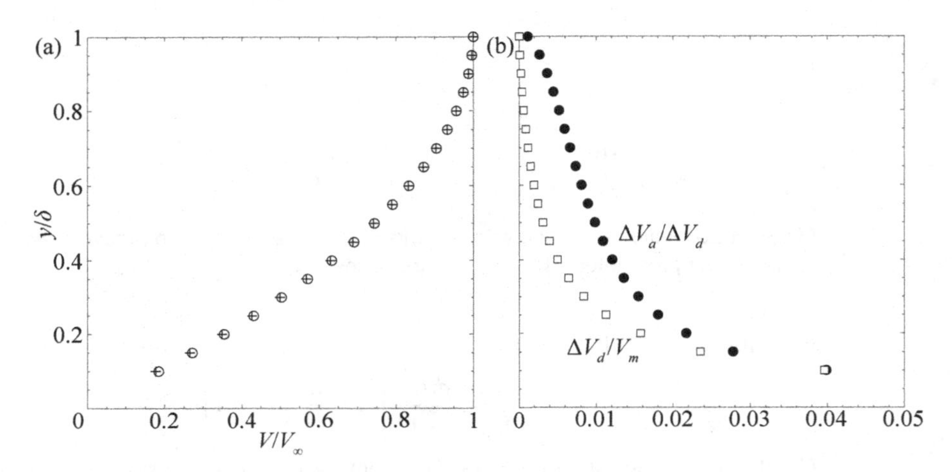

Figure 12.4 (a) Measured (○) and corrected (+) boundary layer velocity profiles; (b) relative velocity correction $\Delta V_d/V_m$ due to displacement (□) and relative velocity correction $\Delta V_a/\Delta V_d$ due to pressure averaging (•).

Answer

The measured velocities are shown in Fig. 12.4a. We first find approximate values of the shear parameter α from the expression that was fitted to the measurements. We find that $0.0003 \leq \alpha \leq 0.1453$, with the largest value corresponding to $y = 0.1\delta$ ($y = 1$ mm), which is the measurement location closest to the wall. The velocity corrections $\Delta V_d = (\mathrm{d}V / \mathrm{d}y)\delta y$, due to displacement, and $\Delta V_a = (\mathrm{d}V / \mathrm{d}y)\delta y'$, due to total pressure averaging, are shown in Fig. 12.4b. We note that the largest velocity correction due to displacement is at $y = 0.1\delta$ ($y = 1$ mm) and is roughly 4% of the measured velocity, that is to say that the true velocity at $y = 1$ mm is 4% lower than the measured value. The figure also highlights that the velocity correction due to pressure averaging is roughly 4% of the correction for displacement, or roughly $0.16\% V_m$, at this location. Therefore, we can neglect the changes due to averaging and apply the correction due to displacement only. The corrected velocity profile, shown in Fig. 12.4a, has moved to lower values than the measured ones.

Wall proximity effect: When a pressure tube is positioned near a wall, it blocks the flow in its vicinity, displacing streamlines away from the wall, thus, towards higher ve-

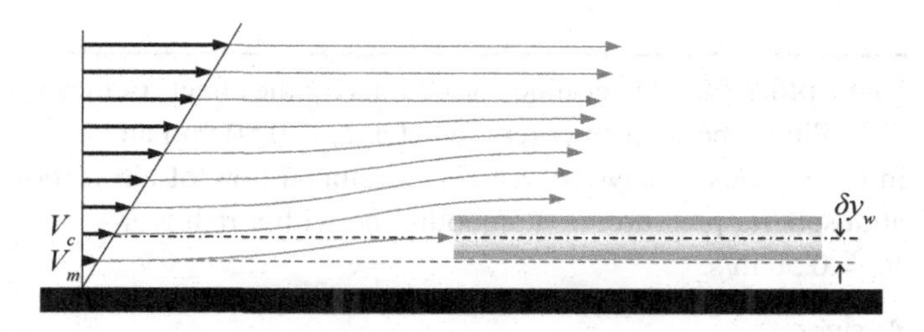

Figure 12.5 Streamline displacement in flow around a Pitot tube positioned close to a wall.

locities, which is the opposite of the displacement effect due to shear. Consequently, the tube reading would correspond to a velocity that is lower than the far-upstream velocity at the level of the tube axis, which may be interpreted as the effect of streamline displacement by δy_w away from the wall (see Fig. 12.5). Circular tubes in a moderate Reynolds number range would be free of wall displacement as long as their axis is positioned more than about $2d_o$ away from the wall [79]; for flattened tubes, it seems that a similar bound is valid, provided that the width (largest dimension) of the tube cross-section is used instead of d_o [93], but the available evidence is not conclusive. For distances from the wall $y \geq 2d_o$, one may use the shear displacement correction, Eq. (12.5). For $y < 2d_o$, one may use the empirically derived wall displacement correction [81]

$$\frac{\delta y_w}{d_o} = \begin{cases} 0.150\,, & \text{for } d^+ < 8\,, \\ 0.120\,, & \text{for } 8 \leq d^+ < 110\,, \\ 0.085\,, & \text{for } 110 \leq d^+ < 1{,}600\,. \end{cases} \tag{12.8}$$

where $d^+ = d_o u_\tau / \nu$ is a non-dimensional diameter, based on the skin friction velocity u_τ and the fluid kinematic viscosity ν. Consequently, the velocity V_m measured by the tube would be lower than the velocity V_c at the extension of its axis, such that

$$V_m = V_c - \frac{\mathrm{d}V}{\mathrm{d}y}\delta y_w\,. \tag{12.9}$$

A comparison of Eqs. (12.7) and (12.9) shows that errors due to shear and wall proximity have opposite signs and, therefore, the two effects may partially compensate for each other in the vicinity of the boundary between the two domains.

As will be discussed in the next chapter (see Section 13.3), thin tubes in contact with the wall serve mainly for the measurement of wall shear stress, in which case they are referred to as *Preston tubes* (see Section 13.3.2). It should be noted that the previous correction requires the value of u_τ. In fully developed pipe flows, u_τ can be conveniently determined from pressure drop measurements along the pipe, but in free boundary layers, developing pipe flows and complex wall-bounded flows, an independent measurement of u_τ may be quite difficult.

Example 12.2 Determine corrections for the effects of mean shear and wall proximity for a Pitot tube with an outer diameter $d_o = 0.80$ mm and a diameter ratio $d_i/d_o = 0.60$ in the viscous sublayer of a fully developed flow of air at room temperature and near atmospheric pressure, in a smooth pipe with a radius $R = 0.50$ m and a bulk velocity $V_b = 0.50$ m/s.

Answer

We may take the density of air as $\rho \approx 1.2$ kg/m^3 and its kinematic viscosity as $\nu \approx 15 \times 10^{-6}$ m^2/s. The bulk Reynolds number for this flow is $\mathrm{Re}_b = 2RV_b/\nu \approx 33{,}300$, which corresponds to a fully turbulent pipe flow. We may estimate the wall shear stress from the empirical expression [38]

$$\tau_w \approx 0.0332\rho V_b^2 \left(\frac{\nu}{RV_b}\right)^{0.25} = 0.88 \times 10^{-3}\ \mathrm{N/m^2}\ ,$$

which gives the friction velocity as $u_\tau = \sqrt{\tau_w/\rho} = 0.027$ m/s. Then, the viscous length is found as $\nu/u_\tau = 0.56$ mm and the conventional thickness of the viscous sublayer as $5\nu/u_\tau \approx 2.8$ mm. Within the viscous sublayer, the time averaged velocity $\overline{V}$ changes linearly with distance y from the wall, as $\overline{V}/u_\tau = u_\tau y/\nu$. Consequently, the mean shear would be

$$\frac{\mathrm{d}\overline{V}}{\mathrm{d}y} = \frac{u_\tau^2}{\nu} = 48.7\ \mathrm{s}^{-1}$$

and the local mean velocity, in mm/s, can be found as

$$\overline{V} = \frac{\mathrm{d}\overline{V}}{\mathrm{d}y} y = 48.7y\ .$$

where y is in mm. The mean velocity at the edge of the viscous sublayer would be about 136 mm/s, which corresponds to a total pressure that is far too small to be measured with a manometer. For the purposes of this example, we shall assume that we find a way to measure the total pressure accurately and that the static pressure at the wall is also estimated accurately.

When the tube axis is at $y = 2d_o = 1.6$ mm, the local mean speed is 77.9 mm/s and the error due to streamline displacement by shear may be estimated from Eqs. (12.3) and (12.5) to be 5.5 mm/s, or about 7% of the local mean velocity.

Now, let us estimate the wall proximity effect for the same tube at the same location. The tube dimensionless diameter is $d^+ = d_o u_\tau/\nu \approx 1440$. Equation (12.8) gives a streamline displacement $\delta y_w = 0.085 d_o = 0.076$ mm, which, when substituted into Eq. (12.3), gives an error of -3.3 mm/s, or about -4% of the local velocity.

In closing, we may note that, because the mean shear is constant within the viscous sublayer, both percentage errors are independent of any flow specifications and depend only on the value of $y^+ = y u_\tau/\nu$, increasing as y^+ decreases.

Viscous effect: When the Reynolds number based on the tube diameter is sufficiently small, the tube response no longer follows the inviscid flow relationship and the pressure coefficient $(p_o - p)/\frac{1}{2}\rho V^2$ would be greater than 1. Viscous effects are significant when $\mathrm{Re}_{d_i} \lesssim 50$ or $\mathrm{Re}_{d_o} \lesssim 100$. To account for viscous effects on a circular tube response at low Reynolds numbers, one may use the following expressions [115, 130] to compute the corrected velocity V from the measured velocity $V_m = [2(p_o - p)/\rho]^{1/2}$:

$$V = \begin{cases} V_m(4.1/\mathrm{Re}_{d_i})^{-1/2}, & \text{for } \mathrm{Re}_{d_i} < 0.7, \\ V_m(1 + 2.8/\mathrm{Re}_{d_i}^{1.6})^{-1/2}, & \text{for } 0.7 < \mathrm{Re}_{d_i} < 18, \\ V_m(1 + 10/\mathrm{Re}_{d_o}^{1.5})^{-1/2}, & \text{for } 30 < \mathrm{Re}_{d_o} < 100. \end{cases} \tag{12.10}$$

Viscous effects may be significant when thin tubes are used in highly viscous fluids or in low-speed flows, including wall regions. One may note that the use of the expressions in Eq. (12.10) may introduce systematic errors, because these expressions may correspond to configurations that are different from the one of interest here. It is hard to estimate a general-purpose uncertainty level, but perhaps a value about 10–20% of the correction may be appropriate.

Example 12.3 Determine a viscous correction for the measured boundary layer velocity profile described in Example 12.1. Compare the magnitudes of this correction and the shear correction calculated in Example 12.1.

Answer

Considering that the kinematic viscosity of air at 20°C is $\nu = 0.000015\ \mathrm{m^2/s}$, we find that the Reynolds number Re_{d_o}, based on the outer diameter of the tube, varies between 37 and 200 across the boundary layer. Only the four measurements closest to the wall require viscous correction, as their Re_{d_o} values are, respectively, 37.0, 54.4, 70.8, and 86.2, thus within the range 30 to 100, as specified in Eq. (12.10). This equation provides the corresponding relative velocity corrections due to viscous effects as $\Delta V_v/V_m = (V_m - V)/V_m = 0.0215, 0.0122, 0.0083$, and 0.0062. The magnitudes of these corrections are roughly half of those due to displacement effects calculated in the previous example; hence, neither of these corrections are negligible. This example raises a practical question: which of these two corrections should be applied first? The matter shall be discussed in a following paragraph of this section.

Turbulence effect: Consider that a pressure tube is inserted in a turbulent flow with a time-averaged velocity $\overline{V}$ and turbulent velocity component variances $\overline{u_1^2}$, $\overline{u_2^2}$, $\overline{u_3^2}$, where the axis x_1 and the tube axis are aligned with the direction of the mean velocity vector. The strength of turbulence is measured by the turbulent kinetic energy per unit mass, defined as $k = (\overline{u_1^2} + \overline{u_2^2} + \overline{u_3^2})/2$, but it is usually expressed in terms of the streamwise turbulence intensity $u_1'/\overline{V}$, where $u_1' = \overline{u_1^2}^{1/2}$. The ratio $\beta = k/u_1'^2$ depends on the turbulence structure: in isotropic turbulence, $u_1' = u_2' = u_3'$, so that $\beta = 3/2$, whereas, in shear flows, $u_1' > u_2', u_3'$, so that $\beta < 3/2$; representatively, in uniformly sheared flows and in outer regions of turbulent boundary layers, $\beta \approx 1$. For turbulence intensities lower than about 10%, one may neglect their effects on the response of Pitot and Pitot-static tubes, whereas, for greater intensities, especially ones than exceed 20%, such effects are likely to be significant and should not be ignored. Turbulence effects have found a rather limited treatment in the literature and they are hard to isolate from mean shear, wall proximity and viscous effects. It is clear that turbulence effects would depend not only on the turbulence intensity, but also on its length scale and the turbulence structure, which is characterised by the turbulent stress tensor anisotropy.

Velocity fluctuations always tend to produce values of both the time-averaged total pressure $\overline{p}_{om}$, read by a Pitot tube, and the time-average static pressure $\overline{p}_m$, read by a static-pressure tube or a wall pressure tap, which are larger than the corresponding values $\overline{p}_o$ and $\overline{p}$ in a steady laminar flow with a velocity that is equal to $\overline{V}$. Unfortunately, the two errors would not cancel each other, if one were to compute the flow velocity as $\overline{V}_m \approx [2(\overline{p}_{om} - \overline{p}_m)/\rho]^{1/2}$.

Theoretical analysis [43] has shown that the total pressure indicated by a Pitot tube that is aligned with the mean flow would be

$$\overline{p}_{om} = \overline{p} + \frac{1}{2}\rho\left(\overline{V}^2 + 2k\right) = \overline{p} + \frac{1}{2}\rho\overline{V}^2\left[1 + 2\beta\left(\frac{u_1'}{\overline{V}}\right)^2\right]. \tag{12.11}$$

Given an accurate value of $\overline{p}$, measured or computed by independent means, and a fair estimate of the turbulence intensity and anisotropy, one may use Eq. (12.11) to determine $\overline{V}$ as

$$\overline{V} \approx \overline{V}_m\left[1 - 2\beta\left(\frac{u_1'}{\overline{V}}\right)^2\right]^{1/2}. \tag{12.12}$$

In practice, however, turbulence also introduces an error in the measurement of $\overline{p}$, which would be different for static holes in Pitot-static tubes and wall pressure taps. An analytical prediction of this error has been based on assumptions, which are not of general validity, particularly the assumption of isotropic turbulence ($\beta = 1$). Under such assumptions, this analysis provides a rough estimate of the corrected Pitot-static tube reading as [91]

$$\overline{V} \approx \overline{V}_m\left[1 - \gamma\left(\frac{u_1'}{\overline{V}}\right)^2\right]^{1/2}, \tag{12.13}$$

where the value $\gamma = 1$ corresponds to a turbulence length scale that is small compared to the distance between the total and static pressure orifices (so that the turbulence effects on the total and static pressure measurements are independent from each other), whereas the value $\gamma = 5$ corresponds to a relatively large turbulence scale (so that each turbulent impacts simultaneously on both the total and the static pressure holes). The value $\gamma = 1$ also applies to Pitot-tube–static-pressure-tap measurements [81].

Vibration effect: The effects of vibration are similar to turbulence effects, as they result in the tube being exposed to pressure fluctuations. If the tube has a right-angle bend, vortex shedding from the tube stem may introduce transverse vibrations, whereas mechanical vibrations transmitted to the tube from the apparatus wall are likely to be aligned roughly with the tube axis. A correction for the effects of axial vibration with frequency f (in Hz) and amplitude a on the reading of a Pitot tube may be made with the use of the expression [91]

$$\overline{V} \approx \overline{V}_m \left[1 - \left(\frac{2\pi a f}{\overline{V}_m} \right)^2 \right]^{1/2} . \tag{12.14}$$

This expression would be inaccurate when tube vibrations are more complex than simple axial oscillations. It is evident that one should strive to reduce and, if possible, eliminate tube vibrations by stiffening or bracing the tube or detaching the tube mount from vibrating surfaces.

Compressibility effect: In compressible flow, Eq. (12.1) is no longer valid and one may instead use an isentropic expression along a streamline as

$$C_P = \frac{p_o - p}{\frac{1}{2}\rho V^2} = \left[\left(1 + \frac{\gamma - 1}{2} \mathrm{M}^2 \right)^{\frac{\gamma}{\gamma-1}} - 1 \right] \frac{2}{\gamma \mathrm{M}^2} . \tag{12.15}$$

In the low subsonic regime, the previous expression may be approximated through a binomial expansion as

$$C_P \approx 1 + \frac{\mathrm{M}^2}{4} + \frac{(2-\gamma)\,\mathrm{M}^4}{24} . \tag{12.16}$$

Compared to the more accurate Eq. (12.15), the incompressible expression, Eq. (12.1), would give an error of less than about 0.5% for $\mathrm{M} < 0.15$, whereas the first two terms in Eq. (12.16) would give an error of less than 5% for $\mathrm{M} < 0.7$. Clearly, Eq. (12.15) would not hold in transonic and supersonic flows, in which the presence of the tube may introduce shock waves and other discontinuities. It is reminded that the total pressure decreases across a normal shock, according to the *Rankine–Hugoniot relationship* (Eq. (9.6)). Conical-nose tubes are preferable to blunt-nose tubes in supersonic flows, particularly for static-pressure measurement. Compressibility effects are coupled with other sources of error, such as displacement and misalignment effects, introducing a large uncertainty into any correction procedure [20].

Combination of different effects: In many experiments, errors may occur as the result of the simultaneous presence of more than one among the previously mentioned effects. For example, measurements by small Pitot tubes in the inner region of a turbulent boundary layer would likely be subject to errors due to viscous effects, turbulence, shear and/or wall proximity. As it does not seem possible to develop a general strategy for decoupling such effects, one is advised to follow an iterative and approximate procedure. First, one may consider each effect in isolation from others and use previously presented relationships to estimate the corresponding correction. Then, one may apply sequentially such corrections to the measurements and possibly iterate until some convergence is achieved. For the inner boundary layer example, one may first apply an approximate viscous-effect correction, based on Eq. (12.10); then, apply an approximate turbulence correction, based on Eq. (12.12); then, apply a shear-effect correction, based on Eq. (12.7); and, finally, use the corrected results as input to the different correction expressions and repeat this procedure, if a significant difference is detected. It is obvious

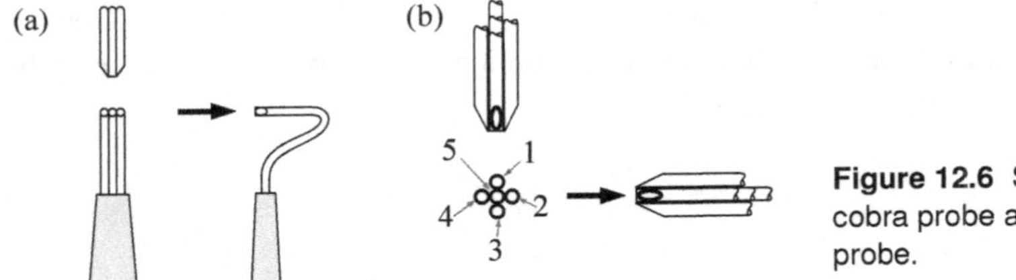

Figure 12.6 Sketches of (a) a cobra probe and (b) a five-hole probe.

that this approach may be subject to a large uncertainty and, so, if one finds multiple corrections to be large, one is advised to incorporate these values into the uncertainty estimates.

Other sources of uncertainty: Besides the previously discussed types of possible error in Pitot tube measurements, there could be additional ones, which may contribute to the overall measurement uncertainty. A common source of error in near-wall velocity profiles is the possible error in the distance of the probe from the wall. This error may be hard to determine accurately, especially if the experimental set-up does not allow optical access. One is also cautioned that fine tubes may be deformed during attempts to bring them in contact with the wall as a reference position. In order to apply some of the previous corrections, one needs to fit smooth functions to measurements, for example, for estimating the shear. Curve fitting introduces both systematic and random uncertainty. The random uncertainty (precision limit) may be estimated roughly (95% confidence limit) as the standard deviation of the differences between the measured velocity values and corresponding values of the fitted expression. Finally, uncertainty would be introduced by the manometer, pressure transducer, tubing etc.; this type of uncertainty was discussed in Chapter 9.

12.1.2 Multi-Hole Probes

Multi-hole probes are arrangements of two or more orifices at the tip of a probe body with faces inclined with respect to the probe axis, so that they can provide information about the flow direction. In combination with total and static-pressure orifices, they may also measure the flow velocity in magnitude and direction. The simplest type of such probes, called *yawmeters*, consist of two symmetrically inclined, thin tubes. They normally operate in *nulling mode*, which means that the assembly is yawed within the plane containing the axes of the two tubes until the pressure readings of the two tubes match each other, in which case the flow direction should be normal to the line connecting the two orifice centres; for plane flows, this would determine the flow direction. The addition of an upstream-facing orifice between the two inclined ones would provide the total pressure, from which the flow velocity magnitude can be determined; an example of such probes is the *cobra probe* (Fig. 12.6a), in which the tip centre is aligned with the probe axis but the plane of the three orifices is normal to this axis, for convenient insertion and rotation through the apparatus wall. Probes with two pairs of inclined orifices, having faces in a pyramidal arrangement, like the *claw-type* probe, have also been developed for use in three-dimensional flows, in which it would be necessary to determine both the *yaw* and *pitch angles* of the flow with respect to a fixed axis. In many experimental settings, nulling of probes is inconvenient and time consuming and it is preferable to obtain flow information from a fixed probe or one that is traversed parallel to a fixed direction. There is a variety of multiple-hole probes that provide flow direction and velocity. based on analytical expressions or calibration curves. The response of probes with spherical or hemispherical tips may be based on potential flow analysis. Such analysis provides the pressure coefficient on the surface of a sphere as

$$C_P = \frac{p - p_\infty}{\frac{1}{2}\rho V^2} = 1 - \frac{9}{4}\sin^2\theta\,, \tag{12.17}$$

where p_∞ is the undisturbed stream pressure and θ is the angle formed by radial axes passing through the surface point and the stagnation point. Due to imperfections in probe manufacturing, the response of a specific probe may deviate from the theoretical expression and it would be preferable to compute C_P from calibration data. Probe designs with wedge-like, conical, pyramidal, and other shapes have been developed; however, the most popular design, known as the *five-hole probe*, consists of a set of five parallel tubes, a central one leading to an orifice normal to the tube axis at the apex of a pyramid formed by the inclined orifices of four peripheral tubes, arranged symmetrically (Fig. 12.6b). Such probes may be fabricated by a skilled machinist and be made with thicknesses that are as small as 1 mm [73, 96, 116]. For their calibration [13], multi-hole probes are inserted in a uniform stream at different pitch and yaw angles φ and ψ, respectively, and the five pressures p_1, p_3 ('pitch' pressure pair), p_2, p_4 ('yaw' pressure pair) and p_5 ('total' pressure) are recorded. The actual total pressure p_o of the stream is also recorded and an average side-orifice pressure is computed as

$$p_m = \frac{1}{4}(p_1 + p_2 + p_3 + p_4)\,. \tag{12.18}$$

Then, the values of the pitch, yaw, total pressure and dynamic pressure coefficients are computed, respectively, as

$$C_\varphi = \frac{p_3 - p_1}{p_5 - p_m}\,, \tag{12.19}$$

$$C_\psi = \frac{p_4 - p_2}{p_5 - p_m}\,, \tag{12.20}$$

$$C_o = \frac{p_o - p_5}{p_5 - p_m}\,, \tag{12.21}$$

$$C_q = \frac{p_5 - p_m}{\frac{1}{2}\rho V^2} \tag{12.22}$$

and plotted in iso-contour form vs. the two angles φ and ψ. During an experiment, one measures the five pressures, which can be used to compute C_φ and C_ψ. Assuming that each pair of values of C_φ and C_ψ corresponds to a unique pair of φ and ψ, one may then find the orientation of the flow velocity with respect to the probe axis. From the combination of computed values of φ and ψ, one may then find C_o and C_q, from which it is easy to compute the flow velocity V. This process may be automated using multivariate curve-fitting algorithms. The operating range of five-hole probes is typically within a cone of 30° half angle. For applications involving larger flow incidence, probes with seven holes, or even more (up to 19), have been developed, which reportedly perform well within cones with a half angle of up to 75° [35, 95, 105, 106, 131]. Other authors [56, 99, 110, 117] have examined the dependence of multi-hole probe response on Reynolds number, shear, wall proximity, turbulence, compressibility and other factors.

12.1.3 Fast-Response Probes

In most applications, static-pressure tubes are connected to manometers or pressure transducers through tubing, in which case, their frequency response would be relatively slow, typically of the order of 0.1 Hz. Nevertheless, pressure fluctuations have been measured with static tubes in which a fast-response pressure transducer was placed near the static orifices. A much quoted early design [109] not only contained a microphone embedded inside the tube, but also had a nose tip that was mounted to the main probe body through four piezoelectric elements that measured the velocity component normal to the tube axis; analogue circuitry provided compensation to the static-pressure reading for turbulence effects. More recently, in-flow fluctuating pressure has been measured by placing miniature piezoresistive (semiconductor-type) pressure transducers inside pressure tubes and multi-hole probes [5, 53, 54, 55, 69]. Four thin piezoelectric elements are arranged in a Wheatstone bridge configuration and mounted on a pressure-sensing membrane and the sensor output is normally compensated for temperature sensitivity. A different concept is exploited in the *bleed-type pressure tubes* [58, 114]. These tubes are connected to a constant-pressure air supply and the pressure is regulated such as to produce laminar flow through the tube and the static-pressure orifices. The tubes contain a hot-film sensor, which measures flow velocity. The instantaneous pressure differences across the tube is computed through its linear relationship to flow velocity in tube. The probe must be calibrated and its output is compensated to improve its frequency response. All probes described previously are subjected to errors due to vibrations, internal motions and external turbulent fluctuations. If the flow velocity is measured simultaneously by other means, the instantaneous pressure reading may, in principle, be corrected for turbulence effects; the magnitude of such corrections remains an issue of controversy.

12.2 Thermal Anemometry

Thermal anemometers [12, 22, 34, 42, 76, 92] measure the local flow velocity through its relationship to the convective cooling of electrically heated metallic sensors. These include *hot wires*, usable in clean air or other gas flows, and *hot films*, usable in both gas and liquid flows. A related method, *pulsed-wire anemometry* is, in a strict sense, a thermal tracer method, but it will be included in this section as it utilises instrumentation similar to that used by the other thermal anemometers.

Until the development of laser Doppler velocimetry, in the 1970s, thermal anemometry was essentially the only method having space, time and amplitude resolutions that are sufficiently high for the measurement of turbulent characteristics down to the smallest scales of dynamic interest. Although the use of hot wires as velocity sensors was anticipated by some work in the late nineteenth century, the seminal paper that clarified the relationship between heat transfer from a heated cylinder and flow speed was written by King in 1914 [63]. The resulting equation, known as *King's law*, is still commonly used. Early hot wires were relatively thick and operated in the constant current mode, which severely limited their frequency response. Turbulence measurement was made possible only following the use of frequency compensation circuits [29]. The next major development in thermal anemometry was the design of a workable constant temperature electronic circuit [67, 90].

12.2.1 Sensor Materials and Mounting Procedures

Although platinum, nickel and various alloys have been used as hot-wire sensors, the materials that are almost exclusively used today are platinum alloys, mainly with 20% iridium or 10% rhodium, and tungsten. Pure platinum has excellent thermal properties but relatively low mechanical strength, which would result in frequent breakage caused by vibrations or aerodynamic loading. The platinum alloys have improved mechanical strength, although reduced thermal resistivity. Tungsten has sufficient mechanical strength, high thermal resistivity and a high melting temperature, but it oxidises at about 350°C, so that it may not be used at extremely high overheats or in high temperature flows. The length of sensing elements is usually 0.8 to 1.5 mm, while their diameter is commonly 5 μm; thicker sensors (7.5 μm) are used for higher strength in high speed flows, while thinner sensors (2.5 μm) are used for the measurement of the turbulence fine structure.

A hot wire is mounted at its two ends on thin metallic *prongs*, usually tapered and having diameters of less than 1 mm, as shown in Fig. 12.7. Tungsten wire is platinum-plated and spot-welded on the prongs, by discharging a capacitor through a needle-type electrode. Platinum alloys can be spot-welded as well. Spot-welded wires have the disadvantage of been active over their entire length and thus permitting temperature non-uniformity due to heat conduction to the prongs; they also suffer from increased aerodynamic interference of the prongs. These problems may be overcome by using

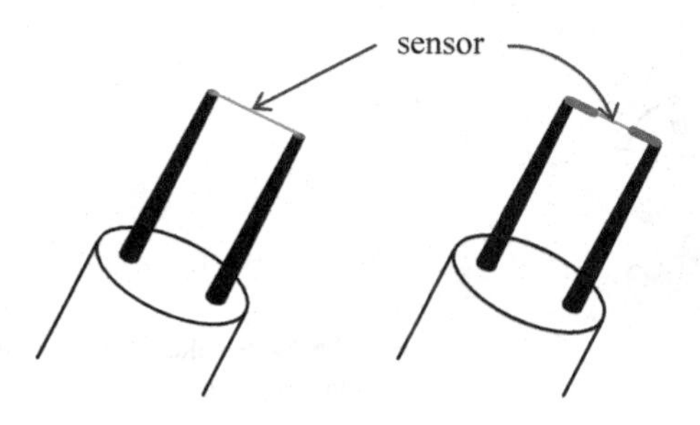

Figure 12.7 Sketches of a spot-welded single hot-wire sensor (left) and a Wollaston or gold-plated single hot-wire sensor (right).

a wire that is only active in its middle region, while the two end regions are much thicker and thus not subjected to appreciable Joule heating. This may be achieved by gold plating the ends of the wire in a special bath, or by etching the middle portion of a *Wollaston* wire (consisting of a Pt or Pt-alloy core and a silver coating, about 25 μm thick) in a thin jet of nitric acid at about 30% concentration. Then, the gold-plated or silver-coated ends of the wire are soft-soldered on the prongs. Sensor material of all kinds as well as probe repair services are available commercially, although a laboratory that uses hot wires routinely should be equipped with probe repairing apparatus.

With advances in microelectromechanical (MEMS) manufacturing techniques, a range of nanoscale thermal anemometer probes (NSTAP)[36], including designs capable of measuring two velocity components [39], have been developed in recent years. The manufacturing process consists of the sequential deposition of material and subsequent etching to produce the desired shape of the sensors [8, 68]. A representative probe of this type has a prismatic platinum sensing element with a length of $l = 60$ μm, a width of 2 μm and a frontal height of 100 nm, allowing for turbulence measurements with significantly higher spatial and temporal resolutions than those of conventional hot-wire sensors.

Hot-film sensors are manufactured by depositing a thin (about 0.1 μm thickness) film of platinum or nickel on an insulating substrate, usually made of quartz or mica; then, a layer of quartz, of about 1 μm thickness, is added on the outside to protect and insulate the active film. Hot-film probes usually have cylindrical sensors, mounted on prongs like hot wires but having a much larger diameter (typically 50 μm); probes with other shapes, including conical, wedge-type and spherical shapes, are also available.

12.2.2 Heat Transfer Characteristics

First, consider a cylindrical heater with a diameter d, a length l, and at a uniform temperature T_w, immersed in uniform, *steady* flow of a gas with a far-field temperature $T < T_w$ and a free-stream velocity V, inclined with respect to a plane normal to the cylinder axis by an angle φ (see Fig. 12.8). An important operating parameter is the *temperature overheat ratio*, defined as

$$a_T = \frac{T_w - T}{T} \,. \tag{12.23}$$

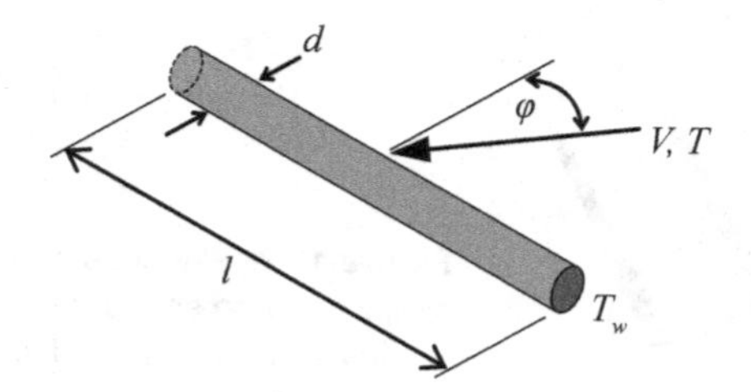

Figure 12.8 Sketch of a heated cylinder in uniform flow.

Heat transfer from the cylinder to the gas will, in general, take place in all possible modes: convection (natural, forced or mixed), conduction to supports at the end of the cylinder, and radiation. For the usual hot-wire anemometry temperature range of a few hundred degrees centigrade, radiation losses are less than 0.1% of the convective losses and may be disregarded (exceptions include the study of combustion with the use of hot wires heated to very high temperatures and measurements in very low density flows, in which convection is low). The gas temperature in the vicinity of the cylinder will have values between T and T_w, and the gas flow properties will vary accordingly. It is a customary convention to define a *film temperature* as $T_f = \frac{1}{2}(T + T_w)$, and evaluate the gas thermal conductivity k, density ρ, kinematic viscosity ν, and thermal diffusivity γ at the film temperature. If $\dot{q}$ is the heat transfer rate from the wire, one may define the dimensionless heat transfer rate, called the *Nusselt number*, as

$$\mathrm{Nu} = \frac{\dot{q}}{\pi l k (T_w - T)} . \tag{12.24}$$

In the general case, the heat transfer will depend on several factors, which, in dimensionless form, may be expressed as

$$\mathrm{Nu} = \mathrm{Nu}(\mathrm{Re}, \mathrm{Pr}, \mathrm{Gr}, \mathrm{M}, \mathrm{Kn}, a_T, l/d, \varphi) , \tag{12.25}$$

where $\mathrm{Re} = Vd/\nu$ is the Reynolds number, $\mathrm{Pr} = \nu/\gamma$ is the Prandtl number, $\mathrm{Gr} = g\beta(T_w - T)d^3/\nu^2$ is the Grashof number ($\beta = 1/T$ is the volume coefficient of expansion), $\mathrm{M} = c/V$ is the Mach number (c is the speed of sound), and $\mathrm{Kn} = \lambda/d$ is the Knudsen number (λ is the molecular mean free path; $\mathrm{Kn} = \sqrt{\frac{1}{2}\pi c_p/c_\upsilon} \mathrm{M}/\mathrm{Re}$). Restricting the analysis to air flows, one may disregard the dependence on Pr, which will be essentially constant. Most hot-wire applications satisfy the criterion $\mathrm{Gr} < \mathrm{Re}^{1/3}$ for the effects of natural convection to be negligible; this assumptions is expected to become inaccurate for $\mathrm{Re} < 0.2$. Further considering only low-speed flows that may be treated as incompressible ($V < 100$ m/s; $\mathrm{M} < 0.3$), the dependence on M may also be neglected. Finally, hot wires usually operate in the 'continuum regime' ($\mathrm{Kn} < 0.01$), or very close to the lower boundary of the 'slip flow regime' ($0.01 < \mathrm{Kn} < 1$), which also essentially eliminates their dependence on Kn. Under such conditions, the previous expression is reduced to

$$\mathrm{Nu} = \mathrm{Nu}(\mathrm{Re}, a_T, l/d, \varphi) . \tag{12.26}$$

Further considering cylinders with a large aspect ratio l/d (e.g., $l/d > 100$), so that end conduction effects may also be neglected, and with their axes normal to the flow ($\varphi = 0$), one may derive the considerably simplified relationship

$$\mathrm{Nu} = \mathrm{Nu}(\mathrm{Re}, a_T) , \tag{12.27}$$

which permits relatively easy experimental determination. Following King's analysis, one may anticipate a relationship of the type

$$\mathrm{Nu} = (A + B\,\mathrm{Re}^n)\left(1 + \frac{1}{2}a_T\right)^m , \tag{12.28}$$

(notice that $1 + \frac{1}{2}a_T = T_f/T$) and, indeed, the well-known empirical response equation by Collis and Williams [21] gives, for the vortex shedding regime,

$$\mathrm{Nu} = (0.24 + 0.56\,\mathrm{Re}^{0.45})(1 + \tfrac{1}{2}a_T)^{0.17}, \quad \text{for } 44 < \mathrm{Re} < 140 \tag{12.29}$$

and, for the attached-flow regime,

$$\mathrm{Nu} = 0.48\,\mathrm{Re}^{0.51}(1 + \tfrac{1}{2}a_T)^{0.17}, \quad \text{for } 0.02 < \mathrm{Re} < 44\,. \tag{12.30}$$

These response equations appear to hold universally for all hot wires satisfying the previous assumptions. In practice, however, the uncertainty in the determination of geometrical and material properties and the interference of many other factors make it necessary for each sensor to require individual calibration and its response to be expressed in terms of practical parameters, rather than dimensionless groups.

During an experiment using hot-wire anemometry, the main measurand, namely, the flow velocity V, needs to be determined from the voltage output E of the measuring system, which includes the hot-wire operating bridge, an amplifier, a signal conditioner, a filter and an analogue-to-digital converter. Therefore, one needs to know the hot-wire response in the form of a relationship between E and V. Such a relationship can be obtained by calibrating the probe under conditions that are identical to those during the experiment, in particular, using the same overheat ratio and the same flow temperature T. This may be possible by calibrating the hot wire in situ, in the same wind tunnel or other flow apparatus as the one used for the experiment and measuring the voltage output for different flow velocities, while maintaining the flow temperature constant. It is assumed that the flow velocity is measured by other means, for example, the pressure difference across a contraction or the output of a Pitot-static tube. Then, the obtained set of pairs of values V, E can be fitted by a suitable algebraic relationship, which, for the usual hot-wire operating conditions, describes both the static and the dynamic response of the hot wire.

The two most common types of such relationships are the *modified King's law*, and a fourth-order polynomial fit. The modified King's law is

$$E^2 = A + BV^n\,, \tag{12.31}$$

where the constants A, B and n are determined by optimal fitting of this expression to calibration results. The usual range of the exponent n is between 0.40 and 0.55, which includes the original value 0.50 proposed by King, based on convective heat transfer analysis. Equation (12.31) can be easily solved for V in terms of a measured voltage E. The polynomial fit is simply

$$V = a_0 + a_1E + a_2E^2 + a_3E^3 + a_4E^4\,, \tag{12.32}$$

where the coefficients $a_i, i = 0, ..., 4$ are obtained by curve fitting.

The response of hot-film sensors, both cylindrical and non-cylindrical, is more complex than that of hot wires, especially due to their small aspect ratio and to heat conduction to the substrate. However, such effects may be largely absorbed into the empirical coefficients in the calibration relationship, which is, thus, applicable to all thermal sensors.

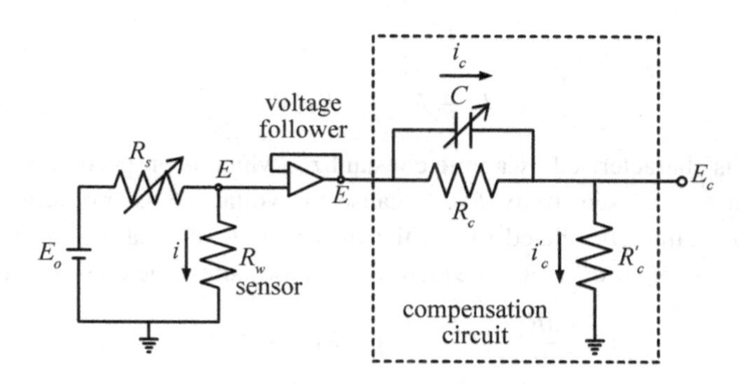

Figure 12.9 Sketch of a simple constant current circuit with a simple frequency compensation circuit; the voltage follower has been added to prevent loading of the sensor by the compensation circuit.

Now consider unsteady flow, in which the flow speed varies with time, but, for simplicity, let the flow temperature be constant. Let i be the electric current through the sensor. The power input to the wire is the Joule heating power i^2R_w. Neglecting heat conduction and radiation, the heat losses from the wire consists only of the convective heat transfer rate $\dot{q}$, which depends on the flow velocity and the sensor temperature, according to the relationships presented earlier. Then, the energy equation applied to the sensor gives

$$mc\frac{\mathrm{d}T_w}{\mathrm{d}t} = i^2R_w - \dot{q}(V, T_w) \tag{12.33}$$

where m is the mass of the sensor and c is its specific heat. Clearly, this equation cannot be solved because it contains three unknowns, T_w, i and V (R_w is directly related to T_w). To render this equation solvable, one must keep either the current or the sensor temperature constant, which may be achieved with the use of suitable electric circuits. The corresponding methods are, respectively, known as *constant current anemometry* (CCA) and *constant temperature anemometry* (CTA). In the constant temperature mode, the unsteady term in the previous equation vanishes and the unsteady response equation becomes identical to the steady response equation, so that a static calibration is sufficient. In the constant current mode, the sensor temperature is permitted to vary, and is, therefore, subjected to 'thermal inertia' effects, so that the anemometer will have a limited frequency response. For this reason, it is the constant temperature mode that is used almost universally. The constant current mode is used rarely, mostly in high-speed or high-temperature flows.

12.2.3 Constant Current Anemometry

An approximately constant current i may be supplied to a sensor with resistance R_w by simply connecting it, in series with a large ballast resistor $R_s \gg R_w$, to a constant-voltage source E_o such that $i = E_o/\,(R_s + R_w) \approx E_o/R_s \approx$ const. (Fig. 12.9). The voltage output will be $E = iR_w$.

The unsteady energy equation, Eq. (12.33), is highly non-linear. When linearised in the vicinity of an operating point, namely at a particular flow speed V_{op} and sensor temperature $T_{w_{op}}$, it leads to the first-order differential equation

$$\tau_w \frac{\mathrm{d}T_w}{\mathrm{d}t} + \left(T_w - T_{w_{op}}\right) = K_T(V - V_{op}) \ . \tag{12.34}$$

This equation is characterised by a time constant τ_w, which is proportional to the overheat ratio, and a static sensitivity K_T. Because the voltage E is proportional to R_W, which, in turn, is linearly related to T_w, the linearised E–V relationship will also be governed by a first-order differential equation, with the same time constant, as

$$\tau_w \frac{\mathrm{d}E}{\mathrm{d}t} + \left(E - E_{op}\right) = K(V - V_{op}) \, . \tag{12.35}$$

A change in the flow velocity (assuming a constant flow temperature, for simplicity), will affect the convective cooling of the sensor, whose temperature T_w will change, according to a first-order system response. Typical values of τ_w, measured by superimposing a low-amplitude square-wave-type voltage on E_o, are of the order of 1 ms for thin hot wires and 10 ms for slim cylindrical hot films. Such values are far too large for the wire to resolve turbulent motions. In a flow with variable speed or temperature, the overheat ratio a_T, and thus τ_w, will vary as well. For i =const., increased overheating at low speeds will result in burnout, whereas, at high speeds, reduced overheating will result in low static sensitivity K of the sensor.

To improve the frequency response of a constant current anemometer, one may use a *frequency compensation circuit*, like the simple R–C circuit shown in Fig. 12.9. To analyse this circuit, notice that $i_c + i_{R_c} = i_{R'_c}$, which leads to the linear differential equation

$$R_c C \frac{\mathrm{d}(E - E_c)}{\mathrm{d}t} + (E - E_c) = E_c \frac{R_c}{R'_c} \ . \tag{12.36}$$

To derive the response equation of the entire compensated anemometer system, one may combine Eq. (12.35) and Eq. (12.36). By selecting the compensation circuit components such that $R_c C = \tau_w$ and $R'_c \ll R_c$, the response equation simplifies to

$$\tau_w \frac{R'_c}{R_c} \frac{\mathrm{d}E_c}{\mathrm{d}t} + \left(E_c - E_{op} \frac{R'_c}{R_c}\right) = K \frac{R'_c}{R_c} (V - V_{op}) \ . \tag{12.37}$$

This equation shows that the compensated system is, approximately, a first-order system. Its time constant is $\tau_w R'_c / R_c \ll \tau_w$, and, thus, it has a far better frequency response than the uncompensated system. However, because τ_w depends on the flow velocity, in order to achieve the previous response, one must readjust the R–C circuit (e.g., by adjusting a variable capacitor) at each speed of operation, which makes the use of this anemometer in shear flows quite cumbersome.

12.2.4 Constant Temperature Anemometry

In this mode, the current through the sensor is continuously adjusted, through an electronic feedback system and in response to changes in convective cooling, such that the sensor's resistance, and, consequently, its temperature, remain essentially constant. As a result, the unsteady heat transfer equation is identical to its steady form and the dynamic

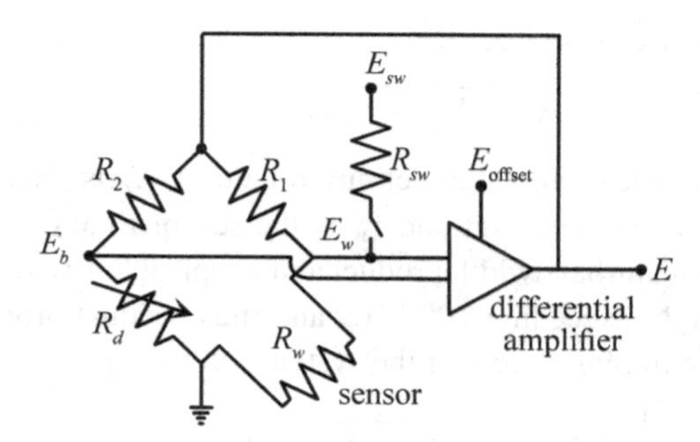

Figure 12.10 Sketch of a constant temperature circuit.

response of the anemometer is the same as its static response within a wide frequency range. By proper selection and adjustment of the electronic feedback system, it is possible to extend the frequency response of the unit beyond the usual ranges of even the most demanding turbulent and transient flows.

For an explanation of the system's operation, consider the typical constant temperature anemometer sketched in Fig. 12.10. The sensor R_w comprises one arm of a Wheatstone bridge, the opposite arm of which consists of an adjustable decade resistor array R_d. In order to supply most of the available power to the sensor, and not to the passive arm, the *CTA bridge ratio* $k_{\text{CTA}} = R_2/R_1$ is fixed at a relatively large value, 10 or 20, although other values in the range 1 to 50 have also been used for specific purposes (a lower bridge ratio results in higher frequency response). Note that the CTA bridge ratio is, by definition, different from the bridge ratio of generic Wheatstone bridges (Section 2.2.3). The two midpoints of the bridge are connected to the inputs of a high-gain, low-noise differential amplifier, the output of which is fed back to the top of the bridge. If $R_2/R_d = R_1/R_w$, then $E_B - E_w = 0$, and the amplifier output will be zero. However, if R_d is increased to a value R'_d, the resulting bridge imbalance will generate an input imbalance of the amplifier. Then the amplifier will produce an output voltage, which will create some current though both branches of the bridge. The additional current through the hot wire will create additional Joule heating, which will tend to increase its temperature and, thus, its resistance, until the resistance increases sufficiently to balance the bridge once more. A problem with this method is that, with a balanced bridge, the amplifier's output voltage would be zero and no current would flow through the system at the operating condition. To avoid this situation, the amplifier's input is offset by a voltage E_{offset}, so that some current flows through the bridge at all times. This produces an imbalance of the bridge, such that the sensor's resistance appears to be $R_w(1 + \delta)$, namely higher than its actual value by a factor $\delta \ll 1$, called the *imbalance parameter.*

Dynamic response: The dynamic response of the constant temperature anemometer is non-linear. Under certain assumptions, one may formulate a third-order linear model, which has been studied to some extent. However, the operation of this system is usually analysed and optimised within the context of a second-order model, which is approximately valid under common operating conditions. According to this model, the natural frequency ω_o of the system is expressed as

$$\omega_o = \sqrt{2a_R K_o/(\tau_a \tau_w)}\,, \tag{12.38}$$

where a_R is the resistance overheat ratio (see definition below), K_o is the system gain parameter, τ_a is the amplifier time constant and τ_w is the sensor time constant. For a modern amplifier with a high gain-bandwidth product and a typical hot-wire sensor, this natural frequency is quite high, exceeding 100 kHz, and thus adequate for practically any flow measurement. The damping ratio ζ of the system can be expressed as

$$\zeta = \frac{1}{2}(1 + K_o\delta)\sqrt{\tau_w/(2K_o\tau_a a_R)}\,. \tag{12.39}$$

To avoid oscillations of the output, which may cause burnout of the sensor, while maintaining the widest possible frequency response, it is recommended to use a critically damped ($\zeta = 1$), or a very slightly underdamped system. Overdamping introduces nonlinearities, which further complicate the dynamic response and introduce measurement errors. For a given sensor, overheat ratio, and amplifier, an adjustment of the damping ratio may only be accomplished by adjusting the imbalance parameter δ, which is possible with the use of additional controls associated with the amplifier's offset.

Operating procedure: The operating procedure of a constant temperature anemometer typically consists of the following steps:

- *Measurement of the 'cold' resistance* R: The sensor, connected to the bridge as described previously, is inserted into the flow stream, so that it takes the flow temperature T, at which its resistance is R. This resistance is measured by disconnecting the feedback ('open loop' configuration), feeding the bridge with some external small voltage, and adjusting the decade resistor R_d until the bridge is balanced.
- *Application of an overheat*: The value of the decade resistor is increased to a value $R'_d = R_d(1 + a_R) = R_d R_w/R$, where the *resistance overheat ratio* $a_R = (R_w - R)/R$ is related to, but distinct from, the temperature overheat ratio $a_T = (T_w - T)/T$. Typical values for a_T are 0.5 to 0.9 for hot wires and 0.1 to 0.2 for hot films; higher values may result to damage to the sensor (e.g., oxidisation of tungsten sensors at 350°C), whereas lower values will produce low velocity sensitivity.
- *Adjustment of the anemometer's dynamic response*: The feedback is connected again ('closed loop' configuration). From now on, the circuitry will strive to maintain the sensor's resistance at the value R_w, irrespective of any changes in the flow velocity. The adjustment of the damping ratio is accomplished by performing the square-wave test. For this test, the sensor is inserted into a stream with a speed comparable to the average speed of the experiment, but free of velocity fluctuations. A square-wave type voltage E_{sw}, fed to the sensor through a large resistor R_{sw}, produces a weak current, with an amplitude equal to about 20% of the sensor's current and superimposed to it. The voltage output of the system is observed on an oscilloscope and some control is adjusted finely until the output oscillations cease (critically damped system) or nearly cease (slightly underdamped system).
- *Calibration of the anemometer*: The sensor is inserted in a low turbulence calibration facility, usually a jet or a wind tunnel, and its output is recorded for different values

of the flow velocity, while the flow temperature is also monitored. A response equation, usually the modified King's law or a fourth-order polynomial, is fitted to the calibration data.

- *Operation of the anemometer*: Finally, the calibrated probe is carefully inserted into the flow of interest for the collection of the experimental results. Because the frequency response of the constant temperature anemometer is high, the voltage output will contain a significant amount of electronic noise, which must be reduced by low-pass filtering the output signal before processing it.

12.2.5 Various Effects and Error Sources

Velocity direction effect: In the previous analysis of the cylindrical heated sensor response, it has been assumed that the flow velocity has a direction normal to the sensor axis. For a general velocity orientation, one may define an *effective cooling velocity* V_{eff}, as the magnitude of the (hypothetical) velocity normal to the sensor that would produce the same convective heat transfer as the actual velocity vector. First, consider a sensor long enough for the effects of prong interference to be negligible and cooled by a flow with a velocity vector with magnitude V and direction forming an angle φ with the normal to the sensor axis (Fig. 12.8). If one assumes that cooling of the cylinder is due to the normal component only, then,

$$V_{eff} = V \cos\varphi \ . \tag{12.40}$$

However, in reality, there would be some cooling even if the flow velocity were tangential to the sensor. To account for tangential cooling, the previous relationship can be modified as

$$V_{eff} = V \sqrt{\cos^2\varphi + k^2 \sin^2\varphi} \ , \tag{12.41}$$

where the empirical coefficient k^2 has to be determined by calibration of each particular probe, exposed to a flow with a constant magnitude and changing direction. Typical values of k^2 for hot-wire probes are between 0.05 and 0.20. Because of the small value of k^2, and considering also the small value of $\sin\varphi$ at small angles, tangential cooling effects are negligible for small flow–probe axis misalignment.

Prong interference effects: Interference of the prongs and the probe body produce additional complications of the heat transfer characteristics. For example, for a sensor mounted to two prongs with identical lengths and its axis normal to the axis of the probe body, a stream in the binormal direction (namely a direction normal to both the sensor axis and the probe body axis) will produce a higher rate of cooling than a stream with the same velocity magnitude but in the normal direction (namely normal to the sensor and parallel to the probe axis). Again, the only practical way to account for this effect is to perform a direct calibration of each particular probe by varying its orientation within a three-dimensional envelope. The usual relationship that is fitted to such calibration data is

$$V_{eff} = \sqrt{V_N^2 + k^2 V_T^2 + h^2 V_B^2} \ , \tag{12.42}$$

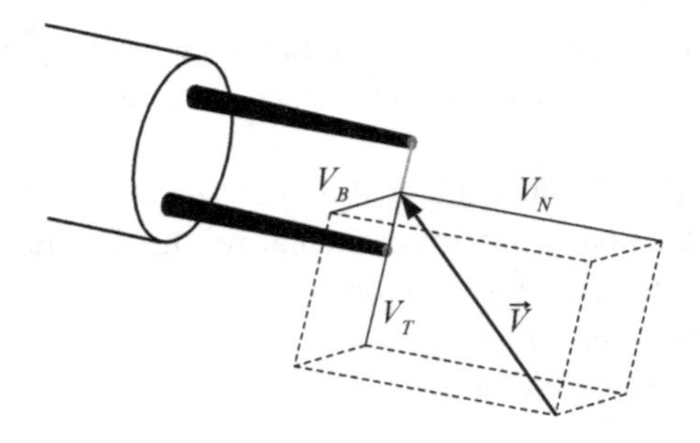

Figure 12.11 Three-dimensional flow past a hot-wire probe.

where V_N, V_T and V_B are, respectively, the normal, tangential and binormal velocity components, as defined previously and illustrated in Fig. 12.11. Typical values of h^2 for hot-wire probes are between 1.1 and 1.2. To minimise prong and probe body interference effects, it is recommended to utilise long and thin (preferably with a diameter less than $50d$) prongs, spaced as widely as possible, and the thinnest possible probe body. Tapered prongs and probe bodies are also recommended. Prong and probe interference, as well as axial conduction effects (see next paragraph) are lower for Wollaston-type or gold-plated sensors than for bare sensors. When orientation effects are significant, one must take particular care to maintain the same relative probe axis–velocity orientation during calibration and during measurement.

Heat conduction effects: The previous analysis of the thermal anemometer response was based on the assumption that the contribution of heat conduction to the total heat transfer was negligible. For cylindrical sensors, this requires a very large aspect ratio l/d, ideally $l/d \to \infty$. The effect of end conduction has been studied analytically by various authors. An often quoted analytical solution, derived on the assumption of axial conduction alone, provides the sensor temperature $T_w(z/l)$ at a relative distance z/l from the middle of the sensor, as

$$\frac{T_w(z/l) - T}{T_{w\infty} - T} = 1 - \frac{\cosh(z/l_c)}{\cosh(l/2l_c)}, \tag{12.43}$$

where T is the flow temperature, assumed to be uniform, and $T_{w\infty}$ is the (uniform) temperature of a sensor with the same material and diameter d, but infinite length, subject to the same electric current as the finite sensor; the latter condition implies that this analysis applies to CCA and not to CTA. The *cold length* l_c is defined as

$$l_c = d\sqrt{\frac{1}{4}(k_w/k)(1 + a_R)/\mathrm{Nu}} \tag{12.44}$$

and, thus, depends on the sensor operating conditions (k_w is the thermal conductivity of the sensor and k is the thermal conductivity of the fluid). The dimensionless temperature difference along a heated cylindrical sensor has been plotted in Fig. 12.12. It may be seen that a finite sensor would have an identifiable section with a nearly uniform temperature only when $l/2l_c \gtrsim 7$. For shorter sensors, the temperature along the sensor will vary continuously between the flow temperature value, at the two ends, and a maximum temperature, in the middle. Equation (12.44) indicates that the cold length decreases

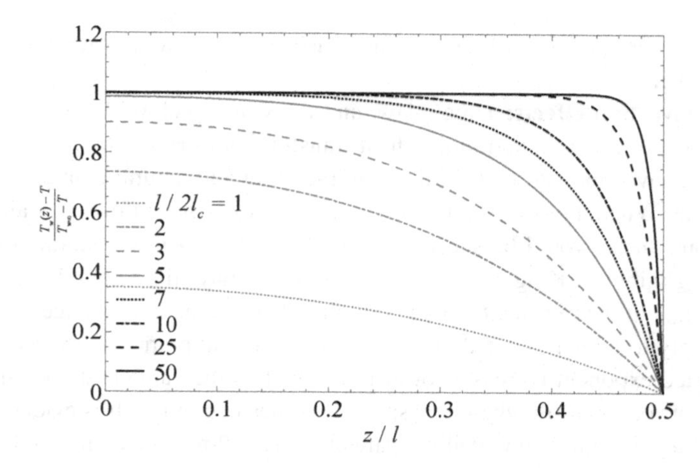

Figure 12.12 Dependence of the dimensionless temperature difference along a heated cylindrical sensor on the ratio $l/2l_c$.

with decreasing thermal conductivity of the wire; therefore, from the viewpoint of minimising heat conduction effects, it would be preferable to use a sensor material with as low heat conductivity as possible, while also maintaining a good mechanical strength and a relatively high electric conductivity. Among the three most popular hot-wire materials, namely tungsten, platinum and platinum–10% rhodium, it is the latter that has the lowest k_w and which should be preferred when it is necessary to use short sensors.

In the usual 'constant temperature' operating mode, the bridge provides a current that raises the sensor resistance to a design value R_w, which is computed under the assumption that the entire sensor has a design temperature $T_{w,d}$. When, however, the sensor temperature $T_w(z)$ is non-uniform, R_w corresponds to the average sensor temperature $T_{w,av}$. The fact that the maximum temperature at mid-sensor ($z = 0$) would be higher than $T_{w,d}$ raises the possibility that the central part of the sensor may be overheated and possibly fail.

Conduction losses could be reduced by increasing the aspect ratio l/d, either by increasing l or by decreasing d. Both approaches would result in lower mechanical strength of the sensor, while an increase in l would degrade the sensor's spatial resolution. Obviously, the optimal l/d ratio for a particular flow would be achieved as a compromise among several conflicting requirements.

Hot-film sensors would be subject to lower axial conduction heat losses than hot wires with the same length, because of their much smaller thickness. On the other hand, the radial conduction losses to the substrate are usually very significant and, for this reason, heat transfer from hot films is more complex than heat transfer from hot wires. Although it has been recognised that heat conduction effects are not negligible in thermal anemometry, the common approach (with the exception of a few very sophisticated studies) has been to disregard the explicit influence of heat conduction in analytical expressions, and to absorb its effects empirically into the coefficients of the calibration equation (e.g., the modified King's law, particularly the exponent n). End conduction effects are expected to decrease significantly as the Reynolds number based on sensor diameter increases [72]. This is of considerable benefit in turbulence research, as the need for higher spatial resolution of the sensor becomes more pressing at higher Reynolds numbers.

Natural convection effects: King's law, and the Collis and Williams expressions, were derived under the assumption that heat transfer from the sensor was exclusively due to forced convection and that conduction, radiation and natural convection effects were negligible. When, however, a hot wire is used in low-speed flows (for air at standard conditions, these would be speeds that are lower than 4 m/s) natural convection may not be negligible. A King's law type of expression may still be fitted to low-speed calibration data, but the exponent would be higher than values in the range from 0.4 to 0.5, which apply to higher-speed flows. As the range of calibration is shifted to lower values, the fitted exponent increases towards 1, which is the value that corresponds to pure natural convection [46], when flow speed becomes negligible. It is evident that the coefficients of a polynomial fit will also be affected as the flow speed range is decreased; it is therefore essential that one calibrates the probe over a velocity range that matches, or preferably exceeds, the expected operating range of the experiment.

Compressibility effects: As the flow speed increases, roughly for M > 0.6, the velocity and temperature fields around the sensor become quite complex and simple relationships, such as King's law, are not sufficient to describe the hot-wire response. Compressible air flows may be characterised by three independent variables, for example, the flow velocity V, the density ρ and the total (or stagnation) temperature T_o, while the pressure is related to the other properties by the perfect gas law. In consequence, the hot-wire output is sensitive to variations of all three parameters. For simplicity, one may linearise the hot-wire response equation to express the output voltage fluctuations as the sum of contributions of fluctuations in V, ρ and T_o, each multiplied by the corresponding sensitivities, S_V, S_ρ or S_{T_o}. Then, in general, it would be impossible to measure velocity fluctuations from a single hot wire output. In clearly supersonic flows (M > 1.2), $S_V = S_\rho$, and one may combine the density and velocity contributions into the *mass flow rate* ρV. Then, for a flow with a constant total temperature, one may use an empirical extension of King's law, in the form

$$E^2 = A + B(\rho V)^n \tag{12.45}$$

(with $n \approx 0.55$) to measure the product ρV. Measurement of velocity or density separately is only possible under special conditions, notably when pressure fluctuations can be neglected. In high subsonic (0.7 < M < 0.9) and transonic (0.9 < M < 1.1) flows, all three sensitivities are different from each other and an independent calibration to variations of each of these three parameters becomes necessary [25]. In low density flows, one must also account for the effects of slip, as expressed by the value of the Knudsen number.

Temperature effects: The previously presented thermal anemometer response relationships (Eqs. (12.31) and (12.32)) are based on the assumption that the flow temperatures during the calibration and during the experiment are constant in time and equal to each other. When, however, a sensor that was calibrated in a flow with a temperature T_r, is used in a flow with a different temperature T, the voltages need to be corrected. The correction procedure, which takes into consideration that convective heat transfer is proportional to the difference between the sensor temperature T_w and the flow temperature T, is straightforward when the flow temperature is uniform and contains no fluctuations. T and T_r can be easily measured with an accurate temperature sensor, such as an RTD or a thermistor, but the sensor temperature T_w cannot be measured directly. For metallic sensors and conditions used in hot-wire anemometry, T_w is approximately related to the heated sensor resistance R_w by the linear expression

$$R_w = R_r\,[1 + \alpha_r\,(T_w - T_r)]\ , \tag{12.46}$$

where α_r is the thermal resistivity coefficient at the reference flow temperature T_r. The value of α_r may be provided by the sensor supplier, but is best determined by temperature–resistance calibration in a heated calibration jet. Consequently, the temperature difference $T_w - T$ can be calculated from the measurable resistance difference, as

$$T_w - T = (1/\alpha_r)[R_w - R(T)]\ . \tag{12.47}$$

The temperature-corrected modified King's law is

$$E^2 \frac{T_w - T_r}{T_w - T} = A + BV^n\ , \tag{12.48}$$

which can be also readily solved for the instantaneous local velocity. The temperature-corrected velocity from the polynomial expression, Eq. (12.32), can be found by substituting E by $E\,\sqrt{(T_w - T_r)/(T_w - T)}$ in all terms. In general, no correction would be necessary, when the difference between the calibration and experimental temperatures is lower than 0.5°C.

The correction procedure becomes more involved when the flow temperature varies from one location to another or contains turbulent fluctuations. In such cases, a temperature correction may be applied only if the local instantaneous temperature is measured simultaneously with the velocity. This temperature may be measured by positioning a temperature sensor, for example, a thermistor for slowly varying temperatures or a 'cold wire' for turbulent temperature fields. In non-isothermal turbulent flows, it is preferable to use both a thermistor and a cold wire, positioned close to the hot wire, and decompose the instantaneous temperature as

$$T(t) = \overline{T} + \theta(t)\ , \tag{12.49}$$

where the time-average mean temperature $\overline{T}$ is measured with the thermistor and the mean-free temperature fluctuation is measured with the cold wire.

Compared to the hot-wire sensitivity to velocity fluctuations, its sensitivity to temperature fluctuations decreases as the overheat ratio increases. Therefore, one may reduce or eliminate the need for temperature corrections by operating the sensor at the highest possible overheat. At extremely low overheats, corresponding to a current through the sensor less than 0.3 mA, the thermal anemometer becomes totally insensitive to velocity variations and becomes, instead, a cold wire, namely a *resistance thermometer*.

Composition effects: The composition of the flow also affects the convective heat transfer from a thermal anemometer inasmuch as it affects the heat conductivity of the surrounding fluid. The sensitivity to composition, relative to the velocity sensitivity, depends on the sensor size, shape, material and overheat ratio. In principle, it is possible to measure both velocity and concentration of a species (although not simultaneously) in a binary gas mixture by running the same, properly calibrated, sensor at two significantly different overheats. A more straightforward, although also more cumbersome, procedure is to use two closely positioned sensors with vastly different sensitivities to composition, and then combine their outputs to recover both the velocity and the composition. Three sensors may be used, at least in principle, to measure velocity, temperature and composition at once.

Reverse-flow and high-turbulence effects: A limitation of hot-wire/hot-film sensors is that their output voltage cannot resolve velocity orientation. For example, an infinitely long cylindrical sensor will produce the same output voltage for any orientation of the velocity vector on the normal plane. Thus, forward flow cannot be distinguished from reverse flow of the same magnitude. Furthermore, even in the case of no flow, a thermal anemometer will produce a non-zero output, attributed to cooling by natural convection associated with the rising warm fluid near the sensor. In highly turbulent flow (for turbulence intensities higher than 25%), reverse flow will occur statistically some of the time. This will result in an overestimation of the mean velocity and the velocity variance. Although some correction procedures for such effects have been developed, the use of thermal anemometers in reversing, recirculating or highly turbulent flows is generally discouraged.

12.2.6 Sensor Arrangements

A single cylindrical sensor, mounted perpendicularly to the probe body axis, which has been approximately aligned with the average local flow velocity, will measure mainly the velocity that is parallel to the probe axis, under the assumption that velocity fluctuations are relatively small. Different arrangements of hot-wire sensors are required for measuring additional velocity components.

Slanted sensors: A single sensor, mounted at an inclination to the mean stream, would provide an output corresponding to the effective cooling velocity, which is insufficient to resolve either the magnitude or the direction of the actual flow velocity. An approach, known as *slanted wire anemometry*, is to mount a single cylindrical sensor at an angle (e.g., at 45°) to the probe body axis and to take a number of measurements, one set after the other, by keeping the sensor midpoint fixed and rolling the probe body to obtain different sensor orientations. Each set of readings is analysed to provide equations for those velocity components which affect the corresponding effective cooling velocity and the resulting system of coupled equations is solved to determine all components of the mean velocity as well as some or all of the turbulent stresses. This method is cumbersome and cannot provide simultaneous values. It is mainly used in near-wall studies, where multi-sensor probes would have inadequate spatial resolution.

Two-sensor probes: A very common design, capable of resolving simultaneously two velocity components, is the *cross-wire* (X-wire) *anemometer* (Fig. 12.13), and its variance, the *V-wire anemometer*. This consists of two sensors, each mounted on a sep-

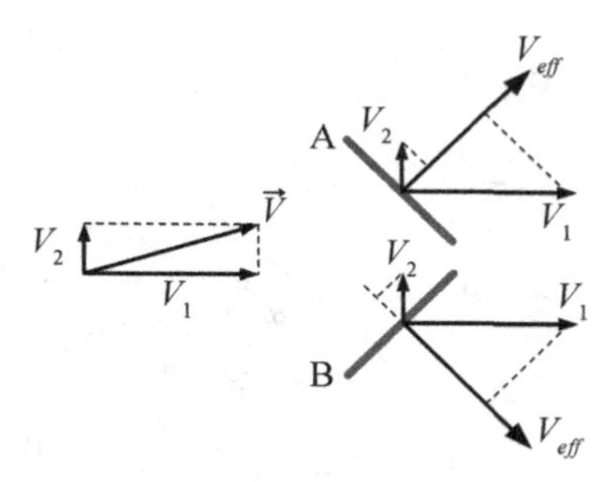

Figure 12.13 Sketch illustrating the effective cooling velocities of the two sensors of a cross-wire probe; for clarity, one sensor has been shifted upwards.

arate pair of prongs and operated by separate circuits, positioned normal to each other and inclined by nominally ±45° with respect to the common probe body axis.

Two common methods are used to calibrate a X-wire probe: the *effective angle method* and the *look-up method* (also known as the *velocity–pitch map* method) [78]. For the effective angle method, the probe is calibrated in a stream parallel to the probe axis, to establish relationships between each voltage output and the corresponding effective cooling velocity. When the probe is put in a stream and aligned with the mean flow direction such that the average binormal velocity is nearly zero, the effective cooling velocities, neglecting tangential cooling and assuming identical sensors inclined at exactly ±45° with respect to the flow direction, would be

$$V_{effA} = \frac{\sqrt{2}}{2}(V_1 + V_2)\,, \tag{12.50}$$

$$V_{effB} = \frac{\sqrt{2}}{2}(V_1 - V_2)\,, \tag{12.51}$$

from which one may calculate the two velocity components as

$$V_1 = \frac{\sqrt{2}}{2}(V_{effA} + V_{effB})\,, \tag{12.52}$$

$$V_2 = \frac{\sqrt{2}}{2}(V_{effA} - V_{effB})\,. \tag{12.53}$$

More accurate values of the effective cooling velocities can be found by angular calibration of the probe, which consists of rotating the probe on a plane parallel to the two sensors and measuring both the tangential cooling coefficients in Eq. (12.41) and the exact angles of inclination of the two sensors.

The look-up method requires fewer assumptions and lesser knowledge of geometric properties of the probe than the effective angle method. The probe is calibrated in a steady stream with a flow velocity $\vec{V}$, in a direction forming a pitch angle φ and a zero yaw angle with the probe body axis (in other words, the probe is allowed to rotate around an axis that is perpendicular to both sensor axes). The calibration procedure starts by setting $\varphi = 0°$ and giving N different values to the velocity magnitude V, while recording the voltages E_1, E_2 of the two sensors. The probe is then rotated to each of M positive or negative pitch angles (see Fig. 12.14a) and the sensor voltages are recorded for the same values of V as in the zero-angle case. This results in a matrix

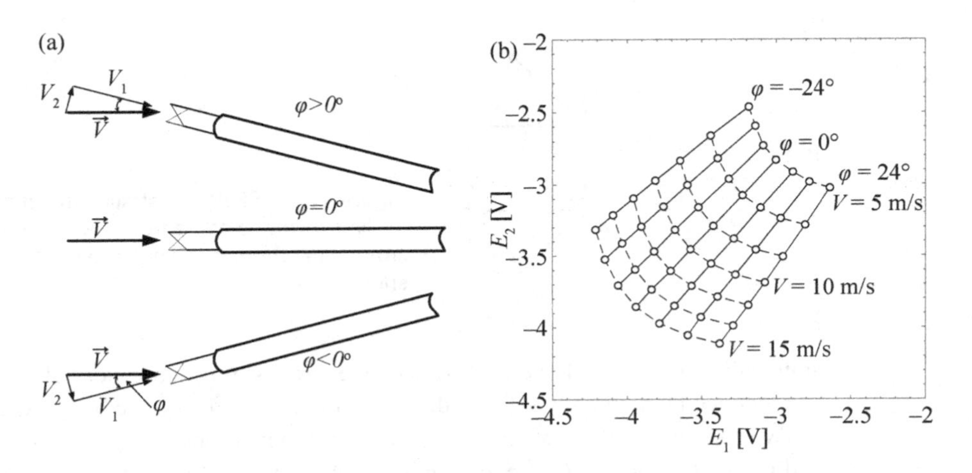

Figure 12.14 (a) X-wire orientated to various pitch angles φ; (b) a sample calibration map for the look-up method.

of $N \times M$ voltage readings for each wire, corresponding to a particular pair of values of the velocity magnitude V and the angle φ. The look-up method is based on the fact that, within certain ranges, the relationships between these two pairs of parameters are unique. Two surface (i.e., two-dimensional) polynomials $V(E_1, E_2), \varphi(E_1, E_2)$, typically of third order, as, for example,

$$V(E_1, E_2) = a_1 E_1^3 + a_2 E_1^2 + a_3 E_1 + a_4 E_2^3 \tag{12.54}$$

$$+ a_5 E_2^2 + a_6 E_2 + a_7 E_1 E_2 + a_8 E_1^2 E_2 + a_9 E_1 E_2^2 + a_{10} \, , \tag{12.55}$$

are then fitted to the calibration data, using least-square regression techniques. These polynomials allow the user to determine, during the experiment, the instantaneous flow velocity projection on a plane parallel to the two sensors, in both magnitude and direction, from a measured pair of voltages E_1, E_2. Then one can obtain the two velocity components as

$$V_1 = V \cos \varphi \text{ and } V_2 = V \sin \varphi \, . \tag{12.56}$$

An example of a calibration map is shown in Fig. 12.14b, where the dashed lines correspond to fixed flow velocities and the solid lines correspond to fixed pitch angles. In this particular example, a calibration was performed for seven equally spaced velocities in the range 5 m/s $\leq V \leq$ 15 m/s, and seven equally spaced pitch angles in the range $-24° \leq \varphi \leq 24°$. For a calibration to be valid for a particular experiment, all of the measured voltages during the experiment must fall within the bounds of the map.

If the calibration of the probe is performed in the same facility as the experiment, the look-up method would have a significant advantage over the effective angle method, because the former would not require a precise initial alignment of the probe body axis and the flow direction. Any error in setting ϕ to zero during calibration would also be present during the experiment and would be accounted for by the fitted polynomials.

Thus one only needs to set accurately the relative change in angle, which can be easily achieved with the use of a small stepper motor. On the other hand, the increased number of calibration points for the look-up method necessitates a longer calibration time (typically four times as long) than that for the effective angle method. Although both methods have their pros and cons, it is generally accepted that the effective angle method should not be used for low velocities and that the look-up method may introduce smaller bias errors for turbulence measurements [15].

Multi-sensor probes: Various geometric configurations of three-sensor [16, 41, 52] or four-sensor [28, 100, 127] probes have also been used to measure simultaneously all three velocity components (see Fig. 12.15). These probes require a three-dimensional calibration and a sophisticated signal analysis. Other multi-sensor probe designs have been used to measure velocity derivatives as well vorticity components [118, 119]. The construction, calibration and use of multi-sensor probes are subject to a number of subtle complications, and should be the task of skillful experimenters.

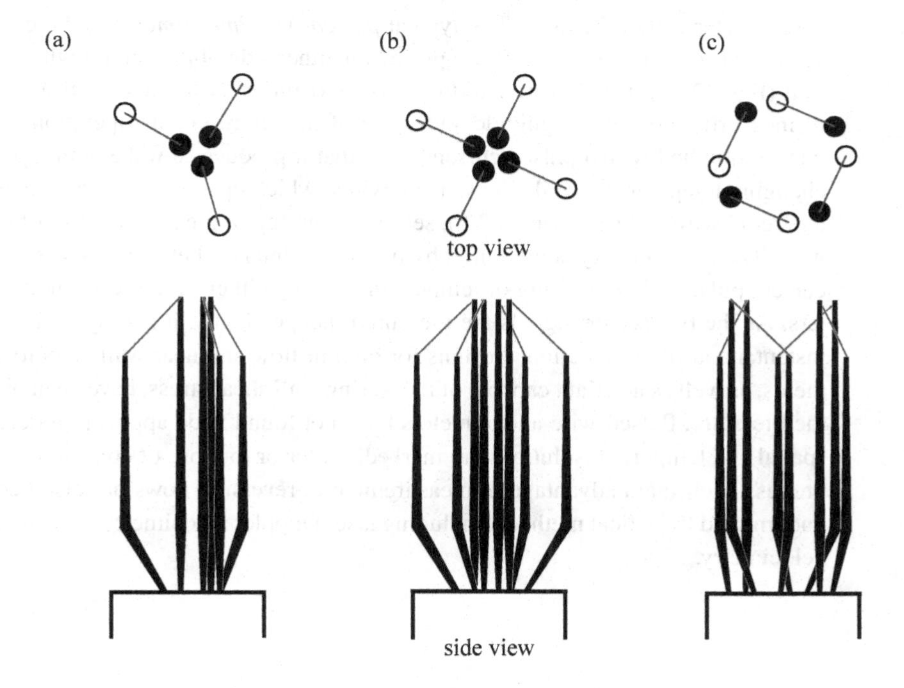

Figure 12.15 Sketches of multi-sensor hot-wire probes for three-dimensional velocity measurement: (a) a three-sensor probe; (b) a four-sensor probe; and (c) an alternative four-sensor probe, which may be also used for streamwise vorticity measurement.

12.2.7 Pulsed-Wire Anemometry

The *pulsed-wire anemometer* [12, 18, 26, 47] has been specifically developed for velocity measurements in high turbulence intensity and reversing flows, in which the hot-wire

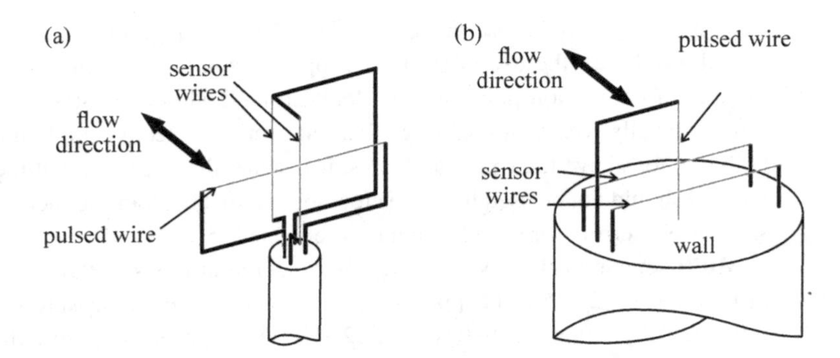

Figure 12.16 Two pulsed-wire anemometer probes: (a) in-flow measurement probe and (b) near-wall measurement probe.

anemometer cannot be used. The typical *pulsed-wire anemometer* probe consists of a central wire and two sensing wires, placed on either side of the central wire and normal to it (Fig. 12.16). It is sensitive to the velocity component normal to all wires, capable of measuring both its magnitude and sense of direction. For its operation, the central wire is supplied with a pulsed current, such that it produces a wake with a periodically changing temperature, while the sensing wires, which operate as resistance thermometers (cold wires; see Section 14.2.6), sense the passing of the heated wake of the central wire. The flow velocity is measured by timing the interval between the application of a current pulse and the sensing of temperature rise by either of the resistance thermometers. As the two sensor signals are measured independently, it is easy to determine the instantaneous flow direction. Designs for both in-flow and near-wall velocity measurements, as well as a variant capable of measuring wall shear stress, have been proposed in the literature. Pulsed-wire anemometers have not found wide application because their spatial and temporal resolutions are markedly inferior to those of conventional hot wire probes. Their main advantage of measurement in reversing flows has also been largely undermined by optical methods, including laser Doppler velocimetry and particle image velocimetry.

12.3 Laser Doppler Velocimetry

Laser Doppler velocimetry (*LDV*), alternately referred to as *laser Doppler anemometry* (*LDA*), was conceptualised in 1964 [129] and witnessed rapid and intensive development to become one of the most popular local velocity measurement methods in the fluid mechanics laboratory during the last quarter of the twentieth century. Although a variety of LDV configurations and a multitude of related components and accessories have been developed, in the present review we shall only discuss the basic principles of LDV operation and some of the most widely used components and techniques [2, 6, 14, 30, 120].

12.3.1 Doppler Shift of Light Frequency

Consider a monochromatic, coherent, linearly polarised and collimated laser beam with a wavelength λ and frequency ν, being transmitted through a fluid containing suspended discrete particles with sizes which are not very small compared to λ and travel with speed $\vec{V}$ (Fig. 12.17). When one of these particles intersects the beam, it will adsorb some of the light, which can be viewed as plane electromagnetic waves, and re-emit it, following Mie scattering laws . At some distance from the particle, the emitted radiation can be viewed as spherical electromagnetic waves. The frequency $\nu + \nu_D$ of these waves as seen by a small photodetector will be different from the incident frequency by an amount ν_D, called the *Doppler frequency*. This phenomenon is known as *Doppler* (or *Doppler–Fizeau*) *phenomenon*. The Doppler frequency depends only on the speed of the particle, the light wavelength and the scattering angle ϕ, as

$$\nu_D = \frac{\vec{V} \cdot \left(\vec{e_s} - \vec{e_i}\right)}{\lambda} = \frac{2 \sin(\phi/2)}{\lambda} V_\phi \,, \tag{12.57}$$

where $\vec{e_i}$ and $\vec{e_s}$ are, respectively, the unit vectors parallel to the incident and scattered beams and V_ϕ is the projection of the particle velocity vector on the direction normal to the bisector of the angle ϕ. Thus, given the parameters λ and ϕ, one may, in principle, determine the velocity V_ϕ from a measurement of the Doppler frequency ν_D, without any need for calibration or other considerations. In practice, because $\nu_D \ll \nu$, it is essentially impossible to measure it accurately as the difference of the incident and scattered light frequencies and some modification of this fundamental configuration is required (a possible exception is the case of extremely high speed flows). Among the few available workable configurations, the one used by most researchers and the usual one in commercial LDV systems is the *dual beam* configuration. Other configurations that have been employed are the *reference beam* and *dual scatter* configurations.

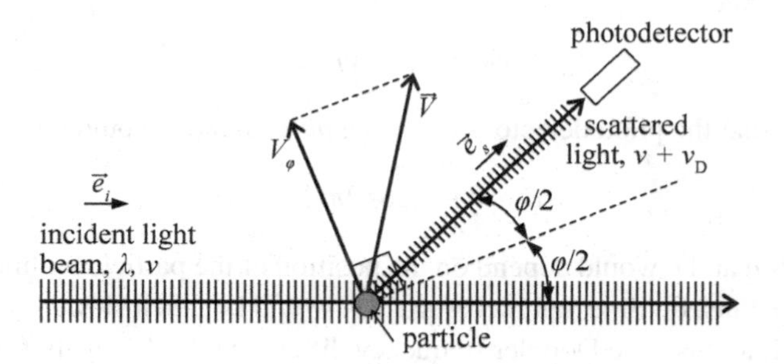

Figure 12.17 Sketch of the light of a laser beam, scattered by a moving particle.

12.3.2 Dual Beam LDV

The basic single-component, forward scatter, dual beam LDV is illustrated schematically in Fig. 12.18. The monochromatic, coherent, linearly polarised beam of a laser is split by a *beam splitter* into two parallel beams, which are focused onto a small control

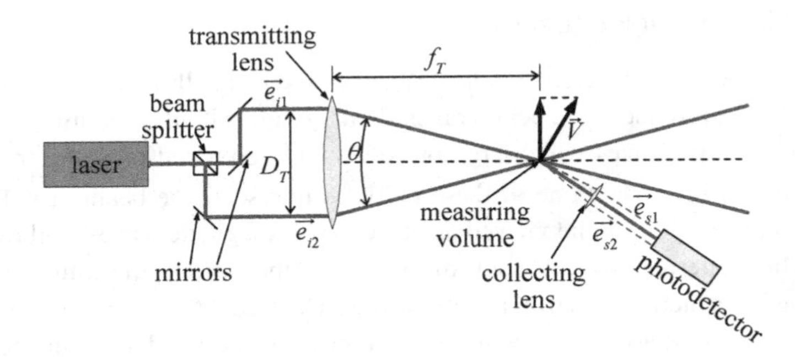

Figure 12.18 Sketch of the dual beam laser Doppler configuration.

volume by the transmitting lens, forming an angle θ with each other. A small particle crossing the intersection volume with velocity $\vec{V}$ scatters the light of both beams, each Doppler shifted according to Eq. (12.57). Both scattered beams are collected by a small-aperture collecting lens and superimposed on a photodetector. Assume first that the intensity of each scattered beam collected by the photodetector varies sinusoidally as $A_i \sin 2\pi [(\nu + \nu_{Di})t]$, $i = 1, 2$. Optical mixing of these beams on the photodetector (*heterodyning process*) produces an output voltage E which is proportional to the square of the combined light intensity, as

$$
\begin{aligned}
E &\propto \{A_1 \sin 2\pi [(\nu + \nu_{D1})t] + A_2 \sin 2\pi [(\nu + \nu_{D2})t]\}^2 \\
&= A_1^2 \sin^2 2\pi [(\nu + \nu_{D1})t] + A_2^2 \sin^2 2\pi [(\nu + \nu_{D2})t] \\
&\quad + A_1 A_2 \{\cos 2\pi [(\nu_{D1} - \nu_{D2})t] - \cos 2\pi [(2\nu + \nu_{D1} + \nu_{D2})t]\} \;. \qquad (12.58)
\end{aligned}
$$

Now, considering that the frequency response of the photodetector is insufficient to resolve any fluctuations with frequency of the order of magnitude of ν (typically, of the order of 10^{14} Hz), but sufficient to resolve fluctuations with the much lower *Doppler frequency difference*

$$\nu'_D = \nu_{D1} - \nu_{D2} \,, \qquad (12.59)$$

it is easy to see that the photodetector output (*Doppler signal*) would be

$$E \propto a + b \cos 2\pi \nu'_D t \,. \qquad (12.60)$$

The coefficients a and b would depend on the position of the particle within the measuring volume and, thus, on time, but their variation would be much slower than the period of Doppler oscillations. The Doppler frequency difference can be easily found as

$$\nu'_D = \frac{\vec{V} \cdot (\vec{e_{s1}} - \vec{e_{i1}})}{\lambda} - \frac{\vec{V} \cdot (\vec{e_{s2}} - \vec{e_{i2}})}{\lambda} = \frac{\vec{V} \cdot (\vec{e_{i2}} - \vec{e_{i1}})}{\lambda} = \frac{2 \sin(\theta/2)}{\lambda} V_\theta \,, \qquad (12.61)$$

where V_θ is the projection of the particle velocity vector on the direction normal to the bisector of the angle θ between the two incident beams. Unlike Eq. (12.57), Eq. (12.61) is independent of the observation angle. The photocurrent generated by the photodetector will be fluctuating with the Doppler frequency ν'_D, which can be measured by

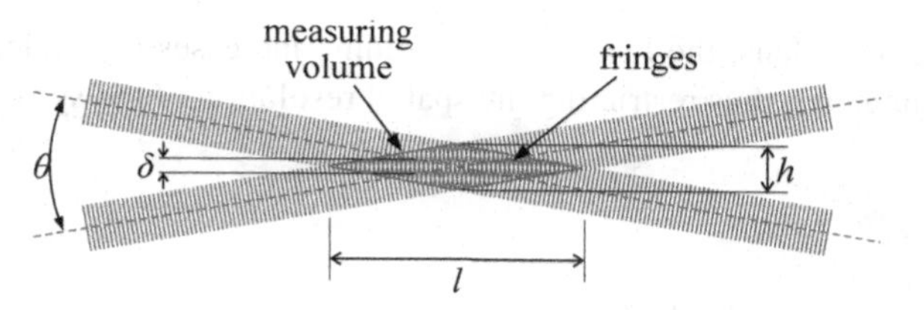

Figure 12.19 Measuring volume and fringe pattern in the dual beam configuration.

various means, thus allowing the direct computation of V_θ as

$$V_\theta = \frac{\lambda}{2\sin(\theta/2)} \nu'_D \,. \tag{12.62}$$

A great advantage of Eq. (12.62) is that it is linear and contains no undetermined constants, thus eliminating the need for calibration and errors due to instrumentation drifting. The proportionality coefficient $\lambda/[2\sin(\theta/2)]$ is called the *calibration factor*, although no calibration is actually necessary.

12.3.3 The Measuring Volume

The laser beam coming out of the laser has a Gaussian distribution of intensity (see Section 7.3.7). The *laser beam diameter* d_e (namely the diameter of the circle within which the light intensity is higher than $1/e^2 \approx 13.5\%$ of the maximum intensity) first diminishes, it reaches a minimum, called the *waist diameter*, and then it increases. The transmitting lens focuses the beam to an even smaller cross section, essentially on its focal plane. Optimal performance, resulting in the smallest possible beam cross section, will be achieved if the beam is focused on its waist. The *focused beam diameter* will be

$$d_{fe} \approx \frac{4 f_T \lambda}{\pi d_{eo}} \,, \tag{12.63}$$

where d_{eo} is the beam diameter at the lens and f_T is the *focal distance* of the lens. When properly positioned, the two focused beams in the dual beam configuration will intersect on their waists to form a volume within which the total local light intensity would be equal to the sum of the two local beam intensities. The *measuring volume*, defined as the space within which the light intensity is higher than $1/e^2$ times its maximum value, has an ellipsoidal shape with the following dimensions (Fig. 12.19)

$$\text{width:} \quad d_{fe} \,, \tag{12.64}$$

$$\text{height:} \quad h = \frac{d_{fe}}{\cos(\theta/2)} \,, \tag{12.65}$$

$$\text{length:} \quad l = \frac{d_{fe}}{\sin(\theta/2)} \,, \tag{12.66}$$

$$\text{volume:} \quad \frac{\pi d_{fe}^3}{6\cos(\theta/2)\sin(\theta/2)} \,. \tag{12.67}$$

The previous relationships show that, for intersection angles $\theta < \pi/2$, the measuring volume length would be its longest dimension. In fact, as θ is reduced to permit measurements away from the lens, the length of the volume increases dramatically, compared to the beam diameter, thus restricting the spatial resolution of the system.

12.3.4 Fringe Model

Considering the two beams in the dual beam configuration as travelling electromagnetic waves with the same frequency and phase, but with directions of propagation that are inclined with respect to each other, one may deduce that their interference within the measuring volume will produce a number of interference surfaces (standing waves). If the beams intersect at their waists, the two travelling waves will be essentially plane and the interference surfaces will be parallel planes, oriented as shown in Fig. 12.19. These surfaces are called *fringes* and appear as bright slabs, flanked by dark slabs. The *fringe spacing* δ is equal to the calibration factor, namely,

$$\delta = \frac{\lambda}{2\sin(\theta/2)}, \tag{12.68}$$

while the *fringe number* within the measuring volume is

$$N = \frac{4}{\pi}\frac{D_T}{d_e}, \tag{12.69}$$

where $D_T = 2f_T \sin(\theta/2)$ is the beam separation before entering the transmitting lens. Although known to lead to certain inconsistencies, the fringe model may be used to explain the dual beam LDV operation. For example, the Doppler signal fluctuations may be viewed as the result of fringe-crossings by the particle. The frequency of fringe-crossings is V_θ/δ, which is identical to the Doppler frequency difference, given by Eq. (12.61).

12.3.5 Beam Expansion

A *beam expander* is a commonly used LDV accessory, whose function is to reduce the size of the measuring volume. This improves the spatial resolution of velocity measurement, while also improving the amplitude resolution, as a result of increased light power density within the measuring volume. The beam expander consists of a diverging lens and a converging lens, in addition to the transmitting lens, which has a larger aperture than that in the basic LDV system (Fig. 12.20). It is characterised by the *beam expansion ratio* $E_x > 1$. This increases the beam diameter from the original value d_e to the expanded beam value $d_{ex} = E_x d_e$ and the beam separation from the original value D_T to

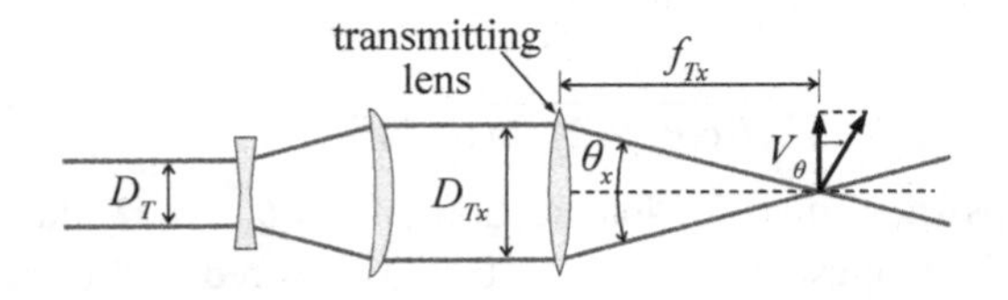

Figure 12.20 Sketch of a beam expander configuration.

the expanded beam value $D_{Tx} = E_x D_T$. The expanded-beam intersection angle θ_x can be found as

$$\theta_x = 2\sin^{-1}\left[\frac{E_x f_T}{f_{Tx}}\sin(\theta/2)\right], \tag{12.70}$$

where f_{Tx} is the focal distance of the transmitting lens of the expanded-beam system. Assuming that the focal distances of the transmitting lenses of the unexpanded and expanded-beam systems are identical, $f_{Tx} = f_T$, one can find that the focused diameter of the expanded beam will be diminished as

$$d_{fex} = \frac{1}{E_x} d_{fe} \approx \frac{4 f_T \lambda}{\pi d_{ex}}. \tag{12.71}$$

The fringe spacing will also be reduced by a factor equal to E_x, while the number of fringes within the measuring volume will remain unaffected. Most importantly, the width of the measuring volume will decrease by a factor of E_x, the height by a factor somewhat less than E_x (due to the increasing value of $\cos(\theta/2)$), while the length, which is normally by far the longest dimension, will decrease by the factor E_x^2. The volume itself will be reduced by a factor somewhat less than E_x^4, which represents a substantial improvement of spatial resolution.

12.3.6 Doppler Signal

First consider that the photodetector is only exposed to light scattered by a single particle traversing the central region of the measuring volume with a constant velocity. The Doppler signal will have a typical appearance shown in Fig. 12.21a. It consists of a slowly varying component, called the *pedestal*, which corresponds to the Gaussian distribution of the light intensity within each beam, and a rapidly fluctuating component, called the *burst*, which is generated by fringe-crossings by the particle and is an amplitude-modulated oscillatory function with frequency ν'_D. The instantaneous value of the burst signal depends on the size and refractive index of the particle and its position in the measuring volume, while the number of cycles within each burst depends

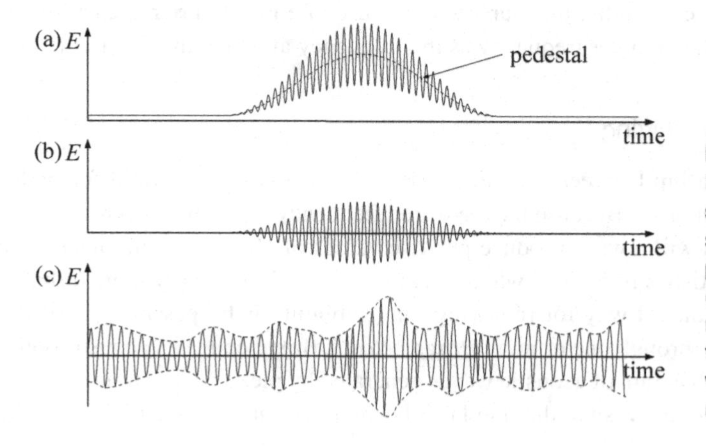

Figure 12.21 (a) Typical Doppler signal of a single particle; (b) the same signal with the pedestal removed (burst); (c) the signal of multiple particles within the measuring volume with the pedestal removed.

on the number of fringes it crosses and, thus, its path through the measuring volume. Particles crossing the volume near its edges will generate bursts with a smaller number of cycles than those crossing it centrally. Also, particles moving normal to the fringes will cross more fringes than particles with inclined paths. The pedestal is usually removed by high-pass filtering, which leaves only the bursts, shown in Fig. 12.21b. The Doppler signal discussed previously appears to be continuous within the duration of each burst, although it vanishes between bursts. This is based on the assumption that the *photon density*, that is, the intensity of light received by the photodetector during each burst is sufficiently high for a continuous output voltage to be generated. At extremely low photon densities, however, the signal consists of a train of pulses corresponding to individual collected photons. We shall not consider this situation any further, as it is normally to be avoided. When more than one particles are simultaneously present in the measuring volume, the Doppler signal gets more complicated and its appearance at any given time depends on the *burst density*, namely the number of particles crossing the measuring volume. Even if all particles present had the same size and the same velocity, amplitude and phase differences in their bursts would be created by differences in paths and times of entry into the measuring volume. In the extreme case in which the particle concentration is high enough for many particles to reside in the measuring volume at all times, the Doppler signal would be the result of superposition of many bursts (*high burst density*), so that it would be continuous over the entire observation period, with a typical appearance as shown in Fig. 12.21c. High burst density signals have an average frequency ν'_D, corresponding approximately to the average particle velocity. However, their instantaneous frequency and phase vary randomly, introducing errors in the velocity measurement, called, respectively, *ambiguity noise* and *phase noise*.

The Doppler signal produced by the photodetector is usually band-pass filtered to remove the pedestal and high-frequency noise, and then processed by a *signal processor*, in order to determine the Doppler frequency. Doppler signal processors have evolved significantly since their early years, following developments in electronics and algorithms. At present, *burst analysers* are the most widely used processors, due to their relatively high sensitivity and accuracy and their ability to operate with low particle densities. They require single-particle bursts for their operation and compute the power spectrum of each individual burst with the use of a Fast Fourier Transform. Then, they determine the Doppler frequency as the frequency at which the spectral peak occurs.

12.3.7 Frequency Shifting

The relationship between V_θ and ν'_D (Eq. (12.62)) is equally valid for both senses of the direction of V_θ. Because the measured frequency ν'_D is always positive, Doppler signal analysis will always produce positive values of V_θ. This would result in erroneous velocity statistics in regions where there is reverse flow or very large turbulence intensity. The standard way for removing this ambiguity is by passing one of the incident laser beams through one or more *Bragg cells*. A Bragg cell is an acousto-optical device, within which a train of acoustic waves generated by piezoelectric transducers modulates the light frequency, such that the laser beam at its output has a frequency shifted with

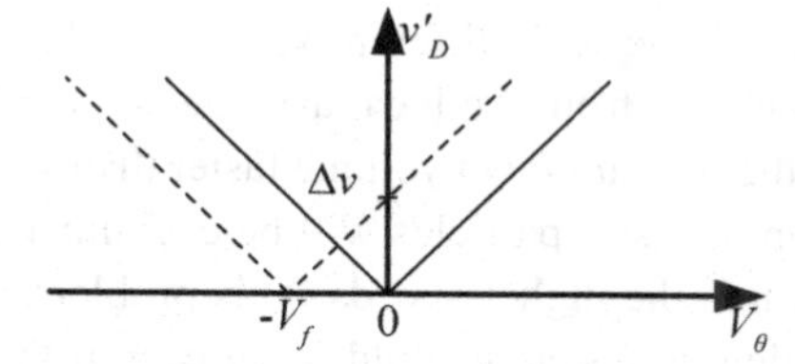

Figure 12.22 Relationship between the Doppler frequency and the flow velocity without (solid lines) and with (dashed lines) frequency shifting.

respect to the incident frequency ν by a fixed amount $\Delta\nu$, equal to about 40 MHz for a single cell. Frequency shifting produces a motion of the interference fringe pattern in the measuring volume with a velocity normal to the fringe plane and equal to

$$V_f = \Delta\nu\delta\,. \tag{12.72}$$

Then, the computed flow velocity would be given by

$$V_\theta = (\nu'_D - \Delta\nu)\,\delta\,, \tag{12.73}$$

which can take zero or negative values and, thus, resolve flow velocity sense, as long as $V_\theta > -V_f$. To illustrate this concept, consider a particle at rest within the control volume. In the absence of frequency shifting, this particle would produce no Doppler signal, as it would not cross any fringes. With frequency shifting, however, the fringes would be convected past the particle, and thus a Doppler signal with frequency $\nu'_D = \Delta\nu$ would be produced and Eq. (12.73) would give the correct value $V_\theta = 0$. The relationship between the Doppler frequency and the flow velocity without and with frequency shifting is illustrated in Fig. 12.22.

An additional advantage of frequency shifting is that it increases the dynamic range of the LDV system by increasing the number of fringes intersected by each particle crossing the control volume. Frequency shifting is necessary for highly turbulent flows but also improves the amplitude resolution of the system, a feature which is useful in measuring low amplitude fluctuations in relatively high speed flows. Frequency shifting by different amounts is also possible by electronic means.

12.3.8 Errors and Uncertainty in LDV Measurements

Previous investigators have identified several sources of measurement error and uncertainty in laser Doppler velocimetry, which, in many cases, can be avoided or reduced by proper designing of the experimental setup. If the beams do not intersect at their waists, the fringes will not be parallel planes, but curved surfaces with spacings that vary across the measuring volume. This would create a measurement uncertainty, called *fringe divergence uncertainty*, as particles with the same speed would appear to have different speeds if they followed different paths through the measuring volume. Another source of uncertainty, which is particularly important for turbulent flows, is due to non-uniform distribution of scattering particles. In such cases, the measured statistical properties would be biased in favour of flow regions that contain more particles.

Even when carefully implemented, laser Doppler velocimetry would be subjected to a number of inherent systematic (bias) errors [31]. The most serious one is *velocity bias*:

when particles cross the measurement volume with different velocities, the arithmetic average of these velocities would be different from the local average velocity. The reason is that faster particles would cross the measurement volume faster than slower ones, with the result that a greater population of faster particles will be encountered and the velocity statistics would be biased towards the higher speeds. This problem is particularly important for high-intensity turbulent flows, and would occur even if the particles were uniformly distributed in the stream. An order-of-magnitude estimate of this bias for the mean streamwise velocity is [83]

$$\frac{\overline{U_m}}{\overline{U}} \approx 1 + \frac{\overline{u^2}}{\overline{U}^2}, \tag{12.74}$$

where $\overline{U_m}$ is the mean velocity measured by the LDV system, $\overline{U}$ is the actual mean velocity and $\overline{u^2}$ is the velocity variance. Among the various methods that have been proposed to correct for this bias, most popular is the so called *residence-time* or *transit-time correction* [14], applicable to burst analysis type processors. Besides the particle velocity, modern burst processors also measure the residence time Δt_i of each particle in the measurement volume, determined as the difference between the times of exit and entry. Then, the velocity statistics are determined by weighing the measured velocity of each particle by the factor $\Delta t_i / \sum_1^J \Delta t_j$, where J is the number of particles considered. A related bias, also due to non-uniform velocities is the *gradient bias*, occurring in shear flows, even laminar ones. In such flows, this bias can be reduced by keeping the measurement volume as small as possible. Near walls, additional errors would be produced by the non-uniform distribution of particles in the flow, as heavier particles would tend to migrate away from the wall. In highly three-dimensional flows, there will also be a *directional bias*, as particles crossing the measurement volume in directions forming small angles with the fringes will not be registered by the processor. Frequency shifting can be used to reduce this bias.

A source of random error, usually called *Doppler ambiguity*, is the broadening of the Doppler frequency f_D due to the finite sampling time, as discussed in Section 5.3. Even if the particle had a constant speed within the measurement volume, the measured spectrum of its Doppler burst would not be a delta-like function but a sinc-like function. One may estimate the uncertainty in the measured Doppler frequency by the half width of its main lobe, which is approximately $1/\Delta t$ (Δt is the residence time), or, in terms of the number of fringes N that the particle intersects, $2f_D/N$ [120]. Obviously, this uncertainty would be reduced by increasing the fringe number that the particle intersects by frequency shifting.

12.3.9 Multi-Component and Multi-Point Systems

The previously described, basic LDV system can measure only one velocity component, the one normal to the bisector of the angle between the intersecting beams. A straightforward extension of this configuration to a *two-component system* is to use two pairs of beams, intersecting within the same measuring volume, but such that the axes of each pair form planes perpendicular to each other, as shown in Fig. 12.23. Thus, the

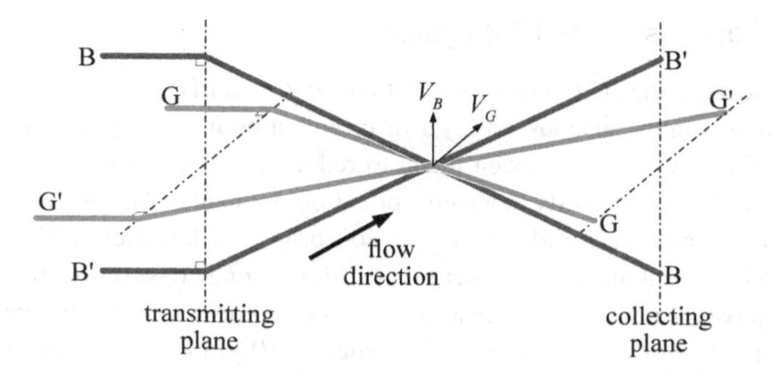

Figure 12.23 Sketch of a two-component LDV configuration using two pairs of beams at orthogonal planes; the letters G and B indicate beams with light at the green and blue spectral peaks of an Ar-Ion laser, as well as the corresponding measured velocity components.

measuring volume will contain two sets of interference fringes and it will become necessary to separate the light emitted by a particle as it crosses one set of fringes from light emitted by the same particle as it crosses the other set. This can be achieved by using light of a different wavelength for each pair of beams and remove, by an optical filter, the light of the other pair before it enters the corresponding photodetector. This is called a *two-colour LDV system.* It can be implemented by combining two lasers that emit at different wavelengths, however, it is far more common to use a single laser with multiple spectral peaks, mostly the Ar-Ion laser at its two strongest spectral peaks in the green (λ = 514.5 nm) and blue (λ = 488 nm) ranges. Because the two velocity components of each particle are measured simultaneously, one may also measure their joint statistical properties, such as their covariance, corresponding to a Reynolds shear stress in turbulent flows. A simplified two-component LDV design utilises three rather than four beams, by combining two beams into one that contains both wavelengths. The two-component system can be extended to a *three-component system* by adding a third pair of beams intersecting on the same measuring volume as the other two, but belonging on a third plane. Unless the facility permits optical access from two orthogonal directions (e.g., one side and the top), the third component will not be orthogonal to the other two, and one needs to perform a coordinate transformation to compute the instantaneous velocity vector. The third pair of beams must also be at a different wavelength. This could be the wavelength of a third spectral peak of an Ar-Ion laser (purple, at λ = 476.5 nm) or light from a different type of laser, for example, a He-Ne laser.

In principle, multi-point measurements by LDV can be performed by using multiple LDV systems with distinct measuring volumes. Compared to multiple hot-wire arrays, they would have the advantage of not interfering with each other. In practice, however, due to the high cost and the need for alignment of many components, this approach is restricted to at most two LDV systems. By traversing one system with respect to the other, one can measure spatial correlations and wavenumber spectra in different directions. A more economical alternative is to scan rapidly the measuring volume [107].

12.3.10 High-Resolution LDV Systems

Because the spatial resolution of conventional LDV systems is insufficient for measurements in the viscous sublayer of turbulent boundary layers at relatively large Reynolds numbers, effort has been made to reduce the size of the LDV measuring volume with the use of carefully selected optical components [24]. A number of authors have devised methods for identifying the location of each particle within the measuring volume, which contains two distinct sets of fringes, one converging and another diverging. This makes it possible to measure the velocity profile within the measuring volume with a spatial resolution that is of the order of 10 μm [23, 108]. Further development of the optical configuration and the signal processing method has allowed the measurement of the full velocity vector with a comparable resolution [77].

12.4 Ultrasonic Doppler Velocimetry

This method is based on the Doppler shift of the frequency of ultrasonic waves scattered by moving particles. This method has been used widely for the measurement of flow rate in pipes, but it has also been extended to the measurement of flow velocity profiles along the path of an ultrasonic beam. The velocity component V_x of a particle in the direction of the beam is calculated as

$$V_x = \frac{f_D}{2f} c \,, \tag{12.75}$$

where f is the frequency of sound emitted by a transmitter, f_D is the Doppler frequency shift due to the particle motion and c is the speed of sound in the medium. The particle position x along the beam path can be calculated as

$$x = \frac{1}{2} c \Delta t \,, \tag{12.76}$$

where Δt is the time delay between emission and reception of sound by the transducer, which alternates as transmitter and receiver. For practical reasons, this method can only be used in liquids. Its advantage over optical methods is that it can be applied to opaque fluids and, for this reason, it has found applications in the food industry, chemical processing, oil refineries, sediment transport and sewer design.

12.5 Particle Tracking

Flow-marker techniques, which were developed for flow visualisation purposes, have been refined to identify the locations of markers at two or more time instances, from which marker displacement, and, hence, velocity can be computed. The simplest such technique is the visual tracking of individual foreign particles (e.g., floating leaves on a flowing river), provided that they are clearly visible, isolated and distinguishable from neighbouring particles. It is remarkable that the earliest documented flow velocity measurement, by Antonie van Leeuwenhoek in 1689, was made by timing the motion of blood cells, serving as tracers. Visual particle tracking may only be possible under particularly favourable conditions and so, in practice, it is the images of particles, recorded by cameras, which are tracked. Besides optical methods, particle tracking can be accomplished with the use of sound waves, electromagnetic fields and other means.

A variant of the particle tracking method is the *line-tracking method*. A line in the fluid transverse to the main flow direction is marked by various means, for example by hydrogen bubbles, wire-generated smoke or an electrochemical or photochemical reaction. Successive images of the line at different times are analysed to produce displacements of points along the line, from which flow velocity profiles can be determined. This approach would not be effective in highly turbulent flows, in which the line would become quickly untraceable due to mixing, or in strongly three-dimensional flows, containing significant velocity components tangential to the line.

All particle tracking methods are non-intrusive, although, at sufficiently high concentrations, foreign particles may cause some loading of a flow, for example, damping of turbulent fluctuations. Another advantage of particle tracking methods is that they provide the velocity directly as the ratio of displacement and time, with the user only needing to specify the spatial resolution of the image and the sampling rate of the camera recording the images. On the other hand, the accuracy of flow velocity measurement depends on the relative velocity of particle with respect to the surrounding fluid, which may be a complicated function of particle properties and flow conditions (see Section 7.6). Although based on a simple principle, current particle tracking methods are quite sophisticated, having undergone spectacular development in recent years by incorporating advances in camera technology, computer capabilities, algorithm design and medical imaging. In view of the existing momentum in the advancement of such methods, one may expect further significant progress in the near future.

The term *particle tracking velocimetry* (*PTV*), also referred to in earlier literature as *chronophotography*, describes techniques that measure the local flow velocity as the ratio of the distance between images of an individual particle, which is transported by the flow and recorded at distinct and known times, and the time step between images [34, 113]. Particle images at multiple positions may be recorded within a single frame or in multiple frames. In single-frame recording, and depending on whether exposure and illumination are continuous or pulsed, the recorded particle trace could be a continuous streak, terminated by the initial and final positions of the particle, or a sequence of distinct images of the particle at different times. In either of these approaches, the sense of particle motion along its path cannot be resolved. Among the techniques that have been suggested to remove this ambiguity is the use of two or more synchronised light sources with different intensities or colours, such that the particle image at a specific time would be distinguishable from those at other times. Another problem that arises when the concentration of particles is relatively high is the overlapping of pathlines and images of different particles. One method of separating pathlines of different particles is to combine continuous illumination at low intensity and interrupted illumination at a higher intensity, which would produce continuous pathlines superimposed on discrete particle positions. Multiple-frame recording, when it is possible, is preferable to single-frame recording, as it can resolve the sense of motion. This requires both short-time exposure, achieved by using high-speed shutters or pulsed illumination, and rapid advancing of the digital frame of the camera. Image improvement, such as removal of blur due to particle movement, can be achieved by image processing methods, which are included in commercial graphics software. Specialised methods of illumination may also assist in the process [66].

Manual analysis of particle tracking images is a very tedious process and, in practice, can only be achieved for a small number of sparsely spaced particles. This necessitated the development of algorithms to identify individual particles and their displacement [57, 64]. It is obvious that the simultaneous tracking of a large number of particles, distributed within the flow domain, would be far more informative than single-particle tracking, as it can provide two- or three-dimensional flow velocity maps, rather than local velocity only. Because, however, iterative algorithms designed to track a large number of particles in 3-D can be computationally prohibitive, modern PTV methods employ specialised data processing techniques [4, 40].

Current particle tracking techniques are most important tools in a rapidly growing arsenal of *volumetric velocimetry* methods, capable of mapping 3-D velocity fields in 3-D flow domains [27]. The time-resolved, simultaneous tracking of closely spaced particles with the use of multiple cameras ideally maps velocity fields in a Lagrangian frame of reference, which is the description of choice for studies of mixing and chemical reactions. Among the most common volumetric velocimetry methods is *tomographic particle image velocimetry* (*tomo-PIV*), which will be discussed in the next section. Related to, but distinct from, this method are the following:

- *scanning tomographic PIV* [17, 70];
- holographic techniques, including *holographic PIV* [9, 49, 50, 61];
- *defocusing PIV* [45];
- *light-field imaging* [10, 84];
- the *shake-the-box* method [89, 104].

12.6 Particle Image Velocimetry

Particle image velocimetry (*PIV*) [1, 2, 3, 44, 80, 85, 94, 111, 123, 124] is currently one of the main measurement techniques in fluid dynamics laboratories. The basic PIV version is *planar PIV*, which provides vector maps of the flow velocity projection on a plane. A more advanced version, *stereoscopic PIV*, provides maps of the three-dimensional velocity vector within a plane area. Even more advanced, but also more complex and expensive, is *tomographic PIV*, which provides maps of the velocity vector within a volume. The spatial resolution of any PIV system depends on the available instrumentation and the settings, but is generally lower than the resolutions of hot-wire anemometers and laser Doppler velocimeters. The temporal resolution of PIV also depends on the capabilities of the available equipment, but, in general, it is also inferior to those of the two previously mentioned methods. The great advantage of PIV by comparison to the two other methods is that it provides velocity maps, rather than merely

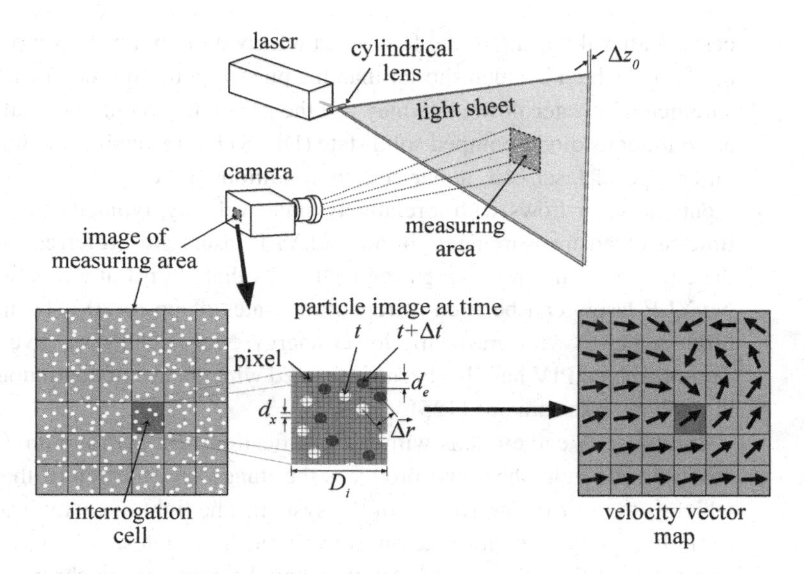

Figure 12.24 Illustration of the components and the principle of operation of a basic particle image velocimetry system.

local values. *Time-resolved PIV* is a very powerful flow measurement tool, permitting the collection of information that was previously inaccessible [11].

12.6.1 Basic Operating Procedure

The basic operating procedure for PIV can be broken down into four key parts: *seeding*, *illumination*, *imaging*, and *processing*. An illustration of the components and principle of operation of a PIV measuring system is shown in Fig. 12.24. A brief description of the PIV measuring procedure is given in the following; details can be found in books dedicated to the topic [3, 94].

Seeding: The working fluid is seeded with particles in sufficient concentration and capable of faithfully tracking the motion of the fluid. The criteria for selection of particles for different working media are discussed in Section 7.6. Critically, one needs to ensure that the Stokes number S, which represents the ratio of the characteristic times of the particle $\tau_p = \rho_p d_p^2/18\mu_f$ (Eq. (7.56)) and the flow, is small. A recommended range for most experiments is $S < 0.05$ [101].

Illumination: A thin volume of the fluid is illuminated by two high power, short duration, light sheets separated by a time increment Δt, such that only particles within this volume are visible. The time increment Δt is referred to as the *interframe time*. Both light sheets are typically produced by a single laser. For measurements in air, the most common type of laser used is a twin-cavity Nd:YAG pulsed laser, which produces pulses with energy in the range 100 to 400 mJ. The rate at which successive pulses can be fired from each cavity, referred to as the *repetition rate* f_r, is typically 30 Hz, with some recent lasers capable of reaching 100 Hz. The second cavity can be fired in quick succession after the firing of the first one, thus giving a time between pulses that is as low as 100 ns, which is much shorter than the pulse repetition time of each cavity. For measurements in water flows, one may use the previously mentioned pulsed lasers, as well as continuous diode-pumped solid state (DPSS) lasers, having an output of 1–5 W. The latter type of laser can, under certain conditions, be used for time-resolved measurements in water flows with a relatively small velocity, typically lower than 1 m/s. For time-resolved measurements in air, Nd:YLF lasers are preferred, however, they produce pulses with lower energy, typically less than 35 mJ at a repetition rate of 1 kHz. Nd:YLF lasers can be fired at repetitions rates of up to 20 kHz, in which case they produce pulses with drastically lower energy. More details are given in Section 7.3.7. Time-resolved PIV has also been performed with the use of continuous-wave lasers and high-frame-rate cameras [125].

Imaging: A camera lens with a magnification factor M and an f-number $f^{\#}$ is focused on the light sheet and produces the image of a portion of the illuminated flow, which is the *measuring volume* of the system. The light intensity is assumed to be uniform throughout the entire measuring volume. Two snapshot images (frames) of the illuminated particles are recorded with a digital camera, in synchronisation with the light sheets, creating *image pairs*. The interframe rate of the PIV measurement is primarily limited by the internal architecture of the camera, with CCD and sCMOS cameras having a higher interframe rate ($\Delta t \sim 100$ ns), compared to CMOS cameras ($\Delta t \sim 500$ ns). The manner in which the two frames are recorded depends on the type of laser used and on whether the measurements are time-resolving or not. In general, the image acquisition mode for a PIV experiment can be classified in one of the following two categories [111]:

- *Frame-straddling mode (FSM)*: In this mode, the first light pulse is triggered just before a frame is advanced to the next one, and the second light pulse is triggered immediately following the frame advance. The time between light pulses is the interframe time Δt, and the successive pairs of light pulses are initiated following a repetition time of $\Delta T = 1/f_r$; for example a laser with a repetition rate of 10 Hz would have a repetition time of 0.1 s. This method is used to achieve the highest possible recording speed with relatively inexpensive digital cameras. The typical timing sequence for CCD and sCMOS cameras is shown in Fig. 12.25a. Note that for CMOS cameras, the duration of the second frame would be the same as the first one, as a result of the faster architecture of these cameras by comparison to CCD and sCMOS cameras.
- *Time-series mode (TSM)*: In this mode, the light pulse and the camera are synchronised to fire at regular intervals, with the light pulse initiated in the middle of the frame. By using both cavities of the laser, one is able to maximise the light intensity. If a continuous laser is used, then one only needs to trigger the camera. In this mode, the interframe time and the repetition time are identical, that is, $\Delta t = \Delta T$, as shown in Fig. 12.25b.

Processing: An example of a PIV image pair is shown in Fig. 12.26 – details of the experiments can be found in [88]. Each frame is divided into subdivisions called *inter-*

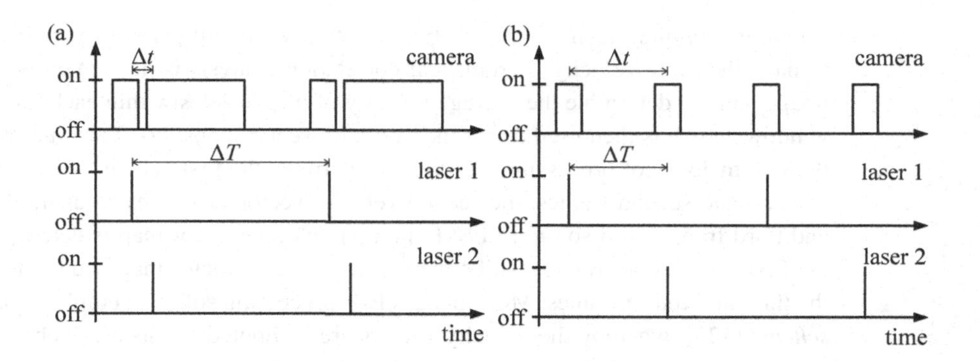

Figure 12.25 Timing diagram for a dual-cavity laser and a camera with image acquisition in (a) a frame-straddling mode and (b) a time-series mode. The interframe time Δt and the repetition time ΔT are also shown.

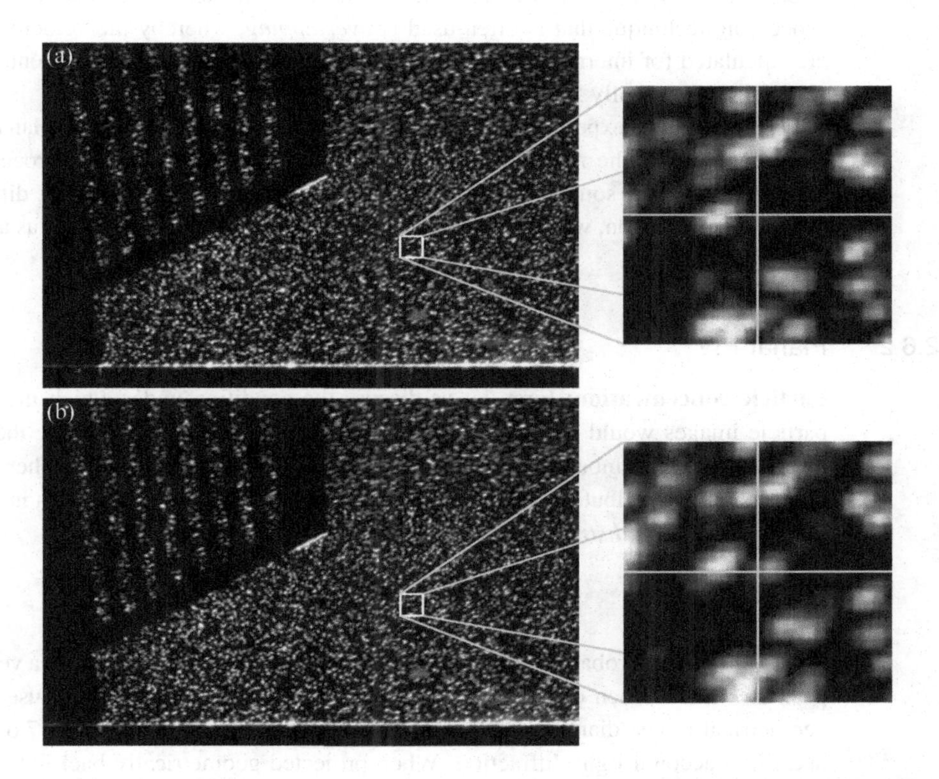

Figure 12.26 Sample planar PIV images downstream of an inclined, perforated plate. The time between the two images (a) and (b) in this image pair is the interframe time Δt. An example of an interrogation window is shown as a white box in both images. Zoomed-in views of the two interrogation windows are shown on the right, with a white cross inserted in each image to facilitate the visual determination of the displacement of the particles from one frame to the other.

rogation windows, having an area D_i^2 and within which all particles may be assumed to have the same velocity. A statistical correlation analysis is performed between the image pairs to determine the average velocity of the particles within each interrogation window, which is then used to produce velocity vector maps. For TSM measurements, the vast majority of processing algorithms determine the first velocity vector map from the first and second frames, the second velocity vector map does so using the second and third frames, and so on. In FSM, the first velocity vector map is determined from the first and second frames, whilst the second velocity vector map is determined from the third and fourth frames. Most modern PIV processing software employ a *multi-pass scheme* [123], whereby the velocity vectors are computed for successively decreasing interrogation windows, while, for each window, using the information from the previous one; for example, the first set of calculations are done for an interrogation window of 64 × 64 pixels (first pass), and the data from that calculation are used to determine the velocity for a subsequently used interrogation window of 32 × 32 pixels (second pass). It is important for the initial window size to be larger than the scale of the pattern of interest, for example, the turbulent integral scale, the vortex core diameter etc.Another processing technique that is often used is *overlapping*, whereby the velocity statistics are calculated for interrogation windows that are shifted by a certain percentage of the original size, typically 50–75% [102].

For a particular experiment and chosen PIV technique, for example, planar, stereo, tomographic etc., the manner in which the illumination, imaging and processing are configured, and the sources of error that arise from them, will of course be different. In the following section, we shall consider the common planar PIV approach as a working example.

12.6.2 Planar PIV

Particle concentration: First, let us discuss the condition under which the recorded particle images would be distinct from those of other particles. Let C be the number density (i.e., the number of particles per unit volume) of monodisperse spherical particles randomly distributed in the flow. The probability of finding k particles in a volume V obeys a *Poisson distribution*, such that

$$P\{k \text{ particles in } V\} = \frac{(CV)^k}{k!} e^{-CV} . \tag{12.77}$$

For $CV \ll 1$, this probability can be approximated as $(CV)^k/k!$, which has a very strong peak at $k = 1$. When considering particle images in PIV, one must not use the ideal geometrical image diameter Md_P, but the parameter d_e, defined by Eq. (7.69), which takes into account light diffraction. When projected geometrically back into the flow, the particle image appears to be produced by a source with diameter d_e/M. The images of two or more particles will overlap if their distances on the image plane are less than d_e. To avoid overlap, there should be at most one particle within the cylindrical volume $\Delta z_o \pi d_e^2/(4M^2)$, which is projected by the lens onto the particle image area $\pi d_e^2/4$. Thus, to ensure non-overlapping particle images, one must select a *particle source density* N_s

such that

$$N_s = C\Delta z_o \frac{\pi d_e^2}{4M^2} \ll 1 \, . \tag{12.78}$$

To proceed, one must consider the resolution of the recorded images, which depends on the recording medium. For digital cameras, the resolution is the size d_x of an individual pixel. Optimal utilisation of the recording system is achieved when $d_e/d_x \approx 2$–4. If the particle size is comparable to the size of the pixel, a bias error known as *peak-locking* will be introduced, whereby the estimate of the displacement will tend towards integer multiples of the pixel size [19, 86]. Assuming that the particle diameter d_P is fixed, one must select the camera lens properties such as to satisfy this condition. In selecting D_i, one needs to ensure that the probability that a significant number of particle images can be found within each interrogation window is high, so that spatial averaging of their displacements would be meaningful. This condition requires one to verify that the number of particles within the *interrogation volume* $V_i = \Delta z_o D_i^2/M^2$ is significant, expressed by the requirement that the *particle image density* N_i is

$$N_i = C\Delta z_o \frac{D_i^2}{M^2} \gg 1 \, . \tag{12.79}$$

Optimal values of the ratio D_i/d_e for resolving particle displacement have been found to be in the range 20–30, which points to optimal interrogation windows of 16×16 or 32×32 pixels. The length D_i is important as it determines the spatial resolution of the PIV system. Whether such resolution is satisfactory, depends on whether the size D_i/M of the window projection into the flow is comparable to or smaller than the smallest length scale of dynamic interest. For turbulent flows, this scale would be the Kolmogorov microscale [51], which is usually too small to be resolved by PIV systems and so some compromise must be tolerated.

Particle velocity calculation: Once the particle images have been recorded and the interrogation window size has been fixed, one needs to determine the average particle displacement $\left\langle \overrightarrow{\Delta r} \right\rangle$ within each window, from which the local flow velocity projected on the plane of the light sheet can be estimated as

$$\vec{U}_p \approx \frac{\left\langle \overrightarrow{\Delta r} \right\rangle}{M \Delta t} \, . \tag{12.80}$$

Among the various methods that have been used to estimate this displacement, the most widely used one is the *cross-correlation method*. Each pixel will produce as output a number, chosen among a small set of permissible values, which represents the intensity of the received light and determines whether a particle image overlaps with the pixel or not. In the cross-correlation method, two images of each window, such as those shown in Fig. 12.26, are processed by an algorithm, which correlate all pixels in the first frame with all pixels in the second one. Figure 12.24 shows an example of a window with both frames superimposed on each other. If the system parameters have been properly selected, the majority of the particles would have both images recorded within the same window and all particle image displacements would be identical within the window.

Then, the cross-correlation would have a peak, whose location with respect to the zero-separation point would correspond to the average particle image displacement on the plane of the image. Interpolation methods are available to determine the peak location at sub-pixel resolution. Shear in the interrogation volume would introduce particle velocity differences, which would cause broadening of the correlation peak and even change of its shape. For broadening effects on the estimate of $\left\langle \overrightarrow{\Delta r} \right\rangle$ to be negligible, the flow velocity differences $\left|\overrightarrow{\Delta U_p}\right|$ within the interrogation volume must satisfy the condition

$$\left|\overrightarrow{\Delta U_p}\right| \ll \frac{d_e}{M\Delta t} . \tag{12.81}$$

It is also evident that correlation distortion would occur by the fact that some particles may only have a single image recorded within a window, because their motion is such that they enter or exit the interrogation volume during Δt. Such errors will be negligible if the in-plane $\overrightarrow{U_p}$ and out-of-plane $\overrightarrow{U_n}$ flow velocity components are such that

$$\frac{\left|\overrightarrow{U_p}\right| M\Delta t}{D_i}, \frac{\left|\overrightarrow{U_n}\right| \Delta t}{\Delta z_o} \ll 1 . \tag{12.82}$$

For example, for a particle density of $N_i = 10$, the displacement of the particles within the interrogation window would ideally be not larger than one quarter of the window width [123]. Note that this requirement is applied to the first pass of the processing algorithm only. Another source of error in PIV, known as *velocity bias*, is that the broadening of the correlation peak results in a systematic underestimation of the average image displacement, which would be relatively small if all particle speeds were identical, but worsen with increasing shear. Displacement errors due to velocity bias are typically of sub-pixel magnitude and can be corrected using methods available in the literature [62, 123].

Once the velocity $\overrightarrow{U_p}$ is determined at all interrogation windows, it can be displayed in the form of a *velocity vector map* (see Fig. 12.24), in which each vector is displayed on the centre of each window, in a manner similar to vector plots in CFD simulations. These velocity values can be used to estimate the vorticity component normal to the light sheet plane and streamline projections on this plane. The experimental design process for a PIV experiment shall be illustrated in the following example, which is based on the experiments of Limbourg and Nedić [74, 75].

Example 12.4 We wish to measure the velocity field during the formation of a vortex ring emanating from a circular orifice with a diameter $d = 50$ mm. Water is pushed along the tube by a piston that is driven by a linear actuator and is issued out of the orifice into a large tank. Based on preliminary visual measurements, the translation speed of the formed vortex is assumed to be roughly 0.14 m/s, although we can expect the speed to be 25% larger in certain regions, based on past experience. We want to obtain time resolved velocity measurements in the tank within a few diameters from the outlet. A high speed camera with a resolution of 2048×2048 pixels, having a size of 10 μm each, is used to record images of the velocity field. After performing a calibration, we find that the physical field of view is 189 mm × 189 mm. A pulsed Nd:YLF laser is available for illumination. Determine an appropriate interframe time for this experiment and the repetition rate for the laser.

Answer

To collect time-resolved measurements, we acquired data in time series mode, and set the interframe time to be equal to the repetition time. In order to achieve a relatively high spatial resolution of the velocity field, we used a two-pass approach. The initial size of the interrogation window was chosen to be smaller than the smallest structure of interest, in the present case the core of the vortex ring. Given that the physical field of view is 189 mm × 189 mm, and that the camera has 2048×2048 pixels, we found that each pixel corresponded to a square area in the flow with a side length of $189/2048 \approx 0.092$ mm. Based on the literature, we expect the diameter of the ring to be comparable to the diameter of the outlet, and the core diameter of the ring to be roughly a quarter of the diameter of the ring [75], namely 12.5 mm, or 135 pixels. We therefore set the initial interrogation window to 64 × 64 pixels, and the final interrogation window to 32 × 32 pixels. The resulting velocity map had $2048/32 = 64$ velocity vectors and a spatial resolution of $189/64 \approx 3.0$ mm.

Assuming that, between consecutive frames, each particle moved by no more than one quarter of the initial interrogation window width, we find that the expected maximum particle displacement between frames would be $0.25 \times 64 = 16$ pixels, which corresponds to a physical displacement of roughly 1.5 mm. From the provided information, we can expect the maximum translation speed of the vortex ring to be around $1.25 \times 0.14 = 0.175$ m/s. Using these two values, we find that the minimum interframe time is $\Delta t \approx 8.5$ ms, which corresponds to a frequency of 118 Hz. We set the repetition rate of the laser to 120 Hz, which ensured that the displacement of the particles satisfied the 'one-quarter rule'.

12.6.3 Stereoscopic PIV

Stereoscopic (*stereo*) *PIV* is an extension of the basic PIV configuration, which is capable of measuring the three-dimensional velocity vector on a plane [7]. This configuration is illustrated in Fig. 12.27a. Like planar PIV, stereo-PIV measures the velocity components on the plane of a planar light sheet, but, in addition to the in-plane particle displacements, it also measures the out-of-plane displacements, from which the map of the full velocity vector on the plane of the sheet is reconstructed. Two cameras are positioned either symmetrically with respect to the normal to the light sheet, or symmetrically with respect to the light sheet plane, with their lenses inclined by an angle α with respect to the sheet plane. Best results, in terms of minimising random errors in the measured velocity components, is obtained for $30° \leq \alpha \leq 45°$ [3]. If the depth of field of the lens is relatively narrow, lens inclination would produce images which would be partially out of focus. This is corrected by tilting the CCD chips of the cameras by an angle β, such that the lens plane and the image plane intersect on the object plane, an

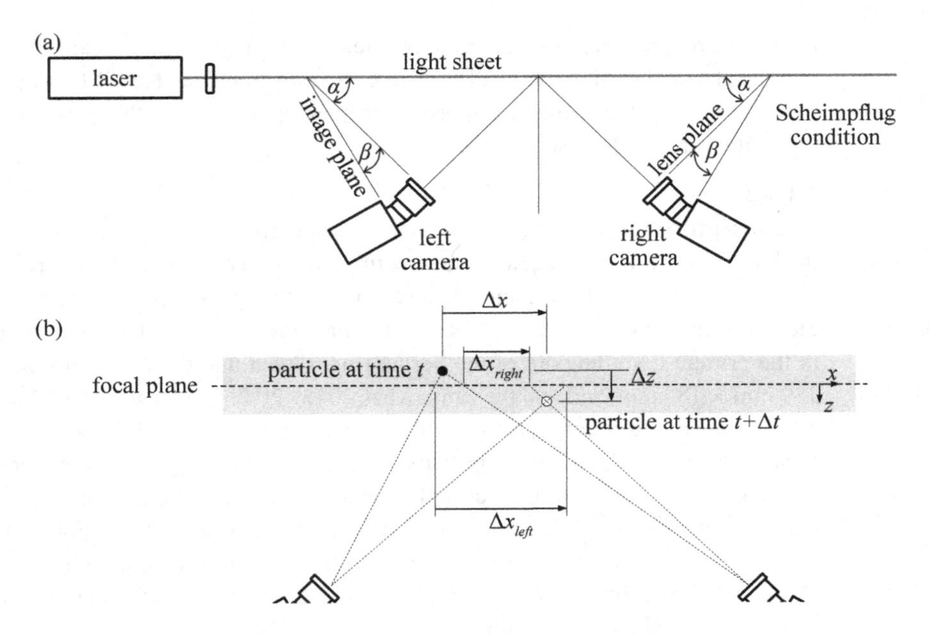

Figure 12.27 (a) Sketch of a stereoscopic PIV set-up; (b) illustration of stereoscopic view of particle displacement.

arrangement known as the *Scheimpflug condition.* Distortion of the object proportions due to image inclination is corrected by calibrating the cameras using a target marked by an array of spots at known positions. Because the motion of particles in the measuring volume is viewed by the two cameras from different angles, the two images would be different, in a manner similar to stereoscopic vision. For example, let us assume that the true in-plane displacement of a particle in the measuring volume during time Δt is Δx, as shown in Fig. 12.27b. Assuming that both lenses are focused on the mid-plane of the light sheet, the left and right cameras will record displacements Δx_{left} and Δx_{right}, which are different from each other and Δx. Analysis of these values can provide both Δx and the out-of-plane displacement Δz [94]. The error in the determination of Δz can be related to the error in Δx by the approximate expression [122]

$$\varepsilon_{\Delta z} = \frac{\varepsilon_{\Delta x}}{\sqrt{2}M \sin\beta} . \tag{12.83}$$

A practical consideration that needs to be taken into account when setting up a stereo-PIV experiment is the change in the refractive index of the fluid between the camera and the focal plane. For example, if a PIV experiment is conducted in water, the cameras would be placed in the air outside the facility and at an angle α relative to the focal plane. Therefore, light rays originating from particles within the focal plane would be refracted as they exit the test section wall and before entering the camera lens, due to a change in refractive index from water to air (see Section 7.2.2); refraction in the test section wall, made of glass or other transparent material, is not a problem, as it merely introduces a small sideways displacement to the beam. Consequently, the particles imaged by the

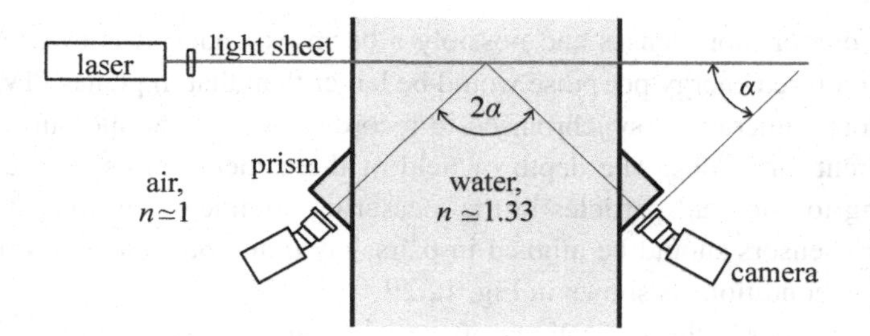

Figure 12.28 Stereo-PIV experimental set-up with liquid prisms placed in front of cameras to reduce astigmatism effects.

camera would appear as streaks (an effect known as *astigmatism*), which would increase the random error. This error can be reduced by placing a prismatic tank containing the same liquid as the one in the test section (thus matching its refractive index), such that one of its faces is in contact with the wall, whereas its other face is perpendicular to the axis of the camera lens, as shown in Fig. 12.28.

An extension of the stereo-PIV is the *dual-plane stereo-PIV*, which provides simultaneous maps of the 3-D velocity vector on two closely spaced, parallel planes [59, 87]. This allows the determination of the full velocity gradient tensor on a plane. Obviously, this technique requires the use of four pulsed-laser light sheets and four synchronised cameras, as well as specially developed data processing algorithms.

12.6.4 Tomographic PIV

Hot-wire anemometry and laser Doppler velocimetry provide point measurements of flow velocity, whereas (planar) PIV provides planar velocity maps. Results from these methods constitute incomplete descriptions of a general flow velocity field, a complete description of which would be achieved only by velocity vector measurements that occupy the entire flow domain of interest, or a substantial part of it, and which are adequately resolved in both space and time. Much effort has been made to devise methods that would ideally provide such information. A direct, and rather obvious, extension of planar PIV is the *scanning light sheet* (*SLS*) method, which is itself a tomographic method and found application in relatively low-speed flows. Nevertheless, the term *tomographic* (*tomo*) *PIV* describes methods that use volume, rather than sheet, illumination, and track and reconstruct the 3-D motion of multiple seed particles using sophisticated interrogation algorithms [32, 103]. When adequately time-resolved, tomo-PIV provides a description of a complex velocity field that is comparable to analytical solutions, or to direct numerical simulations (DNS), even permitting the analytical determination of the in-flow pressure distribution, the measurement of which was previously beyond our experimental capabilities (see Section 9.5.2).

A tomo-PIV system can be based on available planar PIV equipment, but requires a reconfiguration and the addition of components, especially cameras, as shown in Fig. 12.29. The main working principles of tomo-PIV are as follows:

- Pulsed-volume illumination can be provided by a thick laser sheet, produced with the use of one or more lenses and possibly a beam expander. It is evident that the minimum required energy per pulse would be larger than that in planar PIV.

- Four or more cameras are synchronised to record images of the measuring volume from different directions. The depth of field of the camera lenses should be sufficiently long to allow all particles in the measuring volume to be imaged in focus. The camera sensors should be aligned in pairs, say a and b, such as to observe the Scheimpflug condition, as shown in Fig. 12.29.
- Each camera records the projections of particles on its sensor plane. The images of different cameras are analysed by an iterative algebraic reconstruction algorithm, that, following 3-D calibration, provides the 3-D positions of recorded particles at each instant.
- The light intensity distribution in the space of interest is discretised as a 3-D array of cubic *voxels*, from which it is projected onto the pixels on the camera image planes. Mathematical algorithms identify the particle positions within the voxel array.
- The space is divided into cubic interrogation boxes and 3-D correlation algorithms are used to compute flow velocity maps from successive 3-D particle position maps.

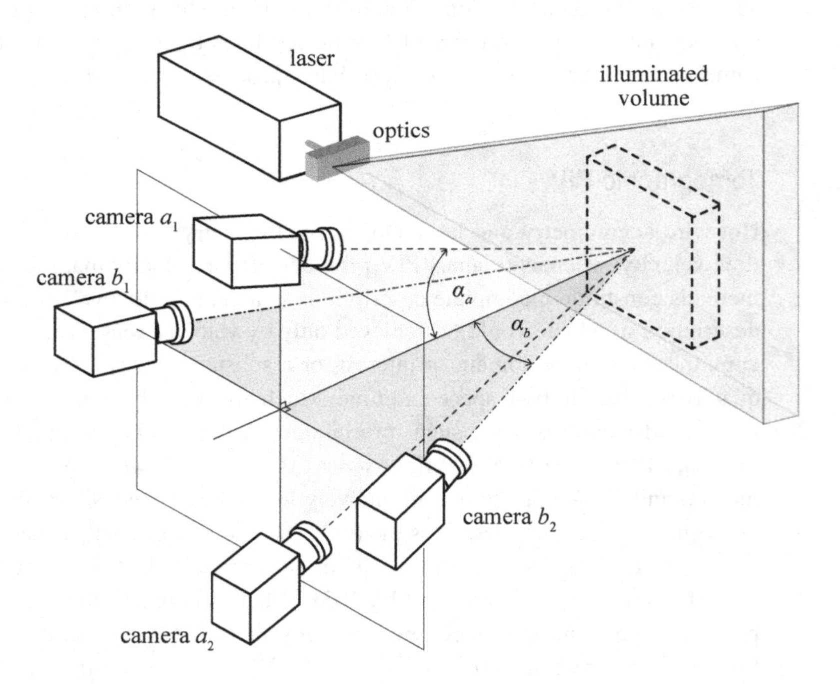

Figure 12.29 Sketch of a tomographic PIV set-up in a cross-like imaging configuration; camera pair a is arranged to have a viewing angle α_a, while camera pair b is arranged to have a viewing angle α_b.

The setting and operation procedures of tomographic PIV systems have been well established. Tomo-PIV has been applied widely for measurements in liquids and gases over a wide range of speeds, including turbulent and supersonic flows. Nevertheless, the in-house development and setting of tomo-PIV systems is beyond the resources and capabilities of typical academic fluid mechanics laboratories.

12.6.5 Micro PIV

In *micro PIV* (*μPIV*) techniques, conventional PIV methods were adapted to allow flow measurements in microdevices, around red blood cells and in other systems having a length scale of the order of a micrometre [121]. A main feature of μPIV techniques is that they replace the camera lens by a microscope lens, which allows focusing on a very small image and a sharply defined objective plane. Rather than using sheet illumination, they usually apply volume illumination, or *evanescent waves*, namely, light that is totally internally reflected at the interface between two media with large differences in their refractive indices. Special particles of sub-micron scale are used for seeding. An additional possible error source in μPIV is the Brownian motion of particles due to random collisions with fluid molecules. μPIV systems can be configured to produce planar velocity maps or three-dimensional ones. The use of long-distance microscopes has permitted measurements with a micrometer-level resolution in the wall region of turbulent boundary layers [60].

12.7 Molecular Tagging Velocimetry

Molecular tagging velocimetry (*MTV*) [65, 112], also referred to as *laser induced photochemical anemometry* and *flow tagging velocimetry*, includes methods which trace the motion of photosensitive molecules in the fluid. When applied to liquids, it has an advantage over PIV and other particle tracking methods, as it requires no seeding, thus being suitable for both single-phase and multi-phase flows. It does require the presence of special materials, however, whose molecules, when activated, will become distinguishable from the background fluid.

The following three types of materials have been used in liquid-flow MTV:

- *Photochromic materials*: These have been introduced in Section 11.2. They produce images with a relatively low contrast, because they are sensitive to changes in light absorbance, unlike more popular MTV methods, which are sensitive to changes in luminescence. Another drawback is that, in general, photochromic materials are not water soluble and require the use of disagreeable liquids, such as kerosene; it must be mentioned, however, that methods capable of producing water-soluble photochromic dyes have also been developed [65].
- *Phosphorescent supramolecules*: These are molecules which, as a result of photon absorption, undergo transition to metastable electronic states and emit radiation at a different wavelength, following a relatively long delay time, which exceeds 0.1 ms (see discussion on phosphorescence in Section 7.4). The emitted radiation is collected by a camera during subsequent 'interrogation' times to locate the tagged fluid and compute its displacement. Recipes for such molecules that are water soluble have

been proposed in the literature [112]. Tagging in the form of single and multiple lines or in meshes of intersecting lines has been employed to map one-, two- and three-dimensional displacement fields [33, 48, 98].

- *Caged dyes*: These are fluorescent dyes, such as fluorescein disodium, the fluorescence of which has been 'caged' by the attachment of a chemical group [71]. Fluorescence is recovered locally, following absorption by the caging group of an ultraviolet photon. This is a permanent, irreversible change and allows the tracing of the tagged fluorescent molecules indefinitely, provided that they get illuminated by suitable laser light. A disadvantage of this method is the high cost of the caged dye material.

The luminescence of specific molecules, which are either generated by internal processes inside a gas mixture or are seeded into it, has been used extensively for flow visualisation and for the measurement of the concentration of a tracer gas. Such techniques are discussed in other chapters of this book, but in this section we will refer to variants of molecular tracing techniques, which provide local measurements of gas velocity. One of these methods utilises fluorescence of strontium ions, seeded into a gas mixture [97]. It requires two different lasers, one to produce a beam along which ionisation occurs and a second one, fired following some specified time delay, to produce a light-sheet, which induces ionic fluorescence. This method has been applied to the measurement of velocity in low-speed burner flames. Several other MTV techniques, having different levels of complexity, have been developed, based on the fluorescence or phosphorescence of excited-state oxygen, ozone, hydroxyl, biacetyl and acetone [65].

12.8 Measurement of Wind Velocity

The measurement of many properties in the Earth's atmosphere is essential for weather prediction, transportation safety, pollution level determination and climate change monitoring. According to a recent, all-encompassing review, 'typical atmospheric measurements include winds, temperature, pressure, humidity, dew point, moisture, radiation, visibility, cloud heights, lightning, gaseous composition, aerosols, and precipitation' [37]. In the present section, however, our interest is restricted to the measurement of wind velocity in the lower planetary boundary layer, typically within 100 m from the ground.

The following are instruments designed to measure mean wind magnitude, wind direction and, to a certain degree, atmospheric turbulence statistics [37, 128]. Such instruments, as well as sensors measuring other atmospheric properties, are typically mounted on dedicated towers.

Cup anemometers: These consist of two, three or more cups, usually of hemispherical shape, mounted on a shaft that is free to rotate, as shown in Fig. 12.30a. A net torque is generated by flow on planes normal to the shaft axis, as a result of higher drag on the concave than the convex side of each cup. The speed of rotation ω of the anemometer

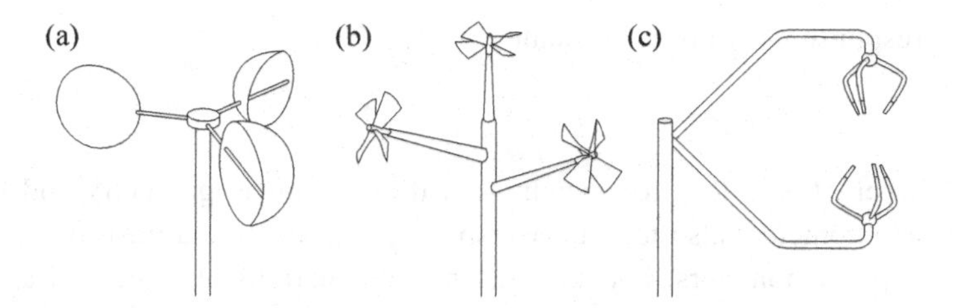

Figure 12.30 Sketches of representative anemometers for wind velocity and turbulence measurement: (a) cup anemometer, (b) three-component Gill anemometer and (c) three-component sonic anemometer.

in steady laminar flow is related to the flow speed V by the linear expression

$$\omega = \frac{k_c V}{r}, \tag{12.84}$$

where r is the mean radius of the cup with respect to the axis of rotation and k_c is an empirical numerical coefficient with a typical value of about 0.3. Cup anemometers typically have a threshold wind speed of about 0.5 m/s and an uncertainty of 0.1 m/s. In turbulent flow, cup anemometers tend to *overspeed*, that is to say they show velocities higher than values calculated from the previous expression. For very small turbulence intensities, the cup anemometer dynamic response can be approximated by a first-order system with a time constant

$$\tau = \frac{I}{\rho r^2 A V}, \tag{12.85}$$

where A is the frontal area and I is the moment of inertia of the cup. More sophisticated, non-linear models have also been formulated. Cup anemometers are usually oriented with their axes of rotation vertical and are meant to measure horizontal wind. When a significant wind component outside the plane of rotation is present, they tend to indicate a speed that is larger than the actual component on that plane, with the error increasing significantly at flow inclinations that are larger than 10° with respect to the rotation plane.

Propeller anemometers: These consist of one or more propeller-type rotors, meant to measure the wind velocity parallel to their axes, and are often mounted on rotating vanes, which orient them into the wind direction. The most common type is called the *Gill anemometer* (Fig. 12.30b), which has four helicoidal blades. The speed of rotation ω of propeller anemometers is, within a range, related linearly to the axial wind speed V, as

$$\omega = k_p V, \tag{12.86}$$

where k_p is a calibration coefficient having a dimension of inverse length. When misaligned with the flow direction, Gill anemometers tend to measure an axial velocity component which is lower than the actual one. The dynamic response of propeller anemometers is non-linear. For rough purposes, one may approximate it by a first-order response with a time constant

$$\tau = \frac{L}{V}, \tag{12.87}$$

where L is a length depending on the instrument geometry and flow conditions. Gill anemometers also tend to overspeed in turbulent and gusty flows, but much less than cup anemometers. For this reason, combinations of three Gill anemometers, together with temperature and humidity sensors, are commonly used for measuring atmospheric turbulence. One must keep in mind, however, that due to their limited frequency response (at most extending to a few hundred Hz), they cannot resolve the turbulence fine structure and may even be missing part of the turbulence kinetic energy.

Vane anemometers: Two types of vanes are used in wind measurement: *rotating vanes* and *fixed vanes*. A main function of rotating vanes is to orient other instruments into the wind direction as well as to indicate this direction by their instantaneous orientation. In high-turbulence flows, vane dynamic response is non-linear. A simplified model gives a second-order response, which means that underdamped vanes are prone to exhibit resonance, with would distort turbulent spectra. Fixed vanes are mounted on strain gauges and measure flow direction from measurements of the applied torque.

Sonic anemometers: Although sonic Doppler anemometers have also been used for atmospheric turbulence measurement, the most common sonic anemometers are of the time-of-flight type, measuring wind speed from the difference in the speed of sound waves propagating in the direction of the flow and against it. A three-axis sonic anemometer is shown in Fig. 12.30c. It consists of three sets of transmitter–receiver combinations. Each set contains a pair of transmitters and a pair of receivers, such that one transmitter and one receiver are on one side, facing the other transmitter and receiver on the opposite side. Both transmitters are triggered simultaneously and the produced sound waves are picked up by the opposite receivers. Let t_1, t_2 be the times it takes for a sound wave emitted by each transmitter in one set to be received by the opposite receiver and V_s be the wind velocity component parallel to the axes of these transducer pairs, defined as positive if it points from transmitter 1 towards receiver 1. Moreover, let s be the distance between each transmitter–receiver pair and c be the local speed of sound. These properties are related as

$$c + V_s = s/t_1 \text{ and } c - V_s = s/t_2, \tag{12.88}$$

from which one can get both V_s and c as

$$V_s = \frac{s}{2}\left(\frac{1}{t_1} - \frac{1}{t_2}\right) \text{ and } c = \frac{s}{2}\left(\frac{1}{t_1} + \frac{1}{t_2}\right). \tag{12.89}$$

If both t_1, t_2 are measured, one may also determine the local *virtual temperature* (namely, the temperature of dry air with the same pressure and specific volume as the moist air between the transducers) as

$$T_v = c^2/(\gamma R), \tag{12.90}$$

where γ is the ratio of specific heats of dry air and R is the gas constant of dry air.

Further considering that $V_s \ll c$, one can approximate the wind speed component as

$$V_s \approx \frac{c^2(t_2 - t_1)}{2s}, \tag{12.91}$$

which only requires the measurement of the time difference in the sound transmission between the two pairs and an estimate of the speed of sound by other means, essentially from a local measurement of temperature and humidity.

In order to achieve a sufficient sensitivity of this instrument, it is necessary to keep the transducer distance s relatively large (e.g., 0.20 m), which limits its spatial resolution. Measurement errors are also introduced by flow blockage by the transducers. Moreover, sonic anemometers cannot be used in dusty or foggy air and when there is precipitation or icing of the unit. The typical frequency response of sonic anemometers is of the order of 10 Hz, which is sufficient for resolving large-scale motions of atmospheric turbulence.

Chapter Digest

12.1 Identify the possible sources of error in measuring flow velocity across a turbulent boundary layer with a Pitot tube.

12.2 How would you use a Pitot tube to measure velocity in a supersonic flow? How would you design such a tube and why?

12.3 If a three-hole probe can measure flow velocity magnitude and direction, why do we have five-hole probes?

12.4 How does the bleed-type pressure tube operate?

12.5 List possible situations in which you would use a hot-film probe, rather than a hot-wire probe.

12.6 What advantages and disadvantages do gold-plated or etched hot-wire sensors have by comparison to spot-welded sensors?

12.7 Why are hot wires unsuitable for measurements in oscillatory flows?

12.8 Why, in your opinion, has constant temperature anemometry prevailed over constant current anemometry?

12.9 Why do we not use hot wires with a relatively small length to diameter ratio?

12.10 Can we use a hot wire to measure flow velocity in a homogeneous gas mixture? In the mixing layer between two different gases?

12.11 Why can we not use laser Doppler velocimeters in the same way we use ultrasonic Doppler velocimeters?

12.12 What are the uses of a beam expander and a Bragg cell in laser Doppler velocimetry?

12.13 Which are the major disadvantages of laser Doppler velocimetry by comparison to particle image velocimetry?

12.14 Comment on the challenges of using current particle tracking techniques in turbulent flows.

12.15 What is the difference between stereoscopic PIV and tomographic PIV?

12.16 What is the Scheimpflug condition and when is it used?

12.17 What is the major advantage of molecular tagging velocimetry techniques, compared to particle tracking methods?

12.18 A hot-wire probe will be used for measuring the turbulent flow velocity profile at the outlet of a heating air duct located on the ceiling of the room. The probe will be calibrated in a heated jet.

(a) Describe how you would measure the instantaneous local flow velocity in the operating duct, whose temperature fluctuates.

(b) Consider describing the hot-wire response by a fitted fourth-order polynomial rather than King's law. Would you recommend this approach for this experiment? Justify your answer.

12.19 Two popular instruments for measuring local flow velocity are the hot-wire anemometer and the laser Doppler velocimeter. Explain which of the two would be preferable in the following situations:

(a) When the highest possible spatial resolution is required.

(b) When the highest possible temporal resolution is required.

(c) When the turbulence intensity exceeds 30%.

(d) When the flow frequently reverses direction.

(e) When the flow is three-dimensional.

(f) When measurements in both gases and liquids are required.

(g) When cost is a determining factor.

(h) When calibration must be avoided.

(i) When multi-point measurements must be conducted simultaneously.

12.20 Compare the ranges of application of laser Doppler velocimetry and particle image velocimetry. Describe comparatively the advantages and disadvantages of these methods. Provide an example of flow measurement for which only LDV would be suitable, and another for which only PIV would be suitable.

12.21 Describe an experiment for which both laser Doppler velocimetry and pulsed-wire anemometry would be suitable but hot-wire anemometry would not. Also give an example of an experiment for which pulsed-wire anemometry would be preferable to laser Doppler velocimetry.

Problems

12.1 Measurements in high-speed flow of helium have provided a stagnation temperature of 500 K, a stagnation pressure of 671 kPa and a static pressure of 400 kPa. Determine the flow velocity and the Mach number. Also estimate the error in velocity, if Bernoulli's equation were used instead of more appropriate relationships.

12.2 A Pitot tube with an inner diameter of 0.85 mm and an outer diameter of 1.22 mm is used to measure total pressure in the wake of a nearly two-dimensional object in air flow at a temperature of 20°C. The difference Δp between the tube reading at different distances y of the tube's centre from the wake axis and the static pressure measured accurately by other means is presented in the following table.

y [mm]	0	1	2	3	4	5
Δp [Pa]	86.4	91.6	96.2	100.4	104.2	107.6
y [mm]	6	7	8	10	12	14
Δp [Pa]	110.7	113.4	115.9	120.0	123.2	125.8
y [mm]	16	18	20	25	30	35
Δp [Pa]	127.8	129.4	130.6	132.6	133.7	134.3

(a) Determine the velocity profile across the wake neglecting the displacement effect.
(b) Estimate appropriate corrections for the displacement effect.
(c) Plot the uncorrected and corrected velocity profiles and discuss possible differences between them.

12.3 Consider a circular Pitot tube with an inner diameter of 0.85 mm and an outer diameter of 1.22 mm inserted in horizontal flow of glycerol at an elevation 200 mm below the free surface, where you may assume that the pressure is hydrostatic. The temperature is 20°C and the atmospheric pressure is standard. Estimate the flow velocity as a function of the total pressure indicated by the tube over the speed range between 10^{-3} and 1 m/s. Plot this estimate in logarithmic axes and fit an appropriate empirical expression to your results to facilitate computing the flow velocity from pressure measurements.

12.4 A hypodermic needle with an external diameter of 0.5 mm and an internal diameter of 0.3 mm is used to measure the total pressure in an open channel flow of a homogeneous mixture of 80% glycerol and 20% water at 20°C. The channel depth is 100 mm. The tube readings p_{om} (gauge total pressure), measured with a micromanometer at different elevations y from the bottom, are presented in the following table. Assume that the static pressure in the channel is hydrostatic. The density and the kinematic viscosity of the mixture can be found from tables as, respectively, 1,208.5 kg/m^3 and 49.57×10^{-6} m^2/s.

y [mm]	0.5	2	6	10	20	30	40
p_o [Pa]	1,179.3	1,167.8	1,137.3	1,106.8	1,087.1	1,141.1	1,409.4
y [mm]	50	60	70	80	90	95	
p_o [Pa]	1,854.7	2,140.4	2,001.3	1,735.2	1,309.3	913.3	

Determine, applying appropriate corrections, if required, the velocity profile across the channel. Discuss the level of confidence in these results and assign uncertainty estimates.

12.5 The thermal resistivity coefficient of annealed tungsten at 20°C is 0.0048 K^{-1}. A hot-wire sensor made of tungsten has a 'cold' resistance of 5.65 Ω, measured at 25°C. Determine the maximum allowable operating resistance R_w of this sensor to avoid oxidisation. The sensor, operating at the resistance R_w as found previously, is inserted in a flow that has a temperature of (a) 25°C and (b) 130°C. Determine the corresponding resistance and temperature overheat ratios. Compare qualitatively the anticipated velocity sensitivities of this sensor in the two flows.

12.6 Consider the constant current anemometer shown in Fig. 12.9, for which $R_c' =$ 100 Ω. The time constant of the sensor was determined by the square wave test to be 7.5 ms. Determine values of the resistor R_c and capacitor C in the R–C circuit that would give a compensated output time constant of 0.05 ms. Plot, in logarithmic coordinates, the amplitude ratios of the uncompensated and compensated outputs vs. frequency. Specify the 3 dB cut-off frequencies of the uncompensated and compensated systems.

12.7 Consider that the response of a hot-wire anemometer follows the modified King's law (Eq. (12.31)), in which the calibration coefficients and their 95% confidence limits at a reference temperature $T_r = 20 \pm 0.2$°C were found to be $A = 2.65 \pm 1\%$, $B = 3.07 \pm 2\%$, $n = 0.48 \pm 0.01$, all in SI units. During an experiment, the following values and their 95% confidence limits were measured: the voltage was $E = 3.673 \pm 0.001$ V; the wire temperature was $T_w = 230 \pm 1$°C (same as during calibration); and the flow temperature was $T = 24 \pm 0.2$°C. Determine the value of the flow velocity and its uncertainty. Explain whether this uncertainty would be acceptable and identify the parameters that contribute most to it. If possible, suggest means to reduce the uncertainty.

12.8 The response of a hot wire, made of platinum-plated tungsten, 5 μm in diameter and 1.25 mm long, is determined by calibration in a heated air jet.

(a) First, the jet speed is kept constant at 8.00 m/s and the resistance of the unheated wire is measured at different flow temperatures. The results are shown in the following table.

T [°C]	25.81	27.71	29.89	32.80
R_w [Ω]	3.44	3.47	3.50	3.54

Determine the thermal resistivity coefficient of the wire at a reference temperature and compare it to values from the literature.

(b) Next, the overheat ratio of the wire is set to different values by setting its operating resistance R_w using a constant-temperature anemometer circuit. The measured circuit output voltage E is related to the current i through the wire by the expression $E = i\,(R_w + R_b)$, where $R_b = 50.30$ Ω is a resistor in series with the hot wire. During these, the flow temperature is kept at 25.80°C and the flow speed is kept at 8.00 m/s. The results are given in the following table.

R_w [Ω]	4.13	4.82	5.50	6.19	5.16	3.78
E [V]	2.448	3.232	3.731	4.098	3.500	1.791

Determine the corresponding resistance and temperature overheat ratios. Then, compute the experimental values of the Nusselt number and compare them to estimates based on the Collis and Williams expressions. Discuss your findings.

12.9 A constant temperature hot-wire anemometer is calibrated in an air jet. The following table provides the anemometer output voltage at different air speeds.

V [m/s]	7.06	8.76	9.48	10.61	11.91	14.06	14.73	15.33	17.08
E [V]	2.894	2.977	3.008	3.054	3.104	3.177	3.197	3.218	3.269

Absorbing the temperature difference into the coefficients of the modified King's law, determine the optimum value of the exponent n in that expression by minimising the mean squared difference between the measured velocities and those calculated from the fitted expression. Further evaluate the sensitivity of this difference to the selected exponent value. Discuss your findings. Repeat the calibration using a fourth-order polynomial, and compare the results.

12.10 A laser Doppler velocimeter is used to measure water flow velocity in a water tunnel having a test section with a width of 200 mm. The LDV system includes a He-Ne laser, a Bragg cell with a frequency shift of ±40 MHz, and three transmitting lenses with focal distances 130, 450 and 800 mm. The beam separation is 30 mm and the beam diameter is 0.8 mm.

(a) Determine the beam intersection half angle in the water; compare the calibration factors in air and in water.

(b) Select the lens that is best suited for measuring the flow velocity in the entire test section; explain your reasoning.

(c) Determine the dimensions of the measuring volume and the number of fringes in the control volume.

(d) Assuming that frequency shifting is used to measure the flow velocity in a recirculating flow region, determine the highest negative velocity that could be measured unambiguously; would you use positive or negative frequency shift?

12.11 We wish to use a planar PIV system to measure the mean velocity of an axisymmetric jet with a Reynolds number of 10,000, based on the jet diameter. The outlet diameter of the jet is $d = 10$ mm, and the desired field of view is $10d$ downstream from the outlet. It is known that the velocity profile is approximately uniform, within $1D$ of the outlet, and approaches an approximately Gaussian shape with a spreading angle of approximately 12 degrees, as the downstream distance increases. We have at our disposal a pulsed Nd:YLF laser and a high-speed camera with a resolution of 2048×2048 pixels, having a size of 10 μm each. Determine the interframe time of the camera and the repetition rate of the laser, assuming that the working fluid is (a) water at room temperature and (b) air at room temperature. How would the setting change, if we account for a turbulence intensity of 20%?

References

[1] R. Adrian. Particle-imaging techniques for experimental fluid mechanics. *Annu. Rev. Fluid Mech.*, 23:261–304, 1991.

[2] R.J. Adrian. Laser velocimetry. In R.J. Goldstein, editor, *Fluid Mechanics Measurements*, pages 175–299. Taylor & Francis Publishers, Washington DC, 1996.

[3] R.J. Adrian and J. Westerweel. *Particle Image Velocimetry.* Cambridge University Press, Cambridge, UK, 2011.

[4] N. Agüera, G. Cafiero, T. Astarita, and S. Discetti. Ensemble 3D PTV for high resolution turbulent statistics. *Meas. Sci. Technol.*, 27:124011 (12pp), 2016.

[5] R.W. Ainsworth, R.J. Miller, R.W. Moss, and S.J. Thorpe. Unsteady pressure measurement. *Meas. Sci. Technol.*, 11:1055–1076, 2000.

[6] Anonymous. Laser doppler anemometry: An introduction to the basic principles. Technical Report 1203E, DISA Elektronik A/S, Skovlunde, Denmark, 1981.

[7] M.P. Arroyo and C.A. Greated. Stereoscopic particle image velocimetry. *Meas. Sci. Technol.*, 2:1181–1186, 1991.

[8] S.C.C. Bailey, G.J. Kunkel, M. Hultmark, M. Vallikivi, J.P. Hill, K.A. Meyer, C. Tsay, C.B. Arnold, and A.J. Smits. Turbulence measurements using a nanoscale thermal anemometry probe. *J. Fluid Mech.*, 663:160, 2010.

[9] D.H. Barnhart, R.J. Adrian, and G.C. Papen. Phase-conjugate holographic system for high resolution particle image velocimetry. *Appl. Opt.*, 33:7159–7170, 1995.

[10] J. Belden, T.T. Truscott, M.C. Axiak, and A.H. Techet. Three-dimensional synthetic aperture particle image velocimetry. *Meas. Sci. Technol.*, 21:125403, 2010.

[11] S.J. Beresh. Time-resolved particle image velocimetry. *Meas. Sci. Technol.*, 32:102003, 2021.

[12] H.H. Bruun. *Hot-Wire Anemometry*. Oxford University Press, Oxford, 1995.

[13] D.W. Bryer and R.C. Pankhurst. *Pressure-Probe Methods for Determining Wind Speed and Flow Direction*. National Physical Laboratory, London, 1971.

[14] P. Buchhave, W.K. George, and J.L. Lumley. The measurement of turbulence with the laser doppler anemometer. *Ann. Rev. Fluid mech.*, 11:443–503, 1979.

[15] P. Burattini and R.A. Antonia. The effect of different x-wire calibration schemes on some turbulence statistics. *Exp. Fluids*, 38(1):80–89, 2005.

[16] T.L. Butler and J.W. Wagner. Application of a three-sensor hot-wire probe for incompressible flow. *AIAA J.*, 21:726–732, 1983.

[17] T.A. Casey, J. Sakakibara, and S.T. Thoroddsen. Scanning tomographic particle image velocimetry applied to a turbulent jet. *Phys. Fluids*, 25:025102, 2013.

[18] I.P. Castro and M. Dianat. Pulsed wire velocity anemometry near walls. *Exp. Fluids*, 8:343–352, 1990.

[19] K.T. Christensen. The influence of peak-locking errors on turbulence statistics computed from piv ensembles. *Exp. Fluids*, 36:484–497, 2004.

[20] S.H. Chue. Pressure probes for fluid measurement. *Prog. Aerosp. Sci.*, 16(2):147–223, 1975.

[21] DC Collis and M.J. Williams. Two-dimensional convection from heated wires at low Reynolds numbers. *J.Fluid Mech.*, 6:357–384, 1959.

[22] G. Comte-Bellot. Hot-wire anemometry. *Ann. Rev. Fluid Mech.*, 8:209–231, 1976.

[23] J. Czarske, L. Büttner, and T. Razik. Boundary layer velocity measurements by a laser-doppler profile sensor with micrometre spatial resolution. *Meas. Sci. Technol.*, 13:1979–1989, 2002.

[24] D.B. DeGraaff and J. K. Eaton. A high-resolution laser Doppler anemometer: design, qualification and uncertainty. *Exp. Fluids*, 30:522–530, 2001.

[25] F. DeSouza and S. Tavoularis. Hot-wire response in compressible subsonic flow. *AIAA J.*, 58(8):3332–3338, 2020.

[26] W.J. Devenport, G.P. Evans, and E.P. Sutton. A traversing pulsed-wire probe for velocity measurements near a wall. *Exp. Fluids*, 8:336–342, 1990.

[27] S. Discetti and F. Coletti. Volumetric velocimetry for fluid flows. *Meas. Sci. Technol.*, 29:042001, 2018.

[28] K. Doebbeling, B. Lenze, and W. Leuchel. Basic considerations concerning the construction and usage of multiple hot-wire probes for highly turbulent three-dimensional flows. *Meas. Sci. Technol.*, 1:924–933, 1990.

[29] H.L. Dryden and A.M. Kuethe. The measurement of fluctuations of air speed by the hot-wire anemometer. Technical Report NACA 581, National Advisory Committee for Aeronautics, Washington DC, 1929.

[30] F. Durst, A. Melling, and J.H. Whitelaw. *Principles and Practice of Laser Doppler Anemometry*. Academic Press, New York, 1976.

[31] R.V. Edwards. Report of the special panel on statistical particle bias problems in laser anemometry. *J. Fluids Eng.*, 109:89–93, 1987.

[32] G.E. Elsinga, F. Scarano, B. Wieneke, and B.W. van Oudheusden. Tomographic particle image velocimetry. *Exp. Fluids*, 41:933–947, 2006.

[33] J. R. Elsnab, J. P. Monty, C.M. White, M. Koochesfahani, and J.C. Klewicki. Efficacy of single-component MTV to measure turbulent wall-flow velocity derivative profiles at high resolution. *Exp. Fluids*, 58:128, 2017.

[34] R.J. Emrich, editor. *Methods of Experimental Physics, Vol. 18A and B: Fluid Dynamics*. Academic Press, New York, 1981.

[35] K.N. Everett, A.A. Gerner, and D.A. Durston. Seven-hole cone probes for high angle flow measurement: Theory and calibration. *AIAA J.*, 21:992–998, 1983.

[36] Y Fan, G Arwatz, TW Van Buren, DE Hoffman, and M Hultmark. Nanoscale sensing devices for turbulence measurements. *Experiments in Fluids*, 56(7):1–13, 2015.

[37] H.J.S. Fernando, M. Princevac, and R.J. Calhoun. Atmospheric measurements (chapter 17). In C. Tropea, A.L. Yarin, and J.F. Foss, editors, *Springer Handbook of Experimental Fluid Mechanics*. Springer, Berlin, 2007.

[38] R.W. Fox, A.T. McDonald, and J.W. Mitchell. *Fox and McDonald's Introduction to Fluid Mechanics (10th Edition)*. Wiley, New York, 2019.

[39] M.K. Fu, Y. Fan, and M. Hultmark. Design and validation of a nanoscale cross-wire probe (X-NSTAP). *Exp. Fluids*, 60(6):1–14, 2019.

[40] T. Fuchs, R. Hain, and C.J. Kähler. Non-iterative double-frame 2D/3D particle tracking velocimetry. *Exp. Fluids*, 58:119, 2017.

[41] T.J. Gieseke and Y.G. Guezennec. An experimental approach to the calibration and use of triple hot-wire probes. *Exp. Fluids*, 14:305–315, 1993.

[42] R.J. Goldstein. *Fluid Mechanics Measurements (2nd Edition)*. Taylor & Francis, Washington DC, 1996.

[43] S. Goldstein. A note on the measurement of total head and static pressure in a turbulent stream. *Proc. Roy. Soc. A*, 155:570–575, 1936.

[44] I. Grant. Particle image velocimetry: A review. *Proc. Instn. Mech. Engrs. Part C*, 211:55–76, 1997.

[45] R.L. Grothe and D. Dabiri. An improved three-dimensional characterization of defocusing digital particle image velocimetry (DDPIV) based on a new imaging volume definition. *Meas. Sci. Technol.*, 19:065402, 2008.

[46] M.S. Guellouz and S. Tavoularis. A simple pendulum technique for the calibration of hot-wire anemometers over low-velocity ranges. *Exp. Fluids*, 18:199–203, 1995.

[47] P.M. Handford and P. Bradshaw. The pulsed-wire anemometer. *Exp. Fluids*, 7:125–132, 1989.

[48] R.B. Hill and J.C. Klewicki. Data reduction methods for flow tagging velocity measurements. *Exp. Fluids*, 20:142–152, 1996.

[49] K.D. Hinsch. Three-dimensional particle velocimetry. *Meas. Sci. Technol.*, 6:742–753, 1995.

[50] K.D. Hinsch. Holographic particle image velocimetry. *Meas. Sci. Technol.*, 13:R61–72, 2002.

[51] J.O. Hinze. *Turbulence (2nd Edition)*. McGraw-Hill, New York, 1975.

[52] G.D. Huffman. Calibration of tri-axial hot-wire probes using a numerical search algorithm. *J. Phys. E: Sci. Instrum.*, 13:1177–1182, 1980.

[53] E.S. Johansen and O.K. Rediniotis. Unsteady calibration of fast-response pressure probes, part 1: Theoretical studies. *AIAA J.*, 43:816–826, 2005.

[54] E.S. Johansen and O.K. Rediniotis. Unsteady calibration of fast-response pressure probes, part 2: Water-tunnel experiments. *AIAA J.*, 43:827–834, 2005.

[55] E.S. Johansen and O.K. Rediniotis. Unsteady calibration of fast-response pressure probes, part 3: Air-jet experiments. *AIAA J.*, 43:835–845, 2005.

[56] E.S. Johansen, O.K. Rediniotis, and G. Jones. The compressible calibration of miniature multi-hole probes. *J. Fluids Eng.*, 123:128–138, 2001.

[57] P.R. Jonas and P.M. Kent. Two-dimensional velocity measurement by automatic analysis of trace particle motion. *J. Phys. E: Sci. Instrum.*, 12:604–609, 1979.

[58] B.G. Jones. A bleed-type pressure transducer for in-stream fluctuating static pressure sensing. *TSI Quarterly*, 7 (2):5–11, April-June 1981.

[59] C.J. Kähler and J. Kompenhans. Fundamentals of multiple plane stereo particle image velocimetry. *Exp. Fluids*, 29:S70–7, 2000.

[60] C.J. Kähler, U. Scholz, and J. Otmanns. Wall-shear-stress and near-wall turbulence measurements up to single pixel resolution by means of long-distance micro-PIV. *Exp. Fluids*, 41:327–341, 2006.

[61] J. Katz and J. Sheng. Applications of holography in fluid mechanics and particle dynamics. *Annu. Rev. Fluid Mech.*, 42:531–555, 2010.

[62] R.D. Keane and R.J. Adrian. Theory of cross-correlation analysis of PIV images. *Appl. Sci. Res.*, 49:191–215, 1992.

[63] L.V. King. On the convection of heat from small cylinders in a stream of fluid: Determination of the convection constants of small platinum wires with application to hot-wire anemometry. *Phil. Trans. Roy. Soc. London*, A214:373–432, 1914.

[64] T. Kobayashi, T. Ishihara, and N. Sasaki. Automatic analysis of photographs of trace particles by microcomputer system. In W.J. Yang, editor, *Flow Visualization III*, pages 231–235. Hemisphere Publishing Corp., Washington DC, 1985.

[65] M.M. Koochesfahani. Molecular tagging velocimetry (MTV): Progress and applications. In *Proc. 30th AIAA Fluid Dynamics Conference, 28 June--1 July 1999, Norfolk, VA, AIAA 99-3786*, 1999.

[66] G.J. Kostrzewsky and R.D. Flack. Simple system for fluid flow visualization and measurement using a chronophotographic technique. *Rev. Sci. Instrum.*, 57 (12):3066–3074, 1986.

[67] L.S.G. Kovasznay. Simple analysis of the constant temperature feedback hot-wire anemometer. Technical Report AERO/JHU CM-478, The Johns Hopkins University, Baltimore, MD, 1948.

[68] G. Kunkel, C. Arnold, and A.J. Smits. Development of NSTAP: Nanoscale thermal anemometry probe. In *36th AIAA Fluid Dynamics Conference and Exhibit*, page 3718, 2006.

[69] P. Kupferschmied, P. Koeppel, W. Gizzi, C. Roduner, and G. Gyarmathy. Time-resolved flow measurements with fast-response aerodynamic probes in turbomachines. *Meas. Sci. Technol.*, 11:1036–1054, 2000.

[70] J.M. Lawson and J.R. Dawson. A scanning PIV method for fine-scale turbulence measurements. *Exp. Fluids*, 55:1857, 2014.

[71] W.R. Lempert, K. Magee, P. Ronney, K.R. Gee, and R.P. Haughland. Flow tagging velocimetry in incompressible flow using photo-activated nonintrusive tracking of molecular motion (PHANTOMM). *Exp. Fluids*, 18:249–257, 1995.

[72] J.D. Li, B.J. McKeon, W. Jiang, J.F. Morrison, and A.J. Smits. The response of hot wires in high Reynolds-number turbulent pipe flow. *Meas. Sci. Technol.*, 15:789–798, 2004.

[73] P.M. Ligrani, B.A. Singer, and L.R. Baun. Miniature five-hole pressure probe for measurement of three mean velocity components in low-speed flows. *J. Phys. E: Sci. Instrum.*, 22:868–876, 1989.

[74] R. Limbourg and J. Nedić. An extension to the universal time scale for vortex ring formation. *Journal of Fluid Mechanics*, 915:A46, 2021.

[75] R. Limbourg and J. Nedić. Formation of an orifice-generated vortex ring. *Journal of Fluid Mechanics*, 913:A29, 2021.

[76] C.G. Lomas. *Fundamentals of Hot Wire Anemometry*. Cambridge University Press, Cambridge, UK, 1986.

[77] K.T. Lowe and R.L. Simpson. An advanced laser-Doppler velocimeter for full-vector particle position and velocity measurements. *Meas. Sci. Tech.*, 20:045402, 2009.

[78] R.M. Lueptow, K.S. Breuer, and J.H. Haritonidis. Computer-aided calibration of x-probes using a look-up table. *Exp. Fluids*, 6(2):115–118, 2004.

[79] F.A. Macmillan. Experiments on pitot tubes in shear flow. Technical Report Reports and Memoranda No. 3028, Aeronautical Research Council, London, 1957.

[80] F. Mayinger and O. Feldmann, editors. *Optical Measurements (2nd Edition)*. Springer, Berlin, 2001.

[81] B.J. McKeon, J. Li, W. Jiang, J.F. Morrison, and A.J. Smits. Pitot probe corrections in fully developed turbulent pipe flow. *Meas. Sci. Technol.*, 14(8):1449, 2003.

[82] B.J. McKeon et al. Velocity, vorticity and Mach number (chapter 5). In C. Tropea, A.L. Yarin, and J.F. Foss, editors, *Springer Handbook of Experimental Fluid Mechanics*. Springer, Berlin, 2007.

[83] D.K. McLaughlin and W.G. Tiedermann. Biasing correction for individual realization of laser anemometer measurements in turbulent flow. *Phys. Fluids*, 16:2082–2088, 1973.

[84] L Mendelson and A.H. Techet. Quantitative wake analysis of a freely swimming fish using 3D synthetic aperture PIV. *Exp. Fluids*, 56:135, 2015.

[85] C. Mercer. *Optical Metrology for Fluids, Combustion and Solids*. Kluwer Academic Publishers, Dordrecht, 2003.

[86] D. Michaelis, D. R. Neal, and B. Wieneke. Peak-locking reduction for particle image velocimetry. *Meas. Sci. Technol.*, 27:104005, 2016.

[87] J. A. Mullin and W. J. A. Dahm. Dual-plane stereo particle image velocimetry (DSPIV) for measuring velocity gradient fields at intermediate and small scales of turbulent flows. *Exp. Fluids*, 38:185–196, 2005.

[88] J. Nedić, B. Ganapathisubramani, J.C. Vassilicos, J. Borée, L-E. Brizzi, and A Spohn. Aeroacoustic performance of fractal spoilers. *AIAA J.*, 50:2695–2710, 2012.

[89] M. Novara, D. Schanz, N. Reuther, C. J. Kähler, and A. Schröder. Lagrangian 3D particle tracking in high-speed flows: Shake-The-Box for multi-pulse systems. *Exp. Fluids*, 57:128:1–20, 2016.

[90] E. Ossofsky. Constant temperature operation of the hot-wire anemometer at high frequency. *Rev. Sci. Instrum.*, 19:881–889, 1948.

[91] E. Ower and R.C. Pankhurst. *The Measurement of Air Flow*. Pergamon Press, Oxford, 1977.

[92] A.E. Perry. *Hot-Wire Anemometry*. Clarendon Press, Oxford, 1982.

[93] A. Quarmby and H.K. Das. Displacement effects on pitot tubes with rectangular mouths. *Aeron. Quart.*, 20:129–139, 1969.

[94] M. Raffel, C.E. Willert, F. Scarano, C.J. Kähler, S.T. Wereley, and J. Kompenhans. *Particle Image Velocimetry: A Practical Guide (3rd Edition)*. Springer, 2018.

[95] V. Ramakrishman and O.K. Rediniotis. Development of a 12-hole onmidirectional flow-velocity measurement probe. *AIAA J.*, 45:1430–1432, 2007.

[96] O.K. Rediniotis and R. Vijayagopal. Miniature multihole pressure probes and their neural-network-based calibration. *AIAA J.*, 37:666–674, 1999.

[97] H. Rubinsztein-Dunlop, B. Littleton, P. Barker, P. Ljungberg, and Y. Malmsten. Ionic strontium fluorescence as a method for flow tagging velocimetry. *Exp. Fluids*, 30:36–42, 2001.

[98] R. Sadr and J.C. Klewicki. A spline-based technique for estimating flow velocities using two-camera multi-line MTV. *Exp. Fluids*, 35:257–261, 2003.

[99] M. Samet and S. Einav. Directional pressure probe. *Rev. Sci. Instrum.*, 55:582–588, 1984.

[100] M. Samet and S. Einav. A hot-wire technique for simultaneous measurement of instantaneous velocities in 3D flows. *J. Phys. E: Sci. Instrum.*, 20:683–690, 1987.

[101] M. Samimy and S. Lele. Motion of particles with inertia in a compressible free shear layer. *Phys. Fluids A*, 3:1915–1923, 1991.

[102] F. Scarano. Iterative image deformation methods in PIV. *Meas. Sci. Technol.*, 13:R1–R19, 2002.

[103] F. Scarano. Tomographic PIV: principles and practice. *Meas. Sci. Technol.*, 24:012001, 2013.

[104] D. Schanz, S. Gesemann, and A. Schröder. Shake-The-Box: Lagrangian particle tracking at high particle image densities. *Exp. Fluids*, 57:70:1–27, 2016.

[105] S. Shaw-Ward, S.C. McParlin, P. Nathan, and D.M. Birch. Optimal calibration of directional velocity probes. *AIAA J.*, 56:2594–2603, 2018.

[106] S. Shaw-Ward, A. Titchmarsh, and D.M. Birch. The calibration and use of n-hole velocity probes. *AIAA J.*, 53:336–346, 2015.

[107] K.A. Shinpaugh and R.L. Simpson. A rapidly scanning two-velocity component laser doppler velocimeter. *Meas. Sci. Technol.*, 6:690–701, 1995.

[108] K. Shirai, T. Pfister, L. Büttner, J. Czarske, H. Müller, S. Becker, H. Lienhart, and F. Durst. Highly spatially resolved velocity measurements of a turbulent channel flow by a fiber-optic heterodyne laser-Doppler velocity-profile sensor. *Exp. Fluids*, 40:473–481, 2006.

[109] T.E. Siddon. On the response of pressure measuring instrumentation in unsteady flow. Technical Report 136, University of Toronto Institute for Aerospace Studies, Toronto, 1969.

[110] N. Sitaram, B. Lakshminarayana, and A. Ravindranath. Conventional probes for the relative flow measurement in a turbomachinery rotor blade passage. *J. Eng. for Power*, 103:406–414, 1981.

[111] B.L. Smith and D.R. Neal. Particle image velocimetry (chapter 48). In R.W. Johnson, editor, *Handbook of Fluid Dynamics (2nd Edition)*. CRC Press, Boca Raton, FL, 2016.

[112] A.J. Smits and T.T. Lim, editors. *Flow Visualization Techniques and Examples*. Imperial College Press, London, 2000.

[113] E.F.C. Somerscales. Fluid velocity measurement by particle tracking. In R.B. Dowdell, editor, *Flow – Its Measurement and Control in Science and Industry*, pages 795–808. Instrument Society of America, Pittsburgh, PA, 1974.

[114] B.W. Spencer and B.G. Jones. A bleed-type pressure transducer for in-stream measurement of static pressure fluctuations. *Rev. Sci. Instru.*, 42 (4):450–454, 1971.

[115] S. Tavoularis. Techniques for turbulence measurement. In N.P. Cheremisinoff, editor, *Encyclopedia of Fluid Mechanics*, volume 1, pages 1207–1255. Gulf Publishing Company, Houston, TX, 1986.

[116] A.L. Treaster and H.E. Houtz. Fabricating and calibrating five-hole probes. In R.A. Bajura and M.L. Billet, editors, *Proc. Fluid Measurements and Instrumentation Forum, AIAA/ASME 4th Fluid Mechanics, Plasma Dynamics and Lasers Conference*, pages 1–4, Atlanta, GA, 1986.

[117] A.L. Treaster and A.M. Yocum. The calibration and application of five-hole probes. *ISA Trans.*, 18 (3):23–34, 1979.

[118] P. Vukoslavcevic and J.M. Wallace. A 12-sensor hot-wire probe to measure the velocity and vorticity vectors in turbulent flow. *Meas. Sci. Technol.*, 7:1451–1461, 1996.

[119] J.M. Wallace and J. F. Foss. The measurement of vorticity in turbulent flows. *Annu. Rev. Fluid Mech.*, 27:469–514, 1995.

[120] B.M. Watrasiewitz and M.J. Rudd. *Laser Doppler Measurements*. Butterworths, Sydney, 1976.

[121] S.T. Wereley and C.D. Meinhart. Recent advances in micro-particle image velocimetry. *Annu. Rev. Fluid Mech.*, 42:557–576, 2010.

[122] M.P. Wernet. Stereo viewing 3-component, planar PIV utilizing fuzzy inference, 1006. AIAA-96-2268, June 17–20.

[123] J. Westerweel. Fundamentals of digital particle image velocimetry. *Meas. Sci. Technol.*, 8:1379–1392, 1997.

[124] J. Westerweel, G.E. Elsinga, and R.J. Adrian. Particle image velocimetry for complex and turbulent flows. *Annu. Rev. Fluid Mech.*, 45:409–436, 2013.

[125] C.E. Willert. High-speed particle image velocimetry for the efficient measurement of turbulence statistics. *Exp. Fluids*, 56, 2015.

[126] F.A.L. Winternitz. Simple shielded total-pressure probes. *Aircraft Eng.*, pages 313–317, October 1958.

[127] K.S. Wittmer, W.J. Devenport, and J.S. Zsoldos. A four-sensor hot-wire probe system for three-component velocity measurement. *Exp. Fluids*, 24:416–423, 1998.

[128] J.C. Wyngaard. Cup, propeller, vane and sonic anemometers in turbulence research. *Ann. Rev. Fluid Mech.*, 13:399–423, 1981.

[129] H. Yeh and H.Z. Cummins. Localized fluid flow measurements with a He-Ne spectrometer. *Appl. Phys. Lett.*, 4:178, 1964.

[130] M.V. Zagarola and A.J. Smits. Mean-flow scaling of turbulent pipe flow. *J. Fluid Mech.*, 373:33–79, 1998.

[131] G.G. Zilliac. Calibration of seven-hole pressure probes for use in fluid flows with large angularity. Technical Report NASA Technical Memorandum 102200, NASA Ames Research Center, Moffet Field, CA, 1989.

13 Measurement of Forces and Wall Shear Stress

Forces and moments applied on an object or a surface by a fluid are collectively referred to as *loads* and are of great importance in many engineering applications. Examples of loading are the hydrostatic force on a dam wall, the lift and drag force on moving vehicles and other immersed objects, and the aerodynamic moments on maneuvering aircraft. This chapter discusses devices and measurement techniques that have been developed to determine these loads, either by measuring the load itself directly or by determining the load indirectly from its relationship to pressure or velocity measurements. In addition, it discusses the various available methods for the measurement of wall shear stress, from which one can determine the friction factor for internal flows and the skin friction coefficient for external ones.

13.1 Strain Gauges

13.1.1 Strain Gauge Description

The strain gauge is a versatile sensor, which, when bonded to a solid object, is able to sense strain induced by tensile, compressive, bending or torsional loads [3, 13, 29, 61]. In simple terms, a strain gauge converts a strain in its axial direction to a change in its electrical resistance. Strain gauges are the main components of force and torque transducers, which shall be discussed in the following section.

A strain gauge consists of a long and thin electrical conductor, folded over several times in a 'grid' pattern, as shown in Fig. 13.1. Compared to a single, straight conductor length with the same resistance, the folded gauge is much shorter, and therefore has a much higher spatial resolution of strain measurement. Strain gauges are available in the form of foils, with the conductor grid attached to a *carrier matrix*, which is bonded with an adhesive upon the object, the strain of which they measure. The most common strain gauge materials are metallic (e.g., a chrome-nickel alloy), but there are also semiconductor (e.g., silicon or germanium) sensors. Typical values of the unloaded resistance of strain gauges are in the range between 120 and 1,000 Ω (with the values 120 and 350 Ω being the most common) for metallic sensors and in the range between 1,000 and 5,000 Ω for semiconductor sensors. Strain gauges are available with lengths between 0.2 and 100 mm. Semiconductor strain gauges have a much higher sensitivity than metallic ones and a much higher frequency response (up to several hundreds of Hz), so that they are more suitable for measuring dynamic loads. Adversely, semiconductor strain

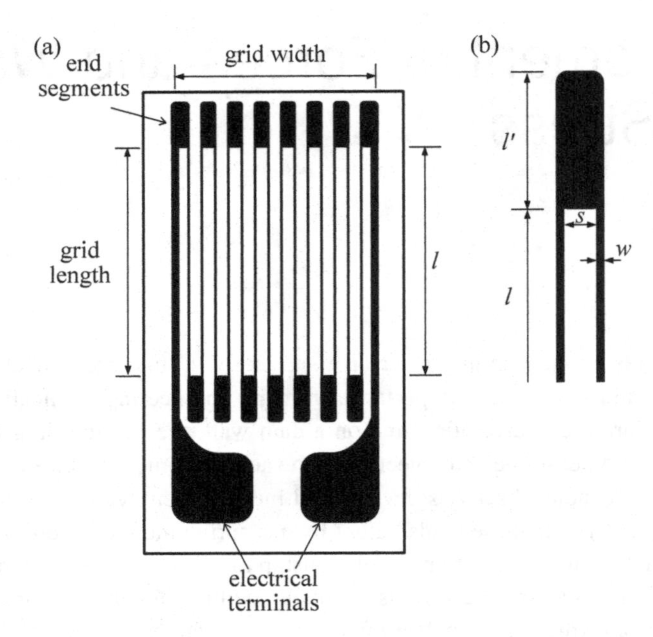

Figure 13.1 Sketches (not to scale) of a bonded, grid-type, strain gauge; the thickness of the active conductor segments (normal to the shown plane) is h.

gauges are highly sensitive to temperature, have a non-linear response and are difficult to compensate.

An important parameter that characterises strain gauges is the *gauge factor*, defined as the ratio of the relative change of its resistance $\delta R/R$ and the strain ε_A along its longitudinal axis, namely, as

$$G = \frac{\delta R/R}{\varepsilon_A} . \tag{13.1}$$

The static sensitivity of a system or device was defined in Section 2.3.2 to be the slope of the output–input relationship. Considering that the input of a strain gauge is the strain and its output is the change in resistance, one finds that the gauge factor of a strain gauge is proportional to its static sensitivity. For metallic strain gauges, $G \approx 2$, whereas, for semiconductor strain gauges, G is of the order of 100.

13.1.2 Principle of Operation

Straight conductor under axial loading: Consider an electrically conducting object with a length L and a rectangular cross section having a width w and a height h, thus a cross-sectional area $A = wh$, as shown in Fig. 13.2. When this object is loaded by a purely axial force F_x, it will have a purely axial stress $\sigma_x = F_x/A$ and will deform, in the elastic range, shown as dashed lines in Fig. 13.2. First of all, its length will change by δL, which may be expressed in terms of the *axial strain* ε_x (or ε_A) as

$$\delta L = \varepsilon_x L . \tag{13.2}$$

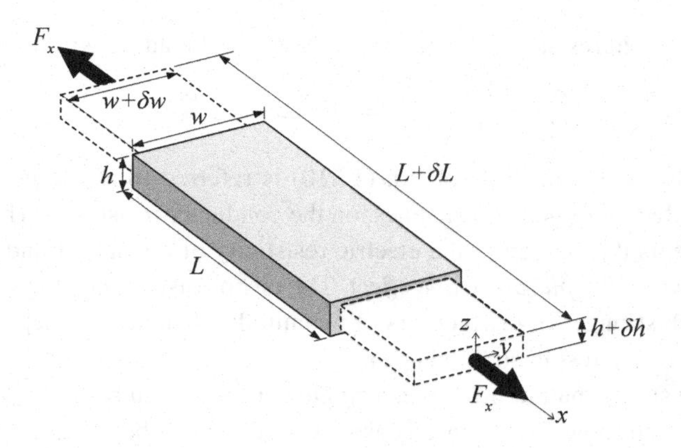

Figure 13.2 An electrical conductor with a length L, a width w and a height h, outlined by solid lines, and its deformed shape under an axial load F_x, outlined by dashed lines.

The axial strain is related to the axial stress, thus, the axial load, as

$$\varepsilon_x = \sigma_x/E = F_x/(EA)\,, \tag{13.3}$$

where E is *Young's modulus of elasticity*. Axial loading will also result in a transverse deformation of the object (*Poisson effect*), so that its width and height will change by, respectively, δw and δh, which may be expressed in terms of the corresponding *transverse strains* ε_y and ε_z (equal to each other and both denoted by ε_T), as

$$\delta w = \varepsilon_y w \quad\text{and}\quad \delta h = \varepsilon_z h\,. \tag{13.4}$$

The *Poisson ratio*

$$\nu = -\varepsilon_T/\varepsilon_A \tag{13.5}$$

is a property of the material and is readily available for common materials (e.g., $\nu \approx 0.31$ for nickel and $\nu \approx 0.44$ for silicon). We may assume that the deformation is very small, so that $\delta L/L, \delta w/w, \delta h/h \ll 1$. Consequently, the cross-sectional area of the object will deform by an amount

$$\delta A = (w - \delta w)(h - \delta h) - wh \approx -\delta w\, h - \delta h\, w\,, \tag{13.6}$$

and the relative change in cross-sectional area will be

$$\frac{\delta A}{A} \approx \frac{\delta w}{w} + \frac{\delta h}{h} = 2\varepsilon_T = -2\nu\varepsilon_A\,. \tag{13.7}$$

The electric resistance of this object is given by *Pouillet's law* as

$$R = \frac{\rho L}{A}\,, \tag{13.8}$$

where ρ is the *electric resistivity* (also referred to as *specific electric resistance*) of its material. The deformation of the object would also cause a change in electrical resistivity, which would similarly be small, that is, $\delta\rho \ll \rho$. The resistance change of this object will be

$$\delta R = \frac{(\rho + \delta\rho)(L + \delta L)}{A + \delta A} - \frac{\rho L}{A} \approx \frac{A\rho\delta L - L\rho\delta A + LA\delta\rho}{A^2} \tag{13.9}$$

and the relative change in resistance of the object under axial loading will be

$$\frac{\delta R}{R} \approx \left(\frac{\delta L}{L} - \frac{\delta A}{A}\right) + \frac{\delta\rho}{\rho} = (1 + 2\nu)\varepsilon_A + \frac{\delta\rho}{\rho} . \tag{13.10}$$

The first term on the right-hand side of Eq. (13.10) is referred to as the *geometrical effect*, namely the effect of dimension changes, on the conductor resistance. The second term, which represents the change in the electric resistivity of a material under an applied strain, is known as the *piezoresistive effect.* The piezoresistive effect is very weak for metals, but much stronger (by two orders of magnitude) than the geometrical effect for semiconductors of interest in strain gauges.

Equation (13.10) shows that, in principle, a single conductor can be used to measure axial strain. In practice, however, such a sensor would need to be extremely long, if it were to have a resistance change that is within a measurable range. To increase the conductor resistance, commercially available strain gauges have a grid arrangement, usually manufactured by a printed-circuit procedure.

Object under three-dimensional loading: Now assume that the loading of the object is three-dimensional, so that it has three normal stresses $\sigma_x, \sigma_y, \sigma_z$. Then, it will also have three strains, which, in the elastic range, will be related to the stresses by the expressions

$$\begin{aligned}
\varepsilon_x &= \frac{1}{E}[\sigma_x - \nu(\sigma_y + \sigma_z)], \\
\varepsilon_y &= \frac{1}{E}[\sigma_y - \nu(\sigma_x + \sigma_z)], \\
\varepsilon_z &= \frac{1}{E}[\sigma_z - \nu(\sigma_x + \sigma_y)] .
\end{aligned} \tag{13.11}$$

Therefore, the load can be determined from measurements of the three strains.

Bonded strain gauge: A representative, bonded-foil type, strain gauge is shown in Fig. 13.1. Its conducting element consists of four parts: (a) a grid of $n - 2$ straight, inner parallel segments, each having a length l, a width w and separated from each other by a distance s; for the device shown in the figure, $n = 16$; (b) two outer parallel segments, having a width $w' > w$; (c) $n - 1$ end segments, which connect the ends of the parallel segments in an in-series configuration; the end segments are made so that they have a length (in the direction of current) $2w + s \ll l$ and a width $l' > w$; and (d) two short and wide end pieces that connect the outer parallel segments to the electrical terminals. The total resistance R of the conducting material between the two terminals consists of four parts, corresponding to the previously identified segments of the stain gauge conductor. However, under the made assumptions concerning the dimensions of the different segments, one may, at first pass, approximate the total resistance by the resistance of the $n - 2$ inner parallel segments, as if the total length of the conductor were $L \approx (n - 2)l$. Then, we may approximate the relative change of the strain gauge resistance using Eq. (13.10).

The literature on strain gauges does not contain a clear and universally accepted description of the physical mechanism of transmission of the strains in a specimen to the conductor in a strain gauge. There are concerns that such a mechanism is affected by differences in the mechanical properties of the adhesive, the matrix and the conductor. The limitation of a theoretical understanding of this mechanism is not detrimental to the practical use of strain gauges, because the response of specific sensors can be determined by calibration. It has been established that strain gauges of the type shown in Fig. 13.1 are mainly sensitive to a specimen strain in the direction of the parallel segments, which we referred to as the axial strain ε_A. This strain can be determined from a measured resistance change with the use of Eq. (13.1), where the value of the gauge factor G is specified by the manufacturer of the particular product. Besides being sensitive to the axial strain, strain gauges may also be sensitive to the transverse strain, albeit usually by a much smaller amount. Transverse strain sensitivity is expressed by the *transverse gauge factor*

$$G_T = \frac{(\delta R/R)_T}{\varepsilon_T} . \qquad (13.12)$$

The ratio G_T/G, referred to as the *transverse sensitivity factor*, is small, typically in the range from 0 to 10%, and may also be specified by the manufacturer. Methods to correct strain-gauge measurements for transverse strain sensitivity are available [59].

13.1.3 Strain Gauge Configurations and Sensitivity

Connection to the bridge: Strain gauges are customarily connected to Wheatstone bridges (see Section 2.2.3), which convert a change in resistance to a voltage signal. Typically, a strain gauge is placed in one or more of the four arms of a classical Wheatstone bridge. If only one arm is occupied by a strain gauge, the set-up is said to have a *quarter-bridge configuration*, whereas, if all four arms are occupied by strain gauges, the set-up is said to have a *full-bridge configuration*.

Arrangement on the specimen: Strain gauges are also arranged upon the test specimen in different configurations, each of which will make the set-up sensitive to loads acting in particular directions. Moreover, the order in which the strain gauges are connected to the Wheatstone bridge also plays an important role in the directional sensitivity of the configuration, as well as the bridge sensitivity. Representative cases of both the strain gauge arrangement on a specimen and their connection to the bridge are shown in Fig. 13.3. According to the definition of static sensitivity in Section 2.3.2, the static sensitivity of a strain gauge bridge configuration is $\mathrm{d}V_o/\mathrm{d}\varepsilon$. Instead of this property, however, it is customary to use the *normalised bridge sensitivity*, defined as

$$S_B = \left| \frac{\mathrm{d}V_o}{\mathrm{d}\varepsilon} \frac{1}{GV} \right| , \qquad (13.13)$$

where V is the voltage supplied to the top of the bridge. The load directions in which the configurations shown in Fig. 13.3 are sensitive, and the corresponding normalised bridge sensitivities, are listed in Table 13.1. A load applied in the x direction would produce tension or compression, whilst a load in the z direction would produce bending. In general, the larger the number of strain gauges is, the higher the sensitivity of the configuration will be. A method by which the theoretical bridge sensitivity of the set-up can be determined is presented in the following example.

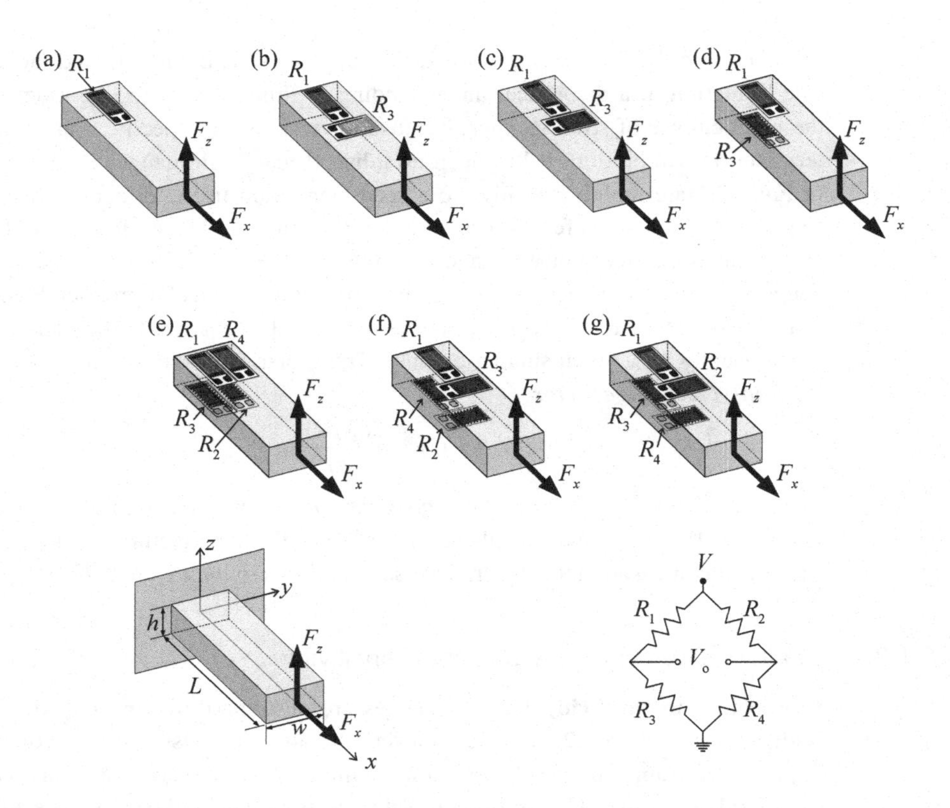

Figure 13.3 (a)–(g) Common strain gauge arrangements on a specimen. Dashed contours mark strain gauges that are connected to the bottom surface. Grey fill in (b) denotes dummy gauges, which are placed on the specimen, but are not bonded to it. The bottom left sketch shows a beam attached at its base on a wall, the coordinate system and the beam dimensions, all applicable to all configurations. The bottom right sketch shows a standard Wheatstone bridge with strain gauges and/or other resistors connected to its arms.

Table 13.1 Load direction and normalised bridge sensitivity for the strain gauge configurations shown in Fig. 13.3

Configuration	F_x	F_z	Temperature compensation	Normalised bridge sensitivity S_B
a	✓	✓	No	0.25
b	✓	✓	Yes	0.25
c	✓	✓	Yes	$0.25(1+\nu)$
d	X	✓	Yes	0.5
e	X	✓	Yes	1
f	X	✓	Yes	$0.5(1+\nu)$
g	✓	X	Yes	$0.5(1+\nu)$

Example 13.1 Determine the voltage output and the sensitivity of the strain gauge set-up shown in Fig. 13.3d, when the test specimen is subjected to loads in the x and z

directions. Assume that, in the absence of a load, all resistances in the bridge, including those of the two strain gauges, are equal to R, so that the unloaded bridge ratio is $k_B = 1$.

Answer

We first recall from Section 2.2.3 that the voltage output of the Wheatstone bridge shown in Fig. 13.3 is

$$V_o = V\left(\frac{R_3}{R_1 + R_3} - \frac{R_4}{R_2 + R_4}\right) = V\left(\frac{R_3}{R_1 + R_3} - \frac{1}{2}\right) .$$

When the specimen is loaded, the resistances of the two strain gauges will change, respectively, by δR_1 and δR_3, which will be positive, when the corresponding strain gauge is under tension, and negative, when it is under compression. Let us first assume that a force is applied in the positive x direction. Both strain gauges will experience a tensile strain, and their resistances would increase by equal amounts δR. We can, therefore, write that $R_1 = R_3 = R + \delta R$. Hence, the first term on the right hand side is also equal to $1/2$, and, therefore, the voltage output will be zero. The same would be true, if a compressive load in the x direction were applied. Therefore, this configuration is insensitive to forces applied in the x direction.

Now, consider that a load is applied in the positive z direction, causing the specimen to bend. The strain gauge at the top would experience a compressive strain, so that $R_1 = R - \delta R$, whilst the strain gauge at the bottom would experience a tensile strain, so that $R_3 = R + \delta R$. The voltage output would, therefore, be

$$\frac{V_o}{V} = \frac{R + \delta R}{R + \delta R + R - \delta R} - \frac{1}{2} = \frac{1}{2}\frac{\delta R}{R} ,$$

from which, with the use of Eq. (13.1), one finds

$$V_o = \frac{1}{2} G \varepsilon_A V .$$

The normalised bridge sensitivity for a strain gauge with a gauge factor G would, therefore, be

$$S_B = \left|\frac{\mathrm{d}V_o}{\mathrm{d}\varepsilon_A}\frac{1}{GV}\right| = \frac{1}{2} .$$

Temperature compensation: A common source of error in strain gauge measurements is a change in temperature, because this would cause a change δR_Θ in the resistance of the strain gauge, in addition to the change δR caused by the strain. For small temperature differences $\delta T = T - T_r$ from a reference temperature T_r, we may estimate the relative resistance change at a temperature T as

$$\delta R_\Theta / R \approx \alpha_r \delta T , \tag{13.14}$$

where α_r is the thermal resistivity coefficient of the strain gauge material at the reference temperature.

The quarter-bridge configuration is particularly vulnerable to temperature changes. Temperature effects can be compensated for, at least partially, with the use of an even number of identical strain gauges, as indicated in Table 13.1. As an example, consider the two temperature-compensated configurations shown in Fig. 13.3b and c. In both configurations, the first gauge is bonded to the specimen and connected to the bridge as R_1. This gauge experiences strain, as well as having a sensitivity to temperature. In Fig. 13.3b, a second, identical, gauge, which is connected as R_3, is only bonded to the specimen at the edge of the matrix, so that the active part is loose. The second gauge would experience the same change in resistance, as the first one, due to a temperature change, but no change in resistance due to strain. When used in this manner, the strain gauge is referred to as a *dummy gauge*. Connection to the bridge compensates the temperature sensitivity of the first gauge and measures strain only. If the second strain gauge is also fully bonded, as is the case in Fig. 13.3c, then one obtains increased bridge sensitivity, as well as temperature compensation. It is noted that the use of a dummy gauge in the two previous configurations is rather obsolete. However, there are other configurations, in which the use of dummy gauges allows the user to compensate for temperature effects, as well as allow for the configuration to be sensitive to loads in certain directions; an example of such a configuration is the one shown in Fig. 13.3e, with R_2 and R_3 replaced by dummy gauges.

Example 13.2 Show that the strain gauge configuration shown in Fig. 13.3c has a normalised bridge sensitivity $S_B = 0.25(1 + \nu)$, and that it compensates for temperature effects. Assume that all resistances in the bridge, with the unloaded specimen at the reference temperature, are equal to R, so that the reference bridge ratio is $k_B = 1$.

Answer

Let us consider the case of an axial load on the specimen, which means that a force F_x is applied in the x direction; this force has a positive value, if tensile, and a negative value, if compressive. The strain gauge connected at location R_1 would experience a change in resistance, as the result of an axial strain ε_A and a change in temperature δT, so that

$$R_1 = R(1 + G\varepsilon_A + \alpha_r \delta T) .$$

The strain gauge connected at location R_3 is mounted such that its axial direction is perpendicular to the direction of the applied load, and is hence sensitive to the transverse strain $\varepsilon_T = -\nu\varepsilon_A$, which is created by the applied axial load; in addition, it is sensitive to the temperature change, so that

$$R_3 = R(1 - \nu G\varepsilon_A + \alpha_r \delta T) .$$

Assuming that all resistance changes are very small, one may determine the voltage output of the Wheatstone bridge as

$$V_o \approx \left(\frac{R_3}{R_1 + R_3} - \frac{1}{2} \right) V = \frac{-G\varepsilon_A(1+\nu)V}{4 + 4\alpha_r\delta T + 2k\varepsilon_A(1-\nu)} \approx -0.25G\varepsilon_A(1+\nu)V ,$$

which shows that the bridge sensitivity is $S_B \approx 0.25(1 + \nu)$. It is important to note that, although the resistances of both strain gauges remain sensitive to temperature changes, the bridge output does not depend on the temperature change, which proves that this configuration has a temperature compensation.

13.2 Load Cells and Torque Transducers

The measurement of forces and torques on laboratory models is essential in a wide variety of engineering applications, including the estimation of aerodynamic forces on road vehicles and aircraft and the torque produced by a wind turbine. The present section discusses the main conceptual and practical considerations associated with measurements of forces and torques in wind and water tunnels.

13.2.1 Load Cells

The term *load cell* describes a family of transducers that convert an applied force into a voltage signal, typically through an elastic mechanical deformation of an element. Load cells may be configured in different designs, with a main design objective being to create the largest possible deformation for a given load, which, in turn, would produce the largest load cell sensitivity $\mathrm{d}V_o/\mathrm{d}F$. They are typically characterised by the mechanism that converts deformation into a voltage signal. Strain gauge, capacitive and pneumatic/hydraulic load cells are the most common types.

Stain gauge load cells: These have a relatively low cost and find extensive use as both commercial and custom-built load cells. Custom-built load cells in a research laboratory commonly consist of a cantilever beam onto which a test model, for example, a wing, is attached; following application of a force on the model, strain gauges, mounted onto the sides of the beam, respond to the beam deformation (i.e., strain) and, when connected to a bridge, produce a convenient voltage output. Typical strain gauge load cell arrangements on beams are shown in Fig. 13.3.

The commercial version of the cantilever beam load cell (see Fig. 13.3, bottom left sketch) is known as a *bending beam load cell.* Two other common types of commercial load cells are shown in Fig. 13.4. *Compressive/tensile load cells*, shown schematically in Fig. 13.4a, consist of two parallel mounting plates that are connected by a column, which is either solid (with a square or circular cross section) or, more commonly, a hollow tube. When a force is applied on a plate, the column deforms, resulting in strain, which produces a voltage output via the strain gauges connected to a bridge. As the name suggests, this type of load cell is only capable of measuring compressive and tensile loads. The strain of the column is $\varepsilon = F_x/(EA)$, where F_x is the applied axial force component, E is Young's modulus and A is the solid part of the cross-sectional area of the column. An alternative design of compressive/tensile load cells contains a diaphragm supported around its edge, onto which strain gauges are attached; when a compressive/tensile force is applied, the diaphragm gets deflected, resulting in a measurable strain.

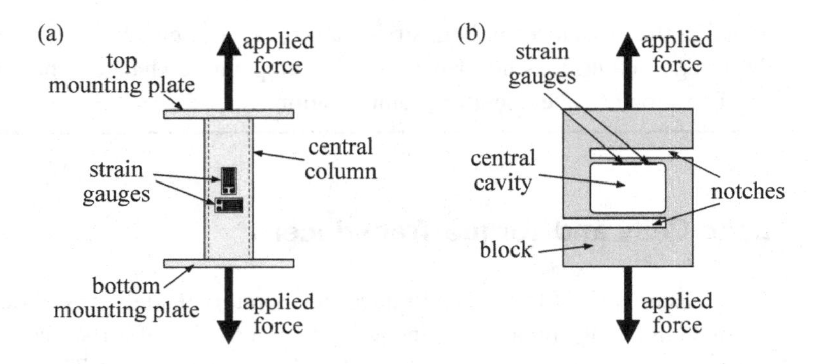

Figure 13.4 Sketches of (a) a compressive/tensile load cell and (b) an S-type load cell.

The *S-type* load cell consists of a metallic block, from which material has been removed to form two transverse notches and a central cavity, as shown in Fig. 13.4b. Under a given load, an S-type column would have a smaller stiffness and a larger axial deformation than a solid block. The stiffness of a block is defined as $k = F_x/\delta L = F_x/(L\varepsilon)$, where L is the length of the block. Strain gauges are placed in the central cavity of the block, where the strain due to an applied load is largest. The central cavity is covered, so that the strain gauges are protected.

Example 13.3 Determine the relationship between the voltage output V_o of the cantilever beam strain gauge shown in Fig. 13.3d and the transverse point force F_z at the free end of the beam. Assume that the bridge ratio is $k_B = 1$.

Answer

Assume that the beam has a width w, a thickness h, a length L, a Young's modulus E and a moment of inertia I. Further assume that the beam is initially straight and then a transverse force F_z is applied at its free end. From elementary solid mechanics, we know that the maximum axial stress is at the base of the beam and is equal to $\sigma_{x,\max} = hLF_z/(2I)$. As the only normal stress experienced by the beam is the axial one, Eq. (13.11) shows that the maximum axial strain on the beam is $\varepsilon_{x,\max} = hLF_z/(2EI)$. The strain decreases linearly from the base of the beam towards the free end, as $\varepsilon_x(x) = \varepsilon_{x,\max}(1 - x/L)$. Therefore, a strain gauge, placed with its axis parallel to the beam axis, would undergo a change in resistance that is proportional to the average axial strain along its active length l. This average strain is

$$\varepsilon_{x,\mathrm{av}} = \frac{1}{l}\int_{x_0-l/2}^{x_0+l/2} \varepsilon_x(x)\mathrm{d}x = \varepsilon_{x,\max}(1 - x_o/L)\,,$$

where x_o is the distance along the beam where the gauge midpoint is placed, assuming that the origin of the coordinate system is at the base on the beam. Given that $k_B = 1$, we use the result presented in Example 13.1, and find that the voltage output is proportional to the applied transverse force, namely, that

$$V_o = V\left(\frac{hL}{4EI}\right)\left(1 - \frac{x_o}{L}\right)GF_z .$$

The smallest value x_o can take is $l' + l/2$ (see Fig. 13.1), which would occur if the edge of the strain gauge were mounted flush with the base of the beam. For a given applied transverse force, the voltage output would increase with decreasing strain gauge size.

Example 13.4 Select dimensions and material properties for the cantilever beam load cell discussed in Example 13.3, so that its static sensitivity is as large as possible.

Answer

The sensitivity of the cantilever beam load cell discussed in Example 13.3 is $\mathrm{d}V_o/\mathrm{d}F_z = VhLG(1 - x_o/L)/(4EI)$. For a beam with a rectangular cross section, having a width w and a height h, as shown in Fig. 13.2, one gets $I = w^3h/12$, and, therefore, $\mathrm{d}V_o/\mathrm{d}F_z = 3VLG(1 - x_o/L)/(Ew^3)$, which demonstrates that the load cell sensitivity is independent of the beam thickness, but depends strongly (following an inverse cubic power law) on the beam width. For a given power supply voltage V on the bridge, and with the strain gauge mounted as closely to the base as possible, one may increase the sensitivity by increasing the value of the parameter $L/(Ew^3)$. First of all, it is evident that one must choose a beam material with as small a Young's modulus as possible. The remaining choices concern only the geometrical dimensions L, w of the beam. These should be chosen as to conform with practical considerations related to a specific experimental set-up, as well as to meet a material constraint: the axial stress at the base of the beam under the largest foreseeable load $F_{z,\mathrm{max}}$ should remain sufficiently smaller than the yield stress σ_{ys} of the material, namely, $\sigma_{\mathrm{max}} = hLF_{z,\mathrm{max}}/(2I) = 6LF_{z,\mathrm{max}}/w^3 < \sigma_{ys}$. We may then choose a value of the beam length L that is compatible with practical considerations, which leaves us with a lower bound for the beam width, which should not be smaller than the width of the strain gauge. The lower bound of the beam thickness h can be found by considering that the maximum shear stress on the beam, which is $\tau_{zx,\mathrm{max}} = 3F_{z,\mathrm{max}}/(2hw)$, must be sufficiently lower than the corresponding shear yield stress, which, for rough purposes, may be approximated as $\sigma_{ys}/2$; this gives us a rough lower bound of h as $w^2/2L$.

Capacitive load cells: These load cells consist of a capacitor, namely, two conducting parallel plates, called the *electrodes*, which are separated by a dielectric material. The capacitance of this load cell is inversely proportional to the distance between the two electrodes (see Eq. (2.6)). When a load is applied to one of the electrodes, the distance between them would decrease and the capacitance would increase. Capacitive load cells have a higher sensitivity than strain gauge load cells. On the other hand, the operation of the former is complicated by the fact that their response is non-linear.

Pneumatic and hydraulic load cells: These load cells contain a gas or a liquid, and are essentially deadweight gauges, which have been discussed in Section 9.2.2. The fluid pressure is measured by an electrical pressure transducer and, through calibration, is converted into a force reading. Although these devices are cumbersome to operate, they are exceptionally accurate. By modifying the size of the fluid chamber and the density of the fluid used within it, one may easily scale-up these load cells, both in physical size and in operating range. Consequently, such devices are particularly suitable for accurate, large-scale, wind-tunnel model testing.

13.2.2 Torque Transducers

The measurement of torques (or moments) on a stationary sting is often part of wind-tunnel testing of wings and vehicle models, but is also used in robotics and other engineering applications. Devices that measure this torque are known as *reaction torque transducers*. The torque on a shaft is determined from the measurement of the shaft shear strain γ via the use of strain gauges, placed on the surface of the shaft. Strain gauges are placed on the surface of the shaft, with their axes aligned with a principal stress axis, which, for a shaft under pure torsion, is at a 45° angle with the shaft axis, as shown in Fig. 13.5a. Because the strain gauges are aligned in this way, the strain they experience is half the shear strain, that is, $\varepsilon_A = \gamma/2$. The measured strain is, therefore, $\varepsilon_A = Td(1+\nu)/(2JE)$, where T is the torque, d is the shaft diameter, J is the shaft *polar moment of inertia*, E is Young's modulus and ν is the Poisson ratio. Two pairs of strain gauges are typically used, one pair being under tension and the other under compression, as shown in Fig. 13.5a. The connection of four strain gauges obviously requires a full-bridge configuration.

Performance analysis of rotating machinery, such as pumps, turbines and propellers, requires the measurement of torque on a rotating shaft. Devices that measure this torque are known as *rotary torque transducers*. The mechanical power produced or consumed by the machine and transmitted by a rotating shaft is the product of measured torque and rotation speed, which necessitates the measurement of the rotational speed. The rotational speed is measured with *tachometers*, the most common types of which are the *magnetic pickup tachometer* and the *laser tachometer* [19, 61]. As for the reaction torque transducers, the torque is measured by pairs of strain gauges bonded onto a shaft. Slip rings are used to transmit the signal from the strain gauges on the rotating shaft to stationary terminals via brushes, which are in contact with the slip rings, as shown in Fig. 13.5b. The rotating machine may be attached to one end of the shaft via a shaft coupling, whilst the other end is attached to a device that produces/consumes the power of the rotating machine; for example, a wind turbine can be added to one end and an electric generator to the other.

13.2.3 Multi-Axis Measurements

Multiple force and moment components on a stationary shaft are measured with the use of multiple strain gauges that are sensitive to deformations along different axes. For example, a two-component load cell can be created by placing strain gauges on all four

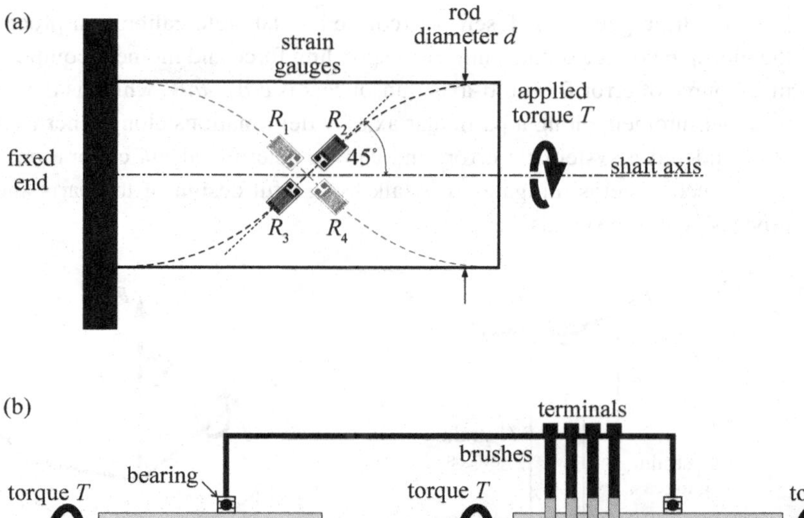

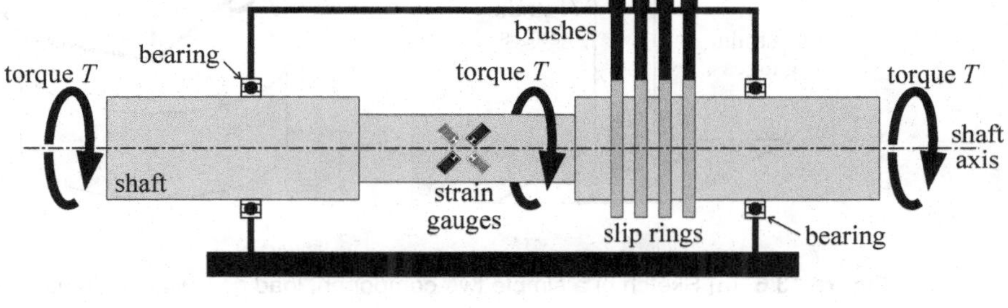

Figure 13.5 Sketches of (a) a reaction torque transducer, in which dashed lines are tangent to a local principal axis, and (b) a rotary torque transducer. Black strain gauges R_2, R_3 are in tension, when the applied torque has the direction shown in the figure, whereas grey strain gauges R_1, R_4 are under compression under the same torque.

faces of a square column, as shown in Fig. 13.6a. In general, the minimum number of strain gauges required for a multi-axis transducer, also referred to as *load balance*, or simply *balance* for short, is equal to twice the number of axes that are being considered; therefore, a three-component balance would have six strain gauges. An essential requirement in the design of multi-component devices is that the strain gauges produce an output only when a load is applied in a particular direction and are insensitive to loads in other directions. For example, the strain gauge arrangement in Fig. 13.6a is such that the F_z strain gauges are sensitive only to the force component in the z-direction and do not produce an output when a force component in the y-direction is applied. Commercial devices that are able to measure both forces and moments are commonly known as *force/torque* (F/T) sensors; note that the term transducer is more appropriate, because, in these devices, strain gauges not only sense mechanical deformation due to a load, but also convert it to a change in electric resistance and, ultimately, to an electric voltage. Most commercial F/T sensors are six-axis types, measuring all three force components and all three moment components, as shown in Fig. 13.6b; note, however, that manufacturers may use different symbols for the coordinate axes. Such devices contain at least 12 strain gauges. F/T sensors require an elaborate calibration process to convert the multiple voltage outputs into corresponding force and moment components. A common source of error for multi-axis transducers is *cross-talk*, which is the contamination of a measurement along a particular axis by deformations along other axes. Cross-talk is considered a systematic error, and can be determined via calibration. Manufacturers of such systems mitigate cross-talk by careful design of the parts and specialised processing of the signals.

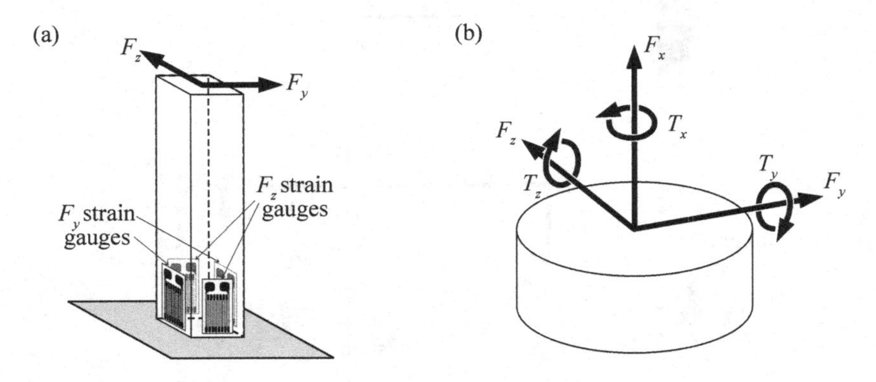

Figure 13.6 (a) Sketch of a simple two-component load cell, with strain gauges located on all four faces of a square bar; (b) coordinate system for a six-axis F/T sensor.

13.2.4 Dynamic Response

In the absence of external models or parts attached to them, force and torque transducers would act as second-order dynamic systems (see Section 2.4.5). The deformation of the load cell under an applied load is inherently linked to the stiffness of the material from which it is constructed; the less stiff the material is, the more more spring-like its behaviour would be. Moreover, the damping of any motion imposed by the load would decrease as the stiffness of the material increases. A useful analogy for the behaviour of load cells is the lumped-parameter, spring–mass–damper model, which allows one to estimate the natural frequency ω_n and damping ratio ζ as

$$\omega_n^2 = \frac{k}{M}, \tag{13.15}$$

$$\zeta = \frac{c}{2M\omega_n} = \frac{c}{2\sqrt{kM}}, \tag{13.16}$$

where c is the *damping coefficient* and M is the mass of the transducer. The damping coefficient depends on the design, the geometrical and material properties of the solid parts of the device, as well as the fluid properties, in the case of pneumatic and hydraulic load cells. Manufacturers typically only specify the natural frequency of the load cell. Equation (13.16) shows that, when the natural frequency is sufficiently large, the damping ratio would be very small and, hence, the system would be underdamped. This has important implications for the measurement of dynamic loads on a model, as shall be elaborated in the following example.

The mounting of a model onto a load cell would generally increase the stiffness of the combined system, but the stiffness of the load cell would generally remain the dominant factor. The main effect of the model would be to increase the mass of the combined system and one may substitute the transducer damping ratio and the total mass into Eqs. (13.15) and (13.16) to approximate the natural frequency and the damping ratio of the combined system. One may determine the damping ratio of the combined system by performing a step-change test, but such test may not be easy to perform for practical reasons. Instead, one may use a rough estimate of the damping ratio, which is based on past experience with load cells. For experiments in air, the system is likely to be strongly underdamped, with typical damping ratio values in the range $0.1 \leq \zeta_{air} \leq 0.2$. A lower bound for the frequency bandwidth (see Example 13.5) of the system can be obtained by assuming that the system is undamped ($\zeta_{air} = 0$). If, on the other hand, the experiment were conducted in water, the system would have a notably larger damping ratio, and, most likely, would be overdamped. For experiments in water, therefore, one may assume that the system is critically damped ($\zeta_{water} \approx 1/\sqrt{2}$); unlike the damping ratio estimate for air, the estimate for water is not necessarily conservative.

Example 13.5 We wish to measure the periodic force due to vortex shedding on a cylinder using a commercial load cell. The cylinder has a mass $M_c = 50$ g and the manufacturer specifies that the load cell has a mass $M_l = 100$ g and a natural frequency $\omega_n = 6{,}500$ rad/s. Estimate the output frequency bandwidth within which the error in the load measurement would be lower than 1%.

Answer

We may assume that the stiffness of the load cell with the cylinder mounted on it is approximately the same as that of the load cell itself. Using Eq. (13.15), we first find the stiffness of the load cell to be $k = \omega_n^2 M_l = 4.2 \times 10^6$ N/m. We now use this stiffness to calculate the natural frequency of the load cell–cylinder system as $\omega_{n,s} = \sqrt{4.2 \times 10^6/(0.10 + 0.05)} \approx 5300$ rad/s. In the absence of a means to determine the damping coefficient of the measuring system, we will find the lower bound of its bandwidth by assuming that the system is undamped ($\zeta = 0$). As shown in Example 2.23, the measurement error of an undamped system under a periodic load would be lower than 1% within a bandwidth $0 \leq \omega \leq 0.10\omega_{n,s}$. Therefore, the bandwidth of the present measuring system would be at least $0 \leq \omega \leq 530$ rad/s, or $0 \leq f \leq 84$ Hz, which implies that the frequency of vortex shedding should be less than 84 Hz for the force measurement error to be lower than 1%. In fact, the actual cut-off frequency of the present measuring system would be somewhat higher, as this system would certainly have a non-zero damping ratio. This example clearly demonstrates that, for the transducer to have a high frequency response, its natural frequency should be as large as possible.

13.3 Measurement of Wall Shear Stress

The wall shear stress is of great importance in fluid mechanics research and applications, as it represents the local tangential force by the fluid on a surface in contact with it. By integrating the wall shear stress over the surface, one can compute its contributions to the lift and drag on immersed objects and the pressure drop in internal flows. Even in a qualitative sense, shear stress is important, as an indicator of flow separation, flow instability and transition to turbulence. In engineering applications involving stationary turbulent flows, it is the time-averaged wall shear stress that is usually of interest, and so the temporal resolution requirement is often relaxed. There are other applications, however, which require high temporal and spatial resolutions of wall shear stress measurement. In particular, measurements of wall shear stress fluctuations are required for the validation of turbulence theories and numerical models and as input to flow control procedures, aimed at reducing skin friction drag. Measurements of the fluctuating wall shear stress in cardiovascular vessels and prosthetic devices are valuable for the detection and control of blood flow patterns that may lead to abnormal operation or disease. A great number of wall shear stress measurement techniques have been developed. Classical techniques, as well as more recent ones, have been described in several thorough reviews [18, 24, 25, 37, 40, 42, 62].

13.3.1 Estimates from Measured Velocity Profiles

For Newtonian fluids, the wall shear stress τ_w is related to the velocity derivative normal to the wall as

$$\tau_w = \mu \frac{\partial U}{\partial y}\bigg|_{y=0}, \tag{13.17}$$

where μ is the fluid viscosity. In laminar boundary layers, the velocity changes gradually and it is usually possible to fit a smooth curve to near-wall measurements, from which τ_w can be obtained through Eq. (13.17). Unfortunately, this is not a common case for turbulent boundary layers, in which the velocity changes dramatically over a short distance from the wall. The resolution and accuracy of conventional instrumentation, such as Pitot tubes, hot wires, LDV and PIV systems, become inadequate as the wall is approached, and even the distance from the wall cannot always be determined very precisely.

In the near-wall region of turbulent boundary layers, the relevant velocity scale is not the *free-stream velocity* U_e, but the *friction velocity*

$$u_\tau = \sqrt{\frac{\tau_w}{\rho}}. \tag{13.18}$$

Hence, the near-wall mean velocity is presented in dimensionless form as

$$u^+ = \frac{\overline{U}}{u_\tau}. \tag{13.19}$$

Similarly, the relevant near-wall length scale is the *viscous length* ν/u_τ and the dimensionless distance from the wall is expressed as

$$y^+ = \frac{y}{\nu/u_\tau} \,. \tag{13.20}$$

In turbulent flows over smooth walls, the flow region closest to the wall, called the *viscous* or *laminar sublayer*, is dominated by viscous forces. Within this sublayer, which is assumed to extend in the range $0 \leq y^+ \lesssim 5 - 7$, the mean velocity varies linearly, as

$$u^+ = y^+ \,. \tag{13.21}$$

Thus, if reliable mean velocity measurements within the viscous sublayer were available, linear least square fitting, combined with Eq. (13.17), would easily provide the mean wall shear stress. In practice, however, such measurements may be subject to significant systematic and random errors, which may be apparent in the form of profile non-linearity or scatter, but may also include subtle systematic effects [15, 30]. Thus, it is often necessary to estimate wall shear stress from velocity measurements outside the viscous sublayer, which have much lower uncertainty.

A suitable region beyond the viscous sublayer is the *inertial or logarithmic sublayer*, in which the velocity is commonly described by the *law of the wall*

$$u^+ = \frac{1}{\kappa} \ln y^+ + B \,, \tag{13.22}$$

where κ (called the *von Kármán constant*) and B are empirical constants. This layer starts at $y^+ \approx 30$, but its outer bound cannot be specified, because it increases with increasing Reynolds number. The region $5 - 7 \lesssim y^+ \lesssim 30$ is called the *buffer sublayer*, where the velocity is described by an interpolation expression. Although the universality of both Eq. (13.22) and the values of the constants have been topics of debate among researchers, one may plausibly adopt this expression with the typical values $\kappa \approx 0.39$ and $B \approx 5.0$. Equation (13.22) is often written as

$$u^+ = A \log y^+ + B \,, \tag{13.23}$$

where, obviously,

$$A = \frac{\ln 10}{\kappa} \approx 5.9 \,. \tag{13.24}$$

A traditional approach to estimate u_τ from velocity measurements within the logarithmic sublayer is to use the *Clauser chart* [8], which consists of plots of the function

$$f\left(y, c_f\right) = \sqrt{\frac{c_f}{2}} \left(A \log \frac{y U_e}{\nu} + A \log \sqrt{\frac{c_f}{2}} + B \right) \tag{13.25}$$

vs. $\log \frac{y U_e}{\nu}$, with the *skin friction coefficient*

$$c_f \equiv \frac{\tau_W}{\frac{1}{2}\rho U_e^2} = 2 \left(\frac{u_\tau}{U_e} \right)^2 \tag{13.26}$$

as a parameter (Fig. 13.7).

To use the Clauser chart, one plots velocity measurements, normalised as $\overline{U}/U_e$,

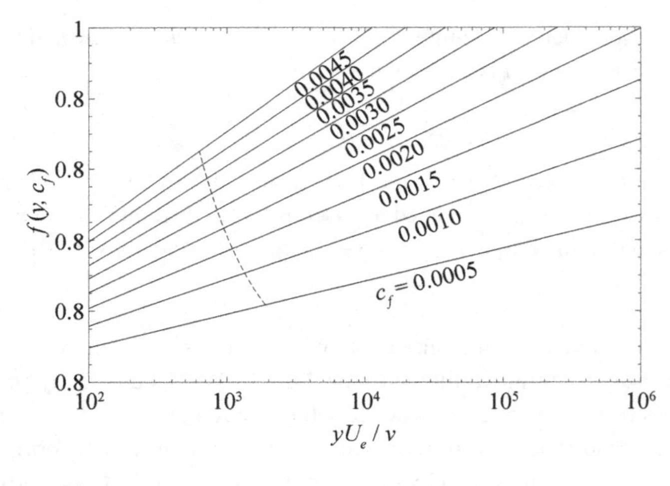

Figure 13.7 Clauser chart for $A = 5.9$ and $B = 5.0$; the dashed line indicates the conventional edge of the logarithmic sublayer, at $y^+ = 30$.

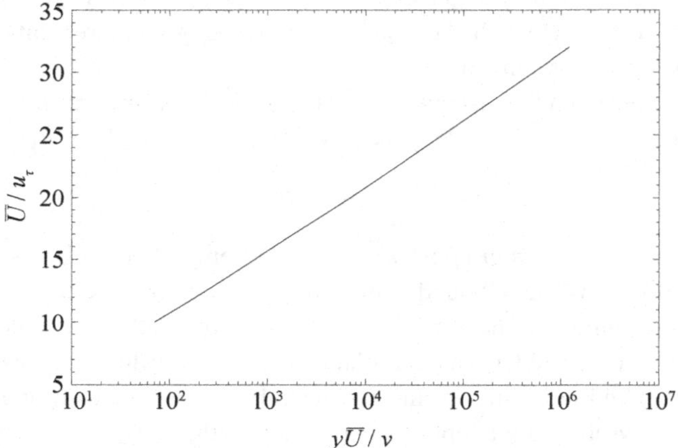

Figure 13.8 Chart for the direct determination of friction velocity from a single measurement within the logarithmic sublayer, according to Eq. (13.27) with $A = 5.9$ and $B = 5.0$.

vs. log $\frac{yU_e}{\nu}$ in this chart. The value of c_f that is best suited to these data is chosen to be the one corresponding to the f-curve that is closest to the measurements. Once c_f is determined, one can also easily determine the wall shear stress τ_w from Eq. (13.26). To reduce the subjectiveness of the Clauser chart method, one may compute u_τ directly by solving the law of the wall, rewritten in the form

$$\log \frac{y\overline{U}}{\nu} = \frac{1}{A}\left(\frac{\overline{U}}{u_\tau} - B\right) + \log \frac{\overline{U}}{u_\tau}, \qquad (13.27)$$

for a selected pair of measurements $\left(y, \overline{U}\right)$ within the logarithmic sublayer. This equation cannot be solved analytically, but its solution can be determined graphically [43] or numerically (see Fig. 13.8). The random uncertainty of u_τ can be reduced by averaging estimates determined from several pairs of values $\left(y, \overline{U}\right)$ within the logarithmic sublayer.

An alternative approach allows for a partly optimal fitting of the law of the wall to a specific available set of measurements, by assuming a value for the constant A only, rather than for both A and B. For this approach, it is more convenient to rewrite the law of the wall as

$$\frac{\overline{U}}{U_e} = \sqrt{\frac{c_f}{2}} A \log \frac{yU_e}{\nu} + \sqrt{\frac{c_f}{2}} \left(\log \sqrt{\frac{c_f}{2}} + B \right) . \tag{13.28}$$

Given a set of measured pairs of values $\left(y, \overline{U}\right)$, plots of $\overline{U}/U_e$ vs. $\log(yU_e/\nu)$ can be fitted by a straight line, extending only in the data range in which linear fitting seems appropriate. The optimal value of c_f can be determined from the slope $\sqrt{c_f/2}A$ of the fitted line, if one assumes a value for A, for instance the value 5.9, corresponding to $\kappa \approx 0.39$. The optimal value of the coefficient B can be determined by extrapolating the fitted line to $\overline{U}/U_e = 0$. Repeating the procedure with minor adjustments in the boundaries of the 'linear' range and the chosen value of A would provide a range of values of c_f, from which one can estimate its uncertainty.

Generalised expressions that contain the law of the wall, but are also valid in the buffer, viscous and/or outer sublayers, have also been used for the calculation of wall shear stress [25, 62]. All such methods depend on the applicability of the assumed universal expression to the particular set of measurements at hand. Universal laws of the wall describe well two-dimensional boundary layer profiles at relatively large Reynolds numbers and away from their origin, but their applicability to low Reynolds number boundary layers, and flows in their early stages of development, over highly curved and three-dimensional surfaces, or near separation locations, cannot be taken for granted.

13.3.2 Estimates from Pressure Differences

The wall shear stress is intimately related to static-pressure losses in pipes and channels and to pressure differences generated by obstructions at or near the wall. In general, however, the wall shear stress also depends on other factors, so that it cannot be determined from a single measured pressure difference and a simple theoretical expression. The notable exception is fully developed flow in horizontal circular pipes, in which streamwise momentum balance gives the (uniform) wall shear stress in terms of the pipe diameter D and the streamwise wall pressure gradient, as

$$\tau_w = \frac{D}{4} \left(-\frac{\mathrm{d}p}{\mathrm{d}x} \right) . \tag{13.29}$$

This expression is valid for both laminar and turbulent flows, if mean values are considered for the latter case. The entrance length L for flow development can be typically estimated as $L/D \approx 0.06\mathrm{Re}$ for laminar flows and $L/D \approx 25$–40 for turbulent ones. Although fully developed pipe flows are rarely encountered in most applications of interest, such a flow produced in the laboratory would be suitable for the calibration of wall shear stress measurements by other methods. Early techniques using empirical correlations between the wall shear stress and pressure differences include the use of two wall taps of different sizes, inclined taps [41], taps distorted to resemble forward- or, backward-facing steps, and others. Such techniques have been of limited application due to their low sensitivity and the difficulty in being reproduced accurately [62]. In the following, we shall discuss three related techniques, which are relatively easy to apply and have enjoyed some popularity.

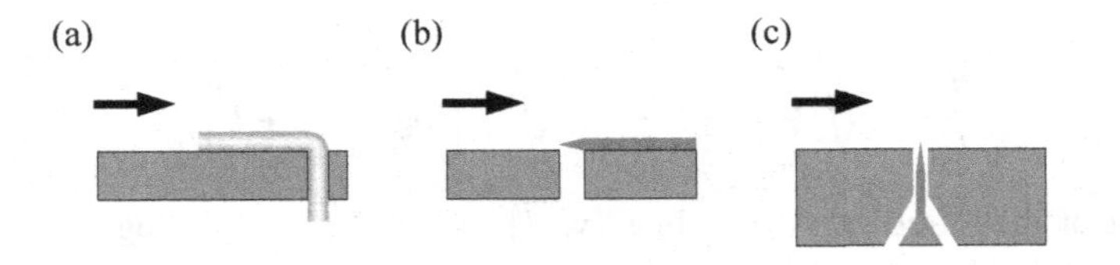

Figure 13.9 Schematic illustrations of (a) a Preston tube, (b) a Stanton gauge and (c) a sublayer fence.

Preston tubes: The *Preston tube* is essentially a Pitot tube (see Section 9.5), resting on the wall (Fig. 13.9a). This technique presumes universality of an inner boundary layer scaling law, although not any specific theoretical relationship. It measures the shear stress τ_w, through calibration, from the difference Δp between the total pressure measured by the tube and the local static pressure measured by a nearby wall tap, with the two parameters represented in dimensionless forms as $\tau_w / \left(\rho v^2 / d^2\right)$ and $\Delta p / \left(\rho v^2 / d^2\right)$, where d is the external tube diameter, ρ is the density and v is the kinematic viscosity. Rather than Preston's [47] original calibration expression, most researchers use the one due to Patel [46], which correlates the parameters

$$x^* = \log \frac{\Delta p}{4\rho v^2 / d^2} \quad \text{and} \quad y^* = \log \frac{\tau_w}{4\rho v^2 / d^2} = 2 \log (d^+ / 2) \ , \tag{13.30}$$

where $d^+ = d u_\tau / v$ is the dimensionless external tube diameter, which may also be viewed as the tube Reynolds number. Patel's empirical relationships are

$$y^* = 0.037 + 0.50 x^* \ , \qquad \begin{cases} 0 \le x^* \le 2.9 \ , \\ 0 \le y^* \le 1.5 \ , \\ 0 \le d^+ \le 11.2 \ , \end{cases}$$

$$y^* = 0.8287 - 0.1381 x^* + 0.1437 x^{*2} - 0.0060 x^{*3} \ , \qquad \begin{cases} 2.9 \le x^* \le 5.6 \ , \\ 1.5 \le y^* \le 3.5 \ , \\ 11.2 \le d^+ \le 110 \ , \end{cases}$$

$$y^* = -0.9654 + 0.718 x^* + 0.0175 x^{*2} - 0.0005 x^{*3} \ , \qquad \begin{cases} 5.6 \le x^* \le 7.6 \ , \\ 3.5 \le y^* \le 5.3 \ , \\ 110 \le d^+ \le 1{,}600 \ . \end{cases} \tag{13.31}$$

It is noted that the latter of the previous three expressions is a polynomial fit to Patel's data, which were originally fitted by a different expression. The three ranges in Eq. (13.31) roughly correspond to the viscous, buffer and logarithmic sublayers. Uncertainties in the expressions for the latter two layers have been cited as 1% and 1.5%, respectively. For higher Reynolds numbers, one may use the following polynomial fit to available data [63]

$$y^* = -1.1649 + 0.784x^* + 0.0104x^{*2} - 0.000235x^{*3}\,, \begin{cases} 6.4 < x^* < 11.3\,, \\ 4.3 < y^* < 8.7\,, \\ 280 < d^+ < 45{,}000\,. \end{cases} \tag{13.32}$$

A compound plot of $\tau_w/\left(4\rho\nu^2/d^2\right)$ vs. $\Delta p/\left(4\rho\nu^2/d^2\right)$, based on all previous expressions and also showing the variation of d^+, is presented in Fig. 13.10.

The previous expressions apply to flows with zero or negligible streamwise pressure gradient. It is known that Preston tubes would generally overestimate the wall shear stress, when inserted in flows with either favourable ($\mathrm{d}p/\mathrm{d}x < 0$) or adverse ($\mathrm{d}p/\mathrm{d}x > 0$) pressure gradient [46, 62]. An error bound of 3% is to be expected for flows in which the pressure gradient parameter $\Delta = \left(\nu/\rho u_\tau^3\right)(\mathrm{d}p/\mathrm{d}x)$ is in the range $-0.005 < \Delta < 0.01$, provided that $d^+ \leq 200$ and the flow is not in the process of relaminarisation ($\mathrm{d}\Delta/\mathrm{d}x < 0$, when $\Delta < 0$) [46]. Besides applying to smooth walls, the previous expressions also apply to *hydraulically smooth* walls, namely, walls having roughness elements with a height k that is smaller than the laminar sublayer thickness, that is, $ku_\tau/\nu < 5$. Correction factors to Patel's equations for rough walls have been presented in the literature but are known to have considerable uncertainty [28]. Additional corrections for effects of turbulence and compressibility are also available [24, 62].

The accuracy of Preston tube measurements also depends on practical details. To keep flow disturbance low, it is recommended to insert the tube through the wall or fasten it on the wall, rather than introduce it, like a Pitot tube, through the flow. To maintain the tube tip within the inner boundary layer, while also ensuring a fairly fast response, one may use a compound tube, consisting of a tip with a smaller diameter and progressively larger sections downstream. The use of flattened tubes has also been suggested, although this is expected to increase measurement uncertainty, unless a specific calibration is performed. Static-pressure taps should be of good quality, have a small diameter and be placed at the same streamwise location as the tube tip. They should be close to the

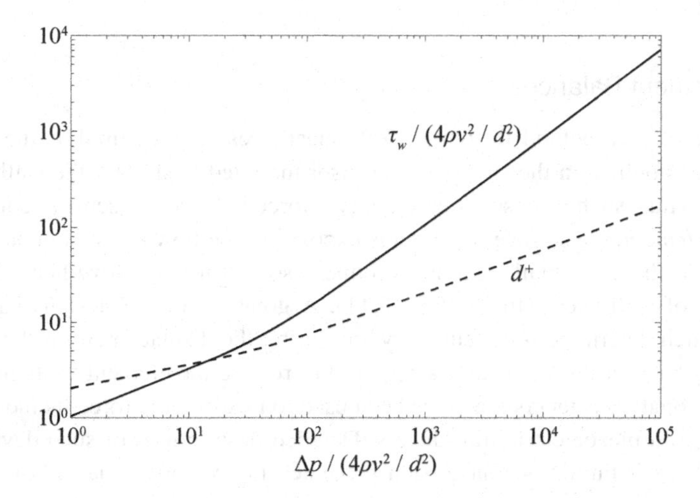

Figure 13.10 Preston tube calibration chart.

tube side, but far enough to avoid blockage effects (this can be tested by comparing measured static pressures with and without the tube). Averaging the readings of two static taps on either side of the tube might also be of some advantage. Preston tube readings do not appear to be particularly sensitive to the inner-to-outer diameter ratio, but it seems prudent to keep this near the main value of 0.6 that was used by Patel. A few variants of the basic configuration, involving tubes with slanted tips and multiple tubes have also been employed to a limited extent [62].

Razor blade technique: This approach originated with an arrangement known as *Stanton tube* [56], which consisted of a rectangular tube resting on the wall and sufficiently thin to be immersed within the viscous sublayer. The most widely used variant, sometimes referred to as the *Stanton gauge* (Fig. 13.9b), employs a thin blade, firmly attached to the wall magnetically or with the use of an adhesive, which blocks partially the opening of a static-pressure tap. The blade thickness must also be smaller than the viscous sublayer thickness and the wall shear stress is determined, through calibration, from the difference between the pressure in the blocked tap and the local static pressure. Empirical expressions, similar to those for Preston tubes, have been proposed by different authors [16, 45], but they are very sensitive to geometrical particulars, thus necessitating the individual calibration of each device. Uncertainty estimates for calibrated Stanton gauges are typically 3% and increase dramatically in flows with a significant pressure gradient.

Sublayer fence: This device consists of a wall tap with a thin blade mounted inside it, such that it partitions the tap into upstream and downstream halves (Fig. 13.9c) [18, 26, 58]. The tip of the blade must extend slightly into the flow, but must remain immersed within the viscous sublayer, as the name *sublayer fence* indicates. The difference in pressures in the two halves of the tap is linearly related to the wall shear stress, with the proportionality coefficient determined through calibration. Sublayer fences have a higher sensitivity than Stanton tubes. Sublayer fences fabricated with the use of micro-electro-mechanical (MEMS) technology are bidirectional (namely, can measure wall shear stress in both forward and reverse flows) and have a frequency response of up to 1 kHz [53].

13.3.3 Floating Element Balances

The principle of this method is simple: wall shear stress is measured as the ratio of the shear force applied on the surface of a sensor mounted flush with the wall and the sensing area. Thus, such a sensor is essentially a force balance. Its sensing component is a *floating element*, the surface of which is exposed to the flow and is mounted inside a wall cavity with some small clearance around it so that it may move laterally under the influence of wall shear stress (Fig. 13.11). A great variety of designs have been proposed, which determine force either by measuring the displacement of the floating element or by measuring the feedback required to restore the element to its null position. Conventional transducers that have been used to measure the force include LVDT, strain gauges and piezoelectric transducers. Detailed descriptions of such devices and discussions of their limitations have been presented in previous reviews [24, 25, 62].

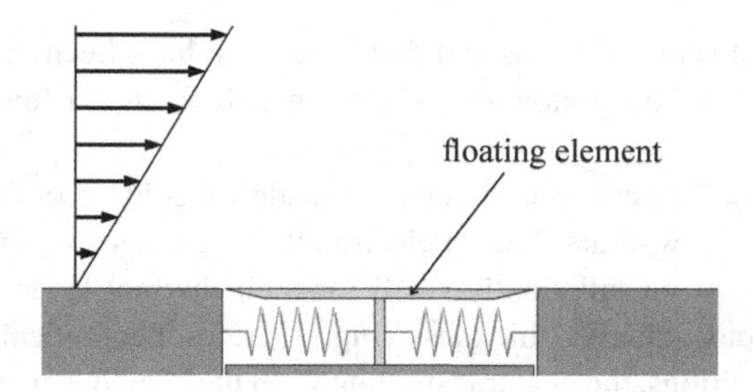

Figure 13.11 Schematic illustration of a floating element balance.

Limitations include relatively low spatial and temporal resolutions and sensitivity to floating element misalignment with the surface, surrounding clearance size, temperature changes, acceleration, vibrations, and pressure gradients. Although their output is directly proportional to the shear force on the wall and their response does not require the use of any analogy or theory, wall shear stress balances require calibration under conditions comparable to those in the experiment. Floating elements are used for the measurement of skin friction drag on sections of both smooth and rough walls [1, 2].

Floating element sensors have also been constructed using silicon micromachining technology (MEMS devices) [33, 40]. Besides the fact that they are much smaller than conventional sensors, MEMS sensors also have the advantage of lower misalignment errors, because of their integrated fabrication. In electrical MEMS sensors, the sensing element is tethered to its support by semiconductor links, which act as elastic springs, and its displacement is detected by capacitive or piezoelectric components. In an optical-detection design [44], the floating element is mounted over two photodiodes, which measure displacement by collecting the light of a laser beam, blocked partially by the element. This device is suitable for measuring shear stress fluctuations up to a frequency of 10 kHz, with a resolution of 0.003 Pa.

13.3.4 Micro-Pillars

Micro-pillars are small cylinders, made of an elastomer (namely, a rubber-like material with elastic properties) and attached to the wall, while being totally immersed in the viscous sublayer. When exposed to a flow, they deform and one can determine the local, time-dependent wall shear stress in magnitude and direction by analysing their deflection, recorded with a high-resolution camera [6, 22, 23].

13.3.5 Thermal Techniques

Thermal sensors for wall shear stress measurement are very similar to those used in hot-wire/hot-film and pulsed-wire anemometry (see Section 12.2). Their main difference is that the thermal fields of the former must be immersed in the viscous sublayer, in which the local flow velocity is proportional to the wall shear stress, according to Eq. (13.21). Like the corresponding velocity thermal sensors, they are divided into two categories: those measuring wall shear stress through its analogy to heat flux from the heated element and those that measure wall shear stress from the time of flight of heated spots. Details on the different sensor designs and their operation have been included in previous reviews [7, 18, 62], so the following discussion will focus on fundamental concepts.

Flush-mounted hot films: These are film sensors, made of gold or other metals, which are mounted flush on the wall and heated electrically by a constant temperature anemometer circuit. They must be sufficiently small for their thermal layer thickness to be smaller than the viscous sublayer thickness (Fig. 13.12a). Theoretical analysis predicts that, under such conditions, the heat transfer rate from the heated element would be proportional to the 1/3 power of the local wall shear stress [32]. To account for small differences among individual sensors, their response is described by an empirical expression analogous to the modified King's law in hot-wire anemometry (Eq. (12.31)), as

$$\frac{E^2}{T_w - T_f} = A + B\tau_w^{1/3}, \tag{13.33}$$

where E is the supplied voltage, T_w is the film temperature, T_f is the bulk fluid temperature, and the coefficients A and B are determined by calibration vs. a Preston tube or other standard. Because of their small size, hot film sensors are sensitive to wall shear stress fluctuations and may capture unsteady phenomena and turbulent fluctuations, although not necessarily the fine structure, at least in high-speed air flows. Like all thermal sensors, flush-mounted films cannot discriminate between forward and reverse flows. Nevertheless, if mounted on a turntable, they can identify the near-wall mean flow direction, which will be parallel to their long axes at an orientation such that the heat transfer rate is at a minimum.

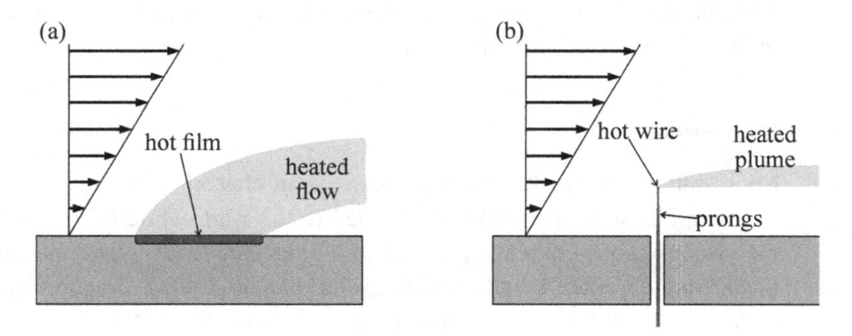

Figure 13.12 Schematic illustrations of (a) a flush-mounted hot film and (b) a wall hot wire.

Hot film arrays: Arrays of co-planar, rectangular, thin metallic films are available commercially and may be also fabricated using integrated circuit technology. They are epoxied on the wall surface with the film long axes normal to the flow direction. Each film must be supplied by a separate anemometer circuit in the constant temperature mode. In principle, they can be calibrated so that each sensor would measure the local wall shear stress, however, the responses of individual sensors would be complicated by interference from the thermal fields of other sensors. In practice, they are only used for qualitative purposes, mainly to identify the location of flow separation, transition to turbulence, shock waves and other discontinuities near the wall. For this reason, they may be powered by relatively simple and inexpensive circuitry, which is an essential requirement, when considering that such arrays often contain a large number of elements. Their frequency response may be sufficiently high to detect transient and oscillatory phenomena. Micro-hot-film arrays with large numbers of active elements (in the hundreds or even thousands) can be manufactured using MEMS technology [27].

Wall hot wires: These are conventional single-wire sensors, which are mounted on prongs that protrude slightly through the wall while remaining within the viscous sublayer, and measure wall shear stress through its proportionality to local flow velocity (Fig. 13.12b). Their voltage output is related to wall shear stress by a modified King's law, including a correction for wall effects on the sensor heat transfer [52]. The sensitivity of wall hot wires improves with increasing distance from the wall and their frequency response is higher than that of other wall shear stress sensors.

Pulsed hot films and hot wires: These transducers consist of a central heated element, which is supplied with a current pulsed at some appropriate frequency, and two temperature-sensing elements, on either side of the central one, which detect the convection of heated parcels of fluid. Both flush-mounted pulsed film arrangements and near-wall mounted pulsed wire arrangements have been used. The main advantage of these sensors over other techniques is that they can identify the instantaneous flow direction and measure wall shear stress fluctuations in separated and reverse flows.

Infrared thermography: In this approach, a laser beam is used to heat a spot on the surface of an immersed object, which is made of a material with a low thermal conductivity and a high emissivity [36]. The spot temperature is continuously monitored with an infrared camera. When a steady temperature is reached, the laser is turned off and the spot temperature time history is recorded. This time history is used as input to a numerical algorithm that calculates heat transfer rates and, through analogy, wall shear stress. The uncertainty of this method is relatively high (about 10%).

13.3.6 Electrochemical Method

The *electrochemical*, or *electrodiffusion*, method measures wall shear stress from the current that must be supplied to an electrode, which is flush with the wall, in order to sustain a chemical reaction in the flowing fluid [24]. An advantage of this method over previously discussed ones is that its response can be derived from theoretical considerations, thus, eliminating the need for calibration, although calibration would reduce errors due to the uncertainty of dimensions and material properties. When used in unsteady or turbulent flows, this method has an advantage over thermal methods, the frequency response of which is limited by heat conduction to the substrate.

In a typical set-up, the sensor is a small, thin platinum film, flush with the wall (similar to the hot film shown in Fig. 13.11a), serving as the cathode, while the anode is a second, much larger, electrode, made of stainless steel or nickel and usually positioned downstream. The method is used in water that contains, as active elctrolytes, potassium ferricyanide ($K_3Fe(CN)_6$) and potassium ferrocyanide ($K_4Fe(CN)_6$), each mixed with the water at a precisely controlled bulk concentration C_b. A supporting, inert electrolyte, commonly sodium hydroxide ($NaOH$) or potassium sulphate (K_2SO_4), is also added in excess to the mixture, in order to increase the conductivity of the solution and to ensure that the reaction at the probe is governed by diffusion mass transfer. When a voltage is applied between the cathode and the anode, free electrons released at the cathode reduce ferricyanide ions into ferrocyanide ions and the reaction is reversed at the anode, as

$$[\mathrm{Fe(CN)_6}]^{3-} + \mathrm{e}^- \underset{\text{cathode}}{\overset{\text{anode}}{\leftrightarrows}} [\mathrm{Fe(CN)_6}]^{4-} . \tag{13.34}$$

Because the molecular diffusivity γ_c of ions is three orders of magnitude smaller than the kinematic viscosity of water, the mass diffusion layer (namely, the layer within which the ion concentration increases from a zero value at the wall to the bulk concentration in the mixture) is much thinner than the hydrodynamic boundary layer. For laminar flows, the velocity profile in the diffusion layer may be approximated by a linear one and, for turbulent flows, the diffusion layer would be immersed within the viscous sublayer, where the flow velocity is proportional to the distance from the wall.

The output of the electrochemical measuring system is the current that is supplied to the cathode. As the applied voltage is increased, the current increases, until it reaches a saturated value and remains nearly constant within a range of voltages. This is called the *limiting current* i, for which the electrochemical reaction occurs at its maximum rate, and at which the measuring system is operated. By analogy to heat transfer from a heated element at the wall (see Eq. (13.33)), the mass transfer from the electrode would be proportional to the 1/3 power of the wall shear stress. In steady or quasi-steady flows over a rectangular cathode, having an exposed area A and a length L in the streamwise direction, the wall shear stress can be calculated as

$$\tau_w = \frac{\mu L I^3}{0.807^3 C_b^3 \gamma_c^2 n^3 A^3 \mathcal{F}^3} , \tag{13.35}$$

where n is the number of electrons involved in the reaction (in this case, $n = 1$) and $\mathcal{F} = 94{,}484.6$ C/mol is Faraday's constant. The same relationship applies to circular electrodes, provided that the length L is replaced by the effective length $0.814d$, where d is the electrode diameter. The uncertainty of measurements in steady flows with a calibrated electrochemical sensor is about 3–4%.

Equation (13.35) is inaccurate in unsteady and turbulent flows, but several extensions of the electrochemical method have been developed, making it possible to measure the time-dependent wall shear stress. In the *spectral method*, the frequency response of the probe is determined first and then it is used to correct measured spectra [9, 10, 12]. This method has a high frequency response, but is only suitable for relatively small (up to typically 30%) fluctuation levels. In the *Sobolík method*, the quasi-steady value is corrected by a term that is proportional to its first-order derivative, accounting for capacitive effects in the diffusion layer [55]. This method can be used for large fluctuation amplitudes but cannot resolve reversing flows. The most accurate approach is the inverse method, by which one solves iteratively the advection–diffusion differential equation for the ion concentration until its solution matches the measurements [35, 50].

This method can also resolve the sense of the wall shear stress (namely, whether the flow is forward or reverse). Besides single cathodes, *segmented cathodes*, with two or three segments, have also been used by several authors [31, 60]. These can resolve both the direction and the magnitude of wall shear stress. The spectral method has been applied to segmented probes for measuring wall shear stress fluctuations with relatively small amplitudes [11], whereas the inverse method has been applied to flows with strong wall shear stress fluctuations [31].

13.3.7 Optical Techniques

Optical techniques are generally attractive for being non-intrusive and having a response that is based on theoretical relationships, thus eliminating the need for calibration. On the other hand, they are subject to other limitations, such as the need for flow seeding or the use of special coatings. In the following, we shall briefly discuss three optical techniques, which find increasing application. We should also mention that any non-intrusive, high-resolution velocity measurement method, including particle image velocimetry and molecular tagging velocimetry, can, at least in principle, be used for measuring near-wall velocity profiles, from which the wall shear stress can be determined with the use of Eq. (13.17).

Laser Doppler technique: This technique is a variant of laser Doppler velocimetry, adapted to the measurement of wall shear stress [20, 39]. A laser beam is passed through a diffractive lens and enters the test section through two narrow, closely spaced, parallel slits, etched on a chromium coating of a glass window, as shown in Fig. 13.13. Interference of the light diffracted by the two slits creates interference fringes (see Section 7.3.8) in the wall region, with a spacing δ that increases linearly with $y - y_0$. For slits with a very close spacing S, one may assume that $y_0 \ll y$, which provides the fringe spacing as

$$\delta = \frac{\lambda}{S} y \,, \tag{13.36}$$

where λ is the laser light wavelength. Particles seeded into the flow scatter light, as they cross the fringes, at the Doppler frequency (see Section 12.3)

$$f_D = \frac{U}{\delta} \,. \tag{13.37}$$

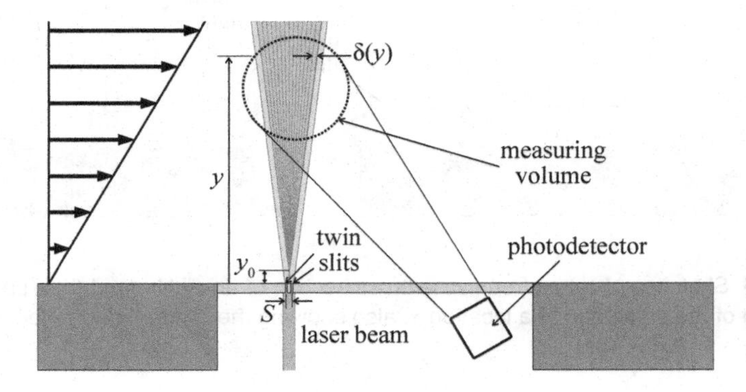

Figure 13.13 Sketch of a laser Doppler wall shear stress measurement set-up.

A receiving lens collects the light from a measuring volume and projects it on a photodetector. If the measuring volume is in the viscous sublayer, where the velocity is proportional to the distance from the wall (i.e., $U/u_\tau = yu_\tau/\nu$), then all particles will scatter light at the same Doppler frequency and the wall shear stress can be determined as

$$\tau_w = \frac{\mu\lambda}{S} f_D \,. \tag{13.38}$$

This technique has the advantages of not requiring calibration and providing the instantaneous value of τ_w, from which statistical properties can be determined. With frequency shifting of light transmitted through one of the slits, it can also determine flow reversal. Its disadvantage is that it requires seeding, thus it is subject to the limitations of particle response near the wall.

Besides the original version of this technique, which used an Ar-Ion laser and conventional optical components [39], a miniaturised version has also been developed, using a small laser diode, a photodiode, optical fibres and MOEMS (micro-optical-electrical mechanical systems) technology [20].

Oil film interferometry: This method, which is an extension of the oil streak technique used for flow visualisation, has found considerable application in wind-tunnel flows past models. It measures wall shear stress through its relationship to the thickness of an oil film in contact with the wall. To demonstrate this principle, consider a small oil droplet applied on a smooth wall. When exposed to a flow, the droplet deforms into a wedge-like film, the local thickness of which diminishes continually as time passes (Fig. 13.14).

For simplicity, assume that the oil flow is two-dimensional and that the shear stress is uniform on the surface. For small film thicknesses, one may neglect the effects of gravity, pressure gradient and surface tension and use lubrication theory to derive the following relationship between wall shear stress, local thickness h and time t

$$\tau_w = \frac{\mu_{\text{oil}} x}{h(x,t)t} \,. \tag{13.39}$$

In practice, the oil thickness is of the order of tens of μm and is most conveniently

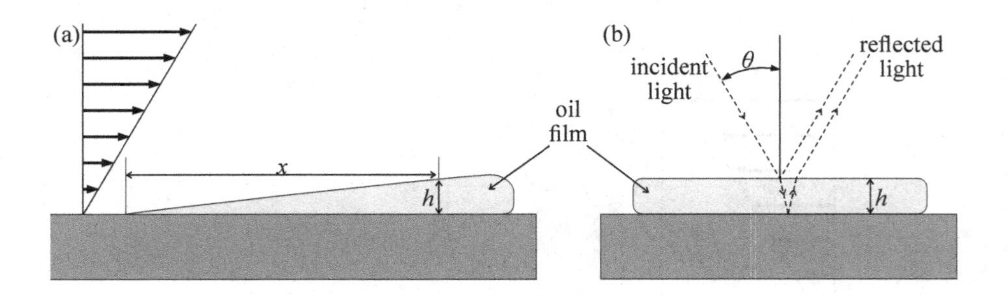

Figure 13.14 Sketches of (a) a streamwise cross section of an oil film and (b) a spanwise cross-section of the same film at a location x, also showing the incident and reflected light beams.

measured by an interferometric set-up. Although white light has also been used [34], in the most common version of this method, monochromatic light from a sodium or mercury lamp is directed on the film at an angle of incidence θ. Part of this light is reflected off the oil surface, while another part enters the oil. Part of the latter is reflected off the wall material (commonly used materials include glass, polished steel and mylar) and passes for a second time through the film to enter the air stream again. The two reflected beams interfere with each other, producing dark fringes on the oil surface, when their phase shift ϕ is an odd multiple of 2π, and bright fringes, when the phase shift is an even multiple of 2π. The local film thickness is related to the phase shift as

$$h(x,t) = \frac{\lambda\phi}{4\pi\sqrt{n_{oil}^2 - n_{air}^2 \sin^2\theta}}, \tag{13.40}$$

where n_{oil} and n_{air} are, respectively the refractive indices of oil and air and λ is the wavelength of the incident light. Obviously, the fringes will shift downstream with time, as the film thickness diminishes. Images of the film are used to identify the locations and orders of the various fringes, from which the wall shear stress can be determined using Eqs. (13.39) and (13.40). Theoretical treatments and practical details on the method can be found in the literature [5, 14, 17, 18, 37, 38, 51, 57]. Oil film interferometry is reputedly the most accurate wall shear stress measurement method, with an uncertainty that is as low as 1%. It is suitable for steady laminar flows and for the measurement of mean wall shear stress in stationary turbulent flows, but cannot measure wall shear stress fluctuations.

Liquid crystal technique: Liquid crystal coatings are frequently used for flow visualisation and surface temperature mapping, but have also found application for the measurement of wall shear stress [4, 21, 40, 48, 49, 54, 64]. Among the different variants of this approach, one that has quantitative capabilities is as follows. A thin coating of liquid crystals is applied on a surface and illuminated by white light, in a direction normal to the surface, within ±15°. The choice of crystals among the available commercial products is such that their appearance in the absence of shear stress is in the red-orange range. When viewed by an observer positioned upstream, the crystal colour changes towards the blue, as the shear stress increases. In contrast, an observer positioned downstream would see no colour change. This happens because the birefringence of liquid crystals depends on the applied shear stress, resulting in changes of the spectral content of light reflected by the crystal. The perceived colour depends both on the magnitude and the orientation of the shear stress. To resolve both, several images of the area of interest are taken by digital cameras from different directions. The coating has to be replaced after each use. This technique is fairly cumbersome and requires attention to many details.

Chapter Digest

13.1 What are wall shear stress measurements used for?

13.2 How does the stiffness of a load cell affect its sensitivity?

13.3 Why is it beneficial to have at least two strain gauges in a single-axis load cell?

13.4 What is the difference between reaction and rotary torque transducers? List possible applications of each type.

13.5 What are the main differences in the velocity profiles of laminar and turbulent boundary layers? On the same plot, sketch such profiles, normalised by the free-stream velocity, and with the distance from the wall normalised by the physical boundary layer thickness (i.e., the distance at which the velocity is equal to 0.99 times the free-stream value).

13.6 Identify the different sublayers of a velocity profile in a turbulent boundary layer.

13.7 What is a Clauser chart? Is there a better alternative?

13.8 How do you determine wall shear stress in fully developed pipe flows from pressure measurements? Does this method also apply to the development region? Justify your answer.

13.9 What is the difference between a Pitot tube and a Preston tube? Can you use the Pitot tube response expression for Preston tube measurements? If not, why?

13.10 Can you use a Preston tube in separated flows?

13.11 What is a Stanton tube and how does it work?

13.12 Which are the main advantages and disadvantages of floating element balances, as wall shear stress measuring devices?

13.13 Which are the two main methods of operation of thermal sensors, as wall shear stress measuring devices? Can these be used in separated flows?

13.14 Can electrochemical methods be used in a wind tunnel? A water tunnel? What would it take to do so?

13.15 Which are the main advantages and disadvantages of optical techniques for measuring wall shear stress?

13.16 Which are the flow properties on which liquid crystal response depends, in general?

Problems

13.1 Derive the normalised bridge sensitivities, as shown in Table 13.1, for the strain gauge configurations shown in Fig. 13.3.

13.2 Consider the strain gauge configuration shown in Fig. 13.3e, except this time with dummy strain gauges at R_2 and R_3. Show that the voltage output is temperature compensated. Compare the voltage sensitivity, and sensitive loading directions, to the configuration in the previous example.

13.3 A force $F_z = 100$ N is applied on the free end of the cantilever beam shown in Fig. 13.3e. The beam is made of aluminium, it is 100 mm long and it has a thickness of 10 mm and a width of 50 mm. Determine the voltage reading, if the bridge is powered by a 5 V DC power supply and the strain gauges have a gauge factor of 2.

13.4 A voltmeter with a resolution of 10 mV is used to measure the force in the previous question. Determine the force resolution of the load cell. What design changes can you make to this load cell to increase its resolution?

13.5 A wing model with a mass of 200 g is mounted onto a two-axis load cell in a wind tunnel, and an experiment is performed with the objective of measuring the lift and drag force on the model as a function of air velocity. The manufacturer states that the load cell has a mass of 50 g and a natural frequency of 3,000 Hz, and we found, using a step-change test, that the damping ratio is approximately 0.1.

(a) How long should we wait before taking a measurement, after a sudden wind speed change from 10 to 15 m/s?

(b) Estimate the bandwidth of the measuring system. To solve this problem, use material presented in Chapter 2.

13.6 The simplest form of a compressive/tensile load cell is that of a single column of material. If one end of the column is fixed and the other end is free and has an axial force applied to it, show that the natural frequency of the load cell is $\omega_n^2 = EA/(LM)$, where M is the mass of the load cell. Discuss how the cross-sectional area and length of the cell affect the natural frequency.

13.7 A compressive/tensile load cell is to be used to measure the down-force created by the rear wing (spoiler) of a racing car in a wind tunnel. How should the strain gauges be attached on the load cell and on the bridge in order to measure this force? Derive a relationship between the voltage output and the applied force.

13.8 A compressive/tensile load cell is made out of a stainless steel rod with a diameter of 6 mm and a length of 25 mm. A model of a car, with a mass of 1 kg and a height of 100 mm, is mounted on top of the load cell. Calculate the natural frequency of the load cell and the natural frequency of the entire measuring system. Compare the two values.

13.9 The frequency of the periodic thrust generated by a propeller is directly related to its rotational speed, which implies that the rotational speed of the propeller can be obtained from measurements of the periodic thrust.

(a) Explain how the rotational speed can be obtained from time histories of the measured force.

(b) How does the frequency bandwidth of the load cell affect your ability to measure the time varying force and the rotational speed?

(c) The combined mass of the propeller, motor and shaft is 253 g. From previous tests, we know that the thrust generated by this particular propeller, as a function of its rotational speed ω, is $F_t = 17\times10^{-6}\omega^2$. Moreover, we have found that, under such loading, the compression of the entire measuring system was $\delta L = 34 \times 10^{-12}\omega^2$. Assuming that the damping ratio of the measuring system is $\zeta = 0.12$, calculate the maximum rotational speed of the propeller for the error in the periodic thrust measurement to be less than 5%; repeat this for an 1% error.

13.10 The frequency of the periodic drag force of wind on a cylindrical chimney may be approximated as $f_v = 0.2U/d$, where U is the wind speed and d is the diameter of the chimney. A scaled model of a particular chimney with a diameter of 35 mm is placed on a load cell in a water tunnel. The model has a mass of 350 g. The supplier of the load cell specifies that the load cell has a mass of 55 g and a natural natural frequency of 7500 rad/s. Estimate the maximum speed we can run the water tunnel for the error in the measurement of the oscillatory force amplitude to be lower than 1%.

13.11 The torque transducer shown in Fig. 13.5 has strain gauges mounted onto a shaft, which can be either a solid cylinder or a tube.

(a) Derive an expression for the shear strain for each of these two cases.

(b) Based on this expression, explain why the rotary torque transducer has a step change in shaft diameter where the strain gauges are placed.

(c) For a given shaft diameter, which cross section type is preferable and why?

13.12 Assume that the resistances of the strain gauges shown in Fig. 13.5 correspond to the resistances in the Wheatstone bridge shown in Fig. 13.3.

(a) Starting from the resistances on the bridge, derive an expression that relates the bridge output to the applied torque.

(b) The rotary torque transducer has four slip rings, each of which transmits a signal from a strain gauge. Sketch how you would wire up the strain gauges to the slip rings in order to measure the voltage output; explain your reasoning.

13.13 Laser Doppler velocimetry measurements in a low Reynolds number turbulent boundary layer, generated in a water tunnel at a temperature of 20°C, consist of the following pairs of distance y from the wall (in mm) and mean streamwise velocity $\overline{U}$ (in mm/s):

y	$\overline{U}$	y	$\overline{U}$	y	$\overline{U}$	y	$\overline{U}$	y	$\overline{U}$
0.2	42	1.8	196	8	256	26	307	55	340
0.3	65	2	205	9	258	28	312	60	341
0.4	85	2.5	208	10	264	30	315	70	342
0.5	106	3	220	12	273	32	318	80	342
0.6	121	3.5	225	14	278	34	320	90	342
0.8	143	4	232	16	282	36	323	100	341
1	162	4.5	234	18	289	38	324	110	341
1.2	174	5	239	20	295	40	329	120	341
1.4	182	6	245	22	300	45	334	130	342
1.6	191	7	249	24	303	50	337	150	341

(a) Determine the wall shear stress by assuming that a number of the measurements closest to the wall are in the viscous sublayer; verify the validity of this assumption.

(b) Determine the wall shear stress using a Clauser chart.

(c) Determine the wall shear stress by solving Eq. (13.27) for several pairs of measurements within the logarithmic sublayer; compare these values and repeat the procedure for different combinations of values of the coefficients A and B.

(d) Determine the wall shear stress from Eq. (13.28) and the associated procedure; determine the optimal value of B and compare it with those used in questions (b) and (c); repeat the procedure for different values of A.

(e) Compare the results of the previous calculations and discuss the accuracy of these procedures.

13.14 Assume that a Preston tube is used to measure the wall shear stress in the boundary layer discussed in Problem 13.13. Estimate the pressure difference between the Preston tube and a nearby wall tap, when the tube diameter is 0.5, 1, 3 and 10 mm.

13.15 A Preston tube with a diameter of 1.2 mm reads a pressure of 3.9 kPa above static in air flow in a duct. Determine the local wall shear stress.

13.16 Assume that a flush-mounted hot film that has been calibrated in steady flow is used in a pulsatile flow in which the wall shear stress varies sinusoidally as

$$\tau_w = \overline{\tau}_w(1 + a \sin \omega t) .$$

Assume that the frequency response of the sensor is sufficient to capture fluctuations without appreciable distortion; derive an expression for the ratio of the estimate of $\overline{\tau}_w$, obtained under the assumption that Eq. (13.33) relates it to the mean voltage $\overline{E}$, and the actual mean wall shear stress $\overline{\tau}_w$; assume that the amplitude a is small compared to 1 and use binomial series expansions to produce a first-order estimate; estimate the limits of validity of this approach, if the acceptable uncertainty must be maintained below 5%.

13.17 Assume that the laser Doppler technique is used to measure the wall shear stress in the boundary layer presented in Problem 13.13. If a He-Ne laser is used, determine the Doppler frequency f_D that would be measured, if the distance S between slits were 10 μm.

References

[1] M. Aguiar Ferreira, E. Rodríguez-López, and B. Ganapathisubramani. An alternative floating element design for skin-friction measurement of turbulent wall flows. *Exp. Fluids*, 59:155, 2018.

[2] W.J. Baars, D.T. Squire, K.M. Talluru, M.R. Abbassi, N. Hutchins, and I. Marusic. Wall-drag measurements of smooth- and rough-wall turbulent boundary layers using a floating element. *Exp. Fluids*, 57:90, 2016.

[3] T.G. Beckwith, R.D. Marangoni, and J.H. Lienhard. *Mechanical Measurements (6th Edition)*. Pearson, Upper Saddle River, NJ, 2007.

[4] P. Bonnett, T.V. Jones, and D.G. Donnell. Shear-stress measurement in aerodynamic testing using cholesteric liquid crystals. *Liquid Cryst.*, 6:271–280, 1989.

[5] H. Bottini, M. Kurita, H. Iijima, and K. Fukagata. Effects of wall temperature on skin-friction measurements by oil-film interferometry. *Meas. Sci. Technol.*, 26:105301, 2015.

[6] C. Brücker, D. Bauer, and Chaves H. Dynamic response of micro-pillar sensors measuring fluctuating wall-shear-stress. *Exp. Fluids*, 42:737–749, 2007.

[7] H.H. Bruun. *Hot-Wire Anemometry*. Oxford University Press, Oxford, 1995.

[8] F.H. Clauser. Turbulent boundary layers in adverse pressure gradients. *J. Aeronaut. Sci.*, 21:91–108, 1954.

[9] C. Deslouis, O. Gil, and B. Tribollet. Frequency response of electro-chemical sensors in a cone- and plate-modulated flow. *Int. J. Heat Mass Transfer*, 33:2525–2532, 1990.

[10] C. Deslouis, O. Gil, and B. Tribollet. Frequency response of electro-chemical sensors to hydrodynamic fluctuations. *J. Fluid Mech.*, 215:85–100, 1990.

[11] C. Deslouis, B. Tribollet, and J. Tihon. Near-wall turbulence in drag reducing flows investigated by the photolithography-electrochemical probes. *J. Non-Newtonian Fluid Mech.*, 123:141–150, 2004.

[12] A. Dib, S. Martemianov, L. Makhloufi, and B. Saidani. Calibration of electrodiffusion probes for turbulent flow measurements. *Flow Measur. Instrum.*, 35:76–83, 2014.

[13] K. Doebbeling, B. Lenze, and W. Leuchel. Basic considerations concerning the construction and usage of multiple hot-wire probes for highly turbulent three-dimensional flows. *Meas. Sci. Technol.*, 1:924–933, 1990.

[14] D.M. Driver. Application of oil film interferometry skin-friction to large wind-tunnels. Technical Report AGARD CP-601, Paper 25, 1997, Advisory Group for Aerospace Research and Development, NATO, Paris.

[15] F. Durst, R. Mueller, and J. Jovanovic. Determination of the measuring position in laser-Doppler anemometry. *Exp. Fluids*, 6:105–110, 1998.

[16] L.F. East. Measurement of skin friction at low subsonic speeds by the razor blade technique. Technical Report 3525, Aeron. Res. Council Research and Memoranda, 1967.

[17] L.B. Esteban, E. Dogan, E. Rodríguez-López, and B. Ganapathisubramani. Skin-friction measurements in a turbulent boundary layer under the influence of free-stream turbulence. *Exp. Fluids*, 58:115, 2017.

[18] H.H. Fernholz, G. Janke, M. Schober, P.M. Wagner, and D. Warnack. New developments and applications of skin-friction measuring techniques. *Meas. Sci. Technol.*, 7:1396–1409, 1996.

[19] R.S. Figliola and D.E. Beasley. *Theory and Design for Mechanical Measurements (7th Edition)*. Wiley, Hoboken, NJ, 2019.

[20] D. Fourguette, D. Modarress, F. Taugwalder, D. Wilson, M. Koochesfahani, and M. Gharib. Miniature and MOEMS flow sensors, AIAA paper 2001-2982. In *Proc. 31st AIAA Fluid Dynamics Conference and Exhibit*, Anaheim, CA, 2001.

[21] N. Fujisawa, A. Aoyama, and S. Kosaka. Measurement of shear-stress distribution over a surface by liquid-crystal coating. *Meas. Sci. Technol.*, 14:1655–1661, 2003.

[22] E.P. Gnanamanickam, H. Nottebrock, Grosse S., J.P. Sullivan, and W. Schröder. Measurement of turbulent wall shear-stress using micro-pillars. *Meas. Sci. Technol.*, 24:1240023, 2013.

[23] S. Grosse and W. Schröder. The micro-pillar shear-stress sensor MPS^3 for turbulent flow. *Sensors*, 9:2222–2251, 2009.

[24] T.J. Hanratty and J.A. Campbell. Measurement of wall shear stress. In R.J. Goldstein, editor, *Fluid Mechanics Measurements (2nd Edition)*, pages 575–648. Taylor & Francis, Washington DC, 1996.

[25] J.H. Haritonidis. The measurement of wall shear stress. In M. Gad-el-Hak, editor, *Advances in Fluid Mechanics Measurements*, pages 229–261. Springer-Verlag, Berlin, 1989.

[26] M.R. Head and I. Rechenberg. The Preston tube as a means of measuring skin friction. *J. Fluid Mech.*, 14:1–17, 1962.

[27] C.M. Ho and Y.C. Tai. Micro-electro-mechanical-systems and fluid flows. *Ann. Rev. Fluid Mech.*, 30:579–612, 1998.

[28] A.B. Hollingshead and N. Rajaratnam. A calibration chart for the Preston tube. *J. Hydr. Res.*, 18:313–326, 1980.

[29] K. Hufnagel and G. Schewe. Force and moment measurement (chapter 8). In C. Tropea, A.L. Yarin, and J.F. Foss, editors, *Springer Handbook of Experimental Fluid Mechanics*. Springer, Berlin, 2007.

[30] N. Hutchins and K.-S. Choi. Accurate measurements of local skin friction coefficient using hot-wire anemometry. *Prog. Aerospace Sci.*, 38:421–446, 2002.

[31] M.-É. Lamarche-Gagnon and J. Vétel. An inverse problem to assess the two-component unsteady wall shear rate. *Intern. J. Thermal Sci.*, 130:278–288, 2018.

[32] H.W. Liepmann and G.T. Skinner. Shearing-stress measurements by use of a heated element. Technical Report Technical Note 3268, National Advisory Committee for Aeronautics, Washington DC, 1954.

[33] L. Loefdahl and M. Gad-el-Hak. MEMS-based pressure and shear stress sensors for turbulent flows. *Meas. Sci. Technol.*, 10:665–686, 1999.

[34] J. Lunte and E. Schülein. Wall shear stress measurements by white-light oil-film interferometry. *Exp. Fluids*, 61:84, 2020.

[35] Z.X. Mao and T.J. Hanratty. Analysis of wall shear stress probes in large amplitude unsteady flows. *Int. J. Heat Mass Transfer*, 34:281–290, 1991.

[36] R. Mayer, R.A.W.M. Henkes, and J.L. Van Ingen. Wall-shear stress measurement with quantitative IR-thermography. AGARD CP-601, Paper 22, 1997.

[37] C. Mercer. *Optical Metrology for Fluids, Combustion and Solids*. Kluwer Academic Publishers, Dordrecht, 2003.

[38] J.D. Murphy and R.V. Westphal. The laser interferometer skin-friction meter: A numerical and experimental study. *J. Phys. E: Sci. Isctrum.*, 19:744–751, 1986.

[39] A.A. Naqwi and W.C. Reynolds. Measurement of turbulent wall velocity gradients using cylindrical waves of laser light. *Exp. Fluids*, 10:257–266, 1991.

[40] J.W. Naughton and M. Sheplak. Modern developments in shear-stress measurement. *Prog. Aerospace Sci.*, 38:515–570, 2002.

[41] G. Onsrud, L.N. Persen, and L.R. Saetran. On the measurement of wall shear stress. *Exp. Fluids*, 5:11–16, 1987.

[42] R. Orlü and R. Vinuesa. Instantaneous wall-shear-stress measurements: advances and application to near-wall extreme events. *Meas. Sci. Technol.*, 31:112001, 2020.

[43] V. Ozarapoglu. *Measurements in Incompressible Turbulent Flows*. PhD thesis, Laval University, Quebec, 1973.

[44] A. Padmanabhan, H.D. Goldberd, M.A. Schmidt, and K.S. Breuer. A wafer-bonded floating-element shear-stress micro-sensor with optical position sensing by photodiodes. *IEEE J. Microelectricalmechanical Syst.*, 5:307–315, 1996.

[45] B.R. Pai and J.H. Whitelaw. Simplification of the razor blade technique and its application to the measurement of wall-shear stress in wall-jet flows. *Aero. Quarterly*, 20:355–364, 1969.

[46] V.C. Patel. Calibration of the Preston tube and limitations on its use in pressure gradients. *J. Fluid Mech.*, 23:185–208, 1965.

[47] J.H. Preston. The determination of turbulent skin friction by means of pitot tubes. *J. Roy. Aeron. Soc.*, 58:109–121, 1954.

[48] D.C. Reda and J.J. Muratore. Measurement of surface shear strress vectors using liquid crystal coatings. *AIAA J.*, 32:1576–1582, 1994.

[49] D.C. Reda, M.C. Wilder, R. Mehta, and G. Zilliac. Measurement of continuous pressure and shear distributions using coatings and imaging techniques. *AIAA J.*, 36:895–899, 1998.

[50] F. Rehimi, F. Aloui, S.B. Nasrallah, L. Doubliez, and J. Legrand. Inverse method for electrodiffusional diagnostics of flows. *Int. J. Heat Mass Transfer*, 49:1242–1254, 2006.

[51] S. Rezaeiravesh, R. Vinuesa, M. Liefvendahl, and P. Schlatter. Assessment of uncertainties in hot-wire anemometry and oil-film interferometry measurements for wall-bounded turbulent flows. *European J. Mech.-B/Fluids*, 72:57–73, 2018.

[52] J.D. Ruedi, H. Nagib, J. Oesterlund, and P.A. Monkewitz. Evaluation of three techniques for wall-shear measurements in three-dimensional flows. *Exp. Fluids*, 35:389–396, 2003.

[53] M. Schober, E. Obermeier, S. Pirskawetz, and H.-H. Fernholz. A MEMS skin-friction sensor for time resolved measurements in separated flows. *Exp. Fluids*, 36:593–599, 2004.

[54] A.J. Smits and T.T. Lim, editors. *Flow Visualization Techniques and Examples.* Imperial College Press, London, 2000.

[55] V. Sobolík, O. Wein, and J. Čermák. Simultaneous measurement of film thickness and wall shear stress in wavy flow of non-Newtonian liquids. *Collect. Czech Chem. Commun.*, 52:913–928, 1987.

[56] T.E. Stanton, D. Marshall, and C.N. Bryant. Om the conditions at the boundary of a fluid in turbulent mótion. *Proc. Roy. Soc. Series A*, 97:413–434, 1920.

[57] L.H. Tanner and L.G. Blows. A study of the motion of oil films on surfaces in air flow, with application to the measurement of skin friction. *J. Phys. E: Sci. Instrum.*, 9:194–202, 1976.

[58] J.B. Vagt and H. Fernholz. Use of surface fences to measure wall shear stress in three-dimensional boundary layers. *Aero. Quarterly*, 2:87–91, 1973.

[59] W.B. Watson. Bonded electrical resistance strain gages. In W.N. Sharpe Jr., editor, *Handbook of Experimental Solid Mechanics*, pages 283–334. Springer, New York, 2008.

[60] O. Wein and V. Sobolík. Theory of direction sensitive probes for electrodiffusion measurement of wall velocity gradients. *Collect. Czech Chem. Commun.*, 52:2169–2180, 1987.

[61] A.J. Wheeler and A.R Ganji. *Introduction to Engineering Experimentation, 3rd Edition.* Pearson, Upper Saddle River, NJ, 2009.

[62] K.G. Winter. An outline of the techniques available for the measurement of skin friction in turbulent boundary layers. *Prog. Aerospace Sci.*, 18:1–57, 1977.

[63] M.V. Zagarola, D.R. Williams, and A.J. Smits. Calibration of the Preston probe for high Reynolds number flows. *Meas. Sci. Technol.*, 12:495–501, 2001.

[64] J. Zhao. Measurement of wall shear stress in high speed air flow using shear-sensitive liquid crystal coating. *Sensors*, 18:1605, 2018.

14 Measurement of Temperature

Unlike velocity, temperature does not have a scientific definition, although it is commonly understood to be a measure of 'hotness' or 'coldness' of an object. The value of temperature is defined with respect to a continuous temperature scale. Two types of such scales exist: (a) *thermodynamic scales*, which are based entirely on thermodynamic considerations, such as the ideal gas law or the thermal noise of electric components; and (b) *practical* or *laboratory scales*, which are constructed by the definition of discrete reference values and interpolation in the intervals between consecutive reference values with the use of consistently performing instruments [35]. Following a brief review of the currently accepted practical temperature scale, this chapter describes the most common temperature sensors and other temperature measurement methods that are used in fluid mechanics research. More details on specific methods can be found in general references [3, 5, 8, 17, 23, 25, 41, 42].

14.1 The International Temperature Scale

According to the International Temperature Scale of 1990 (ITS-90) [32], the unit of *thermodynamic temperature* T is the *kelvin*, denoted as K and defined as the fraction 1/273.16 of the thermodynamic temperature of the *triple point of water* (TPW). The TPW is the unique state at which vapour, liquid and ice coexist in equilibrium, and which occurs when the pressure is 611 Pa; this state can be readily achieved with the use of a water triple-point cell. For historical reasons, the *Celcius temperature* t remains in extensive use, besides the thermodynamic temperature; its unit is *degrees Celsius*, denoted as °C and defined as $t/°\mathrm{C} = T/\mathrm{K} - 273.15$. In this scale, the *ice point of water*, namely the freezing point of pure water at standard atmospheric pressure, is at 0°C and the TPW is at 0.0100°C.

The ITS-90 defines 14 *fixed points* of temperature, corresponding to phase equilibria or phase transitions of water, gases and metals, which can be reproduced consistently. The highest temperature defined by a fixed point in this scale is T_{Cu} = 1,357.77 K (1,084.62°C), corresponding to the freezing point of copper at a pressure of 101,325 Pa, and the lowest fixed point is the triple point of hydrogen, at T_H = 13.80 K (−259.35°C). Between fixed points, and within overlapping ranges, the temperature scale is defined continuously by using the response of standard thermometers. From 3.00 K to 24.56 K a helium glass thermometer is used, whereas, between 13.80 and 1,234.93 K (which is the freezing point of silver), a capsule platinum resistance thermometer is used. Temperatures higher than 1,234.93 K are defined by reference to T_{Cu}, such that the spectral radiance of a black body at the wavelength λ (in m) and at temperature T (in K) in vacuum satisfies *Planck's radiation law* (Eq. (7.27))

$$\frac{L_{e\lambda}(T)}{L_{e\lambda}(T_{Cu})} = \frac{e^{C_2/(\lambda T_{Cu})} - 1}{e^{C_2/(\lambda T)} - 1}, \qquad (14.1)$$

where $C_2 = hc/k_B = 0.014388$ mK and $L_{e\lambda}(T_{Cu})$ is given by a special black body operating at T_{Cu} and used as a standard [46]. Such temperatures are determined with the use of an optical pyrometer (see Section 14.5.1), but they are not very reliable for temperatures that are higher than about 4,000°C. Higher temperatures, up to 10^9 K, such as those occurring in atomic explosions or in the interior of stars, are defined using spectroscopic methods. Temperatures lower than 50 K and approaching absolute zero have been defined using helium vapour pressure thermometers.

14.2 Temperature Sensors

Many different types of temperature sensors are available. The vast majority of them can be classified in one of the following three categories, depending on the principle of their operation:

- *thermal expansion thermometers*, the operation of which utilises the relationship between temperature and volume or length of a material; common types are the *liquid-in-glass thermometers* and the *bimaterial thermometers* ;
- *thermocouples*, the operation of which is based on the *thermoelectric effect*;
- *resistance thermometers*, the operation of which is based on the relationship between temperature and electric resistance; these include *resistance temperature detectors (RTD)*, which are metallic resistance sensors, and *thermistors*, which are semiconductor resistance sensors.

14.2.1 Liquid-in-Glass Thermometers

Liquid-in-glass thermometers [5, 8, 27] consist of an evacuated and sealed glass container, which is filled partially by a liquid, such as alcohol or mercury. The glass container consists of a bulb-type reservoir and a graduated capillary tube (stem). As the temperature of the liquid in the bulb increases, the liquid expands to fill a longer length of the capillary tube. Laboratory-grade thermometers have typical resolutions of about 0.05 to 1 K and can be classified as *complete immersion*, *total immersion* and *partial immersion* types. Complete immersion thermometers must be entirely immersed in the fluid of interest, which may make their reading difficult, at least for liquid flows. Total immersion thermometers must be inserted in the fluid of interest up to the end of the liquid column, while partial immersion thermometers must be inserted in the fluid up to a marked line. Properly used total immersion thermometers are free of errors due to temperature differences between the immersed and non-immersed sections, while partial immersion thermometers suffer from such errors, which may significantly exceed their resolution. The readings of total immersion thermometers used in the partial immersion mode can be corrected by adding a *stem correction*, which is computed from the temperature of the non-immersed portion of the liquid column, measured by another thermometer. Following extensive use, the volume of the bulb might change slightly due to accumulating thermal stresses, and it is advisable to check periodically the reading of accurate liquid-in-glass thermometers versus an ice-point bath or other reference temperature. When inserted into flowing fluids through holes or wells in pipe walls, one must also consider possible errors due to heat conduction through the exposed stem or the well wall. Thermal insulation can be used to reduce such errors. Liquid-in-glass thermometers have poor spatial and temporal resolutions and are unsuitable for measuring temperatures that change rapidly with location or time; however, they are excellent laboratory standards for the calibration of other instruments.

14.2.2 Bimaterial Thermometers

Bimaterial thermometers [8, 41, 42] typically consist of two thin strips of different materials A and B, bonded together tightly at a reference temperature T_r to form an assembly. The two materials are selected such as to have vastly different thermal expansion coefficients α_A and α_B. The lengths L_A, L_B of the two strips at a different temperature T, when they are not bonded to each other, would be

$$L_A(T) = L_A(T_r)[1 + \alpha_A(T - T_r)] , \tag{14.2}$$

$$L_B(T) = L_B(T_r)[1 + \alpha_B(T - T_r)] . \tag{14.3}$$

When the two strips are bonded, the strip with the smaller thermal expansion coefficient will tend to pull the other strip towards itself, as the temperature increases, and so the curvature of the assembly will change. Moreover, the length of the assembly will increase, as a result of thermal expansion of both materials.

Consider a short segment of the assembly of two bonded materials, having thicknesses δ_A, δ_B and thermal expansion coefficients α_A, α_B, such that $\alpha_A < \alpha_B$, as shown in Fig. 14.1a. Let R_{cr} be the mean radius of curvature of the assembly at temperature T_r. When the temperature of the assembly changes, the length of material A will change by a lesser amount than the length of material B, as evident from the previous expressions. Consequently, material A will pull material B, which will result into a change of the radius of curvature of the assembly to a value $R_c(T)$, given by

$$\frac{1}{R_c(T)} - \frac{1}{R_{cr}} = \frac{6(1+m)^2(\alpha_B - \alpha_A)(T - T_r)}{\delta\left[3(1+m)^2 + (1+mn)(m^2 + 1/mn)\right]} , \tag{14.4}$$

where $m = \delta_A/\delta_B$ is the thickness ratio, $\delta = \delta_A + \delta_B$ is the total thickness, and $n = E_A/E_B$ is the ratio of the Young's moduli of elasticity of the two materials. If the two materials have equal thicknesses (i.e., $m = 1$), and, noting that the previous expression is not very

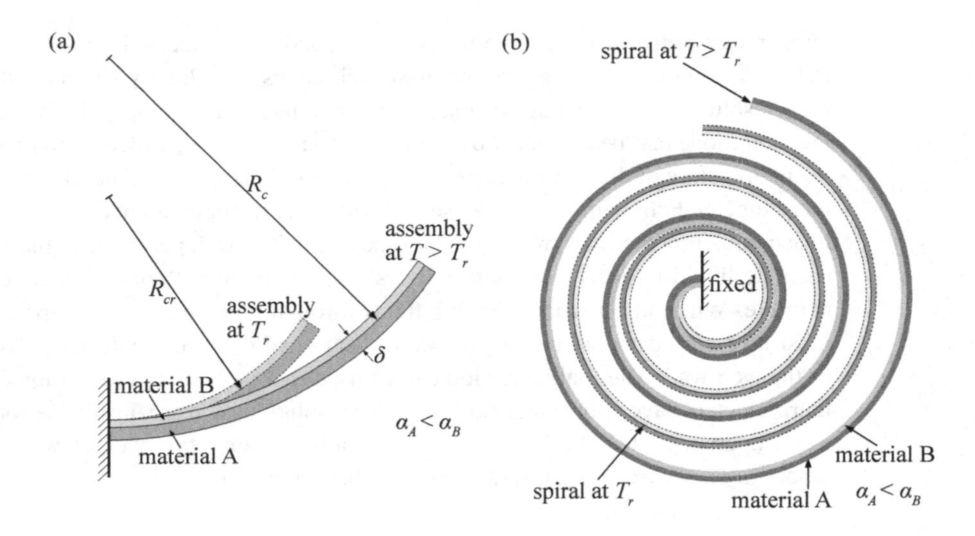

Figure 14.1 (a) A segment of an assembly of two bonded strips of different materials A and B, having one fixed end and a free second end, at two different temperatures T_r and $T > T_r$ ($\alpha_A < \alpha_B$); (b) a schematic diagram of the active element of a spiral bimaterial thermometer.

sensitive to the value of n, we may substitute $n \approx 1$ to derive the simplified expression

$$\frac{1}{R_c(T)} - \frac{1}{R_c} \approx \frac{3(\alpha_B - \alpha_A)(T - T_r)}{2\delta} . \tag{14.5}$$

A bimaterial assembly with its one end fixed and the other end free to move will become an actuator, responding to temperature changes. Analogue bimaterial thermometers usually contain a mechanism that links the free end of a bimaterial assembly to a needle that pivots on an axis, showing temperature on a calibrated scale. The actuation is amplified with the use of a helical or spiral assembly, as shown in Fig. 14.1b. Besides serving as roast temperature indicators, bimaterial thermometers and assemblies are used in a wide range of applications, including many industrial systems and thermostatic controls. In most applications, metals are used for both sides, although non-metallic materials, including carbon and semiconductors, have also been used. The relevant properties of materials used in bimaterial thermometers are listed in Table 14.1 [42]. A favourite choice for one material is Invar, a 36% nickel alloy of steel, which, among available metals and alloys, is the one with the lowest expansion coefficient. Bimaterial thermometers have a typical resolution of about 1% of full scale and a maximum operating temperature of about 500°C.

Example 14.1 A simple bimaterial thermometer is constructed of two strips of copper and nickel, each of which have a thickness of 5 mm and a length of 50 mm. They are bonded together on a flat surface at a temperature of $T_r = 280$ K, with the nickel strip placed on top of the copper strip. One end of the assembly is fixed to a surface,

Table 14.1 Properties of materials used commonly in bimaterial temperature sensors

Material	α [K^{-1}]	E [GPa]
Copper, Cu	383.1 × 10^{-6}	129.8
Aluminium, Al	24 × 10^{-6}	61–71
Nickel, Ni	13.3 × 10^{-6}	199.5
Iron, Fe	12.1 × 10^{-6}	211.4
Titanium, Ti	8.9 × 10^{-6}	120.2
Invar	1.7–2.0 × 10^{-6}	140–150
Silicon, Si	4.7–7.6 × 10^{-6}	113
Silicon Dioxide, SiO_2	0.50 × 10^{-6}	57–85

whereas the other end is free to move, as shown in Fig. 14.2. Determine the approximate vertical deflection, and direction, of the free tip of the assembly, when it is exposed to a temperature $T = 300$ K. Clearly state your assumptions.

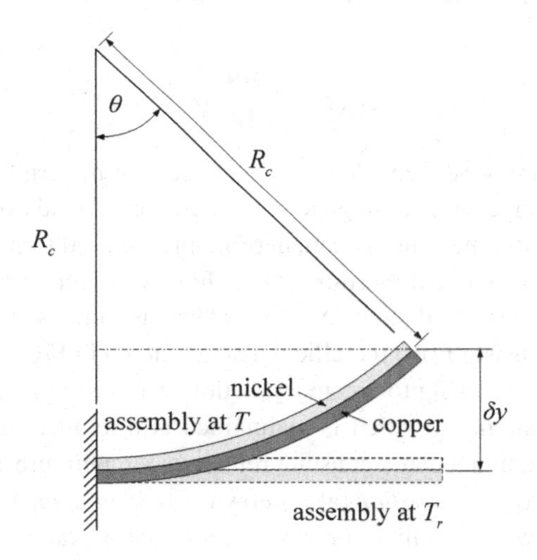

Figure 14.2 Deformation of an initially planar bimaterial assembly, following an increase in temperature.

Answer

Using basic trigonometric relationships, the vertical deflection of the free tip is $\delta y = R_c(1 - \cos\theta)$, where θ is the deflection angle, shown in Fig. 14.2. Because the assembly is flat at T_r, its initial radius of curvature is $R_{cr} \to \infty$. As both strips have the same thickness, and the Young's moduli of the two materials are only moderately different from each other, we may use the simplified expression (Eq. (14.5)) to determine R_c. The deflection angle is found by using the geometrical relationship $\theta = L_{av}/R_c$, where, in this case, the length L_{av} of the circular arc is estimated by averaging the values given by Eqs. (14.2) and (14.3). For the given materials and dimensions, we find $L_{av} = 50.2$ mm, $R_c = 450.7$ mm, and, hence, $\theta = 0.11$ rad $= 6.3°$. The vertical deflection is therefore $\delta y = 2.8$ mm. Considering that $\alpha_{Ni} < \alpha_{Cu}$ and that the nickel strip is placed on top of the copper strip, we expect the assembly to get deflected upwards, as shown in Fig. 14.2.

14.2.3 Thermocouples

Thermocouples (TC) [8, 27, 41, 42] are the most common temperature sensors in engineering applications. They are relatively inexpensive, simple to use, robust, and have an operating range that is wider than the range of any other temperature sensor. Their operation may be briefly explained as follows.

Conversion of mechanical, chemical or thermal energy into electrical energy in a conductor generates an electrical action, called the *electromotive force* (emf) and denoted as $\mathcal{E}$. Consider a thin electrical conductor, made of a homogeneous material, for example, a metallic wire. If the temperature varies along its length, this conductor will also have different local values of emf, which would depend on the local temperature (*Thomson effect*). Therefore, if a conductor has its two ends at two different temperatures, it would generate a voltage difference V_o between its two ends, which is referred to as the *thermoelectric voltage*. For example, consider material A in Fig. 14.3, which has one end at temperature T_s and the other end at temperature T_r. The corresponding thermoelectric voltage would be

$$V_{o,A}(T_r - T_s) \equiv \mathcal{E}_A(T_r) - \mathcal{E}_A(T_s) = \int_{T_s}^{T_r} \sigma_A(T)\mathrm{d}T \,, \tag{14.6}$$

where $\sigma_A(T)$ is the (temperature dependent) *Seebeck coefficient* of material A. In open circuit, the thermoelectric voltage of a homogeneous conductor depends only on the material and the temperature difference, and so can, in principle, be used for temperature measurement. The operation of practical thermocouples, however, is not based on the Thomson effect, but on the related, but distinct, *Seebeck effect*, as follows.

Thermocouples generally consist of two metallic wires, as shown in Fig. 14.3, joined firmly by welding, soldering or pressing to create a junction, referred to as the *sensing junction* (also known as *hot junction*), which is maintained at a temperature T_s. The other ends of both wires are maintained at a constant reference temperature T_r in an arrangement referred to as the *reference junction* (also known as *cold junction*). According to the *law of homogeneous materials*, a thermocouple circuit that is made entirely of a homogeneous material cannot generate a voltage difference across two points that have the same temperature. Consequently, a typical thermocouple circuit consists of two dissimilar wires (e.g., materials A and B in Fig. 14.3), the corresponding ends of which are exposed to a temperature T_s at the sensing junction and a temperature T_r at the reference junction. Then, the measurable voltage difference at the reference junction would be

$$\begin{aligned} V_{o,AB}(T_r - T_s) &= [\mathcal{E}_A(T_r) - \mathcal{E}_A(T_s)] + [\mathcal{E}_B(T_s) - \mathcal{E}_B(T_r)] \\ &= [\mathcal{E}_A(T_r) - \mathcal{E}_B(T_r)] - [\mathcal{E}_A(T_s) - \mathcal{E}_B(T_s)] \\ &= \mathcal{E}_{AB}(T_r) - \mathcal{E}_{AB}(T_s) \,. \end{aligned} \tag{14.7}$$

Practical thermocouple circuits usually contain conductors made of more than two materials, e.g., materials A,B and C in Fig. 14.3. The analysis of such circuits is based on the *law of intermediate materials*, which states that the sum of all emf in a thermocouple circuit using two or more different materials would be zero, if the entire circuit were at the same temperature. A consequence of this law is that, if a third material were

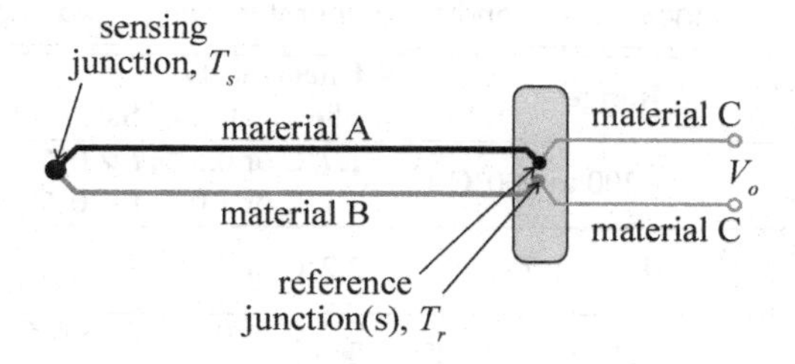

Figure 14.3 Thermocouple circuit made of a sensing junction of two dissimilar wires and a reference junction, which is connected by wires of a third material to a voltmeter.

inserted into a thermocouple circuit, it would not change the measured voltage difference, if the two junctions of the inserted material were at the same temperature. This is readily verified from Eq. (14.6), and is the reason why junctions of two materials can be made using a third material as solder. This concept is illustrated in Fig. 14.3, in which wires of material C, connected to materials A and B within the reference junction, are used to connect the thermocouple circuit to a voltmeter.

Commercially available thermocouples, which conform to National Institute of Standards and Technology (NIST) standards, are designated by a letter, such as J, E and K, which must have specific voltage values at stated temperatures. The operating range and output voltages (relative to a reference junction at 0°C) of selected NIST-standard thermocouples are shown in Fig. 14.4. The properties of some common types are listed in Table 14.2, along with their operating range and sensitivity at 0°C. Typical resolutions of thermocouples are of the order of 1°C. For highest accuracy, it is advisable to calibrate individual sensors separately, fitting the response curve by a low-order polynomial. Good-quality commercial thermocouples are within 0.5 to 3°C of the tabulated values supplied by the manufacturer. Several types of common thermocouples have ranges exceeding 1,000°C, with the highest temperature of 2,930°C reached by the tungsten–

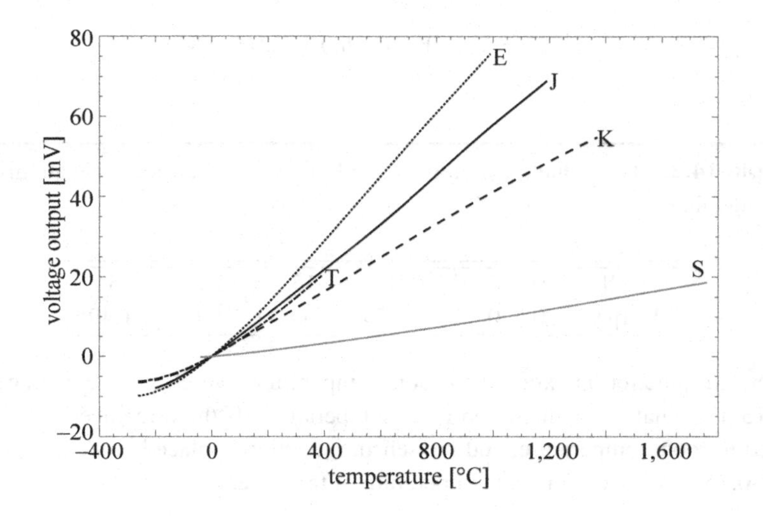

Figure 14.4 Output voltage for several common thermocouple types, with their reference junction temperature at 0°C, as per NIST standards.

Table 14.2 Common types of NIST thermocouples, showing their materials at the positive (+) and negative (−) terminals and their corresponding temperature range and uncertainty

Type	Materials	Range	Uncertainty (the greatest of listed values)
E	(+) chromel (−) constantan	−200 to 900°C	1.7°C or 0.5%, $t > 0$°C 1.7°C or 1.0%, $t < 0$°C
J	(+) iron (−) constantan	0 to 750°C	2.2°C or 0.75%
K	(+) chromel (−) alumel	−200 to 1,250°C	2.2°C or 0.75%, $t > 0$°C 2.2°C or 2.0%, $t < 0$°C
S	(+) platinum/ 10% rhodium (−) platinum	0 to 1,450°C	1.5°C or 0.25%
T	(+) copper (−) constantan	−200 to 350°C	1.0°C or 0.75%, $t > 0$°C 1.0°C or 1.5%, $t < 0$°C

rhenium type. For measuring higher temperatures, up to about 4,000°C, cooled sensor arrangements have been developed, in which two thermocouples are used to measure temperature differences in a cooling stream inside a tube exposed to the hot fluid. The typical size of thermocouple junctions is of the order of 1 mm, with time constants of the order of 1 s. Much smaller micro-junctions are also offered commercially, with time constants of the order of 1 ms, while their frequency response may be further boosted by electronic compensation [38]. The reader is referred to manufacturer catalogues for specialised thermocouple probe and sensor designs and arrangements.

In practice, one may need to determine the response of a thermocouple when it is used with its reference junction at a temperature t_r, based on a calibration with the reference junction at a temperature $t_{rc} \neq t_r$. A similar situation would arise, if $t_r \neq 0$°C, but the thermocouple response is based on NIST tables. The conversion of the thermocouple circuit response to a different reference temperature is based on the *law of intermediate temperatures*, a consequence of which is that

$$V_o(t_s - t_r) = V_o(t_s - t_{rc}) + V_o(t_{rc} - t_r) \,. \tag{14.8}$$

Example 14.2 The voltage readings for a NIST Type E thermocouple are given in the following table.

t [°C]	0	5	10	15	20	25
V_o [μV]	0	0.294	0.591	0.890	1.192	1.495

The sensing junction is exposed to room temperature, which gives a voltage reading of 1.313 μV; what is the measured room temperature? If the reference junction is now exposed to room temperature, and the sensing junction is placed in a cup of water that reads −0.363 μV, determine the temperature of the water.

Answer

Interpolating the given data, we find that a reading of 1.313 μV corresponds to a temperature of 22°C. If the reference junction is now at room temperature, and we obtain a reading of −0.363 μV, then using Eq. (14.7), we convert this reading to a voltage of the NIST tables as $V_O(t-0) = V_O(t-22) + V_O(22-0) = -0.363 + 1.313 = 0.950$ μV, which corresponds to a temperature of 16°C.

14.2.4 Resistance Temperature Detectors

The term resistance temperature detector (RTD) is a commonly used name for pure metallic thermometers. The most popular RTD material is platinum, but nickel, copper and different alloys are also used [5, 8, 27, 41, 42]. In a typical RTD, the sensor is a wire wound over a ceramic substrate in the form of a coil, as shown in Fig. 14.5, or a thin film deposited on the substrate, and the assembly is encased in an electrically insulated metallic (e.g., made of stainless steel) sheath. RTD are very stable and accurate. Common types have a typical resolution of about 0.1 K, and are used extensively in the industry and many technological applications, whereas the Standard Platinum Resistance Thermometer (SPRT), with a resolution of 0.0001 K, is used as a laboratory standard. The sensitivity for a platinum wire is about 0.4% of its resistance per K.

The resistance–temperature relationship of metals over an extensive range of temperatures is non-linear and may be specified in the form of fitted, low-order polynomials. The ratio of the resistance of a platinum sensor is described by the *Callendar–Van Dusen*

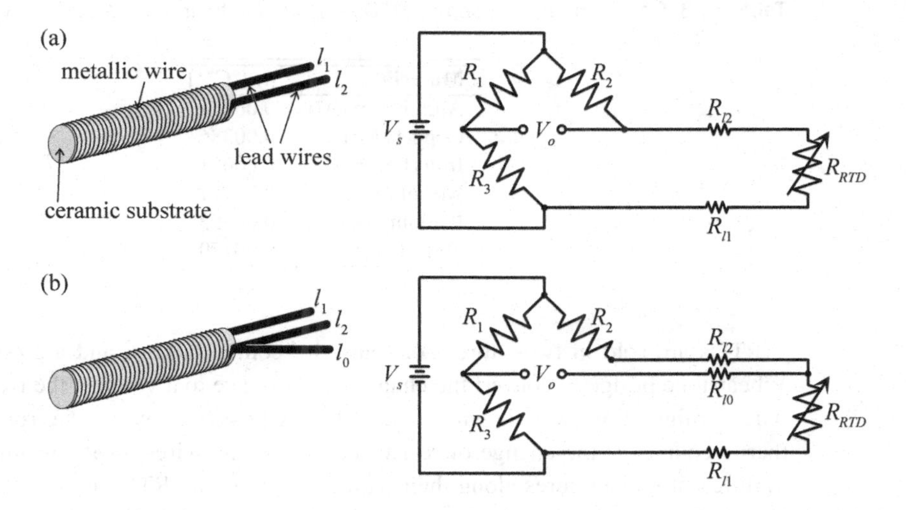

Figure 14.5 RTD assembly and Wheatstone bridge circuit used to measure the voltage produced by a change in temperature for (a) a two-wire configuration and (b) a three-wire configuration. The RTD assembly is comprised of a metallic wire wound around a cylindrical ceramic substrate, and is normally encased in an electrically insulated metallic sheath (not shown in the figure).

equation

$$\frac{R(t)}{R_0} = \begin{cases} 1 + At + Bt^2 + C(t-100)t^3\,, & \text{for } -200°\text{C} \le t < 0°\text{C}\,, \\ 1 + At + Bt^2\,, & \text{for } 0°\text{C} \le t \le 600°\text{C}\,, \end{cases} \tag{14.9}$$

where t is the temperature value in °C and R_0 is the resistance of the platinum RTD at $t = 0°$C. For example, a Pt100 RTD, which is available from different manufacturers, would have a resistance of 100 Ω at $t = 0°$C. Commercially available platinum RTDs, which conform to IEC 60751 standards, have constants $A = 3.9083 \times 10^{-3}\,°\text{C}^{-1}$, $B = -5.7750 \times 10^{-7}\,°\text{C}^{-2}$ and $C = -4.1830 \times 10^{-12}\,°\text{C}^{-4}$.

Within a relatively narrow range of positive temperatures, typically $0°\text{C} \le t \le 100°\text{C}$, the resistance of a metallic sensor can be approximated by the linear expression

$$R(t) \approx R_r[1 + \alpha_r(t - t_r)]\,, \tag{14.10}$$

where R_r is the resistance at a reference temperature t_r and α_r is the thermal resistivity coefficient at t_r. The values of α_r at $t_r = 20°$C for some common metals used as RTD sensors are given in Table 14.3.

Table 14.3 Common metals used as RTD sensors and their thermal resistivity coefficient at $t_r = 20°$C

Material	α_r [°C^{-1}]
Aluminium (Al)	0.00431
Copper (Cu)	0.00386
Iron (Fe)	0.00651
Nickel (Ni)	0.00671
Platinum (Pt)	0.00393
Tungsten (W)	0.00470

RTDs are sold in two-, three- or four-wire configurations, and are connected to a Wheatstone bridge to convert the change in resistance to a voltage; the two- and three-wire configurations are shown in Fig. 14.5. A possible source of error, when using thermocouples is the change of resistance of the lead wires, as the result of exposure to different temperatures along their paths. The two-wire RTD configuration shown in Fig. 14.5a is simple to implement, but it does not compensate the voltage reading for the lead resistance change. The three-wire RTD configuration, shown in Fig. 14.5b, corrects partially for this error, and the four-wire configuration reduces this error further. To achieve the highest possible accuracy of a particular RTD unit, the user is advised to calibrate it against a very accurate standard.

Example 14.3 The normalised voltage output in Fig. 14.5 for two- and three-wire RTDs with lead wires of equal resistance can be shown to be, respectively,

$$\left.\frac{V_o}{V_s}\right|_{2\mathrm{RTD}} = \frac{-[2\frac{R_{l2}}{R_r} + \alpha_r(t - t_r)]}{2[2\frac{R_{l2}}{R_r} + 2 + \alpha_r(t - t_r)]},$$

$$\left.\frac{V_o}{V_s}\right|_{3\mathrm{RTD}} = \frac{-[\alpha_r(t - t_r)]}{2[2\frac{R_{l2}}{R_r} + 2 + \alpha_r(t - t_r)]},$$

under the assumption that $R_1 = R_3$ and that the RTD has a linear response (Eq. (14.10)). The derivation is left as an assignment for the reader.

For a two- and three-wire copper RTD with a reference resistance of 50 Ω and a lead resistance of 5 Ω, both at a reference temperature of $t_r = 20.0°$ C, determine the temperature measurement error, when the RTD is exposed to a temperature $t = 50.0$°C.

Answer

For both the two- and three-wire RTDs, the lead wire resistance will increase due to a temperature change $t - t_r = 50.0 - 20.0 = 30.0$°C. The thermal resistivity coefficient for copper is $\alpha_r = 0.00386°\mathrm{C}^{-1}$ (see Table 14.3), hence the change in resistance of the lead wire, using Eq. (14.10), is $R_{l2} = 5(1 + 0.00386 \times 30) = 5.58$ Ω. The measured normalised voltage output for the two-wire RTD bridge is $\left.\frac{V_o}{V_s}\right|_{2\mathrm{RTD}} = -0.0725$; that is to say, this is the reading we would see when the two-wire copper RTD is exposed to a temperature $t = 50.0$°C. Our interest, however, is to determine the inferred temperature t_i, due to this reading. This is obtained by re-arranging the expression for the two-wire normalised voltage output to the form

$$t_{i,2} - t_r = \frac{-2\left(\left.\frac{R_{l2}}{R_r}\right|_i + 2\left.\frac{V_o}{V_s}\right|_{2\mathrm{RTD}} + 2\left.\frac{V_o}{V_s}\right|_{2\mathrm{RTD}}\left.\frac{R_{l2}}{R_r}\right|_i\right)}{\alpha_r\left(1 + 2\left.\frac{V_o}{V_s}\right|_{2\mathrm{RTD}}\right)},$$

where $\left.\frac{R_{l2}}{R_r}\right|_i = 5/50$ is the inferred, that is, known, initial ratio between the lead wire resistance and the RTD resistance. Using the previous values, we find the inferred temperature for the two-wire RTD configuration to be $t_{i,2} = 56.0$°C, which gives an absolute error of 6.0°C and a relative error of 12%.

In a similar manner, we find the measured normalised voltage output for the three-wire RTD bridge to be $\left.\frac{V_o}{V_s}\right|_{3\mathrm{RTD}} = -0.0248$, and the inferred temperature to be $t_{i,3} = 49.7$°C, which gives a relative error of 0.6%.

14.2.5 Thermistors

Thermistors are semiconductor elements, the resistance of which is a very strong function of temperature [8, 10, 41, 42]. Although thermistors with positive temperature coefficient of resistance (PTC), namely a resistance that increases with increasing temperature, are also available, it is metal oxide thermistors, which have a negative temperature coefficient of resistance (NTC), that are commonly used for temperature measurement in the laboratory. Within their range of operation, thermistors offer several advantages over other temperature sensors, however, they also have important disadvantages. A main advantage is their extremely high sensitivity to temperature for the usual range of laboratory temperatures. As an example, the resistance of a metal oxide thermistor will drop by seven orders of magnitude within the temperature range between −100 and 400°C, in sharp contrast to a platinum RTD, the resistance of which will increase by a single order of magnitude within the previously listed temperature range. In the room temperature range, the sensitivity of a thermistor is typically of the order of 10^5 Ω/K, dropping significantly as temperature increases. Another advantage of thermistors is their insensitivity to aging, electromagnetic fields and radiation. A calibrated thermistor is likely to remain faithful to its calibration relationship over many years. One disadvantage is the strong non-linearity of thermistor response, which may be reduced by operating the sensor in a special bridge configuration. Among the various proposed relationships between thermistor resistance R_{th} and temperature, a fairly accurate one is

$$\frac{1}{T} = A + B \ln R_{th} + C (\ln R_{th})^3 , \tag{14.11}$$

where A, B and C are empirical coefficients specific to each sensor and T is the absolute temperature, measured in K. Thermistors are generally not interchangeable and need to be calibrated individually. Their range of operation is limited to a few hundred degrees C and their time constant typically exceeds 1 s. In order to measure its resistance, one has to supply the thermistor with some current i. If this current is sufficiently large, the dissipated electric power would raise the thermistor temperature above the ambient level (self-heating effect), which would consequently reduce its resistance and generate an error in the voltage output iR_{th}. In fact, if the current exceeds a certain threshold, the voltage output would start decreasing rather than increase with increasing current. Thermistors are available commercially in different forms and shapes, including bare and glass-coated beads, disks, rods, washers and different types of assemblies. Beside their use as thermometers, they find extensive application as parts of thermostatic circuits and other control systems.

14.2.6 Cold Wires

Cold wires are thin metallic resistance thermometers used for measuring temperature fluctuations in turbulent air flows [6]. Their name was introduced to contrast the term hot wire, which applies to similar sensors used for velocity measurement. Like hot wires, cold-wire sensors are typically mounted on two prongs (see Fig. 12.7), either by spotwelding a bare wire or by soldering a Wollaston wire and then etching the silver coating of a central segment. The sensor is powered by a constant current source [12, 29, 39], which permits the measurement of resistance through voltage measurement, following calibration. The current must be sufficiently low to avoid self-heating and sensitivity to velocity fluctuations. Typical current values are 0.5 mA or less, depending on the flow velocity and the intensity of temperature fluctuations. Considering that cold wire signals are generally very weak, one should strive to keep the current as high as possible to reduce the need for high gain amplification, which is accompanied by electronic noise. Common sensor materials are platinum, which has the highest sensitivity to temperature fluctuations, and platinum alloys or tungsten, which have higher mechanical strength. Most researchers have used cold wires with a diameter of 1 μm, but wires with diameters 2.5, 0.6 and 0.25 μm have also been used. It is obvious that the finer the sensor is, the better its frequency response would be; however, choice of diameter is also influenced by the need for mechanical strength and the ease in fabrication and handling. As a rough indication, one may consider that clean cold wires with 1 μm diameter have a high-frequency 3 dB cut-off of about 3 kHz, which increases to about 10 kHz for sensors with 0.25 μm diameters [2, 13, 19, 21]. The frequency response generally increases with increasing flow speed but deteriorates with dirt accumulation on the sensor. Frequent dipping of cold wires in acetone or other solvents to remove coatings is recommended. Frequency compensation or use of more elaborate response relationships can be employed to further improve the temporal resolution of cold wires [21, 40, 43]; such approaches require dynamic calibration of individual sensors, to be discussed separately later in this section. Besides their limited frequency response, cold wires also have limited spatial resolution, especially when used to measure fine structure properties in high Reynolds number turbulence. Theoretical estimates of sensor finite-length effects are available and could be used for estimating the measurement uncertainty and for devising corrections [18, 20, 44]. Two additional problems are related to the interference of the prongs [28, 30]. The first problem is heat conduction from the sensor to the much larger prongs; this results in distorting the sensor's transfer function in the low-frequency range, such that its magnitude attains an intermediate plateau, rather than smoothly approaching a maximum. A second problem may arise when the sensor is immersed in the thermal boundary layers of the prongs; this problem would be accentuated for sensors operating in a gas with a high thermal conductivity, such as helium. Both adverse effects would be reduced, albeit at the expense of spatial resolution, by increasing the sensor's length-to-diameter ratio.

14.3 Dynamic Response of Temperature Sensors

14.3.1 Lumped Parameter Models

When a temperature sensor is immersed in a fluid with a time-dependent temperature field, it will indicate a measured temperature T_m, which, in general, would be different from the actual local temperature T of the fluid. The First Law of Thermodynamics, or Energy Equation, complemented by additional relationships, if necessary, may be used to determine the relationship between T_m and T. In general, heat transfer between the surrounding fluid and the sensor would consist of conduction, convection, and radiation, each of which would depend on the geometry of the configuration and the material properties of the sensor and the fluid, which may not be uniform. Consequently, the heat transfer model would generally consist of one or more three-dimensional, partial differential equations. To simplify the analysis, we often assume that the temperatures of the fluid and the sensor are uniform, which also implies that material properties are uniform as well. This assumption reduces the model to a *lumped parameter model*, which is described by an ordinary differential equation for $T_m(t)$. As mentioned previously, a simple temperature sensor may be modelled as a first-order system, which means that its dynamic response would be fully described by values of its static sensitivity K and time constant τ. Of course, when temperature sensors are parts of electric or electronic circuits or other types of measuring systems, one would have to consider the dynamic response of the entire system and not the thermal response of the sensor alone.

Example 14.4 Derive a model for the dynamic response of a liquid-in-glass thermometer, in which the liquid is contained in a spherical bulb with a diameter d and is immersed in a flowing fluid (see Fig. 14.6). Neglect the thickness of the glass container and assume that heat transfer is dominated by convection around the bulb and that the volume of the fluid in the capillary tube is negligible by comparison to the volume of the bulb.

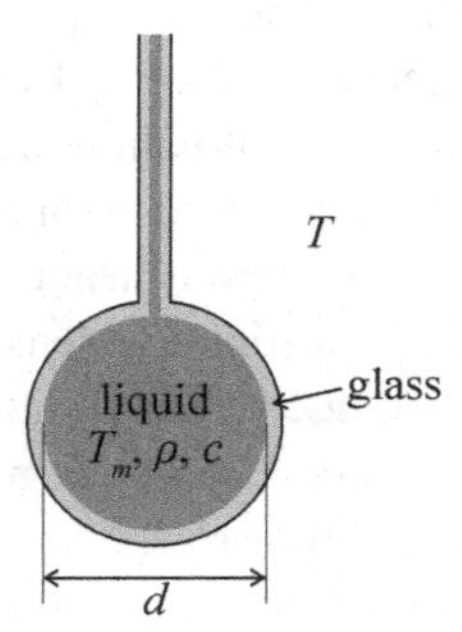

Figure 14.6 Sketch of an idealised liquid-in-glass thermometer.

Answer

We are going to adopt a lump parameter model, assuming that the liquid in the bulb has a uniform temperature T_m and the surrounding fluid has a uniform temperature T. Let $V = \pi d^3/6$ be the volume of the bulb, $A = \pi d^2$ be its area, ρ be the density of the thermometric fluid and c be its specific heat. Further assume that only forced convective heat transfer between the bulb and the flowing fluid takes place and that the convective heat transfer coefficient is h. Then, the Energy Equation, combined with Newton's law of convection, gives the heat transfer during time δt as

$$hA\,(T - T_m)\,\delta t = c\rho V \delta T_m\,,$$

which, in the limit $\delta t \to 0$, leads to the first-order differential equation

$$\frac{c\rho V}{hA}\frac{\mathrm{d}T_m}{\mathrm{d}t} + T_m = T\,. \tag{14.12}$$

Thus, the thermometer would respond to flow temperature changes as a first-order system with a time constant

$$\tau = \frac{c\rho V}{hA} \,. \tag{14.13}$$

Clearly, the time constant depends on the volume of the bulb and the heat transfer coefficient, which generally increases with increasing flow speed. Thus, the dynamic response of a thermometer would be better in fast streams than in slow streams or still fluids, where heat transfer may also take place by conduction or natural convection. The assumption of uniform temperature within the thermometer would be valid only if the thermal conductivity k of the thermometric fluid were sufficiently large for the *Biot number* Bi to be very small [14], namely,

$$\mathrm{Bi} = \frac{hV}{kA} \ll 1 \,.$$

Although a first-order model may be sufficiently accurate for many temperature sensors, in some cases, one may employ an improved lump parameter model of a second or higher order. This would, for example, be appropriate, if the sensor were surrounded by a coating, sheath, well or the like, which would complicate heat transfer. At this time, one is reminded that, although thermal systems are analogous to electrical, hydraulic and pneumatic systems, with which they share the same mathematical models, thermal systems have a distinct peculiarity: heat transfer always takes place from a warmer to a colder object, and, thus, there is no thermal property analogous to mechanical inertia or electrical inductance. Consequently, a temperature sensor will not overshoot above the maximum temperature it has been exposed to. The commonly used term 'thermal inertia' actually indicates a lag in thermal response and it is unrelated to mechanical inertia. It is then evident that a second-order model of a temperature sensor must necessarily be overdamped (this will be illustrated in the following example). Third-order and non-linear models of temperature sensors have also been proposed in the literature [8]; such models may not lead to oscillatory behaviour either.

Example 14.5 Derive a dynamic model for a thermistor embedded in a glass coating and immersed in a fluid stream (see Fig. 14.7). Show that this is an overdamped, second-order model.

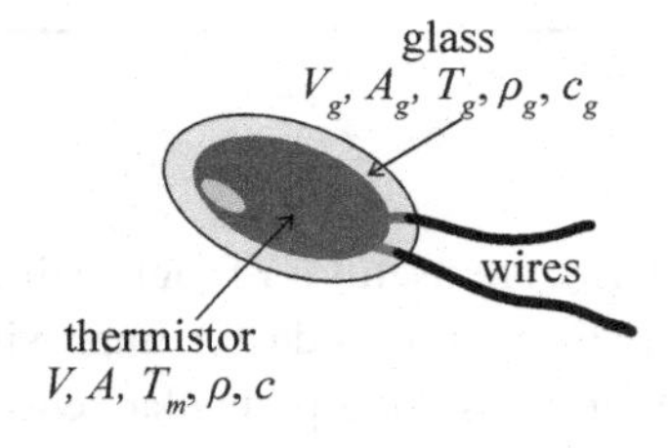

Figure 14.7 Sketch of an idealised glass-coated thermistor bead.

Answer

Consider a thermistor bead with a volume V, a surface area A, a density ρ and a specific heat c. The thermistor is encased within a glass coating, having a volume V_g, a surface area A_g, a density ρ_g and a specific heat c_g. Using the lumped parameter approach, we assume that the thermistor has a temperature T_m, the coating has a temperature T_g and the surrounding fluid has a temperature T. Let the overall heat transfer coefficient from the glass coating to the thermistor be h and the overall heat transfer coefficient from the fluid to the glass coating be h_g. Then, application of the Energy Equation for heat transfer from the flow to the glass/thermistor assembly and from the glass coating to the thermistor leads, respectively, to the equations

$$\frac{c_g \rho_g V_g}{h_g A_g} \frac{\mathrm{d}T_g}{\mathrm{d}t} + T_g + \frac{c\rho V}{h_g A_g} \frac{\mathrm{d}T_m}{\mathrm{d}t} = T \,,$$

$$\frac{c\rho V}{hA} \frac{\mathrm{d}T_m}{\mathrm{d}t} + T_m = T_g \,.$$

Eliminating T_g between these equations, one gets the second-order model [8]

$$\frac{\mathrm{d}^2 T_m}{\mathrm{d}t^2} + \frac{\tau + \tau_g + \tau_c}{\tau \tau_g} \frac{\mathrm{d}T_m}{\mathrm{d}t} + \frac{1}{\tau \tau_g} T_m = \frac{1}{\tau \tau_g} T \,,$$

which, in addition to the thermistor time constant τ, contains the coating time constant

$$\tau_g = \frac{c_g \rho_g V_g}{h_g A_g}$$

and a coupling time

$$\tau_c = \frac{c\rho V}{h_g A_g} \,.$$

The natural frequency and the damping ratio of this model are, respectively,

$$\omega_n = \frac{1}{\sqrt{\tau \tau_g}}$$

and

$$\zeta = \frac{\tau + \tau_g + \tau_c}{2\sqrt{\tau \tau_g}} \,.$$

Letting $\sqrt{\tau/\tau_g} = x$ and noticing that $x + 1/x \geq 2$ for all $x > 0$, it is easy to show that

$$\zeta = \frac{1}{2}\left(\sqrt{\tau/\tau_g} + \frac{1}{\sqrt{\tau/\tau_g}}\right) + \frac{\tau_c}{2\sqrt{\tau\tau_g}} > \frac{1}{2}\left(x + \frac{1}{x}\right) \geq 1 \,,$$

which means that this coated thermistor is an overdamped, second-order system.

14.3.2 Dynamic Calibration

Although analytical expressions, like Eq. (14.13), are useful for rough estimates of the dynamic response of temperature sensors, as well as a guide for improving this response, precise determination of the dynamic response of a particular sensor can only be achieved by dynamic calibration. Dynamic calibration may also reveal possible deviations from first-order response and the need for a more complex model. Among the various sensor heating methods that have been used for dynamic calibration purposes, one could mention the following:

Self-heating technique: This is the easiest method to implement, but it is only suitable for sensors sensing temperature through a change in their electric resistance, which include metallic resistance detectors and thermistors. The sensor is exposed to a flow stream and a low-amplitude square-wave current is superimposed to the steady current that is used to measure its resistance. A sudden change in the Joule heating of the sensor would result in a gradual change of the sensor output, from which the time constant can be determined. Alternately, a sinusoidal current is applied to the sensor and its frequency is varied, while its amplitude is kept constant. The amplitude ratio at a given frequency can be determined as the ratio of the standard deviations of the output and the input and the static sensitivity can be determined as the amplitude ratio at very low frequencies. Either process should be repeated at different flow speeds so that the possible sensitivity of the dynamic characteristics upon the cooling rate can be determined.

External heating technique: This is the most straightforward method and is suitable for all types of sensors. One approach is to immerse the sensor, suddenly or periodically, into a steady heated stream, such as a heated jet, wake or buoyant plume, using a fast-action linear or rotating mechanism, and then determine its step response or frequency response. An alternative approach is to keep the sensor fixed and interrupt the flow (e.g., by interrupting a heated jet by a rotating perforated disk) or its heating (e.g., by putting the sensor in the wake of a thin heating wire, supplied with a pulsed current [2]). Obviously, in situ dynamic calibration would be desirable, but this requires that the time-dependent temperature field be measured by an accurate, fast-response method, which would serve as calibration standard. An example is the dynamic calibration of fine thermocouples in turbulent flows, with a cold wire serving as the standard [38]. In such cases, dynamic calibration may be assisted by the use of cross-spectra, cross-correlations and joint probability densities of the two temperature signals.

Laser heating technique: In this method, the sensor is heated by the beam of a pulsed laser or the chopped beam of a CW laser [7, 43]. Visible or infrared laser radiation can be used and the beam may be focused to achieve its maximum intensity. Because only part of the sensor may be heated, one would have to determine whether complications may arise due to heat conduction within the sensor or to its supports.

14.3.3 Compensation

Quite commonly, the dynamic response of temperature sensors is inadequate to measure accurately the instantaneous temperature in unsteady and turbulent flows. Improved accuracy can be achieved with the use of compensation. For example, if the time constant of a first-order temperature sensor is known, one can estimate the actual flow temperature T from the sensor's reading T_m by using Eq. (14.12). This can be done by analogue components, which include a differentiator to provide the temperature derivative, an amplifier and an adder [43]. Better yet, one can perform these operations digitally, from discrete time histories of the temperature signal, by estimating the derivative as a finite difference, preferably by a second- or higher-order finite difference method [9].

14.4 Thermochromic Materials

Thermochromic are materials which, when illuminated appropriately, emit visible radiation, which changes colour at a certain temperature. Such materials are used for visualisation purposes (see Section 11.2), but, with proper calibration, can also be used for mapping isotherms and measuring local temperature as well. In most applications, the thermochromic material is applied on a surface in the form of a coating, but thermochromic suspensions have also been used for in-flow temperature mapping in liquids. Two main types of thermochromic materials are available: *temperature sensitive paints*, having temperature ranges up to several hundred degrees centigrade, and *liquid crystals*, usable in ranges near room or body temperatures.

14.4.1 Temperature Sensitive Paints

The term *temperature sensitive paints* (*TSP*) describes coatings that contain photoluminescent (fluorescent or phosphorescent) substances, having a luminescence that is sensitive to temperature [17, 22, 25]. Although the use of phosphorescent coatings for the measurement of surface temperature precedes significantly the use of pressure sensitive paints (PSP) [17], modern TSP techniques were developed in parallel with PSP techniques (see Section 9.4), with which they share their operating principle. To minimise sensitivity to pressure by oxygen quenching, TSP utilise coatings which are impenetrable to oxygen. Furthermore, as indicated by Eq. (9.20), to achieve a high temperature sensitivity, one should use materials with a relatively low activation energy ΔE. Some TSP fluoresce at distinct wavelengths, with the energy under each spectral peak depending on temperature. One may measure surface temperature with these substances by comparing the signals of two cameras equipped with optical band-pass filters, so that each only records only the energy under a single spectral peak. Other types of TSP have wide-band fluorescent emission, in which case one may determine temperature by comparing the heated-surface image to a uniform-temperature image of the same set-up. Typical changes in luminescence of TSP are in the range of a few percentage points per degree Kelvin. A number of recipes for luminophore compounds and binder combinations have been proposed in the literature. Collectively, they cover the temperature range between 80 K and 2,000 K, with each compound covering a range of about 100 K.

A different type of thermochromic paints have been used as temperature indicators in aerodynamics research [26], as well as in a variety of industrial and commercial applications [42]. These consist of metallic compound powders, dispersed in an acrylic lacquer, which is applied as a coating to the heated surface. When the paint is exposed to a triggering temperature for a certain time, a chemical reaction occurs, which causes it to change colour. Some materials may actually undergo multiple colour changes as temperature rises. The temperature value can be determined from appropriate calibration in a temperature-controlled oven. Several such materials are available commercially, and their properties are regulated by national and international standards. For some paints, the colour change is permanent, whereas for others it is reversible. The temperature range of some paints reaches 1,300°C and their uncertainty ranges between 1 and 5°C. Besides paints, crayons are also available, which may be easily applied on surfaces. Temperature indicators also include materials that undergo phase change or deform at certain temperatures and are available in a variety of shapes [16, 42].

14.4.2 Liquid Crystals

Liquid crystals (*LC*), or *mesophases* [15, 17, 36], are certain organic compounds, the molecular structure and optical properties of which are intermediate between those of conventional liquids, which are optically isotropic, and crystalline solids, which are anisotropic. Among the different types of available liquid crystals, the ones suitable for temperature measurement are the *thermotropic* or *thermochromic* liquid crystals (*TLC*). These have both a pure crystalline state and an amorphous liquid state, but, within a range of temperatures, they also have an anisotropic liquid state. Mixtures of liquid crystals with other liquid crystals or other materials may also be optically anisotropic. Within their range of operation, liquid crystals have a colour, which changes with temperature and could, therefore, be used for temperature mapping, either applied to a surface as a coating, or suspended in a liquid medium, in which case they can also serve as flow markers. Compared to temperature sensitive paints, liquid crystals have higher amplitude, spatial and temporal resolutions.

Among the three general categories of liquid crystal materials, known as *smectic* (soap-like), *nematic* (thread-like) and *cholesteric* (containing cholesterol) or *chiral nematic*, it is the latter ones, which are thermochromic. Nematic liquid crystals have long molecules, which are oriented in approximately the same direction, although they are not as perfectly structured as crystals. Cholesteric materials consist of parallel nematic layers, about 3 Å thick, with each layer containing molecules oriented at a angle that is slightly different from the angles of the adjacent layers, in a helical fashion. A linearly polarised beam of light propagating through such a material would rotate its plane by as much as 50 rotations per mm thickness. Cholesteric liquid crystals also decompose white light into clockwise- and counterclockwise-rotating components (*dichroism*) and transmit one component, while reflecting the other, with the reflection properties changing with temperature, within a certain range. Therefore, their colour appears to change, at a certain temperature, but the process also depends on the strength of the electromagnetic field, the shear stress, the pressure and the orientations of illumination and observation. Accurate measurement of temperature with TLC requires collimated illumination, which can be achieved with the use of combinations of slits, parabolic mirrors and lenses.

To avoid chemical reactions with the surrounding fluids, liquid crystals are usually enclosed in microcapsules (5–30 μm in diameter), having a gelatinous shell and suspended in a liquid in the form of a slurry. A common approach is to apply the slurry on a surface by a brush or other means and to let it dry. Single layers should have a thickness of about 50 μm, but multiple layers of different liquid crystals can also been applied consecutively. The surface must be painted black before application to ensure sufficient contrast. Liquid crystals are also available in sheet form, about 100 μm thick, which can be bonded to a surface. Finally, liquid crystal slurries can be also suspended in water or other liquids, usually at very small concentrations (typically less than 0.1%).

Each liquid crystal is characterised by its *event temperature range*, namely the temperature range at which it is optically active. The event temperature ranges of different liquid crystals cover temperatures between 0 and 100°C. The width of the event temperature range of a given liquid crystal, called the *ambiguity bandwidth*, typically ranges from 1 to 15°C. As the temperature increases and a liquid crystal spans its event temperature range, its colour changes from the infrared to red, through the visible colours to blue, and then to ultraviolet. A TLC becomes transparent, when its temperature exceeds its *clearing point temperature*, which typically spans the range from −30 to 120°C. In order to obtain a sharp isotherm, one may use monochromatic illumination or white light and band-pass optical filters. The use of multiple filters or the mixing of liquid crystals with different, non-overlapping, event temperature ranges could provide several isotherms at different temperatures. Continuous variation of temperature can be resolved by using liquid crystals with a wide ambiguity band and digital image processing [37]. Although colours appear in brilliant forms, their succession is highly non-linear with temperature. Liquid crystals may be calibrated vs. the readings of thermocouples embedded in the surface underneath them or by exposing the surface to a uniform, constant, or slowly varying, temperature [33]. Although the time constant for colour change of common liquid crystals is relatively small, typically of the order of 0.1 s, the response of applied layers is considerably slower and depends on the layer thickness, the thermophysical properties of the materials and the heat transfer coefficient. Typical cut-off frequencies in practical applications are of the order of 0.1 Hz. When proper care is taken for their calibration and use, the temperature resolution of liquid crystals is very high, allowing the measurement of temperature differences that are as small as 0.1 K. Another advantage of liquid crystals is that their operation is reversible and can be used repeatedly to map temperature fields. In combination with in-flow velocity measurement methods, surface-coated liquid crystals can be used to measure convective heat transfer [31].

14.5 Radiation Emission Methods

14.5.1 Pyrometry

Literally, the term *pyrometry* means measurement in fires. *Optical pyrometers* [3, 5, 8, 23, 25, 41, 42] measure the temperature of objects with the use of thermal radiation laws (see Section 7.3.4). Compared to other temperature measuring systems, they have the advantages of non-contact operation and application to much higher temperatures. A variety of pyrometers are offered by different manufacturers, often employing proprietary designs.

The classical and most accurate pyrometer design is the *disappearing-filament optical pyrometer*, also known as *monochromatic-brightness pyrometer*. In its original version, this instrument required manual adjustments by the user. First, the user observed a small area of the 'target', the temperature of which was to be measured, through an eyepiece. The image of an electrically heated, incandescent tungsten filament was superimposed on the target image and the combined image was viewed through an optical band-pass filter, which removed all radiation but a narrow band in the red part of the spectrum ($\lambda_p = 0.655$ µm). The user adjusted the current through the filament until its image was no longer distinguishable from the background, which indicated that the spectral radiance of the filament matched that of the target. The instrument was calibrated to provide as output the *brightness temperature* T_B, defined as the temperature of a black body with the same spectral radiance as the filament. If the target was not a black body, its temperature T would have to be higher, to compensate for a lower emissivity. Wien's radiation law (Eq. (7.28)) provides the equality

$$L_{e\lambda} = \frac{2\pi hc^2}{\lambda_p^5} e^{-C_2/(\lambda_p T_B)} = \frac{2\pi hc^2 \varepsilon_\lambda}{\lambda_p^5} e^{-C_2/(\lambda_p T)} . \tag{14.14}$$

This leads to a relationship between the two temperatures as

$$\frac{1}{T} = \frac{1}{T_B} + \frac{\lambda_p}{C_2} \ln \varepsilon_\lambda , \tag{14.15}$$

where ε_λ is the monochromatic emissivity of the target for $\lambda_p = 0.655$ µm and $C_2 = 0.01439$ m K. It is noted that the temperature sensitivity to ε_λ is relatively low, because of the latter's appearance in a logarithm. For this reason, an uncertainty in the determination of ε_λ would result in a much lower relative uncertainty in T. Representative values of ε_λ for common materials with unoxidised surfaces are listed in Table 14.4 [5]. For many materials, however, or for different surface conditions, the emissivity is not known and the previous correction method cannot be applied.

Table 14.4 Monochromatic emissivity of some materials

Material	ε_λ
C	0.80–0.93
Cu	0.10 (solid), 0.15 (liquid)
Fe or steel	0.35 (solid), 0.37 (liquid)
Pt	0.30 (solid), 0.38 (liquid)
Ag	0.07 (solid or liquid)

The manual pyrometer has been superseded by automated versions, which employ a photomultiplier tube to sense differences in the brightness of the filament and target images and to adjust the current to enforce equality of these brightnesses. A more recent version is the *two-colour pyrometer*, which measures the ratio of spectral radiances $L_{e\lambda 1}$ and $L_{e\lambda 2}$ at two wavelengths λ_1 and λ_2. Assuming that the corresponding monochromatic emissivities are $\varepsilon_{\lambda 1}$ and $\varepsilon_{\lambda 2}$, one may use Wien's radiation law to estimate the target temperature as

$$\frac{1}{T} = \frac{1}{C_2\left(\frac{1}{\lambda_1} - \frac{1}{\lambda_2}\right)} \left[\ln \frac{L_{e\lambda 1}}{L_{e\lambda 2}} + 5 \ln \frac{\lambda_1}{\lambda_2} + \ln \frac{\varepsilon_{\lambda 1}}{\varepsilon_{\lambda 2}} \right]. \tag{14.16}$$

If the monochromatic emissivities at the two wavelengths can be assumed to be equal, as in the case of grey bodies, then the previous expression would become independent of the emissivity value. Even if this is not the case, one would need only to estimate approximately the emissivity ratio, as the uncertainty in this ratio plays a minor role in the accuracy of the temperature measurement.

When the absorption of radiation by the medium between the pyrometer and the target is not negligible, it would be necessary to replace the monochromatic emissivity in the previous expressions by its product with the *monochromatic transmissivity* $\tau_\lambda \leq 1$, defined as the ratio of the radiation transmitted through the medium at a given wavelength and the emitted radiation.

14.5.2 Infrared Thermography

Infrared thermography [3, 4, 23, 41, 42, 45] is a non-contact technique for the measurement of surface temperature of an object by measuring the infrared radiation emitted by it. Infrared radiation detectors providing point measurements of temperature are available; however, line- and area-scanning instruments are more usable. A common device is the *infrared scanning radiometer* (*IRSR*), which provides two-dimensional images of the surface temperature distribution, called *thermograms*. A typical IRSR is a camera containing two mirrors or prisms, which oscillate on planes normal to each other to scan the image, and an infrared radiation detector, which converts the radiation to an electronic video signal. *Staring-type infrared cameras* are also available. These have no moving parts but contain a two-dimensional array of detectors, which integrate the oncoming radiation over a short time interval and provide this information to a CCD array. In order to convert the received radiation energy to surface temperature, one must know the emissivity of the surface. Further errors would be produced when radiation is absorbed or reflected along its path from the target to the detector, which includes losses in the intervening fluids, test section windows and camera windows. The transmissivity of air is a strong function of radiation wavelength. Two spectral bands have been identified as matching a high air transmissivity and strong response of available infrared detectors: the 3.5 to 5 μm range, called 'short wave band' and the 8 to 12 μm range, called 'long wave band'. These are the two bands which are used in most infrared cameras. Glass and quartz have a low transmissivity to IR radiation and are not suitable as camera windows. Materials with high transmissivity in the ranges of interest include sapphire, germanium and zinc selenide. The temperature resolution of infrared cameras is characterised by two parameters: (a) the *noise equivalent temperature difference*, which is the temperature difference that would create a change in output voltage equal to the peak-to-peak noise voltage; and (b) the minimum resolvable temperature difference, which is the lower limit of temperature difference between a background surface and an array of strips for which the strips can be discriminated visually from the background. Infrared cameras are calibrated vs. a blackbody radiation source at a known temperature. Due to the uncertainty in the values of the surface emissivity and the transmissivity of the radiation path, it is a sound practice to perform in situ calibration, for example, by comparing the camera output to the readings of surface mounted thermocouples. Infrared thermography is used extensively in many industrial applications. In the laboratory, it is a valuable tool for heat transfer experiments and for high-speed flows, especially in the hypersonic range, in which temperature differences on immersed objects are generated spontaneously. For low-speed flows, heating of the surface electrically or by other means can be used to provide thermograms that illustrate surface flow patterns, such as flow separation and transition to turbulence, through their effect on the local heat transfer coefficient [24].

14.5.3 Laser-Induced Photoluminescence

Laser-induced fluorescence (*LIF*) methods are mainly used for the measurement of concentration in liquids and gases (see Section 15.5). Even so, the dependence of fluorescence upon temperature can be used for temperature measurement. This approach has been applied to the measurement of temperature fluctuations in water using Rhodamine B, which has a fluorescence quantum efficiency that decreases, almost linearly, with temperature at a relatively strong rate (2.3% per K at 20°C). The dye is mixed with the water at a uniform concentration, a laser light sheet is used to activate the process, and images are recorded with a digital camera or other suitable means. If the dye concentration is kept at very low levels and the water is free of other substances, light absorption in its path through the water can be neglected and the radiation received by each element of the camera would be proportional to the local quantum efficiency. Because of spatial, and possibly temporal, variation of light intensity on the light sheet, and other disturbances along the light path, pixel-by-pixel calibration would be required. An alternative approach is to monitor undesirable local light intensity variations by measuring simultaneously the local emission by a second fluorescent dye (e.g., Rhodamine 110), which has a negligible fluorescence quantum efficiency sensitivity to temperature [34].

A variant of LIF thermography is a method referred to as *thermographic particle imaging*. By using phosphorescent seeding particles in a fluid, one may measure simultaneously velocity (using standard particle image velocimetry techniques, as outlined in Section 12.6) and temperature planar maps. A key difference with mainstream PIV techniques is that an ultraviolet (UV) laser is used to both illuminate the flow and to excite the phosphorescence of the particles, and two additional cameras are used to image the temperature field [1, 11]. Ideally suited for high-temperature applications, owing to increased phosphorescence of the particles and ease of imaging, this technique has been applied successfully to both laminar and turbulent heated jets.

14.6 Optical Techniques

The optical techniques described in Section 11.3 for flow visualisation purposes are based on the non-uniformity of the refractive index of the fluid, which is directly related to the fluid density non-uniformity. If the experimental conditions are such as that density changes are caused exclusively by temperature changes, one could, in principle, use these optical techniques to measure the local fluid temperature. In practice, however, the resolution of shadowgraphs is insufficient to permit the accurate double integration required to determine temperature from the measured light intensity, so that the shadowgraph technique remains a strictly qualitative one. The resolution of high-quality Schlieren images could be adequate to determine temperature by single integration, followaing proper calibration. Interferograms provide isotherms, from which one could easily extract temperature values, provided that temperature at a reference location is known by other means. For details and references, the reader is referred to Section 11.3.

Chapter Digest

14.1 Can we define temperature in the same way we define velocity or pressure? What do you understand when we say, for example, 'the temperature is twenty degrees'?

14.2 When we give an estimate of the temperature in the Sun, what is the basis for giving this value?

14.3 Identify the principles of operation of the most common temperature sensors. Can you think of any sensors that operate on other principles?

14.4 Based on your general familiarity with scientific instrumentation, which sensor would you consider to be the most and least accurate for measuring room temperature: a liquid-in-glass thermometer, a bimaterial thermometer, a thermocouple, an RTD or a thermistor?

14.5 What do you understand by the term 'thermal inertia' and is this term in any way analogous to the term 'inertia', as used in classical mechanics?

14.6 What would be the consequences of an underdamped second-order model for a thermal measuring system? Are such consequences compatible with logical expectations for the operation of this system?

14.7 Discuss the self-heating effect of resistance thermometers. How would this effect change the reading of a metallic resistance thermometer and a thermistor? What can one do to reduce such errors?

14.8 Which is the reference temperature for NIST standard thermocouples? How would you use the NIST-table voltages to determine the temperature measured by thermocouple that has its reference junction at a temperature that is different from the NIST value?

14.9 Which voltage should a thermocouple have at its reference temperature?

14.10 Can the Callendar–Van Dusen equation (Eq. (14.9)) be used for a platinum RTD with a reference temperature of 20°? Can it be applied to RTD made of other materials?

14.11 Compare the ranges of applications of thermistors, thermocouples and cold wires. Give examples of temperature measurements for which each of the three methods would be suitable, but the other two would not.

14.12 Provide an example of an experiment for which liquid crystals would be preferable to a thermocouple array and another example for the opposite case.

14.13 Provide an example of an experiment for which thermocouples would be preferable to thermistors and another example for the opposite case.

Problems

14.1 A liquid-in-glass thermometer is calibrated in a dual-jet calibration facility such that it is alternately immersed in either of two air streams with different temperatures.. Assume that the air temperature surrounding the thermometer during calibration varies sinusoidally between a minimum of 30°C and a maximum of 50°C with a period of 8 s. The reading of the thermometer fluctuates between a minimum of 35°C and a maximum of 45°C.

(a) Determine the thermometer's time constant.

(b) Determine the time difference between the occurrence of maximum thermometer reading and the occurrence of maximum airstream temperature.

14.2 Repeat Example 14.1 using Eq. (14.4) to determine the radius of curvature, and calculate the percentage difference between this value and the one obtained in Example 14.1. Comment on how the error in the vertical deflection depends on $T - T_r$.

14.3 Consider an initially planar bimaterial thermometer assembly, constructed of two materials with equal thicknesses and comparable Young modulus values. State the assumption(s) that would allow you to approximate the deflection angle as $\theta \approx \frac{3}{2}\frac{L_0}{\delta}(\alpha_B - \alpha_A)(T - T_r)$, where L_0 is the initial length of the strip. For the specifications given in Example 14.1, compute the value of θ and compare it to the more accurate value found in that example. Comment on the general validity of this approximation.

14.4 Derive an expression for the sensitivity of the vertical deflection for a strip bimaterial thermometer. Investigate, and comment, on how the sensitivity of the device can be maximised as a function of design parameters.

14.5 Consider a bimetallic element, 30 mm long and consisting of a plate of copper, 2 mm thick, bonded on a plate of Invar, 1 mm thick. The element is clamped at one end and straight when the ambient temperature is 20°C. Determine the deflection of its free end when the temperature reaches 40°C.

14.6 Figure 14.3 suggests that the listed NIST thermocouples have an approximate linear response for $t \geq 0$°C. Using NIST data sheets, which are freely available online, determine the sensitivities of the thermocouples presented in Table 14.2 for $t \geq 0$°C and the corresponding linearity error (see Section 6.1.1). Comment on the accuracy of the linear approximation for temperature measurements close to 0°C and near the upper ends of the respective NIST ranges.

14.7 A voltmeter with a resolution of 1 mV is used to measure the voltage provided by the NIST standard thermocouples shown in Table 14.2. Determine the corresponding resolutions of temperature measurements, as functions of the measured temperature value. Comment on the suitability of thermocouples for measuring small positive temperatures. If we want to have a temperature resolution of at least 0.1°C for measurements in the range 0°C $\leq t \leq$ 100°C, determine the required voltmeter resolution for each of the listed thermocouples.

14.8 The calibration of a custom-built thermocouple provided the following data, referenced to room temperature, which remained constant during calibration.

t [°C]	0	20	40	60	80	100
V_o [mV]	−1.028	−0.106	0.816	1.738	2.660	3.582

(a) Determine the room temperature during calibration.

(b) Next, let us place the cold junction in an ice bath and the hot junction in a cup of coffee; determine the temperature of the coffee, if the thermocouple reading is 3.319 mV.

14.9 The manufacturer of a thermocouple provided the following data, referenced to a temperature of 0°C.

t [°C]	0	40	80	120	160	200
V_o [mV]	0	5.664	22.656	50.976	90.624	141.600

A number of these thermocouples are used in an experiment to measure the temperature at different locations along the outer wall of a heated tube over an extended period of time. You place the reference junction in an ice bath and verify, with a precision RTD, that the temperature of the ice bath is 0.0°C. Over the course of the experiment, you found that the temperature of the ice bath increased to 2.4°C.

(a) Determine the error in the temperature measurement due to the drift in the reference temperature, when the thermocouples on the wall of the tube read 25, 50, 100 and 175°C.

(b) Based on your findings, would you recommend using this thermocouple to measure small positive temperatures?

14.10 Assume that the calibration of a thermocouple, referenced at a temperature t_{ref}, is of the form $V_o = a(t - t_{\text{ref}})^b$, where a and b are calibration coefficients.

(a) Show that the temperature error that would be caused by a change Δt in the reference temperature is $\varepsilon_t = [(t - t_{\text{ref}})^b - \Delta t^b]^{1/b} - (t - t_{\text{ref}})$.

(b) Specify, with adequate documentation, the conditions under which this error would be reduced for a given reference temperature.

(c) If you want to take measurements at near room temperature, which values of the reference temperature and the exponent b would you recommend, in order to reduce the error in the measurement?

14.11 Compute the slope dR_{pt}/dt of the Callendar–Van Dusen (Eq. (14.9)) at $t = 0$°C. Assuming that a linear expression with the same slope describes the entire resistance-temperature relationship for platinum RTD, compute the error in temperature computed from the linear expression and plot it vs. t for the range from −100 to 700°C.

14.12 A three-wire Pt100 RTD is connected to the bridge circuit shown in Fig. 14.5b. Derive an expression for the voltage output, assuming $R_1 = R_3$ and including the effect of a difference between the resistances of the lead wires. Determine whether the difference between the lead wire resistances or an equal change in the resistances of both lead wires would have a greater effect on the voltage error.

14.13 Determine the temperature error for a three-wire Pt100 with lead wire resistances of 6 Ω, when it is exposed to a temperature of 80°C. Repeat this for a three-wire Pt1000 RTD. Comment on the results.

14.14 Show that the normalised voltage outputs of a two-wire RTD and a three-wire

RTD with lead wires of equal resistance are, respectively,

$$\left.\frac{V_o}{V_s}\right|_{\text{2RTD}} = \frac{-[2\frac{R_{l2}}{R_r} + \alpha_r(t - t_r)]}{2[2\frac{R_{l2}}{R_r} + 2 + \alpha_r(t - t_r)]},$$

$$\left.\frac{V_o}{V_s}\right|_{\text{3RTD}} = \frac{-[\alpha_r(t - t_r)]}{2[2\frac{R_{l2}}{R_r} + 2 + \alpha_r(t - t_r)]},$$

under the assumption that $R_1 = R_3$, and that the RTD has a linear response (Eq. (14.10)).

14.15 An iron RTD with a reference resistance of 100 Ω and lead wire resistances of 2 Ω is connected to the circuits shown in Eq. (14.10). Calculate the temperature measurement error (both absolute and relative) for a temperature range of $-10°\text{C} \leq t \leq 100°\text{C}$ and reference temperatures in the range $0°\text{C} \leq t_r \leq 20°\text{C}$. You may assume that α_r is the same for all reference temperatures. Discuss your findings, in particular, (i) the relative error for both configurations, when $t_r = 0$; (ii) the conditions under which the error is lowest for both the three- and two-wire configurations; and (iii) whether a two-wire configuration can ever give an error that is lower than that of the three-wire configuration.

14.16 You find an RTD in the lab without any specifications and obtain the following data after performing a calibration.

t [°C]	0.1	20.2	99.8
R [Ω]	99.8	107.7	138.8

Determine the reference temperature, resistance and thermal resistivity of this RTD, assuming a linear response (Eq. (14.10)). Next, assume that this RTD follows the Callendar-Van Dusen equation (Eq. (14.9)). Determine the coefficients by curve fitting and compare these values to the IEC 60751 standards. Comment on the suitability of this standard for this device. Determine and compare the RTD temperature values given by the three expressions, if the RTD resistance is measured to be 109.3 Ω.

14.17 Calibration of a thermistor probe provided the following data.

t [°C]	20.0	40.0	60.0
R [Ω]	3,440	1,280	512

Determine the coefficients A, B and C in Eq. (14.11). In semi-logarithmic axes, plot the thermistor resistance vs. temperature for the extended range between −20 and 100°C. Then compute and plot the sensitivities $\mathrm{d}R/\mathrm{d}t$ and $\mathrm{d}t/\mathrm{d}R$ over the extended range. Compare the values of these sensitivities for $t = -20$, 20 and 100°C. Compute the maximum error that would be produced in the range of the calibration data, if a linear expression were fitted to these data.

14.18 Consider that the glass-coated thermistor discussed in Example 14.5 operated inside a smoke plume and was covered by a coat of soot with a uniform thickness. Derive a model for the dynamic response of this thermistor. Investigate whether temperature measured by this thermistor can overshoot, when exposed to a step change in flow temperature.

References

[1] C. Abram, B. Fond, A.L. Heyes, and F. Beyrau. High-speed planar thermometry and velocimetry using thermographic phosphor particles. *Applied Physics B*, 111(2):155–160, 2013.

[2] R.A. Antonia, L.W.B. Browne, and A.J. Chambers. Determination of time constants of cold wires. *Rev. Sci. Instrum.*, 52:1382–1385, 1981.

[3] T. Arts et al. *Measurement Techniques in Fluid Dynamics (2nd Edition).* von Kármán Institute for Fluid Dynamics, Rhode-Saint-Genese, Belgium, 2001.

[4] T. Astarita, G. Cardone, G.M. Carlomagno, and C. Meola. A survey of infrared thermography for convective heat transfer measurements. *Optics and Laser Technol.*, 32:693–610, 2000.

[5] R.P. Benedict. *Fundamentals of Temperature, Pressure and Flow Measurements (2nd Edition).* Wiley Interscience, New York, 1977.

[6] H.H. Bruun. *Hot-Wire Anemometry.* Oxford University Press, Oxford, 1995.

[7] R. Budwig and C. Quijano. A new method for in situ dynamic calibration of temperature sensors. *Rev. Sci. Instrum.*, 60:3717–3720, 1989.

[8] E.O. Doebelin and D.N. Manik. *Doebelin's Measurement Systems (SIE) (7th Edition).* McGraw-Hill, New York, 2019.

[9] M.J. Downs, D.H. Ferriss, and R.E. Ward. Improving the accuracy of the temperature measurement of gases by correction for the response delays in the thermal sensors. *Meas. Sci. Technol.*, 1:717–719, 1990.

[10] Fenwal Electronics. *Thermistor Manual.* Fenwal Electronics, Inc., Milford, MA, 1974.

[11] B. Fond, C. Abram, A.L. Heyes, A.M. Kempf, and F. Beyrau. Simultaneous temperature, mixture fraction and velocity imaging in turbulent flows using thermographic phosphor tracer particles. *Optics Express*, 20(20):22118–22133, 2012.

[12] J. Haugdahl and V. Lienhard. A low-cost, high performance DC cold-wire bridge. *J. Phys. E: Sci. Instrum.*, 21:167–170, 1988.

[13] J. Hojstrup, K. Rasmussen, and S.E. Larsen. Dynamic calibration of temperature wires in still air. *DISA Inf.*, 20:22–30, 1976.

[14] J.P. Holman. *Heat Transfer (9th Edition).* McGraw-Hill, New York, 2002.

[15] N. Kasagi, R.J. Moffat, and M. Hirata. Liquid crystals. In W.-J. Yang, editor, *Handbook of Flow Visualization*, pages 105–116. Taylor & Francis, 1989.

[16] E. Kimmel. Temperature sensitive materials. *Measurements and Control*, October 1979:98–103, 1979.

[17] T.A. Kowalewski, P. Ligrani, A. Dreizler, C. Schulz, U. Fey, and Y. Egami. Temperature and heat flux (chapter 7). In C. Tropea, A.L. Yarin, and J.F. Foss, editors, *Springer Handbook of Experimental Fluid Mechanics*. Springer, Berlin, 2007.

[18] S. Larsen and J. Hojstrup. Spatial and temporal resolution of a thin-wire resistance thermometer. *J. Phys. E: Sci. Instrum.*, 15:471–477, 1982.

[19] J.C. LaRue, T. Deaton, and C.H. Gibson. Measurement of high frequency turbulent temperature. *Rev. Sci. Instrum.*, 46:757–764, 1975.

[20] J.C. Lecordier, A. Dupont, P. Gajan, and P. Paranthoen. Correction of temperature fluctuation measurements using cold wires. *J. Phys. E: Sci. Instrum.*, 17:307–311, 1984.

[21] J. Lemay and A. Benaissa. Improvement of cold-wire response for measurement of temperature dissipation. *Exp. Fluids*, 31:347–356, 2001.

[22] T. Liu, T. Campbell, S. Burns, and J. Sullivan. Temperature and pressure sensitive luminescent paints in aerodynamics. *App. Mech. Rev.*, 50:227–246, 1997.

[23] F. Mayinger and O. Feldmann, editors. *Optical Measurements (2nd Edition).* Springer, Berlin, 2001.

[24] G. Melina, P.J.K. Bruce, J. Nedić, S. Tavoularis, and J.C. Vassilicos. Heat transfer from a flat plate in inhomogeneous regions of grid-generated turbulence. *International Journal of Heat and Mass Transfer*, 123:1068–1086, 2018.

[25] C. Mercer. *Optical Metrology for Fluids, Combustion and Solids.* Kluwer Academic Publishers, Dordrecht, 2003.

[26] W. Merzkirch. *Flow Visualization (2nd Edition).* Academic Press, New York, 1987.

[27] Omega Engineering. *The Temperature Handbook.* Omega Engineering Inc., Stamford, CT, 2001.

[28] P. Paranthoen, C. Petit, and J.C. Lecordier. The effect of thermal prong-wire interaction on the response of a cold wire in gaseous flows (air, argon and helium). *J. Fluid Mech.*, 124:457–473, 1982.

[29] R. Peattie. A simple, low-drift circuit for measuring temperatures in fluids. *J. Phys. E: Sci. Instrum.*, 20:565–567, 1987.

[30] A.E. Perry, A.J. Smits, and M.S. Chong. The effects of certain low frequency phenomena on the calibration of hot wires. *J. Fluid Mech.*, 90:415–431, 1979.

[31] T.J. Praisner, D.R. Sabatino, and Smith C.R. Simultaneously combined liquid crystal surface heat transfer and PIV flow-field measurements. *Exp. Fluids*, 30:1–10, 2000.

[32] H. Preston-Thomas. The International Temperature Scale of 1990 (ITS-90). *Metrologia*, 27:3–10,107, 1990.

[33] D.R. Sabatino, T.J. Praisner, and C.R. Smith. A high-accuracy calibration technique for thermochromic liquid crystal temperature measurements. *Exp. Fluids*, 28:497–505, 2000.

[34] J. Sakakibara and R.J. Adrian. Whole field measurement of temperature in water using two-color laser induced fluorescence. *Exp. Fluids*, 26:7–15, 1999.

[35] J.F. Schooley. *Thermometry.* CRC Press, Boca Raton, FL, 1986.

[36] A.J. Smits and T.T. Lim, editors. *Flow Visualization Techniques and Examples.* Imperial College Press, London, 2000.

[37] J Stasiek. Thermochromic liquid crystals and true color image processing in heat transfer and fluid-flow research. *Heat Mass Transfer*, 33:27–29, 1997.

[38] R. Talby, F. Anselmet, and L. Fulachier. Temperature fluctuation measurements with fine thermocouples. *Exp. Fluids*, 9:115–118, 1990.

[39] S. Tavoularis. A circuit for the measurement of instantaneous temperature in heated turbulent flows. *J. Phys. E: Sci. Instrum.*, 11:21–23, 1977.

[40] P.V. Vukoslavcevic and J.M. Wallace. The simultaneous measurement of velocity and temperature in heated turbulent air flow using thermal anemometry. *Meas. Sci. Technol.*, 13:1615–1624, 2002.

[41] J.G. Webster, editor. *The Measurement, Instrumentation and Sensors Handbook.* CRC Press, Boca Raton, FL, 1999.

[42] J.G. Webster, editor. *Mechanical Variables Measurement: Solid, Fluid and Thermal.* CRC Press, Boca Raton, FL, 2000.

[43] A.R. Weeks, J.K. Beck, and M.L. Joshi. Response and compensation of temperature sensors. *J. Phys. E: Sci. Instrum.*, 21:989–993, 1988.

[44] J.C. Wyngaard. Spatial resolution of a resistance wire temperature sensor. *Phys. Fluids*, 14:2052–2054, 1971.

[45] W.-J. Yang, editor. *Handbook of Flow Visualization, 2nd edition.* Taylor & Francis, 2001.

[46] M.W. Zemansky. *Temperatures Very Low and Very High.* Dover Publications, Inc., New York, 1964.

15 Composition Determination

This chapter describes methods of determining the *composition* of a fluid mixture, namely, identifying the components of the mixture, and measuring the *concentrations* of these components, namely, their relative proportions. Much of the interest focuses on the products of combustion and other chemical reactions, due to our concern for their efficiency and the production of pollutants. Additional topics include the measurement of size and number density (particles per unit volume) of solid and liquid particles suspended in fluids and the measurement of the void fraction, namely, the portion of volume occupied by the gas phase in gas–liquid flows. Only an introductory treatment of measurement methods that are suitable for fluid mixtures can be afforded here, as an in-depth understanding of some of these methods requires an extensive background in chemistry, physics and other fields, which is beyond the scope of the present exposition. Many composition determination methods, especially those based on radiation absorption and emission, were developed for application in chemistry, physics, biochemistry, biology and medical sciences. We are not concerned with such aspects; instead, we shall focus only on aspects that are relevant to fluid mechanics research.

15.1 Sample Analysis

Sampling of a fluid substance is the removal of a volume of the fluid through a *sampling tube* or *sampling probe* for subsequent analysis. It is necessarily intrusive and subject to several limitations. First of all, excessive or inadequate suction applied during sampling may bias the sample in favour of heavier or lighter components, thus distorting the measured composition of mixtures and multi-phase flows. For this reason, it is advisable for sampling to be applied *isokinetically*, which means that the sampling tube should be aligned with the flow direction and the mixture speed inside the tube, corrected for viscous effects and probe wall thickness, should be regulated to match the surrounding flow speed. In the case of reactive flows, one must consider possible changes of composition within the sampling instrumentation, as a result of continuing chemical reactions. Similarly, errors in composition measurement may result from condensation or evaporation of some components within the sampling tubing. When this is a possibility, it may be necessary to control the temperature of the sampling tube and subsequent parts of the measuring system. Other precautions and controls may also be necessary, depending on the particular experimental conditions. In general, sampling is a relatively slow process, with typical response times of the order of 1 s or larger. Much faster sampling rates have been achieved with the use of sampling probes equipped with solenoid valves or piezo-electric transducers [65]. A few representative sample analysis methods for composition determination are described in the following.

Orsat analyser: This is a classical, rather old-fashioned, device used for measuring the combustion efficiency of fossil fuels. A sample of the exhaust gas is injected into a series of graduated glass containers, partly filled with liquid reagents, each of which reacts with a particular gas, for example, oxygen, carbon dioxide or carbon monoxide. Absorption of each gas produces an easily measurable change in volume of the corresponding reagent. This method is simple and inexpensive, but also manual, and, therefore, slow and tedious.

Gas chromatography: This is a method for the separation and analysis of gas and volatile liquid mixtures by *elution* of their components [5, 13, 28, 89]. The main parts of a simple gas chromatograph are illustrated in Fig. 15.1a. A small gas sample is injected into a stream of an inert carrier gas, usually helium or nitrogen, which transports it through a heated column that is packed with a liquid or granular solid sorbent. Each component of the mixture is dissolved and retained by the sorbent for a certain time and exits the column with a certain time lag, depending on the affinity of the component to the sorbent and on the carrier flow rate. Carbon dioxide, unsaturated hydrocarbons, oxygen and carbon monoxide can be eluted sequentially from a certain sorbent. A practical gas chromatograph also contains a detector, which identifies the components. A common detector type is the thermal conductivity detector. It contains two heated elements, one in the analysed stream and one in the pure carrier gas. The presence of a component in the analysed stream is detected from a change in the electrical resistance of the corresponding element; each present component affects the thermal conductivity of the stream, which affects the heat transfer from the element, in turn affecting its temperature and, thus, its resistance. The resistance change is sensed by a bridge, which produces an electric signal output, which is recorded or displayed on a monitor. Other types of detectors have also been used, including flame ionisation and photo-ionisation detectors. A typical *chromatogram*, shown in Fig. 15.1b, appears as a sequence of peaks, each associated with a gas component and separated by specific time intervals. The components of the mixture are identified by the timing of the peaks, while their concentrations can be found by measuring the areas under each peak.

Electronic testers: These are specialised instruments measuring the concentrations of various gases, particularly products of combustion. They contain *electrochemical gas sensors*, each measuring the concentration of a particular type of molecule. For example, a zirconium oxide cell is used to measure oxygen concentration, due to its capacity to develop a voltage difference across its two sides when exposed to differences in oxygen concentration. Testers of carbon monoxide and other gases contain disposable chemical capsules, which change colour when the particular gas is drawn through them.

Continuous emission monitors: Current pollutant emission regulations require continuous monitoring of exhaust gases from the stacks of industrial plants. This is achieved with a variety of electronic instruments, referred to as *continuous emission monitors* (*CEM*). Both extractive and in situ CEM are available. The three main methods of sam-

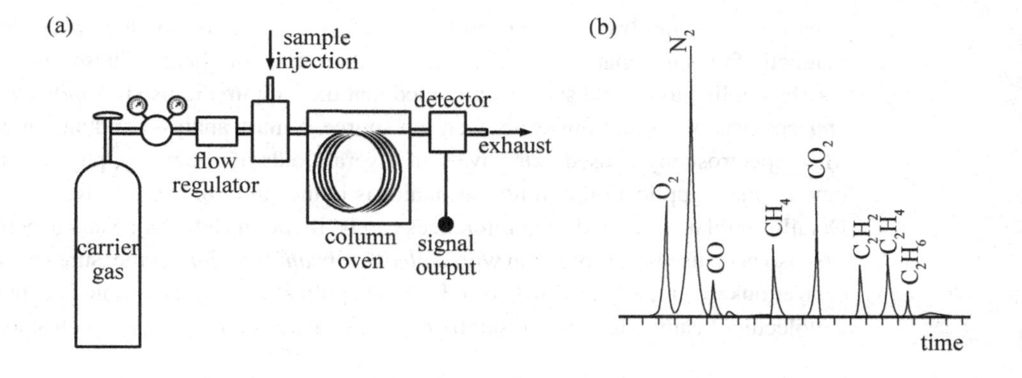

Figure 15.1 (a) Sketch of a gas chromatograph and (b) a typical chromatogram.

ple analysis are absorption spectroscopy, luminescence techniques and electroanalytical methods. Detailed description of CEM can be found in specialised references [44]. In the following, we shall briefly discuss some basic principles of absorption spectroscopy and spectrophotometry.

Absorption spectrophotometry: When radiation passes through a gas, it gets absorbed according to Beer's law (see Section 7.2.5). If the gas consists of a single chemical species, this law can be written in the form [28]

$$A = \log \frac{I_o}{I} = \varepsilon l C \,, \tag{15.1}$$

where A is the *absorbance*, I_o and I are the incident and transmitted radiant intensities (see Section 7.3), respectively, ε is the *molar absorptivity* coefficient of the species, the value of which depends on the radiation wavelength, l is the length of the absorption path in the gas and C is the concentration of the molecules, expressed as moles per unit volume. Thus, by measuring A, it is possible to compute C. In gases containing two or more types of molecules, the total absorbance would be the sum of the individual absorbances, as

$$A = l \sum_{i=1}^{N} \varepsilon_i C_i \,. \tag{15.2}$$

Because each type of molecule preferentially absorbs radiation at particular wavelengths corresponding to its vibrational and rotational energy levels [8, 11, 19, 28] (see Section 7.4), one may analyse the spectrum of the absorbed radiation to identify a particular molecule. Furthermore, by using radiation of different wavelengths, one would, at least in principle, be able to compute the concentrations of the components of a gas mixture by solving a system of equations of the type shown in Eq. (15.2). In practice, the applicability of this method is restricted by various effects, such as deviation from Beer's law due to large species concentrations [28]. A variety of *spectrophotometers* are available, operating in the ultraviolet, visible or infrared ranges.

Mass spectrometry: *Mass spectrometry* refers to methods which employ electric or magnetic fields to separate ions according to their mass or charge. These methods may also be applied to neutral species, provided that they are first ionised. A *mass spectrometer* consists of a sampling system, an ion source, a mass analyser and an ion detector. Mass spectroscopy is used extensively in several fields of chemistry, physics and biology. Its main application in fluid mechanics is in the study of shock waves and flames. Details on related methods and references can be found in Ref. [28]. Mass spectroscopy has also been used in association with *molecular beam sampling* to measure species concentrations in rarefied gas flows [65]. In these applications, a gas sample is converted to a molecular beam, which is ionised before being analysed by a mass spectrometer.

15.2 Thermal Probes

Convective heat transfer from a heated object that is immersed in a fluid mixture depends on the thermal conductivity of the fluid, which, in turn, depends on the mass fractions of its constituents. Thus, the composition of a binary mixture of fluids can, at least in principle, be estimated from heat transfer measurements using thermal sensors. Nevertheless, heat transfer also depends on flow velocity and temperature, among other factors, and one would have to separate the sensor's sensitivity to concentration from other sensitivities. Furthermore, in order to achieve high spatial and temporal resolutions, one would have to use fine thermal sensors, such as hot-wire/hot-film anemometers. This led to the suggestion [20, 21] of a probe consisting of two closely spaced, but non-interfering, hot-wire sensors with substantially different physical characteristics, so that their sensitivities to velocity and concentration would be significantly different. Such probes have been applied [55, 56] to the simultaneous measurement of the fluctuating velocity and concentration in isothermal binary gas mixtures. This approach is, however, difficult to implement, mostly because of inadequate separations of the different sensitivities. More practical, and more extensively used, are two somewhat different thermal techniques, which are, respectively, based on *interfering thermal probes* and *aspirating probes*.

The interfering thermal probe consists of two hot-wire/hot-film sensors with differing materials, sizes and overheats, but positioned such that one of them would be slightly upstream and within the thermal field of the other (Fig. 15.2a). The extent of this thermal field would depend on the thermal conductivity, thus the concentration, of the surrounding fluid. This thermal interference increases dramatically the sensitivity of the upstream probe to concentration. Consequently, calibration of the two-sensor probe in a sequence of flows with different combinations of known velocity and concentration makes it possible to measure both in the flow of interest. Addition of a third sensor would permit measurement of a second velocity component or the temperature. Although tedious and subject to drift and various interferences, this method has been applied successfully to the study of turbulent mixing [39, 51, 74, 76, 85, 86].

The ideal concentration measurement probe would be one that would be entirely insensitive to velocity variations. This has been achieved by enclosing the thermal sensor

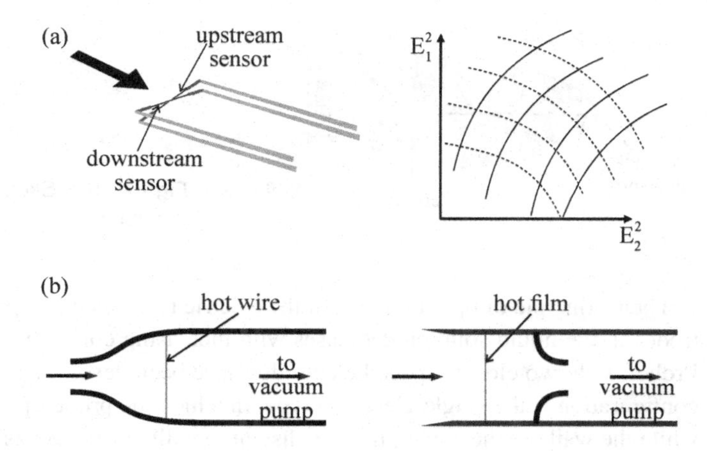

Figure 15.2 (a) Sketch of a two-sensor interfering probe and its calibration plot; solid curves represent constant concentration lines, while dashed curves represent constant velocity lines; E_1 and E_2 are the two hot-wire voltage outputs; (b) sketches of two aspirating probe tips.

inside a small sampling tube, upstream or downstream of a sonic nozzle (Fig. 15.2b). Suction is applied to this tube at a level sufficient for the flow to be 'choked', in which case the flow velocity inside the tube would remain constant, independently of velocity fluctuations in the flow from which the fluid is sampled. This type of aspirating probes, sometimes referred to as *catharometers*, have a strong sensitivity to concentration fluctuations and provide reliable fluctuating concentration measurements [1, 2, 12, 14, 91]. Their outputs are also sensitive to temperature fluctuations; however, the two sensitivities are decoupled, so that one may calibrate the probe separately for the two effects and apply corrections to the concentration measurement, provided that temperature is measured by other means [15]. Finally, these probes also exhibit sensitivity to pressure variations, which would always be present in turbulent flows and may express themselves as an increased output noise level. A variation of the aspirating probe that is based on mass transfer has been applied to the measurement of hydrocarbon concentration in a spark ignition engine [35].

15.3 Electric Conductivity Probes

Electric conductivity probes [4, 18, 25, 31, 49, 53, 60, 61, 64, 83] measure the local electric resistivity of aqueous solutions of electrolytes, such as common salt, from which, through calibration, one can estimate the local concentration of the electrolyte. Such probes have been used for the study of turbulent mixing of passive scalars as well in studies of chemical reactions. Their use for measurement of void fraction will be discussed in Section 15.7.

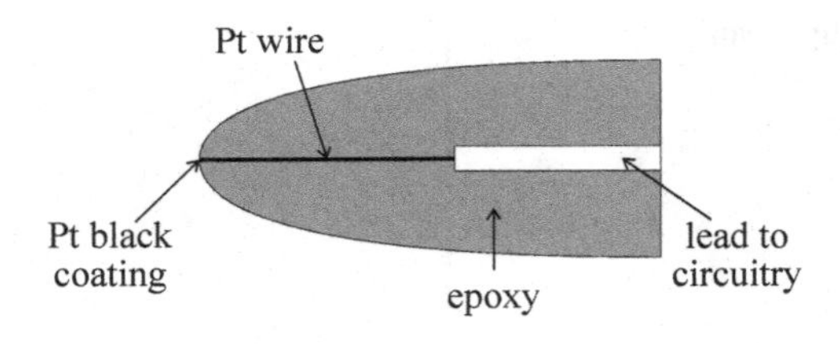

Figure 15.3 Sketch of a conductivity probe tip.

Their principle of operation is that the electric resistance of a path between two electrodes in the liquid solution decreases with increasing concentration of the electrolyte. Probes with two closely spaced electrodes have been developed, but the most common configuration is the single-electrode type, in which the probe tip is the active electrode, while the walls of the apparatus, or a distant metallic piece, act as the second electrode. The probe tip, Fig. 15.3, consists of a short, fine, platinum wire, encased in epoxy, except for the exposed active length, which is coated with platinum black to reduce surface polarisation and noise. The wire is supplied with a relatively high-frequency AC carrier voltage, and the output consists of the low-pass filtered voltage across a fixed resistor in series with the probe. Frequent cleaning, re-plating with platinum black and calibration of the probe in liquid solutions with known uniform concentrations are necessary. The spatial resolution of conductivity probes is remarkably high, with typical characteristic lengths of the order of tens of μm. Such resolution is adequate for the measurement of the fine structure of salinity fluctuations in moderate Reynolds number flows.

15.4 Light Scattering Methods

A variety of light scattering methods have been developed for determining the composition of fluid mixtures and the products of chemical reactions, including combustion processes and flames. Such methods are based on the light scattering phenomena introduced in earlier chapters (Section 7.4 and elsewhere), which include Mie scattering, Rayleigh scattering, Raman scattering and fluorescence. The experimental arrangement for a basic light scattering method is illustrated in Fig. 15.4. A laser beam or other collimated beam of light is focused on a small measuring volume, which can be as small as 1 mm^3 or less. The light scattered from particles within this volume is collected by a collecting lens, separated from other collected radiation with the use of a slit or pinhole acting as a spatial filter, and then projected onto a photodetector for subsequent analysis. In this arrangement, the scattered light is collected at a right angle to the incident beam, in contrast with usual *in situ absorption* and *emission spectroscopic methods* [6], in which the photodetector is aligned with the incident beam (*line-of-sight* methods). Accordingly, unlike line-of-sight methods, which produce a path-average measurement, light scattering methods measure 'local' concentration values. *Planar light scattering methods*, in which light scattered within a volume bounded by two closely spaced planes is collected by a camera, have also become standard tools. This section will briefly discuss some of the major light scattering methods, whereas *laser-induced-fluorescence*

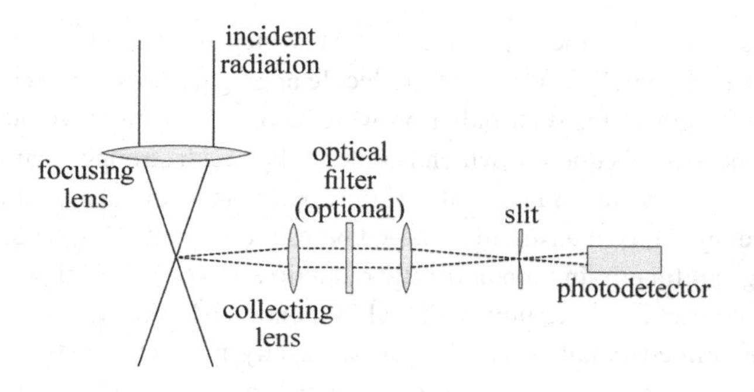

Figure 15.4 Sketch of a basic optical arrangement for the measurement of scattered light.

techniques will be discussed separately in the following section. These topics have been discussed in detail by several thorough reviews [48, 68, 93].

Mie scattering methods: The intensity of light scattered from suspended tracer particles, such as smoke or mists contained in gases, can be used to measure their local concentration; this approach is sometimes referred to as *nephelometry* [9, 10, 73]. The operating principle of this method is quite simple: a laser beam or other strong light is focused in a small measurement volume and a light collection system is also focused on the same volume and leads to a photodetector. Digital camera and photomultiplier tube outputs are linearly related to light intensity over many orders of magnitude and so the theoretical dynamic range of this method is very wide. In practice, however, the accuracy of the method is limited by non-uniformities in particle size (polydispersity) and refractive index, particle coagulation and light absorption. Mie scattering has also been applied to planar measurements (*tomography*) of particle concentrations with the use of laser light sheets [36, 52].

Molecular radiation emission: Far more common than the previous ones, are techniques based on radiation scattering at the molecular level. Like the corresponding radiation absorption methods, radiation emission methods are capable of both identifying the species in a mixture as well as measuring their concentrations. The basis of these methods is the transition of molecules at an energy state n to a state of lower energy m, by emission of photons with energy $h\nu_{nm}$. If N_n is the number of such molecules within the emitting volume, the intensity of spontaneously emitted radiation would be [28]

$$I_{nm} = A_{nm} N_n h\nu_{nm} \,, \tag{15.3}$$

where A_{nm} is the *Einstein transition probability*, which is specific to each transition. Values of A_{nm} are available in specialised sources and may also be determined by calibration. Thus, in principle, one can measure the concentration of such molecules by measuring the intensity of radiation at the specific frequency ν_{nm}. Clearly, radiation emission methods require the use of high-power radiation sources and careful experimental setting.

Rayleigh scattering methods: Consider a pure gas consisting of a single chemical species and such that the mean free path between molecular collisions is relatively long. When illuminated by monochromatic light having a frequency that is much lower than the lowest resonant frequency of its molecules, the gas would emit radiation at the same frequency as that of the incident radiation, which is precisely what we have previously termed as Rayleigh scattering. If molecule aggregations that are comparable in size to the wavelength of incident radiation were also present, they would generate Mie scattering at the same frequency, which is potentially much stronger than Rayleigh scattering. Lenses, mirrors etc. and the walls of the apparatus would also scatter light at the incident frequency. Thus, measurement based on Rayleigh scattering would only be possible following careful elimination or drastic reduction of other scattering effects and the application of corrections to remove residual influences. Moreover, even for single-species gases, the emitted radiation would not be exactly monochromatic, but would have a nearly Gaussian spectral peak, due to *Doppler broadening*, which is caused by thermal motions of the molecules. The width of the spectral peak depends on the temperature, which introduces the possibility of temperature measurement by examination of the spectral shape of the scattered light. An additional limitation is that, when the molecular mean free path is relatively short, the process is contaminated by Doppler shifting caused by the propagation of sound waves in the gas. In such cases, the frequency spectrum of the scattered light would have two peaks, called *Brillouin peaks*, on either side of the incident frequency. This effect and other types of interference make spectral shapes difficult to interpret [28]. In practice, composition measurements based on Rayleigh scattering mainly rely on light intensity monitoring. For a single species, the collected light intensity would be

$$I = AN\sigma_R \,, \tag{15.4}$$

where A is a calibration constant, N is the concentration of the species and σ_R is the Rayleigh scattering cross section (see Section 7.4), which has been tabulated for common species and light wavelengths [30, 40]. Therefore, the gas density can be found as proportional to the scattered light intensity. For a mixture of molecules with relatively low densities, the scattering intensities are cumulative, as

$$I = A \sum N_i \sigma_{Ri} \,. \tag{15.5}$$

This approach has been applied successfully to binary mixtures of gases with substantially different scattering cross sections, such as methane and air (molecular oxygen and nitrogen have nearly the same σ_R) or carbon monoxide and nitrogen, in which the degree of mixedness has been measured with a relatively high-frequency response, in some cases exceeding 10 kHz [32, 41, 62]. For multi-species mixtures, or when the scattering cross sections are not sufficiently separated, Rayleigh-scattered radiation cannot be easily converted to species concentrations. When this method works, its advantage over competing methods is a nearly continuous output signal and a good temporal resolution.

Spontaneous Raman spectroscopy: *Spontaneous Raman spectroscopy* (SRS) employs the spontaneous Raman scattering effect to identify species in fluid mixtures as well as to measure their concentrations. The great advantage of Raman scattering methods, compared to Rayleigh scattering methods, is that the former are not contaminated by incident light or light scattered by impurities in the flow and the walls of the apparatus. All these, as well as Rayleigh-scattered light, can be separated by a spectral filter (e.g., a spectrometer). Its disadvantage is the low intensity of Raman-scattered radiation, which limits the resolution of the method. The collected light intensity obeys the relationship

$$I = AN\sigma_{Rm} , \tag{15.6}$$

where σ_{Rm} is the Raman scattering cross section, available in the literature for common species [87]. This method has been applied successfully to measurements of concentration in high-speed gas flows and in flames [6, 28, 65, 90]. Because each type of molecule has unique vibrational energy levels, it is possible to identify the presence of many different types of molecules, for example, the multiple products of combustion of complex fuels. The number densities and concentrations of different species in the mixture can be found by measuring the areas of the corresponding spectral bands and summing the contributions of bands that belong to the same species. It must be noted that, as the flow temperature increases, the number of vibrational energy levels that contribute to Raman scattering increases. Similarly, the number of rotational energy level increases, with the result of broadening the bands of vibrational levels. Both of these effects increase the probability of spectral band overlapping for different molecules, thus reducing the usability of the method. In general, Raman spectroscopy is relatively easy for mixtures of simple molecules (e.g., O_2 and CO_2), but much more difficult when higher hydrocarbons or other complex molecules are present. High power laser sources are required to overcome the low sensitivity of the method. Pulsed lasers, with typical pulse lengths of 10 ns, are used for measurements in high-speed and rapidly reacting flows and high-power CW lasers can be applied to the study of steady phenomena. Although light in the visible range has been used by some researchers, ultraviolet light is more usable, because some molecules (O_2, N_2 and others) have no visible resonance absorption energy bands. It is also preferable to use radiation with as long wavelength as possible, because the intensity of scattered light increases as the fourth power of wavelength (see Eq. (7.38)). The most commonly used laser for SRS in combustion research is the KrF excimer laser. Typical spatial resolution of SRS systems is about 0.5 mm and its temporal response is of the order of 20 ns [40]. Raman scattering can also be used for measuring temperature, thus permitting the simultaneous measurement of temperature and concentration. Moreover, in combination with laser-Doppler velocimetry or particle image velocimetry, it can be used for measuring simultaneously concentration and flow velocity.

Coherent anti-Stokes Raman spectroscopy: *Coherent anti-Stokes Raman spectroscopy (CARS) methods* are specialised variants of Raman scattering methods [27, 33, 40]. They are generally preferable to SRS for combustion research, because they provide a much stronger signal and are not as sensitive to the presence of soot. The basic CARS configuration uses two monochromatic beams with frequencies ν_1 and $\nu_2 < \nu_1$, respectively. Both beams are focused co-linearly on the same measuring volume within the fluid [79] (see Fig. 15.5). If some molecules within this volume have a vibrational energy state with a frequency $\Delta\nu = \nu_1 - \nu_2$, they will be driven to vibration coherently (namely, in phase), scattering the incoming light into two new co-linear and coherent beams with wavelengths $\nu_1 + \Delta\nu$ (anti-Stokes sideband) and $\nu_2 - \Delta\nu$ (Stokes sideband), respectively. Practical systems use the anti-Stokes band, as this is easier to

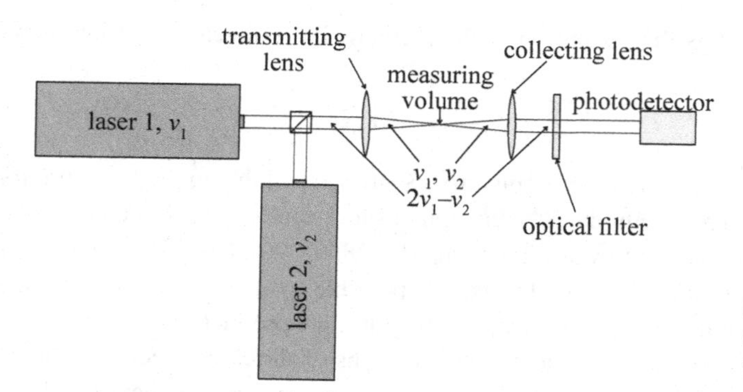

Figure 15.5 Sketch of a CARS system (adapted from [79]).

detect spectroscopically. In order to match the vibrational energies of different species, one of the lasers has to be tunable (i.e., a dye laser), while the other could have a fixed wavelength (commonly used is the frequency-doubled, Nd:YAG laser). By scanning the tunable laser wavelength, one would be able to detect different types of molecules, each of which has distinct vibrational energy levels. Furthermore, the concentration of each species can be found by measuring the intensity of light scattered at a particular spectral band, taking into account that the intensity of scattered radiation is proportional to the square of concentration. Due to the low sensitivity of the method, it is necessary to use high-power, pulsed lasers. Several different variants of CARS have been applied to the measurement of concentration in combustion processes and high-speed aerodynamics. A variant, referred to as *CARS thermometry*, has also been applied to the measurement of temperature in flames [68].

15.5 Laser-Induced Fluorescence

The term *laser-induced fluorescence* (*LIF*) describes a host of experimental methods that are based on the phenomenon of fluorescence of certain molecules (see Section 7.4), when excited by laser light. LIF methods can be applied to both liquids and gases. In all cases, for the method to be effective, the collected light must be passed through an optical band-pass filter, centred around the peak of the fluorescence emission, thus eliminating all interference from the incident radiation, reflections and stray light.

LIF can be applied in a local, a planar or a tomographic measuring configuration. In *local LIF*, the fluid is illuminated by a laser beam, the light of which is collected by a lens focused on a small control volume along the beam and recorded by a photodetector (Fig. 15.6a). In *planar LIF* (*PLIF*), the laser beam is converted into a thin sheet and a two-dimensional image of the concentration field is recorded by a camera (Fig. 15.6b). In *tomographic LIF*, the light sheet is scanned rapidly with the use of rotating mirrors or other means and images are collected with an essentially negligible time delay from each other, so that one can reconstruct the instantaneous 3-D concentration field.

LIF in liquids: The most common compounds used for LIF in water flows, together with representative peaks in their absorption and fluorescence spectra, are listed in Ta-

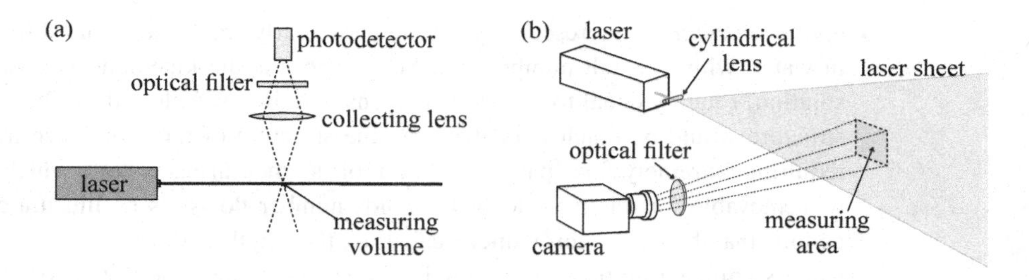

Figure 15.6 Sketches of optical arrangements for (a) local and (b) planar LIF.

ble 15.1. A buffered solution of the dye is injected into the fluid stream and laser light

Table 15.1 Compounds that are commonly used in LIF research

Molecule	Absorption peak	Fluorescence peak
Fluorescein disodium	488 nm	514.5 nm
Rhodamine 6G	530 nm	556 nm
Rhodamine B	542.75 nm	575 nm

is directed into the region of interest. Qualitative LIF for flow visualisation purposes is relatively easy to implement and does not require particular attention to details. On the other hand, the quantitative application of LIF for measuring dye concentration is subject to the following limitations [22, 34, 67, 82, 84, 88]:

- *Depletion* (bleaching): the fluorescence of molecules will diminish with time, when the same fluid volume is steadily exposed to laser radiation; this could happen during calibration of the technique, if a small container is used without mixing, but would not be a problem in a turbulent flow of a large volume of water, in which adequate mixing is provided.
- *Extinction*: for a fixed, relatively low, dye concentration, fluorescence intensity is proportional to the incident light intensity, for at least five orders of magnitude of concentration. This relationship is based on the assumption that absorption of the light would be negligible along its entire path. However, as the concentration increases, the attenuation of the laser beam or sheet along its path will inevitably become significant, causing measuring errors [82].
- *Temperature sensitivity*: the fluorescence quantum efficiency of the substances listed in Table 15.1 decreases with increasing temperature, with the steepest slope exhibited by Rhodamine B (see Section 14.5.3); thus, temperature variations in the flow would introduce uncertainty in concentration measurements; on the other hand, this effect makes LIF suitable also for the measurement of temperature fluctuations in liquids having a uniform dye concentration.
- *pH sensitivity*: within a range, pH affects significantly the extinction of light in the fluorescent medium; for fluorescein disodium, the pH effect is negligible for pH $\geq$ 8.5.

- *Spatial resolution*: fluorescent dyes have particularly small molecular diffusivities γ in water, with Schmidt numbers Sc (Sc $= \nu/\gamma$; ν is the kinematic viscosity of the solution) roughly equal to 2,000; this means that, in a turbulent flow, the scalar fine structure would be much finer than the fine structure of the turbulence itself, thus requiring extremely fine spatial resolution of the measuring system, which may not be achievable for relatively large Reynolds number flows; as an illustration, let us assume that the Kolmogorov microscale (i.e., the length scale of the smallest dynamically significant motions) in a flow is $\eta = 0.2$ mm; then, the Batchelor microscale (i.e., the length scale of the smallest significant concentration differences) would be $\eta_B = \eta \text{Sc}^{-1/2} \approx 4$ μm; this value would most likely much smaller than the typical spatial resolution of LIF systems (for local LIF, this would be the focal volume of the collecting lens, whereas, for planar LIF, this would depend on the laser sheet thickness, the pixel size and the image magnification factor).
- *Non-uniformity of illumination*: as mentioned in Section 7.3.9, laser sheets produced by cylindrical lenses have non-uniform intensities; Powell lenses can be used to produce a nominally uniform intensity across the sheet width, but these cannot eliminate the appearance of dark stripes, which are the expanded images of dust particles in the laser cavity; this effect would create erroneous concentration variations, even in cases with perfectly uniform concentration; the reader is referred to Section 7.3.9 for methods to improve the illumination uniformity and calibration methods of the optical system to correct for such non-uniformity.
- *Secondary fluorescence*: Although, ideally, the camera would only capture light emitted from fluid illuminated by the laser sheet, in practice, it also captures light from its entire range of view. In some cases, such light may be significant, even if its comes from out-of-focus regions [82].

Common sources of illumination for LIF in water are continuous-wave Ar-Ion and He-Ne lasers and pulsed Nd:YAG lasers, which conveniently emit light at wavelengths near the absorption peaks of the compounds listed in Table 15.1.

LIF in gases: LIF has been performed in non-reacting gas flows that contain passive fluorescent seeding materials, such as biacetyl, which emits radiation in the visible range. However, the most extensive LIF application in gases has been in flames, in which fluorescence is usually produced by radicals, such as OH (hydroxyl), CH (methylene), CO (carbon monoxide) and NO (nitric oxide), thus making it possible to map chemical reactions, even in transient states. The hydroxyl radical is the most commonly used species, for both quantitative and qualitative purposes, as it provides a strong signal and its concentration jump can be used to map the interface between burnt and unburnt parts of flames. Flame extinction is usually detected by the concentration of CH. By comparison to other spectroscopic methods, LIF produces the strongest radiation signal from a single species. Together with other optical methods, LIF has become an indispensable tool in combustion research. Excitation of fluorescence is normally provided by tunable, pulsed, dye lasers and excimer lasers, operating in the visible or ultraviolet ranges and pumped by various means. Details of the associated techniques may be found in several excellent reviews [23, 27, 37, 40, 95].

15.6 Particulate Measurement

The term *particulate* covers all sorts of airborne particles, both solid and liquid. Their sources can be natural (e.g., dust, pollen and haze) or industrial/commercial/residential (e.g., soot and smoke). Many types of particulates constitute health and environmental hazards and contribute to pollution. Sprays and aerosols, which are liquid droplet suspensions in gases, are encountered in many industrial systems, such as combustors and paint injectors. From the fluid mechanical viewpoint, the characterisation of particulates and aerosols includes the description of their mass flux, number density, mean size, size distribution, active surface area and shape. Among the common particulate measurement methods are the following.

Gross gravimetric analysis: The simplest approach for obtaining the average mass flux of suspended matter in a fluid stream would be to pass it through a filter of sufficient solidity (i.e., ratio of blocked to total area), which will capture all particles or those larger than a certain size. The average mass flow rate would be equal to the increase in weight of the filter, compared to its clean weight, over a measured time interval. Besides filters, particles may be captured by impacting the stream on surfaces coated with solid or liquid films [7]. When passing the entire stream through a filter is inconvenient or undesirable, one may extract a sample of the fluid through a sampling tube. In such cases, it would be advisable to apply *isokinetic sampling* (i.e., to sample at a speed equal to the local flow speed), in order to avoid biasing the composition of the sample towards heavier or lighter components. Obviously, these approaches cannot separate the different sizes or densities of particles and would be unsuitable for non-stationary flows and for reacting flows, in which the composition of the particulate may vary.

Separation and measurement of sizes: Mixtures of solid particles with linear dimensions greater than a few μm can be separated by size with the use of *sieves*. Samples of captured particles, solid or liquid, can be analysed using optical microscopes, for sizes larger than about 0.5 μm, or electron microscopes, for sizes as small as 10 nm [28, 43]. Separation of particles in a fluid stream according to their inertia can be achieved by passing a fluid sample through a *cascade sampler*, consisting of a set of different nozzles and baffles, or exposing the sample to oscillatory flow [7]. The characterisation of polydisperse particles, particles with different densities and particles with non-spherical shapes requires an unambiguous definition of the *average* (or, more appropriately, *equivalent*) *diameter*. This parameter may be defined in different ways, depending on the measuring method and the intended use of the measurements: the *Sauter mean diameter* (*SMD*) is defined as the diameter of spheres that have the same surface area per unit volume of the suspension, while the *mean aerodynamic diameter* takes into account the drag on the particles, irrespective of their size, shape and density. When analysing particle samples, care must be taken to avoid coalescence, deposition, condensation or evaporation in the sampling tubes; this may require heating or cooling of some components. A number of *particle sizers*, based on different principles, are available commercially and one is advised to contact the manufacturers for details.

Photographic, visual and optical methods: Unlike sampling methods, on-site photographic, visual and optical methods are non-intrusive, thus avoiding many of the problems mentioned above. Conventional photography is not very suitable for particle imaging, because it cannot resolve depth of field; three-dimensional views of particles can be recorded by *stereo-photography* and holographic methods [7]. The various *light scattering methods* described in Section 7.4 may also be used for measuring the local concentration of particulates. Related to these are *obscuration techniques*, which are based on the absorption of light by particles in the path of a light beam, usually transmitted and collected by optical fibres.

Phase-Doppler analysis and related techniques: Among the variety of available optical methods, of particular interest is *phase-Doppler particle analysis* (*PDPA* or *PDA*), also known as *phase-Doppler interferometry* (*PDI*) [3, 54], because it can be combined with laser-Doppler velocimetry (see Section 12.3) to provide simultaneous measurements of flow velocity and size distribution of droplets or other spherical particles. A simplified illustration of the technique is shown in Fig. 15.7. A laser beam is split into two parallel beams, which are made to intersect within a small measuring volume, in which they form interference fringes. A particle passing through this volume scatters light, according to Mie scattering theory. The scattered light is collected by a lens located off-axis and projected onto an array of two or more photodetectors. The output of each photodetector is a Doppler burst, the frequency of which is proportional to the particle velocity, according to LDV analysis. At the same time, light reflected on the surface of the particle or refracted through it will acquire a phase shift, which would depend on the geometrical arrangement of the incident-beam–particle-detector system and the size of the particle. By determining the phase shifts between the Doppler bursts of different detectors, and taking into account the wavelength of light and some optical parameters, one can calculate the particle diameter without a need for calibration. The phase-Doppler technique is not affected by multiple scattering and absorption along the light path. However, it requires the presence of a single particle in the measuring volume, thus limiting its range to relatively low particle concentrations. It is also suitable only for optically homogeneous, spherical particles. Among the techniques which can characterise non-spherical and inhomogeneous particles [24] is the *shadow Doppler technique* [38, 59], which essentially maps the projections of individual moving particles on a large (up to 35) array of photodiodes, from which both the shapes and the velocities of the particles can be extracted. Another method of interest is *global phase-Doppler* (*GDP*) technique [3]. This technique is similar to PDA, with the main difference that the measuring volume is much larger, as the collecting lens is out of focus

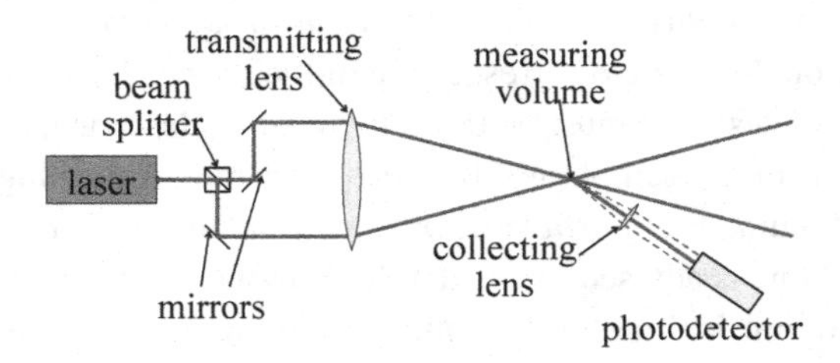

Figure 15.7 Sketch of a phase-Doppler particle analyser.

on the intersection of two light sheets. Particles crossing this volume appear to have interference fringes and the size of particles is determined by counting the number of fringes.

Laser-induced incandescence: The exhaust gases of internal combustion engines, furnaces and other combustion systems contain particulates, produced by an incomplete oxidisation of the fuel and referred to as *soot*. Soot consists of irregularly shaped aggregates of nearly spherical carbon particles of nanoscale dimensions, called the *primary particles*. In addition, the gases may contain liquid droplets, which are the product of vapour condensation, and particles composed of metal sulfates or oxides, which constitute the *ash*. *Laser-induced incandescence* (*LII*) methods [57, 68] apply high-power, short-duration laser radiation pulses to heat the airborne soot particles to sublimation temperatures, causing them to emit radiation, which is collected by photodetectors and analysed to provide the particle size and volume fraction. Because carbon absorbs laser radiation at a much higher rate than liquids or ash, this method is insensitive to the presence of the latter. The size of the soot particles is much smaller than the laser light wavelength, so that the light emission is in the Rayleigh regime. Heat transfer from the hot particles to the surrounding cooler gas takes place mainly by conduction and the particle temperature, following the initial heating stage, decays exponentially, with a time constant that is proportional to the particle size. Thus, assuming that the primary particles are monodisperse, one can estimate their size by analysing the temporal decay of the emitted radiation signal [75, 81]. Mathematical analysis of the LII signal, bandpass-filtered at two wavelengths, may also provide the size distribution of polydisperse particulates [66]. The volume fraction of soot particles can be determined from the intensity of the LII signal, calibrated in a laminar flame containing particles of known size, which has been measured independently by other means; a self-calibrating method providing soot volume fraction has also been proposed [57]. LII, combined with two-colour pyrometry, is capable of both measuring particle size and determining its composition. This method applies not only to soot particles, but also to tungsten, silicon and other materials, and it is based on the measurement of the boiling temperature of the particles [77]. Compared to other optical methods, LII has the advantages of high spatial and temporal resolutions and requires relatively simple means.

15.7 Void Fraction Measurement

An important property of gas–liquid two-phase flows is the *void fraction*, defined as the ratio of the volume occupied by the gaseous phase over the total volume of interest. Depending on the volume considered, one may distinguish between *bulk void fraction*, which is spatially averaged over the entire channel or other large volume of the fluid, a cross section or a line (e.g., a diameter of a pipe), and *local void fraction*, which is (usually) time-averaged at a location in the flow [42]. Besides the void fraction, which is an average property, one may also be interested in determining the shape of the interface between the gas and liquid phases, and, for bubbly flows, the size distribution of bubbles. There is significant interest in such parameters in many applications, ranging from nuclear power generation to physical oceanography. A great variety of methods have been developed for gas–liquid flow measurement, widely varying in resolution and accuracy, and often specialised to the needs and conditions of a particular application. In general, such techniques are specific to the *regime* of the gas–liquid flow, namely the geometry and topology of the phases.

Flow regime identification: Among commonly identified regimes for pipe flows are the *bubbly*, *slug*, *churn*, *annular* and *stratified* regimes [94]. Which regime will occur in a particular channel depends on many factors, including the fluid material, pressure and temperature, the shape, size and orientation of the channel and the phase velocities. For a particular geometry and fluid, the regimes are presented in empirical *regime maps*, in which the two axes are typically measures of the gas and liquid flow rates. As a rule, measurement of the void fraction requires knowledge of the flow regime and so *regime identification* is usually part of the gas–liquid flow measurement method. Because visual identification of the regime is rarely convenient and sufficiently fast, in many cases being impossible, this is accomplished by an automated method that is based on radiation data, pressure difference, change of an electrical property or some other measurand. Machine learning methods (e.g., neural networks and elastic maps) have been applied successfully to regime identification, in combination with algorithms providing measurements of the void fraction and phase velocities [47, 58, 69, 78].

Among the most commonly used void fraction measurement methods are the following:

Quick-closing valve method: This method is suitable for bulk void fraction measurement in stationary flows through closed channels, which can be temporarily interrupted without affecting the operation of the system. A part of the mixture is trapped in a section of the channel by simultaneously closing two solenoid valves, one at at its inlet and another at its outlet. Then, the trapped liquid volume is measured by draining it into a volumetric container [42]. This method is not suitable for high-pressure flows because of valve leakage. It is cumbersome and slow, but it is also robust and requires no calibration or empirical fits. It is used primarily for calibrating other methods.

Sampling methods: A number of sampling methods have been proposed. They all remove continuously samples of one or both phases of the two-phase mixture, separate the phases and analyse them externally. The *wall scoop method* and the *porous wall method* have been used for sampling the liquid film in annular-type flows, whereas *isokinetic sampling probes* have been used to measure the mass flux of liquid droplets in vapour streams [46].

Pressure difference methods: The pressure difference between two closely located points along a vertical pipe depends on the elevation difference, the void fraction and frictional losses and is specific to each flow regime [45, 69, 94]. The measured pressure difference signal can be analysed, most effectively by machine learning algorithms, to determine the void fraction as well as other two-phase flow properties.

Electrical impedance meters: These methods are based on the dependence of the electrical impedance of a multi-phase flow upon the concentration of the different phases. Although various improved designs have been introduced by different authors, the basic *impedance void fraction meter* consists of a number of concentric ring-type electrodes, powered by high-frequency alternating current and surrounding a non-conducting section of a pipe, through which the fluid mixture is passing. This method provides time-dependent, cross-sectional or annular averages of the void fraction with a high-frequency response. However, because the outputs of impedance meters are very sensitive to the phase pattern (i.e., whether it is in the bubbly, annular, slug or mist flow regime), they can be only used if the flow regime is known [16, 42]. Local void fraction impedance probes are also available [46], whose operation is also based on the electrical impedance dependence upon the phase distribution. A commonly used type are the various *conductivity probes* (see Section 15.3), which have a thin electrode (e.g., the tip of an acupuncture needle, with the rest of the needle insulated) in contact with the fluid. Their output signals discriminate between the liquid and gas phases, so that they can be used to measure not only the time-averaged local void fraction, but also bubble size and speed. A variety of designs have been proposed for specific use in studies of boiling, cavitation, fluidised beds and oceanic physics [16, 17, 92].

Wire-mesh sensors: A *wire-mesh sensor* (WMS) consists of two arrays of electrodes, with the axes of one array perpendicular to those of the other and with the planes of these two sets of axes separated by a small distance in the streamwise direction along a pipe [63, 70, 71]. High frequency (up to 10 kHz) electric pulses are passed sequentially through each electrode and an algorithm calculates the conductivity of the gas–liquid mixture that occupies the small volume near the projected nodes of the mesh. Single and dual WMS have been used for tomographic flow visualisation, flow regime identification and the measurement of the void fraction, interfacial velocity and bubble size distribution.

Thermal methods: The constant temperature *hot-film anemometer*, which is normally used as a velocity sensor, can be also used to discriminate between liquid and gas phases in contact with the sensor. The instrument output is proportional to the electrical power required to maintain the sensor in constant temperature; due to the reduced heat transfer in the gas compared to that in the liquid, the anemometer output voltage would be much lower for the gas phase than for the liquid phase. By using an adjustable threshold to separate the signal into low-level and high-level segments, one can measure the fraction of time that the sensor would be immersed in gas, which, ideally, would be equal to the void fraction. With calibration of the probe and application of various corrections, it may be possible to measure both the void fraction and local fluid velocity in one or both phases [46]. A different approach has been to use *microthermocouples* both as local temperature sensors and conductivity sensors to separate the gas and liquid phases [26].

Optical methods: A variety of optical methods have been suggested, mostly utilising the difference in refractive index between the gas and liquid phases. In the *glass-rod method* [46], the tip of a thin glass rod is polished to a 45° angle with respect to its axis. The probe is introduced into the flow and light is transmitted through the glass rod. According to Snell's law (see Section 7.2.2), when the tip is in contact with gas, the light beam will exit the rod, while, when the tip is in contact with liquid, total internal reflection will occur (see Section 7.2.3) and the light will be reflected back into the rod. By collecting the reflected light, it is possible to measure the percent of time that the rod is in contact with the gas, and, thus, the void fraction. Besides the basic glass-rod method, a number of variations and improvements using fibre optics have been developed. A separate approach has been to measure the attenuation or extinction of light transmitted through bubbly flows through reflection, refraction and diffraction on the bubbles. Methods suitable for both relatively high and relatively low concentrations of bubbles are available [50, 72].

Acoustical methods: Void fraction and bubble size distribution have been measured in bubbly flows by utilising the changes in sound (usually ultrasound) propagating through a liquid, when it encounters a bubble. A variety of acoustical methods, based on sound absorption, Doppler shift and transit time, have been applied in both industrial [46] and oceanic applications [80] as well as in microchannels [29].

Radiation methods: Some widely used bulk and local void fraction methods are based on the difference in absorption of gamma and X-rays propagating through the gas and liquid phases [42]. Absorption of a single collimated ray would provide the bulk void fraction along its path. By traversing the ray across the flow or using broad-beam or multi-beam instruments, one may obtain the cross-section-averaged void fraction. The local void fraction can be measured by positioning the radiation detector at an angle with the beam (side-scattering), so that it would detect radiation scattered by a small volume on the beam's path. A related method, based on neutron scattering, has been used in applications in which a particle accelerator is available.

Chapter Digest

15.1 Which is a main general limitation of sampling probes?

15.2 What makes thermal probes sensitive to composition?

15.3 What makes light scattering methods sensitive to composition?

15.4 Discuss the difference between Rayleigh and Raman scattering of light. Identify the main advantages and disadvantages of concentration measuring methods that are based on these two types of scattering.

15.5 When considered in a broad sense, to which fluid properties is laser-induced fluorescence sensitive? How can these sensitivities be separated?

15.6 What is a particulate and why do we need to measure its characteristics?

15.7 What is the void fraction and why do we need to measure it?

Problems

15.1 Consider a homogeneous mixture of two neutral gases with molecular weights M_1 and M_2 and Gladstone–Dale constants K_1 and K_2. Would it be possible to determine its composition from a measurement of its Gladstone–Dale constant? If so, describe the procedure and provide appropriate relationships.

15.2 Consider a stationary turbulent mixing layer between a stream of air and a stream of helium with different speeds and temperatures. Describe in detail the instrumentation and measuring procedures required for the following measurements:

(a) The measurement of the local mean velocity and turbulence statistics across the mixing layer.

(b) The measurement of the local mean and standard deviation of the concentration across the mixing layer.
(c) The determination of instantaneous concentration maps over a cross section of the flow.
(d) The simultaneous measurement of local velocity, temperature and concentration fluctuations with a temporal resolution of at least 1 ms.

References

[1] D. Adler. A hot wire technique for continuous measurement in unsteady concentration fields of binary gaseous mixtures. *J. Phys. E: Sci. Instrum.*, 5:163–169, 1972.

[2] D. Adler and Y. Zvirin. The time response of a hot wire concentration transducer. *J. Phys. E: Sci. Instrum.*, 8:185–188, 1975.

[3] H.-E. Albrecht, M. Borys, N. Damaschke, and C. Tropea. *Laser Doppler and Phase Doppler Measurement Techniques*. Springer-Verlag, Berlin, 2003.

[4] C.V. Alonso. Comparative study of electrical conductivity probes. *J. Hydraulic Res.*, 9:1–10, 1971.

[5] Anonymous. *Elementary Theory of Gas Chromatography with Bibliography and Experiments*. Gow-Mac Instrument Co., Bridgewater, NJ, 1978.

[6] T. Arts et al. *Measurement Techniques in Fluid Dynamics (2nd Edition)*. von Kármán Institute for Fluid Dynamics, Rhode-Saint-Genese, Belgium, 2001.

[7] B.J. Azzopardi. Measurement of drop sizes. *Int. J. Heat Mass Transfer*, 22:1245–1279, 1979.

[8] C.N. Banwell and E.M. McCash. *Fundamentals of Molecular Spectroscopy (4th Edition)*. McGraw-Hill, New York, 1994.

[9] H.A. Becker. Mixing, concentration fluctuations and marker nephelometry. In B.E. Launder, editor, *Studies in Convection*. Academic Press, New York, 1977.

[10] H.A. Becker, H.C. Hottel, and G.C. Williams. On the light-scatter technique for the study of turbulence and mixing. *J. Fluid Mech.*, 30(2):259–284, 1967.

[11] P.F. Bernath. *Spectra of Atoms and Molecules*. Oxford University Press, New York, 1995.

[12] A.D. Birch, D.R. Brown, M.G. Dodson, and F.Swaffield. Aspects of design and calibration of hot-film aspirating probes used for the measurement of gas concentration. *J. Phys. E: Sci. Instrum.*, 9:59–63, 1986.

[13] J.M. Bobbitt, A.E. Schwarting, and R.J. Gritter. *Introduction to Chromatography*. Reinhold, New York, 1968.

[14] G.L. Brown and M.R. Rebollo. A small, fast-response probe to measure composition of a binary gas mixture. *AIAA J.*, 10(5):649–652, 1972.

[15] M. Cabannes, M. Ferchichi, and S. Tavoularis. Temperature variation correction for aspirating concentration probes. *Meas. Sci. Tech.*, 15:1211–1215, 2004.

[16] S.L. Ceccio and D.L. George. A review of electrical impedance techniques for the measurement of multiphase flows. *J. Fluids Eng.*, 118:391–399, 1996.

[17] N.P. Cheremisinoff. *Instrumentation for Complex Fluid Flows*. Technomic Publishing, Inc., Lancaster, PA, 1986.

[18] S.K. Chua, J.W. Cleaver, and A Millard. The measurement of salt concentration in a plume using a conductivity probe. *J. Hydr. Res.*, 24(3):171–178, 1986.

[19] E.R. Cohen, D.R. Lide, and G.L. Trigg. *AIP Physics Desk Reference (3rd Edition)*. Springer, New York, 2003.

[20] S. Corrsin. Extended applications of the hot-wire anemometer. *Rev. Sci. Instrum.*, 18:469–471, 1947.

[21] S. Corrsin. Extended applications of the hot-wire anemometer. Technical Report Technical Note 1864, NACA, Washington DC, April 1949.

[22] J.P. Crimaldi. Planar laser induced fluorescence in aqueous flows. *Exp. Fluids*, 44:851–863, 2008.

[23] J.W. Daily. Laser induced fluorescence spectroscopy in flames. *Prog. Energy Combust. Sci.*, 23:133–199, 1997.

[24] N. Damaschke, G. Gouesbet, G. Grehan, and C. Tropea. Optical techniques for the characterization of non-spherical and non-homogeneous particles. *Meas. Sci. Technol.*, 9:137–140, 1998.

[25] P. Danckwerts. The definition and measurement of some characteristics of mixtures. *Appl. Sci. Res. A*, 3:279–296, 1952.

[26] J.M. Delhay, R. Semeria, and J.C. Flamand. Void fraction, vapor and liquid temperatures: Local measurements in two-phase flow using a microthermocouple. *J. Heat Transfer*, 95C:365–370, 1973.

[27] A.C. Eckbreth. *Laser Diagnostics for Combustion Temperature and Species (2nd Edition)*. Gordon and Breach Publishers, Philadelphia, PA, 1996.

[28] R.J. Emrich, editor. *Methods of Experimental Physics, Vol. 18A and B: Fluid Dynamics*. Academic Press, New York, 1981.

[29] A.R. Gardenghi, E. dos Santos Filho, D.G. Chagas, G. Scagnolatto, R. Monteiro Oliveira, and C.B. Tibiriçá. Overview of void fraction measurement techniques, databases and correlations for two-phase flow in small diameter channels. *Fluids*, 5:216 (26pp), 2020.

[30] W.C. Gardiner, Y. Hidaka, and T. Tanzawa. Refractivity of combustion gases. *Combust. Flame*, 40:213–219, 1981.

[31] C.H. Gibson and W.H. Schwarz. Detection of conductivity fluctuations in a turbulent flow field. *J. Fluid Mech.*, 16 (3):357–364, 1963.

[32] S.C. Graham, A.J. Grant, and J.M. Jones. Transient molecular concentration measurements in turbulent flows using rayleigh light scattering. *AIAA J.*, 12 (8):1140–1142, 1974.

[33] D.A. Greenhalgh. Quantitative CARS spectroscopy. In R.J.H. Clarc and R.E. Hester, editors, *Advances in Nonlinear Spectroscopy*, pages 193–252. Wiley, New York, 1987.

[34] G.G. Guibault. *Practical Fluorescence: Theory, Methods and Techniques*. Dekker, New York, 1973.

[35] P. Guilbert and E. Dicocco. Development of a local continuous sampling probe for the equivalence air-fuel ratio measurement. application to spark ignition engine. *Exp. Fluids*, 32(4):494–505, 2002.

[36] P. Guilbert, M. Durget, and M. Murat. Concentration fields in a confined two-gas mixture and engine in-cylinder flow: Laser tomography measurement by mie scattering. *Exp. Fluids*, 31:630–642, 2001.

[37] R.K. Hanson, J.M. Seitzman, and P.H. Paul. Planar laser-fluorescence imaging of combustion gases. *Appl. Phys. B*, 50:441–454, 1990.

[38] Y. Hardalupas, K. Hishida, M. Maeda, H. Morikita, A.M.K.P. Taylor, and J.H. Whitelaw. Shadow doppler technique for sizing particles of arbitrary shape. *Appl. Opt.*, 33:8417–8426, 1994.

[39] J.L. Harion, M. Favre-Marinet, and B. Camano. An improved method for measuring velocity and concentration by thermo-anemometry in turbulent helium-air mixtures. *Exp. Fluids*, 22:174–182, 1996.

[40] E.P. Hassel and S. Linow. Laser diagnostics for studies of turbulent combustion. *Meas. Sci. Technol.*, 11:R37–R57, 2000.

[41] J. Haumann, G. Wu, and A. Leipertz. Low power laser Rayleigh probe for flow mixing studies. *Exp. Fluids*, 5:230–234, 1987.

[42] G.F. Hewitt. Void fraction. In G. Hestroni, editor, *Handbook of Multiphase Systems*, pages 10.21–10.33. Hemisphere Publishing Corporation, Washington DC, 1982.

[43] R.R. Irani and C.F. Callis. *Particle Size: Measurement, Interpretation, and Application*. Wiley, New York, 1973.

[44] J.A. Jahnke. *Continuous Emission Monitoring (2nd Edition)*. Wiley, New York, 2000.

[45] J. Jia, A. Babatunde, and M. Wang. Void fraction measurement of gas–liquid two-phase flow from differential pressure. *Flow Meas. Instrum.*, 41:75.80, 2015.

[46] O.C. Jones. Two-phase flow measurement techniques in gas-liquid systems. In R.J. Goldstein, editor, *Fluid Mechanics Measurements (1st Edition)*, pages 479–558. Hemisphere Publishing Corporation, Washington DC, 1983.

[47] J. Julia, Y. Liu, S. Paranjape, and M. Ishii. Upward vertical two-phase flow local flow regime identification using neural network techniques. *Nucl. Eng. Des.*, 238:156–169, 2008.

[48] K. Kohse-Höinghaus and J.B. Jeffries, editors. *Applied Combustion Diagnostics*. Taylor & Francis, New York, 2002.

[49] S. Komori, T. Kanzaki, and Y. Murakami. Simultaneous measurement of instantaneous concentrations of two reacting species in a turbulent flow with a rapid reaction. *Phys. Fluids A*, 3(4):507–510, 1991.

[50] D.M. Leppinen and S.B. Dalziel. A light attenuation technique for void fraction measurement of microbubbles. *Exp. Fluids*, 30:214–220, 2001.

[51] P.A. Libby. Studies in variable-density and reacting turbulent shear flows. In B.E. Launder, editor, *Studies in Convection*. Academic Press, New York, 1977.

[52] M.B. Long, B.F. Webber, and R.K. Chang. Instantaneous two-dimensional gas concentration measurements in a jet flow by mie scattering. *Appl. Phys. Lett.*, 34:22–24, 1979.

[53] M. Mahouast. Concentration fluctuations in a stirred reactor. *Exp. Fluids*, 11:153–160, 1991.

[54] F. Mayinger and O. Feldmann, editors. *Optical Measurements (2nd Edition)*. Springer, Berlin, 2001.

[55] J. McQuaid and W. Wright. The response of a hot-wire anemometer in flows of gas mixtures. *Int. J. Heat Mass Transfer*, 16:819–828, 1973.

[56] J. McQuaid and W. Wright. Turbulence measurements with hot-wire anemometry in non-homogeneous jets. *Int. J. Heat Mass Transfer*, 17:341–349, 1974.

[57] C. Mercer. *Optical Metrology for Fluids, Combustion and Solids*. Kluwer Academic Publishers, Dordrecht, 2003.

[58] Y. Mi, M. Ishii, and L. Tsoukalas. Flow regime identification methodology with neural networks and two-phase flow models. *Nucl. Eng. Des.*, 204:87–100, 2001.

[59] H. Morikita and A.M.K.P. Taylor. Application of shadow doppler velocimetry to paint spray: Potential and limitations in sizing optically inhomogeneous droplets. *Meas. Sci. Technol.*, 9:221–231, 1998.

[60] F. Ncube, E.G. Kastrinakis, S.G. Nychas, and K.E. Lavdakis. Drifting behavior of a conductivity probe. *J. Hydr. Res.*, 29(5):643–654, 1991.

[61] S.-R. Park and H. Swerdlow. A miniature electrolytic conductivity probe with a wide linear range. *Electroanalysis*, 19:2294–2300, 2007.

[62] W.M. Pitts and T. Kashiwagi. The application of laser-induced rayleigh light scattering to the study of turbulent mixing. *J. Fluid Mech.*, 141:391, 1094.

[63] H.-M. Prasser, D. Scholz, and C. Zippe. Bubble size measurement using wire-mesh sensors. *Flow Meas. Instrum.*, 12:299–312, 2001.

[64] J.A. Puleo, J. Faries, M. Davidson, and B. Hicks. A conductivity sensor for nearbed sediment concentration profiling. *J. Atmosph. Ocean. Techn.*, 27:397–408, 2010.

[65] B.E. Richards, editor. *Measurement of Unsteady Fluid Dynamic Phenomena*. Hemisphere Publishing Corporation, Washington DC, 1977.

[66] P. Roth and A.V. Filippov. In situ ultrafile particle sizing by a combination of pulsed laser heatup and particle thermal emission. *J. Aerosol Sci.*, 27:95–104, 1996.

[67] P. Sarathi, R. Gurka, G. Kopp, and P. Sullivan. A calibration scheme for quantitative concentration measurements using simultaneous PIV and PLIF. *Exp. Fluids*, 52:247–259, 2012.

[68] C. Schulz, A. Dreizler, V. Ebert, and J. Wolfrum. Combustion diagmostics (chapter 20). In C. Tropea, A.L. Yarin, and J.F. Foss, editors, *Springer Handbook of Experimental Fluid Mechanics*. Springer, Berlin, 2007.

[69] H. Shaban and S. Tavoularis. Identification of flow regime in vertical upward air–water pipe flow using differential pressure signals and elastic maps. *Intern. J. Multiphase Flow*, 61:62–72, 2014.

[70] H. Shaban and S. Tavoularis. The wire-mesh sensor as a two-phase flow meter. *Meas. Sci. Technol.*, 26:015306, 2015.

[71] H. Shaban and S. Tavoularis. On the accuracy of gas flow rate measurements in gas–liquid pipe flows by cross-correlating dual wire-mesh sensor signals. *Intern. J. Multiphase Flow*, 78:70–74, 2016.

[72] B. Shamoun, M. El Beshbeeshy, and R. Bonazza. Light extinction technique for void fraction measurements in bubbly flow. *Exp. Fluids*, 26:16–26, 1999.

[73] E.J. Shaughnessy and J.B. Morton. Laser light-scattering measurements of particle concentration in a turbulent jet. *J. Fluid Mech.*, 80:129, 1977.

[74] A. Sirivat and Z. Warhaft. The mixing of passive helium and temperature fluctuations in grid turbulence. *J. Fluid Mech.*, 120:475–504, 1982.

[75] D.R. Snelling, G.J. Smallwood, I.G. Cambell, J.E. Medlock, and O. Gulder. Development and application of laser-induced incandescence (LII) as a diagnostic for soot particulate measurements. In *AGARD 90th Symposium of the Propulsion and Energetics Panel on Advanced Non-Intrusive Instrumentation for Propulsion Engines*, pages 1–9, AGARD, Brussels, 1979.

[76] R.A. Stanford and P.A. Libby. Further applications of hot-wire anemometry to turbulence measurements in helium-air mixtures. *Phys. Fluids*, 17 (7):1353–1361, 1974.

[77] M. Stephens, N. Turner, and J. Sandberg. Particle identification by laser-induced incandescence in a solid-state laser cavity. *Appl. Optics*, 42:3726–3736, 2003.

[78] T. Tambouratzis and I Pázsit. Non-invasive on-line two-phase flow regime identification employing artificial neural networks. *Annals Nucl. Energy*, 36:464–469, 2009.

[79] J.P. Taran. Molecular diagnostics for rarefied flows. In *AGARD Conference Proceedings CP-601*, Neuilly-sur-Seine, France, September 22-25, 1997. AGARD, 1997.

[80] S. Vagle and D.M. Farmer. A comparison of four methods for bubble size and void fraction measurements. *IEEE J. Ocean. Eng.*, 23 (3):211–222, 1998.

[81] R.L. Vander Wal, T.M. Ticich, and A.B. Stephens. Can soot primary particle size be determined using laser-induced incandescence? *Combust. Flame*, 116:291–296, 1999.

[82] C. Vanderwel and S. Tavoularis. On the accuracy of PLIF measurements in slender plumes. *Exp. Fluids*, 55:1801, 2014.

[83] K. Voloudakis, P. Vrahliotis, E.G. Kastrinakis, and S.G. Nychas. The behaviour of a conductivity probe in electrolytic liquid/solid suspensions. *Meas. Sci. Technol.*, 10:100–105, 1999.

[84] D.A. Walker. A fluorescence technique for measurement of concentration in mixing liquids. *J. Phys. E: Sci. Instrum.*, 20:217–224, 1987.

[85] J. Way and P.A. Libby. Hot-wire probes for measuring velocity and concentration in helium-air mixtures. *AIAA J.*, 8 (5):976–978, 1970.

[86] J. Way and P.A. Libby. Application of hot-wire anemometry and digital techniques to measurements in a turbulent helium jet. *AIAA J.*, 9 (8):1567–1573, 1971.

[87] A. Weber, editor. *Raman Spectroscopy of Gases and Liquids*. Springer, Berlin, 1979.

[88] D. Webster, S. Rahman, and L. Dasi. Laser-induced fluorescence measurements of a turbulent plume. *J. Eng. Mech. ASCE*, 129:1130–1137, 2003.

[89] J.G. Webster, editor. *The Measurement, Instrumentation and Sensors Handbook*. CRC Press, Boca Raton, FL, 1999.

[90] G.F. Widhoff and S. Lederman. Specie concentration measurements utilizing raman scattering of a laser beam. *AIAA J.*, 9(2):309–316, 1971.

[91] D.J. Wilson and D.D.J. Netterville. A fast-response, heated-element concentration detector for wind-tunnel applications. *J. Wind Eng. Industr. Aerodyn.*, 7:55–64, 1981.

[92] C. Wolff, F.U. Briegleb, J. Bader, K. Hektor, and H. Hammer. Effect of suspended solids on the hydrodynamics of bubble columns for application in chemical and biotechnological processes. *Chem. Engng. Technol.*, 13:172–184, 1990.

[93] J. Wolfrum. Lasers in combustion: From basic theory to practical devices. *Proc. Combust. Inst.*, 27:1–42, 1998.

[94] G. Yadigaroglu and F.H. Hewitt. *Introduction to Multiphase Flow: Basic Concepts, Applications and Modelling*. Springer, Cham, Switzerland, 2018.

[95] W.-J. Yang, editor. *Handbook of Flow Visualization, 2nd edition*. Taylor & Francis, 2001.

16 Retrospection and Outlook

In previous chapters, we discussed general experimental concepts and methodologies, as well as classical experimental techniques and some state-of-the-art methods for fluid mechanics measurements. In closing, it seems worthwhile to assess recent progress in the field, to identify areas in which further development is needed and to speculate on the chances of such development being achieved in the near future.

Fluid mechanics has been established as an important branch of physical sciences and a vital engineering discipline, with a wide range of contributions to our quality of life and economic sustainability. Its progress has been based on both theoretical formulation and empirical observation, which, in time, led to prediction and control of physical phenomena and technological processes. As in other scientific fields, analytical, computational and experimental tools used in fluid mechanics research and development have grown in sophistication and complexity, largely in response to ever-expanding human needs, but also because of the challenges and intellectual rewards offered by such activities. On the theoretical side, significant advances in many fluid mechanical applications have followed the development of mathematical methods, such as perturbation analysis and wavelet transforms, while the maturing of computational fluid dynamics and numerical algorithm design has enabled the numerical solution of many problems that could not have been tackled by other means. Computational approaches occupy an important place in fluid mechanics research, but it is also evident that experimentation retains an indispensable role in solving engineering problems and in designing and testing new engineering systems. Moreover, the discovery of new concepts, the understanding of complex physical phenomena and mechanisms, and the validation of theories, models and algorithms still rely heavily on the imaginative conception and meticulous execution of scientific experiments. Based on recent experience and current needs, one may confidently assess that experimental fluid mechanics is at no risk of becoming obsolete.

In 2005, in the first edition of this book, we asked the question 'what would the fluid mechanics laboratory look like in 10, 20 or 30 years?' We now have the opportunity to assess the advances made during the nearly two decades that have passed since then. The following is, in our opinion, a partial list of recent significant advances:

- improvements in illumination with the use of LED and laser diodes and in light recording with the use of CMOS cameras;
- continuing increase of computer speed and power, which allowed the real-time processing of huge volumes of experimental data; a notable example is the calculation of in-flow static pressure by analysis of volumetric velocity measurements;

- general reduction of electronic noise in many devices, with a consequent rise in the corresponding signal-to-noise ratios;
- improvements in the temporal, spatial and amplitude resolutions of many measuring systems, as a result of both technical advances and the development of more powerful data processing algorithms;
- further application of machine learning techniques, notably artificial neural networks, in many calibration and measurement procedures;
- dramatic advancement of methods producing time-resolved, three-dimensional maps of flow properties, including in-flow turbulent static-pressure fluctuations, the hitherto 'holy grail' of experimental fluid mechanics;
- expansion of the experimental fluid mechanics toolbox, in some cases without additional cost, by the broad availability of devices and processes intended for ordinary daily use (e.g., cameras, light sources and software applications in mobile telephone devices);
- increased availability of online scientific information, both under license and through open access; besides its unquestionable benefits, this ease has intensified the onus on the user to assess the credibility of the source and the veracity of online information.

As in the first edition, we will venture to speculate on forthcoming developments. We may reiterate that, because experimental fluid mechanics is a mature and multi-faceted field, its development will probably continue to be steady, arduous and gradual, rather than explosive. Experimental fluid mechanicists have long been exploiting a multitude of mechanical, thermal, acoustic, optical, electrical and electromagnetic principles and phenomena to measure fluid properties. One may plausibly anticipate that new experimental approaches may be devised, based on hereto non-utilised physical concepts. Nevertheless, a more predictable short-term development is that existing techniques will continue being improved, as some of their current limitations are overcome and their ranges of application are expanded.

Our expectation that time-resolving, volumetric measurement methods would be advanced has already materialised and the existing momentum makes one hopeful that further refinement and extension of the range of application of such methods is imminent. Current trends indicate that several powerful methods that are at present successful only for measurements under 'friendly' conditions (e.g., isothermal, single-phase flows at moderate Reynolds numbers) will be extended to higher Reynolds numbers (or wider ranges of other relevant dimensionless groups), multi-phase flows and flows with phase change, corrosive and high-temperature or high-pressure fluids, transonic and hypersonic flows, nano-, pico- and even femto-scale flows, measurements in opaque fluids and measurements in living organisms and biological systems. In particular, new or improved non-optical, non-intrusive methods will hopefully enhance our capabilities for measurements in optically non-accessible apparatus, opaque fluids and multi-phase flows.

Besides technical advances, experimental fluid mechanics, like any other scientific field, will benefit from popularisation. Wider availability of open-access design of instrumentation and data processing algorithms, and lowering of the cost of materials, will make measurement affordable to potential users that operate with limited resources. This will also facilitate the education and training of graduate students and other junior researchers, as well as the enrichment of undergraduate education.

Like any other endeavour into the future, the development of new instrumentation and experimental methods would greatly benefit from the experience of the past. The inventory of any fluid mechanics laboratory will more than likely contain Pitot tubes and liquid manometers, both instruments that have remained essentially unchanged for hundreds of years. What makes such tools persist, while state-of-the-art expensive apparatus has a useful lifetime of just a few years? It is their simplicity of construction and operation, a response based on fundamental laws of mechanics, rather than calibration, and a relative insensitivity to most, although not all, sources of interference. These are qualities that instrumentation of the future must have, if it is to be of lasting service. As a final note, let us once more reconsider the objectives of a 'good' fluid mechanics experiment. It is not to impress with the cost and complexity of apparatus but with the quality of the results. Obviously, the availability of suitable and functioning equipment is a prerequisite for a good experiment. That, however, comes with no guarantee of success. The experimeter is the one to decide how to set up the experiment, what to measure, how to use the equipment properly and how to present the results. Good experimental results can have a lasting significance, much longer than the lifetime of the instruments, and even the persons, that generated them. Figure 16.1 shows the late Stanley Corrsin in action in the GALCIT fluid mechanics laboratory nearly 80 years ago. The equipment he used includes an early hot-wire anemometer, but certainly no data acquisition systems, no computers, not even transistors of any sort. Yet, Corrsin's heated jet measurements of that era are still cited and have not been superseded by more recent ones.

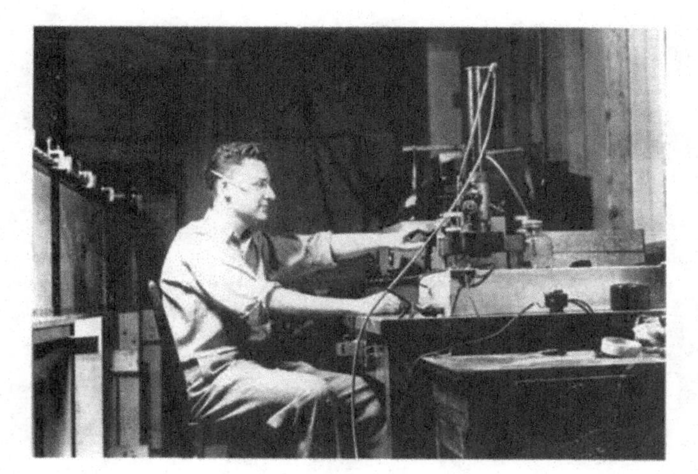

Figure 16.1 Stanley Corrsin operating a hot-wire anemometer at GALCIT *ca.* 1943 (with permission of the California Institute of Technology).

Index